HERBICIDES - CURRENT RESEARCH AND CASE STUDIES IN USE

Edited by **Andrew J. Price**
and **Jessica A. Kelton**

Herbicides - Current Research and Case Studies in Use
http://dx.doi.org/10.5772/56743
Edited by Andrew J. Price and Jessica A. Kelton

Contributors

André Andres, Leandro Galon, Giovani Theisen, Germani Concenco, Timothy Grey, Flavio Martins Garcia Blanco, Sydnei Almeida, Marcus Matallo, Jamal R. Qasem, Tran Dang Khanh, Le Hung Linh, Le Huy Ham, Tran Dang Xuan, Ta Hong Linh, Nguyen Thanh Quan, Do Manh Cuong, Vu Thi Thu Hien, \'Mota Samuel Lesoli, Masibonge Gxasheka, Beyene Solomon, Bethwell Moyo, W. James Grichar, Peter A. Dotray, Jason Woodward, Wendy-Ann Patrice Isaac, Zongjun Gao, Mei Li, Wayne Ganpat, Puran Bridgemohan, Lyn Gettys, William Haller, Greg MacDonald, Ines Santin-Montanya, Rafael Grossi Grossi Botelho, Valdemar Tornisielo, Paulo Alves, Sérgio Henrique Monteiro, Eloana Bonfleur, Gyuhwa Chung, Mona El-Hadary, In-Taek Hwang, Jung-Sup Choi, Dorota Soltys, Agnieszka Gniazdowska, Renata Bogatek, Urszula Krasuska, Carlos Azania, Andrea Azania, Luciana Rossini, Rodrigo Adriano, Dilermando Perecin, Damien A. Devault, Istvan Jablonkai, Jason Ferrell, Brent Sellers, Maria Aparecida Marin-Morales, William Anthony Bailey, Andrew Price, Jessica Kelton, Pill-Soon Song, In-Ja Song, Markkandan Ganesan, Jeong-Il Kim, Hyo-Yeon Lee, Tae-Woong Bae

Published by InTech

Janeza Trdine 9, 51000 Rijeka, Croatia

Edition 2014

Publishing Process Manager Iva Lipovic
Technical Editor InTech DTP team
Cover InTech Design team

Additional hard copies can be obtained from orders@intechopen.com

Herbicides - Current Research and Case Studies in Use, Edited by Andrew J. Price and Jessica A. Kelton
p. cm.
ISBN 978-953-51-1112-2

INTECH
open science | open minds

Contents

Preface

Herbicide use has dramatically increased since the introduction of the first selective herbicides decades ago. Utilization of herbicides for weed control is a crucial aspect of weed management in most crop productions, aquatic systems with invasive weed species, pastures, and non-crop areas such as turf and natural areas. The dynamic nature of weed populations, both in crop and non-crop systems, necessitates continuous adaptations and revisions to weed management strategies in order to ensure effective control of problematic weed species. Furthermore, continuous work is required to detail successful means of integrating innovative weed control tactics into existing management systems.

In recent years, a shift from single weed management practices to multiple, integrated control methods has been the focus for most agricultural and non-agricultural settings. These management practices utilize a number of chemical, mechanical, and biological tools to provide adequate weed control while preserving the efficacy of individual control options and maintaining environmental quality. The value of herbicides in integrated weed management is considerable; however, the overdependence on a single herbicide for weed control can lead to reduced efficacy, herbicide resistance, and potential environmental contamination. Due to these potential risks from herbicide overuse, integrated weed management has become the recommended practice for weed control in most systems.

In this book, chapters explore a wide array of weed control topics in many agricultural and non-crop systems. Authors provide information regarding current weed management practices and potential strategies for future weed control plans. Many chapters focus on the use of integrated control tactics, while other chapters describe individual management practices that can be implemented into existing weed management plans. Topics covered in this book include: integrated weed management in agricultural crops and rangelands, allelopathy and bioherbicides, transgenic crops and herbicide resistance, aquatic herbicide use, and many other subjects related to herbicide use.

The trends and case studies detailed in **Herbicides - Current Research and Case Studies in Use** provide a great deal of information concerning herbicide use in a number of settings. In this regard, the book should be of great benefit to many people that are involved in weed management plan development, herbicide resistance control, education, and technology transfer. It is anticipated that this book will be a useful reference in regards to current herbicide use trends and weed management strategies in both agriculture and non-crop settings.

Andrew J. Price
United States Department of Agriculture, Agricultural Research Service
National Soil Dynamics Laboratory, Auburn, Alabama, USA

Jessica A. Kelton
Auburn University, Auburn, Alabama, USA

Section 1

Row Crop Case Studies

Chapter 1

Weed Resistance to Herbicides in Rice Fields in Southern Brazil

André Andres, Giovani Theisen,
Germani Concenço and Leandro Galon

Additional information is available at the end of the chapter

http://dx.doi.org/10.5772/55947

1. Introduction

Rice (*Oryza sativa* L.) is the main staple food for a great part of the world population, and together with corn and wheat represents most of the cereals produced and grown worldwide [1]. With the growth of the world population, especially in East Asian countries, there are concerns about if rice production will be sufficient to meet the demand in the future [1]. There is the need to increase crop productivity levels, but there are both limitations for the opening of new agricultural areas, and issues regarding environmental pollution and use of natural resources.

The annual rice production in Brazil is 11.6 million tons [2], occupying an average area of 2.43 million ha per year with yields averaging 4.73 t ha^{-1} (Table 1). The southern states of Rio Grande do Sul (RS; 1.05 million ha) and Santa Catarina (SC; 0.15 million ha) contribute with more than 77% of the rice production with about 51% of the cultivated area in Brazil. Average grain yields obtained in the last five years in the RS and SC were around 7.26 t ha^{-1}, almost 55% higher than the national average [2]

The intensification of rice cropping systems in the same area promotes an increase in infestations by weeds. The fields of irrigated rice in southern Brazil provide a special habitat for weeds. During some months of the hot season, in addition to temperature and luminosity suitable for plant growth, there is also abundant soil moisture, which favors the development of weeds. This makes weeds responsible for losses in yield and grain quality, due to the direct interference they cause to the crop [3]. The weeds also cause other indirect negative effects in the production system, such as losses in nutritional value of pastures, interference in cover crops and even depreciating the land value [4-6].

In the fields of southern Brazil, the increase in weed occurrence is well characterized, mainly due to the fact that the irrigated rice was – until recently – almost the only cultivated crop in lowlands. To reduce the impact of weeds in rice, farmers have adopted some technologies. At first, there were modifications in the soil management system, shifting from a conventional plough-and-harrow to other forms of soil cultivation, such as minimum-till, no-till and the water-seeded rice system. Secondly, there was the adoption of ALS-tolerant rice cultivars (Clearfield technology - CL®) and last, the increase in the area of Roundup Ready soybeans in drained lowlands has also contributed to the weed management in rice fields. Herbicides, however, are still heavily used as the main form of weed control in almost all irrigated rice fields in RS state. In complement, organic rice is growing in adoption, but actually is restricted to small fields. The certified organic smallholders account for 400 producers, in an area of about 3,400 ha dispersed in the RS state.

It is known that the average regional yields (7.26 t ha^{-1}) are below those obtained in field trials and in high technology farms. Even though new cultural techniques are often used to control weeds, poor weed control is one to be highlighted among the probable causes of grain yield variability. According to results of [7] and [2] it is estimated that about 1 million tons of rice are lost annually in Brazil, which is roughly equivalent to 8% of the national production of this cereal, even after using all methods available for weed management. This corresponds to an annual loss estimated of about US$ 200 million.

Cropping season		**Area (1000 ha)**			**Yield (kg ha^{-1})**			**Production (1000 t)**	
	Brazil	RS	SC	Brazil	RS	SC	Brazil	RS	SC
2002/03	**3186**	960	145	**3254**	4890	7195	**10367**	4696	1043
2003/04	**3654**	1039	151	**3511**	6064	6630	**12960**	6433	1000
2004/05	**3916**	1050	154	**3377**	5912	6800	**13355**	6333	1050
2005/06	**3018**	1040	156	**3884**	6610	7050	**11722**	6872	1099
2006/07	**2967**	954	156	**3813**	6726	7050	**11316**	6419	1099
2007/08	**2875**	1067	153	**4200**	6902	6650	**12074**	7362	1018
2008/09	**2909**	1106	150	**4332**	7150	6950	**12603**	7905	1040
2009/10	**2765**	1080	150	**4218**	6781	7060	**11661**	7321	1057
2010/11	**2820**	1172	150	**4827**	7600	6625	**13613**	8904	996
2011/12	**2455**	1053	150	**4728**	7350	7180	**11600**	7740	1078

* Source: [2]

Table 1. Historical cultivated area, grain yield and production of rice in Brazil and in the states of Rio Grande do Sul (RS) and Santa Catarina (SC), from 2002 to 2012.

http://dx.doi.org/10.5772/55947

Due to the particular regional characteristics, there are many ways of soil, water and plant management in irrigated rice in southern Brazil. The main system is minimum-till (around 60% of the area) in which the soil is plowed, harrowed, leveled and the levees are done in the autumn, right after the harvest of the summer crop, with chemical desiccation in spring before rice planting, done with a no-till drill in dry soil. Another system is the conventional seeding, where all the tillage is done just before planting rice, in dry soil. Finally, about 20% of the fields are cultivated with the water-seeded system, performed mainly in small farms (up to 30ha) in which rice is sown pre-germinated over a field already flooded (schemes on Figure 1). The system of manual or mechanic transplanting rice seedlings from the nursery to the puddled and flooded field – very common in the Asian paddies – is almost not used in Brazil.

In the last few years, there was a continuous increase in the soybean area in the lowlands of RS, and currently this crop occupies around 250,000ha in rotation with rice (all RS state have approximately 4.19 million hectares of soybean). Probably in the following years, soybean will spread up to 0.5 million hectares in the lowlands of RS, limited by poor soil drainage conditions. Glyphosate-tolerant soybean has changed the scenario of resistant-weeds in rice fields and will be discussed later in this article.

1.1. The main weeds of irrigated rice in southern Brazil

The main weeds in flooded rice fields in Brazil are commonly classified into narrow- and broad-leaved weeds. The major representatives of narrow leaves are weedy rice (*Oryza sativa*), barnyardgrass (*Echinochloa* sp.), the aquatic grasses (*Leersia hexandra* and *Luziola peruviana*), and the sedges (*Cyperus difformis, C. esculentus, C. ferax*, and *C. laetus*).

Recently, there was an increase in the occurrence of Alexander grass (*Brachiaria plantaginea*), crabgrass (*Digitaria horizontalis*) and goosegrass (*Eleusine indica*) in the rice fields. These monocotyledonous weeds, common in dry fields in crops such as corn, sorghum and soybeans, are expanding due to the increase in crop diversification in lowland areas, to the continued use of ALS inhibitors and the abandonment of propanil herbicide in the rice fields. Some places also reported the presence of perennial weeds such as Olive hymenachne (*Hymenachne amplexicaulis*), ribbed murainagrass (*Ischaemum rugosum*), Mexican sprangletop (*Leptochloa uninervia*), Fall panicum (*Panicum dichotomiflorum*), Knotgrass (*Paspalum distichum*) and *Paspalum modestum*. These perennial plants grow in areas with an excess of moisture.

As broadleaved weed representatives, there are the jointvetches (*Aeschynomene* spp.) and in some areas some species of morning glory (*Ipomoea* spp.), water pepper (*Polygonum hydropiperoides*) and alligator weed (*Alternanthera philoxeroides*). The aquatic weeds, associated mainly with fields grown in the water-seeded system (with pre-germinated seeds) are globe fringerush (*Fimbristylis miliacea*), arrowheads (*Sagittaria montevidensis* and *S. guyanensis*), water hyacinth (*Eichornia crassipes*), kidneyleaf mudplantain (*Heteranthera reniformis*) and the Ludwigia complex (*Ludwigia elegans, L. longifolia* and *L. octovalvis*).

Many of these species are difficult to control and severely compete with the crop for resources available in the environment if no control method is adopted. In addition, barnyardgrass,

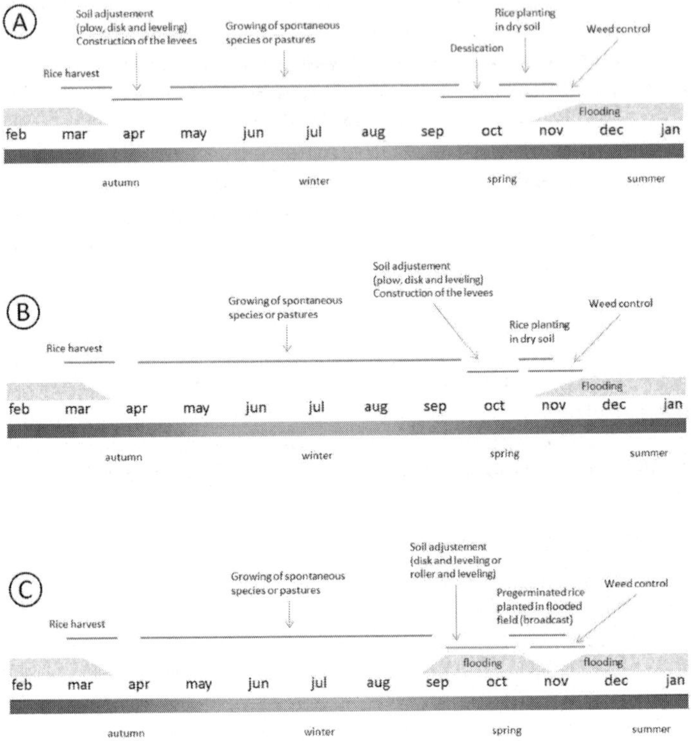

Figure 1. A simplified scheme of the three main production systems of irrigated rice in southern Brazil. (A) Represents the minimum-till system; (B) represents the conventional system; (C) represents the water-seeded system. The schemes illustrate part of a very common two-year sequence of rice cropping.

weedy rice, globe fringerush, arrowhead and some sedges have acquired resistance to herbicides (Table 3).

1.2. How does resistance to herbicides appear in rice fields?

The adoption of herbicide-tolerant rice has increased considerably in the last few years. The results of this unprecedented change in agriculture have been many, but perhaps most dramatic is the simplification of weed-control tactics; growers can now apply a single herbicide group (ALS-inhibitors) at higher rates of active ingredient without concern for injury to the crop. Regardless, the number of chemical groups of herbicides applied has declined, thus increasing the ecological implications such as reducing the biodiversity of arable land, facilitating population shifts in weed communities and the evolution of herbicide-resistant biotypes. Historically, a number of significant changes in agricultural systems have occurred with significant impact on weed communities.

The use of herbicides for weed control has become a common practice in agriculture worldwide. Once, this technology was used mainly by big farmers; it is currently becoming a common practice even among smallholders. Nowadays, weed control in irrigated rice relies almost exclusively on herbicides, mainly because chemical control has been efficient, relatively cheap, readily available and professionally developed. Thus, other methods of control have been left as a second choice or under certain circumstances may present themselves unattractive or unfeasible. It needs to be noted that the strong presence of pesticide suppliers has almost banned the use of other forms of weed management but the herbicides in irrigated rice fields. The widespread and almost exclusive use of the chemical method of weed control in rice promotes changes in the weed flora, from quite easily controlled broadleaved weeds to more hostile grass weeds [8-9]. The recurrent use of herbicides with the same site of action can select individuals that are genetically capable of surviving a dose of a given herbicide which normally would kill or suppress the species [10]. Herbicide resistance is the inherent ability of a species to survive and reproduce following exposure to a dose of herbicide normally lethal to its wild type. Resistance is not directly caused by herbicides, rather, it appears from the selection of natural mutation or minor pre-existing population of herbicide-resistant plants (selection pressure imposed by herbicides) [11] or in rice cases, gene flow from herbicide-resistant to weedy rice [12-14].

As at other places worldwide, in the rice fields of south of Brazil, the continuous use of herbicides has led to the evolution and appearing of herbicide-resistant (HR) weeds, and this is an additional problem in the pest management context. Chemical weed control is used in almost all areas and the scenario in the short-past, at present – and probably to the future – is a continuous intensification of the rice cropping systems. This intensive system, combined with the continued use of herbicides with the same mechanism of action, has resulted in the development of resistant weeds. The resistance of weeds to herbicides in that region was confirmed by several institutions, namely EMBRAPA, EPAGRI, IRGA, UFRGS, UFPEL and UFSM.

2. Main herbicide resistant weeds occurring in rice in Southern Brazil

There are already reported cases of herbicide-resistant biotypes of the main weeds such as *Oryza sativa* (red rice or weedy rice), *Echinochloa* spp., *Cyperus difformis, C. esculentus, C. iria, Fimbristylis miliacea* and *Sagittaria montevidensis.* These weeds are common in almost all rice fields of Southern Brazil and at some places show resistance to ALS-inhibiting herbicides. Some barnyardgrass biotypes resistant to ALS-inhibitors also were resistant to quinclorac herbicide. One of the most important cases of resistant weeds is the occurrence of weedy rice resistant to the ALS-inhibiting herbicides used in the Clearfield® technology [15], because in this particular situation the weedy rice is from the same species as the crop (*Oryza sativa*).

2.1. First cases

Weed resistance to herbicides in rice fields of Southern Brazil was first registered in 1999 [16], with a biotype of arrowhead (*Sagittaria montevidensis*), which evolved resistance to four ALS-inhibiting herbicides. A short time after, other cases of resistance were reported with a new biotype of *Sagittaria* [17]; and also with barnyardgrass (*Echinochloa* spp.) resistant to the herbicide quinclorac [18]. Since then there was an increasing number of reports of weed resistance (Table 3).

2.2. The case of *Echinochloa crus-galli* resistance to the herbicide quinclorac

This species is a monocotyledon that survives in flooded environments, occurring normally in high levels of infestation. It is widely distributed in almost all rice fields of SC and RS. In addition, barnyardgrass presents morpho-physiological similarities with the crop in the early stage of development. The negative effects of its presence in rice include: the high capacity to compete with rice by resources as light and nutrients; the intrinsic difficulties related to control; the increases in the production costs; it causes rice lodging, difficulties in the harvest and depreciation of the product; it is a host of some pests in rice and this species can even decrease the commercial value of arable areas [3,19-20].

Barnyardgrass is also one of the most widely distributed weeds in the grain crops grown in rotation with rice in lowland areas, mainly represented by soybeans, some sorghum [21] and a little portion of areas with maize. In reference [4] reported that many of the ALS-resistant biotypes of *Echinochloa* showed faster initial development compared to susceptible ones. The authors also report that biotypes from different areas are distinct in terms of initial growing speed.

Due to the continuous use of herbicides with the same mode of action, often in the absence of crop rotation and lack of integrated management, barnyardgrass evolved resistance to several herbicides [22] and some biotypes have multiple resistance [23]. In reference [24] reported a biotype of barnyardgrass presenting cross-resistant to quinclorac (auxin-mimic herbicide) and to ALS inhibitors. Herbicides represent the main tool for weed control within the program of integrated management in rice fields of Southern Brazil. Among those used in rice, quinclorac (auxin-mimic) combines flexibility in the application (pre- and post-emergence) and normally offers good efficiency to *Echinochloa crus-galli* and *Aeschynomene rudis* control, low toxicity to humans and animals and high selectivity to rice. This active ingredient started to be used in rice production areas of RS and SC in the early 1990's, being used intensively until mid-1999, when complaints began to emerge about failures in barnyardgrass control. Studies confirmed the occurrence of resistance already in 2000 [18, 25].

2.3. The cases of weed resistance to ALS inhibiting herbicides

Similar to what happened with quinclorac in the past, in more than a half of all cultivated rice areas in RS state, the ALS-inhibitors were (and still are) vastly applied in the fields. This scenario was aggravated by the use of varieties tolerant to the herbicides belonging to this group (CL technology), aiming to achieve efficient control of weedy rice and barnyardgrass. The repetitive use of some ALS-inhibiting herbicides for 4 to 5 years after the launch of the CL technology resulted in resistance of barnyardgrass to the herbicides bispyribac-sodium, penoxsulam, imazethapyr+imazapic and imazapic+imazapyr [26].

2.4. Arrowhead — *Sagittaria montevidensis*

This is an aquatic weed often found in water-seeded or transplanted rice systems. Arrowhead is characterized as a weed that occurs in high levels in most areas of flooded rice in Santa Catarina [27]. This weed presents a low capacity to compete with rice as compared to other species which infest the crop [28]. However, the frequency of high infestations by arrowhead has resulted in increased use of herbicides for its control. In the RS, rice is mainly drill planted in dry soil, and flood irrigation starts about 20-25 days after emergence; in the State of SC almost 100% of its rice area is grown in the water seeded system, which favors arrowhead.

Several biotypes of arrowhead were found to be resistant to ALS inhibitors [29]. In Brazil, populations with cross-resistance to the sulfonylurea and pyrimidinyl thiobenzoates were identified in 1999 in areas treated with these products for about five consecutive years [16]. In reference [30] the authors found that the resistant biotype of arrowhead showed faster emergence, higher seed vigor and absorption of herbicides preferably by shoots instead of roots, when compared to the susceptible population.

Rice areas with arrowhead resistant to ALS inhibitors are common in Brazil due to the extensive and repetitive application of herbicides with this mechanism of action. A recent study revealed the occurrence in SC State of populations of this weed with cross-resistance to several ALS-inhibiting herbicides and multiple resistances to PSII inhibitors [31]. Currently, arrowhead resistant to ALS inhibitors is present in almost all municipalities which grow rice in Santa Catarina State.

In rice fields where the ALS-resistant biotype of arrowhead occurs, the herbicides carfentrazone-ethyl or bentazon can be used as alternatives for chemical control. Both herbicides applied alone or in tank mix allowed control levels of arrowhead superior to 92% at the pre-harvest of water-seeded rice in SC State [32]. It should be emphasized that planting rice at low- or lower-densities that the recommended [33] allows a more favorable environment for aquatic rice weeds, especially arrowhead. In reference [34] it was observed that a strong negative correlation between the planting density of the rice variety BRS 6-Chui and the infestation by arrowhead; in other words, the infestation was more serious as rice density was decreased. According to the authors, this suppression caused by higher rice densities is due to the increased ability of the crop to compete for light, which prevented the weeds from having access to adequate levels of radiation.

2.5. Nutsedges — *Cyperus difformis, C. iria* and globe fringerush — *Fimbristylis miliacea*

Some weed species of the family *Cyperaceae* infest rice fields in the RS and SC states, being responsible for reducing the potential yields of this cereal. *Cyperus difformis* appears as one of the most damaging weeds to rice. This species is distinguished by production of large quantities of seed (50,000 seeds plant^{-1}), promoting rapid infestation with high growth rates. This has, as a consequence, the formation of a large amount of green mass with high competitive potential with rice, especially in the initial stages of development of the crop [19].

The weed control in rice fields can be accomplished with the use of herbicides due to its ease of use and high efficiency. There are difficulties, however, in chemically controlling species of the genus *Cyperus*. Some species of *Cyperus* reproduce both by seeds and vegetatively (tubers and stolons) as in the case of *C. esculentus* and *C. rotundus*. Furthermore, the chemical control of *Cyperus* spp. in pre-emergence is especially problematic due to the scarcity of products to be applied in this modality. For controlling these species, post-emergence herbicides inhibiting the enzyme ALS, as bispyribac-sodium, penoxsulam, pyrazosulfuron-ethyl, ethoxysulfuron, cyclosulfamuron and azimsulfuron, can be applied. It is necessary also to respect the limit of growth stage at the time of application and to use adjuvants specific to each herbicide [33].

The control of *C. difformis* with ALS inhibitors, however, has presented problems due to the development of resistance [35,36]. The authors report that this is mainly due to the intensive cultivation of rice, associated with the use of herbicides with the same mechanism of action for several years, favoring the selection of resistant populations.

From the 1980's, the ALS-inhibiting herbicides have become very important tools for agriculture, and the widespread use of these products was mainly due to its high efficiency at low doses, low toxicity to animals, high selectivity for some crops and reduced environmental impact when compared to other pesticides [37]. These traits contributed to the increased use of these herbicides in various crops. Two years after these products were made available in the market, however, appeared the first case of a weed with resistance to this mechanism of action. Currently, there are 95 resistant species, distributed in 34 countries [38].

Results in reference [36] are shown in Figure 1. One biotype of *C. difformis* presented a high-level of resistance to the herbicide pyrazosulfuron-ethyl (sulfonylurea), and was also cross-resistant to the bispyribac-sodium (pirimidinyl thiobenzoate), both ALS inhibitors. Bentazon is an efficient alternative for the chemical control of the ALS-resistant biotype of *C. difformis* (Figure 2). The same authors point out that, for the management of populations of *C. difformis* resistant to ALS inhibitors in flooded rice areas, it is recommended the adoption of practices such as rotating herbicides with different mechanisms of action and management practices that may restrict the expansion of the resistant populations.

The mechanism involved in the resistance of *C. difformis* to pyrazosulfuron-ethyl is the insensitivity of the enzyme ALS to herbicides, which inhibit this enzyme, conferring high levels of resistance [39]. In [40] tested the herbicides pyrazosulfuron-ethyl, bispyribac-sodium, imazapyr, imazapic and penoxsulam on the species *C. iria* (Table 2), and also proved the resistance of this species to ALS inhibitors due to the low levels of control achieved with all herbicides. In the same study, bentazon (PSII inhibitor) controlled 100% of the biotype. Another study [41] also observed no efficient control of *C. iria* under application of 1x and 2x the recommended dose of pyrazosulfuron-ethyl, imazethapyr, imazapic or ethoxysulfuron.

For rice fields infested with biotypes of weeds resistant to ALS inhibitors, the most effective strategies are pointed out in the following. The application of glyphosate alone or mixed with pendimethalin or clomazone at the so-called "needle point" will ensure that the rice emerges free from the infestation of *Cyperus,* allowing also efficient control of several other weeds. The "needle point" is the rice germination stage immediately prior to the initiation of the emergence, depicted in Figure 3. Usually, when a very few rice seedlings start to emerge in the field indicates the needle point, and the non-selective herbicide should be applied on that day. This

http://dx.doi.org/10.5772/55947

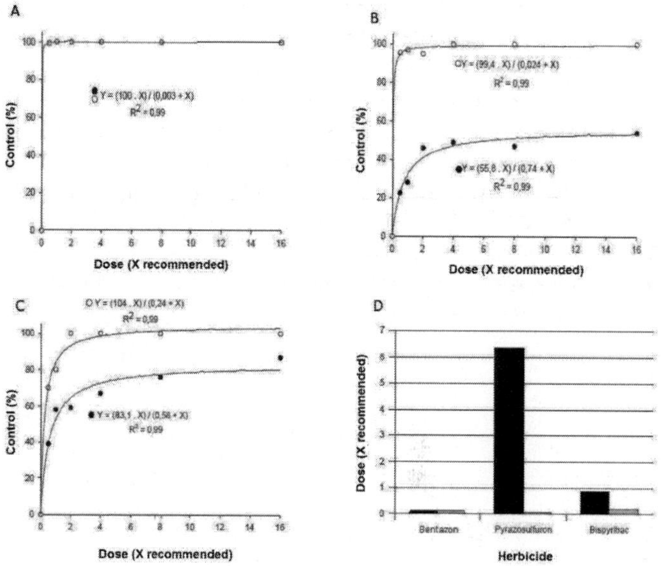

Figure 2. Control (%) of a biotype of *Cyperus difformis* resistant (●) or susceptible (○) to ALS-inhibiting herbicides by using PSII and ALS-inhibiting herbicides as a function of dose. [bentazon (A), pyrazosulfuron-ethyl (B), bispyribac-sodium (C)] In (D) the doses that control 50% of the population (LD_{50}) of the resistant (black bars) and susceptible (grey bars) biotypes are presented. Source: [36]

Herbicide	**Control (%)**		**Dry Mass**
	14 DAH[1]	**28 DAH**	**(g plant^{-1})**
Pyrazosulfuron-ethyl	15 b[2]	6 b	1,59 a
Bispyribac-sodium	6 bc	13 b	2,31 a
Imazapyr + imazapic	2 c	2 b	2,31 a
Penoxsulam	3 c	10 b	1,48 a
No application	0 c	0 b	2,79 a
CV(%)	18,09	38,16	27,41

[1] Days after application of herbicides. Means followed by the same letter, in the column, are not different (Tukey P>0.05). Source: Adapted from [40]

Table 2. Control efficiency and shoot dry mass of *Cyperus iria* as a function of the application of ALS-inhibiting herbicides.

should not affect the stand of rice plants in the field, as the majority of the seedlings will not be emerged on that day. This happens from three to five days after rice planting, depending on environmental conditions (soil moisture and temperature).

Although effective, a delay in the application of glyphosate + pendimethalin or clomazone for a single day from the needle point may cause severe damage to rice. This is particularly a problem if there are frequent rains forecasted for the five days following planting. So, technicians are highly encouraged to evaluate carefully the risk of this practice before recommending it for farmers. In addition, the application of glyphosate should not be done only at the needle point. There is the need for a previous desiccation of the area between 20 and 10 days before planting, which will allow control of the older weed plants.

Another option defined in [40] is the use of PSII inhibitors like bentazon or carfentrazone-ethyl in post-emergence. Carfentrazone, however, may cause severe damage to rice. In addition, both chemicals are contact-only herbicides, which means that a good coverage of the plants by using higher water volumes than the usual followed by flooding on the following day, should allow good results.

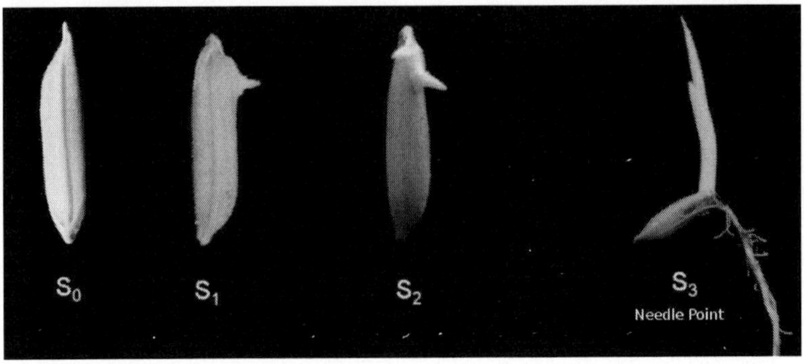

Figure 3. Rice seeds at distinct germination stages, from S0 to S3 (needle point). Source: FREITAS, T. F. S; GROHS, D. (SOSBAI, 2012).

The species *Fimbristylis miliacea,* popularly known as globe fringerush, belongs to the family *Cyperaceae* and is disseminated in various regions of the world. In Brazil it appears to be more common in the Southern coastal region infesting flooded rice [42]. The plant cycle is annual or perennial, depending on the environmental conditions; it presents seed dormancy, and germinates in any season if water is available. In RS and SC, the species is distributed all over the rice producing areas. It is observed that the higher infestations occur generally in areas with no uniform irrigation. The population and crop management determine the potential damage in yields due to globe fringerush, but the average losses can be about 73% under high infestations [27].

There were only three reports of resistance of *F. miliacea* to herbicides in the world, and the first record was in Malaysia in 1989 with biotypes resistant to 2,4-D; the second in 2001, in

Brazil, with biotypes resistant to pyrazosulfuron-ethyl and cyclosulfamuron, and more recently in 2010 in Venezuela, also with resistance to ALS-inhibiting herbicides [38]. It is known that biotypes of this species are resistant to ALS inhibitors in Brazil, especially in SC, but the mechanism of resistance is still unknown.

2.6. The case of weedy rice (red rice) resistant to ALS-inhibiting herbicides

Among the major weeds infesting rice, weedy rice can surely be highlighted as the one which most limits the potential yield of rice [43]. The direct losses resulting from competition exerted by weedy rice in rice paddy fields is estimated at about 20% [43].

There are also several indirect losses, such as raised cost of production, depreciation of the market value of cultivated areas and of the harvested product, equipment damage and reduction in generation of jobs, further reducing the profitability of farming [44]. The degree of interference of weedy rice varies with the level of infestation, soil and climatic conditions, cultivar traits, coexistence period and biotype found in the area [45].

The control of red rice with herbicides has become possible after the development of rice genotypes tolerant to the herbicides from the imidazolinone group (ALS-inhibitors) [46]. The same authors also reported that effective chemical control of red rice is almost impossible using conventional genotypes, because of the morpho-physiological similarity between the cultivated and the weedy rice. Despite the Clearfield® system providing a great advantage in terms of weed control, the adoption of herbicides from the ALS group associated to this technology resulted in selection of resistant genotypes of this weed [15]. Thus it is evident that the continued use of the Clearfield® system in rice areas of Rio Grande do Sul favored the development of populations of red rice resistant to imidazolinone due to its repeated use in the absence of crop rotation or others tools.

The introduction of rice cultivars tolerant to imidazolinone herbicides probably resulted in gene flow of the resistance to wild rice genotypes [47-48]. The occurrence of weedy rice populations resistant to herbicides may be caused by gene flow between cultivated varieties and weedy rice [12,13]. A research at RS has indicated that pollen dispersal occurs between cultivated rice and transgenic rice at levels below [49] and others studies [50-51] showed gene flux between rice CL varieties and weedy rice rates of 0.042 and 0.065%, respectively. It should be noted that even with low rates of gene flow, there might be a considerable increase in the frequency of resistant individuals in the population, given the high degree of infestation of cultivated areas [15]. Another study shows that the gene flow as low as the rate of 0.008% originated 170 individuals of red rice per hectare with resistance [13].

In reference [15] with populations of red rice from the six most rice-producing regions of the RS, the occurrence of resistant biotypes to herbicides from the imidazolinone group (imazethapyr and imazapic) was confirmed to occur in all regions under the Clearfield® system. The predominant mechanism of herbicide resistance in weedy rice in RS and SC is the target site insensitivity due to changes in nucleotide sequence of the ALS enzyme [14,52]. Gene flux was the main origin of imidazolinone herbicide resistance, but independent selection occurred in 1.1 % of the evaluated weedy rice plants [14]. The high frequency of weedy rice resistant plants carrying the $G_{654}E$ mutation, which is the same mutation responsible for the resistance in the rice cultivar largely used in Southern Brazil when the weedy rice plants were collected, suggests that gene flow is occurring from the rice cultivar to weedy rice [52].

2.7. A retrospect of the ALS-inhibiting herbicides and ClearField® technology use in irrigated rice fields of Southern Brazil

The use of Clearfield® (CL) technology in rice areas of southern Brazil began in 2002 with imidazolinone-resistant cultivars. Ten years later, more than 60% of irrigated rice in Rio Grande do Sul State carry the CL technology and are treated with these herbicides. The combined use of imidazolinone-resistant rice cultivars with the correspondent herbicides is often very effective, providing more than 95% of control of weedy rice in most cases [53]. This technology had permitted immediate benefits in terms of efficiency and easiness of weed control, mainly for weedy rice and the *Echinochloa* complex. However, at the beginning of the use of CL rice cultivars there were some difficulties that possibly favored the increasing of the number of the ALS-resistant weeds. First, due to high initial costs of the commercial seeds and of the herbicide, part of the fields was planted with saved-seeds and there were the use of not-registered, illegal herbicides, applied at elevated doses in some fields. Second, the CL rice cultivar was planted repeatedly in areas heavily infested with weedy rice, disregarding the official recommendations for the management, which suggested herbicide rotation, field management rotation and crop rotation in fields of irrigated rice [33].

Even though some weeds presented resistance to ALS-inhibitors before the adoption of the CL technology (Table 3), the selection pressure caused by the increasing use of the ALS-inhibitors should be associated with the emerging of weedy rice (*Oryza sativa*) resistant to ALS-inhibiting herbicides, only four years after the starting of the use of Clearfield® technology in southern Brazil [15] which occurred in USA [53]. The fields infested with these resistant biotypes represent a part of the whole area of rice cultivation, but all regions have dispersed resistant weedy rice and there is an increase in the number of cases of resistance. The farmers and assistants are at the present taking additional management strategies for this weed, as the crop and herbicides rotation to reduce the losses and constrains associated with the weed resistance. In Arkansas [53] after 5 years of imidazolinone-resistant rice technology, crop rotation and use of certified seeds are the main reason for rice fields being free of weedy rice.

2.8. Prevention of herbicide-resistant weeds in irrigated rice of Brazil

An herbicide-resistant weed biotype usually occurs in areas where the common practice for weed control is the repeated use of the same product, or the use of different herbicides but with the same mechanism of action. This is the main scenario at the beginning of the weed-resistance cases in rice fields - the high selection pressure - as reported by [37]. This situation is very common in the RS, where rice still is continuously grown as a mono-crop in most parts of the area. In the state of Santa Catarina, the areas are smaller and the management more varied, with farmers using both herbicides and cultural practices on weed management.

The adoption of best-practices in weed control is one of the main tools to prevent the occurrence of new cases of resistance. In reference [33, 68], some preventative measures to avoid or to

Species	Common name	Active ingredient confirmed	Sources*
Sagittaria montevidensis	arrowhead	Azimsulfuron, bentazon, bispyribac-sodium, cyclosulfamuron, ethoxysulfuron, imazapic+imazethapyr, metsulfuron, penoxsulam, pyrazosulfuron-ethyl	[16, 17, 31]
Echinochloa spp.	barnyardgrass	Quinclorac	[4, 18, 25, 54]
		Bispyribac-sodium, flucarbazone, imazapyr, imazethapyr, imazethapyr+imazapic, imazapyr+imazapic, nicosulfuron, penoxsulam, quinclorac	[55, 56, 60-65]
Cyperus difformis	nutsedges	Azimsulfuron, bispyribac-sodium, cyclosulfamuron, ethoxysulfuron, penoxsulam, pyrazosulfuron-ethyl	[35, 36, 39, 65]
Cyperus iria	nutsedges	Bispiribac-sodium, Ethoxysulfuron, imazapyr+imazapic, imazethapyr+imazapic, penoxsulam, pyrazosulfuron-ethyl	[40, 41, 57, 58]
Fimbristylis miliacea	globe fringerush	Azimsulfuron, bispyribac-sodium, cyclosulfamuron, ethoxysulfuron, penoxsulam, pyrazosulfuron-ethyl	[66-67]
Oryza sativa	weedy rice	Imazethayr +Imazapic Imazapyr	[15, 52]

* The sources lists only a part of the whole studies conducted with the species listed.

Table 3. Herbicide-resistant weeds reported in irrigated rice in Southern Brazil.

minimize the risks are the use of crop rotation, the use of herbicides in the correct time and when necessary; to perform the rotation of herbicides, using those with distinct mechanisms of action; and be aware of the results of herbicide applications, checking for escapes and shifts in weed population. When an escaped plant is observed it must be immediately eliminated, preventing the spread of this suspected resistant biotypes. These recommendations are not always adopted in all fields due to the various difficulties. A good exception is the case of the seed-producers: these farmers really care with the weeds in your fields and adopt the best-management practices in terms of weed control, because there are some weed species whose seeds are expressly prohibited in lots of commercial rice seeds, and its presence would condemn the entire field, preventing it to be sold as seed.

In areas where herbicide resistant weed populations occur, some simple – but important – management strategies are issued. It is recommended not to plant very early in spring, because due to low temperatures, weeds will emerge and grow faster than rice, offering an additional difficultly for control and increased competition. The soil could be prepared, or chemically desiccated, immediately before planting rice to eliminate the weed seedlings already emerged; the machinery should be cleaned when leaving an infested area; the herbicides with proven resistant biotypes should not be used, and resistant escaped plants should not be allowed to produce seeds, by means of the localized chemical desiccation or by manual rouging.

In lowlands of Southern Brazil, rice is the main crop and commonly shares areas with cattle production. The cattle can occupy the fields in winter (between two cycles of rice) and consume cold-adapted grasses and broadleaves belonging to the genus *Lolium, Trifolium, Vicia* and others; or, in summer when the main feed is composed by grasses such as red rice, barnyard-grass, some perennial grasses and others species. Integration crop-livestock in rice fields is an important form of weed management in the production system once they consistently reduce the seed production of some grasses and the number of viable seeds in the soil seed bank will decrease [69].

In recent years, however, soybean has increased in area in the lowlands, also being used as a cash-crop in these fields due to the high prices in the international market. Between one-fourth and one-third of the rice in RS is already rotated with soybeans and this crop is the main – and probably the best – option to the rotational scheme with irrigated rice in terms of increasing the soil fertility and reduction of some pests in rice. Almost all soybeans cultivated in RS are tolerant to glyphosate (Roundup Ready technology) and this herbicide offers very good control of annual grasses such as red rice and barnyardgrass. The consolidation of RR soybean was a step forward in the effectiveness of integrated pest management in irrigated rice in the RS state. The soybean is already used as the main tool of management in the cases of herbicide-resistant weeds occurring in irrigated rice fields, mainly in those well-drained areas. However, there are some concerns about the selective pressure driven by glyphosate, and about the spread of the resistant weeds to glyphosate, such as the Italian ryegrass (*Lolium multiflorum*) and the hairy fleabane (*Conyza* sp.), already present in various places in the south of Brazil.

In terms of herbicide rotation, in the fields with barnyardgrass resistant to ALS-inhibitors and/or auxin-mimic herbicides, the herbicides pendimethalin, trifluralin, thiobencarb, clomazone (in pre-emergence), quinclorac (in pre or post-emergence – avoid it in areas where auxin-mimic resistant biotypes occur), propanil alone or mixed with pendimethalin or clomazone (in early post-emergence) and ACCase inhibitors (in post-emergence of the crop) are good options [33]. There are, however, reports about biotypes of *Echinochloa* with multiple resistances to ALS inhibitors and other chemical groups in several countries of Latin America [70]. As a consequence, no abuses in the chemical control should be allowed, making this weed difficult to be controlled, demanding crop and chemical rotation along the years. It should be highlighted that the use of ACCase inhibiting herbicides in rice fields have promoted efficient control of *Echinochloa* biotypes, but there is the need for rotation of chemical groups to avoid the appearance of biotypes resistant also to this mode of action.

In reference [62] studying methods of application of clomazone and imazapic + imazapyr, reported that the application of clomazone alone or mixed with imazapic + imazapyr in the rice on "needle point" allow efficient control of ALS-resistant *Echinochloa* and the susceptible biotype was efficiently controlled by clomazone alone in the needle point, and by imazapic + imazapyr in all application times.

Several rice farmers use residual herbicides in mixture with glyphosate in the pre-planting desiccation, mainly in areas under minimum- or no-till system (sod seeding) and/or with delayed flooding. In these cases, the elimination of existing weeds is accomplished with glyphosate and the new cohorts of seedlings are controlled by the residual herbicides. One of the most widely used herbicides for this task is clomazone, which presents residual effects over several grasses, especially barnyardgrass [71]. Thus, the use of clomazone with glyphosate, either in the early pre-planting desiccation of sod seeding areas, or in the post-planting on "needle point", is an effective tool for weed suppression. The application of glyphosate in the needle point was previously discussed, being illustrated in Figure 2.

Besides clomazone, pendimethalin may also be used at the "needle point" mixed with glyphosate aiming to suppress the emergence of *Echinochloa* spp. This pre-emergence herbicide plays an important role in the suppression of propanil-resistant junglerice in Central America [72, 73], whose genotypes still were not identified in Brazil. Pendimethalin thus can represent an important herbicide in the management strategy for the Brazilian ALS-inhibiting and Auxin-mimic resistant biotypes. In addition, propanil applied in early post-emergence, mainly mixed with clomazone or pendimethalin, are alternative choices depending on the level of the field infestation and effectiveness of the previously applied treatments [74]. In Brazil there are no reported cases of *Echinochloa* biotypes resistant to ACCase-inhibiting herbicides (Merotto and Noldin, personal information); thus, these herbicides are great options for post-emergence control of biotypes of *Echinochloa* resistant to ALS or Auxin-mimic herbicides. However, herbicides with this mode of action are considered of "high risk" for resistance evolution if not properly managed [70].

Managing herbicides properly within these options will allow farmers to have a 3-year rotation of herbicide, which will reduce both the occurrence of resistant biotypes, and the chance of appearance of a new resistant weed biotype. Farmers should request their technicians to plan the most proper herbicide rotation for every case. Used alone, none of the currently available cultural techniques provides an adequate level of weed control. However, when used in carefully planned combinations, extremely effective barnyardgrass control can be achieved [75].

3. Conclusions

Weeds resistant to herbicides have been of concern for scientists and farmers in the Rio Grande do Sul and Santa Catarina states of Brazil, since most herbicides used for chemical control are no longer effective in many fields. It is noteworthy to mention that the evolution of weeds resistant to herbicides is related to selection pressure, genetic variability of weeds, the number of genes involved, patterns of inheritance, gene flow and dispersal of the propagules. The elucidation of these factors becomes important for future predictions of proportions between resistant, tolerant and susceptible biotypes in the fields, and will require choosing more efficient management methods on these biotypes, aiming also to prevent the multiplication and dissemination of weed-related problems in the area.

In the case of rice, there are some intrinsic difficulties for adoption of full-integrated weed management with crop rotation because the condition of soil, with its susceptibility to be flooded and difficulties for fast drainage. The weed resistance to herbicides may cause losses to the rice production in many regions of Southern Brazil. Without the introduction of new herbicide mechanisms of action or better herbicide-resistance management, a technology that has allowed increases in agricultural productivity is at risk [76]. Despite the success attained in some cases, more research and investments must be directed to this field of study in irrigated rice in Brazil, especially in the Southern region, which is the main producer, so that the problem can be more understood and specific strategies to manage this problem can be established and applied by the farmers.

Author details

André Andres[1*], Giovani Theisen[1], Germani Concenço[2] and Leandro Galon[3]

*Address all correspondence to: andre.andres@embrapa.br

1 Embrapa Temperate Agriculture, Pelotas, Brazil

2 Embrapa Western Agriculture, Brazil

3 Federal University of the Southern Border, Erechim, Brazil

References

[1] FAO. Food and Agriculture Organization. http://www.fao.org (accessed 22 October 2012).

[2] CONAB. Companhia Nacional de Abastecimento. Arroz - Brasil. Série Histórica de: área, produtividade e produção. http://www.conab.gov.br (accessed 22 October 2012).

[3] Agostinetto D, Galon L, Silva JMBV, Tironi SP, Andres A. Interference and economic weed threshold (Ewt) of barnyardgrass on rice as a function of crop plant arrangement. Planta Daninha 2010; 28(special issue) 993-1003.

[4] Andres A, Concenço G, Melo PTBS, Schmidt M, Resende RG. Detection of *Echinochloa* sp. Resistance to quinclorac in rice fields in southern Brazil. Planta Daninha 2007; 25(1) 221-226.

[5] Galon L, Agostinetto D, Moraes PVD, Dal Magro T, Panozzo LE, Brandolt RR, Santos LS. Economic threshold level for barnyardgrass *(Echinochloa* spp.*)* control decision in flooded rice *(Oryza sativa)*. Planta Daninha 2007; 25(4) 709-718.

[6] Galon L, Agostinetto D. Comparison of empirical models for predicting yield loss of irrigated rice (*Oryza sativa*) mixed with *Echinochloa* spp. Crop Protection 2009; 28(10) 825-830.

[7] Oerke EC. Crop losses to pests. The Journal of Agricultural Science 2006; 144(1) 31-43.

[8] Mortimer AM, Hill JE. Weed species shifts in response to broad-spectrum herbicides in sub-tropical and tropical crops. Brighton Crop Protection Conference 1999; 2(1) 425-437.

[9] Olofsdotter M, Navarez D, Rebulanan M, Streibig JC. Weed-suppressing rice cultivars - does allelopathy play a role? Weed Research 1999; 39(6) 441-454.

[10] Gressel J, Segel LA. Interrelating factors controlling the appearance of resistance: The outlook on the future. In: LeBaron, H.L. and Gressel, J. (eds.) Herbicide Resistance in Plants. New York: Wiley; 1982. p325-347.

[11] Prather TS, Ditomaso JM, Holt JM. Herbicide Resistance: Definition and Management Strategies. University of California, Division of Agriculture and Natural Resources, Publication 8012. http://anrcatalog.ucdavis.edu/pdf/8012.pdf (accessed 10 October 2012).

[12] Gealy DR, Mitten DH, Rutger JN. Gene flow between red rice (*Oryza sativa*) and herbicide-resistant rice (*O. sativa*): implications for weed management. Weed Technology 2003; 17(3) 627–645.

[13] Shivrain VK, Burgos NR, Anders MM, Rajguru SN, Moore J, Sales MA. Gene flow between Clearfield rice and red rice. Crop Protection 2007; 26(3) 349–356.

[14] Goulart ICG, Pacheco MT, Nunes AL, Merotto Jr. A. Identification of origin and analysis of population structure of field-selected imidazolinone-herbicide resistant red rice (*Oryza sativa*). Euphytica 2012; 187(3) 437-447.

[15] Menezes VG, Mariot CHP, Kalsing A, Goulart ICGR. Red rice (*Oryza sativa*) resistant to the herbicides imidazolinones.Planta Daninha 2009; 27(special issue) 1047-1052.

[16] Noldin JA, Eberhardt DS, Knoublauch R. Resistência de *Saggitaria montevidensis* a herbicidas: primeiras evidências. In: Embrapa Clima Temperado (ed.): proceedings of the I Congresso Brasileiro de Arroz Irrigado, 1-4 August 1999, Pelotas, Brazil, Pelotas: Embrapa Clima Temperado; 1999.

[17] Noldin JA, Eberhardt DS, Knoublauch R. Sagitária resistente a herbicidas inibidores da enzima ALS, In: SBCPD (ed.): proceedings of the Congresso Brasileiro da Ciência das Plantas Daninhas, 6-11 June 2000, Foz do Iguaçú, Brazil, Foz do Iguaçú: SBCD: 2000.

[18] Merotto Jr A, Vidal RA, Fleck NG, Reis B, Andres A. Resistência de *Echinochloa* sp à quinclorac. In: SBCPD (ed.): proceedings of the Congresso Brasileiro da Ciência das Plantas Daninhas, 6-11 June 2000, Foz do Iguaçú, Brazil, Foz do Iguaçú: SBCPD: 2000.

[19] Kissmann KG, Groth D. Plantas infestantes e nocivas. São Paulo: BASF; 1997.

[20] Lopez-Martinez N, Salva AP, Finch RP, Marshall G, Prado RD. Molecular markers indicate intraspecific variation in the control of *Echinochloa* spp. with quinclorac. Weed Science 1999; 47(3) 310-315.

[21] Andres A, Concenço G, Theisen G, Galon L, Tesio F. Management of red rice *(Oryza sativa)* and barnyardgrass (*Echinochloa crus-galli*) grown with sorghum with reduced rate of atrazine and mechanical methods. Experimental Agricultural 2012; 48(4) 587–596.

[22] Ruiz-Santaella JP, Fischer AJ, De Prado R. Alternative control of two biotypes of *Echinochloa phyllopogon* susceptible and resistant to fenoxaprop-ethyl. Communications in Agricultural and Applied Biological Sciences 2003; 68(4) 403-407.

[23] Lopez-Martinez N, Marshall G, De Prado R. Resistance of barnyardgrass (*Echinochloa crus-galli*) to atrazine and quinclorac. Pesticide Science 1997; 51(2) 171-175.

[24] Mariot CHP, Menezes VG, Souza PA. Resistência múltipla e cruzada de capim-arroz a herbicidas na cultura de arroz irrigado no Rio Grande do Sul. In: SBCPD (ed.): proceedings of the Congresso Brasileiro da Ciência das Plantas Daninhas, 19-23 July 2010, Ribeirão Preto, Brazil. Londrina: SBCPD; 2010.

[25] Eberhardt DS, Noldin JA, Gutierez M, Dittrich RC. Resistência de capim-arroz (*Echinochloa crusgalli*) ao herbicida quinclorac. In: SBCPD (ed.): proceedings of the Congresso Brasileiro da Ciência das Plantas Daninhas, 6-11 June 2000, Foz do Iguaçú, Brazil, Foz do Iguaçú: SBCD: 2000.

[26] Mariot CHP, Rubin R, Celmer A, Tormen N. Controle de capim-arroz resistente a imidazolinonas com a associação de Ricer + Clincher em arroz irrigado no Rio Grande do Sul. In: Epagri/SOSBAI (eds.): proceedings of the VII Congresso Brasileiro de Arroz Irrigado, August 2011. Balneário Camboriú, Brazil. Itajaí: Epagri/SOSBAI; 2011.

[27] Noldin JA, Eberhardt DS, Rampelotti FT, Zunino J, Concenço G. Freqüência de plantas de *Sagittaria montevidensis* resistentes ao herbicida Only. In: SBCPD (ed.): proceedings of the Congresso Brasileiro da Ciência das Plantas Daninhas, 24-28 May 2004, São Pedro, Brazil. Londrina: SBCPD; 2004.

[28] Gibson KD, Breen JL, Hill JE. California arrowhead is a weak competitor in water-seeded rice. Weed Science 2001; 49(3) 381-384.

[29] Merotto Jr A, Kupas V, Nunes AL, Goulart ICGR. Isolamento do gene ALS e investigação do mecanismo de resistência a herbicidas em *Sagittaria montevidensis*. Ciência Rural 2010; 40(11) 2381-2384.

[30] Concenco G, Noldin JA, Lopes N F, Comiotto A. Aspectos da resistência de *Sagittaria montevidensis* ao herbicida pirazosulfuron-ethyl inibidor da ALS. *Planta Daninha* 2007; 25 (1)187-194. http://www.scielo.br/scielo.php?script=sci_arttext&pid=S0100-83582007000100021&lng=en&nrm=iso (access on 22 October 2012).

[31] Eberhardt DS, Noldin JA. Multiple Herbicide-Resistant *Sagittaria montevidensis* Population in Santa Catarina State (Brazil) Rice Fields. WSSA Abstracts, 2011. http://

www.weedscience.org/Case/Reference.asp?ReferenceID=1166 (accessed 22 October 2012).

[32] Menezes VG, Kalsing A, Felin JP. Controle químico de *Sagittaria montevidensis* (SAG-MO) em áreas de arroz cultivadas no sistema pré-germinado. In: Epagri/SOSBAI (eds.): proceedings of the VII Congresso Brasileiro de Arroz Irrigado, 9-12 August 2011. Balneário Camboriú, Brazil. Itajaí: Epagri/SOSBAI; 2011.

[33] Sociedade Sul-Brasileira de Arroz Irrigado – SOSBAI. Arroz irrigado: recomendações técnicas da pesquisa para o Sul do Brasil. Porto Alegre: SOSBAI, 2012. http://www.sosbai.com.br/BoletimRecomendacoesTecnicas_2012.zip (accessed 22 October 2012).

[34] Ferreira FB, Pinto JJO, Sperandio CA, Lamego FP, Resende AL, Lazaroto CA, Galon L. Influência da população de arroz na infestação de sagitária. In: SBCPD/Embrapa Clima Temperado (eds.): proceedings of the Congresso Brasileiro da Ciência das Plantas Daninhas, 29 July – 01 August 2002, Gramado, Brazil. Pelotas: SBCPD/Embrapa Clima Temperado; 2002.

[35] Noldin JA, Eberhardt DS, Rampelotti FT. *Cyperus difformis* L. resistente a herbicidas inibidores da ALS em Santa Catarina. In: SBCPD/Embrapa Clima Temperado (eds.): proceedings of the Congresso Brasileiro da Ciência das Plantas Daninhas, 29 July – 01 August 2002, Gramado, Brazil. Pelotas: SBCPD/Embrapa Clima Temperado; 2002.

[36] Galon L, Panozzo LE, Noldin JA, Concenço G, Tarouco CP, Ferreira EA, Agostinetto D, Silva AA, Ferreira FA. Herbicide Resistance of *Cyperus difformis* to ALS-Inhibitors in Paddy Rice of Santa Catarina. Planta Daninha 2008; 26(2) 419-427. http://www.scielo.br/scielo.php?script=sci_arttext&pid=S0100-83582008000200019&lng=en&nrm=iso (accessed 25 October 2012)

[37] Saari LL, Cotterman JC, Thill DC. Resistance to acetolactate synthase inhibiting herbicides. In: Powles SB, Holtur, JAM. (eds.) Herbicide resistance in plants: biology and biochemistry. Boca Raton: Lewis; 1994. p83-139.

[38] Heap, I. International Survey of Herbicide Resistant Weeds. http://www.weedscience.org (accessed 22 October 2012).

[39] Dal Magro T, Santos L, Schaedler CE, Agostinetto D, Vargas L, Noldin JA. Dose resposta de pyrazosulfuron-ethyl em biotipos de *Cyperus difformis* L. resistente e suscetível. In: SOSBAI/IRGA (eds.): proceedings of the VI Congresso Brasileiro de Arroz Irrigado, 11-14 August 2009. Porto Alegre, Brazil. Porto Alegre: SOSBAI/IRGA; 2009.

[40] Ulguim AR, Agostinetto D, Vargas L, Manica-Berto R, Westendorff N, Rubin R, Danielowski H. Resistência de *Cyperus iria* l. (CYPIR) aos inibidores de acetolactato sintase (ALS) no Rio Grande do Sul. In: Epagri/SOSBAI (eds.): proceedings of the VII Congresso Brasileiro de Arroz Irrigado, 9 – 12 August 2011. Balneário Camboriú, Brazil. Itajaí: Epagri/SOSBAI; 2011.

[41] Scherer MB, Dornelles, SHB, Sanchotene, DM, Macedo LCP de, Espíndola EFS, Cirolini AN. Manejo químico alternativo de *Cyperus iria* resistente aos herbicidas inibidores da enzima ALS. In: Epagri/SOSBAI (eds.): proceedings of the VII Congresso Brasileiro de Arroz Irrigado, 9 – 12 August 2011. Balneário Camboriú, Brazil. Itajaí: Epagri/SOSBAI; 2011.

[42] Kissmann KG. (ed.). Plantas infestantes e nocivas. São Paulo: BASF Brasileira S.A.; 2007.

[43] Fleck NG, Agostinetto D, Galon L, Schaedler CE. Relative competitivity among flooded rice cultivars and a red rice biotype. Planta Daninha 2008; 26(1)101–111

[44] Menezes VG, Silva PRF. Manejo de arroz-vermelho através do tipo e arranjo de plantas em arroz irrigado. Planta Daninha 1998; 16(1) 45-57.

[45] Agostinetto D, Fleck NG, Rizzardi, MA, Merotto Jr A, Vidal RA. Red rice: Ecophysiology and Strategies of control. Ciência Rural 2001; 31(2) 341-349.

[46] Croughan TP. Application of tissue culture techniques to the development of herbicide-resistant rice. Louisiana Agriculture 1994; 37(3) 25–26.

[47] Zhang W, Linscombe SD, Webster E, Tan S, Oard J. Risk assessment of the transfer of imazethapyr herbicide tolerance from Clearfield rice to red rice (*Oryza sativa*). Euphytica 2006; 152(1) 75-86.

[48] Shivrain VK, Burgos NR, Sales MA, Mauromoustakos A, Gealy DR, Smith KL, Black HL, Jia M. Factors affecting the outcrossing rate between ClearfieldTM rice and red rice (*Oryza sativa*). Weed Science 2009; 57(4):394–403.

[49] Magalhães Jr AM, Franco DF, Andres A, Antunes P, Luzzardi R, Dode LB, Tillmann MAA, Silva MP. Método para identificação de sementes de arroz transgênico resistente ao herbicida glufosinato de amônio. Agropecuária Clima Temperado 2000; 3(1) 31-38.

[50] Ramírez HB. Polinização cruzada em arroz irrigado. Doctoral thesis. Universidade Federal de Pelotas, Brazil; 2003.

[51] Villa SCC, Marchezan E, Avila LA, Massoni PFS, Telo GM, Machado SLO, Camargo ER. Arroz tolerante a imidazolinonas: controle do arroz-vermelho, fluxo gênico e efeito residual do herbicida em culturas sucessoras não-tolerantes. Planta Daninha 2006; 24(4) 761-768.

[52] Roso AC, Merotto Jr A, Delatorre A, Menezes VG. Regional scale distribution of imidazolinone herbicide-resistant alleles in red rice (*Oryza sativa* L.) determined through SNP markers. Field Crops Research 2010; 119(1) 175-182.

[53] Burgos NR, Norsworthy JK, Scott RC, Smith KL. Red rice (*Oryza sativa*) status after 5 years of imidazolinone resistant rice technology in Arkansas. Weed Technology 2008; 22() 200–208.

[54] Menezes VG, Ramirez H. Resistance *Echinochloa crus-galli* L. to quinclorac in flooded rice in southern Brazil. In: IWSC (ed.) proceedings of the III International Weed Science Congress, 6 - 11 June 2000, Foz do Iguaçu, Brazil. Corvalis: IWSC; 2000.

[55] Menezes VG, Mariot CHP, Oliveira CAO, Kalsing A, Soares DC. Resistência de capim-arroz a herbicidas do grupo químico das imidazolinonas no sul do Brasil. In: SOSBAI/IRGA (eds.): proceedings of the VI Congresso Brasileiro de Arroz Irrigado, 11-14 August 2009. Porto Alegre, Brazil. Porto Alegre: SOSBAI/IRGA; 2009.

[56] Ulguim AR, Westendorf N, Noldin JA, Agostinetto D, Manica-Berto R, Ludtke R, Thurmer L. Resposta de biótipos de *Echinochloa crusgalli* (L.) Beauv. resistentes e suscetível aos inibidores de ALS. In: Epagri/SOSBAI (eds.): proceedings of the VII Congresso Brasileiro de Arroz Irrigado, 9 – 12 August 2011. Balneário Camboriú, Brazil. Itajaí: Epagri/SOSBAI; 2011.

[57] Dornelles SHB, et al. Controle pré-emergente de *Cyperus iria* resistente a herbicidas inibidores da enzima ALS. In: Epagri/SOSBAI (eds.): proceedings of the VII Congresso Brasileiro de Arroz Irrigado, 9 – 12 August 2011. Balneário Camboriú, Brazil. Itajaí: Epagri/SOSBAI; 2011.

[58] Dornelles SHB, et al. *Cyperus iria* resistente a herbicidas inibidores da enzima Aceto Lactato Sintase. In: Epagri/SOSBAI (eds.): proceedings of the VII Congresso Brasileiro de Arroz Irrigado, 9 – 12 August 2011. Balneário Camboriú, Brazil. Itajaí: Epagri/SOSBAI; 2011.

[59] Concenço G, Melo PTBS, Ferreira EA, Silva AF, Aspiazú I, Galon L, Ferreira FA, Silva AA, Noldin JA. Competitividade de biótipos de capim-arroz resistente e suscetível ao quinclorac. Planta Daninha 2008; 26(1) 195-202.

[60] Mariot CHP, et al. Resistência múltipla e cruzada de capim-arroz a herbicidas na cultura do arroz irrigado no Rio Grande do Sul. In: SBCPD (ed.): proceedings of the Congresso Brasileiro da Ciência das Plantas Daninhas, 19-23 July 2010, Ribeirão Preto, Brazil. Londrina: SBCPD; 2010. [CD-ROM].

[61] Merotto Jr A, Kupas V, Nunes AL, Costa RF. Resistência de Capim-arroz (*Echinochloa crusgalli*) aos herbicidas inibidores da enzima ALS. In: SOSBAI/IRGA (eds.): proceedings of the VI Congresso Brasileiro de Arroz Irrigado, 11-14 August 2009. Porto Alegre, Brazil. Porto Alegre: SOSBAI/IRGA; 2009. http://www.sosbai.com.br/admin/artigos/bk20100528133117.pdf (accessed 12 October 2012).

[62] Perboni LT, et al. Controle de capim-arroz resistente e suscetível à ALS pela aplicação de herbicidas em diferentes épocas. In: Epagri/SOSBAI (eds.): proceedings of the VII Congresso Brasileiro de Arroz Irrigado, 9 – 12 August 2011. Balneário Camboriú, Brazil. Itajaí: Epagri/SOSBAI; 2011.

[63] Noldin JA, Eberhardt DS, Andrade S, Pinheiro GF. Capim-arroz com resistência múltipla a herbicidas em Santa Catarina. In: SOSBAI/IRGA (eds.): proceedings of the VI Congresso Brasileiro de Arroz Irrigado, 11-14 August 2009. Porto Alegre, Brazil. Por-

to Algre: SOSBAI/IRGA; 2009. http://www.sosbai.com.br/admin/artigos/bk20100528133117.pdf (accessed 12 October 2012).

[64] Pinto JJO, Noldin JA, Donida A da C, Piveta LB, Pinho CF de, Pohlmann TS, Batista DD. Controle de capim-arroz em áreas de arroz com suspeita da presença de biótipos resistentes a herbicidas inibidores da ALS. In: SOSBAI/IRGA (eds.): proceedings of the VI Congresso Brasileiro de Arroz Irrigado, 11-14 August 2009. Porto Alegre, Brazil. Porto Alegre: SOSBAI/IRGA; 2009.

[65] Theisen G, Andres A. Tolerância de capim-arroz (*Echinochloa crus-galli* spp.) ao herbicida imazetapir + imazapic em arrozais da região Sudeste do RS. Pelotas, Brazil: Embrapa Clima Temperado; 2010. http://www.cpact.embrapa.br/publicacoes/download/comunicados/comunicado_253.pdf (accessed 22 October 2012).

[66] Rampelotti FT, et al. Monitoramento da resistência de *Cyperus difformis* e *Fimbristylis miliacea* aos herbicidas inibidores de ALS em Santa Catarina. In: SBCPD (ed.): proceedings of the Congresso Brasileiro da Ciência das Plantas Daninhas, 24-28 May 2004, São Pedro, Brazil. Londrina: SBCPD; 2004.

[67] Noldin JA, Eberhardt DS, Rampelotti FT. *Fimbristylis miliacea* (L.) Vahl resistente a herbicidas inibidores da ALS em Santa Catarina. In: SBCPD/Embrapa Clima Temperado (eds.): proceedings of the Congresso Brasileiro da Ciência das Plantas Daninhas, 29 July – 01 August 2002, Gramado, Brazil. Pelotas: SBCPD/Embrapa Clima Temperado; 2002.

[68] Empresa de Pesquisa Agropecuária e Extensão Rural de Santa Catarina – EPAGRI. Sistema de produção de arroz irrigado em Santa Catarina (pré-germinado). Florianópolis, Brazil: EPAGRI; 2010.

[69] Marchezan E, Oliveira APBB, Avila LA, Bundt ALP. Red rice seed bank dynamics affected by cattle trampling and fallow duration. Planta Daninha 2003; 21(1) 55-62.

[70] Valverde BE. Status and management of grass-weed herbicide resistance in Latin America. Weed Technology 2007; 21(2) 310-323.

[71] Chaves ICPV, Garcia L. Avaliação da combinação de Aurora 400 CE + Gamit 500 CE, aplicada em mistura com glifosato, na dessecação de erva-de-bicho (*Polygonum persicaria*) e seu efeito residual no controle de capim-arroz (*Echinochloa* sp.). In: SOSBAI (ed.): proceedings of the IV Congresso Brasileiro de Arroz Irrigado and XXVI Reunião da Cultura do Arroz Irrigado, 9 – 12 August 2005. Santa Maria, Brazil. Santa Maria: Orium/SOSBAI; 2005. [CD-ROM].

[72] Riches CR, Knights JS, Chaves L, Caseley JC, Valverde BE. The role of pendimethalin in the integrated management of propanil-resistant Echinochloa colona in Central America. Pesticide Science 1997; 51 (3) 341-346.

[73] Valverde BE, Riches CR, Caseley JC. Prevention and management of herbicide-resistant weeds in rice: experiences from Central America with *Echinochloa colona*. Costa Rica: Cámara de Insumos Agropecuarios; 2000.

[74] Andres A, Machado SLO. Plantas daninhas em arroz irrigado. In: Gomes AS, Magalhães Jr AM (eds.) Arroz irrigado no sul do Brasil. Brasília: Embrapa; 2004. p.611-726.

[75] Gill GS, Holmes JE. Efficacy of Cultural Control Methods for Combating Herbicide-Resistant *Lolium rigidum*. Pesticide Science 1997; 51(3) 352-358.

[76] Vencill W, Grey T, Culpepper S. Resistance of Weeds to Herbicides, Herbicides and Environment, Dr Andreas Kortekamp (Ed.). Rijeka: InTech; 2011. http://www.intechopen.com/books/herbicides-and-environment/resistance-of-weeds-to-herbicides (accessed 22 October 2012).

Chapter 2

Cotton (*Gossypium hirsutum* L.) Response to Pendimethalin Formulation, Timing, and Method of Application

Timothy Grey and Theodore Webster

Additional information is available at the end of the chapter

http://dx.doi.org/10.5772/56184

1. Introduction

The introduction of glyphosate-resistant cotton for production in the southeast United States changed herbicide application strategies and increased the profitability of no-tillage and strip-tillage techniques. Glyphosate (*N*-[phosphonomethyl]-glycine) is a highly effective herbicide that controls a broad spectrum of annual and perennial grass and broadleaf weeds in cotton [3, 37]. When glyphosate-resistant cotton varieties were first introduced, glyphosate was applied two to four times on most fields and may have been the only herbicide used [4, 5]. In Georgia, 93% of the cotton acres received at least one glyphosate application in 2005 [3]. The technology allowed growers to reduce or eliminate soil-applied herbicides, allowing them to abandon cultivation and make the transition to conservation tillage, which promotes soil conservation and compliance with USDA Federal regulations. Greater than 50% of Georgia cotton was produced using no-tillage or strip-tillage techniques in 2007, a strategy that has been affected by glyphosate weed control [1, 11].

2. Importance

With the elimination of cultivation as a control tactic in conservation tillage systems, herbicides were the primary and often only method used for weed control [24]. However, the incidence of herbicide-tolerant or resistant weeds emerging in the southeast United States [33, 34] has increased the need for multiple herbicide modes of action in both conservation tillage and conventional tillage weed management systems [3, 5, 16]. In Georgia, there are populations of Palmer amaranth (*Amaranthus palmeri* S. Wats.) (Figure 1) with resistance to glyphosate, ALS, triazines, dinitroanilines, with some populations demonstrating resistance to multiple mechanisms of action [5, 26, 31, 38]. While glyphosate- and ALS-resistant Palmer amaranth is widespread in Georgia, the frequency and distribution of triazine- and dinitroaniline-resistant has not been characterized in Georgia. With the potential mobility of herbicide resistance traits, through movement of pollen [27, 28] or seed [18] and/or potentially high levels of naturally occurring mutations conferring resistance, cotton production in the region is threatened by herbicide resistant weeds.

Figure 1. Glyphosate resistant Palmer amaranth in conventional upland cotton in Georgia.

The increased occurrence of herbicide-resistant weeds necessitates the search for alternative control tactics. For instance, metolachlor had not been traditionally used in cotton because of excessive crop injury when applied preemergence after planting. However, changing its use pattern to be applied after cotton emergence avoided crop injury, while controlling an exotic weed that had become troublesome [4]. This technology and new mechanism of action has been instrumental in current management of glyphosate-resistant Palmer amaranth. Research on a new use pattern for pendimethalin may provide an additional tool for weed management at different times in the growing season.

3. Background information on soil applied herbicides

Herbicides with soil persistence and weed control activity were extensively used for pre-emergence weed control in cotton until the commercial release of herbicide-resistant cotton in 1997. Cotton herbicides with soil residual properties included cyanazine (2-((4-chloro-6-(ethylamino)-1,3,5-triazin-2-yl]amino]-2-methylpropanenitrile), diuron (*N*'-(3,4-dichloro-phenyl)-*N,N*-dimethylurea), flumeturon (*N,N*-dimethyl-*N'*(3-(trifluoromethyl)phenyl]urea), pendimethalin (N-(1-ethylpropyl)-3,4-dimethyl-2,6-dinitrobenzenamine), trifluralin (2,6-dinitro-N,N-dipropyl-4-(trifluoromethyl(benzenamine), and others. Pendimethalin was registered for cotton in 1975 [22]. These herbicides were applied pre-plant soil incorporated (PPI), pre-emergence (before cotton and weed emergence) and/or post-directed (where applications are directed to the soil and bottom portion of the stems of mature cotton plants). Cotton in the southeastern U.S. has a growing season that can extend to over 150 days ranging from late March to early November. Growers can PRE apply pendimethalin but have to PPI trifluralin. This allows conservation tillage cotton growers an option to use a dinitroaniline herbicide for grass and small seeded broadleaf weed control. A weakness in weed efficacy of these residual herbicides was the lack of extended weed control due to dissipation of the herbicide in the soil. With the introduction and high rate of adoption of glyphosate-resistant cotton varieties and almost exclusive use of glyphosate for weed control, the herbicides with soil residual activity was reduced in favor of total post-emergence weed control programs. The cotton registration for cyanazine was eventually canceled in 2002 in the United States. However, even with increased herbicide-resistant weeds in growers' fields in the first decade of the 2000's, diruon, flumeturon, and pendimethalin use did not increase, even though residual herbicides could improve weed control (Figure 2). Diuron and flumeturon are widely applied to cotton as post-directed sprays in this region. However, growers using conservation tillage practices in cotton often rely on pendimethalin for early season residual weed control with preemergence applications either sprayed or impregnated on fertilizers.

3.1. Pendimethalin

Pendimethalin is a member of the dinitroanaline family of herbicides. Pendimethalin prevents plant cell growth by inhibiting spindle formation during cell division [6]. Pendimethalin is applied PRE to the soil surface, with or without incorporation into the soil, to approximately 37% of Georgia cotton [17] for control of grasses and small-seeded broadleaf weed species [2]. Pendimethalin inhibits mitotic cell division in susceptible plants [30], while tolerant crops grow through, or are planted below, the treated zone [13, 14]. Among the dinitroanaline herbicides, pendimethalin has greater water solubility of 0.275 ug mL^{-1} and less volatility at 9.4×10^{-6} mm Hg at 25 C [22], allowing it to be applied to the soil surface rather than needing mechanical incorporation [35, 36]. However, pendimethalin still requires moisture in the form of rainfall or irrigation in order to move it into the active zone of weed germination. Cotton selectivity of pendimethalin pre-emergence is due to differences in metabolism and sequestration of pendimethalin in the lysigenous glands [25]. Pendimethalin is registered for PRE application up to 2 days after cotton planting. However, delayed application in combination with excessive moisture (rainfall or irrigation) can result in injury to seedling cotton. Pendimethalin injury to

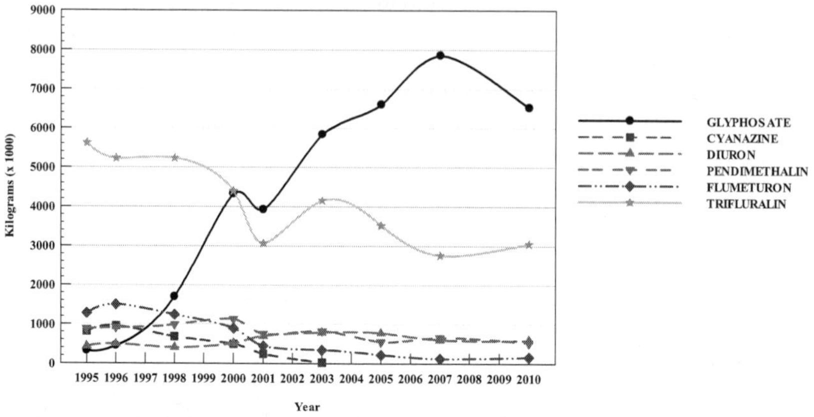

Figure 2. Residual cotton herbicides use as compared to glyphosate in United States cotton production since the advent of glyphosate resistant cotton [17].

cotton seedlings results in delayed hypocotyl development and can also cause abnormal root growth. This injury is commonly associated with enlarged lower stems and 'bottle brush' root development. Microbial decomposition is the main method of pendimethalin dissipation [19, 32]. While pendimethalin has a reported soil half-life of 74 to 114 days [30], surface applied half-lives of 4 to 6 days can occur due to volatilization, photo-chemical, and other degradation processes [21]. Additionally, increased degradation can occur with no-tillage application [9].

3.2. Pendimethalin weed control

Pendimethalin is often used in cotton to supplement control of grass weeds and small-seeded broadleaf weed species. According to the University of Georgia Extension recommendations, pendimethalin provides excellent (90%) control of crabgrass (*Digitaria sanguinalis* (L.) Scop.), crowfootgrass (*Dactyloctenium aegyptium* (L.) Willd.), foxtails (*Setaria* species), goosegrass (*Eleusine indica* (L.) Gaertn.), seedling johnsongrass (*Sorghum halepense* (L.) Pers.), and sandbur (*Cenchrus echinatus* L.); good control (80-90%) of fall panicum (*Panicum dichotomiflorum* Michx.) and Texas millet (*Urochloa texana* (Buckl.) R. Webster). Pendimethalin also provides excellent (90%) to good (80-90%) control of the broadleaf species Florida pusley (*Richardia scabra* L.), pigweeds (*Amaranthus* species), lambsquarters (*Chenopodium album* L.), and pink purslane (*Portulaca pilosa* L.); and fair to good (60-90%) control of Palmer amaranth.

3.3. Pendimethalin formulation

There are two liquid formulations of pendimethalin registered for cotton in the United States. One contains 37.4% pendimethalin (0.41 kg ai/L) formulated with aromatic naphtha as an emulsifiable concentrate (EC), and the other contains 38.7% pendimethalin (0.47 kg ai/L) formulated as a microencapsulated (ME) aqueous capsule suspension [12] (Figure 3). One potential method of obtaining extended weed control to apply pendimethalin as an in-season application, i.e. from emergence to when the cotton crop has up to six leaves, or just prior to canopy formation. However, injury to cotton from the EC formulation has prevented topical applications in the past.

(Photo courtesy Sidney Cromer, University of Georgia).

Figure 3. Pendimethalin microencapsulated aqueous capsule suspension (left) and pendimethalin emulsifiable concentrate (right)

3.4. Research

Cotton response to pendimethalin ME applied at different growth stages is less injurious to cotton because of its formulation. An alternative method of application is to impregnate pendimethalin onto fertilizer for in-season application to extend residual weed control, reducing the number of herbicide applications [15, 20], and minimizing potential crop injury. Crop injury has been noted with pendimethalin EC and ME when applied topically to cotton at the 4th leaf growth stage [7] and its effects on cotton nutrient uptake [10]. Weed control for comparing pendimethalin EC to ME in cotton have been made using spray applications [11]. Florida pusley and Texas millet control were similar and consistent for PRE applied EC and ME formulations (Table 1). While weed control has been evaluated, cotton crop response to applications made PRE up to the 6th leaf growth stage comparing season- long factors is also needed. Therefore, this chapter will emphasize pendimethalin use, formulation (EC and ME), and cotton response. Additionally, this chapter will focus on pendimethalin formulations when applied as an aqueous solution in water or impregnated on fertilizers [15].

Formulation	Application method	Timing	Texas millet	Florida pusley
			%	
Pendimethalin EC	Spray	PRE	75	66
Pendimethalin ME	Spray	PRE	75	68

[a]Abbreviations: EC, emulsifiable concentrate; ME, microencapsulated; PRE, prior to plant emergence.

Table 1. Weed control in Georgia cotton with pendimethalin EC[a] and ME[a] formulations applied at planting.

4. Studies

4.1. Field studies

Field trials were conducted in 2005, 2006, and 2007 at the University of Georgia Ponder Research Station near Ty Ty, Georgia. Soil was Tifton loamy sand (fine-loamy, kaolinitic, thermic Plinthic Kandiadults) with 83% sand, 12% silt, 5% clay, organic matter content of 1 to 1.8%, and pH of 5.6 to 6.1. Conventional tillage was used during all three years of the study to obtain optimal herbicide/soil contact, since pendimethalin has been observed to adsorb to cover crop residue [9]. Delta and Pineland 555 BG/RR was planted in 2005 and Delta and Pineland Flex 445 BG/RR in 2006 and 2007 using a Monosem precision vacuum planter set to deliver 14 seeds per linear meter of row with 0.9 m between row centers. The experimental design was a two factor randomized complete block with treatments replicated four times. Plots were 1.8 m (two rows) wide by 8 m long. Four different methods of pendimethalin application were made at four different timings during the growing season. All herbicide treatments consisted of 1.1 kg active ingredient/ha of pendimethalin EC or ME. Only the method or time of application varied. Treatments were pendimethalin EC or ME applied as either an aqueous solution in water, or impregnated on fertilizer (10-10-10) that was applied at 280 kg ha^{-1} with a Gandy fertilizer applicator (Figure 4). All herbicide spray treatments were made with a CO_2-pressurized backpack sprayer using Teejet 11002 flat fan nozzles, which delivered 140 L/ha of water at 130 kPa. For the fertilizer treatment, pendimethalin EC or ME at 1.1 kg active ingredient ha^{-1} was impregnated on fertilizer using a CO_2–pressurized sprayer with a Teejet 8002 flat fan nozzle at 130 kPa. Fertilizer was rotated at a constant speed of 12 meter minute^{-1} using a rotating steel drum. The drum freely rotated on a twin roller rod system set at a 30º angle, powered by an electric motor, with speed adjusted by a rheostat (Figure 5).

http://dx.doi.org/10.5772/56184

Figure 4. Pendimethalin impregnated fertilizer treatment on soil surface (left) and application (right).

Figure 5. Fertilizer prior to (left) and after (right) treatment with pendimethalin formulation Prowl 3.3EC.

All plots received the same fertilizer rates to ensure no variability for fertility. Plots were then irrigated the day after treatments were applied. Treatments were made at four different application timings, at planting prior to plant emergence (PRE), at seedling emergence (AE), to 3rd leaf, or to 6th leaf cotton. A non-treated control was included for comparison for a total of 17 treatments. All plots were maintained weed free by hand pulling weed escapes and treatments with glyphosate. Other cultural and pest management practices were based upon recommendations by the Georgia Cooperative Extension Service. Supplemental overhead sprinkler irrigation was applied as needed. Cotton injury ratings were evaluated after applications using a scale of 0 (no injury) to 100 % (plant death) [8]. Cotton height measures were made up to five times in 2005, 2006 and 2007. Both rows of each plot were harvested with a spindle picker, and seed cotton yield was quantified. Data were subjected to mixed model ANOVA using Proc Mixed in SAS 9.1, with random effects of years and replications. Mean separation was determined using the PDMIX800 macro. Regression analysis was performed using Sigmaplot 12 nonlinear regression. The intent was to determine if the response could be described by using the exponential growth, Stirling Model.

$$y = y0 + \frac{a(e^{bx}) - 1]}{b} \quad (1)$$

Where y is the response variable of treatment, $y0$ is the value of the response variable (y) when X is equal to zero, a is the rate of growth, and X is time in days. Data for growth were analyzed by ANOVA under the general linear models procedure and used mean separation of 95% asymptotic confidence intervals for comparison of parameter estimates.

4.2. Laboratory studies

Fertilizer samples were taken prior to and after treatment with EC and ME pendimethalin. Samples were viewed at ×125 and ×200 magnification with a light microscope. Images were captured with a digital camera with image analysis software. Figure 6 notes the smooth surface for the EC formulations verses the course texture of the ME formulation alone and when impregnated on fertilizer.

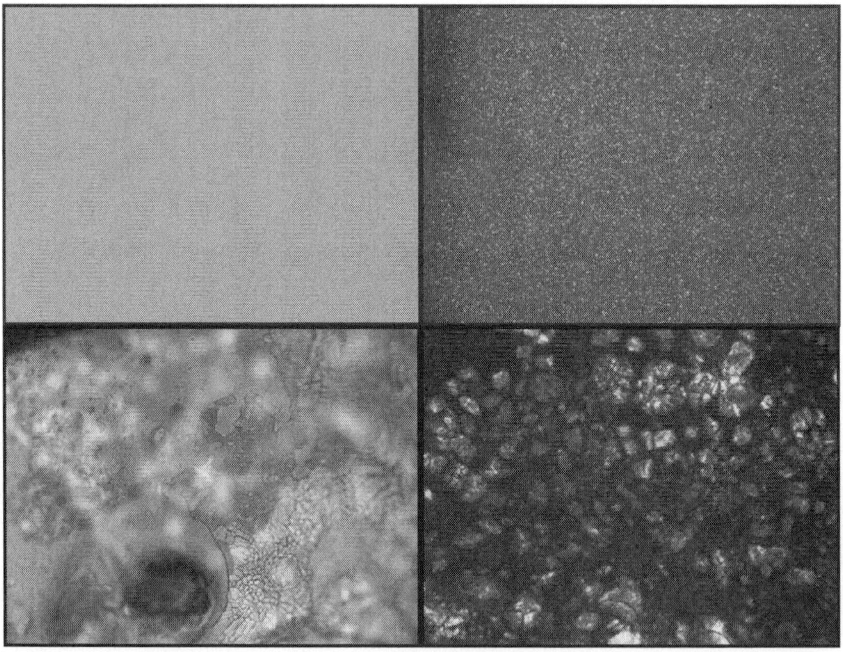

Figure 6. Pendimethalin EC (top left) and ME (top right) formulations alone (x125 light microscope magnification), and EC (bottom left) and ME (bottom right) impregnated on fertilizers (x200).

5. Cotton response

There were significant formulation by application method, application method by timing, and formulation by timing interactions for cotton plant injury and cotton yield. Since the non-treated control had no associated timing effects and did not differ significantly in cotton yield or injury from the PRE applications (Table 2), comparisons of injury and yield included only the treated plots to simplify the model.

5.1. Cotton injury

Spray applications of pendimethalin EC resulted in greater crop injury (27%) than when pendimethalin EC was applied with fertilizer (12%) or both application methods of pendimethalin ME (≤12%) (Table 2). Pendimethalin on fertilizer applied at the 3rd leaf stage and both application methods applied PRE or the 6th leaf stage of cotton had lower levels (≤7%) of cotton injury than all other treatments. For PRE applications, pendimethalin injury in the form of stunting, leaf curl, leathery cotyledons, swollen hypocotyl, and intense green color were observed, but this did not affect plant establishment, confirming previous results [14]. There was similar and significant injury when pendimethalin (Figure 7) was applied as cotton emerged (AE) with both the fertilizer (27%) and spray (42%) application and when sprayed at the 3rd leaf stage (27%). Previous reports of cotton injury resulting from a topical application of pendimethalin ME at the 4th leaf growth stage (≤20%) was lower than that from pendimentalin EC (≤33%) [7]. When averaged over application method, there was minimal cotton injury when either pendimethalin formulation was applied PRE or at the 6th leaf stage. Greatest injury occurred when pendimethalin EC was applied AE (47%). At both the AE and 3rd leaf stage timings, pendimethalin ME caused less cotton injury than pendimethalin EC.

Figure 7. Cotton injury from pendimethalin EC (110) as compared to pendimethalin ME (112). Both rates were 1.1 kg active ingredient/ha at cotton emergence (AE) applied.

			Injury			LSD[a]
Formulation	Application	Timing	%			
Pendimethalin EC[bc]	Spray		27	a[d]	(4)[d]	7
	Fertilizer[f]		12	b	(4)	
Pendimethalin ME	Spray		12	b	(4)	
	Fertilizer		8	b	(4)	
	Spray	PRE	7	c	(5)	10
	Fertilizer	PRE	6	c	(5)	
	Spray	AE	42	a	(5)	
	Fertilizer	AE	27	b	(5)	
	Spray	3LF	27	b	(5)	
	Fertilizer	3LF	5	c	(5)	
	Spray	6LF	1	c	(5)	
	Fertilizer	6LF	3	c	(5)	
Pendimethalin EC		PRE	7	c	(5)	10
Pendimethalin ME		PRE	6	c	(5)	
Pendimethalin EC		AE	47	a	(5)	
Pendimethalin ME		AE	21	b	(5)	
Pendimethalin EC		3LF	24	b	(5)	
Pendimethalin ME		3LF	8	c	(5)	
Pendimethalin EC		6LF	3	c	(5)	
Pendimethalin ME		6LF	7	c	(5)	

[a]Because *proc Mixed* measures pair-wise differences, multiple LSDs may be obtained. In these cases, the LSD (α=0.05] included is the mean LSD for all treatments.

[b]Pendimethalin rates were 1.1 kg ai/ha for the EC and ME formulations.

[c] Abbreviations: EC, emulsifiable concentrate (0.41 kg ai/L); ME, microencapsulated (0.47 kg ai/L); PRE, prior to plant emergence; AE, at seedling emergence; 3LF, to 3rd leaf cotton; 6LF, 6th leaf cotton

[d]Means within a variable followed by the same letter are not significantly different using Fisher's protected $LSD_{(P=0.05)}$. Standard error of the mean for that treatment enclosed in ().

[f]Fertilizer [10 -10-10] rate was 280 kg/ha, with all plots equally treated. Pendimethalin EC and ME were spray impregnated.

Table 2. Interaction effects between pendimethalin formulation, application method, and application timing for injury in conventional tillage cotton.

5.2. Cotton height

There were no significant effects on cotton height during the year regardless of the pendimethalin formulation or application type (Figures 8 to 10). The pendimethalin EC formulation (Figure 8) and spray application (Figure 9) did reduce height at 45 days after planting, but this was not significant and was not observed by 75 days after planting for either scenario. Cotton height was reflected in the injury for the timing of application (Figure 10). No differences were noted in height for the 6th leaf treatment timings. While there was cotton injury and height reduction when pendimethalin EC was spray applied at the AE or 3rd leaf timings, cotton recovered and height measures were equivalent by the end of the season. Utilizing exponential growth Stirling model, all curves converged with the analysis at no greater than 14 iterations (data not presented) with no differences for parameter estimates (Tables 3, 4 and 5). The long growing season in tandem with cotton's physiological ability to compensate for early season injury essentially explains why growth models can be effectively used to predict the lack of net negative effects from early season injury from pendimethalin applications.

	Rate of cotton growth[b]					
Herbicide	***a*[c]**		**95% CL**	**b**		**95% CL**
Pendimethalin EC	0.0537	a	±0.0179	0.0513	a	±0.00555
Pendimethalin ME	0.0516	a	±0.0173	0.0514	a	±0.0056
Nontreated	0.0669	a	±0.0558	0.0471	a	±0.0140

[a]Each herbicide for first-order rate constants for each column followed by the same letter are not significantly different according to Fisher's protected LSD test (P≤0.05). General linear models procedures were used for mean separation with 95% asymptotic confidence intervals.

[b]Rates of cotton growth were calculated by nonlinear regression of the herbicide treatments with respect to time in days after planting.

[c]Abbreviations: *a*, rate of cotton growth; CL, confidence limit.

Table 3. Rate of cotton growth (*a*) as a response to pendimethalin formulation.[a]

	Rate of cotton growth[b]					
Application method	***a*[c]**		**95% CL**	**b**		**95% CL**
Fertilizer	0.0653	a	±0.0208	0.0485	a	±0.0053
Spray	0.0418	a	±0.0144	0.0545	a	±0.0057
Nontreated	0.0689	a	±0.0594	0.0463	a	±0.0145

[a]Each application method for first-order rate constants for each column followed by the same letter are not significantly different according to Fisher's protected LSD test (P≤0.05). General linear models procedures were used for mean separation with 95% asymptotic confidence intervals.

[b]Rates of cotton growth were calculated by nonlinear regression of the herbicide treatments with respect to time in days after planting.

[c]Abbreviations: *a*, rate of cotton growth; CL, confidence limit.

Table 4. Rate of cotton growth (*a*) as a response to method of pendimethalin application.[a]

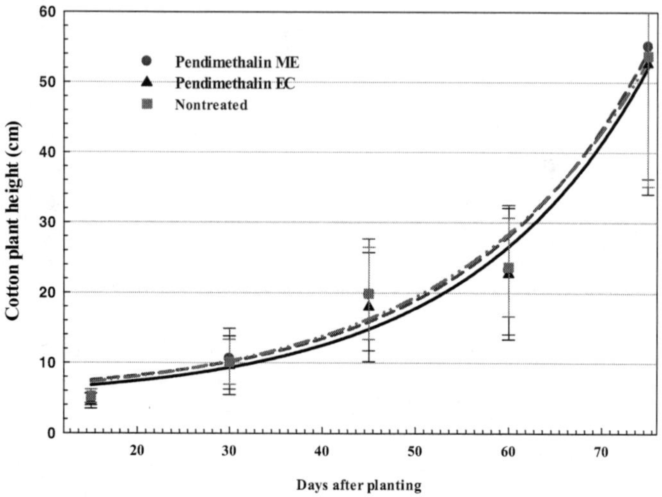

Figure 8. Cotton growth response as affected by pendimethalin formulation. The line represents the first-order regression equation. Data points are the means of replications with bars indicating the standard error of the mean:

$$\text{Fertilizer applied } y = 6.31 + \frac{0.0537\left[\left(e^{0.05137x}\right) - 1\right]}{0.05137} \quad P < 0.0001$$

$$\text{Spray applied } y = 5.62 + \frac{0.0516\left[\left(e^{0.0514x}\right) - 1\right]}{0.0514} \quad P < 0.0001$$

$$\text{Nontreated } y = 5.83 + \frac{0.0669\left[\left(e^{0.0471x}\right) - 1\right]}{0.0471} \quad P < 0.0001$$

	Rate of cotton growth[b]					
Application timing	***a*[c]**		**95% CL**	**b**		**95% CL**
Preemergence	0.1104	a	±0.0621	0.0395	a	±0.0067
At cotton emergence	0.0649	a	±0.0384	0.0488	a	±0.0099
3rd leaf cotton	0.0550	a	±0.0372	0.0512	a	±0.0112
6th leaf cotton	0.0415	a	±0.0319	0.0559	a	±0.0126
Nontreated	0.0689	a	±0.0594	0.0463	a	±0.0145

[a]Each application timing for first-order rate constants for each column followed by the same letter are not significantly different according to Fisher's protected LSD test (P≤0.05). General linear models procedures were used for mean separation with 95% asymptotic confidence intervals.

[b]Rates of cotton growth were calculated by nonlinear regression of the herbicide treatments with respect to time in days after planting.

[c]Abbreviations: *a*, rate of cotton growth; CL, confidence limit.

Table 5. Rate of cotton growth (*a*) as a response to timing of pendimethalin application.[a]

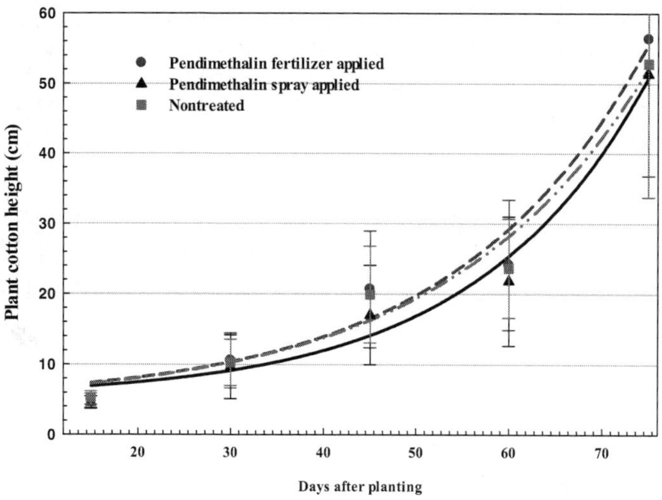

Figure 9. Cotton growth response as affected by application method. The line represents the first-order regression equation. Data points are the means of replications with bars indicating the standard error of the mean:

$$\text{Pendimethalin ME } y = 5.93 + \frac{0.0653\left[\left(e^{0.0485x}\right) - 1\right]}{0.0485} \quad P < 0.0001$$

$$\text{Pendimethalin EC } y = 5.95 + \frac{0.0418\left[\left(e^{0.0545x}\right) - 1\right]}{0.0545} \quad P < 0.0001$$

$$\text{Nontreated } y = 5.78 + \frac{0.0689\left[\left(e^{0.0463x}\right) - 1\right]}{0.0463} \quad P < 0.0001$$

5.3. Cotton yield

Cotton yields reflected the trends initially revealed with cotton injury. Pendimethalin EC spray applied (3,610 kg ha^{-1}) had lower cotton yield than pendimethalin EC applied on fertilizer (4,010 kg ha^{-1}) and both pendimethalin ME treatments (≥4,000 kg ha^{-1}) (Table 6). The treatments that caused the greatest cotton injury for application method by timing interaction had the lowest yields, included both spray AE and 3rd leaf stage of cotton applications. Application timing of pendimethalin on fertilizer did not affect cotton yield. When averaged over application method, cotton yield for the pendimethalin ME treatments had equivalent cotton yields across all application timings. Only pendimethalin EC applied AE or 3rd leaf stage cotton lower yields compared to the typical PRE use-pattern.

None of the PRE or 6th leaf application treatments displayed crop injury, significant decreased growth, or significant yield loss. The AE and 3rd leaf application treatments resulted in significant cotton crop injury and decreased yield, with pendimethalin EC treatments having greater injury than the pendimethalin ME, with spray applications exhibiting more injury than the fertilizer-applied treatments. The fertilizer application of pendimethalin at 3rd leaf did not significantly enhance crop injury, but did enhance injury at the AE application timing. Based on injury, subsequent height, and final yield measurements, pendimethalin ME caused less injury than pendimethalin EC, and fertilizer application of both formulations was less injurious than spray application. The AE application timing was prone to greater injury by any formulation or application method and should be avoided. The 3rd leaf appears to be more prone to spray injury than fertilizer injury.

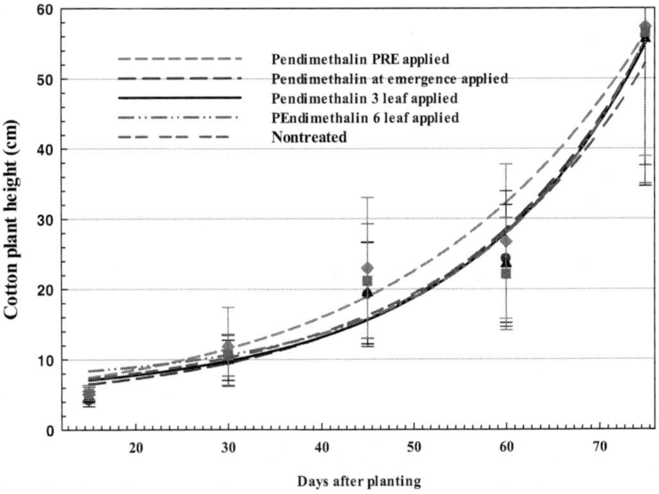

Figure 10. Cotton growth response as affected by application timing. The line represents the first-order regression equation. Data points are the means of replications with bars indicating the standard error of the mean:

$$\text{Pendimethalin PRE applied } y = 5.18 + \frac{0.1104\left[\left(e^{0.0395x}\right) - 1\right]}{0.0395} \quad P < 0.0001$$

$$\text{Pendimethalin AE applied } y = 5.08 + \frac{0.0649\left[\left(e^{0.0488x}\right) - 1\right]}{0.0545} \quad P < 0.0001$$

http://dx.doi.org/10.5772/56184

Pendmethalin 3 leaf applied $y = 5.87 + \frac{0.0550\left[\left(e^{0.0512x}\right) - 1\right]}{0.0512}$ $P < 0.0001$

Pendimethalin 6 leaf applied $y = 7.42 + \frac{0.0450\left[\left(e^{0.00559x}\right) - 1\right]}{0.0559}$ $P < 0.0001$

Nontreated $y = 5.78 + \frac{0.0689\left[\left(e^{0.0463x}\right) - 1\right]}{0.0463}$ $P < 0.0001$

			Yield			LSD
Formulation	**Application**	**Timing**	**kg/ha**			
Pendimethalin EC	Spray		3610	b	(145)	252
	Fertilizer		4010	a	(149)	
Pendimethalin ME	Spray		4000	a	(149)	
	Fertilizer		4230	a	(154)	
	Spray	PRE	4130	a	(176)	370
	Fertilizer	PRE	4260	a	(180)	
	Spray	AE	3570	b	(172)	
	Fertilizer	AE	4050	a	(175)	
	Spray	3LF	3450	b	(170)	
	Fertilizer	3LF	4070	a	(176)	
	Spray	6LF	4080	a	(188)	
	Fertilizer	6LF	4110	a	(182)	
Pendimethalin EC		PRE	4140	a	(181)	369
Pendimethalin ME		PRE	4250	a	(178)	
Pendimethalin EC		AE	3630	bc	(169)	
Pendimethalin ME		AE	3980	ab	(175)	
Pendimethalin EC		3LF	3510	c	(169)	
Pendimethalin ME		3LF	4000	ab	(174)	
Pendimethalin EC		6LF	3960	ab	(185)	
Pendimethalin ME		6LF	4230	a	(185)	

[a]Because *proc Mixed* measures pair-wise differences, multiple LSDs may be obtained. In these cases, the LSD (α=0.05) included is the mean LSD for all treatments.

[b]Pendimethalin rates were 1.1 kg ai/ha for the EC and ME formulations.

[c] Abbreviations: EC, emulsifiable concentrate (0.41 kg ai/L); ME, microencapsulated (0.47 kg ai/L); PRE, prior to plant emergence; AE, at seedling emergence; 3LF, to 3-leaf cotton; 6LF, 6-leaf cotton

[d]Means within a variable followed by the same letter are not significantly different using Fisher's protected $LSD_{(P=0.05)}$. Standard error of the mean for that treatment enclosed in ().

[f]Fertilizer (10-10-10) rate was 280 kg/ha, with all plots equally treated. Pendimethalin EC and ME were spray impregnated.

Table 6. Interaction effects between pendimethalin formulation, application method, and application timing for yield in conventional tillage cotton.

6. Discussion

Comparing the EC to ME pendimethalin formulations, when either spray or fertilizer impregnated applied, indicated the ME formulation consistently reduced cotton injury. The reason for the reduced cotton injury from the ME as compared to the EC-pendimethalin formulation is due to the microencapsulation. This has been observed with another ME formulated herbicide, alachlor [29]. While pendimethalin has lower volatilization than other dinitroanaline herbicides such as trifluralin [21], the ME formulation decreases volatilization and provides extended activity. As previously noted, pendimethalin half-lives of 74 to 114 days in soil have been reported [30], surface applied half-lives of 4 to 6 days can occur due to volatilization, photo-chemical, and other degradation processes with EC formulation [21]. By utilizing the ME formulation, supplementing, or even delaying pendimethalin application to in-season timings impregnated on fertilizer, growers could extend residual weed control until cotton can canopy and suppress weed growth. Our recommendation would be to utilize pendimethalin as a PRE application followed by an in-season application impregnated on prilled fertilizers to extend weed control. Total seasonal pendimethalin applications in cotton are up to 2.24 kg ha^{-1}. Cotton fertility recommendations for the southeast include in-season nitrogen applications which could be pendimethalin impregnated. Given advanced global positioning systems (GPS) used for accurate fertilizer applications, even greater precision for pesticide applications can now be achieved in tandem with these advanced technologies. These data indicate that cotton growers can successfully incorporate in-season pendimethalin application into their cotton production programs with minimal potential for cotton injury, while supplementing weed control with a residual herbicide.

Author details

Timothy Grey[1*] and Theodore Webster[2]

*Address all correspondence to: tgrey@uga.edu

1 Crop and Soil Sciences Department, University of Georgia, Tifton Georgia, USA

2 Crop Protection and Management Research Unit, USDA-ARS, Tifton Georgia, USA

References

[1] AnonymousConservation Technology Information Center. (2005). National crop residue management survey. Available at www.ctic.purdue.edu/CTIC/CRM.htmlaccessed 12 Mar. 2009). CTIC, West Lafayette, IN. online].

[2] Byrd JrJ.D. and A.C. York. (1987). Annual grass control in cotton with fluazifop, sethoxydim, and selected dinitroaniline herbicides. Weed Sci. , 35, 388-394.

[3] Culpepper, A. S. (2007). Cotton weed control. Georgia pest control handbook. Coop. Ext. Serv. The Univ. of Georgia College of Agr. and Environ. Sci., Athens, GA.

[4] Culpepper, A. S, Flanders, J. T, York, A. C, & Webster, T. M. (2004). Tropical spiderwort (Commelina benghalensis) control in glyphosate-resistant cotton. Weed Technology , 18, 432-436.

[5] Culpepper, A. S, Grey, T. L, Vencill, W. K, Kichler, J. M, Webster, T. M, Brown, S. M, York, A. C, Davis, J. W, & Hanna, W. M. (2006). Glyphosate-resistant Palmer amaranth (Amaranthus palmeri) confirmed in Georgia. Weed Sci. DOI:WS-06-001R.1., 54, 620-626.

[6] Devine, M. D, Duke, S. O, & Fedtke, C. (1993). Physiology of Herbicide Action. Englewood Cliffs, New Jersey: Prentice Hall. 441 p.

[7] Dodds, D. M, Reynolds, D. B, Huff, J. A, & Irby, J. T. (2010). Effect of pendimethalin formulation and application rate on cotton fruit partitioning. Weed Technology , 24, 77-84.

[8] Frans, R. E, Talbert, R, Marx, D, & Crowley, H. (1986). Experiment design and techniques for measuring and analyzing plant responses to weed control practices. In D. Camper, ed. Research Methods in Weed Science. 3rd ed. Champaign, IL: Southern Weed Sci. Soc. , 29-46.

[9] Gaston, L. A, Boquet, D. J, & Bosch, M. A. (2003). Pendimethalin wash-off from cover crop residues and degradation in a loess soil. Communications in Soil Sci. and Plant Analysis DOI:10.1081/CSS-120024783 , 34, 2515-2527.

[10] Gordon, J. A, & Green, C. J. (1999). Comparative field and greenhouse studies of trifluralin and pendimethalin on cotton growth, development, and nutrient uptake. In Proc. Beltwide Cotton Conf., Orlando, FL, Natl. Cotton Counc. Am. Memphis, TN., 536-539.

[11] Grey, T. L, Webster, T. M, & Culpepper, A. S. (2008). Weed control as affected by pendimethalin timing and application method in conservation tillage cotton (Gossypium hirsutum). J. Cotton Sci. , 12, 318-324.

[12] Hatzinikolaou, A. S, Eleftherohorinos, I. G, & Vasilakoglou, I. B. (2004). Influence of formulation on the activity and persistence of pendimethalin. Weed Technol , 18, 397-403.

[13] Keeling, J. W, & Abernathy, J. R. (1989). Response of cotton (Gossypium hirsutum) to repeated application of dinitroaniline herbicides. Weed Technol. , 3, 527-530.

[14] Keeling, J. W, Dotray, P. A, & Abernathy, J. R. (1996). Effects of repeated applications of trifluralin and pendimethalin on cotton (Gossypium hirsutum). Weed Technol. , 10, 295-298.

[15] Martens, A. R, Burnside, O. C, & Cramer, G. L. (1978). Compatibility and phytotoxicity of herbicide-fertilizer. Agron. J. , 70, 1089-1098.

[16] Mueller, T. C, Mitchell, P. D, Yound, B. G, & Culpepper, A. S. (2005). Proactive Versus Reactive Management of glyphosate-resistant or-tolerant weeds. Weed Technol. , 19, 924-933.

[17] National Agricultural Statistics Service (NASS) (2010). National Agricultural Statistics Service U.S. Dept. of Agri.. Published Estimates Database. NASS-USDA, Washington, DC. http://www.nass.usda.gov/Statistics_by_Subject/Environmental/index.asp

[18] Norsworthy, J. K, Smith, K. L, Steckel, L. E, & Koger, C. H. (2009). Weed seed contamination of cotton gin trash. Weed Technol. , 23, 574-580.

[19] Parochetti, J. V, & Dec, G. W. Jr. (1978). Photodecomposition of eleven dinitroaniline herbicides. Weed Sci. , 26, 153-156.

[20] Rabaey, T. L, & Harvey, R. G. (1994). Efficacy of corn (Zea mays) herbicides applied at reduced rates impregnated in dry fertilizer. Weed Technol. , 8, 830-835.

[21] Savage, K. E, & Jordan, T. N. (1980). Persistence of three dinitroaniline herbicides on the soil surface. Weed Sci. , 28, 105-110.

[22] Senseman, S. A. (2007). Weed Science Society of America Herbicide Handbook, 9th ed. Lawrence, KS. , 283-285.

[23] Shaner, D. L. (2000). The impact of glyphosate-tolerant crops on the use of other herbicides on resistance management. Pest Manag. Sci. , 56, 320-326.

[24] Shaner, D. L. (2000). The impact of glyphosate-tolerant crops on the use of other herbicides on resistance management. Pest Manag. Sci. , 56, 320-326.

[25] Shaner, D. L, Tecle, B, & Johnson, D. H. (1998). Mechanisms of selectivity of pendimethalin and trifluralin in cotton (Gossypium hirsutum) and weeds. In Proc. Beltwide Cotton Conf., San Diego, CA, Natl. Cotton Counc. Am. Memphis, TN., 1399-51402.

[26] Sosnoskie, L. M, Kichler, J. M, Wallace, R. D, & Culpepper, A. S. (2011). Multiple resistance in Palmer amaranth to glyphosate and pyrithiobac confirmed in Georgia. Weed Sci. , 59, 321-325.

[27] Sosnoskie, L. M, Webster, T. M, & Culpepper, A. S. (2012). Estimates of Palmer amaranth (Amaranthus palmeri) seedbank longevity and potential post-dispersal herbivory. Weed Sci. (submitted).

[28] Sosnoskie, L. M, Webster, T. M, Dales, D, Rains, G. C, Grey, T. L, & Culpepper, A. S. (2009). Pollen grain size, density, and settling velocity for Palmer amaranth (Amaranthus palmeri). Weed Sci. Walker, A. and W. Bond. 1977. Persistence of the herbicide AC92,553,N-(1-ethylpropyl)-2,6 dinitro-3,4-xylidine in soils. Pestic. Sci. 8:359-365., 57, 404-409.

[29] Vasilakoglou, I. B, & Eleftherohorinos, I. G. (1997). Activity, adsorption, mobility, efficacy, and persistence of alachlor as influenced by formulation. Weed Sci. , 45, 579-585.

[30] Vencill, W. K. (2002). Weed Science Society of America Herbicide Handbook, 8th ed. Lawrence, KS. , 231-234.

[31] Vencill, W. K, Grey, T. L, Culpepper, A. S, Gaines, C, & Westra, R. (2008). Herbicide-resistance in the Amaranthaceae. J. Plant Dis. Prot. Special Iss. XXI: , 41-44.

[32] Weber, J. B. (1990). Behavior of dinitroanaline herbicides in soils. Weed Technol. , 4, 394-406.

[33] Webster, T. M, & Nichols, R. L. (2012). Changes in the prevalence of weed species in the major agronomic crops of the Southern United States: 1994/1995 to 2008/2009. Weed Sci. , 60, 145-157.

[34] Webster, T. M, & Sosnoskie, L. M. (2010). The loss of glyphosate efficacy: a changing weed spectrum in Georgia cotton. Weed Sci. , 58, 73-79.

[35] Wilcut, J. W, Wehtje, G. R, & Hicks, T. V. (1990). Evaluation of herbicide systems in minimum-and conventional tillage peanuts (Arachis hypogaea). Weed Sci. , 38, 243-248.

[36] Wilcut, J. W, Patterson, M. G, Wehtje, G. R, & Whitwell, T. (1988). Efficacy and economics of pendimethalin herbicide combinations for weed control in cotton (Gossypium hirsutum). App. Ag. Res. , 3, 203-208.

[37] Wilcut, J. W, Coble, H. D, York, A. C, & Monks, D. W. (1996). The niche for herbicide-resistant crops in U.S. agriculture. In S.O. Duke (ed.) Herbicide-resistant crops: Agricultural, environmental, economic, regulatory, and technical aspects. CRC Press, Boca Raton, FL., 213-230.

[38] Wise, A. M, Grey, T. L, Prostko, E. P, Vencill, W. K, & Webster, T. M. Establishing the geographical distribution and level of acetolactate synthase resistance of Palmer

amaranth (Amaranthus palmeri) accessions in Georgia. Weed Tech. Weed Technol. , 23, 214-220.

Chapter 3

Herbicide — Soil Interactions, Applied to Maize Crop Under Brazilian Conditions

Flavio Martins Garcia Blanco,
Sydnei Dionisio Batista de Almeida and
Marcus Barifouse Matallo

Additional information is available at the end of the chapter

http://dx.doi.org/10.5772/56006

1. Introduction

This chapter discusses the behavior of herbicides in soil cultivated with maize crop in Brazilian conditions, reporting case studies of herbicide use in different periods, from the earliest to the present time, covering ecotoxicological aspects and reflections on the future of the use of the technology in herbicide-resistant transgenic maize.

Maize (*Zea mays* L.) is an annual herbaceous plant adapted to the most diverse ecological conditions. It is an economically important crop in tropical, subtropical and temperate climates, as well as in extreme altitudes, allowing its worldwide presence in several continents.

Brazilian maize production is third in the world ranked behind United States and China. Currently, maize is one of the main crops in Brazil with annual grain yields around 57.5 million tons over a large area of production (13.8 million hectares). It is the most consumed cereal in the country under a variety of forms, in nature and processed food. The exportation volume estimate for 2012 is around 14 million tons, which corresponds to US $ 2,766 billion income for Brazil [1].

Since the late 1970's maize has been cropped in two distinct yearly periods, in the main Brazilian producing regions: one, called "full-season harvest" corn, sowed in the beginning of the rainy season (September, spring); and the other, called "safrinha" or "little harvest" or fall-corn cropping, sowed in the end of this rainy season (from January to April). Usually, fall corn is sowed after soybeans or common-beans harvest, in the same area where these crops had been previously grown, mainly in the South-Central Brazilian region, involving the States of Paraná, São Paulo, Minas Gerais, Goiás, Mato Grosso and Mato Grosso do Sul.

Therefore, the maize crop system adopted by Brazilian farmers has evolved from subsistence agriculture to technical agriculture by using improved adapted cultivars for each edaphoclimatic situation and pest management. Currently, maize cropping has shown expressive productivity increases, due to the modern crop production systems and top cultivars obtained via biotechnology. Farm unities with average grain yields above 7 ton per ha are commonly found in those regions.

Despite the fact that fall corn is subjected to higher production risks during the dry season, there is an economical compensation, due to the new market situation (better grain prices) after the full-season harvest offer. Additionally, there are lower production costs because farmers usually use second generation seeds from the hybrid full-season harvest and grow plants only with the residual fertilizers and herbicides, without any extra management.

This type of crop management has contributed to improved corn production in Brazil during the last 30 years: the production area increased from 11.6 to 13 million ha; annual grain yield increased from 19 to 54.1 thousand tons and average productivity from 1.6 to 4.1 kg ha^{-1}. It is important to emphasize the small production area increase (10.7%) compared to the significant increases in grain yields (184%) and crop productivity (156%). Evidently, such increases, besides the two harvest seasons per year, were mainly due to research improvements in crop management, plant breeding and biotechnology areas.

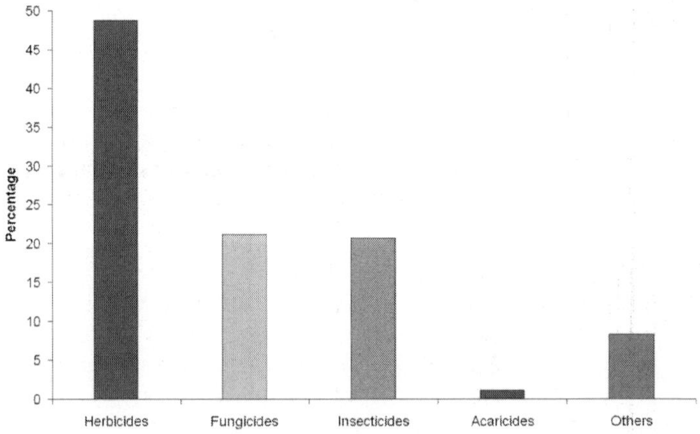

Figure 1. Agricultural pesticides sales (%) in Brazil (2011). SINDAG [6]

Concerning the research results on maize/ weed cohabitation under Brazilian conditions, classic papers [2-5] have demonstrated yield losses between 22% and 83% due to weed competition with the maize crop within a critical period between 15 and 45 days after seedling emergence. There is evidence that weeds prevent maize plants from expressing their maximum production potential, impairing grain yields, even using top maize cultivars obtained through-

http://dx.doi.org/10.5772/56006

out the highly technical breeding programs associated with biotechnology and modern crop management procedures.

Inasmuch, weed control is the prevalent factor to an economically successful maize crop, and for that, herbicide use is required.

Data from the National Association for Plant Defense (ANDEF) and the National Syndicate of Industrial Products for Agriculture Defense (SINDAG) indicated that Brazil is the world's largest pesticide market, and this industrial business mobilized US $ 14.1 billion in 2010, divided into the classes described in Figure 1.

The maize crop was the third largest consumer of herbicides in 2011, ranked behind soybeans and sugarcane crops. Therefore, the knowledge on herbicide-soil interaction processes, applied to control weeds, is highly relevant to understanding the herbicide ecotoxicological effects on the maize crop.

2. Herbicide use in maize crop — A retrospective

Since the first synthetic herbicide release of 2,4-D, a selective herbicide for Gramineae (Poaceae), in 1946, a revolution has occurred in the field for the crops of the Gramineae family, such as the cereals (wheat, rice, maize, barley and oats).

Concerning the maize crop, [7], cited by [8], reported grain yield increases of 25 thousand tons in a cropping area of 7,000 ha, due only to the use of 2,4-D, just after its release in EUA.

Furthermore, another positive aspect provided by herbicide use in the maize crop was the lower spacing among plant rows adopted, with consequent early shading. The old cultivators that required larger spacing among rows to cultivate the soil and eliminate emerging weeds were not necessary anymore. Therefore, [7] affirmed that only this row spacing change in the field allowed increasing plant population from 30 (in 1950) to 50 thousand plants per hectare (in 1970).

Another factor contributing to farmers' fast adoption of herbicides in maize cropping was the fact that women and children were set free of the hard work of hand-weeding. Weeds were removed by hand among plants within rows because cultivators would remove weeds between rows but not between plants in the row. In hand-weeded maize cropping, a man would spend 12 hours per hectare and the whole crop cycle would require three to five hand-weeding procedures [8, 9]. [10] calculated the human workpower necessary just to maintain the same level of annual production at that time and concluded that 18 million men would be required only to hand-weed maize cropping. Considering the work-hour average cost increase from US $ 0.50 (in 1950) to the current US $ 7.50, the choice for herbicide use is almost obligatory economic success in maize.

The advent of s-triazines started in 1952 by researchers from the J. R. Geigy Ltd. Enterprise, in Basel, Swiss: the first patent was obtained in 1954 for the 2-chloro-4,6-bis (alkylamine)-s-triazines; 2-metoxi and 2-methylthio-4,6 bis (alkylamine)-s-triazines. The triazine selectivity description for maize was published in 1955 [11, 12]. The first assays with triazines began in 1952 with the chlorazine molecule. In the following years, so many molecules were synthesized in the same chemical group that a specific symposium was organized in Riverside, California, in 1969 [13].

Since then, herbicide use in the maize increased significantly, because s-triazines were more selective than 2,4-D, which were more phytotoxic to several maize genotypes than s-triazines. Atrazine, specifically, showed low phytotoxicity to maize plants and could also control some dicotyledonous weeds, a distinct property not shown by 2,4-D that is a specific graminicide herbicide. Therefore, important competitive weeds to the maize crop could then be controlled, such as *Bidens pilosa, Emilia sonchifolia, Amaranthus sp, Euphorbia heterophylla, Portulaca oleraceae,* and *Sonchus oleraceae*, representing an advance in weed control management in maize.

Extensive literature concerning s-triazines interactions in the soil can be found because they are among the soil applied herbicides most used worldwide, making it difficult to present a complete review on this subject. A significant number of international reports about atrazine and simazine are available about the most used triazines in maize, but little literature on the environmental toxicology area for Brazilian conditions is available.

Among the herbicides of the s-triazine group used in maize is atrazine; since its release up to now, it has been considered an excellent herbicide due to its selectivity, range of weed control and safety, not causing phytotoxicity for successive crops. At present, it is estimated that 75% of maize-cropped area in the USA is treated with atrazine.

Atrazine ($C_8H_{14}ClN_5$) properties are: chemical name (IUPAC) 6-chloro-N^2-ethyl-N^4-isopropyl-1,3,5-triazine-2,4-diamine; fusion point = 175ºC; solubility in $H_2O_{(20º)}$ = 33 mg kg^{-1}; vapor pressure = 3.0 10^{-7}; $pK_{(21º)}$ = 1.68 and Log $Kow_{(25º)}$ = 2.61, [12].

In Brazil, atrazine is largely used and registered for pineapple, sugarcane, pine, rubber-tree, sorghum and maize.

Atrazine is mainly taken up by roots and also through leaves of plants. When absorbed by roots, it is rapidly transported upwards via xylem and accumulated in the meristems; its movement in the phloem is restricted. Atrazine functions through photosynthesis inhibition by impairing the Hill reaction in the photosystem II, leading to death of susceptible plants. In tolerant plants, like maize and sorghum, atrazine is bound to glutathione (GHS), blocking the atrazine molecule herbicide action [14].

At first, atrazine and other s-triazines were recommended only as pre-emergent herbicides, that is, applied directly on the soil or incorporated just after sowing. However, in the early 1990's, farmers faced climate difficulties that impeded application as recommended, because the pre-emergent application would require a dry period without rain just after sowing to put the implements in the field; in many cases, the dry period would not occur and when the climate conditions were favorable, both the maize seeds and weeds had already emerged their second leaves, characteristic of the first emergence flow. Then, farmers did not have other options than that of applying the herbicide over the plants in the initial stage of development, characterizing a post-emergent herbicide application. From then on and to date, atrazine has

http://dx.doi.org/10.5772/56006

been observed to efficiently control weeds and not cause any toxicity to maize plants, allowing its recommendation also as a post-emergent herbicide.

This occurred at the same time that new post-emergent herbicides were released for maize including: nicosulfuron, isoxaflutole, foramsulfuron + iodosulfuron-methyl, mesotrione and tembotrione. The main advantage of the post-emergent procedure is to better adjust the herbicide dose to control the emergent weed flora, avoiding excessive rates, saving money and decreasing environmental impact. However, there are toxicity risks mainly concerning maize or other more susceptible crops in succession to maize due to residual herbicide effects in the soil. Since the herbicides are indicated for post-emergent application and they are applied over plants, at first, it might erroneously suggest that such chemical products do not persist in the soil or show low persistence.

It is important to highlight that many of those new herbicides have been indicated for agricultural use as components of mixtures with atrazine, similar to the usual recommendation for metolachlor (chloroacetamides group). The herbicide action of atrazine + metolachlor mixture consists of the inhibition of weed cellular division mainly in plants from Poaceae (Gramineae group), complementing the broadleaf weed atrazine action (dicotyledonous plants). Therefore, this herbicide mixture has a wider range of action over weed species which explains its commercial success; up to now, it is considered the best standard herbicide mixture for maize.

Now, research work must focus on these two herbicides in studies concerning the herbicide-soil interactions, applied to maize, since herbicide residues might persist in soils for longer periods than expected, causing phytotoxicity to the next season crop in succession or rotation practices. The knowledge on herbicide persistence in soils is critically important for the adequate use of such products in sustainable environmental systems.

3. Herbicide interaction in soils

In agricultural systems, the soil represents the final destination of large numbers of herbicides applied directly on the soil or over the plant shoots [15].

Herbicides interact with the environment throughout three main routes: (1) physical processes such as soil desorption, volatilization, lixiviation (by water) and erosion together with soil (by wind and water); (2) chemical processes such as photodecomposition, adsorption, reaction with soil components; and (3) biological processes such as molecule decomposition by microorganisms and absorption by plants [16, 17].

According to [18-22], all those processes are dependent on the soil chemical and physical characteristics (soil texture, structure, colloid nature and concentration, pH, etc.) and climatic conditions, particularly, the soil temperature and moisture. On the other hand, the herbicide chemical characteristics depend on the molecular structure, ionization, water solubility, liposolubility, polarization and volatization.

Different external factors exert important roles on herbicide-soil interactions, such as the herbicide formula, rate and mode of application, which are illustrated in Figure 2.

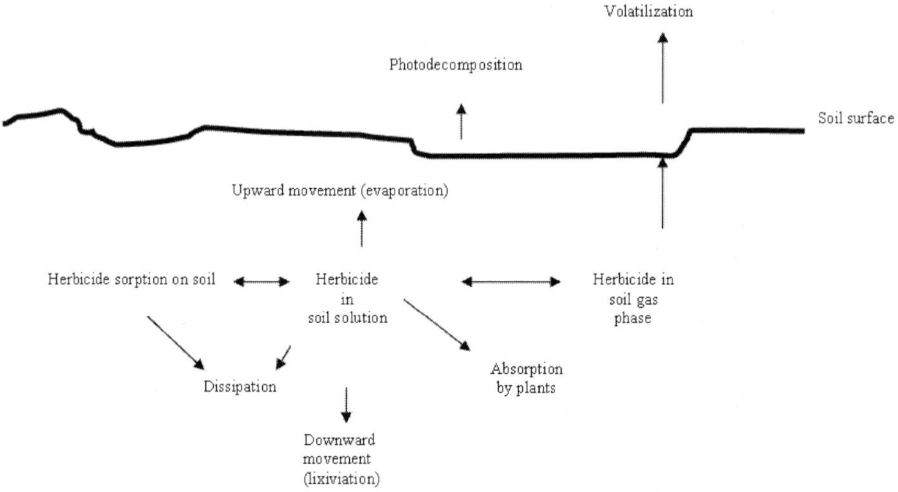

Figure 2. Diagram of the main herbicide-soil interaction processes, adapted from [20].

The processes of soil colloid adsorption-desorption of herbicide molecules greatly influence herbicide movement and transformation in the environment.

4. Soil sorption (adsorption) and desorption of herbicide molecules

The soil adsorption process is understood as the adherence of a molecule, an ion or other particle on the soil surface, as a result of the interaction between both the adsorbing (clay, organic matter) and adsorbed surfaces' field strengths (in this case, the herbicide). Herbicide particles may also be *absorbed* (taken up) by soil colloids. [23], discussed about the difficulty in differentiating between the absorption and adsorption phenomena, suggesting the term "sorption" to express both processes.

Soil sorption is generally a reversible phenomenon (sorption/desorption) and an equilibrium is reached between the adsorbed (sorbed) herbicide concentration on clay/organic matter and the herbicide concentration in the soil solution [24, 25]. The soil sorption is generally a physical process in which attraction forces are involved: such as the van der Waals strengths, bipolar particle interactions, hydrogen bridges and hydrophobic binds. When soil sorption is resultant from chemical processes, an irreversible chemical reaction (herbicide-soil colloid interaction) might occur and a third compound or a stable complex molecule is formed. Soil sorption is a fundamental factor in environmental toxicological studies, because it is determinant to other processes like lixiviation and microbial decomposition [26, 27].

Soil sorption is affected by the involved molecule size. [24] demonstrated that large organic molecules, like herbicides, when adsorbed in montmorillonite clay type, can hardly be substituted by small ions. However, the authors affirmed that the most influential property of clay is the charge type. A stronger electrical charge is generated from dissociation and a weaker charge is resultant from a non-uniform electron distribution on the molecule surface, causing weak polarity.

Briggs [18, 19], affirmed that the extension and intensity of processes involved in the sorption/desorption phenomena are greatly dependent on the herbicide molecular properties and soil temperature and moisture. Similarly, [15] cited the importance of herbicide physical-chemical properties, as well as of soil pH, soil colloid type and soil cation retention.

Velini [22], evidenced that the knowledge on how much the sorption process influences the herbicide sorption on soil colloids is fundamental to define the herbicide application rate to efficiently control weeds.

According to [28], the sorption/desorption processes are highly influenced by the soil colloid type because the larger the specific surface (organic matter, 2:1 clay type), the larger the sorption on soil colloid. Soil moisture also significantly affects the process, once the higher the moisture the lower the sorption. This is due to the fact that H^+ ions, with concentrations dependent on the soil moisture content, compete with the herbicide molecules for the sorption sites at the soil colloids' surface. Therefore, higher herbicide sorption occurs under a water deficit.

The acidic and alkaline compounds' ionization is dependent on the soil pH and herbicide pK. In the case of atrazine, $pK_{(21^\circ)} = 1.68$; and under Brazilian soil pH conditions (pH>5.5), most molecules would be in the molecular form[1], subject to the non-ionic sorption processes, such as hydrogen bridges and van der Waals forces.

In accordance [50], the lack of knowledge about the sorption/desorption phenomena is not surprising because the soil is a highly complex biological and chemical medium, making it difficult to completely understand the interdependent relationships and interactions among the several components involved which certainly affect the herbicide sorption/desorption processes in the soil.

5. Herbicide movement in the soil

Pre-emergent herbicides are expected to present a certain movement in the soil and to be taken up by weed seedling roots because such movement provides an important soil surface

---

$$1\%ionization(base) = \frac{100}{1 + anti\log(pH - pK)}$$

incorporation, allowing a better herbicide contact with greater number of weed seeds or seedlings and maximizing weed control.

Herbicides applied to the soil might move in all directions and phases - gaseous or liquid phases – in areas exposed to intense winds during specific year periods, which could transport considerable herbicide amounts [28], but the vertical descendent route is the most significant movement, characterizing herbicide lixiviation [29, 15].

The herbicide lixiviation in soils is a relevant factor affecting herbicide persistence in the environment. Herbicide lixiviation is dependent on several factors related to the herbicide molecule properties (as intrinsic molecule unity - volatilization, ionization capacity, water solubility, molecular size and weight and lipophilicity) and edaphoclimatic factors (soil type, organic matter content, relief, rainfall and temperature), as well as the herbicide application method. All these factors will determine an herbicide immobilization rate by soil sorption and will influence the herbicide lixiviation. When the herbicide is dragged into deeper soil layers by lixiviation, it persists for longer periods in the environment due to the absence or lower number of microorganisms responsible for the molecule decomposition [30].

The knowledge on herbicide movement routes in soils is essential to a better herbicide/weed management (dose and application method) as well as to understand the potential contamination risks to the environment. The possible herbicide routes that might severely contaminate environment resources include lixiviation to underground waters, superficial molecular movement in solution or suspension (erosion) to water flows, volatilization (air contamination), and removal by live organisms [30].

The higher the herbicide lipophilicity, the higher the tendency to be sorbed on soil colloids, and consequently, lower herbicide lixiviation would be expected. On the other hand, high hydrophilic herbicides would be expected to show lower soil sorption and higher lixiviation rates.

Besides the herbicide vertical movement in the soil profile being an important indicator of its potential contamination risk to underground water and deeper layers, it is also an indicator of herbicide persistence and potential contamination risks to plants with deeper root systems.

Herbicide soil persistence can be determined by biological methods (using bio-indicators) and by chemical or radiometric methods. Both methodologies have advantages and disadvantages and allow assessing the period of herbicide presence in the soil within the detection limits of each method used [31].

Several research works of environment monitoring for potentially toxic residues have been carried out during the last decades, mainly in developed countries. For instance, [32] reported the soil analysis results of 130 different pesticide and herbicide residues applied to agricultural soils in annual crops of 43 USA states. Among the 130 chemical products, only 24 were found in soils during the harvest period: 6 herbicides, 5 phosphorous pesticides, 11 chlorinated pesticides and 2 arsenium pesticides. Among the herbicides, atrazine and simazine, both from the triazine group, persisted in the soil for periods of 12 and 10 months, respectively. [33], when monitoring more than 2200 wells in areas of irrigated maize, detected the presence of several herbicides, such as atrazine, simazine, propazine, prometon and ametrine, and also traces of metolachlor in several well-water samplings.

In Brazil, significant amounts of atrazine and simazine residues were detected in artesian well-waters in the recharge area of Guarani aquifer, an important underwater natural resource [34].

Almeida [35], carried out detailed s-triazine sorption/ desorption studies on different soils from the region of Ubatuba municipal district, State of São Paulo, Brazil. The authors reported that potential herbicide lixiviation and/or superficial runoff would depend on the intrinsic soil characteristics and that the herbicide recommendation must be supported and evaluated based on such soil attributes. Furthermore, the authors observed high s-triazine sorption and consequently lower lixiviation potential in high organic-C content soils; and low herbicide sorption in low C-content soils, favoring desorption and increasing the potential risk of subsoil contamination. They concluded that the soil organic-C content is directly related to the s-triazine sorption and it might be an important indicator of herbicide lixiviation potential, corroborating the results observed in [36].

6. Determination of herbicide persistence and lixiviation: Simazine, atrazine and metolachlor

Blanco [37], determined herbicide persistence and lixiviation up to 50 cm soil depth, using gas-chromatography, in a field experiment with simazine applied as a pre-emergent herbicide in maize at the rate of 3 kg ha^{-1} (a.i. = active ingredient), in the State of São Paulo, Brazil. The results obtained are described as follows:

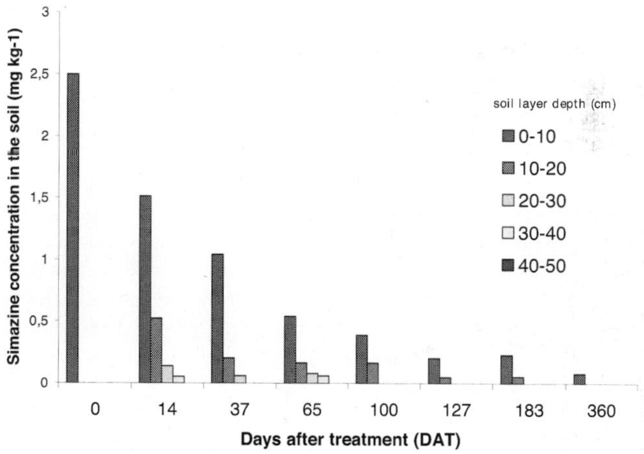

Figure 3. Simazine residue means determined in several soil depth samples (layers from 0-10 until 40-50 cm), collected in different sampling dates [37].

The highest simazine concentration was found in the 0-10 cm superficial soil layer (Figure 3) and decreasing simazine concentrations were found in deeper layers (30-40 cm), 14 until 65 days after treatment (DAT), but at levels near the method detection limit (0.05 mg.kg^{-1}). No simazine residue was found in the 20-30 and 30-40 cm layers, 100 DAT. Simazine persisted in the 0-10 cm layer until 360 DAT in concentrations near the method detection limit. At 10-20 cm layer, simazine persisted until 100 DAT, and afterwards (127 and 183 DAT), only residues near the detection limit were found. The simazine persistence curve was obtained by regression analysis, considering the total residue data in the soil profile (0-50 cm depth, Figure 4).

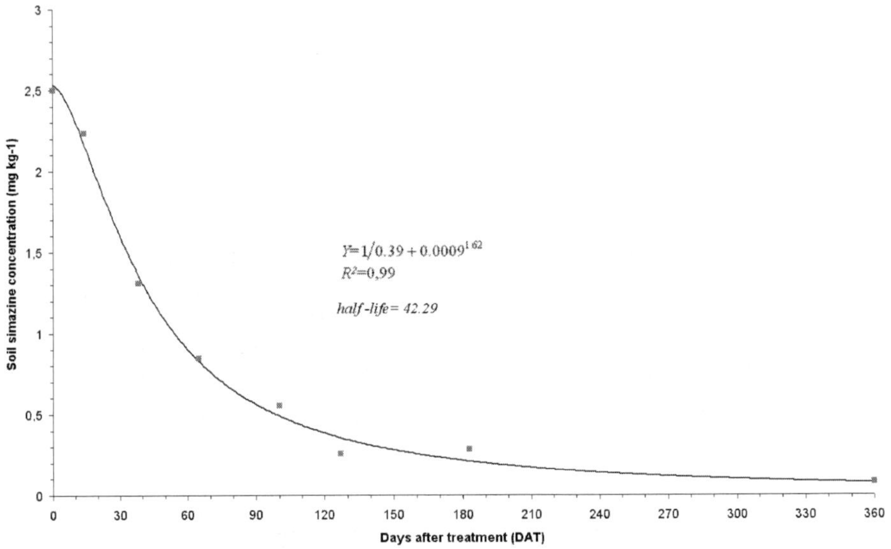

Figure 4. Simazine persistence curve determined until 360 days after treatment, applied as pre-emergent herbicide in maize crop [37].

The persistence curve (Figure 4) fitted an exponential equation, showing a fast decrease of soil simazine concentration until 120 DAT; and afterwards, a slower decreasing slope was observed until 360 DAT. These results might be explained by the rainfall distribution (Figure 3), because a dry period occurred between 112 and 215 DAT, causing adverse conditions to microbial development with consequent increased herbicide molecule adsorption and decreased availability/ dissipation. After 230 days, frequent rainfalls and high soil moisture favored dissipation by biotic agents and soil desorption, once the higher the soil moisture the higher molecule availability and dissipation; other dissipation types might also occur.

Researchers tended to confirm these results, that is, under different Brazilian conditions, simazine remained in the superficial soil layer [38 – 44]. Nevertheless, such affirmations must be supported with information on soil conditions and the triazine group that the studied molecule belongs to as well as on the field experiment local climate where the results were obtained.

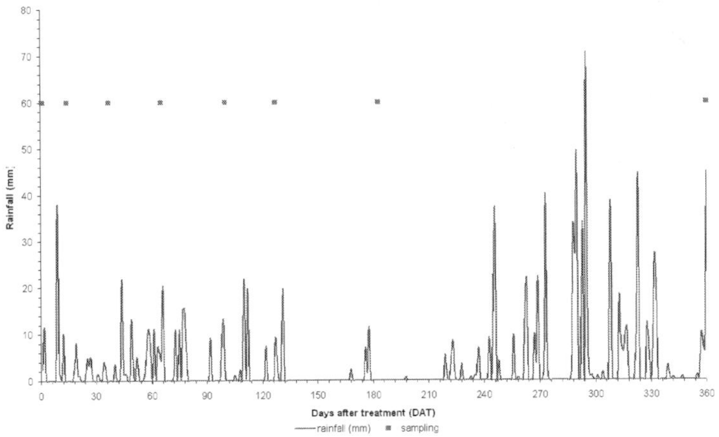

Figure 5. Daily rainfall occurred during the experiment period, January 14th, 1992 to January 8th, 1993, [37].

Dawson [41], found simazine residues one year after the last annual application (of a total of six applications) in the 10-20 cm depth layer; Albers and Homburg cited by [44], also found simazine movement below 15 cm depth for a six-month period in several soil types.

In this experiment, [37] observed that the soil solution pH varied from 6.8 to 5.3 in soil samples; and the simazine $pK_{(21°C)}$ = 1.7 indicated that, under these conditions and despite simazine being a weak base, most herbicide molecules would be in the molecular form (only 0.006% would be ionized). According to [45, 46], simazine presents log Kow = 1.51 characterizing its lipophilic property, and increasing the chances of herbicide molecule sorption by soil colloids.

Gast [13] cited by [30], demonstrated that simazine mobility is affected by soil organic matter (OM). In soil columns with 27 to 30% of OM, the herbicide did not percolate, but in sandy soils, herbicide lixiviation occurred until 17.50 cm depth. In field experiments, [39] found higher simazine concentrations below 30 cm than in the first 15 cm above, 16 months after application. These authors' results were related with the low soil OM (0-10 cm layer = 0.60% OM; and 40-50 cm layer = 0.19% OM) what might explain the simazine lixiviation until deeper soil layers (40 cm).

In another research, following the same procedures, [47] evaluated the atrazine and metolachlor herbicide persistence and lixiviation, applied pre-emergence to maize, as the commercial product Primestra at the rate of 8.0 $L.ha^{-1}$ (1600 g ai atrazine + 2400 g ai metolachlor). The highest atrazine herbicide concentration was found restricted to the superficial soil layer (0-10 cm) (Figure 6). In the 10-20 cm depth layer, the herbicide was found only 15 DAT; and no residue was found in deeper layers until 380 DAT. Atrazine persistence was detected until 184 DAT.

The atrazine persistence curve fitted an exponential equation obtained by regression analysis of the total residue data from 0 to 50 cm depth layers (Figure 7). The initial atrazine residue level was depleted very rapidly from soil and tended to stabilize 100 DAT, remaining constant

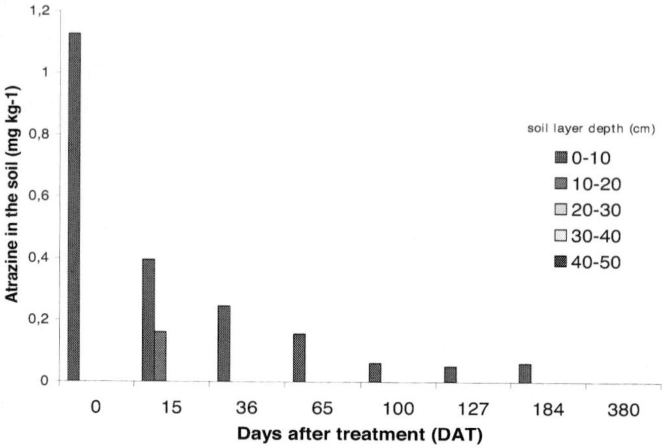

Figure 6. Average atrazine values found in different soil depth layers and sampling dates, in maize crop [47].

until 184 DAT. Afterwards, it tended to zero, and it was not detected at any other sampling dates.

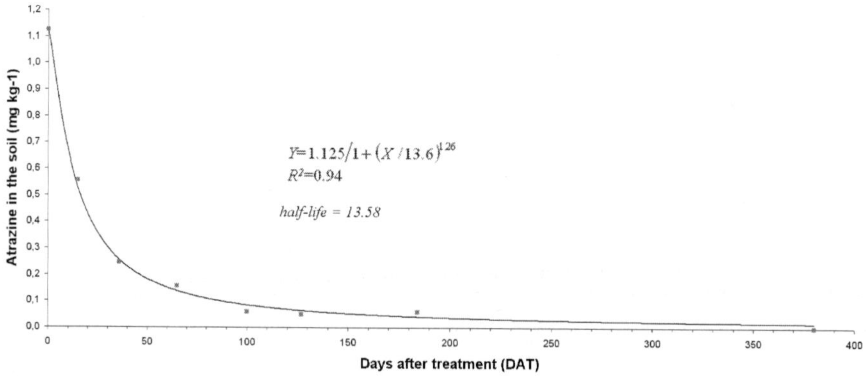

Figure 7. Atrazine herbicide persistence, applied in pre-emergence to maize crop [47].

The metolachlor persistence curve (Figure 8) was similar to the atrazine curve, showing higher residue concentration in the 0-10 cm depth layer. However, metolachlor persistence differed from atrazine due to the fact that it was detected until 380 DAT, and also, because metolachlor lixiviated until 20-30 cm and 10-20 cm depth, 15 and 100 DAT, respectively.

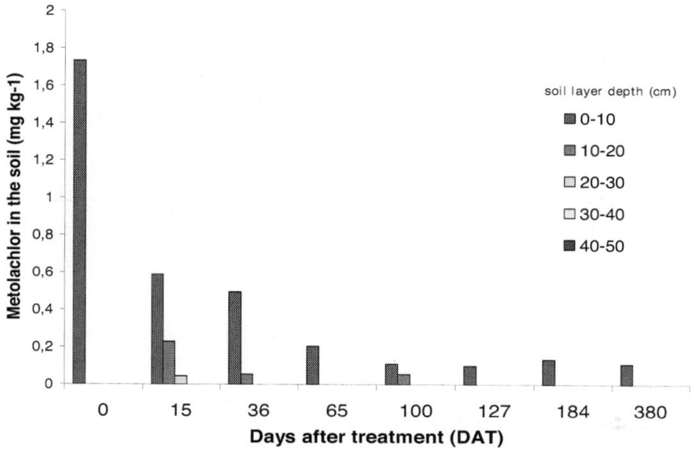

Figure 8. Average metolachlor values found in different soil depth layers and sampling dates, in maize crop [47].

Regression analysis was applied to the total residue data up to the 50 cm depth and the persistence curve fitted an exponential equation (Figure 9).

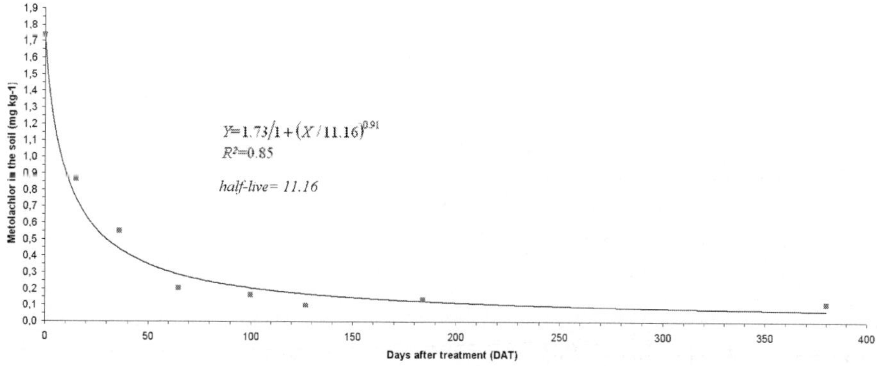

Figure 9. Metolachlor herbicide persistence, applied pre-emergence to maize [47].

A rapid decrease of metolachlor residue concentrations were observed until 100 DAT (0.20 mg kg^{-1}), and afterwards it tended to stabilize reaching the method detection limit (0.05 mg kg^{-1}) 380 DAT.

The rainfall regime is presented in Figure 10. In the experiment beginning, rainfalls were not abundant favoring metolachlor sorption and immobilization on soil colloids, and thus, reducing the dissipation factors. The frequent and abundant rainfalls observed 220 DAT favored metolachlor desorption; its molecules were released in the soil solution and entered through dissipation processes and, consequently, the metolachlor concentration was fast depleted from the soil solution.

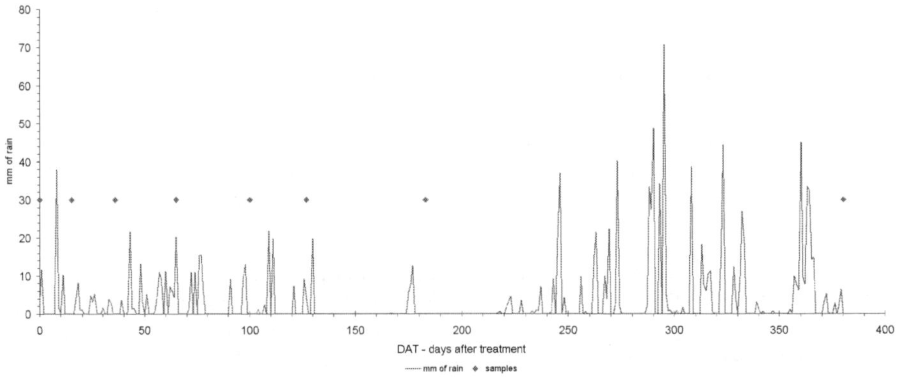

Figure 10. Daily rainfall occurred during the experiment period, January 14th, 1992 to January 20th, 1993 [47].

Blanco et. al [37, 47] carried out field experiments in the same period and edaphoclimatic conditions, allowing the comparison between simazine and metolachlor results: metolachlor presented higher persistence (380 DAT) than atrazine (184 DAT), and this latter showed lesser lixiviation (until 20 cm depth), once metolachlor and simazine lixiviated until 30 and 40 cm depth, respectively.

From these results it might also be inferred that when these herbicides are sequentially used in successive crops, undesirable product residue amount might accumulate in the soil. For instance, at maize harvest (100-120 days after sowing), significant herbicide residual concentrations might be left until the next crop sowing date, causing damage to the environment and plants as well as promoting plant resistance to herbicides.

In the research described in item [56] used bioassay methods to evaluate the atrazine persistence applied as post-emergent herbicide in maize at rates of 1000 and 2000 g ha^{-1} and observed that persistence ended 56 DAT, independently of the atrazine rate. Although comparing different methods, the above result was similar to the one presented in Figure 7, where the end of atrazine persistence was found 83 DAT, determined by gas-chromatography.

The edaphoclimatic condition effects on herbicide persistence in soils are well-known. Soil and climate conditions may directly alter the herbicide persistence or impair different degradation routes in many ways, whatever biotic (caused by microorganisms) or abiotic processes occur. Little Brazilian literature is found concerning the environmental impact caused by herbicide persistence, dissipation and lixiviation, when compared to foreign literature, especially about atrazine that is largely studied in foreign countries. However, data from foreign countries cannot be extrapolated to Brazilian soil and climate conditions.

For example, [48] studied atrazine dissipation in a clayey loam type soil cropped with maize in England (with atrazine rates of 1.10 and 3.30 kg ha^{-1}) and observed an exponential atrazine dissipation, but longer half-life (3 to 3.6 months), much different from the half-life found by [45] under Brazilian conditions for the same herbicide (13.58 days) (Figure 7).

It is described in the paper [49], studying atrazine degradation in field soils of Spain, found a half-life of 30 days, and demonstrated the microbial and chemical nature of atrazine degradation, corroborating the results reported by [11]. The latter authors affirmed there is a strong relationship between s-triazines' inactivation and optimal conditions for microbial community growth. Nevertheless, several soil factors such as increasing soil temperature and moisture, low pH and high soil organic matter content usually favor triazine chemical degradation by hydrolysis [11].

The foreign literature has cited metolachlor (acetanilide group) as the most persistent herbicide in soils, superior to propachlor and alachlor. [50] and [28] reported metalochlor half-life of 33 and 15 days for sandy loam soils and clayey loam soils, respectively, both soils under 80% of water field capacity. Results reported in [47], found similar metalochlor half-life (11.16 days) in clayey loam soil (Figure 9).

Many authors reported metolachlor degradation as an essentially microbial degradation type [51-54]; the soil organic matter is preponderant to metolachlor dissipation because this herbicide shows lipophilic molecule characteristics (log Kow > 3) and it is strongly adsorbed in high OM-soils [45, 55]; thus, explaining the metalochlor lixiviation observed in low OM-soil until 30 cm depth reported by [47] (Figure 8).

When herbicide persistence is determined through biological methods, a specific susceptible test plant is used as an indicator. For that, the test plant is submitted to herbicide residue and the time period of herbicide bioactivity is evaluated as well as its molecule impact on the environment. Since test plants are more susceptible than crop plants, it is possible to estimate the time period that an herbicide is potentially active in the soil to cause damage to susceptible crop plants in succession to the previously treated crop [31].

7. Determination of mesotrione and tembotrione herbicide persistence in soils

Results described in [56], evaluated the tembotrione and mesotrione persistences, applied at two rates, as post-emergent herbicides to maize under different planting systems, using beetroot (*Beta vulgaris*, Early Wonder cv) as a test plant. Soil samples from the field experiment were collected at predetermined dates and used to grow potted test plants under growth-chamber conditions (Conviron phytotron, PVG386 model). After 14 days, plants were cut above the soil and evaluated for shoot fresh matter. The treatment means with and without (control) herbicide were compared by *t* test at 0.05 ($t_{5\%}$).

In this way, bioassays with test plants were used to determine tembotrione and mesotrione soil persistence during four consecutive field experiments.

Tembotrione (2-{2-chloro-4-mesil-3-[(2,2,2-trifluoroetoxi) metil]benzoil} ciclohexane -1,3-dione), solubility = 28 mg L^{-1}, pKa= 4.22, and *mesotrione* (2-(4-mesyl-2-nitrobenzoyl)cyclohexane-1,3-dione), solubility = 168.7 mg L^{-1}, pKa = 3.07, is an herbicide from the triketone group. The herbicide mechanism of action is the inhibition of the hydroxyphenylpyruvate-dioxygenase enzyme impairing carotenoids biosynthesis and destroying cellular membranes, leading to the death of susceptible plants.

In Brazil, they are indicated as post-emergent herbicides in both maize growth periods (full-season harvest and little harvest or "safrinha" corn).

8. Mesotrione and tembotrione herbicide persistence determination in "safrinha" corn

The bioassay was carried out in potted test plants grown in medium texture soil (pH 4.9 and 3% OM), treated with two mesotrione rates (192 and 384 g ha^{-1}). The results (Figure 9) showed that beetroot plants were able to grow and develop only after the sixth soil sampling (84 DAT), with the lower herbicide rate. With the double rate, the test plants grew only 114 DAT. From then on, plants showed increasing shoot fresh matter yields in both treatments, until the moment when no significant differences were found between the control and treated plants (H0 - null hypothesis accepted). Therefore, the end of mesotrione persistence was determined 114 DAT and 177 DAT for the first and second rates (192 and 384 g ha^{-1}), respectively.

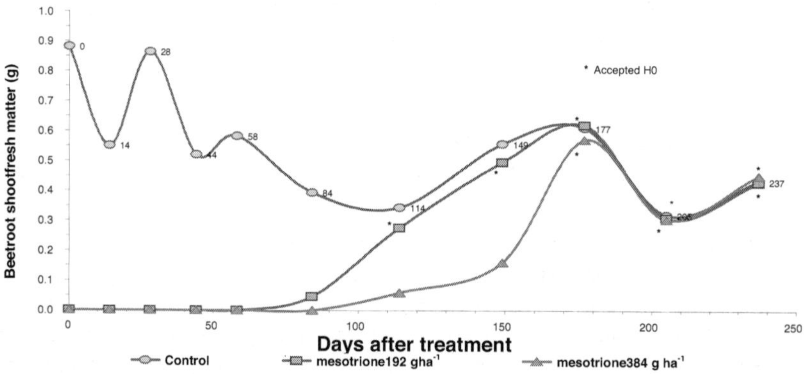

Figure 11. Residual effects of mesotrione herbicides on potted beetroot plants, used as susceptible test plants, grown under growth-chamber conditions. Data is referred to shoot fresh matter (g) and means were compared by t test (0.05). [56].

The rainfall regime during the field experiment is shown in Figure 12. In the beginning of the experiment less frequent and little rainfall occurred. Despite the considerable 140 mm rainfall peak volume between 30 and 40 DAT, the favorable situation did not persist and it was followed by a 130 day-dry period. This fact favored herbicide sorption to soil colloids making it unavailable in the soil solution and restricting the dissipation processes.

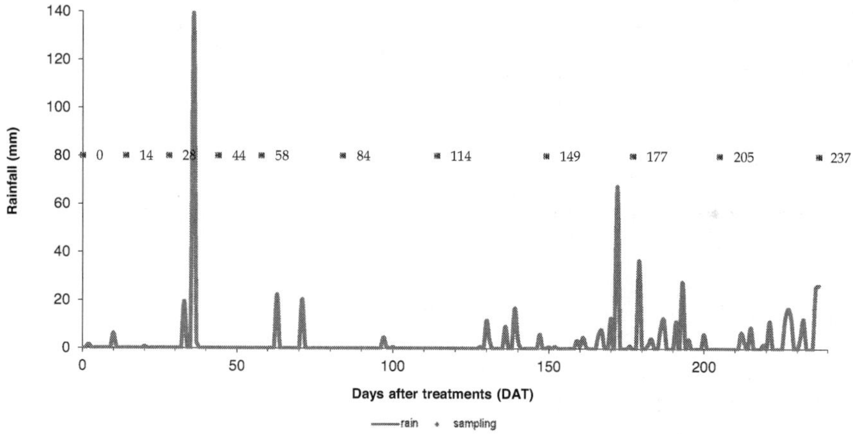

Figure 12. Rainfall distribution observed during the experiment period described in Figure 11. [56]

Such rainfall distribution explained the slow mesotrione dissipation during the initial dry period because the herbicide final persistence was only found 114 DAT, for the lower treatment rate. When rainfall started 130 DAT, mesotrione desorption was favored, increasing its availability, followed by its depletion from the soil solution through dissipation processes. At this point, the test plants were not affected anymore (177 DAT), indicating the mesotrione persistence end for the second rate (384 g ha^{-1}).

The results obtained with test plants grown in soil samples with residual tembotrione from the "safrinha" field corn crop (medium texture soil, pH 5.1 and OM = 1.1%) are presented in Figure 13.

The beetroot shoot fresh matter for different soil sampling periods (Figure 13) showed that these susceptible plants started to grow after the third soil sampling (32 DAT). This means that the null hypothesis (H0) for the rate of 100.8 g ha^{-1} was rejected at this time (significant differences between treatment and control means were found by t test 0.05). The null hypothesis (H0) was only accepted between 55 and 120 DAT, meaning that no significant differences in plant growth among treatments and control were observed at this time, and that the tembotrione persistence ended 55 DAT for the first rate (100.8 g ha^{-1}). For the second rate (201.6 g ha^{-1}), the test plant growth occurred between 75 and 120 DAT, evidencing the end of tembotrione persistence 75 DAT.

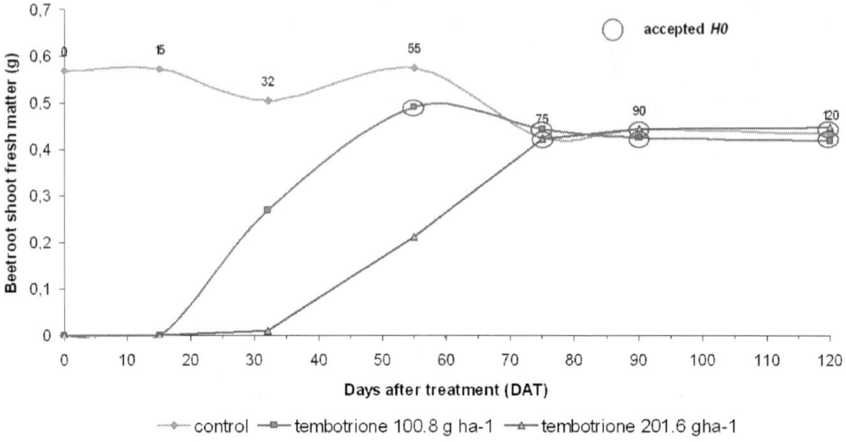

Figure 13. Residual effects of tembotrione herbicides on potted beetroot plants, used as susceptible test plants, grown under growth-chamber conditions. Data is referred to shoot fresh matter (g) and means were compared by t test (0.05). [56]

The rainfall regime during the experiment (Figure 14) showed frequent and abundant rains in the period beginning, and consequently, the high soil moisture favored herbicide release in the soil solution and its subsequent rapid dissipation. This fact explains the low soil herbicide persistence - 55 and 75 DAT - obtained for the first (100.8 g ha^{-1}) and second doses (201.6 g ha^{-1}), respectively. The soil moisture favorable conditions persisted until 70 DAT, almost coincident with the second rate persistence end (75 DAT).

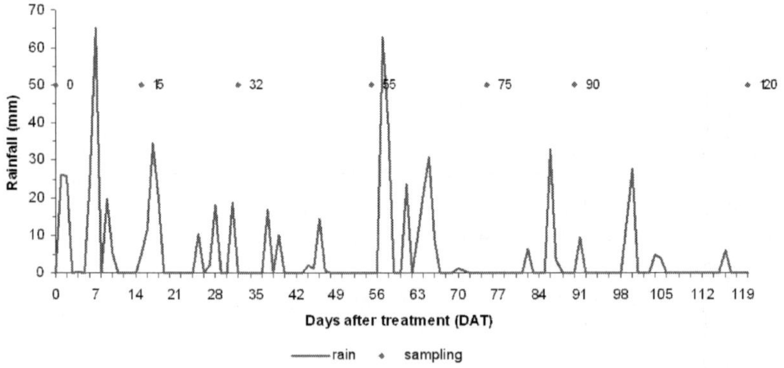

Figure 14. Rainfall regime during the experiment period described in Figure 13. [56]

Although both experiments were carried out under "safrinha" fall conditions (dry season), the mesotrione and tembotrione persistence results (Figures 11 and 13) cannot be compared to each other, because different rainfall regimes occurred during the two field experiments (Figures 12 and 14). Therefore, a third experiment was carried out in a medium texture soil (pH 6.6 and OM = 3%), also under fall conditions ("safrinha" corn) as presented in Figure 15.

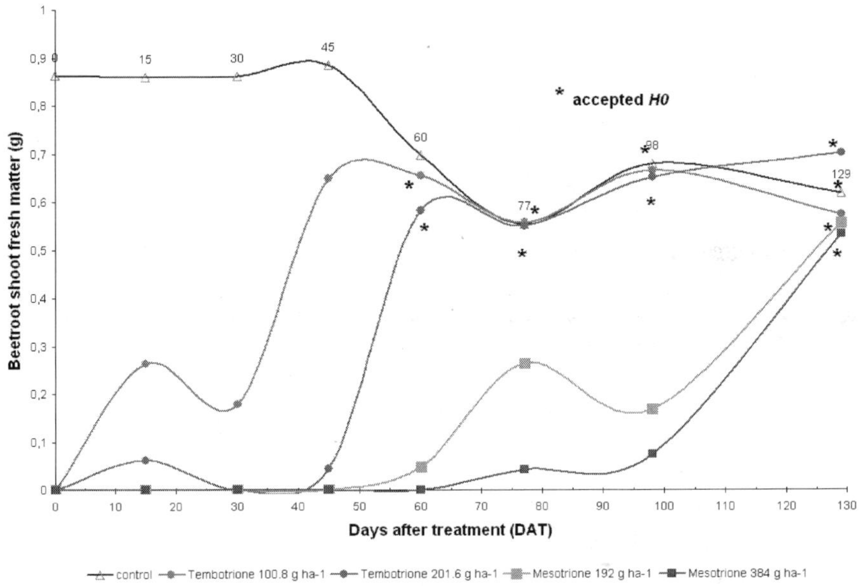

Figure 15. Residual effects of mesotrione and tembotrione herbicides on potted beetroot plants, used as susceptible test plants, grown under growth-chamber conditions. Data is referred to shoot fresh matter (g) and means were compared by t test (0.05). [56].

The herbicides tembotrione and mesotrione were observed to differently affect the beetroot shoot fresh matter during the experiment time period (Figure 15).

The mesotrione residual effect of the first rate (192 g ha^{-1}) actually restricted the susceptible plant growth until 45 DAT, meanwhile the tembotrione residues did not restrict plant growth, except for the period 0-30 DAT at the second rate (201.6 g ha^{-1}). These differences were pointed out by the null hypothesis analysis (Figure 15), which indicated significant contrast differences between the control and the herbicide treatments. The analysis also indicated that tembotrione persistence ended 60 DAT and mesotrione persistence ended 129 DAT, independently of both herbicide rates.

The rainfall regime during the experiment period (Figure 16) showed that the rain volume and intensity until 75 DAT favored tembotrione dissipation that was completely depleted from soil solution 60 DAT, and from then on, did not affect beetroot plant growth (Figure 15). A dry period occurred from 75 to 115 DAT, which restricted mesotrione dissipation and favored its molecule sorption on soil colloids. After that period, new rainfalls caused fast mesotrione dissipation, evidenced by the persistence end 129 DAT.

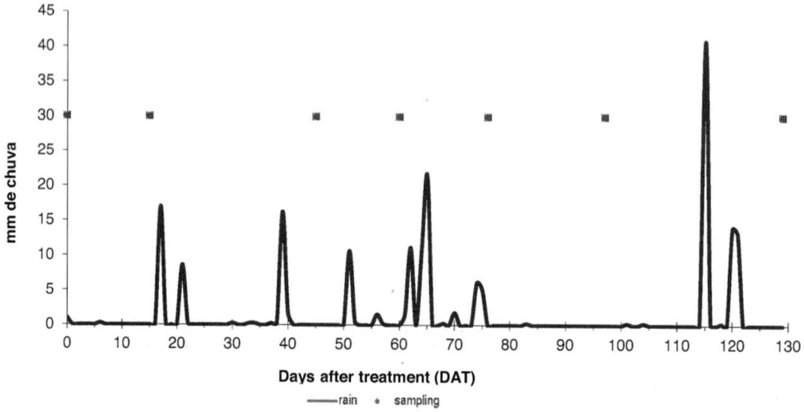

Figure 16. Rainfall regime occurred during the experiment period described in Figure 15. [56]

9. Herbicide persistence determination under full-season cropping conditions

Bailey & Coffey [54] complemented the research work and carried out the same trials during the full-season maize crop, occurred during spring/summer seasons.

The results obtained with test plants grown in soil samples with residual mesotrione and tembotrione (applied to a full-season maize crop, grown in a medium texture soil, pH 5.9 and OM = 2.5%) are presented in Figure 17.

The results indicated that beetroot plants (shoot fresh matter) grown in soil samples with tembotrione residues did not differ from the control plants 56 DAT, independently of the herbicide rate, until the end of the experiment (132 DAT). Plants grown in soil samples with mesotrione residues (first rate = 144 g ha^{-1}) did not differ from the control plants 83 DAT until the end of experiment (132 DAT). However, plants grown in soil samples treated with the second rate (288 g ha^{-1}) were severely affected, except for the last sampling period, when test plants did not show any phytotoxicity symptoms (132 DAT).

Therefore, tembotrione persistence ended 56 DAT independently of the field rate applied; mesotrione persistence ended 83 and 132 DAT for the first and second rates applied to the field (144 and 288 g ha^{-1}), respectively.

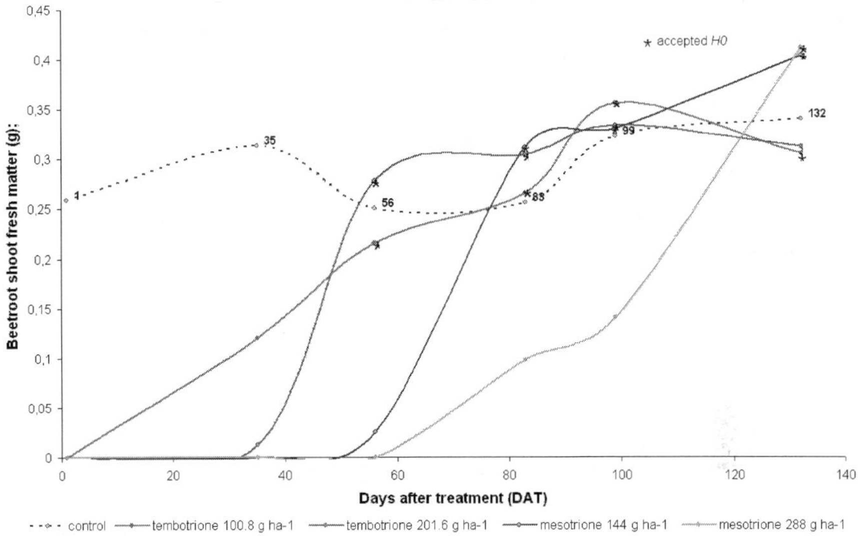

Figure 17. Residual effects of mesotrione and tembotrione herbicides on potted beetroot plants, used as susceptible test plants, grown under growth-chamber conditions. Data is referred to shoot fresh matter (g) and means were compared by t test (0.05). [56]

The rainfall regime during the experiment period (Figure 18) indicated a typical condition observed during full-season maize crop in the State of São Paulo, Brazil, with frequent and abundant rains.

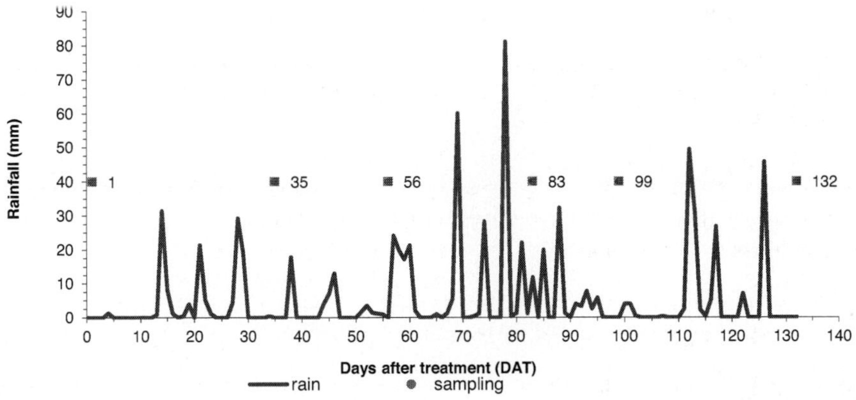

Figure 18. Rainfall regime observed during the experiment period described in Figure 17 [56].

The rainfall distribution favored lower herbicide sorption in soil colloids and higher availability in the soil solution, and thus, the herbicides were easily subject to biological and/or chemical dissipation processes. This explains the shorter tembotrione persistence (56 DAT). However, longer mesotrione persistence was observed (83 DAT) for the first rate (144 g ha^{-1}), which required one more rain period to be dissipated; and still longer (132 DAT) for the second rate (288 g ha^{-1}) that required even another period of rain to be dissipated.

Beetroot plants showed similar susceptibility to both herbicides, but it is possible to infer that tembotrione has a shorter persistence in the soil than mesotrione, independently of the crop season, and that tembotrione provides less environmental impact and toxicity risk to successive crops.

10. Current scenario in Brazil

During the human evolution process, since the beginning of agriculture 7,000 years ago, man has continuously developed technology for that activity, which nowadays, is a highly technical agriculture. However, certain facts have made us think about Carl Gustav Jung's (1875-1961) citation: *"knowledge does not mean wisdom"*.

When mesotrione and tembotrione herbicides, among others, were released in the market, one of the main highlighted advantages at that time was the lower rates required to control weeds, which would result in significantly less environmental impact and phytotoxicity risks to crops in rotation.

Nowadays, such advantage is being revealed, because with the advent of glyphosate-resistant transgenic soybeans, the first resistant weed biotypes started to appear. Currently, there is an increasing concern about glyphosate-resistant weed biotypes after transgenic maize release, which is also resistant to glyphosate. For this reason, a glyphosate mixture with other herbicides, or else a sequential application, has been recommended justified by the need for herbicide rotation or another option to control weeds.

Actually, there has been a tendency to go back to the past with all the old misconceptions and misdirections, that is, to recommend the use of residual herbicides with high environmental impact and risk to successive crops.

It seems that it would be imperative to put in practice not only an herbicide rotation, but also a rotation between transgenic and conventional crops in order to decrease the selection pressure over weed communities, attempting to more efficiently postpone the appearance of herbicide resistant weed biotypes.

This subject will certainly be the new challenge for weed science.

Author details

Flavio Martins Garcia Blanco, Sydnei Dionisio Batista de Almeida and Marcus Barifouse Matallo

Instituto Biológico de São Paulo, Centro Experimental, Campinas, Brazil

References

[1] Companhia Nacional de Abastecimento. CONAB: Acompanhamento da Safra Brasileira de Grãos:http://www.conab.gov.br/OlalaCMS/uploads/arquivos/12_08_27_09_50_57_boletim_portugues_agosto_2012.pdf (accessed 10 October 2012)

[2] Blanco, H. G.; Oliveira, D. A.; Araujo, J. B. M. Estudo sobre a competição das plantas daninhas na cultura do milho (Zea mays L.). I – Experimento para verificar onde realizar o controle do mato. Arquivos do Instituto Biológico, São Paulo 1973; 40(4) 309-320.

[3] Blanco, H. G.; Haag, H. P.; Oliveira, D. A. Estudo sobre a competição das plantas daninhas na cultura do milho (Zea mays L.). II – Influência do mato na nutrição do milho. Arquivos do Instituto Biológico São Paulo 1974; 41(1) 5-14.

[4] Blanco, H. G.; Araujo, J. B. M.; Oliveira, D. A. Estudo sobre a competição das plantas daninhas na cultura do milho (Zea mays L.). IV – Determinação do período de competição. Arquivos do Instituto Biológico, São Paulo 1976; 45(3-4) 105-114.

[5] Blanco, H. G.; Oliveira, D. A.; Araujo, J. B. M. Estudo sobre a competição das plantas daninhas na cultura do milho (Zea mays L.). III – Controle do mato em faixas sobre a linha da cultura. Arquivos do Instituto Biológico, São Paulo 1976; 43(1-2) 3-8.

[6] Sindicato Nacional da Indústria de Produtos para Defesa Agrícola. SINDAG: http://www.sindag.com.br/noticia.php?News_ID=2256 (accessed 10 October 2012)

[7] Hanson, N. S.; Past, present, and future in the North Central Weed Control: conference. proceedings of the 4th Annual Meeting of the North Central Weed Control conference, 8-12 1947.

[8] Gianssi, L. & N. Reigner, N. The Value of Herbicides in U.S. Crop Production, Weed Technology 2007; 21(1) 559–566.

[9] Grubinger, V. P.; Sustainable Vegetable Production From Start-Up To Market. Ithaca, NY, 1999. Cooperative Extension NRAES-104.

[10] Nalewaja, J. D. Herbicidal weed control uses energy efficiently. Weeds Today, Fall 1975. 10-12.

[11] Esser, H.O., Dupuis, G., Ebert, E., Vogel, C., Marco, G.J. S-triazines. In: P.C. Kearney & D.D.Kaufamn, ed. Herbicides: Chemistry, Degradation and Mode of Action, N.Y. 1975; 1(2) 129-208.

[12] Worthing, C. R. The Pestcide Manual. 7 ed. Croydon: The British Crop Council, 1983.

[13] Gast, A. Use and performance of triazines herbicides on major crops and major weeds throughout the world. Residue Rev. 1970; 32(1) 11-8.

[14] Dan Hess, F. Herbicide effects on plant structure, physiology, and biochemistry. In: Pesticide interactions in crop production. CRC Press Inc. 1993. p13-34.

[15] Walker, A. Evaluation of simulation model for prediction of herbicide movement and persistence in soil. Weed Res. 1987; 27(1) 143-152.

[16] Kearney, P. C., Sheets, P. J., Smith, J. W. Volatility of seven s-triazines. Weed 1964; 12(1) 83-7.

[17] Blanco, H. G. Destino, comportamento e resíduos de herbicidas no solo. O Biológico 1979; 45(11-12) 225-48.

[18] Briggs, G. G. Degradation in soil.In: Persistence of insecticides and herbicides. The British Crop Council 1976. 17(1) 41-54.

[19] Briggs, G. G. Molecular structure of herbicide and their sortion by soil. Nature 1969; 223(1) 288.

[20] Walker, A. The fate and significance of herbicide residue in soil. In: Scientific horticulture 1983; 34(1) 35-47.

[21] Walker, A.; Allen, J. G. Influence of soil and environmental factors on pesticide. Soil and Crop Protection Chem. 1984; 27(1) 27.

[22] Velini, E. D. Comportamento de herbicidas no solo. In: Simpósio Nacional Sobre Manejo Integrado de Plantas Daninhas em Hortaliças. FCA-UNESP, Botucatu, 1992.

[23] Harper, S. S. Sortion-desorption and herbicide behavior in soil. Rev. Weed Sci. 1994; 6(1) 207-225.

[24] Bailey, G. W., White, J. L. Factors influencing the adsorption, desorption, and movement of pesticides in soil. Residue Rev. 1970; 32(1) 29-92.

[25] Hayes, M. H. B Adsorption of triazine herbicides on soil organic matter, including a short review on soil organic matter chemistry. Residue Reviews 1970; 32(1) 131-174.

[26] Weber, J. B., Weed, J. B., Ward, T. M. Adsorption of s-triazines by soil organic matter. Weed Sci. 1969; 17(1) 417-421.

[27] Weber, J. B. Mechanism of adsorption of s-triazines by clay colloids and factors affecting plant availability. Residue Rev. 1970; 32(1) 93-130.

[28] Walker, A. & Brown, P.A. The relative persistence in soil of acetanilide herbicides. Bull. Environ. Contam. Toxicol. 1985; 134(1) 143-149.

[29] Riley, D. Physical loss and redistribution of pesticides in the liquid phase. The British Crop Protection Council 1978; 17(1) 109-116.

[30] Helling, G.S. Movement of s-triazine herbicides in soils. Residue Reviews 1970; 32(1) 175-210.

[31] Blanco, F. M. G.; Velini E. D.; Batista Filho, A. Persistence of Herbicide Sulfentrazone in Soil Cultivated with Sugarcane and Soy and Effect on Crop Rotation, Herbicides - Properties, Synthesis and Control of Weeds, Mohammed Naguib Abd El-Ghany Hasaneen (Ed.), ISBN:978-953-307-803-8,InTech;2012.Availablefrom: http://www.intechopen.com/books/herbicides-properties-synthesis-and-control-of-weeds/persistence-of-herbicide-sulfentrazone-in-soil-cultivated-with-sugarcane-and-soy-and-effect-on-crop-

[32] Kenaka, E. E. Evaluation of the harzard of pesticides residues in the environment. In: Watson, D. L., Brown, A. W. A. (ed) Pesticides management and insecticides resistance. New York: Academic Press 1977. p51-95.

[33] Spalding, R. F., Burbach, M. E., Exner, M. E. Pesticides in Nebraska's ground water. Grond Water Monitoring Rev. 1989; 9(4) p126-133.

[34] Cerdeira, A.L.; Santos, N. A. G.; Ueta, J.; Shuhama, I. K.; Pessoa, M.C.P.Y.; Smith JR, S.; Lanchote, V. L. Atrazine in Water and Biodegradation in a Recharge Area of Guarany Aquifer in Brazil. Bulletin of Environmental Contamination and Toxicology 2004; 73(1) 117-124.

[35] Almeida, S.D.B.; Costa, E.; Gomes, M.A.F.; Luchini, L.; Spadotto, C.; Matallo, M. B.Sorção de Triazinas em Solos Tropicais. I. Pré seleção para recomendação de uso na região de Ubatuba, São Paulo, Brasil. In: IV Congreso Iberoamericano de Física Y Química Ambiental 2006, Cáceres. MEDIOAMBIENTE EN IBEROAMERICA - Visión desde la Física y La Química en los albores del siglo XXI 2006. 2(1) 17-24.

[36] Piccolo, A. & Conte, P.. Advances in nuclear magnetic resonance and infrared spectroscopies of soil organic particles. In: Structure and Surface Reactions of Soil Particles (eds P.M. Huang, N. Senesi & J. Buffle), Wiley-Interscience, New York, 1998, p375–435.

[37] Blanco, F. M. G.; Blanco, H. G.; Machado, T. R. Persistência e lixiviação do herbicida simazine em solo barrento cultivado com milho. Planta Daninha, 1997; 15(1) 130-140.

[38] Bouchet, F. Estude de l'influence de la nature du sol sur l'action herbicide de la simazine. Weed Res. 1967; 7(1) 102-116.

[39] Burnside, O. C.; Fenster, C. R.; Wicks, G. A. Dissipation and leaching of monuron, simazine, and atrazine in Nebraska soils. Weed 1963; 11(1) 209-213.

[40] Clay. D.V., Mckone, C.E. The persistence of chlorthiamid, lenacil and simazine in uncropped soil. British Weed Control Conference, 9, Brigthon, England 1968. Proceedings; 2(1) 933-938.

[41] Dawson, J.H., Bruns, V.F., Clore, W.J. Residual monuron, diuron and simazine in a vineyard soil. Weed Sci. 1968; 16(1) 63-65.

[42] Kozlowski, T.T., Kuntz, J.E. Effects of simazine, atrazine, propazine, and Eptan on growth and developement of pine seedlings. Soil Sci. 1963; 95(1) 164.

[43] Kozaczenko, H. Factores affecting the efficiency of herbicides. Biul. Warzyw. 8 (31). 1965. Apud. Weed Abstracts, 1966; 15 (1), 1780 p.

[44] Sheets, T.J. The comparative toxicities of monuron and simazine in soil. Weeds 1959; 7(1) 189-219.

[45] Briggs, G.G. Theoretical and experimental relationships between soil adsorption, octanol - water partition coefficients, water solubilities, bioconcentration factors, and the parachor. J. Agric. Food Chem. 1981; 29(1) 1050-1059.

[46] Briggs, G.G. Factors affecting the uptake of soil-applied chemicals by plants and other organisms., Proceedings, symposium on soil and crop protection chemicals 1984; p35-47.

[47] Blanco, F. M. G.; Machado, T. R. Persistence and leaching of atrazine and metolachlor in soil under corn. In: Third International Weed Science Congress. Fox do Iguassu, 2000, Abstrats, p 90.

[48] Frank, R. & Sirons, G. J. Dissipation of atrazine residues from soils. Bulletin of Environmental Contamination and Toxicology 1985; 34(4) 541-48.

[49] Durand, G. & Barcelo, D. Environmental degradation of atrazine, linuron and fenitrothion in soil samples. Toxicological and Environmental Chemistry 1992; 36(3-4) 225-34.

[50] Zindahl, R. L. & Clark, S. K. Degradation of three acetanilide herbicides in soils. Weed Sci. 1982; 30(1) 545-548.

[51] Beestman, G. B. & Deming, J. M. Dissipation of acetanilide herbicides from soil. Agron. J. 1974; 66(1) 308-311.

[52] Mcgahen, L. L. & Tidge, J. M. Metabolism of two new acetanilide herbicides, Antor herbicides (H-22234) and Dual (metolachlor) by the soils fungus Chaetomiun globosum. J. Agric. Food Chem. 1978; 26(1) 414-419.

[53] Dermont, C. B.; Lavy, T. L.; Marx, D. B. Rate of metribuzin, metolachlor and fluometron in soil. Weed Sci. 1982; 30(1) 629-632.

[54] Bailey, A. M. & Coffey, M. D. Characterization of microorganisms involved in accelerated biodegradation of metalaxyl and metolachlor in soils. Can. J. Microbiol. 1986; 32(1) 562-569.

[55] Peter, C. J. & Weber, J. B. Adsorption, mobility and efficacy of alachlor and metolachlor as influenced by soil properties. Weed Sci. 1985; 33(1) 874-81.

[56] Blanco, F. M. G.; Franco, G. V.; Ramos, Y. G. Persistência dos herbicidas Tembotrione, mesotrione e atrazina aplicados na cultura do milho. In: 27º Congresso Brasileiro da Ciência das plantas daninhas. Ribeirão Preto, SP, 2010. p1738-1742

Chapter 4

Integration of Allelopathy to Control Weeds in Rice

T.D. Khanh, L.H. Linh, T.H. Linh, N.T. Quan,
D.M. Cuong, V.T.T. Hien, L.H. Ham, K.H. Trung and
T.D. Xuan

Additional information is available at the end of the chapter

http://dx.doi.org/10.5772/56035

1. Introduction

Rice (*Oryza sative* L.) is the main food crop in Asia and the staple food of the majority of the population in many regions of the world. The population pressure in rice-consuming countries demands that more attention be directed towards new approaches to sustainable rice production. Improvement of both crop quality and yield is an urgent task [1]. Optimally, rice yield improvement must be sought through agronomic approaches that are environmentally safe [2]. Weed management using allelopathy may effect a yield improvement without environmental cost, which is one of the most important considerations for worldwide scientists working to secure the world's food supply for future generations.

Allelopathy is described as the ability of plants to inhibit or stimulate growth of other plants in the environment by exuding chemicals. The concept of allelopathy was first raised by Hans Molisch to describe both the beneficial and the detrimental chemical interactions of plants and microorganisms [3]. Since then, the term "allelopathy" has undergone several changes and it has been described as any direct or indirect harmful or beneficial effects of one plant on another through the production of chemical compounds that it releases into the environment [4]. The subject of allelopathy currently receives much attention from scientists; the increasing interest in allelopathy in recent years has been stimulated by the recognition that agro-ecological applications of allelopathy may provide alternatives to synthetic herbicides for weed management [5] and with the evidence that allelopathy has the potential for weed control [6-7].

The overuse of agrochemicals has caused environmental degradation, pest tolerance and human health concerns. Agriculture worldwide is currently using about 3 million tons of herbicides annually, and herbicide-resistant weeds have become more prolific, which has further expanded the use of herbicides [8]. To solve these problems, it is necessary to develop sustainable weed management systems that may reduce both herbicide dependency and the burden of manual weeding. With attempts to exploit rice's allelopathic properties for weed control in rice growing, research into rice allelopathy was begun in the early 1970s and has been widely studied in the USA, Europe, Japan, Korea, India and China. If the allelopathic property of crops can be improved, it implies that the competitive ability of crops against weeds can be strengthened, the amount of applied herbicides lowered and environmental risks reduced. Improved crops' allelopathic potential may be useful for rice and all other crops [9]. Crop allelopathy may be a successful tool to manage weed infestations in agricultural production, if it can be exploited appropriately in a rotational cropping system [10]. However, in the case of rice, it is difficult to rotate different crops in a paddy field; therefore, enhancing weed suppression by rice itself may be among the most feasible means of controlling weeds. The isolation and identification of allelochemicals responsible for weed suppression by rice plants may be helpful for understanding the chemical interactions of rice. The introduction of allelopathic traits into cultivated rice via a breeding program may give the possibility of utilizing rice allelopathy in agricultural production.

The aims of this chapter are to present some aspects of integration of allelopathy to control weeds in rice that is pertinent to sustainable agriculture. The following points are discussed: (i) role of allelopathy in weed management; (ii) methodology of allelopathy utilization in rice; (iii) incorporation of higher plants with strong allelopathy to control weeds; (iv) developing allelochemicals and their derivatives for weed management; (v) effort to utilize rice allelopathy for rice weed control; (vi) benefits from allelopathy integrated into sustainable agriculture.

2. Rice weeds

Weeds cause major yield losses in crops and also reduce their quality. Without weed management, rice yield may be reduced by 16 to 86%, or even 100% [11]. Worldwide more than 1000 weed species have been reported in rice [12]. However, 13 species are the most serious weeds spp.: *Cyperus rotundus* L. (purple nutsedge), *Cynodon dactylon* (L.) Pers. (Bermunda grass), *Echinochloa crus-galli* (L.) Beauv (barnyardgrass), *Echinochloa colonum* (L.) Link. (jungle rice), *Eleusine indica* (L.) Gaertner (goosegrass), *Eichhornia crassipes* (Mart.) Solms (water hyacinth), *Portulaca oleracea* L. (purslane), *Chenopodium album* L. (lambsquarter), *Digitaria arvensis* L. (field bindweed), *Sorghum halepense* (L.) Pers. (Johnson grass), *Imperata cylindrical* (L.) Beauv. (spear grass), *Avena fatua* L. (wild oat), and *Amaranthus retroflexus* L. (redroot pigweed) [13-14]. The type of weed species to infest mainly depends on weather, temperature and latitude, and where the rice crop is grown. For instance in Australia, *Cyperus difformis* L. (dirty dora), *Elatine gratioloides* (waterwort), *D. minus* (starfruit) and *E. crus-galli* (L.) Beauv. (barnyardgrass) are major noxious weeds [15] (Table 1). The overuse of herbicides results in herbicide resistance in weeds, which cause more difficulties in weed management. Approximately 200 weed biotypes from 125 different species worldwide have become resistant to herbicides [16]. Traditional weed management in rice was dependent on weather, water coverage and hand weeding. These methods are time-consuming and labor intensive, hence, current weed control depends on synthetic herbicides, but these are harmful to the environment and humans. Therefore, a new strategy for biological weed management in sustainable agriculture should be developed.

Botanical and common name	
Ammannia spp (Redstems)	*Jussiaea decurrens* Walt.(Winged waterprimrose)
Brachiaria mutica Forssk(Bufallo grass)	*Marsilea quadrifolia* L. (Waterclover)
Bacopa spp.(Waterhyssops)	*Monochoria vaginalis* Burm.f.(Monochoria)
Cyperus iria. L (Ricefield flatsedge)	*Murdannia nudiflora* L.(Nakedstem dewflower)
Cyperus difformis.L (Dirty-dora)	*Murdannia keisak* Hassk. (Wartremoving herb)
Commelina diffusa. Burm (Dayflower)	*Ischaemum rugosum* Salisb (Wrinkle grass)
Dopatrium junceum Roxb Hamilt (Horsefly's eye)	*Lindernia pyxidaria* L. (Lindern)
Dactyloctenium aegyptium L. Beauv (Crowfoot grass)	*Leptochloa chinensis* L. Nees (Red spangletop)
Echinochloa colonum L. Link (Shama millet)	*Paspalum distichum* L.(Knotgrass)
Echinochloa crus-galli L. Beauv (Barnyardgrass)	*Leptochloa fascicularis* Lam.(Sprangletop)
Eleocharis acicularis L. Roemer (Needle spikerush)	*Rotala indica* Wild. (India toothcup)
Elatine triandra Schkuhr (Waterwort)	*Sagittaria longiloba* Engelm. (Arrow head)
Fimbristylis dichotoma L. Vahl (Forked fimbry)	*Sphenoclea zeylanica* Gaertin (Gooseweed)
Fimbristylis miliacea L. Vahl (Grasslike fimbry)	*Scirpus mucronatus* L.(Bulrush)
Isachne globosa Thumb (Chigozasa)	*Salvinia molesta* Mitchel.(Kariba weed)
Heteranthera limosa Sw.Willd (Ducksalad)	*Scirpus juncoides* Ferm(Weakstalk bulrush)

Table 1. List of major rice weeds in paddy field

3. Role of allelopathy in weed management

Agriculture worldwide has struggled to control weed interference and the appearance of herbicide-resistant weeds that require the development of new herbicides, and increasing doses of synthetic herbicides in practice. There are about 30000 species of weeds affecting food crops, which cause great losses of crop yields worldwide [17]. In the USA alone, about $20 billion worth of crops are lost each year, accounting for 10% of production [18]. Many high-yield crops have been bred, but this simultaneously increases the heavy dependence on agrochemicals. The desire for safer control of weeds with less environmental impact has become a worldwide concern. In this regard, integrating allelopathy can be a source of new methods for sustainability of agriculture systems.

4. Weed control by allelopathy

Weeds compete with crops for nutrients, water, space, and requirements for photosynthesis, which reduces crop yield. Synthetic herbicides can control weeds effectively and reduce labor in weeding but can cause numerous detriments to the environment and humans, and increase the occurrence of herbicide-resistant weeds. Since it is known that plants can self-regulate their densities and distribution in nature via allelopathic interactions, scientists have attempted to exploit these characteristics of crops and weeds in agriculture. The use of allelopathy for biological control of weeds in agriculture practice has attracted the interest of many agronomic scientists [1].

One approach of utilizing the allelopathic property of crops is to screen accessions to examine their potential for weed suppression [11, 19]. To place crops in a more favourable competitive position in relation to allelopathy over weeds is important for the establishment of sustainable agriculture [20]. The strategy for using allelopathy for weed management could be either through directly exploiting natural allelopathic interactions, especially of crop plants, or applying allelochemicals as a source of natural herbicides. Derivatives of allelochemicals from plants used as herbicides with environmental properties include mesotrione [21-22] and citronella and bilanaphos oil [21]. Several microbial allelochemical products are marketed worldwide, such as glufosinate and bialaphos.

5. Methodology of allelopathy utilization

5.1. Crop rotation

Crop rotation is one of the traditional practices whereby some crops, particularly leguminous species, are grown in short rotation with the main crops [1]. Crop rotation implies growing different crops in systematic and recurring sequence on the same land. This rotational system can help minimize the interference of weeds, fungi, pathogens, insects, and nematodes, and improve soil physical properties, fertility, and organic matter content and reduce soil erosion and heal soil sickness, and crop yields are therefore increased. Allelopathy and crop selection may play a key role in management strategies of weeds and pests. Use of allelopathy in a cropping system relies on better knowledge of the chemicals involved and their behaviour in the agro-ecosystem [23]. Lampkin, 1994 [24] suggested that the principles of selecting crops for rotational sequences should be: (i) alternating between autumn and spring germinating crops, (ii) rotating between annual and perennial crops, (iii) replacing between closed and dense crops, which shade out weeds and open crops such as maize (*Z. mays*), which encourage weeds, and (iv) cutting or topping operations (in particular the traditional cleaning crops, leys, and green manures). Some reports indicated that rotation of maize–cowpea and maize–soybean gave higher yield than monoculture, and the nutrient status of soil was also improved [25-26]. Rotating tobacco–rye grass–maize could minimize the root rot diseases caused by a soil-borne pathogen [27]. This may be the result of the fungitoxins produced by rye grass that inhibited the germination of conidia or chlamydospores of *Thielaviopsis basicola* [28]. Johnson, 1985 [29] conducted a series of exhaustive field trials to determine the suitability of various non-host/poor-host plants for various cropping systems of sweet corn–soybean–wheat–soybean–spinach (*Spinacia oleracea*) that showed significant control of Meloidogyne incognita infestation. Furthermore, Rizvi and Rizvi, 1992 [30] demonstrated that some food crops such as wheat, barley, rye, maize, and triticale (*Triticosecale wittmack*) with high concentrations of gramine or hyroxamic acids were useful for controlling fields with high aphid populations. Allelochemical interactions of plants–plants, plants–soils, plants-micro-organisms, and plant residues from a crop rotation play an active role in enhancing crop yields. Those allelochemicals released from rotated crops then interacted with many physiological processes, which could help promote the growth and yield of crops. If plants used in a rotational system can be determined appropriately, the amount of chemical nitrogenous fertilizers is lowered and environmental hazard is reduced, whereas the sustainability of agriculture by substituting them with biologically fixed nitrogen from legumes is enhanced [31]. However, at present, negligible work has been done on the mode of action of allelochemicals in crop rotation, maybe due to their complicated transformation in nature. Moreover, Chou et al 1980 [77] reported 25% reduction in rice yield of second crop in Taiwan due to the phytotoxins produced during the decomposition of rice residues of first crop left in the soil. The phytotoxic effects of decomposing rice residues in the soil on the succeeding crop are problematic in some countries. In Southeast Asia, rotational systems give greater rice yield than rice monoculture and use of appropriate crops can also minimize the weed biomass significantly. In general, legume crops are preferred as preceding crops to suppress the weeds in succeeding rice crops [1].

5.2. Cover crops, green manure, mulch and intercropping

The term 'cover crop' is defined as crops cultivated with regular cropping for soil and moisture conservation, promotion of nutrient recycling, biomass production, temperature lowering, nuisance weed inhibition, and forage supply [32, 33, 34]. Cover crops may be referred to as either green manure crops or sometimes implied catch crops [35]. Popular allelopathic crops used as cover crops are: barley (*Hordeum vulgare*), sorghum (*Sorghum* spp.), maize (*Z. mays*), wheat (*T. aestivum*), rye (*S. cereale*), buckwheat. (*Fagoprum esculentum*), velvetbean (*M. pruriens*), crimson clover (*Trifolium incarnatum*), subterranean clover (*Trifolium subterraneum*), hairy vetch (*Vicia vilosa*) sweet potato (*I. batatas*), and convolvulaceae (*Tricolor batatas*) [32]. These allelopathic plants exhibited significant weed reduction [36-37]. Excluding phytotoxins released from cover crops into soil, shading effects of the cover crops as well as their thick and dense population, and fast growth could effectively suppress weeds [38]. Legume species and some cruciferous plants could improve soil fertility contributing organic matter and nitrogen to the soil. Successfully established cover crops can develop sufficiently dense canopies in the autumn to interfere with growth of perennial and winter annual weeds [39]. Application of green manure crops can enhance soil organic matter and reduce weed growth. Some plants are used as green manures, including: *Mucuna* spp., *Canavalia* spp., *Trifolium* spp., *Brassica* spp., and *Ipomoea* spp. [32]. Several non-leguminous plants belonging to the family of Brassicaceae, such as field mustard (*Brassica campestris*), white or yellow mustard (*Brassica hirta*), brown/ Indian mustard (*Brassica nigra*), rapeseed/soilseed rape/canola (*B. napus*), black mustard (*B. nigra*), and garden cress (*L. sativum*), were promising sources of green manure and significantly reduced weed biomass [40-41]. Among crops used for covering and green manure, leguminous species should be given priority as they provide rich nutrients including nitrogen to soil [42].When bracken fern (*Pteridium aquilinum*) was used as a green manure, it showed significant herbicidal and fungitoxic activities [43]. The integration of a cover crop into a cropping system by relay cropping, over- seeding, inter-seeding, and double cropping may be useful to supply nitrogen for grain crops and reduce soil erosion and interference of weeds [44]. Some secondary metabolites from cover crops such as volatile glucosinolates and the breakdown isothiocyanates, nitriles, epithinitriles, and ionic thiocyanates were responsible for weed and fungi inhibitory activities [45]. When plants with different growth habits and morphology are intercropped, weed biomass can be lowered. For instance, in maize, mung bean provides more weed suppression than peanut [46]. Barley, rye, and *Vicia faba* were planted in monoculture after the harvest of summer crop [47]. Barley+ V. faba and rye+ V. faba showed effective weed suppression. This was explained by the release of allelochemicals from root exudates during crop growth and from decomposing crop residues [47].

6. Incorporation of higher plants with strong allelopathy to control weeds in rice

6.1. Direct use of plant materials in rice fields

Many plants in the plant ecosystem exert significant allelopathic potential, and when they were incorporated into paddy fields, it resulted in excellent weed reduction. Our research, conducted during 1999–2006, was mainly exploring allelopathic potential of plants in Southeast Asia and Japan for paddy weed control. The preliminary screening for the allelopathic potential of plants in the plant ecosystem should be made with the following requirements: (i) an assessment of their invasiveness and area in the plant ecosystem; (ii) ensuring the plants have less natural weed density in their canopy and surroundings than other plants in their ecosystem; and (iii) using those are traditionally used as green manure, weed or pest management by local farmers [48-49]. Minimizing the hazardous impacts of pesticides (herbicides, insecticides, nematicides and fungicides) in agriculture is the current trend in modern agriculture. Many plants with strong allelopathic properties inhibited the growth of indicator test plants in our laboratory and greenhouse studies. Afterwards, plant species with strong weed suppression were examined against weeds grown in paddy fields. The direct incorporation of allelopathic plant materials into rice fields remarkably reduced the weed interference [48-49].

Southeast Asia has a rich diversity in plant ecosystems; hence, we tested a few hundred plants. More than 30 species including crops strongly inhibited the emergence of pathogens and weeds. In a preliminary investigation, we separated leaves, stems and roots of plants to test their effects on germination and growth of indicator plants (lettuce, radish) and noxious weeds in paddy fields [*E. crus-galli* (barnyardgrass) and *Monochoria vaginalis* (monochoria)] in bioassays and in greenhouse trials. In field trials, some plant species reduced weeds and increased the rice yield (Table 1). We suggested that these plants could be used as source of natural herbicides.

Plant species	Weed reduction (%)	Increased in rice yield ton/ha^{-1}
Ageratum conyzoides L. (billy goat weed)	80.8	20.9
Alocasia cucullata (Chinese taro)	78.4	17.0
Azadirachta indica A.Juss (neem)	91.0*	–
Bidens pilosa L.(Beggar tick)	81.8	23.3
Blechnum orientale L.(White fern)	74.7	23.3
Eupatorium canabium L.(Fragrant thoughoutwork)	75.8	23.3
Euphobia hirta L. (Asthma weed)	87.9	23.3
Helianthus tuberosus (Jerusalem artichoke)	77.8	17.0
Galactia pendula Pers (Galactia)	84.8	7.0
Fagopyrum esculentum Moench (Buckwheat) Pellets	70.0	–
Leucaena glauca L.(White lead-tree)	85.9	23.3
Melia azedarach L.(Chinaberry)	86.9	4.7
Nerium odeander (Oleander)	74.5	19.5
Medicago sativa L. (Alfalfa) Pellets	70.0	–
cv. Rasen	80.0	80.6
cv. Yuba	65.0	29.0
Morus alba L. (Mulberry)	72.7	23.3
O. sativa L. (Rice)		
Hulls	51.7	19.4
Bran	25.1	-6.5*
Hulls +Rasen	88.3	77.4
Bran+Yuba	53.1	29.0
Piper methysticum (Kava)	86.3*	–
Passiflora incarnate (Passionflower)	75.1	21.5
Passiflora edulis (Passionflower)	72.7	34.5
Sophora japonica (Japanese pagoda tree)	84.1	9.9
Stylosanthes guianensis (Stylo)	72.0	25.8
Tephrosia candela L. (White tephrosia)	91.9	23.3
Herbicide (5L ha-1)**	77.8	11.6
Hand weeding	71.7	25.6

(-) Calculation was not conducted; Inhibited compared with the control, applied dose: 1-2 tons ha-1; *: only greenhouse trial was conducted; ** : active ingredients in herbicides: pyributicard, bromobutide, butanamide, benzofenap [Shizetto furoaburu (5 L ha-1), Sankyo Ltd., Japan], and butachlor (600 g L-1 (Butataf, Monsato company, UK). Source: [48, 50].

Table 2. Allelopathic plants inhibitory to paddy weeds and stimulatory to rice yields over their control

6.2. Dose of application

The application of 1-2 tons ha^{-1} biomass of alfalfa (*Medicago sativa*), buckwheat (*Fagopyrum esculentum*), kava (*Piper methysticum*), neem (*Azadirachta indica*), leucaena (*Leucaena glauca*), billy goat weed (*Ageratum conyzoides*), galactia (*Galactia pendula*), chinaberry (*Melia azedarach*), frangrant thoroughwort (*Eupatorium canabium*) and passion fruit (*Passiflora edulis*), strongly reduced the growth of major paddy weeds including *E. crus-galli, M. vaginalis, Rotala indica, Cyperus difformis, Digitaria ciliaris* [50-54]. Plant species exhibiting suppression > 20% were selected for weed control. Plant materials applied < 1 ton ha^{-1} suppresses only weed emergence. The application of alfalfa plants and its pellets or buckwheat pellets at 1-2 tons ha^{-1} caused significant reduction in weeds. The magnitude of weed reduction in rice fields was proportional to the applied dose of plant materials. However, it should not exceed 2 tons ha^{-1}, because application of higher rates causes practical problems for its application, etc. [48]. Despite drastic suppression of paddy weed biomass, the allelopathic plants did not injure the rice plants, rather enhanced their yields by 20% (Table 2). The magnitude of weed inhibition depended on applied plant species. The nutrients released from the plants applied to paddy fields increased the rice yields.

6.3. Methods of application

The ability of allelopathic plants to reduce weeds in paddy fields depends on the treatment method. The plants with strong weed suppressing ability in the screening should be exploited for paddy weed control [51, 53-54]. The leaves of the screened plants are commonly used to provide a large biomass; however, their nutrient contents should be monitored before conducting field trials. Spreading plant materials evenly on the surface of paddy field, 1-5 days after saturating with water at 1 ton ha^{-1} causes greatest weed biomass reduction. Application of allelopathic materials in fields, 7 days after adding water did not influence paddy weed emergence. Major paddy weeds (*E. crus-galli* and *R. indica*) re-emerged in treatments with alfalfa pellets, alfalfa plants, rice hulls and rice bran [52, 55]. A sequential application of biomass was also studied. In the first application, 1 ton ha^{-1} allelopathic material was added 1-2 days after irrigating the paddy soils. In the second and third applications, the same doses were added at 10 days intervals. Each application caused an additional 10-15% inhibition of weeds. However, a greater amount of plant material was needed, which requires more fieldwork, hence, becomes costly [48, 53, 56-57].

7. Developing allelochemicals and their derivatives to control weeds in rice

7.1. Role of allelochemicals in paddy fields

The allelochemicals released from the plants incorporated into paddy soil play a crucial role in inhibiting the paddy weed growth. Many weed growth inhibitors identified from *M. sativa, Piper methysticum, A. indica* (neem), *A. conyzoides, O. sativa*, and *B. pilosa* belong to phenolic acids [52,56, 58-63], fatty acids [56], lactones [62-63], and amino acids [64]. These compounds inhibit the paddy weed growth at low concentrations in bioassays. However, the evidence of how these growth inhibitors act in paddy field conditions has remained unclear. We also examined the correlation of inhibitory potential of plant materials [alfalfa (*M. sativa*) and kava (*P. methysticum*)] incorporated in paddy soil against weeds [60]. Both alfalfa and kava strongly inhibited barnyardgrass and monochoria (*M. vaginalis*) growth up to 10 days after incorporation (80-100% weed control) and suppression persisted for 20-25 days (50% weed control). Many phenolic acids were found in the soil even after 50 days in low concentration, but their concentrations was maximized at 10-15 days and were efficacious until 20-25 days after incorporation. Some growth inhibitors found in the kava treatment showed strong inhibition until 25 days after application, these may be lactones (major constituents in kava roots) and are plant and fungal growth inhibitors [63].

Observations from laboratory, greenhouse and field trials showed that the effects of plant materials on weed species are selective [48]. Different plant materials may possess different quantities and types of toxins, of which the amount released into soil after incorporation, is also species dependent. Despite the identification of many growth inhibitors, their fates after penetrating the soil, how they accumulate at phytotoxic levels and influence the weed growth, the interaction of these compounds with soil factors such as nutrients, pH, minerals and soil microbes, have not yet been fully understood. Even though these issues are complex, we need to understand the actual mode of action of allelochemicals in the environment, so that their efficacies can be increased and become more helpful to develop novel bioactive herbicides.

7.2. Syntheses of novel compounds

Searching the growth inhibitors from plants and testing their efficacies against weeds in the laboratory, greenhouse and fields are just the initial steps to developing bioactive herbicides. However, it is necessary to develop bioactive herbicides, because: (i) direct use of allelochemicals as herbicides is not successful as these compounds are degraded in nature, before reaching the targets, (ii) to isolate allelochemicals from plants is complex, promising compounds for weed suppression exist in low quantities in plants, hence, it is too costly to use as herbicides and (iii) despite the promising weed reduction by direct application of plant materials to paddy soils, it requires a very high amount of plant biomass, therefore, does not meet the current requirements of trend in agricultural production in many countries. However, despite obtaining numerous compounds with herbicidal activities, very few constituents from plants have been marketed as herbicides than from bacteria and fungi [65]. Further, most reported secondary metabolites with strong herbicidal activity have complex chemical structures, hence, may not be processed as novel herbicides, because of difficulties in their synthesis and thus become costly. Thus searching for compounds having a simple form with strong herbicidal activities should be a priority. The synthesis of compounds derived from allelochemicals, attached with further functional groups and possessing herbicidal activities, is indispensable to developing novel bioactive herbicides.

Dihydro-5,6-dhydrokawain (DDK) (Figure 1) is a major compound in all parts of Alpinia (*Alpinia zerumbet*), a plant distributed widely in the subtropics and tropics. Besides many promising pharmaceutical efficacies, DDK exerts herbicidal and antifungal activities in bioassay trials. Our team has synthesized numerous DDK derivatives (Figure 1) [66] and tested for their influences against indicator plant and plant fungi. The derivative dimethyl phosphorothionate exhibited maximum antifungal activity of 91% and 72% against *Corticium rolfsii* and *Pythium* spp., respectively [67]. Twenty-four kinds of esters were made from cinnamic acid, p-coumaric acid and ferulic acid, alcohols and the components of Alpinia [68]. Among these derivatives, isopropyl 4-hydroxycinnamate and butyl 4-hydroxy-cinnamate were fungitoxic to *Pythium* spp. at 10 ppm. Further syntheses of DDK derivatives are being carried out in our laboratory.

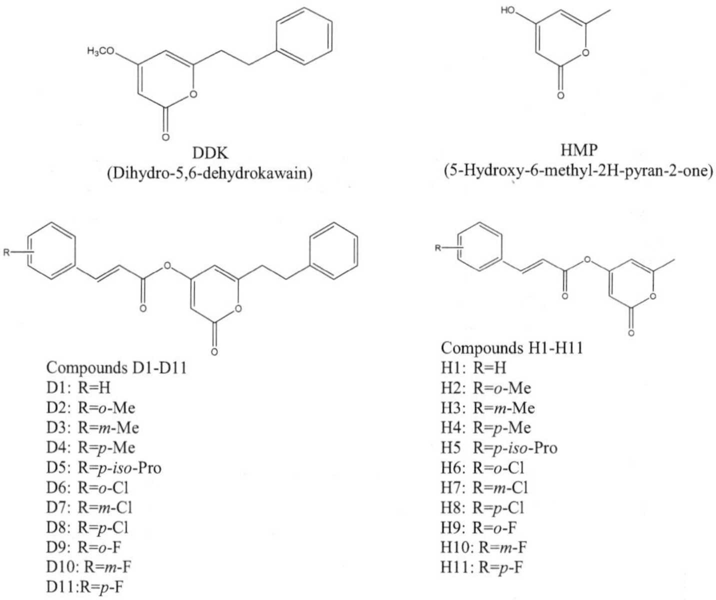

Figure 1. Structures of DDK, HMP and the Pyranyl - substituted Cinnamates. [11, 66]

8. Effort to utilize rice allelopathy for rice weed control

Reducing weed infestation by exploiting the allelopathic properties of rice may be the most important goal of research involved in rice allelopathy and has been a hope of many agronomists. The direct use of rice residues and genetic control of rice allelopathy via breeding programmes to enhance weed suppression may be the most feasible strategy.

Allelopathic activity has been shown to be variety-dependent and origin-dependent, where Japonica rice shows greater allelopathic activity than Indica and Japonica-Indica hybrid.

Extensive efforts of researchers worldwide to clarify allelopathic activities among rice cultivars have been made. They provided important information for further work such as genetic analyses, gene mapping of allelopathic characteristics and breeding new rice cultivars with strengthened weed suppression ability [10].

8.1. Rice residue

Utilization of rice residues in paddy fields has long been recognised as an important source to improve the organic matter status of soil and was also reported to reduce the emergence of weeds. In Asia, farmers are often left with a large amount of rice residues, leaves, stubble and straw in the paddy fields after harvest. Incorporating the residues of rice with high allelopathic activity minimised rice flatsedge (*Cyperus iria* L.) growth to a similar degree as achieved by the application of propanil and bentazon herbicides [69]. Furthermore, another trial showed residues of rice (var. Sarjoo 52) blended into the soil (5–6 cm in depth, 5 tons ha^{-1}) suppressed jungle rice [*Echinochloa colona* (L.) Link], monarch redstem (*Ammania baccifera* L.), Ammania multiflora Roxb., and gulf leaf flower (*Phyllanthus fraternus Webster*) [70]. Other experiments reported that rice straw suppressed the germination of oat (*Avena sativa*) and wheat (*Triticum aestivum*), Lens sp., *Convolvulus arvensis* L., *Avena ludoviciana* and *Phalaris minor Retz* [71-72].

To date, decomposition of rice straw and stubble has reduced the occurrence of both broad-leaved and grassy weeds [73]. Leaf plus straw and hulls of some rice cultivars with strong allelopathic property dramatically inhibited weed interference about 60–95% [74]. A pot study of soil incorporation of a mixture of stubble and straw in 15 cm of soil in the pots (7.4 tons ha^{-1} of blended stubble and straw) revealed inhibition on growth of weed density and decrease of the dry biomass of weeds [73]. Straw, leaves and hulls of some rice cultivars suppressed the germination of field bind weed (*Convolvulus arvensis*) and little seed canary grass (*Phularis minor*) [71-72, 75-76]. Similarly, Pheng et al (2010) [77] suggested that if the rice residue incorporation was suspended for 2 weeks or only a proportion of the residue was incorporated, the rice crop could withstand the growth-suppressive effect. This research suggests that rice possessing high allelopathy can control some weeds in rice and can be integrated with existing weed management practice. Residues of rice allelopathy may be helpful for weed control, but they sometimes cause trouble with rice autotoxicity. From the residual effects of decomposing rice plant materials, the rice plant may obtain adaptive mechanisms to prevent a severe autotoxic effect. For instance, Chou, (1980) [78] reported in Taiwan that decomposed rice residues left on the paddy field soil persisted into the next crop season and could reduce the rice yield by up to 25% compared with that of the first crop. Such a reduction was suggested to be primarily attributable to the phytotoxins produced, which inhibited paddy weed growth and minimised rice yield. Singh et al. (1999) [79] reported that autotoxicity in rice could provide an adaptive strategy to plants because they are grown in adequately water-logged soils sufficient in oxygen and thus develop a negative redox potential in soil because of decomposing rice residues. This induced the inhibition of root growth of rice plants accompanied by swelling of root cells in order to capture more oxygen [80]. Rice hulls and bran were reported to suppress paddy weeds and could be exploited for weed management [81]. Xuan et al. (2003) [52] noted that rice hulls and bran each at 1 ton ha^{-1} reduced paddy weed biomass by about 25% and 50%, respectively. The combination of rice by-products and alfalfa strengthened weed suppression by 70–80% and controlled more weed species and increased rice yield more than the incorporation of single rice by-products.

8.2. Molecular research in rice allelopathy and breeding

Allelopathy is one of the last areas of plant science to use molecular biology as a tool in understanding the phenomena. Allelopathic competition, which may be defined as the unequal sharing of resources such as nutrition, light and water, is dependent on several physiological and phenological traits, and its allelopathy is polygenic and quantitatively inherited [82-84]. To be able to breed a more competitive crop with strong allelopathic potential, it is crucial to know which genes are involved in crop competitiveness and allelopathic potential. Molecular marker-aided genetics is presently the best tool for identifying quantitative traits, mapping the genes involved onto the chromosomes with a reasonable level of precision and analysing the relationship between the traits of interest and other important agronomic traits [82]. Allelopathic activity in rice has demonstrated to be a polygenic trait that is only slightly correlated with yield or other agronomic features. The quantitative inheritance of rice allelopathy curbed the breeding of allelopathic rice cultivars against paddy weeds under varying environmental condition [84, 86-87]. Recent research of Xu et al 2012 [85] has provided the evidence that diterpenoid momilactones (allelochemical) isolated from a rice cultivar plays a novel genetic for natural product-mediated allelopathy and furnished a molecular target for breeding and metabolic engineering of a rice cultivar. The selection of rice cultivars with strong weed suppression ability through transgenic and breeding programmes may successfully utilise rice allelopathy for weed control. Allelopathic activity of rice varies among cultivars and origins and correlates with some growth characteristics; therefore, the existence of genes determining rice allelopathy is presumed and should be detected. It was proposed that allelopathic activity may be a polygenic trait slightly correlated with yield or other agronomic features. Allelopathic potential in rice was demonstrated to be quantitatively inherited, but the allelopathic traits were not identified [83].

8.3. Genomic analysis and gene mapping

Despite research on rice allelopathy beginning in the early 1970s, the genetic allelopathy control programme started only in 1996 [88]. Dilday et al. (1998) [89] crossed the allelopathic rice cultivar PI312777 (PI) with another non-allelopathic rice cultivar Lemont and noted that the F2 was allelopathic against *Heteranthera limosa* and was quantitatively inherited. Jensen et al. (2001) [90] studied quantitative trait loci (QTLs) mapping using a population of 142 recombinant inbred lines (RILs) derived from a cross between IAC 165 (Japonica upland cultivar) and CO 39 (Indica irrigated cultivar). Four main QTLs located on three chromosomes, 2, 3 and 8, were identified and claimed 35% of the total phenotypic variation of the allelopathic activity against barnyardgrass. Okuno & Ebana (2003) [91] identified seven QTLs controlling rice allelopathy on chromosomes 1, 3, 5, 6, 7, 11 and 12. Digenic interactions in five pairs among the seven QTLs were detected. This study showed 125 out of 215 restriction fragment length polymorphism (RFLP) generated polymorphic bands between PI312777 and Rexmont under QTL analysis. A map of 12 linkage groups was constructed and covered a genetic distance of 1336.2 cM. The total number of probes ranged from 12.7% to 76.4% among 12 chromosomes. With RFLP marker loci to the allelopathic QTLs at all pinpoints, the PI312777 alleles were more suppressive against lettuce than the Rexmont alleles. The positive allelopathic effect was shown by QTL located on chromosome 7 that suppressed root growth and necrosis on lettuce [92]. Zeng et al. (2003) [93] used a double-haploid population derived from ZYQ8/JX17, a typical Indica and Japonica hybrid. Four QTLs correlated to allelopathy belonging to chromosomes 3, 9, 10 and 12 were detected and their logarithm of odds scores were 3.40, 2.68, 2.75 and 3.08, respectively. Among them, additive effects of the QTLs on chromosomes 3 and 10 were 1.65 and 1.43 and on chromosomes 9 and 12 were –1.44 and –1.58, respectively. Recently, Lee et al. (2005) [94] identified nine QTLs controlling allelopathic effects of rice on *E. crusgalli* on chromosomes 1, 2, 3, 4, 5, 8, 9 and 12. Of these, QTLs on chromosomes 1 and 5 were the most allelopathic and explained 36.5% of total phenotypic variation. Lin et al. (2005) [95] used the inter-simple sequence repeat approach to detect the genetic diversity of allelopathic potential in 57 rice cultivars. Thirty-four polymorphic bands were generated, and the percentage of polymorphic bands was 53.0%. Rice from the same geographical location and those cultivars with higher allelopathic potential could be clustered into each group, implying that the genes conferring allelopathy in rice might be isolocus. However, some cultivars of rice with markedly different allelopathic potential clustered into a group with a lower level of genetic polymorphism, and this might be attributed to selection oriented for high-yielding traits in breeding. More recent advances in rice genome research have provided a powerful tool for the genetic analysis of quantitative traits. The use of high density genetic linkage maps and DNA markers mapped onto rice chromosomes may enable the identification of the QTLs controlling the allelopathic effect of rice on weeds [96]. QTL analysis is the initial step in rice genetic analysis. Identification of QTLs from close linkage of a DNA marker to the QTL would be useful for producing near-isogenic lines. Application of DNA marker-assisted selection, map based cloning of allelopathic QTLs and a near-isogenic line may help to determine allelopathy-correlated genes in rice. Nine possible differently expressed genes 1, 4, 5, 7, 8 and 9 involved in allelopathic potential of Indica type rice variety, namely Sathoi, capable of producing nicotianamine against growth of barnyardgrass indicated higher while three differentially expressed genes 2, 3 and 6 showed low expression. It implies that these genes were found to be homologous to other genes [96-98]. To date, under low-nitrogen stress, rice cultivar PI exhibited increased allelopathic activity. Nine genes involved in phenylpropanoid metabolism, including phenylalanine ammonia-lyase (PAL), became up regulated and the content of phenolic compounds in rice was enhanced [98-99]. Song et al. (2008) [101] reported that the intensification of allelochemical biosynthesis in rice grown under stress nutrition (i.e., low levels of nitrogen) disclosed the overexpression of genes that encode for PAL (phenylalanine ammonia-lyase), O-methyltransferase, triosephosphate isomerise and P450-all related to the synthesis of phenolic compounds and detoxification. Furthermore, a proteomic analysis of rice growing with barnyardgrass revealed the induction of the following proteins: PAL, a thioredoxin and 3-hydroxy-3-methilglutaril-coenzyme a reductase 3 (HMGR) [102]. On the other hand, the differential proteomic analyses have validated that enhanced allelopathic potential in rice exposed to stress is due to increased expression of enzyme genes involved in the biosynthesis of phenolic compounds and reduced expression of enzyme genes associated with terpenoid biosynthesis [103]. The identification of these genes and proteins shows different signs, plant-environment interactions or plant-plant communication triggering the biosynthesis of phenolic compounds that are also known to be related with plant defence processes [102,104]. Moreover, allelopathic enhancement of allelopathic rice cultivars in the vicinity of barnyard-grass was due to improvement in carbon assimilation deriving from the regulation of photosynthesis genes and the activation of the enzyme system [103, 105].

8.4. Breeding new rice allelopathic cultivars

To breed new rice cultivars having strong competitiveness against weeds may bring important benefits to farmers in rice-cultivating nations. In the breeding programme, both traditionally bred and hybrid rice with allelopathy may be feasible. Courtois & Olofsdotter (1998) [88] indicated that if a high number of QTLs with low effect are involved, a traditional breeding method can be a reasonable alternative, in which two parents with contrasting behaviour are crossed and RILs are derived through the single seeded descent method (SSD). Kim & Shin (2003) [106] crossed Donginbyeo (a non-allelopathic cultivar, but a high yielding rice of good quality) and Kouketsumochi (an allelopathic cultivar, close to a wild type) and advanced by SSD breeding method. The F5 of this cross exhibited allelopathic potential in bioassays and was continuously examined under field conditions. The three-line hybrid rice widely cultivated in China may be a good source because of its rapid and profuse vegetative growth in comparison with an inbred line [106]. Lin et al. (2000) [107] applied a simultaneous backcrossing and self-breeding method to develop a hybrid rice with allelopathic activity and, its counter-part, an isogenic hybrid rice with no allelopathic effect on weeds. Three lines of rice Kouketsumochi, Rexmont and IR24 were used as the allelopathic donors, non-allelopathic and restoring genes, respectively. The selected restorer lines were crossed with cytoplasm-sterile lines and tested for the outcross rate. This work illustrated a scheme for developing hybrid rice having allelopathic potential. On the other hand, the heterotic effect on rice allelopathy was positively significant, showing higher heterosis over the mid-parent. This specific hybrid rice showed a suppressive effect on barnyardgrass, exhibiting a large deviation from the resource competition curve [107]. Hybrid rice with stronger weed suppression ability could be bred, but the quality factors associated with rice allelopathy should be carefully considered in the breeding programme as an important standard for the new cultivars. A newly bred rice, namely K21 showed highly allelopathic and agronomically fit. This cultivar inherited its good agronomic performance from the female parents (Dongjibyeo) and attained its potent allelopathic potential from male parent (Koutetsumochi) [108-109]. Moreover, Kim and Shin, 2008 [108] suggested that identified allelochemicals and genes which responsible for allelopathic activity can further be incorporated into the cultivars via breeding or genetic engineering. For instance, the diterpenoid momilactones and phenolics in rice work as the major inhibitor substances to suppress weeds, which are able to be produced in a conventional rice cultivar by inserting the genes CA4H and OsDTS2 for *p*-coumatic acid and momilactone, respectively through genetic engineering or even conventional breeding [108, 85, 103]. Also, Kong et al. 2011 [84] has successfully developed commercially acceptable allelopathic rice cultivars via crosses between allelopathic rice variety PI12777 and commercial cultivars. The bred Huagan-3 showed 80% inhibition on noxious barnyardgrass and 30-50% of a total reduction in paddy weeds. However, it should be noted that developing allelopathic rice cultivars must therefore be accompanied with an evaluation of the cultural practices required for consistent suppression under variable environmental conditions [84, 86]. On the other hand, before starting any plant breeding program to enhance allelopathic activity, it is important to utilize a practical effective screening method in both controlled and natural conditions for measurement of allelopathic potential. It is hoped that with assistance of modern genetic techniques, new rice cultivars with strong weed suppression ability and acceptable for cultivation by farmers will hopefully appear very soon.

9. Benefits from allelopathy integrated into sustainable agriculture

If allelopathy can be integrated into sustainable agriculture appropriately, the heavy dependence on synthetic pesticides and other agrochemicals can be significantly minimized. Mono culture has caused imbalances in agricultural production, and this would be replaced by a more ecological and sustainable cropping system. In modern agriculture with its shortage of labour, it is difficult to completely alter the use of agrochemicals, but the biological characteristics of crops including allelopathy and strength of competition should be exploited to reduce the amount of pesticides and agrochemicals used. Furthermore, unsafe pesticides and agrochemicals must be replaced by safer bioactive products, which are derived from living organisms such as plants, fungi, bacteria, and micro-organisms. The detrimental effects from allelopathy integration into agricultural production should also be noted, as only their benefits have been detailed [37]. The competition and chemical interaction of crops can effectively inhibit weeds and other pests, but they may also have harmful effects for crops in the next cropping seasons. Allelochemicals released from living plants and decomposition includes many toxins, which may suppress growth of useful bacteria, fungi, and micro-organisms, but they may cause problems to mineralization and nitrification in soils. This issue can be excluded with common crops, but should be examined when plant (other than common crops and legumes) materials are incorporated into soils. This style of application is still useful in many developing countries, in which a major proportion of the population is still involved in agricultural production. The modes of action of allelochemicals need further research to exploit novel allelochemicals and their derivatives in the development of bioactive pesticides. However, in addition, the extent to which they cause detrimental effects to crops and soils needs careful examination. Despite the fact that many hypotheses have been developed and discussed, and many experiments have been carried out to test them, the actual modes of action of allelopathy in nature are still somewhat unclear, unlike the allelopathic phenomena that we could easily observe. The allelopathic characteristics of plants have been known for centuries, and extensive research worldwide has been conducted for more than 40 years to elucidate the mode of allelopathy as well as efforts to utilize allelopathy more effectively in agricultural production. However, it can be said that farmers have not yet received much efficacy from what has been observed and reported. Much knowledge on plant allelopathy has been documented, but few approaches have already been successfully applied in agricultural practice. There is no doubt that organic and sustainable agricultural practices are indispensable forms of resource management, with the source of knowledge being traditional agriculture throughout the world [37,110]. What we have researched and discussed about multiple cropping, the use of cover crops, organic compost, and biological controls of pests has been traditionally conducted by farmers without knowledge of allelopathy. Therefore, our achievements on allelopathy should be carefully incorporated with the traditional practices of farmers to create sustainable agriculture integrated with allelopathy. Otherwise, this system will never be feasible for farmers to adopt for economic reasons and in the complex ecological conditions of the tropics, these practices would be inappropriate [110]. In our modern agriculture, ecological and sustainable factors are indispensable. Therefore, what crop species are used and how they are applied in the cropping system are important. Of which, both crop allelopathy and nutrient cycle should be further studied to enhance biological characteristics of crops in the agricultural production. The establishment of allelopathy-integrated sustainable agriculture is obviously varied among cultivating regions, of which opinions of farmers regarding traditional cropping system should be referred, and should be carefully examined and repeated before introducing to farmers for agricultural practices. An agricultural production that is sustainable, economical, less labour-intensive, can be easily implemented by farmers, and supported by local authorities could be helpful for farmers in developing countries to eliminate poverty. To date, a number of phytotoxins involved in the allelopathic activities of worldwide rice cultivars have been identified and isolated, and the fate of these compounds in the environment has been gradually understood, and mode of allelopathy is therefore much clearer. Many novel secondary metabolites have been synthesized and marketed as bioactive pesticides, which effectively aid the integration of sustainable agriculture with allelopathy. The use of allelopathy as a tool for a more bio-rational management of natural resources is not a simple panacea for the solution of ecological problems in agro-ecosystems or in natural ecosystems. It is necessary to develop a scientific approach based on the disciplines of botany, ecology, chemistry, microbiology, agronomy, entomology, and biochemistry, and to work together to clarify these bio-chemical interactions from a holistic point of view, as well as utilize them for beneficial purposes in the management of natural resources in agro-ecosystems [37, 110]. The application of crop rotation, cover crop, mulch, green manure, and incorporation of plant materials with strong allelopathic potential may be more effective in the agricultural practice. The integration of allelopathy via breeding and/or genetic manipulation in rice cultivars may clearly provide specific opportunities for successful implementation of alternative weed management systems [111]. However, knowledge about allelopathy for weed and pest management and establishment of sustainable agriculture integrated with allelopathy should be further introduced to local extension workers and farmers. The modification of allelopathy-integrated sustainable agriculture is needed to allow it to be suitable for different regions. Undoubtedly, the integration of allelopathy in rice will benefit from worldwide collaboration with ecologists, plant breeders, and molecular biologists leading to the successful utilization of new tools for selection of rice cultivars with weed-suppressive traits.

Author details

T.D. Khanh[1], L.H. Linh[1], T.H. Linh[1], N.T. Quan[1], D.M. Cuong[1], V.T.T. Hien[1], L.H. Ham[1], K.H. Trung[1] and T.D. Xuan[2*]

*Address all correspondence to: tdxuan@hiroshima-u.ac.jp; khanhkonkuk@gmail.com

1 Department of Molecular Biology, Agricultural Genetics Institute, Vietnam

2 Graduate School for International Development and Cooperation (IDEC), Hiroshima University, Japan

References

[1] Khanh TD, Chung IM, Xuan TD, Tawata S. The exploitation of crop allelopathy in sustainable agricultural production. Journal of Agronomy and Crop Science 2005; 191 172–184.

[2] Olofsdotter M, Navarez D, Moody K. Allelopathy potential in rice (*Oryza sativa* L.) germplasm. Annals of Applied Biology 1995; 127 543–560.

[3] Molisch H. Der Einfluss einer Pflanze auf die andere—Allelopathie. Jena, Germany: Gustav Fischer; 1937.

[4] Rice EL. Allelopathy. Physiological Ecology. New York, NY: Academic Press; 1974.

[5] Romeo JT, Weidenhamer JD. Bioassays for allelopathy in terrestrial plants. In: Eds. Haynes KF and Millar JG. (eds.) Methods in Chemical Ecology. MA: Kluwer Academic Publishing; 1999. p179–211.

[6] Rice EL. Allelopathy. Physiological Ecology. Orlando, FL: Academic Press; 1984

[7] An M, Pratley JE, Haig T, Jellett P. Genotypic variation of plant species to the allelopathic effect of vulpia residues. Australian Journal of Experimental Agriculture 1997; 37 647–660.

[8] Shibayama H. Weeds and weed management in rice production in Japan. Weed Biology and Management 2001 1 53–60.

[9] Olofsdotter M, Navarez D, Rebulanan M, Streibig JC. Weed suppressing rice cultivars—does allelopathy play a role? Weed Research 1999; 39 441–454.

[10] Khanh TD, Xuan TD, Chung IM. Rice allelopathy and the possibility for weed management. Annals of Applied Biology 2007; 151 324-339.

[11] Khanh TD, Elzaawely AA, Chung IM, Ahn JK, Tawata S, Xuan TD. Role of allelochemicals for weed management in rice. Allelopathy Journal 2007; 19 85-96.

[12] Baltaza AM and Dedatta SK. Weed management in rice. Weed abstract 1992. 41 495-497.

[13] Aliotta G, Cafiero G, Otero AM. Weed germination, seedling growth and their lesson from allelopathy in agricultural. In: Reigosa MJ, Redrol N, Gonzales L. (eds.) Allelopathy: A Physiological Process with Ecological Implications. Dordrecht, the Netherlands: Springer Publisher; 2006. p 285-299.

[14] Holm LG, Plucknett DL, Pancho JV, Herberger JP. The World's Word weeds distribution and biology. The University Press of Hawaii: Honodulu; 1991. p 609.

[15] McIntyre S, Finlayson CM, Ladiges PY, Mitchell DS. Weed community composition and rice husbandry practices in New South Wales, Australia. Agriculture, Ecosystems and Environment 1991; 34 27-45.

[16] Heap IM. The occurrence of herbicides-resistant weeds worldwide. Pesticide Science 1997; 51 235-243.

[17] Anaya AL. Allelopathic organisms and molecules: promising bio-regulators for the control of plant diseases, weeds, and other pests. In: Inderjit, Mukerji KG. (eds.) Allelochemicals: Biological Control of Plant Pathogens and Diseases. Springer Dordrecht: The Netherlands; 2006. p 31–79.

[18] IFIC. International Food Information Council. Agriculture and Food Production, Background on Agriculture and Food Production; 2007. http://www.ific.org/ food/agriculture/index.cfm (accessed 25 November 2012).

[19] Kohli RK, Batish D, Singh HP. Allelopathy and its implication in agroecosystems. Journal of Crop Production 1998; 1 169–202.

[20] Worsham AD. Current and potential techniques using allelopathy as an aid in weed management. In: Chou CH, Waller GR. (eds.) Phytochemical Ecology: Allelochemical, Mycotoxins and Insect Pheromones and Allomones. Monograph Series, No. 9, Institute of Botany, Academia Sinica, Taipei, ROC; 1989. p 275–89.

[21] Duke SO, Dayan FE, Rimando AM. Natural products and herbicide discovery. In: Cobb HS, Kirkwood RC. (eds.) Herbicides and their Mechanism of Action. Sheffield Academic Press, Sheffield: UK; 2000. p 105–33.

[22] Mitchell G, Bartlett DW, Fraser TEM, Hawkes TR, Holt DC, Townson JK et al. Mesotrione: a new selective herbicide for use in maize. Pest Management Science 2001; 57 20–28.

[23] Mamolos AP, Kalburtji KL. Significance of allelopathy in crop rotation. Journal of Crop Production 2001; 4 197–218.

[24] Lampkin L. Organic Farming. Farming Press, Ltd, Ipswich: UK; 1994.

[25] Horst WJ, Haerdter R. Rotation of maize with cowpea improves yield and nutrition use of maize compared to maize monocropping in an alfalfa soil in Northern Guinea Savanna of Ghana. Plant and Soil 1994; 160 171–183.

[26] Kessavalou A, Walters DT. Winter rye as a cover crop following soybean under conservation tillage. Agronomy Journal 1997; 89 68–74.

[27] Patrick ZA, Koch LW. 1963 The adverse influence of phytotoxic substances from decomposing plant residues on resistance of tobacco to black root rot. Canadian Journal of Botany 1963; 41 747–758.

[28] Chou CH. Allelopathy in relation to agricultural productivity in Taiwan: problems and prospects. In: Rizvi SJH, Rizvi V. (eds.) Allelopathy: Basic and Applied Aspects. Chapman and Hall: London; 1992. p 179–204.

[29] Johnson AW. Specific crop rotation effects combined with cultural practices and nematicides. In: Sasser JN, Carter CC. (eds.) An Advanced Treatise on Meloidogyne. North Carolina State University Press: Raleigh NC; 1985. p283–301.

[30] Rizvi SJH, Rizvi V. Exploitation of allelochemicals in improving crop productivity. In: Rizvi SJH, Rizvi V. (eds.) Allelopathy, Basic and Applied Aspects. Chapman and Hall: London; 1992. p 443–73.

[31] Narwal SS. Allelopathy in ecological sustainable agriculture. In: Reigosa MJ, Pedrol N, Gonzalez L. (eds) Allelopathy – A Physiological Process with Ecological Implications. Springer Dordrecht: The Netherlands; 2006.p 512–37.

[32] Batish DR, Singh HP, Kohli RK, Kaur S. Crop allelopathy and its role in ecological agriculture. Journal of Crop Production 2001; 4 121–62.

[33] Swanton CJ, Murphy SD. Weed science beyond the weeds: the role of integrated management (IWM) in agroecosystem health. Weed Science 1996; 44 437–445.

[34] Gallandt ER, Liebman M, Huggins DR. Improving soil quality: implications for weed management. Journal of Crop Production 1999; 2 95–121.

[35] Fageria NK, Baligar VC, Bailey BA. Role of cover crop in improving soil and row crop productivity. Communications in Soil Science and Plant Analysis 2005; 26 2733–2757.

[36] Worsham AD, Blum U. Allelopathic cover crops to reduce herbicide inputs in cropping systems. In: Richardson RG. (ed.) Proceedings of the First International. Weed Control Congress. Weed Science Society of Victoria : Melbourne,VIC Australia; 1992. p 577–579.

[37] Khanh TD, Chung IM, Tawata S, Xuan TD. Allelopathy for weed management in sustainable agriculture. CAB Reviews: Perspectives in Agriculture, Veterinary Science, Nutrition and Natural Resources 2007; 2 (034).

[38] Foley MC. Genetic approach to the development of cover crops for weed management. Journal of Crop Production 1999; 2(1) 77–93.

[39] Frick B, Johnson E. Using allelopathic and cover crops to suppress weeds. Research Report – Scott Research Farm; 2002. p. 125–126.

[40] Boydston RA, Al-Khatib K, Hang A, Krishnan G, Nissen S. Weed control with rapeseed (Brassica napus) and white mustard (*Brassica hirta*) as green manure crops. Weed Science Society of American, Abstract 1994; 34 89.

[41] Al-Khatib K, Boydston R. Weed control with Brassica green manure crop. In: Narwal SS. (ed.) Allelopathy Update, Vol. 2. Basic and Applied Aspects. Oxford and IBH: New Delhi, India; 1999. p 255–227.

[42] Kohli RK, Batish DR, Singh HP. Allelopathic interactions in agroecosystems. In: Reigosa MJ, Pedrol N, Gonzalez L. (eds.) A Physiological Process with Ecological Implication. Springer; 2006. p 465–493.

[43] Nava RV, Fernandez LE, Del Amo RS. Allelopathic effects of green fronds of *Pteridium aquilinum* on cultivated plant, weeds, phytopathogenic fungi and bacteria. Agriculture, Ecosystems and Environment 1987;18 357–379.

[44] Hartwig NL, Hoffman LD. Suppression of perennial legume and grass cover crop for no-tillage corn. Proceedings of the Northeastern Weed Science Society 1975; 29 82–88.

[45] Vaughan SF, Boydston RA. Volatile allelochemicals released by crucifer green manures. Journal of Chemical Ecology 1997;23 2107–2116.

[46] Bantilan RT, Palada MC, Harwood RK. Integrated weed management. I. Key factors affecting crop weed balance. Philippines Weed Science Bulletin 1974; 1 14–36.

[47] Gliessman SR. Allelopathy and agricultural sustainability. In: Chou CH, Walter GR. (eds.) Phytochemical Ecology: Allelochemicals, Mycotoxins, Insect Pheromones and Allomones. Monograph No. 9. Institute of Botany Academia Sinica: Taipei, Taiwan; 1989. p 69–80.

[48] Xuan TD, Tawata S, Khanh TD, Chung IM. Biological control of weeds and plant pathogens in paddy rice by exploiting plant allelopathy: an overview. Crop Protection 2005;24 197–206.

[49] Khanh TD, Xuan TD, Chin DV, Chung IM, Elzaawely AA, Tawata S. Current status of biological control of paddy weeds in Vietnam. Weed Biology and Management 2006; 6 1–9.

[50] Hong NH, Xuan TD, Tsuzuki E, Khanh TD. Paddy weed control by higher plants from South East Asia. Crop Protection 2004; 23 255-261.

[51] Khanh TD, Hong NH, Xuan TD, Chung IM. Paddy weed control by medicinal leguminous plants from Southeast Asia. Crop Protection 2005; 24 421-431.

[52] Xuan TD, Tsuzuki E, Terao H, Matsuo M, Khanh TD. Alfalfa, rice by-products, and their incorporation for weed control in rice. Weed Biology and Management 2003; 3 137-144.

[53] Xuan TD, Tsuzuki E, Terao H, Matsuo M, Khanh TD, Chung IM. Evaluation on phytotoxicity of neem (*Azadirachta indica.* A. Juss) to crops and weeds. Crop Protection 2004; 23 335-345.

[54] Xuan TD, Tsuziki E, Tawata S, Khanh TD. Methods to determine allelopathic potential of crop plants for weed control. Allelopathy Journal 2004; 13 149-164.

[55] Xuan TD , Tsuzuki E. Effects of application of alfalfa pellet on germination and growth of weeds. Journal of Crop Production 2001; 4 303-312.

[56] Khanh TD, Chung IM, Tawata S, Xuan TD. Weed suppression by *Passiflora edulis* and its potential allelochemicals. Weed Research 2006; 46 296-303.

[57] Xuan TD, Tawata S, Hong NH, Khanh TD, Chung IM. Assessment of phytotoxic action of *Ageratum conyzoides* L. (billy goat weed) on weeds. Crop Protection 2004; 23 335-345.

[58] Deba F, Xuan T D, Yasuda M, Tawata S. Herbicidal and fungicidal activities and identification of potential phytotoxins from *Bidens pilosa.* var. *Radiata.* Weed Biology and Management 2007; 7 77-83.

[59] Xuan TD, Tsuzuki E, Matsuo M, Khanh TD. Correlation between inhibitory exhibition and suspected allelochemicals in alfalfa (*Medicago sativa* L.). Plant Production Science 2003; 6 165-171.

[60] Xuan TD, Tawata S, Khanh TD, Chung IM. Decomposition of allelopathic plants in soil. Journal of Agronomy and Crop Science 2005; 191 162-171.

[61] Khanh TD, Hong NH, Nhan DQ, Kim SL, Chung IM, Xuan TD. Herbicidal activity of *Stylosanthes guianensis* and its phytotoxic components. Journal of Agronomy and Crop Science 2006; 192 427-433.

[62] Chung IM, Kim JT, Kim SH. Evaluation of allelopathic potential and quantification of momilactone A, B from rice hull extracts and assessment of inhibitory bioactivity on paddy field weeds. Journal of Agricultural and Food Chemistry 2006; 54 2527-2536.

[63] Xuan TD, Elzaawely AA, Fukuta M, Tawata S. Herbicidal and antifungal activities of lactones in kava (*Piper methysticum*). Journal of Agricultural and Food Chemistry 2006; 54 720-725.

[64] Xuan TD, Elzaawely AA, Deba F, Fukuta M, Tawata S. Mimosine as a potent herbicide. Agronomy for Sustainable Development 2006; 26 89-97.

[65] Duke SO, Dayan FE, Romagni JG, Rimando AM. Natural products as sources of herbicides: current status and future trends. Weed Research 2000; 40 99-111.

[66] Zhu J, Majikina M, Tawata S. Syntheses and biological activities of pyranyl-substituted cinnamates. Bioscience Biotechnology and Biochemistry 2001; 65 161-163.

[67] Tawata S, Taira S, Kobamoto S, Ishihara M, Toyama S. Syntheses and biological activities of dihydro-5,6-dehydrokawain derivatives. Bioscience Biotechnology and Biochemistry 1996; 60 1643-1645.

[68] Tawata, S., Taira, S., Kobamoto, N., Zhu, J., Ishihara, M. and Toyama, S. (1996). Synthesis and antifungal activity of cinnamic acid esters. Bioscience Biotechnology and Biochemistry 60: 909-910.

[69] Lin J Jr, Smith RJ, Dilday RH. Allelopathic activity of rice germplasm on weed. Proceedings in Southern Weed Science Society; 1992 45 99.

[70] Khan AH, Vaishya RD. Allelopathic effects of different crop residues on germination and growth of weeds. In: Tauro P, Narwal SS.(eds.) Proceedings of National Symposium on Allelopathy in Agroecosystem, 12–14 February 1992, CCS Haryana Agricultural University. Hisar, India: Indian Society of Allelopathy; 1992.p 59–60.

[71] Young CC, Zhu C, Throne LR, Waller GR. Phytotoxic potential of soils and wheat straw in rice rotation cropping systems of subtropical Taiwan. Plant Physiology 1989; 120 95–101.

[72] Tamak JC, Narwal SS, Singh L, Singh I. Effect of aqueous extract of rice stubble and straw + stubble on the germination and seedling growth of wheat, oat, berseem and lentil. Crop Research 1994; 8 180–185.

[73] Narwal SS. Weed management in rice: wheat rotation by allelopathy. Critical Reviews in Plant Science 2000; 19 249–266.

[74] Jung WS, Kim KH, Ahn JK, Hahn SJ, Chung IM. Allelopathic potential of rice (*Oryza sativa* L.) residues against *Echinochloa crus-galli*. Crop Protection 2004; 23 211–218.

[75] Ahn JK, Chung IM. Allelopathic potential of rice hulls on germination and seedling growth of barnyard grass. Agronomy Journal 2000; 92 1162–1167.

[76] Inderjit, Rawat D, Foy CL. Multifaceted approach to determine rice straw phytotoxicity. Canadian Journal of Botany 2004; 82 168–176.

[77] Pheng S, Olofsdotter M, Jahn G, Adkins S. Use of phytotoxic crop residues for weed management. Weed Biology and Management 10; 176-184

[78] Chou CH. Allelopathic researches in subtropical vegetation in Taiwan. Comparative Physiology and Ecology 1980; 5: 222–234.

[79] Singh HP, Daizy R, Batish DR, Kohli RK. Autotoxicity: concept, organism, and ecological significance. Critical Reviews in Plant Science 1999; 18 757–772.

[80] Chou CH. Allelopathy and sustainable agriculture. In: Inderjit, Dakshini KMM, Einhellig FA. (eds.) Allelopathy: Organisms, Processes and Application . ASC Symposium Series No. 582. Washington, DC: American Chemistry Society; 1995. p 211–223.

[81] Kuk YI. Evaluation of rice by-products for weed control. Weed Science 2001; 49 141–147.

[82] Olofsdotter M. Getting closer to breeding for competitive ability and the role of allelopathy—an example from rice (*Oryza sativa*). Weed Technology 2001; 15 798–806.

[83] Olofsdotter M. Rice-a step toward use of allelopathy. Agronomy Journal 2001; 93 3–8.

[84] Kong CH, Chen XH, Hu F, Zhang SZ. Breeding of commercially acceptable allelopathic rice cultivars in China. Pest Management Science 2011; 67 1100-1106.

[85] Xu M, Galhano R, Wiemann P, Bueno E, Tiernan M, Wu W, Chung IM, Gershenzon J, Tudzynski B, Sesma A, Peter RJ. Genetic evidence for natural product-mediated plant-plant allelopathy in rice (*Oryza sativa*). New Phytologist 2012; 193 570-575.

[86] Belz RG. Allelopathy in crop/weed interactions – an update. Pest Management Science 2007; 63:308–326.

[87] Dilday RH, Mattice JD, Moldenhauer AK. An overview of rice allelopathy in the USA. In: Kim KU, Shin DH. (eds.) Rice Allelopathy. Kyungpook National University, Taegu, Korea; 2000. p 15–26.

[88] Courtois B, Olofsdotter M. (1998) Incorporating the allelopathy trait in upland rice breeding program. In Pro-ceedings of Workshop on Allelopathy in Rice, 25–27 November 1996, Makati City, pp. 57–67. Ed M. Olofsdotter. Manila, The Philippines. International Rice Research Institute.

[89] Dilday RH, Yan WG, Moldenhauer AK, Gravois KA. Allelopathic activity in rice for controlling major aquatic weeds. In: Olofsdotter M. (ed.) Proceedings of the Workshop on Allelopathy in Rice, 25–27 November 1996, Makati City, Manila, The Philippines: International Rice Research Institute; 1998.p 7–26.

[90] Jensen LB, Courtois B, Shen L, Li Z, Olofsdotter M, Mauleon RP. Locating genes controlling allelo-pathic effects against barnyardgrass in upland rice. Agronomy Journal 2001; 93 16–21.

[91] Okuno K, Ebana K. Identification of QTL controlling allelopathic effects in rice: genetic approaches to biological control of weeds. Japan Agricultural Research Quarterly 2003; 37 77–81.

[92] Okuno K, Ebana K, Hegab M. Challenges for biological weed control using genetic diversity of rice-QTL and candidate compounds associated with allelopathic effect. CS2–S1, 5th International Crop Science Congress and Exhibition (ICSC 2008); 2008.

[93] Zeng DL, Qian Q, Teng S, Dong GJ, Fujimoto H, Yasufumi K, Zhu LH. Genetic analysis of rice allelopathy. Chinese Science Bulletin 2003; 48 265–268.

[94] Lee SB, Seo KI, Koo JH, Hur HS, Shin JC. QTLs and molecular markers associated with rice allelopathy. In: Harper JDI, An M, Wu H, and Kent JH. (eds.) Fourth World Congress on Allelopathy "Establishing the Scientific Base"Wagga Wagga, Australia: CharlesSturt University2005. p 505–507

[95] Lin WX, He HQ, Chen XX, Song BQ, Liang YY, Liang KJ. Use of ISSR molecular markers approach to estimate genetic diversity in rice and barley allelopathy. In: Harper JDI, An M, Wu H, Kent JH. (eds). Proceedings of World Fourth Congress on Allelopathy. Wagga Wagga, Australia: Charles Sturt University; 2005.p 168–174.

[96] Harushima Y, Yano M, Shomura A, Sato M, Shimano T, Kuboki Y, Yamamoto T, Lin SY, Antonio BA, Parco A, Kajiya H, Huang N, Yamamoto K, Nagamura Y, Kurata N, Khush GS, Sasaki T. A high-density rice genetic linkage map with 2275 markers using a single F2 population. Genetics 1998; 148 479–494.

[97] Junaedi A, Jung WS, Chung IM, Kim KH. Differentially expressed genes of potentially allelopathic rice in response against barnyardgrass. Journal of Crop Science and Biotechnology 2008; 10 231–236.

[98] Maqbool N, Wahid A, Farooq M, Cheema ZA, Siddique KHM. Allelopathy and abiotic stress interaction in crop plants. In: Allelopathy, Cheema ZA et al. (eds.) Springer-Verlag Berlin: Heidelberg 2013. P 451-468.

[99] Xiong J, Wang HB, Qiu L, Wu HW, Chen RS, He HB, et al. qRT-PCR analysis of key enzymatic genes related to phenolic acid metabolism in rice accessions (*Oryza Sativa* L.) exposed to low nitrogen treatment. Allelopathy Journal 2010; 25 345–356.

[100] Wang HB, He HB, Ye CY, Lu JC, Chen RS, Liu CH, et al. Molecular physiological mechanism of increased weed suppression ability of allelopathic rice mediated by low phosphorus stress. Allelopathy Journal 2010; 25 239–248.

[101] Song B, Xiong J, Fang C, Qiu L, Lin R, Liang Y, Lin W. Allelopathic enhancement and diferential gene expression in rice under low nitrogen treatment, Journal of Chemical Ecology 2008; 34 688–695.

[102] Lin W, He H, Shen L, Chen X, Ke Y, Guo Y, He H. A proteomic approach to analysing rice allelopathy on barnyard grass (*Echinochloa crus-galli* L.) Proceedings of the 4th International Crop Science Congress, Brisbane: Australia; 2004 http://www.cropscience.org.au/icsc2004/poster/2/4/1/1414_xionglw.htm (accessed 26 November 2012).

[103] Jabran K, Farooq M. Implication of potential allelopathic crops in agricultural systems. In: Cheema ZA et al. (eds.) Allelopathy. Springer-Verlag : Berlin Heidelberg; 2013. p 349-385.

[104] De Albuquerque MB, Dos Santos RC, Lima LM, Melo Filho PA, Nogueira RJMC, Da Camara CAG, Ramos AR. Allelopathy, an alternative tool to improve cropping systems. A review. Agronomy for Sustainable Development 2011; 31 379-395.

[105] He HQ, Shen LH, Xiong J, Jia XL, Lin WX, Wu H. Conditional genetic effect of allelopathy in rice (*Oryza sativa* L.) under different environmental conditions. Plant Growth Regul 2004; 44 211–221.

[106] Kim KU, Shin DH. The importance of allelopathy in breeding new cultivars. In: Labrada R. (ed.) Weed Management for Developing Countries. Addendum 1—FAO Plant Production and Protection Paper 120, Rome, Italy: Food Agriculture Organization of the United Nations; 2003. p 290.

[107] Lin W, Kim KU, Liang K, Guo Y. Hybrid rice with allelopathy. In Kim KU, Shin DH. (eds.) Rice Allelopathy, Proceedings of the International Workshop in Rice Allelopathy, 17–19 August 2000, Institute of Agricultural Science and Technology, Kyungpook National University, Taegu, Korea: Chan-Suk Park; 2000. p 49–56.

[108] Kim KU, Shin DH. Progress and prospect of rice allelopathy research. In: Zeng RS, Malik AU, Luo SM (eds.) Allelopathy in sustainable agriculture and forestry. Springer: The Netherlands; 2008. p 189–213.

[109] Ma HJ, Shin DH, Lee IJ, Koh JC, Park SK, Kim KU. Allelopathic K21 selected as promising allelopathic rice. Weed Biology and Management 2006; 6 189–196.

[110] Anaya AL. Allelopathy as a tool in management of biotic resources. Critical Reviews in Plant Sciences 1999;18 697–39.

[111] Bertin C, Weston LA, Kaur H. Allelopathic crop development: Molecular and traditional plant breeding approaches. In: Janick J. (ed.) Plant Breeding Review; 2008. p 231-254.

Chapter 5

Weed and Disease Control and Peanut Response Following Post—Emergence Herbicide and Fungicide Combinations

W. James Grichar, Peter A. Dotray and
Jason E. Woodward

Additional information is available at the end of the chapter

http://dx.doi.org/10.5772/55949

1. Introduction

Peanut, or groundnut (*Arachis hypogaea* L.), is a species in the legume or "bean" family (Fabaceae). Hypogaea means "under the earth" [1]. Peanuts are known by many other local names such as earthnuts, goober peas, monkey nuts, pygmy nuts and pig nuts [2,3]. Peanut was probably first domesticated and cultivated in the valleys of Paraguay [3].

The domesticated peanut is an amphidiploid or allotetraploid, meaning that it has two sets of chromosomes from two different species, thought to be *A. duranensis* and *A. ipaensis*. These likely combined in the wild to form the tetraploid species, *A. monticola*, which gave rise to the domesticated peanut [4,5]. This domestication might have taken place in Paraguay or Bolivia, where the wildest strains are found today. Archeologists have dated the oldest specimens to about 7,600 years, found in Peru [3,4]. Cultivation spread as far as Mesoamerica where the Spanish conquistadors found the tlalcacahuatl (Nahuatl = "peanut", whence Mexican Spanish, cacahuate and French, cacahuète) being offered for sale in the marketplace of Tenochtitlan (Mexico City). The plant was later spread worldwide by European traders [3].

Peanuts grow best in light, sandy loam soil. They require 120 to 150 days of warm weather, and an annual rainfall of 380 to 650 mm or the equivalent in irrigation water [6]. It is an annual herbaceous plant growing 30 to 50 cm tall. The leaves are opposite, pinnate with four leaflets (two opposite pairs; no terminal leaflet), each leaflet 1 to 7 cm long and 1 to 3 cm wide. The orange-veined, yellow-petaled, pea-like flower (2 to 4 cm across) of *A. hypogaea* is borne in axillary clusters above ground. Following self-pollination, the flowers fade and wither. The stalk at the base of the ovary, called the pedicel, elongates rapidly, and turns downward.

Continued stalk growth pushes the ovary underground where the mature fruit develops into a legume pod (the peanut). The fruits or pods have wrinkled shells that are constricted between pairs of the one to four (usually two) seeds per pod [4,5].

Harvesting occurs in two stages [6]. In modern, mechanized systems, a machine called a digger is used to cut off the main or tap root of the peanut plant by cutting through the soil just below the level of the peanut pods. The machine lifts the plant from the ground, shakes and inverts the plant, leaving it upside down on the ground to keep the peanut pods out of the soil. This allows the peanuts to dry slowly to a bit less than a third of their original moisture level over a period of three to four days depending on weather conditions [6]. Prior to mechanization, peanuts were pulled and inverted by hand [6]. The second stage consists of the use of a combine to remove peanuts from the vine.

World peanut production totals approximately 34 million metric tons per year (Table 1). China leads in production of peanuts, having a share of about 46% of overall world production, followed by India (17%), and the United States (6 %) [7]. The United States is one of the world's leading exporters, with average annual exports of between 200,000 and 250,000 metric tons. Argentina and China are other significant exporters [7].

Peanut production requires the use of a wide range of agrichemical products to control weed and diseases and optimize crop growth and development [8-10]. Peanut has several unique features that contribute to challenging weed management [10]. Peanut cultivars grown in the United States require a fairly long growing season (140 to 160 days), depending on cultivar and geographical region [10,11]. Consequently, soil-applied herbicides may not provide season-long control and mid-to-late season weed emergence can occur. Peanut also has a prostrate growth habit, a relatively shallow canopy, and is slow to shade inter-rows allowing weeds to be more competitive [10,12]. Additionally, peanut fruit develop underground on pegs originating from branches that grow along the soil surface. This prostrate growth habit and pattern of fruit development restricts cultivation to an early season control option [10,13]. With conventional row spacing (91 to 102 cm), complete ground cover may not be attained until 8 to 10 weeks after planting. In some areas of the United States peanut growing region, complete canopy closure may never occur.

Pigweed (*Amaranthus* spp.) is listed as one of the ten most common weeds in most peanut-growing states in the United States, with Palmer amaranth (*Amaranthus palmeri* S. Wats) ranked as the fourth most common weed in South Carolina [14]. Palmer amaranth is not generally ranked as a troublesome weed in most crops in the United States; however, it is a common weed in many crops produced around the world. Palmer amaranth is currently found in the southern half of the United States [15] while in Texas, Palmer amaranth can be found in all areas of the state [16] and is a severe problem in many peanut fields when not properly controlled [17]. Texas millet (*Urochloa texana* (Buckley) R. D. Webster) is a large seeded, vigorous, fast growing annual grass commonly found in peanut fields in parts of Florida, South Carolina, Oklahoma, and Texas [14]. It is listed as one of the most trouble-some weeds in all peanut growing states except Alabama and Georgia [14]. During the digging operation, the peanut plant is lifted out of the ground and inverted. A heavy stand of Palmer amaranth or Texas millet can reduce the effectiveness of the process. The tight fibrous root system becomes intertwined with the peanut plant, causing peanut pods to be stripped from the vine during digging. Peanuts that become detached from the plant remain unharvested in or on the soil [18].

Rank	Country	Production (Million metric tons)
1	China	15.64
2	India	5.85
3	United States	1.89
4	Nigeria	1.55
5	Senegal	1.29
6	Indonesia	1.25
7	Burma	1.14
8	Argentina	1.05
9	Sudan	0.85
10	Chad	0.47
11	Ghana	0.44
12	Vietnam	0.44
13	Congo Kinshasa	0.37
14	Burkino Faso	0.35
15	Mali	0.28
16	Malawi	0.27
17	Guinea	0.26
18	Cameroon	0.24
19	Brazil	0.23
20	Egypt	0.19
	Total	34.05

Source: USDA Foreign Agricultural Service; Table 13 Peanut Area, Yield, and Production (Created 8/10/2012)

Table 1. Worldwide peanut production.

The dinitroaniline herbicides are registered for use in over forty crops [19]. These herbicides provide excellent control of annual grasses [10,18,20] and are the only soil-applied herbicides registered for use in peanut that will provide full-season control of Texas millet [10,21,22]. Peanut tolerance to the dinitroaniline herbicides has been questioned previously [23,24,25]. Greenhouse studies showed that ethalfluralin inhibited seedling growth more than pendimethalin at equivalent rates applied preplant incorporated; however, injury by these herbicides following preemergence applications were similar [26]. In runner peanuts, which are more prone to peg injury compared to Spanish peanut [27], proper herbicide incorporation was needed to prevent injury [28]. Merkle [27] stated that sporadic injury to runner peanut from trifluralin was due to the failure to properly incorporate the herbicide. No differences were observed in a study examining peanut growth, yield, and grade effects with ethalfluralin, pendimethalin, or trifluralin in two different studies [24,29]. In Florida, ethalfluralin did not cause peanut injury at any rate or application timing [23]. Dinitroaniline injury on peanut includes swollen hypocotyl, abnormal lateral root growth, and stunted plants [18,28].

Metolachlor is commonly used in peanut for control of small-seeded broadleaf weeds, some annual grasses, and yellow nutsedge [30]. *S*-metolachlor is labeled for either preplant incorporated (PPI), POST-plant incorporated, preemergence (PRE), postemergence (POST), or layby in peanut [31]. The registered rate for the southwest United States is 1.1 to 1.4 kg/ha [31]. However, many growers have reported peanut stunting when soil applications of metolachlor have been followed by rain [30]. Grichar et al. [30] reported that POST applications of metolachlor followed by (fb) irrigation within 24 hour could be effective for yellow nutsedge control and reduce the chance of peanut injury from soil applications of metolachlor. Combinations of factors, such as herbicide rate, moisture conditions at planting, soil organic matter, and pH may affect peanut injury by chloroacetamide herbicides such as *S*-metolachlor [32-35]. Cardina and Swann [32] reported that metolachlor often delayed peanut emergence and reduced peanut growth when irrigation followed planting. However, yield loss was observed only when metolachlor was applied at a 3X rate.

Several postemergence herbicides are used to control weed escapes. Imazethapyr and imazapic are imidazolinone herbicides registered for use in peanut. Imazethapyr may be applied PPI, PRE, ground cracking (GC), or POST for effective weed control [10]. Imazethapyr applied PPI or PRE controls many troublesome weeds such as coffee senna (*Cassia occidentalis* L.), common lambsquarters (*Chenopodium album* L.), morningglory species (*Ipomoea* spp.), pigweed species (*Amaranthus* spp.) including Palmer amaranth, prickly sida (*Sida spinosa* L.), purple and yellow nutsedge (*Cyperus rotundus* L. and *C. esculentus* L., respectively), spurred anoda [*Anoda cristata* (L.) Schlecht.], and wild poinsettia (*Euphorbia heterophylla* L.) [29,36-39].

Imazethapyr applied POST provides broad spectrum and most consistent control when applied within 10 days of weed emergence [37,40,41]. Imazethapyr and imazapic are the only POST herbicides to effectively control both yellow and purple nutsedge [29,42]. Control is most effective when imazethapyr is applied to the soil or to yellow nutsedge that is no more than 12 cm tall [10,42,43].

Imazapic is similar to imazethapyr and controls all the weeds controlled by imazethapyr [10,44-46]. In addition, imazapic provides control and suppression of Florida beggarweed [*Desmodium tortuosum* (S.W.) D.C.] and sicklepod [*Senna obtusifolia* (L.) Irwin & Barneby), which are not adequately controlled by imazethapyr [47]. Imazethapyr provides consistent control of many broadleaf and sedge species if applied within 10 days after emergence, but imazapic has a longer effectiveness period when applied POST [10,42,46,48]. Imazapic also is effective for control of rhizome and seedling johnsongrass [*Sorghum halepense* (L.) Pers.], Texas millet, large crabgrass [*Digitaria sanguinalis* (L.) Scop.], southern crabgrass [*Digitaria ciliaris* (Retz.) Koel.], and broadleaf signalgrass [*Brachiaria platyphylla* (Griseb.) Nash] [46].

Peanut is susceptible to numerous fungal diseases caused by foliar and soilborne pathogens. Chlorothalonil has been the most widely used fungicide in the United States peanut production areas for control of early leaf spot caused by *Cercospora arachidicola* S. Hori, late leaf spot caused by *Cercosporidium personatum* Berk. & M.A. Curtis, and rust caused by *Puccinia arachidis* Speg. for over 30 years [49,50]. Despite its widespread use across the peanut belt, chlorothalonil continues to provide effective control of foliar diseases [50,51]; however, it has no activity against any of the soilborne diseases such as southern stem rot caused by *Sclerotium rolfsii* Sacc. or Rhizoctonia pod or stem rot caused by *Rhizoctonia solani* Kühn [49,52,53]. Within the past 15 years, several fungicides, including the sterol biosynthesis inhibitor fungicide, tebuconazole, along with the strobilurin fungicides azoxystrobin and pyraclostrobin have been registered for use in peanut for control of both leaf spot and soilborne diseases [49,52-55].

Depending on the fungicide, the calendar spray regime in the southeastern United States may result in seven applications [50,52] while in the southwest United States peanut growing region a maximum of five fungicide applications may be applied during the growing season [53,56]. Chlorothalonil is used to fill the remaining treatment slots in an azoxystrobin, pyraclostrobin, tebuconazole program to minimize the risk of fungal pathogens developing resistance to triazole or strobilurin fungicides [57].

Prothioconazole is a sterol biosynthesis inhibitor fungicide in the new triazolinthione class of fungicides [58] that has shown activity against the leaf spot pathogens, *C. arachidicola* and *C. personatum,* as well as the soilborne pathogens *S. rolfsii* and *R. solani* [59]. Prothioconazole has shown promise for control of cereal diseases in Europe when applied alone or in combination with strobilurin fungicides [58]. In addition, the activity of this fungicide on foliar diseases is of special interest because populations of both *C. arachidicola* and *C. personatum* have displayed reduced sensitivity to tebuconazole and noticeable reductions in efficacy of that fungicide [59]. Prothioconazole plus tebuconazole received registration for use in peanut during the 2008 growing season [60].

Management strategies to protect peanut from various weeds, insects, and fungi require multiple applications of herbicides, insecticides, or fungicides. Timing of application of herbicides and fungicides may coincide during the growing season, and co-application of these pesticides is desirable if herbicide or fungicide performance and peanut tolerance are not compromised [61]. Potential interactions related to physiological effects on plants and other organisms, application variables such as adjuvant, water quality, commercial formulation, and environmental stress can affect pesticide compatibility [61].

2. Research needs

Considerable research has been conducted over the past several years to define interactions among pesticides including interactions of herbicides in mixture with other herbicides and fungicides [62-65]. Peanut fungicides are applied beginning approximately 30 to 60 days after planting and can be applied until a few weeks prior to digging. Efficacy of clethodim and sethoxydim can be reduced by co-application with copper-containing fungicides or azoxystrobin, chlorothalonil, and pyraclostrobin [8,66,67]. Fluazinam and tebuconazole did not reduce grass control compared with graminicides applied alone [8,9,66]. Efficacy of herbicides that control dicotyledonous weeds and sedges are not generally affected by fungicides [66]. Weed species and size, and plant stress can affect the magnitude of interactions between herbicides and fungicides [66].

Additional research was conducted to define potential interactions of various postemergence herbicides and fungicides when used in combination on peanut for control of various broadleaf weeds and annual grasses. Therefore, the purpose of this research was to determine interactions of postemergence grass (clethodim and sethoxydim) and broadleaf herbicides (lactofen, imazethapyr, imazapic, aciflurofen, and 2,4-DB) with commonly used peanut fungicides (boscalid, fluazinam, pyraclostrobin, tebuconazole, or prothioconazole plus tebuconazole) for annual grass and broadleaf weed control in peanut as well as the response to foliar and soilborne disease development.

3. Research methods with tank-mix combinations for weed and disease control

3.1. Weed control with tank-mix combinations

Field studies were conducted in two different peanut growing regions of Texas from 2007 through 2010 to determine weed efficacy and peanut response to applications of herbicides and fungicides applied alone and in combination. Field studies at south Texas were conducted at the Texas A&M AgriLife Research site near Yoakum and on the Texas Southern High Plains at Lamesa or Halfway. Soils at the Yoakum site were a Tremona loamy fine sand (thermic Aquic Arenic Paleustalfs) with less than 1% organic matter and pH 7.0 to 7.2. The location near Lamesa was at the Agricultural Complex for Research and Extension Center (AG-CARES) on a Amarillo fine sandy loam (fine-loamy, mixed, superactive, thermic Aridic Paleustalf) with 0.4% organic matter and pH 7.8. The Halfway location was located west of Plainview at the Texas A&M AgriLife Research and Extension Center on a Acuff clay loam (fine-loamy, mixed, thermic Aridic Paleustolls) with less than 1.0% organic matter and pH 7.9.

The experimental design was a randomized complete block with a factorial arrangement of two grass or five broadleaf herbicides by three fungicides with three replications. All studies included a non-treated control. Each plot consisted of two rows spaced 97 or 101 cm apart and 7.6 m long.

3.1.1. Weed efficacy studies

Weed efficacy studies were divided into two groups: 1) a grass herbicide study and 2) a broadleaf weed study. The grass herbicide study included clethodim at 0.14 kg ai/ha or sethoxydim at 0.21 kg ai/ha while the broadleaf weed study included aciflurofen at 0.42 kg ai/ha, imazapic at 0.07 kg ai/ha, imazethapyr at 0.07 kg ai/ha, lactofen at 0.22 kg ai/ha, or 2,4-DB at 0.42 kg ai/ha. Fungicides evaluated included pyraclostrobin at 0.27 kg ai/ha, tebuconazole at 0.23 kg ai/ha, and the premix of prothioconazole at 0.084 kg ai/ha plus tebuconazole at 0.168 kg ai/ha.

Herbicides and fungicides were applied alone and in combination to determine efficacy against various weeds. A crop oil concentrate (Agri-Dex, a blend of 83% paraffin-based petroleum oil and 17% surfactant) at 2.3 L/ha was added to each treatment except in 2007 at Yoakum where a non-ionic surfactant (X-77, 90% nonionic surfactant) at 0.25% v/v was added. Herbicide and fungicides at Yoakum were applied with a CO_2-pressurized backpack sprayer equipped with TeeJet 11002 DG flat fan spray tips (Spraying Systems Company, P.O. Box 7900, North Avenue, Wheaton, IL 60188) that delivered a spray volume of 190 L/ha at 180 kPa while on the Texas High Plains locations, fungicides and herbicides were applied with a CO_2 pressurized backpack sprayer using TeeJet 110015 TT flat fan nozzles calibrated to deliver a spray volume of 94 L/ha at 207 kPa. At Yoakum, the peanut variety Tamrun OL02 [68] was planted in each year at a seeding rate of 112 kg/ha. At the Texas High Plains locations, Flavor Runner 458 [69] was planted at the rate of 100 kg/ha.

Texas millet and southern crabgrass were present at Yoakum in 2007 and 2009 while broadleaf signalgrass was present in 2008. Texas millet was present at Lamesa in 2007. Palmer amaranth was present at Yoakum in 2007, 2008, and 2009, Lamesa in 2007, and Halfway in 2008 and 2009. Smellmelon (*Cucumis melo* L. var. Dudaim Naud.) was present at Yoakum in 2007, 2008, and 2009 while horse purslane (*Trianthema portulacastrum* L.) was present at Yoakum only in 2009. When present, all field plots were naturally infested with dense populations of Texas millet and broadleaf signalgrass at 4 to 6 plants/m^2, southern crabgrass at 6 to 8 plants/m^2, horse purslane at 6 to 8 plants/m^2, smellmelon at 6 to 8 plants/m^2, or Palmer amaranth at 4 to 6 plants/m^2. Typically, treatments were applied when annual grasses were 10 to 26 cm tall, Palmer amaranth was 15 to 30 cm tall, horse purslane was 10 to 20 cm tall, and smellmelon was 15 to 30 cm in length. No attempt was made to harvest peanut in the efficacy studies due to the difficulty in digging weedy plots [10,13,17].

3.1.2. *Weed-free studies*

Studies also were conducted under weed-free conditions at the Lamesa and Halfway in 2008 and 2009. Plots were maintained weed-free with ethalfluralin (Sonalan HFP®, Dow Agro-Sciences, 9330 Zionsville Road, Indianapolis, IN 46268) at 0.84 kg/ha applied preplant incorporated. At Lamesa, Flavor Runner 458 was planted in 2008 while Tamrun OL02 was planted in 2009; at Halfway, the Spanish market type, OLin [70] was planted both years of the study. Seeding rate for the runner market cultivars (Flavor Runner 458, Tamrun OL02) was 90 kg/ha while OLin was planted at 100 kg/ha. Peanut phytotoxicity ratings were recorded throughout the growing season and peanut yield was obtained by digging each plot separately, air-drying in the field for 4 to 7 days, and harvesting pods from each plot with a combine. Weights were recorded after soil and trash were removed from plot samples were adjusted to 10% moisture.

Weed control and peanut phytotoxicity, expressed as chlorosis and necrosis of leaf tissue, was visually estimated on a scale of 0 to 100 (0 indicating no weed kill or leaf chlorosis or necrosis and 100 indicating complete weed or peanut kill), relative to the non-treated control. Weed control was recorded approximately four weeks after POST herbicide applications while peanut phytotoxicity was recorded 5 to 14 days after herbicide application.

3.1.3. Data analysis

Weed control and peanut injury data were transformed to the arcsine square root prior to analysis of variance, but are expressed in their original form for clarity because the transformation did not alter interpretation. Visual estimates of weed control and peanut injury, and yield were subjected to analysis of variance to test effects of POST herbicide and fungicide. Means were compared with the appropriate Fisher's Protected LSD test at the 5% probability level. The non-treated was not included in weed efficacy or peanut injury analysis but was included in peanut yield analysis.

3.2. Disease control with tank-mix combinations

Studies were conducted in two different peanut growing regions of Texas to determine disease control and peanut response to applications of herbicides and fungicides applied alone and in combination. Field studies at south Texas were conducted at the Texas A&M AgriLife Research site near Yoakum while the central Texas studies were conducted at the Texas A&M AgriLife Research and Extension Center near Stephenville. Soils at the Yoakum site were described previously. This site has been in continuous peanut for over forty years so there was a high concentration of soil-borne and foliar disease inoculum. The soil at the Stephenville site was a Windthorst loamy sand (fine mixed thermic Udic Paleustalfs) with less than 1% organic matter and pH 7.6.

3.2.1. Disease efficacy studies

Studies in south Texas were conducted from 2008 to 2010 on early leaf spot and southern blight. These studies included the fungicides pyraclostrobin at 0.27 kg ai/ha, tebuconazole at 0.23 kg ai/ha, and the premix of prothioconazole at 0.084 kg ai/ha plus tebuconazole at 0.168 kg ai/ha and the herbicides aciflurofen at 0.42 kg ai/ha, clethodim at 0.14 kg ai/ha, imazapic at 0.07 kg ai/ha, imazethapyr at 0.07 kg ai/ha, lactofen at 0.22 kg ai/ha, sethoxydim at 0.21 kg ai/ha, or 2,4-DB at 0.42 kg ai/ha. Fungicides and herbicides were applied alone and in combination to determine efficacy against foliar and soilborne diseases. No adjuvant was included in these studies in 2008 or 2009; however, in 2010 a crop oil concentrate (Agri-Dex, a blend of 83% paraffin-based petroleum oil and 17% surfactant) at 2.3 L/ha was added to each treatment.

Fungicides and herbicides alone and in combination were applied with a CO2-pressurized backpack sprayer equipped with three D2-23 hollow-cone spray nozzles per row in 140 L of water/ha at a pressure of 504 kPa. The experimental design was a randomized complete block with a factorial arrangement of seven herbicides by three fungicides. All studies included a non-treated control. Each plot consisted of four rows spaced 97 cm apart and 6.3 m long. The variety Tamrun OL02 [68] was planted in 2008 and 2009 while Florida 07 [71] was planted in 2010 at the rate of 112 kg/ha. Planting dates were June 16, 2008, July 1, 2009, and May 24, 2010.

Studies conducted in central Texas focused on early leaf spot and Sclerotinia blight caused by *Sclerotina minor* Jagger. These studies included the herbicides clethodim at 0.14 kg ai/ha and sethoxydim at 0.21 kg ai/ha and the fungicides boscalid at 0.49 kg ai/ha and fluazinam at 0.88 kg ai/ha. Agridex at 1.0% v/v was included in all treatments. Each plot consisted of two rows spaced 91 cm apart and 7.9 m long. Fungicides and herbicides were applied alone and in combination with a CO2-pressurized backpack sprayer equipped with two 8002VS flat fan spray nozzles per row in 140 L of water/ha at a pressure of 134 kPa. The runner-type variety Flavor Runner 458 [69] was planted each year of the study at 95 kg/ha.

Typical peanut injury resulted in rapid damage to plant tissue after application and manifested as small necrotic lesions. The visible injury on leaflets with 2,4-DB was common and consisted of typical 2,4-DB damage which consisted of elongated leaflets with a slightly faded appearance [10]. This symptomology was not visible on new growth and remained visible on lower leaves throughout the growing season. Peanut phytotoxicity ratings were recorded 7 days after treatment at Yoakum. Peanut injury was estimated visually on a scale of 0 to 100 (0 indicating no leaf chlorosis or necrosis and 100 indicating complete peanut kill), relative to the non-treated control. Severity of leaf spot was rated in the center two rows using the Florida leaf spot scoring system where 1 = no leaf spot, and 10 = plants completely defoliated and dead because of leaf spot [49,59]. Values of 1 through 4 on the scale reflect increasing incidence of leaflets with spots, and occurrence of spots in lower versus upper canopy of the plots; whereas values 4 through 10 reflect increasing levels of defoliation [51]. The leaf spot rating was recorded immediately prior to peanut digging.

Loci of southern stem rot or Sclerotinia blight (where applicable) were counted immediately after peanut plants were inverted. A locus represented 31 cm or less of linear row with one or more plants infected with *S. rolfsii* or *S. minor* [72]. Plots were harvested in south Texas in 2008 and 2010, but not in 2009 due to extremely wet conditions which persisted during late October and November and prevented digging of individual plots (Table 1). Plots were harvested in 2008 and 2009 in central Texas.

All test areas were maintained weed-free with a preemergence tank-mix application of pendimethalin at 1.06 kg ai/ha plus *S*-metolachlor at 1.42 kg ai/ha. Overhead sprinkler irrigation was applied on a 1- to 2-week schedule throughout the growing season as needed.

3.2.2. *Data analysis*

Peanut yields were obtained by digging each plot separately, air-drying in the field for 4 to 7 days, and harvesting pods from each plot with a combine. Weights were recorded after soil and trash were removed from plot samples were adjusted to 10% moisture. Leaf spot ratings and incidence of soilborne disease development were used for comparison of tank-mix combinations. Data were analyzed using PROC GLM with SAS (SAS Institute, Inc., Cary, NC) and a model statement appropriate for a factorial design. Treatments means were separated by Fisher's protected least significant difference test at $P \leq 0.05$.

4. Effects of tank mix combinations on weed control, peanut phytotoxicity, and peanut yield

4.1. Weed efficacy with combinations of herbicides plus fungicides.

There was no herbicide by fungicide by year interaction for Texas millet, southern crabgrass, or broadleaf signalgrass control; therefore, that data are combined over clethodim and sethoxydim herbicides.

4.1.1. Annual grass control

No differences in broadleaf signalgrass, Texas millet, or southern crabgrass control were noted between clethodim or sethoxydim when applied alone or in combination with any of the fungicides (Table 2). Grichar [73] reported that clethodim and sethoxydim controlled 3 to 10 cm tall Texas millet and southern crabgrass at least 85%. Clethodim applied to 15 to 25 cm tall Texas millet or southern crabgrass provided no better than 89% Texas millet control while southern crabgrass control varied from 51 to 95% [73]. Sethoxydim applied to the same height Texas millet or southern crabgrass controlled Texas millet 79 to 87% and southern crabgrass control was no better than 76% [73].

Herbicide	Rate	Texas[c] millet	Southern crabgrass	Broadleaf signalgrass
	kg ai/ha		%	
Clethodim	0.14	96	96	98
Sethoxydim	0.21	95	96	98
LSD (0.05)		NS[d]	NS	NS

[a] Herbicides and rates included clethodim at 0.14 kg ai/ha and sethoxydim at 0.21 kg ai/ha. Fungicides and rates included pyraclostrobin at 0.27 kg ai/ha, tebuconazole at 0.23 kg ai/ha, and the premix of prothioconazole at 0.084 kg ai/ha + tebuconazole at 0.168 kg ai/ha. Data were combined over fungicides due to a lack of interaction.
[b] Texas millet present in south Texas in 2007 and 2009 and at Lamesa in 2007. Southern crabgrass present in south Texas in 2007 and 2009. Broadleaf signalgrass present in south Texas in 2008.
[c] Texas millet, *Urochloa texana* (Buckley) R. D. Webster; Southern crabgrass, *Digitaria ciliaris* (Retz.) Koeler; broadleaf signalgrass, *Brachiaria platyphylla* (Griseb.) Nash.
[d] NS, not significant at the 5% level of probability.

Table 2. Annual grass control with clethodim and sethoxydim.[a,b]

Lancaster et al. [8,9] reported large crabgrass control was reduced with clethodim when applied with pyraclostrobin, chlorothalonil, and azoxystrobin; however, fluazinam, propiconazole plus trifloxystrobin, and tebuconazole did not reduce large crabgrass control by clethodim. Similarly, Jordan et al. [66] reported that azoxystrobin and chlorothalonil, but not tebuconazole, reduced annual grass control by clethodim. Also Lancaster et al. [8,9] reported that large crabgrass control was reduced when sethoxydim was applied with azoxystrobin or pyraclostrobin, but not fluazinam, propiconazole plus trifloxystrobin, or tebuconazole.

4.1.2. Palmer amaranth control

At Yoakum in 2007 and 2008 and Halfway in 2008 there was an herbicide by fungicide interaction; therefore, those data are presented as a 2-way interaction of broadleaf herbicide by fungicide (Table 3). However, only herbicide effects were significant at Lamesa in 2007 and Halfway and Yoakum in 2009 (Table 4).

In 2007 at Yoakum, lactofen, aciflurofen, and imazapic alone controlled Palmer amaranth at least 91% while 2,4-DB and imazethapyr alone provided 83% and 68% control, respectively (Table 3). Lactofen plus tebuconazole and aciflurofen plus either the premix of prothioconazole plus tebuconazole or tebuconazole reduced Palmer amaranth control over each respective herbicide applied alone. In 2008 at Yoakum, all herbicides alone controlled Palmer amaranth at least 92%. Reduced control from each respective herbicide alone was noted with acifluorfen plus either pyraclostrobin or tebuconazole and imazethapyr or 2,4-DB plus pyraclostrobin. At the Halfway location, lactofen and aciflurofen alone provided poor control (≤ 25%) of Palmer amaranth while imazethapyr, imazapic, and 2,4-DB controlled Palmer amaranth at least 77% (Table 3). Only the combination of 2,4-DB plus the premix of prothioconazole plus tebuconazole reduced control when compared to 2,4-DB alone.

At Lamesa, all herbicides controlled Palmer amaranth less than 60% while at Yoakum there was no difference in Palmer amaranth control following all herbicide treatments (Table 4). At the Halfway location, lactofen, imazapic, and imazethapyr controlled Palmer amaranth at least 98% while 2,4-DB and aciflurofen controlled this weed 75% and 54%, respectively.

Grichar [74] reported that imazapic at 0.04 to 0.07 kg/ha controlled Palmer amaranth at least 95% when applied to weeds that were less than 15 cm tall while imazethapyr provided at least 90% control in 2 of 3 years. In other research, Jordan et al. [66] reported that smooth pigweed (*A. hybridus* L.) control by imazethapyr was reduced by tank mixing with fungicides.

4.1.3. Horse purslane control

There was an herbicide by fungicide interaction for horse purslane in 2009. Lactofen and 2,4-DB alone and in combination with fungicides provided almost complete control of horse purslane (Table 3). Aciflurofen alone controlled 97% horse purslane while antagonism was noted with acifluorfen plus the premix of prothioconazole plus tebuconazole combinations. All imazethapyr plus fungicide combinations reduced horse purslane control compared to imazethapyr alone. Imazapic alone or in combination failed to control horse purslane.

Horse purslane can be a stronger competitor with peanut early in the growing season than common purslane due to a more upright growth than that of common purslane [75]. Grichar [75] reported that aciflurofen and lactofen alone or combinations of these herbicides with 2,4-DB controlled horse purslane at least 70% when evaluated 21 days after treatment (DAT), but no greater than 75% control was observed when rated up to 115 DAT. In later work, Grichar

		Palmer amaranth[c]			Horse purslane[d]
		2007	2008		
		Yoakum	Yoakum	Halfway	Yoakum
Herbicide	Fungicide		%		
Lactofen	-	93	92	22	100
Lactofen	Pyraclostrobin	100	100	18	100
Lactofen	Prothioconazole + tebuconazole	78	93	17	100
Lactofen	Tebuconazole	70	93	17	99
Acifluorfen	-	91	97	25	97
Acifluorfen	Pyraclostrobin	73	80	18	80
Acifluorfen	Prothioconazole + tebuconazole	57	97	30	58
Acifluorfen	Tebuconazole	60	85	18	99
Imazethapyr	-	68	100	77	80
Imazethapyr	Pyraclostrobin	88	86	75	0
	Prothioconazole		98		
Imazethapyr	Prothioconazole + tebuconazole	82	98	82	25
Imazethapyr	Tebuconazole	87	93	80	10
Imazapic	-	94	97	96	13
Imazapic	Pyraclostrobin	94	99	94	0
Imazapic	Prothioconazole + tebuconazole	86	99	94	7
Imazapic	Tebuconazole	97	93	94	0
2,4-DB	-	83	96	88	100
2,4-DB	Pyraclostrobin	67	87	87	100
2,4-DB	Prothioconazole + tebuconazole	98	100	28	100
2,4-DB	Tebuconazole	100	97	83	99
LSD (0.05)		20	9	18	32

[a] Agri-Dex at 2.3 L/ha was added to each treatment except in 2007 at Yoakum where X-77 at 0.25% v/v was added.
[b] Herbicides and rates included aciflurofen at 0.42 kg ai/ha, imazapic at 0.07 kg ai/ha, imazethapyr at 0.07 kg ai/ha, lactofen at 0.22 kg ai/ha, or 2,4-DB at 0.42 kg ai/ha. Fungicides and rates included pyraclostrobin at 0.27 kg ai/ha, tebuconazole at 0.23 kg ai/ha, and the premix of prothioconazole at 0.084 kg ai/ha + tebuconazole at 0.168 kg ai/ha.
[c] Palmer amaranth, *Amaranthus palmeri* S. Wats.; horse purslane, *Trianthema portulacastrum* L.
[d] Present only in 2009.

Table 3. Palmer amaranth and horse purslane control with herbicide-fungicide combinations.[a,b]

[76] reported that in one year, lactofen applied to horse purslane less than 15 cm tall controlled this weed 93% while in another year, lactofen applied to horse purslane less than 15 cm tall or 20 to 30 cm tall provided at least 93% control while acifluorfen applied to horse purslane less than 15 cm tall controlled this weed 77%.

4.1.4. Smellmelon control

Only herbicides were significant for smellmelon control (Table 4). No difference in smellmelon control was noted with any herbicides in 2007, while in 2008 imazethapyr produced the worst control. In 2009, lactofen controlled less smellmelon than imazapic. Grichar [76] reported that imazapic provided the most consistent control of smellmelon while acifluorfen, imazethapyr, imazapic, and lactofen controlled at least 80% smellmelon in some years but in other years control was less than 70%. Imazapic at 0.04 to 0.07 kg/ha controlled smellmelon greater than 90% in corn (*Zea mays* L.) regardless whether applied PRE, early POST, or late POST [77]. Grichar [78] reported that imazapic provided consistent control (> 85%) of citronmelon (*Citrullus lanatus* var. *citroides*) in peanut. Typically, season-long smellmelon control with 2,4-DB is poor. This can be attributed to lack of any residual activity of 2,4-DB and continued germination of seed and smellmelon growth [76].

4.1.5. Peanut phytotoxicity with tank mix combinations

4.1.5.1. Clethodim/sethoxydim plus fungicide combinations

No peanut phytotoxicity was noted with any graminicide by fungicide combinations at Yoakum or Halfway (data not shown); however, at Lamesa there was a treatment by year interaction.

	Palmer amaranth[c]			Smellmelon		
	Lamesa	Yoakum	Halfway	2007	2008	2009
			%			
Lactofen	49	94	98	93	99	89
Aciflurofen	38	90	54	88	99	96
Imazethapyr	28	88	99	88	91	95
Imazapic	25	90	98	99	98	98
2,4-DB	59	96	75	93	99	96
LSD (0.05)	6	NS[d]	12	NS	4	9

[a] Data are pooled over herbicides due to a lack of interaction. Herbicides and rates included aciflurofen at 0.42 kg ai/ha, imazapic at 0.07 kg ai/ha, imazethapyr at 0.07 kg ai/ha, lactofen at 0.22 kg ai/ha, or 2,4-DB at 0.42 kg ai/ha. Fungicides and rates included pyraclostrobin at 0.27 kg ai/ha, tebuconazole at 0.23 kg ai/ha, and the premix of prothioconazole at 0.084 kg ai/ha + tebuconazole at 0.168 kg ai/ha.

[b] Palmer amaranth present at Lamesa in 2007, Yoakum in 2009, and Halfway in 2009.

[c] Palmer amaranth, *Amaranthus palmeri* S. Wats.; smellmelon, *Cucumis melo* L. var. Dudaim Naud.

[d] NS, not significant at the 5% level of probability.

Table 4. Weed control with various postemergence herbicides.[a,b]

In 2007 (with Texas millet pressure) and in 2009 (weed-free), peanut phytotoxicity (up to 12%) was evident with clethodim and sethoxydim combinations with either pyraclostrobin, tebuconazole, and the premix of prothioconazole + tebuconazole up to two weeks after application (Table 5). In 2007, clethodim, sethoxydim, or tebuconazole alone or clethodim or sethoxydim in combination with tebuconazole caused no phytotoxicity. All other combinations resulted in at least 3% phytotoxicity. Either graminicide in combination with prothioconazole plus tebuconazole or prothioconazole plus tebuconazole alone caused the greatest phytotoxicity. In 2009, similar results were noted; however, pyraclostrobin alone or in combination with either graminicide caused the greatest injury (Table 5). Subsequent new growth did not exhibit adverse effects of any tank-mix combination and was 2% or less, four weeks after application (data not shown).

Herbicide	Fungicide	2007	2009
		%	
Clethodim	-	2	0
Clethodim	Pyraclostrobin	3	13
Clethodim	Tebuconazole	0	0
Clethodim	Prothioconazole + tebuconazole	10	8
Sethoxydim	-	2	0
Sethoxydim	Pyraclostrobin	5	12
Sethoxydim	Tebuconazole	0	0
Sethoxydim	Prothioconazole + tebuconazole	12	3
-	Pyraclostrobin	4	11
-	Tebuconazole	0	0
-	Prothioconazole + tebuconazole	8	0
LSD (0.05)		3	3

[a] Herbicides and rates included clethodim at 0.14 kg ai/ha and sethoxydim at 0.21 kg ai/ha. Fungicides and rates included pyraclostrobin at 0.27 kg ai/ha, tebuconazole at 0.23 kg ai/ha, and the premix of prothioconazole at 0.084 kg ai/ha + tebuconazole at 0.168 kg ai/ha.

Table 5. Peanut phytotoxicity with graminicide plus fungicide combinations at Lamesa in 2007 and 2009.[a]

4.1.5.2. Aciflurofen, imazapic, imazethapyr, lactofen, or 2,4-DB plus fungicide combinations

Phytotoxicity observations were not recorded in the weed efficacy studies with the exception of Yoakum in 2008; however, phytotoxicity ratings were recorded in the weed-free studies conducted at Lemasa in 2008 and 2009 and Halfway in 2009. In these studies, there was a significant herbicide by fungicide interaction; therefore, data are presented separately by location. Phytotoxicity varied across locations and treatments but in most instances was greater with the use of aciflurofen or lactofen.

In 2008 at Yoakum, lactofen alone and in combination with prothioconazole plus tebuconazole or tebuconazole alone caused at least 10% peanut phytotoxicity while aciflurofen alone or in combination with any of the fungicides caused 4 to 7% phytotoxicity. At Lamesa, combinations with aciflurofen, lactofen, and 2,4-DB caused the greatest injury (Table 6). Imazethapyr or imazapic alone or in combination with pyraclostrobin resulted in no injury. Imazethapyr plus tebuconazole caused no injury while imazapic plus tebuconazole resulted in 10% injury.

In 2009 at Lamesa, imazapic, imazethapyr, and 2,4-DB alone resulted in no injury; however, imazapic plus either pyraclostrobin or prothioconazole plus tebuconazole, imazapic plus pyraclostrobin, and 2,4-DB plus any fungicide resulted in 5 to 15% phytotoxicity (Table 6). Slight peanut phytotoxicity was also noted with the fungicides pyraclostrobin and tebuconazole. At Halfway, peanut injury with aciflurofen or lactofen was greater than at Lamesa with the exception of lactofen plus pyraclostrobin which caused 9 to 10% injury at both locations (Table 6).

4.1.6. Peanut yield as influenced by tank mix combinations

Under weed-free conditions, when using either the grass or broadleaf herbicides with fungicides, no negative response with respect to peanut yield was noted when compared with the non-treated control for either runner or Spanish market types (data not shown). Most studies conducted on herbicide-fungicide interactions on peanut have focused on either weed efficacy or disease control and few have reported on effect on peanut yield. No studies could be found that reported any peanut yield reductions with clethodim or sethoxydim under weed-free conditions. Although lactofen at 0.22 kg/ha caused peanut leaf bronzing and spotting [74], lactofen produced a similar yield when compared to the untreated, weed-free control [79]. Richburg et al. [80] reported no yield differences with runner, Spanish, or Virginia peanut cultivars with imazethapyr at 0.07 kg/ha in Georgia or Texas. No reduction in peanut grade or yield following imazapic treatments have been observed in several studies [76,81,82]. Grichar et al. [83] reported that single and multiple applications of 2,4-DB at 0.45 kg/ha did not affect runner-type yield.

		Phytotoxicty[c]			
		2008		2009	
		Yoakum	Lamesa	Lamesa	Halfway
Herbicide	**Fungicide**	%			
-	-	0	0	0	0
Lactofen	-	10	10	13	23
Lactofen	Pyraclostrobin	2	4	9	10
Lactofen	Prothioconazole + tebuconazole	12	10	8	22

		Phytotoxicty[c]			
		2008		2009	
		Yoakum	Lamesa	Lamesa	Halfway
Lactofen	Tebuconazole	12	12	7	25
Acifluorfen	-	5	5	5	20
Acifluorfen	Pyraclostrobin	4	5	5	17
Acifluorfen	Prothioconazole + tebuconazole	7	9	5	23
Acifluorfen	Tebuconazole	6	9	5	22
Imazethapyr	-	0	0	0	0
Imazethapyr	Pyraclostrobin	2	0	7	0
Imazethapyr	Prothioconazole + tebuconazole	0	4	5	12
Imazethapyr	Tebuconazole	0	0	0	0
Imazapic	-	0	0	0	0
Imazapic	Pyraclostrobin	0	0	8	0
Imazapic	Prothioconazole + tebuconazole	0	10	0	3
Imazapic	Tebuconazole	0	10	0	3
2,4-DB	-	0	5	0	10
2,4-DB	Pyraclostrobin	3	12	15	18
2,4-DB	Prothioconazole + tebuconazole	1	12	10	20
2,4-DB	Tebuconazole	0	6	5	5
-	Pyraclostrobin	0	0	10	3
-	Prothioconazole + tebuconazole	0	1	0	13
-	Tebuconazole	0	0	5	0
LSD (0.05)		3	1	2	5

[a] Agri-Dex at 2.3 L/ha was added to each treatment.
[b] Herbicides and rates included aciflurofen at 0.42 kg ai/ha, imazapic at 0.07 kg ai/ha, imazethapyr at 0.07 kg ai/ha, lactofen at 0.22 kg ai/ha, or 2,4-DB at 0.42 kg ai/ha. Fungicides and rates included pyraclostrobin at 0.27 kg ai/ha, tebuconazole at 0.23 kg ai/ha, and the premix of prothioconazole at 0.084 kg ai/ha + tebuconazole at 0.168 kg ai/ha.
[c] Rating index: 0=no leaf chlorosis or necrosis, 100=plants completely dead.

Table 6. Peanut phytotoxicity with herbicide-fungicide combinations when rated 12 to 15 days after treatment.[a,b]

http://dx.doi.org/10.5772/55949

5. Effects of tank mix combinations on foliar and soilborne disease control, peanut phytotoxicity, and peanut yield

5.1. Disease control with tank mix combinations

Rainfall in south Texas was below average in 2008 and the early to mid-part of the 2009 peanut growing season (May through August); however, rainfall amounts were above average for the latter portion of the 2009 season (September through November). Rainfall amounts for 2010 were above average for May, July, August, and September (Table 7). In central Texas, rainfall amounts in 2008 were below average for all months (May through November) with the exception of July which was slightly above average while in 2009 rainfall was below average for all months with the exception of July and October (Table 7).

	South Texas				Central Texas		
Month	2008	2009	2010	60-yr avg	2008	2009	30-yr avg
				mm			
May	1.3	16.3	118.4	112.2	76.5	65.5	117.6
June	65.3	3.8	95.0	109.2	30.5	8.6	100.0
July	54.9	5.3	200.7	65.8	47.2	79.0	34.7
August	57.9	42.7	89.4	78.7	50.3	2.0	58.3
September	2.5	114.0	223.3	102.6	55.9	10.6	70.5
October	14.2	352.6	0	94.5	32.5	127.3	72.3
November	25.9	111.3	71.1	75.4	40.4	25.9	54.5
Total	222.0	646.0	797.9	638.4	333.3	318.9	507.9

Table 7. Rainfall amounts in south Texas and central Texas from 2008 through 2010

5.1.1. Early leaf spot control in South Texas

There was an herbicide by fungicide interaction for early leaf spot control in 2008 and 2009. In 2010, the main plots of herbicide and fungicide were significant for early leaf spot control; therefore, that data were averaged over herbicides and fungicides only. Foliar disease development was moderate in 2008 due to extreme drought and hot conditions that persisted throughout the 2008 and the early portion of the 2009 growing seasons. Typically, early leaf spot epidemics are favored by temperatures of approximately 16 to 250 C and long periods of high relative humidity are required for infections to occur [84]. All herbicides alone, with the exception of sethoxydim and lactofen, were not different from the non-treated control with respect to early leaf spot development in 2008 (Table 8). All fungicides alone or in combination with any of the herbicides produced leaf spot levels that were less than the non-treated control. When individual fungicides were compared with the respective fungicide plus herbicide treatments some differences were noted. Pyraclostrobin alone resulted in less early leaf spot than pyraclostrobin plus either imazapic, lactofen, or sethoxydim. No differences were noted between tebuconazole alone or in combination with any herbicide. Prothioconazole plus tebuconazole alone resulted in less early leaf spot than prothioconazole plus tebuconazole in combination with acifluorfen (Table 8).

		Leaf spot[b]		Southern blight[c]	Phytotoxicity[d]		Yield
Fungicide	Herbicide	2008	2009	2010	2009	2010	2008
		Florida scale		% Incidence	%		kg/ha
-	-	6.8	9.4	37	0	0	1860
-	Clethodim	6.3	9.3	24	0	0	1680
-	Sethoxydim	5.7	9.3	89	0	0	1510
-	Lactofen	5.5	9.6	45	11	11	1860
-	Aciflurofen	6.9	9.4	61	5	7	2320
-	Imazethapyr	6.0	9.3	69	0	0	1500
-	Imazapic	6.3	9.2	50	0	0	1810
-	2,4-DB	7.0	9.2	21	0	3	1550
Pyraclostrobin	-	2.5	5.6	29	0	6	2670
Pyraclostrobin	Clethodim	3.0	5.7	21	0	8	2470
Pyraclostrobin	Sethoxydim	3.5	5.8	27	0	10	1630
Pyraclostrobin	Lactofen	3.5	5.8	13	4	7	2440
Pyraclostrobin	Acifluorfen	3.2	5.9	21	1	7	1830
Pyraclostrobin	Imazethapyr	3.0	5.6	29	0	8	1550
Pyraclostrobin	Imazapic	3.8	6.6	18	0	7	2060
Pyraclostrobin	2,4-DB	3.0	5.7	17	1	10	1560
Tebuconazole	-	3.7	7.0	10	0	0	1780
Tebuconazole	Clethodim	4.0	7.8	24	0	0	1870
Tebuconazole	Sethoxydim	4.0	7.2	24	0	0	1970
Tebuconazole	Lactofen	4.5	6.3	19	8	8	1890
Tebuconazole	Acifluorfen	4.0	8.4	27	4	8	1720
Tebuconazole	Imazethapyr	4.0	7.2	35	0	0	1970
Tebuconazole	Imazapic	3.3	7.7	10	0	0	1450
Tebuconazole	2,4-DB	4.0	7.1	21	0	4	2670
Prothioconazole	-	3.0	6.8	37	0	0	2080

		Leaf spot [b]		Southern blight [c]	Phytotoxicity [d]		Yield
Fungicide	Herbicide	2008	2009	2010	2009	2010	2008
		Florida scale		% Incidence	%		kg/ha
+ tebuconazole							
Prothioconazole + tebuconazole	Clethodim	3.7	6.7	55	0	1	1890
Prothioconazole + tebuconazole	Sethoxydim	3.8	6.7	21	0	6	2380
Prothioconazole + tebuconazole	Lactofen	3.3	6.5	31	10	7	2470
Prothioconazole + tebuconazole	Acifluorfen	4.2	7.3	21	5	8	2020
Prothioconazole + tebuconazole	Imazethapyr	3.0	6.8	39	0	0	1500
Prothioconazole + tebuconazole	Imazapic	3.5	6.9	17	0	1	2080
Prothioconazole + tebuconazole	2,4-DB	3.5	7.0	16	1	9	1510
LSD (0.05)		1.0	0.6	31	1	2	780

[a] Fungicides and rates: pyraclostrobin at 0.27 kg ai/ha, tebuconazole at 0.23 kg ai/ha, and the premix of prothioconazole at 0.084 kg ai/ha + tebuconazole at 0.168 kg ai/ha. Herbicides and rates included clethodim at 0.14 kg ai/ha, sethoxydim at 0.21 kg ai/ha, aciflurofen at 0.42 kg ai/ha, imazapic at 0.07 kg ai/ha, imazethapyr at 0.07 kg ai/ha, lactofen at 0.22 kg ai/ha, or 2,4-DB at 0.42 kg ai/ha.

[b] Florida leaf spot scoring system where 1 = no leaf spot, and 10 = plants completely defoliated and dead because of leaf spot. Values of 1 through 4 on the scale reflect increasing incidence of leaflets with spots, and occurrence of spots in lower versus upper canopy of the plots. Values 4 through 10 reflect increasing levels of defoliation.

[c] Loci of southern stem rot were counted immediately after peanut plants were inverted. A locus represented 31 cm or less of linear row with one or more plants infected with *S. rolfsii*. Percent incidence based on number of loci/12.7 m rows.

[d] Peanut phytotoxicity ratings (leaf chlorosis and necrosis) ratings were taken 7 days after treatment. Peanut injury was visually estimated on a scale of 0 to 100 (0 indicating no leaf chlorosis or necrosis and 100 indicating complete peanut kill), relative to the non-treated control.

Table 8. Disease control and peanut response to fungicide-herbicide combinations in south Texas.[a]

Although early-season rainfall was below normal in 2009, September rainfall was above normal leading to conditions for late-season development of high levels of foliar diseases. No differences were noted between the non-treated control and any herbicide with respect to early leaf spot control (Table 8). All fungicides alone or in combination with herbicides resulted in less early leaf spot than the non-treated control. When fungicides were compared alone or in combination, pyraclostrobin alone resulted in less early leaf spot than the combination of

	Southern blight [b]	Leaf spot [c]	Yield
Herbicide	**2008**	**2010**	**2010**
	% Incidence	**Florida scale**	**Kg/ha**
Herbicide			
No herbicide	24	6.5	3635
Aciflurofen	13	7.6	3047
Clethodim	16	6.6	3302
Imazapic	13	6.8	3581
Imazethapyr	15	7.4	3387
Lactofen	17	7.3	3048
Sethoxydim	15	6.8	3240
2,4-DB	28	6.8	3461
LSD (0.05)	NS [c]	0.5	NS
Fungicide			
No fungicide	22	8.8	2834
Pyraclostrobin	16	6.1	3490
Tebuconazole	17	6.5	3401
Prothioconazole + tebuconazole	16	6.5	3627
LSD (0.05)	NS [d]	0.5	419

[a] Fungicides and rates: pyraclostrobin at 0.27 kg ai/ha, tebuconazole at 0.23 kg ai/ha, and the premix of prothioconazole at 0.084 kg ai/ha plus tebuconazole at 0.168 kg ai/ha. Herbicides and rates included clethodim at 0.14 kg ai/ha, sethoxydim at 0.21 kg ai/ha, aciflurofen at 0.42 kg ai/ha, imazapic at 0.07 kg ai/ha, imazethapyr at 0.07 kg ai/ha, lactofen at 0.22 kg ai/ha, or 2,4-DB at 0.42 kg ai/ha.

[b] Loci of southern stem rot were counted immediately after peanut plants were inverted. A locus represented 31 cm or less of linear row with one or more plants infected with *S. rolfsii*. Percent incidence based on number of loci/12.7 m rows.

[c] Florida leaf spot scoring system where 1 = no leaf spot, and 10 = plants completely defoliated and dead because of leaf spot. Values of 1 through 4 on the scale reflect increasing incidence of leaflets with spots, and occurrence of spots in lower versus upper canopy of the plots. Values 4 through 10 reflect increasing levels of defoliation.

[d] Abbreviation: NS, not significant at the 5% level of significance.

Table 9. Disease control and peanut response to herbicides and fungicides in south Texas.[a]

pyraclostrobin plus imazapic while tebuconazole alone resulted in less leaf spot than tebuconazole plus either imazapic or aciflurofen. No differences were noted between prothioconazole plus tebuconazole alone or in combination with any herbicides.

Weather conditions in 2010 were conducive for development of early leaf spot (Table 7). When herbicides were compared, aciflurofen, imazethapyr, and lactofen resulted in greater early leaf spot than where no herbicide was used (Table 9). All fungicides resulted in less early leaf spot than where no fungicide was used.

5.1.2. Early leaf spot control in central Texas

Early leaf spot data was collected only in 2008 and neither fungicide nor herbicide effects were significant. Due to dry conditions, early leaf spot pressure was moderate and there were no differences with any factors (Table 10). Management of early and late leaf spot of peanut is essential for peanut production in most areas of the world [59]. In the southeastern United States, control of these diseases is heavily reliant upon multiple fungicide applications [59,84] while far fewer applications are necessary in the southwestern United States [53,56,85].

5.1.3. Southern blight control

Control of southern blight was not significant for any factor in 2008; however, in 2010 there was a fungicide by herbicide interaction. Since peanut were not dug in 2009, no southern blight ratings were taken. In 2008, no differences were noted with respect to development of southern blight (Table 8). In 2010, under low to moderate pressure, sethoxydim alone produced the highest levels of southern blight with over 85% disease incidence (Table 9). No differences were noted between fungicides alone or the combinations of a fungicide with a herbicide.

5.1.4.Sclerotinia blight control

Sclerotinia blight control was significant for both fungicides in both years; whereas herbicides did not impact disease control. Sclerotinia blight pressure was moderate to heavy in each year (Table 10). In 2008, fluazinam provided the best control of Sclerotinia blight compared with the non-treated control while both boscalid and fluazinam reduced Sclerotinia blight compared to the non-treated control in 2009. Fluazinam has provided good to excellent disease control depending on the rate applied [86-88]. Smith et al. [89] reported in field studies that the application of boscalid or fluazinam that preceded the largest incremental increase in disease incidence provided the best control of disease or increased yield. They advised that disease advisories or intensive scouting should be used to determine when epidemics initiate so that a fungicide can be applied prior to infection.

5.2. Peanut phytotoxicity with tank mix combinations

In south Texas, peanut phytotoxicity ratings were recorded in 2009 and 2010 and an herbicide by fungicide interaction was observed in each year. In 2009, lactofen alone or in combination with any fungicide resulted in the greatest amount of foliar chlorosis or necrosis (Table 8). The addition of a fungicide to lactofen reduced phytotoxicity 10 to 64% compared with lactofen alone. Lactofen is classified as a diphenyl ether (cell membrane disruptor), which interferes with protoporphyrinogen IX oxidase and causes accumulation of protoporphyrin IX [90]. Protoporphyrinogen IX is a potent photosensitizer that generates high levels of singlet oxygen

Fungicide	Leaf spot[b]	Sclerotinia blight[c]	
		2008	2009
	Florida scale	%	
None	5.4	31.9	39.3
Boscalid	5.1	24.3	18.5
Fluazinam	5.6	16.6	12.7
LSD (0.05)	NS[d]	13.9	8.0

[a] Fungicides and rates: boscalid at 0.49 kg ai/ha and fluazinam at 0.88 kg ai/ha. Herbicides and rates included clethodim at 0.14 kg ai/ha, sethoxydim at 0.21 kg ai/ha, acifluorfen at 0.42 kg ai/ha, imazapic at 0.07 kg ai/ha, imazethapyr at 0.07 kg ai/ha, lactofen at 0.22 kg ai/ha, or 2,4-DB at 0.42 kg ai/ha. Data combined over fungicides due to a lack of interaction.
[b] Leaf spot assessed using the Florida 1-10 scale where 1=no disease and 10=completely dead. Leaf spot present only in 2008.
[c] Loci of Sclerotinia blight were counted just prior to peanut plants being inverted. A locus represents 31 cm or less of linear row with one or more plants exhibiting disease symptoms or signs of *S. minor*.
[d] NS, not significant at the 5% level of probability.

Table 10. Foliar disease and Sclerotinia blight control with fungicides in central Texas.[a]

in the presence of molecular oxygen and light, leading to light-induced oxidative breakdown of cell constituents [90]. Aciflurofen, also a diphenyl ether herbicide, caused injury similar to lactofen; however, this injury was not as great as that observed with lactofen (Table 8). Peanut and soybean (*Glycine max* L.) tolerance to aciflurofen and lactofen is based on metabolism, which often results in some leaf bronzing and spotting of leaves and plant growth can be temporarily reduced [79,91].

In 2010, aciflurofen and lactofen exhibited similar phytotoxicity symptoms as exhibited in 2009; however, more phytotoxicity overall was noted with other fungicide-herbicide combinations than was seen in 2009. This increase in phytotoxicity was probably due to the addition of Agridex to all treatments in 2010, which was not added in 2008 or 2009. Phytotoxicity was noted with pyraclostrobin, which is never seen (authors personal observations). Pyraclostrobin and prothioconazole plus tebuconazole combinations with herbicides were more phytotoxic than tebuconazole combinations with herbicides. With tebuconazole, other than aciflurofen or lactofen, only the combination of tebuconazole plus 2,4-DB resulted in observed phytotoxcity. However, with pyraclostrobin or prothioconazole plus tebuconazole, phytotoxicity resulted from combinations with either clethodim, sethoxydim, imazethapyr, or imazapic in addition to aciflurofen or lactofen (Table 8).

5.3. Peanut yield with tank mix combinations

In south Texas, there was a fungicide by herbicide interaction for peanut yield in 2008; therefore, data are presented as an interaction while in 2010 only fungicide treatment was significant. In 2008, no treatments affected peanut yield when compared with the non-treated control (Table 8). Only pyraclostrobin alone or tebuconazole plus 2,4-DB resulted in an increase in yield over the non-treated control. The lack of response to fungicides is probably related to

http://dx.doi.org/10.5772/55949

Herbicide	Fungicide	Yield	
		2008	2009
		Kg/ha	
-	-	2720	1985
Clethodim	-	2713	2099
Clethodim	Fluazinam	3408	3337
Clethodim	Boscalid	2973	3060
Sethoxydim	-	2930	2351
Sethoxydim	Fluazinam	2778	2930
Sethoxydim	Boscalid	2865	4240
-	Fluazinam	3060	2865
-	Boscalid	3971	4402
LSD (0.05)		NS [b]	855

[a] Fungicides and rates: boscalid at 0.49 kg ai/ha and fluazinam at 0.88 kg ai/ha. Herbicides and rates included clethodim at 0.14 kg ai/ha and sethoxydim at 0.21 kg ai/ha.
[b] NS, Not significant at the 5% level.

Table 11. Peanut yield as influenced by fungicide and herbicide alone and in combinations in central Texas.[a]

the hot, dry conditions during the growing season and relatively low disease pressure. In 2010, all fungicides improved peanut yield over the non-treated control (Table 9).

In central Texas, there was no difference with any factor in 2008; however, a significant fungicide by herbicide interaction was observed in 2009. In 2008, there were no differences with any factor for yield while in 2009 there was a fungicide by herbicide interaction; however, yields were extremely variable (Table 11). Damicone and Jackson [92] reported that yield reductions of over 50% can occur following severe outbreaks of Sclerotinia blight. All boscalid or fluazinam treatments improved peanut yield over the non-treated control . Boscalid alone or in combination with sethoxydim produced greater yield than fluazinam alone or fluazinam in combination with sethoxydim. This agrees with the results of Smith et al. [89] who reported that in both field and greenhouse studies, boscalid performed marginally better than fluazinam.

6. Conclusions of using tank mix combinations on weed efficacy and peanut response

Adding fungicides to either clethodim or sethoxydim did not have an effect on annual grass efficacy. No phytotoxicity was noted on peanut and yield was not affected with any graminicide -fungicide combinations. Lancaster et al. [8] reported that pyraclostrobin and tebuconazole did not reduce the amount of ^{14}C-labled clethodim or sethoxydim absorbed in large crabgrass. Although tebuconazole did not reduce efficacy of either graminicide in the field, pyraclostrobin reduced efficacy of clethodim and sethoxydim in some instances. They concluded that reduced absorption was not the mechanism for reduced large crabgrass control but may be the result of a biological response or a chemical interaction. Pyraclostrobin is a strobilurin fungicide which inhibits fungal respiration and acts systemically within the plant [93]. Therefore, the formulated product is not likely to remain on leaf surfaces and interfere with herbicide absorption [8,9]. With Palmer amaranth, antagonism was noted 33% of the time with aciflurofen plus either pyraclostrobin or tebuconazole and 2,4-DB plus pyraclostrobin. Horse purslane also exhibited reduced control with herbicide-fungicides while smellmelon showed no effects of these combinations. Peanut leaf phytotoxicity was most evident with combinations that included aciflurofen or lactofen but this is to be expected since these two herbicides can cause bronzing and leaf spotting when applied alone.

7. Conclusion of tank mix combinations on disease control and peanut response

Control of early leaf spot was reduced with pyraclostrobin plus imazapic combinations compared with pyraclostrobin alone in two of three years while pyraclostrobin plus either sethoxydim or lactofen, tebuconazole plus either clethodim or aciflurofen or the premix of prothioconazole plus tebuconazole in combination with aciflurofen reduced leaf spot control over the respective fungicide in one of three years. Fungicide-herbicide combinations did not affect southern blight or Sclerotinia blight disease development over the respective fungicide alone. Peanut phytotoxicity was greatest with aciflurofen or lactofen combinations. Under early leaf spot and southern blight or Sclerotinia blight disease pressure, no negative response was noted for peanut yield with any fungicide-herbicide combinations over the respective fungicide alone.

Many variables can affect interactions of herbicides with fungicides. Adjuvant selection, herbicide and fungicide rate, commercial formulation, active ingredient, spray volume, water quality, and environmental conditions can affect interactions [61]. Applying a higher rate of the herbicide that may be adversely affected can compensate for interactions [94-96]. Applying ammonium sulfate with bentazon reduced the negative effect of adding bentazon to clethodim or sethoxydim [97,98,99]. Differential response to clethodim has been noted when applied with different formulations of chlorothalonil [66]. Applying graminicides in higher spray volumes can hasten the negative influence of herbicides and fungicides on weed control by graminicides [66,100,101]. Environmental conditions that affect plant response to herbicides or fungicides can influence the magnitude of interactions. Negative effects of interactions associated with the efficacy of systemic herbicides, especially graminicides, are increased when grasses are stressed and the physiological processes that reduce absorption and translocation occur [63,102-105].

http://dx.doi.org/10.5772/55949

Acknowledgements

The National Peanut Board through the Texas Peanut Producers Board provided funds for this research. Kevin Brewer, Dwayne Drozd, Lyndell Gilbert, Bill Klesel, and Ira Yates provided technical assistance.

Author details

W. James Grichar[1*], Peter A. Dotray[2] and Jason E. Woodward[3]

*Address all correspondence to: w-grichar@tamu.edu

1 Texas A&M AgriLife Research, Corpus Christi, TX, USA

2 Texas Tech University, Texas A&M AgriLife Research, Lubbock,TX, USA

3 Texas Tech University, Texas A&M AgriLife Extension Service, Lubbock, TX, USA

References

[1] Anonymous. World Geography of the Peanut, University of Georgia. 2004-01-02. http://www.lanra.uga.edu/peanut/knowledgebase/ (accessed 22 Aug 2012.

[2] Seijo G, Lavia GI, Fernandez A, Krapovickas A, Ducasse DA, Bertioli DJ, Moscone EA. Genomic relationships between the cultivated peanut (Arachis hypogaea, Leguminosae) and its close relatives revealed by double GISH. American Journal of Botany. 2007;94(12) 1963–1971.

[3] Hammons R. Early History and Origin of the Peanut. In: Tripp L. (ed.) Peanuts-Culture and Uses. Roanoke: American Peanut Research and Education Association; 1973. p17-45.

[4] Gregory WC, Gregory MP, Krapovickas A, Smith BW, Yarbrough JA. Structures and Genetic Resources of Peanuts. In: Tripp L. (ed.) Peanuts-Culture and Uses. Roanoke: American Peanut Research and Education Association; 1973. p47-134.

[5] Hammons R. Genetics of Arachis hypogaea. In: Tripp L. (ed.) Peanuts-Culture and Uses. Roanoke: American Peanut Research and Education Association; 1973. p135-173.

[6] Sturke DG, Buchanan GA. Cultural Practices. In: Tripp L. (ed.) Peanuts-Culture and Uses. Roanoke: American Peanut Research and Education Association; 1973. p299-326.

[7] Soyatech. Peanut facts. http://www.soyatech.com/peanutfacts.htm (accessed 27 August, 2012.

[8] Lancaster SH, Jordan DL, York AC, Burke IC, Corbin FT, Sheldon YS, Wilcut JW, Monks DW. Influence of selected fungicides on efficacy of clethodim and sethoxydim. Weed Technology 2005;19 397-403.

[9] Lancaster SH, Jordan DL, York AC, Wilcut JW, Monks DW, Brandenburg RL. Interactions of clethodim and sethoxydim with selected agrichemicals applied to peanut. Weed Technology 2005;19 456-461.

[10] Wilcut JW, York AC, Grichar WJ, Wehtje GR. The Biology and Management of Weeds in Peanut (Arachis hypogaea). In: Pattee HE and Stalker HT. (eds.) Advances in Peanut Science. Stillwater: American Peanut Research and Education Society; 1995. p207-244.

[11] Henning RJ, Allison AH, Tripp LD. Cultural Practices. In Pattee HE and Young CT. (eds.) Peanut Science and Technology. Yoakum: American Peanut Research and Education Society; 1982. p123-138.

[12] Walker RH, Wells LW, McGuire JA. Bristly starbur (Acanthospermum hispidum) interference in peanuts (Arachis hypogaea). Weed Science. 1989;37 196-200.

[13] Brecke BJ, Colvin DL. Weed Management in Peanuts. In Pimentel D. (ed.) CRC Handbook of Pest Management in Agriculture. Boca Raton: CRC Press; 1991. p239-251.

[14] Webster TM. Weed Survey-Southern States. Proceedings of Southern Weed Science Society 2005;58 291-306.

[15] Anonymous. Weed Identification Guide. Southern Weed Science Society, Champaign, IL.1999.

[16] Correll DS, Johnson, MC. Manual of the Vascular Plants of Texas. University of Texas at Dallas. Richardson, TX;1979 p555-556.

[17] Grichar WJ, Besler BA, Lemon RG, Brewer KD. Weed management and net returns using soil-applied and postemergence herbicide programs in peanut (Arachis hypogaea L.). Peanut Science 2005;32 25-31.

[18] Buchanan GA, Murray DS, Hauser EW. Weeds and Their Control in Peanuts. In; Pattee HE, Young CT. (eds.) Peanut Science and Technology. Yoakum: American Peanut Research and Education Society, Incorporated; 1982. p206-249.

[19] Anonymous. Crop Protection Reference. 26th Edition. Chemical & Pharmaceutical Press, Inc. New York. http://www.greenbook.net. (accessed 28 September 2012).

[20] Chamblee RW, Thompson L Jr, Bunn TM. Management of broadleaf signalgrass (Brachiaria platyphylla) in peanuts (Arachis hypogaea). Weed Science 1982;30 40-44.

[21] Wilcut JW, Wehtje GR, Walker RH. Economics of weed control in peanuts (Arachis hypogaea) with herbicides and cultivations. Weed Science 1987;35 711-715.

[22] Wilcut JW, Wehtje GR, Patterson MG. Economic assessment of weed control systems for peanuts (Arachis hypogaea). Weed Science 1987;35 433-437.

[23] Brecke BJ, Currey WL. Weed control in peanuts with ethalfluralin. Peanut Science 1980;7 124-127.

[24] Dotray PA, Keeling JW, Grichar WJ, Prostko EP, Lemon RG. Peanut response to ethalfluralin, pendimethalin, and trifluralin preplant incorporated. Peanut Science 2004;30 34-37.

[25] Prostko EP, Johnson WC III, Mullinix BG Jr. Annual grass control with preplant incorporaed and preemeergence applications of ethalfluralin and pendimethalin in peanut (Arachis hypogaea). Weed Technology 1999;15 36-41.

[26] Johnson WC III, Mullinix BG Jr. Peanut seedling response to dinitroaniline herbicides applied preplant incorporated and preemergence. Peanut Science 1999;26 28-32.

[27] Merkle MG. Weed control. In: Peanut Production in Texas. College Station: Texas Agricultural Experiment Station; 1975. p50-52.

[28] Greer HA, Tripp LD, Santleman PW. The influence of environmental conditions on weed control and Spanish peanut injury by herbicides. In: Proceedings Southern Weed Science Society. 1969;22 145-149.

[29] Grichar WJ, Colburn AE. Effect of dinitroaniline herbicides upon yield and grade of five runner cultivars. Peanut Science 1993;20 126-128.

[30] Grichar WJ, Colburn AE, Baughman PA. Yellow nutsedge (Cyperus esculentus) control in peanut (Arachis hypogaea) as influenced by method of metolachlor application. Weed Technology 1996;10 278-281.

[31] Anonymous. Dual Magnum label SCP 816A-L1N 0403. Syngenta Crop Protection. Greensboro 2004.

[32] Cardina J, Swann CW. Metolachlor effects on peanut growth and Development. Peanut Science 1988;15 57-60.

[33] Wehtje G, Wilcut JW, Hicks TV, McGuire J. Relative tolerance of peanuts to alachlor and metolachlor. Peanut Science 1988;15 53-56.

[34] Osborne BT, Shaw DR, Ratliff RL. Response of selected soybean (Glycine max) cultivars to dimethenamid and metolachlor in hydroponic conditions. Weed Technology 1995;9 178-181.

[35] Mueller TC, Shaw DR, Witt WW. Relative dissipation of acetochlor, alachlor, metolachlor, and SAN 582 from three surface soils. Weed Technology 1999;13 341-346.

[36] Cole TA, Wehtje GR, Wilcut JW, Hicks TV. Behavior of imazethapyr in soybeans (Glycine max), peanuts (Arachis hypogaea), and selected weeds. Weed Science 1989;37 639-644.

[37] Wilcut JW, Walls FR Jr, Norton DN. Imazethapyr for broadleaf weed control in peanuts (Arachis hypogaea). Peanut Science 1991;18 26-30.

[38] Wilcut JW, Walls FR Jr, Norton DN. Weed control, yield, and net returns using imazethapyr in peanuts (Arachis hypogaea). Weed Science 1991;39 238-242.

[39] York AC, Wilcut JW. Potential for Cadre and Pursuit applied to peanuts to carryover to cotton in North Carolina and Georgia. Proceedings Beltwide Cotton Conference; 1995. p.602.

[40] Wilcut JW, Richburg JS III, Eastin EF, Wiley GR, Walls FR Jr, Newell S. Imazethapyr and paraquat systems for weed management in peanut (Arachis hypogaea). Weed Science 1994;42 601-607.

[41] Wilcut JW, Richburg JS III, Wiley G, Walls FR Jr., Jones SR, Iverson MJ. Imidazolinone herbicide systems for peanut (Arachis hypogaea L.). Peanut Science 1994;21 23-28.

[42] Richburg JS III, Wilcut JW, Wehtje GR. Toxicity of foliar and/or soils applied AC 263,222 to purple (Cyperus rotundus) and yellow nutsedge (C. esculentus). Weed Science 1993;42 398-402.

[43] Wilcut JW, York AC, Wehtje GR. The control and interaction of weeds in peanut (Arachis hypogaea). Review of Weed Science 1994;6 177-205.

[44] Nester PR, Grichar WJ. Cadre combinations for broadleaf weed control in peanut. Proceedings Southern Weed Science Society 1993;46 317.

[45] Grichar WJ, Colburn AE, Nester PR. Weed control in Texas peanut with Cadre. Proceedings American Peanut Research and Education Society 1994;26 70.

[46] Wilcut JW, Eastin EF, Richburg JS III, Vencil WK, Wells FR, Wiley G. Imidazolinone systems for southern weed management in resistant corn. Proceedings Weed Science Society America 1993;33 5.

[47] Grey TL, Bridges DC, Prostko EP, Eastin EF, Johnson WC III, Vencil WK, Brecke BJ, MacDonald GE, Tredaway Ducar JA, Everest JW, Wehtje GR, Wilcut JW. Residual weed control with imazapic, diclosulam, and flumioxazin in southeastern peanut (Arachis hypogaea). Peanut Science 2003;30 22-27.

[48] Richburg JS III, Wilcut JW, Colvin DL, Wiley GR. Weed management in southeastern peanut (Arachis hypogaea) with AC 263,222. Weed Technology 1996;10 145-152.

[49] Culbreath AK, Brenneman TB, Bondari K, Reynolds KL, McLean HS. Late leaf spot, southern stem rot, and peanut yield responses to rates of cyproconazole and chlorothalonil applied alone and in combination. Plant Disease 1995;79 1121-1124.

[50] Hagan AK, Rivas-Davila ME, Bowen KL, Wells L. Comparison of fungicide programs for the control of early leaf spot and southern stem rot on selected peanut cultivars. Peanut Science 2004;31 22-27.

[51] Chiteka ZA, Gorbet DW, Shokes FM, Kucharek TA, Knauft DA. Components of resistance to late leafspot in peanut. I. Levels and variability-implications for selection. Peanut Science 1988;15 25-30.

[52] Brenneman TB, Murthy AP, Csinos AS. Activity of tebuconazole on Sclerotium rolfsii and Rhizoctonia solani, two soilborne pathogens of peanut. Plant Disease 1991;75 744-747.

[53] Grichar WJ, Besler BA, Jaks AJ. Use of azoxystrobin for disease control on Texas peanut. Peanut Science 2000;27 83-87.

[54] Bowen KL, Hagan AK, Weeks JR. Number of tebuconazole applications for maximizing disease control and yield of peanut in grower's fields in Alabama. Plant Disease 1997;81 927-931.

[55] Branch WD, Brenneman TB. Pod yield and stem rot evaluation of peanut cultivars treated with tebuconazole. Agronomy Journal 1996;88 933-936.

[56] Besler BA, Grichar WJ, Brewer KD, Baring MR. Assessment of six peanut cultivars for control of Rhizoctonia pod rot when sprayed with azoxystrobin or tebuconazole. Peanut Science 2003;30 49-52.

[57] Culbreath AK, Brenneman TB, Bondari K, Reynolds KL, McLean HS. Late leaf spot, southern stem rot, and peanut yield responses to rates of cyproconazole and chlorothalonil applied alone and in combination. Plant Disease 1995;79 1121-1124.

[58] Dutzmann S, Suty-Heinze A. Prothioconazole: a broad spectrum demethylation inhibitor (DMI) for arable crops. Pflanzenschutz-Nachr Bayer 2004;57 249-264.

[59] Culbreath AK, Kemerait, RC Jr, Brenneman TB. Management of leaf spot diseases of peanut with prothioconazole applied alone or in combination with tebuconazole or trifloxystrobin. Peanut Science 2008;35 149-158.

[60] Anonymous. Variety Guide. In: The Peanut Grower February 2010; p10-18.

[61] Jordan DL, Gurinderbir SC, Lancaster SH, Beam JB, York AC. Defining Interactions of Herbicides with Other Agrochemicals Applied to Peanut. In Soloneski S., Larramendy M. (eds.) Herbicides, Theory, and Applications. Rijeka: InTech; 2011. p73-92.

[62] Barret M. Interactions of Herbicides and other Agrochemicals in Plants: Interactions in Mixtures with other Herbicides and with Safeners, Fungicides, Insecticides, and Nematicides. In Altman NJ (ed.) Pesticide Interactions in Crop Production: Beneficial and Deleterious Effects. Boca Raton: CRC Press; 1993. p113-132.

[63] Green MJ. Herbicide antagonism at the whole plant level. Weed Technology 1989;3 217-226.

[64] Hatzois KK, Penner D. Interaction of herbicides with other agricultural chemicals in higher plants. Review of Weed Science 1985;1 1-64.

[65] Putnam AR, Penner D. Pesticide interactions in higher plants. Residue Review 1974;50 73-110.

[66] Jordan DL, Culpepper AS, Grichar WJ, Ducar JT, Brecke BJ, York AC. Weed control with combinations of selected fungicides and herbicides applied postemergence to peanut (Arachis hypogaea L.). Peanut Science 2003;30 1-7.

[67] Moore JD, Banks PA. Interactions of foliarly applied herbicides on three weed species in peanut (Arachis hypogaea). Weed Science 1991;39 614-621.

[68] Simpson CE, Baring MR, Schubert AM, Black MC, Melouk HA, Lopez Y. Registration of 'Tamrun OL 02' peanut. Crop Science 2006;46 1813-1814.

[69] Beasley J, Baldwin J. Peanut cultivar options and descriptions. 2009 http://www.uga/commodities/fieldcrops/peanuts/production/cultivardescription.html. (accessed 21 September 2012.

[70] Simpson CE, Baring MR, Schubert AM, Melouk HA, Lopez Y, Kirby JS. Registration of 'Olin' peanut. Crop Science 2003;43 1880-1881.

[71] Gorbet DW, Tillman BL. Registration of 'Florida 07' peanut. Journal of Plant Registration 2009;3 14-18.

[72] Rodriquez-Kabana R, Backman PA, Williams JC. Determination of yield losses to Sclerotium rolfsii in peanut fields. Plant Disease Reporter 1975;59 855-858.

[73] Grichar WJ. Control of Texas panicum (Panicum texanum) and southern crabgrass (Digitaria ciliaris) in peanuts (Arachis hypogaea) with postemergence herbicides. Peanut Science 1991;18 6-9.

[74] Grichar WJ. Control of Palmer amaranth (Amaranthus palmeri) in peanut (Arachis hypogaea) with postemergence herbicides. Weed Technology 1997;11 739-743.

[75] Grichar WJ. Horse purslane (Trianthema portulacastrum) control in peanut (Arachis hypogaea). Weed Technology 1997;7 199-202.

[76] Grichar WJ. Horse purslane (Trianthema portulacastrum), smellmelon (Cucumis melo), and Palmer amaranth (Amaranthus palmeri) control in peanut with postemergence herbicides. Weed Technology 2007;21 688-691.

[77] Thompson AM, Rosales-Robles E, Chandler JM, Nester PR, Tingle CH. Crop tolerance and weed management systems in imidazolinone-tolerant corn (Zea mays L.). Weed Technology 2005;19 1037-1044.

[78] Grichar WJ. Citronmelon (Citrullus lanatus var. citroides) control in Texas peanut (Arachis hypogaea) using postemergence herbicides. Weed Technology 2001;15 481-484.

[79] Dotray PA, Grichar WJ, Baughman TA, Prostko EP, Grey TL, Gilbert LV. Peanut (Arachis hypogaea L.) response to lactofen at various postemergence timings. Peanut Science 2012;39 9-14.

[80] Richburg JS III, Wilcut JW, Grichar WJ. Response of runner, Spanish, and Virginia peanut ciultivars to imazethapyr. Peanut Science 2006;33 47-52.

[81] Dotray PA, Baughman TA, Keeling JW, Grichar WJ, Lemon RG. Effect of imazapic application timing on Texas peanut (Arachis hypogaea). Peanut Science.2001;15 26-29.

[82] Wilcut JW, Richburg JS III, Wiley GL, Walls FR Jr. Postemergence AC 263,222 systems for weed control in peanut (Arachis hypogaea). Weed Science 1996;44 615-621.

[83] Grichar WJ, Sestak DC, Besler BA. Effects of various timings of 2,4-DB on runner-type peanut development and yield. Peanut Science 1997;24 105-106.

[84] Nutter FW Jr., Shokes FM. Management of Foliar Diseases Caused by Fungi. In Melouk HA, Shokes FM. (eds.) Peanut Health Management. St. Paul: The American Phytopathological Society; 1995. p65-73.

[85] Grichar WJ, Jaks AJ, Woodward J. Using prothioconazole plus tebuconazole for foliar and soilborne disease control in Texas peanut. Crop Management 2010 doi:10.1094/CM-2010-0405-02-RS.

[86] Damicone JP, Jackson KE. Disease and yield responses to fungicides among peanut cultivars differing in reaction to Sclerotinia blight. Peanut Science 1996;23 81-85.

[87] Smith FD, Phipps PM, Stipes RJ. Agar plate, soil plate, and field evaluation of fluazinam and other fungicides for control of Sclerotinia minor on peanut. Plant Disease 1991;75 1138-1143.

[88] Smith FD, Phipps PM, Stipes RJ. Fluazinam: A new fungicide for control of Sclerotinia blight and other soilborne pathogens of peanut. Peanut Science 1992;19 115-120.

[89] Smith DL, Garrison MC, Hollowell JE, Isleib TG, Shew BB. Evaluation of application timing and efficacy of the fungicides fluazinam and boscalid for control of Sclerotinia blight of peanut. Crop Protection 2008;27 823-833.

[90] Duke SO, Lydon J, Becerril JM, Sherman TD, Lehnen LP Jr., Matsumoto H. Protoporphyrinogen oxidase-inhibiting herbicides. Weed Science 1991;39 465-473.

[91] Harris JR, Gossett BJ, Murphy TR, Toler JE. Response of broadleaf weeds and soybeans to the diphenyl ether herbicides. Journal of Production Agriculture 1991;4 407-411.

[92] Damicone JP, Jackson KE. Effects of application method and rate on control of Sclerotinia blight of peanut with iprodione and fluazinam. Peanut Science 2001;28 28-33.

[93] Clough JM, Godfrey CRA. The Strobilurin Fungicides. In: Hutson D, Miyamoto J. (eds.) Fungicidal Activity Chemical and Biological Approaches to Plant Protection. New York: J. Wiley; 1998. p109-148.

[94] Chernicky JP, Slife FW. Effects of sublethal concentrations of bentazon, fluazifop, haloxyfop, and sethoxydim on corn (Zea mays). Weed Science 1986;34 171-174.

[95] Rhodes GN Jr, Coble HD. Influence of application variables on antagonism between sethoxydim and bentazon. Weed Science 1984;32 436-441.

[96] Rhodes GN Jr, Coble HD. Influence of bentazon on absorption and translocation of sethoxydim in goosegrass. Weed Science 1984;32 595-597.

[97] Penner, D. The impact of adjuvants on herbicide antagonism. Weed Technology 1989;3 227-231.

[98] Jordan DL. Influence of adjuvants on the antagonism of gramancides by broadleaf herbicides. Weed Technology 1995;9 741-747.

[99] Jordan DL, York AC. Effects of ammonium fertilizers and BCH 81508 S on antagonism with sethoxydim plus bentazon mixtures. Weed Technology 1989;3 450-454.

[100] Buhler DD, Burnside OC. Effect of application factors on the phytotoxicity of fluazifop-butyl, haloxyfop-methyl, and sethoxydim. Weed Science 1984;32 574-583.

[101] Kells JJ, Wanamarta G. Effect of adjuvant and spray volume on quackgrass (Agropyron repens) control with selective postemergence herbicides. Weed Technology 1987;1 129-132.

[102] Burke IC, Price AJ, Wilcut JW, Jordan DL, Culpepper AS, Ducar JT. Annual grass control in peanut (Arachis hypogaea) with clethodim and imazapic. Weed Technology 2004;18 88-92.

[103] Burke IC, Wilcut JW. Physiological basis for antagonism of clethodim by imazapic on goosegrass [Eleusine indica (L.) Gaertn.]. Pesticide Biochemistry Physiology 2003;76 37-45.

[104] Wanamarta G, Penner D. Identification of efficacious adjuvants for sethoxydim and bentazon. Weed Technology 1989;3 60-66.

[105] Wanamarta G, Penner D, Kells JJ. The basis of bentazon antagonism on sethoxydim absorption and activity. Weed Science 1989;37 400-404.

Chapter 6

Weed Management in Cereals in Semi-Arid Environments: A Review

Inés Santín-Montanyá, Encarnación Zambrana-Quesada and José Luis Tenorio-Pasamón

Additional information is available at the end of the chapter

http://dx.doi.org/10.5772/55970

1. Introduction

1.1. The weed problem on cereal arable fields

With growing concern about the environment, and the increased public interest in environmental conservation, traditional agriculture has led to profound changes in in recent years. Cereals are the most important crop in dry-land areas of southern Europe. In Spain, nearly 5.5 million ha of winter cereals are sown each year [1]. Research in agriculture has undergone a paradigm shift, favoring systems aimed at improving the performance of cropping systems without deleterious effects to the environment. To achieve this, weed managers continually develop comprehensive programs for crop protection, in which an essential component is the use of crops more competitive with weeds [2], in order to maintain the stability of agricultural production.

The selection of a crop is not an easy task and it involves the consideration of numerous environmental and socioeconomic factors. Additionally, in any cropping system, we always can observe the presence of weeds that invade, persist and survive. They are unwanted and we refer to them as plants "out of place". There are numerous definitions of a weed: a plant that is out of place and not intentionally sown; a plant that grows where it is not wanted or welcomed; a plant whose virtues have not yet been discovered; a plant that is competitive, persistent, pernicious, and interferes negatively with human activity. Weeds possess one or more of the following characteristics that allow them to survive and increase in nature: abundant seed production; rapid population establishment; seed dormancy; long-term survival of buried seed; adaptation for spread; presence of vegetative reproductive structures and ability to occupy sites disturbed by humans.

Therefore, to control effectively we should ask: *why do weeds emerge;* and *what factors limit their development?*. There is abundant evidence that the presence of weeds reduce crop yields; weeds compete for environmental resources, especially water, light and soil nutrients, resulting in decreased crop yield or reducing the crops quality by contaminating the commodity, interfering with harvest, serving as hosts for crop diseases or providing shelter for insects to overwinter, limiting the choice of crop rotation sequences and cultural practices. The most important parameters that characterize the infestation of weeds in a crop and that determine the competitive relationships between them are the density and time of weed competition. Their competitive ability is associated with the establishment of a dense infestation, and is caused by the different habits of growth of weeds and crops. Weeds have developed a number of features that allow them to survive and even dominate in adverse environmental conditions. Also, to learn more about competition exerted by weeds is necessary to know their life cycle, and we can observe three major life cycle groups in cereal arable fields:

Annuals

Summer annuals germinate in the spring, mature, produce seed, and die in one growing season.

Winter annuals germinate in late summer or fall, mature, produce seed, and then die the following spring or summer.

Biennials

Weeds grow from seed anytime during the growing season. They normally produce a rosette of leaves close to the soil surface the first year, then flower, mature, and die during the second year. A true biennial never produces flowers or seeds the first year. There are relatively few biennial weeds.

Perennials

Simple perennials form a deep taproot and spread primarily by seed dispersal.

Creeping perennials may be either herbaceous or woody and can spread by both vegetative structures as well as by seed.

When we study the competition process between species, we must consider what resources are limiting in the environment, which will account for more competition. Since weeds are so prevalent in many areas of the landscape, management techniques are necessary to maintain order. Weed management is most successful when it involves an integrated approach using a variety of methods. The common methods used to manage weeds include prevention and cultural, mechanical, biological, and chemical means.

Herbicides remain the predominant weed management tool with the greatest influence on weed selection over the last 60 years [3]. Reliance on chemicals for weed control has increased significantly in the last decades [4]. However, herbicide use also carries risks that include environmental, ecological, and human health effects. It is important to understand both the benefits and disadvantages associated with chemical weed control before selecting the appropriate control. Many factors determine when, where, and how a particular herbicide can

http://dx.doi.org/10.5772/55970

be used most effectively. Understanding some of these factors enables you to use herbicides to their maximum advantage. Urzúa [5] recorded the following precepts:

1. When any plant is established and persists in a given area, it is likely to have established a presence of seeds, tubers, rhizomes or other means propagative in the place; that environmental conditions are favorable for reproductive success; and competes successfully with established plant populations. Furthermore, morphological and physiological differences between plants being constantly selected will likely be the most suited to climate, soil and agricultural management, for their establishment and persistence and will likely dominate [6]. Yenish [7] pointed out that it is not economical nor practical to try to eradicate the most problematic species already established, when the presence of them is high in the soil seed bank; in most cases, they can be kept under control with the application of herbicides. In a period of about five years we may reduce the seed bank to less than 5%, but we should also consider that in a single year without control, their seed production may be sufficient to exceed 50% of the original population [7].

2. The weed composition in different communities is not always the same, and it changes over time; this has been called succession. According to this theory, when the habitat remains relatively constant, we do not record considerable changes in the community. When the conditions are modified, the species adapted to the "original conditions" are replaced by those that the new environment is more conducive for their development. At the same time, the presence of new species modifies the new environmental conditions and favors the establishment of other species [8]. In agricultural land the succession process is different than in natural areas since agricultural practices constantly disrupt natural succession process, and the dynamic successional cycle begins. With the suspension of agricultural operations, successional processes in vegetative populations are restored [9].

3. The practices used by the farmers to produce their crops each year favor the development of certain species of weeds so that populations that occur in different plots reflect agricultural management provided to crops that year and previous years.

4. The competitive damage to the crop depends on the species, the density of each range, the proximity in which it is growing when they emerge to the crop plant and the duration of the competition. There are many species that do become problematic during a crop cycle in a particular field, depending on crop. However, it has been found that the early stages of crop development are more sensitive to competition by weeds.

5. Herbicides are available in the market, which when selected appropriately for each particular problem, can efficiently control weeds. To succeed, it is not enough to acquire and apply herbicides recommended for cultivation, it is necessary to take into account the factors that affect the efficiency of action of these herbicides, such as:

- In post-emergence applications, the species present, their size, age, growth rate and environmental conditions.
- In pre-emergence applications, soil type (texture, pH and organic matter content), soil moisture at the time of application and weed species to be controlled.

In addition, the selected herbicide must fulfill other requisites about their mode of action, which are:

- Control weeds with a sufficient dose.
- Penetrate into the weed.
- Move to where conduct its physiological action.
- Affect any vital function.

Herbicides provide a convenient, economical, and effective way to help manage weeds. They allow fields to be planted with less tillage, allow earlier planting dates, and provide additional time to perform the other tasks that farm or personal life require. However, if herbicides are not applied in a timely and appropriate manner in terms of dosage and coverage, or resistant weeds are present, they can have ineffective control.

In this context, long-term experiments, carried out for decades, are considered very important in agricultural research when evaluating the sustainability of crop systems in which are being developed programs of integrated crop protection, in order to maintain stability of agricultural production. The weed vegetation in an agricultural area can change quickly and vary greatly among fields and regions. The factors that influence the weed community are numerous and are difficult to evaluate each factor independently, in a culture system (Figure 1): climatic factors relevant to the persistence of plants, soil factors, which involved the physical and chemical properties of soil, human factors, which are involved in various legislative measures and the use and farm management and technological factors, where one is constantly innovating and researching systems tillage, crop rotation, herbicides, fertilization, and irrigation.

Intensification of land use has also been identified as a major cause of the current biodiversity decline in agro ecosystems [10, 11]. For instance, arable weeds have suffered a severe decline over all Europe, which has developed concerns over the sustainability and environmental consequences of the intensification of land use in agricultural systems [12]. Plant diversity in dry land Mediterranean cereal fields is affected by agricultural intensification at any of these abovementioned scales, as reflected by a decrease of plant species richness and changes in species composition [13]. But the ecological implications of these changes still remain uncertain, because in such agro systems there is a high variability in the local occurrence of plant species [14, 15].

Historically in central semiarid Spain, arable fields have been dominated by cereal production. In this region, tillage intensity has markedly decreased in order to decrease soil loss. There has been an increasing trend towards utilizing conservation tillage systems and the use of herbicides in winter cereals holds a prominent place in the overall use of pesticides in Spain. However, in recent years, climate change, grain prices, cost of

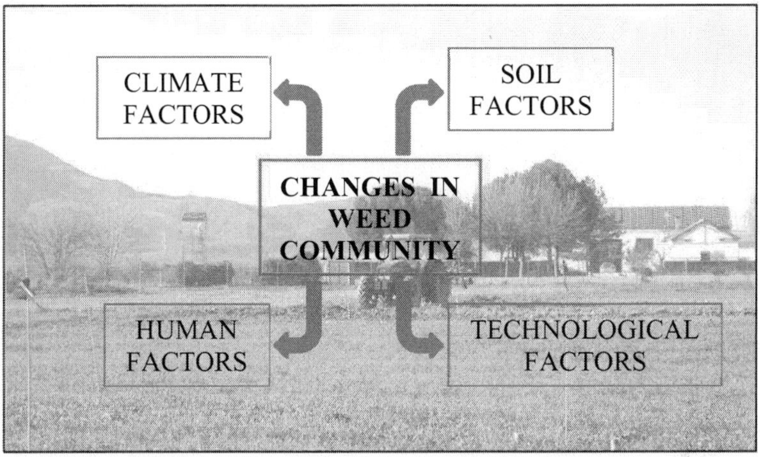

Figure 1. Factors involved in changes of weed community present in a field.

herbicides and the development of resistant weeds has led to seek integrated weed management systems more. Integrated weed management requires more knowledge on how weed community compositions respond to changing agronomic practices after one crop rotation cycle with different practices. Gerard [16] observed that the prediction of the distribution and abundance of weed infestations likely in each field could help to plan and carry out timely control measures in an efficient and economical manner, in accordance with ecology and the interests of the society. The above statement is framed within what is known as "integrated management of weeds", where the main objective is to cause displacement of species difficult to control, by others less problematic and / or reduce the density of populations of noxious weeds at levels that do not cause damage. Therefore, such rationalization goes through the realization of a good diagnosis of the situation, by using a series of agronomic practices that hinder the development of weed populations most problematic and the use of clear decision criteria based on scientific knowledge.

1.1.1. Weed ecology in dry land cereal agriculture

Cereals are the most important crop in dry-land areas of southern Europe. In Spain, nearly 5.5 million ha of winter cereals are sown each year [1]. In Mediterranean areas, weed species are adapted to crops and to management techniques like soil disturbance by tillage. However, the agricultural intensification in the last decades is a process occurring at different scales, which reduces biodiversity, simplifies communities, leads to a loss of ecosystem services [17- 19] and reduces species richness [20]. At the landscape scale, farming intensification has caused the replacement of most natural habitats with arable fields [21], which leads to large, uniformly cropped areas with low spatial heterogeneity [22, 23]. At the field scale, intensification is related to the farming practices performed: i.e., high amount of external inputs (mainly chemical fertilizers and herbicides), low complexity of crop-rotational schemes and improvements in seed-cleaning techniques [24].

In this sense, the patterns of weed species composition in cereal fields are often attributable to a complex number of interacting factors and multivariate analysis has been used in many studies to discuss them. The selection of weeds is constantly evolving in response to crop management practices; therefore, these practices have an important role in the flora composition and its fluctuations in the short and long term at the field level. Management practices, geographical gradients and climatic factors have been found to be the driving factors to explain weed species composition and richness in Northern Europe [25] and in Central Europe [26-29]. Thus, changes in flora may be the result, among other factors, of complex interactions between agronomic practices (choice of species, tillage systems, and strategies for weed control) and environmental factors (soil quality, temperature, and rainfall). It is well known that sometimes, the use of some methods of control, or changes in them by others, causes a change in the composition of the flora, and we can say that weed communities are not static, producing the phenomenon known as *Flora Inversion*.

Although major weeds can be quite different from one region to another, from one farm to another and even between different locations of the same farm, we can select a few species that are widespread throughout the Spanish geography which represent a serious threat by the competitiveness, by the difficulty of control and by the rapid expansion of their populations. Among them we can mention four annual grasses:

Avena sterilis L. and *A. fatua* L. ("Wild oats"), these weeds are found throughout the peninsula, has an almost identical cycle of cereals, germinating simultaneously with them and for a fairly long period of time and matures at the same time as grain crops. These attributes, combined with its ability to emerge from depths relatively high (up to 25 cm) and the prolonged persistence of seeds in the soil (over 3 years) facilitates the development and presence in tilled fields. However, the main reason for its spread is its ability to cause high losses in cereal yields.

Lolium rigidum Gaudin and *Lolium multiflorum* L. ("Ryegrass") are widespread geographically, being especially prevalent in cereals. These species germinate with the first rains of autumn, usually beginning their nascence before sowing of cereal. If the first plants were not completely destroyed by seedbed preparation tillage or pre-plant herbicides they can become great competitors with the crop. Most seeds germinate the following year of their production, making containment or eradication of their population easier than in the case of the other grass.

Bromus spp. and Phalaris spp. were a very common species in the margins of roads and cultivated fields until the arrival of conservation agriculture. With tillage reduction or elimination, they have been introduced in the fields quickly causing major problems. These species are well adapted to emerge from the soil surface zone. Its emergence period is very short, beginning with the first rains of autumn, and almost all seeds germinate the following year.

Besides the grasses mentioned above, there are some dicotyledonous annuals that are harmful, either because of their abundance, their competitiveness or difficulties involved in their control. In the case of the "poppy" (Papaver spp. and Hypecoum spp.) the problem is more

http://dx.doi.org/10.5772/55970

due to their abundance (associated with a huge seed production and a high persistence of these seed on the soil surface) than competitiveness with the crop (relatively low). Similarly, the Cruciferae family (*Sinapis arvensis* L., *Diplotaxis erucoides* L. and Raphanus spp.) produces high numbers of seeds although the competition with the crop can be quite high. These species had adapted to conventional tillage, but the increasing use of herbicides has reduced their populations while favoring the presence of other species: "Cleavers" (*Galium aparine* L.), which are fast growing and can outcompete almost completely the cereal plants; "Speedwell" (*Veronica hederifolia* L. and *Veronica persica* L.), "Chamomile" (*Matricaria chamomilla* L.), *Polygonum aviculare* L., etc. Other species of the genus Tussilago, Epilobium, Conyza, Artemisia, Lactuca, etc., have problems in soils subjected to periodic disturbances, and they have also adapted to no-tillage fields. Also, the species Chenopodium spp., Amaranthus spp., Salsola spp., etc. can invade the cereal fallows, which can require investment in specific herbicides for control. Finally, perennial weeds base their success on their bodies' underground reserves that enable rapid development at the beginning of spring. They are represented by the bindweed (Convolvulus spp.) and several thistles (Cirsium spp.).

In this paper, we will not create a weed inventory or abundance, but focus on identifying the most significant risks to which crop will face during its development. Before herbicide treatment, it is imperative to carry out a diagnosis as accurate as possible of the weed situation. This idea is according to the National Academy of Sciences (1980): "to induce population changes in response to agricultural management, it is necessary to know the biology of the species involved and environmental modifications that causes each agronomic practice". This requires knowledge of the dynamics of weed populations that cause *a favorable succession* and it is necessary to know the majority of weed species present in the plots treated. In this sense, decisions regarding herbicide treatments should be based on four main points:

First, it is necessary to select the most appropriate treatments taking into account the efficacy and selectivity of the products available on the market. In view of the problems identified in each field, we will need to find which products adequately control all high-risk species. In Spain there are over 30 different active ingredients for use in cereal crops and over two times that many commercial products (with various formulations and/or combinations of active substances). The selection of products to be used will be dictated by the timing of treatments. Table 1 lists some of the most widely used herbicides in cereals and their application times. We should note that the application of these products is not always carried out in isolation, so it is important to know if there is a problem of incompatibility between products (relatively frequent event). There would also be possible to find problems of sensitivity of crops because not all products are equally safe for barley and wheat, and even within the same crop, there are differences in sensitivity in some varieties.

In the case of herbicides used in pre-emergence, this decision will have to be made based on the problems identified in previous years. In that sense, it is highly desirable to have some information about the history of the field, i.e. crops that were planted, cultural practices, herbicides used, and what kind of weed problems developed. This information will help us to identify the type and severity of the problem to be faced in the coming season. Since weed

Timing of herbicide application	Controlled weeds	Active substances
Pre emergence	Dicotyledonous	clortoluron, isoproturon, trifluralina, clorsulfuron, linuron, bifenox,pendimetalina, triasulfuron.
	Grass	isoproturon
Early post emergence	Dicotyledonous	clortoluron, diclofop-metil, fenoxaprop-etil, 2.4-D, MCPA, fluroxipir, bentazona, tifensulfuron-metil, tribenuron-metil.
	Grass	iodosulfuron-metil-sodio, pendimetalina, tralkoxidim.
Late post emergence	Dicotyledonous	clodinafop, tralkoxidim
	Grass	Fenoxaprop-p-etil,

Table 1. Herbicides used in cereal crops depending on the timing and type of weed.

infestations are often not distributed evenly throughout the field, it will also be useful to know the location of problematic weeds populations and if they are particularly aggressive species or found in very high densities. Pre-emergent herbicides act upon weed seeds, seedlings or form a barrier in the soil to prevent weed seed germination or establishment. These herbicides are usually used in the spring to prevent seeds establishing when the soil temperatures begin to warm up and a properly timed application can provide control for several months.

In the case of herbicides used in post-emergence (the most common use), it is desirable to perform the evaluation of the main weeds that are invading each field as soon as the cereal is established. This assessment should be made as soon as possible in order to plan and carry out early treatment, which is recommended due to their greater efficiency. Post-emergent herbicides work on actively growing weeds and can be further broken down into two categories:

- Selective herbicides can be applied to an area and target weeds (i.e. dicots or monocots) while having little or no effect on the crop or non-target weeds. Some products may require repeated applications for effective control.
- Non-selective herbicides kill all susceptible plants they come into contact with. The most used non-selective herbicide is glyphosate.

After choosing the herbicide, it is necessary to decide the dose to apply. Typically there is a relatively large dose range according to what weeds dominate; what is the stage of development (the higher development, the greater the dose needed to control them) and what is the texture and the organic matter content of soil (in cases of pre-sowing applications or pre-emergence, the higher the content of clay and organic matter, the greater the dose).

Second, one must consider the costs of treatments considered. There are large differences between the costs of different products. For example, while the cost of treatment with hormonal herbicides (2.4-D, MCPA, etc.) for overall control of dicots is almost negligible, the use of specific herbicides against Galium spp. or Avena spp. may be a considerable investment.

Third, it is necessary to estimate the economic benefits of a treatment application. This involves estimating the expected yields in the crop (and its sale value) and the losses that would be avoided by such treatment. In this sense, while the application of herbicides in areas of high productivity (yields higher than 4 t/ha) is usually economically profitable in more marginal areas (with an income below 2 t/ha) these benefits are rather dubious. Similarly, in meteorologically favorable years higher investments in inputs may provide higher profits. In relation to avoidable losses, we should consider the competitiveness of the dominant species (it´s not the same having a plot infested by Avena spp. or it infested by Papaver spp.), and the level of weed infestation of plot.

Finally, we must consider the potential side effects arising from the application of such treatment. This section is not only to consider the effects on the environment (pollution of waterways, loss of biodiversity) but also the risk of resistance. The emergence of resistance as a result of poor practices is increasingly common. Continued application of the same product (or products belonging to the same chemical family or families with the same mode of action) over a certain period of time leads, sooner or later, to the emergence of resistant weed biotypes. The best strategy to prevent the emergence of weed resistance is the integrated use of prevention and control of many methods as possible:

- Use of crop rotations, using spring crops needed to eliminate resistant biotypes before planting or use alternative herbicides not applicable in cereal crops.
- Employment of fallow and mechanical control practices.
- Avoid seeds with resistance movement from one field to another, carefully cleaning tillage and harvesting equipment.
- Using appropriate densities for a competitive cultivation.
- Herbicide use only when necessary, alternating herbicides belonging to different groups according to their mode of action.

1.1.2. The climate influence in an agro system with a semi-arid environment

The climatic factors more relevant to the persistence of the plants are: light, temperature, water, wind and seasonal characteristics of these factors:

The intensity, quality and duration of **light** are important for determining the growth, reproduction and distribution of such plants. Light governs the photoperiodic response and determines the flowering time of seed maturation; therefore, it determines the latitudinal distribution limits of species.

The air and soil **temperature** and the duration of the frost, are important limits on the distribution of weeds. The soil temperature is directly related to the seed germination, and a drop in temperature will influence the same seed dormancy and survival of their underground parts. Therefore, temperature is a critical factor for the persistence and adaptation of annual and perennial weeds.

Water is the most important environmental factor in the habitat, with a marked morphological expression in the plant. The total water available in a location is related to both the initial supply with losses by runoff, evaporation and transpiration. The seasonal distribution of water is a key factor, since sometimes its scarcity at critical stages of the plant leads to lack of reproduction and survival.

The speed, **wind** direction and wind frequency defines the presence of all plants, including weeds. Also, it can produce transpiration losses of plants.

In summary, the weeds are primarily affected by the same factors as the crop: water, and the factors related to their availability (insolation and transpiration) and nutrients. If these parameters are not restricted, the weed growth will be higher than the crop.

On the other hand, when conditions are not suitable, the agronomic practices may be ineffective in inducing seed germination. In this sense, one of chief limiting factor of crop yield in cereal agro systems with a semi-arid environment is the scarce irregular rainfall distribution. For this reason, we initiated a field experiment, at the experimental farm of INIA "La Canaleja", located in Alcala de Henares (Madrid). The field trials were located in a semi-arid agro system of central Spain, with an average total annual rainfall of 470 mm, and rainfall distribution registered over fifteen years were used to assess the effects of environmental conditions on weed community.

Our results showed that seasonal distribution of rainfall did restrict the effectiveness of the weed management practices and it affected the weed density. In 2000-2001 and 2010-2011, it we recorded higher annual rainfall than the average for this area, and in accordance with the increase of water availability, the weed density, measured by sampling (size of each sample of 0,125 m^2), increased considerably. Between years 1995 and 2011 herbicides controlling dicotyledonous and / or against grass were used to control the weed community present in the field. In this situation, total weed density was maintained except in the 2009-2010 period, when weed density was large though the annual rainfall was below normal; this was mainly due to herbicides not being used in this period favoring the weed competition with the crops (Table 2 & Figure 2).

The community of weeds present in the field differed with the annual distribution of rainfall and may limit the effectiveness of the system used to control weeds, leading the specialization of some species under certain crop conditions. We observed in our field, that high rainfall occurring in the spring favored early-emergence weeds, such as *Papaver roheas* L. and high rainfall occurring in autumn favored late-emergence weeds such us *Lolium rigidum* Gaud. and *Hypecoum procumbens* L.; and weeds with extended patterns of emergence such as *Anacyclus clavatus* L. and Veronica spp.; or perennial weeds (Cardaria spp. and Convolvulus spp.) were favored by a general increase of annual rainfall in the area. Furthermore, increasing knowledge of how plants respond to different environmental conditions and the application of this knowledge allows more effective and efficient use of available tillage tools in combination with other weed control practices.

http://dx.doi.org/10.5772/55970

Year	J	F	M	A	My	J	Jy	A	S	O	N	D	Annual
1994	18,4	33,4	9	18,2	76,8	10,8	3,6	0,8	40,6	52,2	26,8	4,6	295,2
1995	18,2	28,4	0,8	10,8	33,4	56,2	7,6	3,0	11,4	5,6	45,2	71,2	291,8
1996	84,6	24,2	16,4	9,4	94,4	5,8	1,4	6,8	12,4	4,0	58,4	86,6	404,4
1997	76,2	2,8	0,0	63,2	47,2	6,8	23,6	32,0	28,0	17,6	145,4	53,4	496,2
1998	28,5	45,1	15,9	27,6	14,2	1,6	2,0	0,4	0,2	1,0	29,6	20,0	186,1
1999	36,0	14,8	18,0	55,3	65,8	22,6	4,0	3,0	57,0	96,7	45,8	0,0	419,0
2000	64,0	0,8	30,0	103,8	69,6	25,0	10,0	0,0	18,2	26,5	70,0	129,7	547,6
2001	108,8	20,7	60,1	26,5	37,7	7,2	4,3	7,0	14,0	79,4	4,9	0,0	370,6
2002	54,0	5,7	46,2	41,5	76,7	12,3	12,0	5,1	32,5	50,6	86,0	48,2	470,8
2003	51,2	52,3	39,5	57,2	22,2	1,8	1,2	2,6	9,4	99,5	67,6	35,5	440,0
2004	4,2	74,7	55,1	43,7	102,2	5,3	46,1	18,5	5,4	99,4	20,4	17,8	492,8
2005	0,0	15,5	11,3	7,2	7,1	1,3	0,0	3,2	12,2	86,3	66,0	27,0	237,1
2006	40,2	45,5	20,0	37,0	14,0	34,8	1,5	5,7	15,9	84,6	98,3	20,5	418,0
2007	7,8	43,8	13,6	104,5	95,7	37,0	0,0	6,8	10,3	30,8	30,3	4,0	384,6
2008	27,6	32,6	2,1	80,1	106,0	36,9	0,0	0,0	22,2	56,7	25,2	42,0	431,4
2009	38,7	43,9	11,2	31,4	8,0	17,3	0,0	20,7	8,6	23,2	12,8	111,7	327,5
2010	70,2	84,8	51,3	47,5	33,5	58,1	17,5	2,8	40,6	31,0	41,0	62,9	541,2
2011	44,0	30,0	46,7	65,0	168,5	24,5	1,0	24,0	1,6	33,2	49,5	6,0	494,0
Historical Average	42,92	33,28	24,84	46,11	59,61	20,29	7,54	7,91	18,92	48,79	51,29	41,17	402,68

Table 2. Annual distribution of rainfall (mm) and historical average during years object of study.

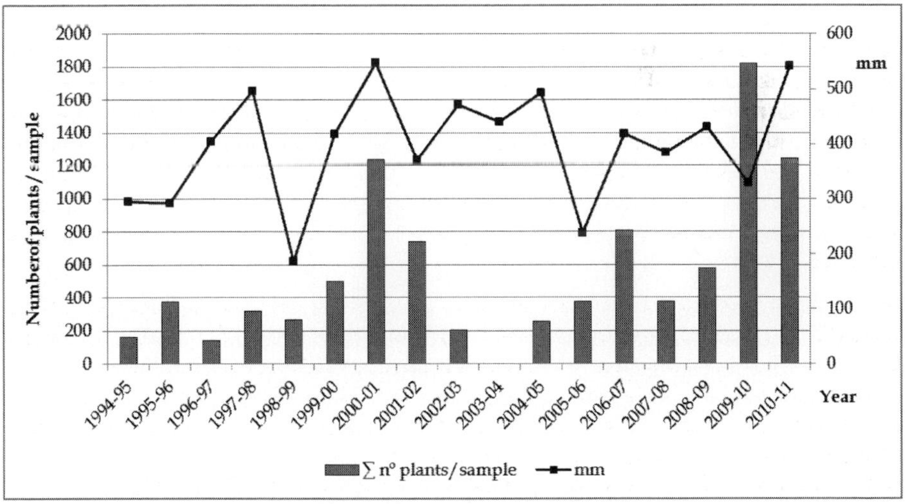

Figure 2. Total number of plants recorded per sample (0,125 m^2) and annual rainfall (mm) from 1995 to 2011.

2. The adoption of conservation tillage systems

The European agricultural situation is modifying quickly due to the pressure of economic factors and to the increased sensitivity of environmental problems. Nowadays, integrated weed management could be a possible solution to rationalize the inputs of herbicides and to increase the use of complementary methods of weed control forming an integral component of sustainable agriculture [30]. However, the adoption of these practices have a considerable impact on communities of weeds, and therefore, their management should be different from that undertaken in a conventional system [31-33]. The benefits of conservation tillage include reducing soil erosion, increasing organic matter, improving soil structure, and reducing fuel consumption and some tillage machinery. As a result of these agronomic and economic incentives, direct seeding practices have been adopted in many regions. Weed control is often cited as the main challenge in minimum tillage systems and no-tillage, and often leads to increased herbicide use, so we must pay special attention to this system. Otherwise, conservation tillage systems are believed to worsen weed problems with higher weed emergence promoted by higher concentrations of seed in the surface soil and shifts of the weed community towards increased abundance of troublesome species, e.g. grasses and perennials [34].

In summary, minimum tillage, particularly no-tillage, may favor a relative emergence of weeds over crops. Moreover, the increase of prior crop residues in these systems can alter the competitive ability of crops with weeds at early stages, increasing production losses thereof [35]. Thus, it appears that common tasks tend to select annual weeds and little work allows the dominance of perennial or biennial species. However, these predictions are strongly influenced by cultural practices and environmental conditions used in a specific area. Currently, insufficient information exists about the processes associated with changes in weed communities; such information is crucial in managing weeds. As a means of control it is necessary to assess the presence of weeds, setting thresholds for treatment of major species in crops and the adequate product selection, dose and time of application best suited among those authorized, while taking into account the environmental conditions.

Development of improved weed management systems requires more knowledge on how weed species respond to changing agronomic practices. In order to monitor weed development subjected to different agronomic practices, one experiment was conducted to determine weed population response to various tillage intensities in a cereal agro system in central Spain (Figure 3A). Field trials under a cold semi-arid environment were conducted in successive growing seasons from 1995 to 2011, to assess the effects of management practices on the weed community with three tillage systems: (1) conventional tillage (CT); (2) minimum tillage (MT) and (3) no-tillage system (NT). The experiment consisted of a field divided in four randomized complete blocks with three different tillage systems and four replications. To study the effectiveness of different managements, we performed a first identification of the flora present in the field where the experiment was developed.

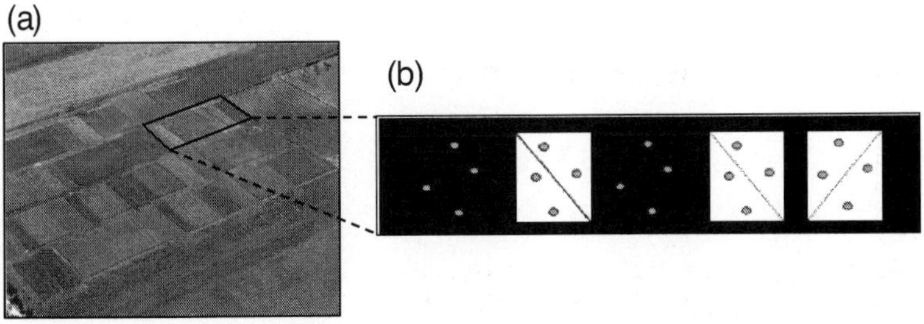

Figure 3. A) Field trials in the experimental farm "La Canaleja", and B) Weeds sampling scheme realized it in each tillage system.

The natural community of weeds present in the assay is comprised by dicotyledonous weed species and grasses, annual and perennial species typical of crop fields in the area (Table 3). Later, during all years of the study, weeds were counted by species with a similar methodology based on the selection of four random samples in the field with a quadrant of 0,125m^2, taken in zigzag on the diagonal of a rectangle defined in each sub-plot (Figure 3B), which were identified and quantified in situ, the weed species present. Total density of weeds referred to the unit area (1m^2).

Adonis annua	Chondrilia juncea	Hordeum murimus	Scabiosa spp.
Amaranthus albus	Chrozopera tinctoria	Hypecoum pendulum	Scorzonera laciniata
Amaranthus blitoides	Cichorium intybus	Hypecoum procumbens	Senecio vulagaris
Amaranthus retroflexus	Cirsium arvense	Lactuca serriola	Setaria viridis
Anacyclus clavatus	Cnicus benedictus	Lamium amplexicaule	Silybum marianum
Anchusa azurea	Convulvulus arvensis	Lavatera spp.	Sisymbrium iria
Andryala integrifolia	Conyza spp.	Linaria micranha	Sisymbrium orientale
Asperugo procumbens	Datura stramonium	Lolium rigidum	Solanun rigidum
Avena spp.	Descurania Sophia	Medicago spp.	Sonchus spp.
Belladia trixago	Diplotaxis erucoides	Melilotus spp.	Stellaria media
Biscutella auriculata	Echallium elaterium	Papaver hybridum	Torilis nodosa
Bromus rigidus	Echium spp.	Papaver rhoeas	Tragopogum psp.
Bromus rubens	Epilolium brachycarpum	Plantago spp.	Trifolium angustifolium
Buglossoides arvensis	Eruca vessicaria	Polygonum aviculare	Trigonella polyceratia
Campanula erimus	Eryngium spp.	Portulaca aleracea	Veccaria pyramidata
Capsella burs-pastori	Filago spp.	Rapistrum rugosum	Veronica hederifolia
Cardaria draba	Fumaria officinalis	Reseda phyteuma	Vicia spp.
Centaurea aspera	Galium murale	Roemeria híbrida	Xanthium spinosum
Chenopodium album	Heliotroium europaeum	Salsola kali	

Table 3. Initial weed community in the farm "La Canaleja".

The herbicides employed in the trials were post emergence against dicotyledonous weeds from 1994 to 2000; against dicotyledonous and grasses from 2004 to 2009; in 2009 we did not employ any herbicide and afterward, we used post emergence herbicide against dicotyledonous weeds. Also, in the NT system, the crops were seeded each year after an application of glyphosate at 2 $l.ha^{-1}$. Within the time frame of this research, weed density and species composition were affected by year, which differs in environmental conditions, and by tillage intensity, indicating fluctuations in changes of weed community composition associated with changes in agronomic practices and environmental conditions are complex and difficult to predict, especially in semiarid regions with low and / or irregular rainfall.

Specific research regarding the impact of crop production systems on weed communities is lacking and currently, there is not a common position among authors about which system produces the best weed control. Several researchers have described the effect of the tillage system on weed flora composition and valued the long term dependence on the crop system used and their studies showed changes in weed species composition as a consequence of tillage practices [36]. According with this idea, we observed that the community of weeds present in a field differs with the tillage system employed (Figure 4). Minimum tillage systems (MT) and no-tillage (NT) showed higher weed densities compared to conventional tillage (CT).

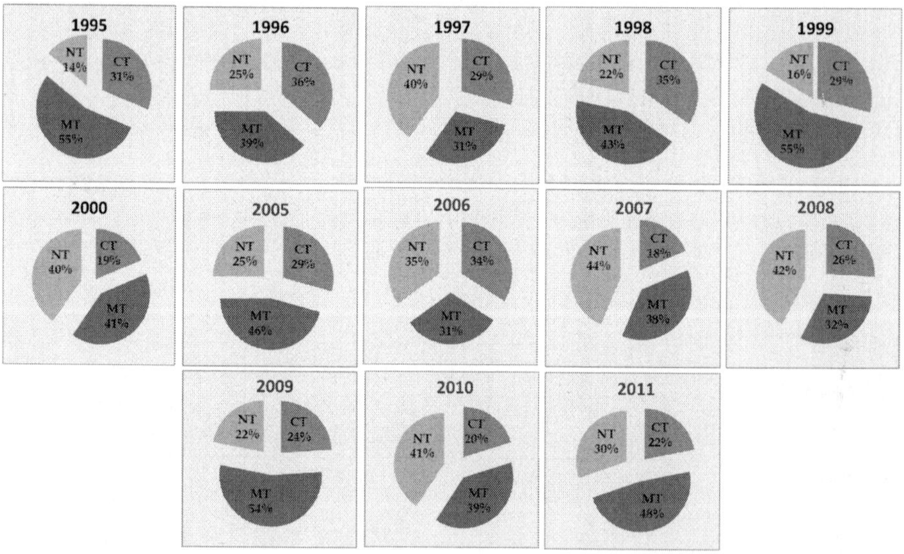

Figure 4. Percentage of total weeds observed in each tillage system studied from 1995 to 2011.

Other researchers have described the predominant weeds of different tillage systems, such as Lolium spp. in minimum tillage system [37]; Poligonum spp. in conventional tillage [38, 39], or *Fumaria officinalis* L. and *Lamium amplexicaule* L., also favored in conventional tillage [40].

Some species display greater capacity of infestation when the intensity of tillage is reduced [41-44].These species shifts generally resulted in the emergence of species tolerant to existing weed management practices [45, 46]. In this sense, Froud-Williams [47] also predicted that annual and perennial grasses, perennial dicotyledonous species, wind-disseminated species, and volunteer crops would increase and annual dicotyledonous weeds would decrease in association with MT systems; although, these predictions were strongly influenced by the agronomic practices employed within a specific study; and Liebman & Davis [48] suggested a possible solution for weed problems would be the combination of different soil tillage systems. Nevertheless, other authors have suggested that tillage did not produce any selective effect in the composition of weed flora [49].

In this context, it is very important to identify which are the most troublesome weeds, because they are the most difficult to control. Also, we should follow those species maintained in the seed bank of soil without an initial risk because they present low density, but one change in the crop system and /or the environmental conditions can favor their propagation and convert them into a dominant species of the field. The specific objectives of the work reported here were to determine if decreasing tillage is accompanied by a predicted increase in the presence of annual and perennial grasses, perennial dicot species, wind disseminated species, and volunteer crops, but a decrease in annual dicotyledonous weeds.

In order to realize the following of several weed species along the year's object of study, we determined the relative weed density in the field each five years for representative species (Figure 5). In general, years with high rainfall in fall, 1995 and 2005, favoured later-emergence weeds and perennial species to escape suppression by the crops. Many weeds had patterns of emergence that peak in October and November such as *Fumaria officinalis* L.; *Lolium rigidum* Gaudin and *Hypecoum procumbens* L., as well as the perennial weed *Cardaria draba* L. Desraux, which increased the year where higher than average rainfall was received in fall. However, years with high rainfall occurring in April and May, 2000 and 2010, they favoured early-emergence weeds such as Papaver spp. At the same time, in our experiment, we noted a reduction of dicotyledonous weeds *Cardaria draba* L. and *Fumaria officinalis* L., and the increase of Papaver spp., *Lamium amplexicaule* L. and Veronica spp. in sub-plots with NT system. Also, we could observe a clear tendency of increasing of *Lolium rigidum* L. and *Hypecoum procumbens* L. density in MT sub-plots and another perennial species such as Cirsium spp. and Convolvulus spp., which we typically found in field margins, appeared frequently within NT sub-plots.

The decrease in soil water evaporation due to the residual cover in both NT and MT could have increased the soil water content compared with CT, and this could be one of the reasons for the increase in the density of weeds within these systems [50]. Also, the annual distribution of rainfall may limit the effectiveness of the system used to control weeds, predisposing the specialization of some species under certain crop conditions. Generally, no-till systems can be difficult to maintain over a long period of time without adequate weed management, and knowledge of the emergence process of weeds will increase the effectiveness of a post-emergence herbicide, assuming an important qualitative advance in the integrated control of weed populations.

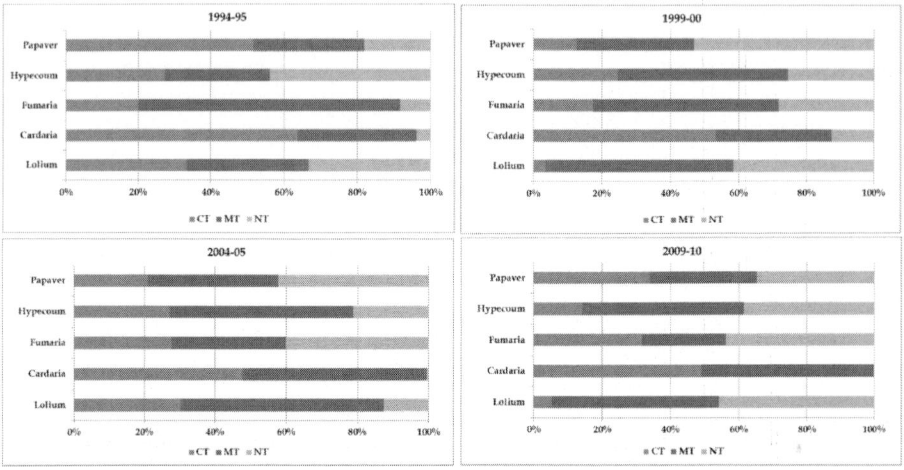

Figure 5. Relative density of weed species more representative in the field object of study.

3. Conclusion

At the moment, sustainable agriculture is being promoted in Europe, and its industrialization using technologies that help to increase crop production should be designed in order to protect the environment. In this context, the increasing awareness of the farmers requires the adoption and adaptation of techniques that, without undermining the economic benefit of farms, could be also accepted by the environment.

Sometimes we ignore the ecological processes that occur in agro systems, and weed control problems associated with herbicide selectivity and changes occurring in weed communities within MT and NT systems have been reported by numerous authors. In this sense, changes in agricultural technologies, such as the employment of selective herbicides, require reevaluation of assumptions regarding the nature of weed communities in MT and NT systems and the information on the association of weeds species with tillage systems and herbicides are key in determining directions of future research in weed management.

Acknowledgements

We appreciate the funds received from different Ministries for the realisation of this long-term experiment. This work has been funded by projects: INIA SC94-005-C2-2; SC94-003-C3-2; SC98-020-C4-2; MCYT-INIA RTA-02-058-C3-2; MEC-INIA RTA2006-00121-C03-02 and MICINN-INIA RTA2010-0006-C03-02.

We are grateful to all members of the experimental farm "La Canaleja" for helping managing the experiment.

Author details

Inés Santín-Montanyá, Encarnación Zambrana-Quesada and José Luis Tenorio-Pasamón

Instituto Nacional de Investigación y Tecnología Agraria y Alimentaria (INIA), Spain

References

[1] MARM (Ministerio de Agricultura. Alimentación y Medioambiente). Anuario de Estadística Agroalimentaria. http://www.marm.es/es/agricultura/estadisticasaccessed 19 January (2012).

[2] Korres, N. E, & Froud-williams, R. J. Effects of winter wheat cultivars and seed rate on the biological characteristics of naturally occurring weed flora, Weed research (2002)., 42, 417-428.

[3] Hass, H, & Streibig, J. C. (1982). Changing patterns of weed distribution as a result of herbicide use and other agronomic factors. In: LeBaron, H.M. & Gressel, J. (eds.) Herbicide Resistance in Plants. John Wiley & Sons, New York. 1982., 57-79.

[4] Thill, D, & Lemerle, D. Resistance Management in Wheat-dominated Agro-ecoystems, In: Powles, S. & Shaner, D. (eds.) Herbicide Resistance Management and World Grain, CRC Press. Boca Raton, London, New York, Washington DC (2001)., 165-194.

[5] Urzúa, S. F. Pruebas de efectividad biológica de herbicidas. In: Mota, S.D., Rodríguez, J.C., Sánchez, M.A.H. & Lagunas, T.A. (eds.) Aprobación en estudios de efectividad biológica de plaguicidas. Colegio de Postgraduados, México. (1995)., 260-282.

[6] Daubenmire, R. Citation classic-a canopy-coverage method of vegetational analysis. Current Contens/Agriculture Biology & Environmental Sciences (1984).

[7] Yenish, J. P, Doll, J. D, & Buhler, D. D. Effects of tillage on vertical-distribution and viability of weed seed in soil. Weed Science (1992)., 40(3), 429-433.

[8] Odum, E. P, Park, T. Y, & Hutcheson, K. Comparison of the weedy vegetation in old-fields and crop fields on the same site reveals that fallowing crop fields does not result in seed bank build-up of agricultural weed. Agriculture, Ecosystems & Environment (1994)., 49, 247-252.

[9] Oppong, F. K, & Sagar, G. R. Degradation of triasulfuron in soil under laboratory conditions. Weed Research (1992)., 32(3), 167-173.

[10] Reidsma, P, & Tekelenburg, T. van den Berg, M. & Alkemade, R. Impacts of land-use change on biodiversity: An assessment of agricultural biodiversity in the European Union. Agriculture, Ecosystems & Environment (2006)., 86-102.

[11] Robinson, R. A, & Sutherland, W. J. Post-war changes in arable farming and biodiversity in Great Britain. Journal of Applied Ecology (2002). , 39, 157-176.

[12] Petit, S, Firbank, R, Wyatt, B, & Howard, D. MIRABEL: models for integrated review and assessment of biodiversity in European landscapes. Ambio (2001). , 30, 81-88.

[13] Jose-maria, L, Armengot, L, Blanco-moreno, J. M, Bassa, M, & Sans, F. X. Effects of agricultural intensification on plant diversity in Mediterranean dry land cereal fields. Journal of Applied Ecology (2010). , 47, 832-840.

[14] Diaz, S, & Cabido, M. Plant functional types and ecosystem function in relation to global change. Journal of Vegetation Science (1997). , 8(4), 463-474.

[15] Diaz, S, & Cabido, M. Vive la différence: plant functional diversity matters to ecosystem processes. Trends in Ecology & Evolution (2001). , 16(11), 646-655.

[16] Gerard, C. J, Sexton, P, & Shaw, G. Physical Factors Influencing Soil Strength and Root Growth. Agronomy Journal (1982). , 74(5), 875-879.

[17] Green, V. S, Cavigelli, M. A, Dao, T. H, & Flanagan, D. C. Soil physical properties and aggregate-associated C, N, and P distributions in organic and conventional cropping systems. Soil Science (2005). , 170(10), 822-831.

[18] Matson, P. A, & Vitousek, P. M. Agricultural intensification: Will land spared from farming be land spared for nature? Conservation Biology (2006). , 20(3), 709-710.

[19] Tilman, D, Fargione, J, Wolff, B, Antonio, D, Dobson, C, Howarth, A, Schindler, R, Schlesinger, D, Simberloff, W. H, & Swackhamer, D. D. Forecasting agriculturally driven global environmental change. Science (2001). , 292, 281-284.

[20] Roschewitz, I, Thies, C, & Tscharntke, T. Are landscape complexity and farm specialisation related to land-use intensity of annual crop fields?. Agriculture Ecosystems & Environment (2005). , 105, 87-99.

[21] Klejin, D, & Sutherland, W. J. How effective are European agri-environment schemes in conserving and promoting biodiversity? Journal of Applied Ecology (2003). , 40, 947-969.

[22] Gabriel, D, Roschewitz, I, Tscharntke, T, & Thies, C. Beta diversity at differential spatial scales: plant communities in organic and conventional agriculture. Ecological Applications (2006). , 16, 2011-2021.

[23] Tscharntke, T, Klein, A. M, Kruess, A, Steffan-dewenter, I, & Thies, C. Landscape perspectives on agricultural intensification and biodiversity-ecosystem service management. Ecology Letters (2005). , 8, 857-874.

[24] Sutherland, W. J. Restoring a sustainable countryside. Trends in Ecology and Evolution (2002). , 17, 148-150.

[25] Salonen, J. Weed infestation and factors affecting weed incidence in spring cereals in Finland-a multivariate approach. Agriculture Science of Finland (1993). , 2, 525-536.

[26] Pysek, P, Jarosik, V, Kropac, Z, Chytrý, M, Wild, J, & Tichý, L. Effects of abiotic factors on species richness and cover in Central European weed communities. Agriculture, Ecosystems & Environment (2005). , 109, 1-8.

[27] Lososová, Z, Chytrý, M, Cimalová, S, Kropác, Z, Otýpková, Z, Pysek, P, & Tichý, L. Weed vegetation of arable land in Central Europe: Gradients of diversity and species composition. Journal of Vegetation Science (2004). , 15, 415-422.

[28] Cimalova, S, & Lososova, Z. Arable weed vegetation of the northeastern part of the Czech Republic: effects of the environmental factors on species composition. Plant Ecology (2009). , 203(1), 45-57.

[29] Fried, G, Norton, L. R, & Reboud, X. Environmental and management factors determining weed species composition and diversity in France. Agriculture, Ecosystems & Environment (2008). , 128, 68-76.

[30] Ball, D. A, & Miller, S. D. Cropping History, Tillage, and Herbicide Effects on Weed Flora Composition in Irrigated Corn. Agronomy Journal (1993). , 85, 817-821.

[31] Hammond, C. M, Luschei, E. C, Boerboom, C. M, & Nowak, P. J. Adoption of integrated pest management tactics by Wisconsin farmers. Weed Technology (2006). , 20(3), 756-767.

[32] Rew, L. J, Maxwell, B. D, & Aspinall, R. Predicting the occurrence of nonindigenous species using environmental and remotely sensed data. Weed Science (2005). , 53(2), 236-241.

[33] Smith, R. G, & Gross, K. L. Rapid changes in the germinable fraction of the weed seed bank in crop rotations. Weed Science (2006). , 54, 1094-1100.

[34] Bàberi, P, & Locascio, B. Long-term tillage and crop rotation effects on weed seedbank size and composition. Weed Research (2001). , 41, 325-340.

[35] Blackshaw, R. E, & Brandt, R. N. Nitrogen Fertilizer Rate Effects on Weed Competitiveness is Species Dependent. Weed Science (2008). , 56(5), 743-747.

[36] Cussans, G. W. Weed control in reduced cultivation and direct-drilling systems. Outlook Agriculture (1975). , 8, 240-242.

[37] Bàrberi, P, Bonari, E, & Manzzoncini, M. Weed density and decomposition in winter wheat as influenced by tillage systems. In: L. García-Torres, L., Benites, J. & Martínez-Vilela, A. (eds.) Conservation Agriculture, a worldwide challenge. (2001). Vol. II, , 451-455.

[38] Legere, A, Samson, N, Rioux, R, Angers, D. A, & Simard, R. R. Response of spring barley to crop rotation, conventional tillage, and weed management intensity. Agronomy Journal (1997). , 89, 628-638.

[39] Liebman, M, Drummond, F. A, Corson, S, & Zhang, J. Tillage and rotation crop effects on weed dynamics in potato production system. Agronomy Journal (1996). , 88, 18-26.

[40] Navarrete MartínezL., Fernández-Quintanilla, C., Hernanz Martos, J.L. & Sánchez-Girón Renedo "El Encín" Clima, Suelo y Vegetación. In: Dirección General de Educación y Promoción Ambiental de la C.A.M. (eds.) Clima, Suelo y Vegetación (2000). , 73-89.

[41] Buhler, D. D, Soltemberg, D. E, Becker, R. L, & Gunsolus, J. L. Perennial weed populations after 14 years of variable tillage and cropping practices. Weed Science (1994). , 42, 205-209.

[42] Derksen, D. A, Lafond, G. P, Thomas, A. G, Loepky, H. A, & Swanton, C. J. Impact of agronomic practices on weed communities: Tillage Systems. Weed Science (1994). , 33, 176-181.

[43] Legere, A, Samson, N, & Rioux, R. Perennial weeds in conservation tillage systems: More of an issue than in conventional tillage systems? In: Br. Crop Prot. Council, (eds.) Weeds Proceedings Brighton Crop Protection. Farnham, U.K. (1993). , 747-752.

[44] Catalán, G, Hervella, A, De Andrés, F. E, Sánchez, F. J, Eyerbe, L, & Tenorio, J. L. Distintos laboreos para la rotación cebada-veza en la España semi-árida fría. In: XLI Workshop S.E.E.P. Biodiversidad en pastos (2001). , 563-568.

[45] Wrucke, M. A, & Arnold, W. E. Weed species distributions as influenced by tillage and herbicides. Weed Science (1985). , 33, 853-856.

[46] Radosevich, S. R, & Holt, J. S. Implications for vegetation management. John Wiley & Sons, New York (eds.) In: Weed ecology (1984). , 198-203.

[47] Froud-williams, R. J, Chancellor, R. J, & Drennan, D. S. H. Potential changes in weed flora associated with reduce-tillage cultivation systems for cereal production in temperate regions. Weed Research (1981). , 21, 99-109.

[48] Liebman, M, & Davis, A. S. Integration of soil, crop and weed management in low-external input farming systems. Weed Research (2000). , 40, 27-47.

[49] Roberts, H. A. Emergence and longevity in cultivated soil of seeds of some annual weeds. Weed Research (1964). , 4, 296-307.

[50] Santín-montanyá, I. Tenorio Pasamón, J.L. & García Baudín, J.M. Changes in Weed Community as affected by Tillage Systems in a Semi-Arid Environment. Italian Journal of Agronomy (2008).

Chapter 7

The Use of Glyphosate in Sugarcane: A Brazilian Experience

Carlos Alberto Mathias Azania,
Luciana Rossini Pinto, Rodrigo Cabral Adriano,
Dilermando Perecin and Andréa Padua Azania

Additional information is available at the end of the chapter

http://dx.doi.org/10.5772/54958

1. Introduction

In Brazil, sugarcane (*Saccharum* spp.) fields are renewed in intervals of five to six profitable crops on average. With each harvest, sugarcane displays a decrease in productivity due to diverse factors. Genetic, phytosanitary and edaphoclimatic issues are the main factors contributing to the degeneration that necessitates the renewal of sugarcane fields with more productive cultivars. After the last economical harvest, the ratoon crop is destroyed using mechanical or chemical processes or a combination of both. Chemical destruction is more practical and causes less impact on soil structure and quality due to less soil disturbance. Glyphosate is the most widely used non-selective herbicide in the chemical eradication of ratoon crops because there is a broad spectrum of plants susceptible to glyphosate. Glyphosate's mechanism of action is through inhibition of the 5-enolpyruvylshikimate-3-phosphate synthase (EPSPS), a precursor of the aromatic amino acids phenylalanine, tyrosine and tryptophan, which are essential for protein synthesis. The recommended dosage of glyphosate for the eradication of plants is 1440 to 2880 g acid equivalents (a. e.) ha^{-1} [1,2]. However, sugarcane cultivars present varying degrees of susceptibility and require different amounts of herbicide for the complete death of the plant.

In Brazil, sugarcane cultivars commercially released by genetic improvement programmes are not characterised in terms of their susceptibility to glyphosate. Nevertheless, knowledge of the degree of cultivar tolerance to glyphosate can generate savings for producers and benefit to the environment through the reduction of the quantity of applied herbicide. Literature studies of cultivar responses to herbicides, especially glyphosate, are supported solely by phytotechnical observations, such as plant height, girth and mass gain. However, these characteristics are greatly affected by the environment and require longer periods of evaluation and data collection, such as the 12-month studies of [3] and [4].

The use of isoenzymatic markers allows for the prompt analysis of results with larger sample numbers while using a relatively simple and inexpensive technique that can substitute or reduce field experimentation. However, the choice of the correct enzymes to analyse is critical to the success of this technique and obtaining robust results.

2. Sugarcane crops

Sugarcane probably originated in New Guinea, and from there, it was introduced to India, where the oldest evidence of its existence has been recorded [5]. Officially, Martins Afonso de Souza brought the first sugarcane plant to Brazil in 1532 and started its cultivation in the Captaincy of São Vicente (Capitania de São Vicente). This transfer was the beginning of an industry that found in Brazil, among other nations that would later initiate production, its most fertile ground for rapid expansion and perpetuation for an almost uninterrupted 500 years. Starting in the 1970s, sugarcane farming became increasingly important for Brazil as the agro-industrial sector was tapped to contribute to a solution to the emerging energy crisis because of the potential for energy production from sugarcane as a renewable source [6]. Growth in the sugarcane-ethanol sector is important for the Brazilian economy in that the sector's growth entails both the creation of jobs and of 100% national renewable energy.

According to taxonomic classification, sugarcane belongs to the *Poaceae* family and the *Saccharum* genus. Sugarcane is a semi-perennial plant requiring a tropical or sub-tropical climate [7]. With a C4 metabolism, sugarcane is classified as having among the highest rates of photosynthetic efficiency and a high efficiency for water usage [8]. The sugarcane plant is divided into aerial (culm, leaves and inflorescences) and underground parts (roots and rhizomes). The culms are cylindrical and are composed of nodes and internodes; these parts are defined as the aboveground portion that supports the leaves and inflorescences [9]. According to [10], each node has one alternating bud and a root system. Inflorescences are panicles with a hermaphrodite flower containing one ovule; the pistils terminate in purple or reddish stigmae that characterise the flower's plumose panicle [9]. The root system is fasciculated and serves to support, as well as to absorb and transport water and nutrients [8]. Sugarcane tillering influences the sugarcane handling system because each tiller behaves as an independent plant with individual organs, such as roots, leaves and fruits [11].

The most appropriate agricultural conditions for sugarcane propagation are found between the 30° north and 30° south latitudes, which are characteristic of tropical and subtropical regions. Outside of these latitudes, lower temperatures limit the growth and development of the plant [12]. According to [10], the optimal temperature range for the growth of this crop is between 20 and 35 °C with an ideal photoperiod of 10 to 14 hours [12] and an annual rainfall ranging between 1,000 and 1,600 mm, preferentially with abundant rain during the vegetative

http://dx.doi.org/10.5772/54958

growth period and a dry period during maturation, which favours increased sucrose accumulation [13].

In Brazilian regions where it is traditional to grow sugarcane, planting may occur at different times of the year, as long as the producer possesses an irrigation system and cultivars that are adapted to each season [14]. Traditionally, in south-central Brazil, there are two cycles for planting: "cane of the year" and "year and a half". In "cane of the year", planting is performed between September and November, and the cane is harvested after 12 months [14]. This type of plantation addresses the demand for raw materials in the spring cycle (at the end of the harvest). In "year and a half" cane, planting is performed between January and April-May. In contrast to cane of year, this cycle allows for harvest during the autumn season (the beginning of harvest). Additionally, several producing units have practiced winter planting, particularly June through July, using rescue irrigation, and these units have obtained great productivity compared to "year and a half" cane planting.

Currently, Brazil is the largest producer of sugarcane in the world followed by India, China and Thailand [15]. The national production is estimated as 641.982 million tons with an average productivity of approximately 76.4 t ha^{-1} [16].The national sugar-energy industry sector accounted for 1,283,258 jobs up to 2008 with 37.5% occupied by plant growth, 44.8% in the production and refining of sugar and 17.7% in the production of ethanol. This sector also accounted for approximately 3.85 million people that are employed indirectly [17]. The production and processing of sugarcane is currently managed by the private sector in Brazil, which achieves the lowest cost for production worldwide for both sugar and ethanol, emerging as a highly competitive segment in international markets [18].

3. Characteristics of the herbicide glyphosate

Glyphosate was commercially released in 1974 under the trade name Roundup initially in the USA for industrial purposes, in the United Kingdom for use in wheat crops and in Malaysia for use in rubber trees. Currently, the molecule is registered in more than 130 countries for the control of over 300 species of weeds in over 100 types of crops [19], making it the most widely used herbicide [20]. Worldwide, there are numerous registered trademarks of the herbicide, which, according to [50], number more than 150. In Brazil, glyphosate is also registered for the eradication of sugarcane ratoon crops [21].

The molecule belongs to the glycine-derived chemical group (Group G), and its mechanism of action consists of inhibiting the 5-enolpyruvylshikimate-3-phosphate synthase (EPSPS) in the shikimic acid pathway [22], which is found only in microorganisms and plants [23]. The molecular formula of glyphosate is $HO_2CCH_2NHCH_2PO(OH)_2$. Glyphosate has a solubility in water of 15,700 mg L^{-1} at 25 °C and pH 7, a density of 1.74 g mL^{-1}, a vapour pressure of 2.45 x 10^{-8} Pa (45 °C), pK_a values of 2.6, 5.6 and 10.3 (acid) and a k_{ow} between 0.0006 and 0.0017. In the soil, glyphosate is strongly adsorbed to colloids, and its leachability is notably low. The compound has an average K_{oc} of 24,000 mL g^{-1}, and its volatilisation and photodegradation are negligible [22]. The half-life of the molecule in the environment depends on the surrounding soil texture and microbial activity and may vary from a few days to several years [24]. In roots, the absorption is slow due to the low diffusion and high adsorption to the soil, which also favours microbial action in the transformation of the molecule into its main metabolite, aminomethylphosphonic acid [25; 26]. Glyphosate is absorbed through leaf cuticles, and its translocation occurs mainly via the cellular symplast to the leaves and apical meristem, as well as to underground organs [22]. According to [27], glyphosate absorption depends on such factors as the age of the plant, environmental conditions, surfactants and herbicide concentration in the soil milieu.

As an herbicide, glyphosate is among the less hazardous agro-toxins used in agriculture [28]. Glyphosate-based herbicides, when used according to their respective guidelines, display low toxicity and are safe to humans [30].

Often, the glyphosate molecule is not efficient in penetrating waxy cuticles. Therefore, commercial formulations contain surfactants capable of reducing surface tension in herbicide droplets, thus increasing their penetration in leaves [20]. However, these surfactants are more toxic than the glyphosate molecule [29]. For example, polyoxyethylene amine, the predominant surfactant in Roundup® [30], has been classified as moderately to highly toxic in laboratory tests [31]. Glyphosate is a unique molecule, and although it is considered to be of low toxicity, its unrestrained use can affect the environment through direct or indirect effects on non-target organisms [32].

In plants, the EPSPS enzyme catalyses a reaction between shikimate-3-phosphate (S3P) and phosphoenolpyruvate (PEP) to produce 5-enolpyruvylshikimate-3-phosphate and inorganic phosphate. Glyphosate binds to the catalytic site of EPSPS and to the S3P substrate to form the EPSP synthase-S3P-glyphosate complex [33]. The relevance of the shikimate pathway is such that approximately 35% of all plant mass is related to derivatives from this pathway; moreover, 20% of all of the carbon fixed during photosynthesis also travels through this metabolic pathway [34].

According [35], the inhibition of amino acids compromises the production of carotenoids and chlorophyll, thus causing irreversible cellular damage. Therefore, the translocation of the herbicide throughout the entirety of the plant causes plant death in a few days or weeks (Figure 1). The inhibition of EPSPS leads to the accumulation of high levels of shikimate in vacuoles, which is intensified by the loss of control of the carbon flow across this pathway [36]. Thus, there is an obstruction in the production of the aromatic amino acids phenylalanine, tyrosine and tryptophan, which are essential for protein synthesis and serve as precursors for secondary metabolites that are important for plant growth [37], resulting in the slow development of symptoms [33].

4. Reforming sugarcane fields and the use of glyphosate

With each harvest cut, sugarcane sprouts new tillers that develop into culms [38]. Nevertheless, ratoon-crop productivity gradually diminishes with an increasing number of cuts [39], thereby requiring the renewal of the field. In the state of São Paulo, the average productivity of sugarcane fields is approximately 80 to 85 t ha^{-1}, considering the longevity of ratoons to be between five and six cuts [40]. However, in the region of Ribeirão Preto, SP, ratoon crops after the sixth cut are no longer economically viable, and renewal of the sugarcane field is necessary [41]. Degeneration after successive years of production makes the renewal of sugarcane fields essential. The causes of degeneration are diverse and involve a combination of genetic, physiological, phytosanitary, edaphoclimatic and phytotechnical factors. The factors impacting degeneration may also be linked to characteristics of the growing environment, such as a decrease in soil fertility [42]. Another cause for degeneration can be soil compaction and consequent difficulties in root development, as proposed by [43]. The authors note that compacted soil still presents difficulties for root development, even if its humidity levels are close to the soil's capacity.

[44] found that degeneration is linked to the health of the plants. Sugarcane-field longevity may also be affected by competition with weeds [45], nematode infestations [46] and uprooting of tufts during mechanical harvesting [47].

The fact that various cuts are performed from a single plantation allows for the formation of a significant number of root systems, which often make the elimination of the ratoon crop difficult, especially if the eradication is performed mechanically, which may also compromise the settlement of the next plantation. At the time of crop renewal, the ratoon crop is first eliminated through desiccating herbicides, specifically glyphosate, and after plant death, eradication is later completed using mechanical destruction of the crop [48].

Glyphosate is the most widely used herbicide for the chemical eradication of sugarcane ratoons due to its ease of use, low cost and absence of residual effects on the soil, which allows for repeated plantation in the same area, as is often practiced by farmers [49, 50]. Tolerance to glyphosate is highly prevalent in cultivars, and while certain cultivars are eradicated with a dose of 1080 g a. e. ha^{-1}, others require a dose of 2520 g a. e. ha^{-1}. According to [1,2], the minimal lethal dose for sugarcane is 1440 g a. e. ha^{-1}.

The progression of symptoms caused by a glyphosate application occurs in a gradual fashion (Figure 2) until the eradication of the plants [23, 36]. The authors noted that glyphosate-induced damage develops slowly until complete death in contrast to the effects of other herbicides. According to these authors, molecular stability inside the plant allows for the occurrence of irreversible effects on processes that control both annual and perennial plants (Figures 3 and 4).

After the application of glyphosate in the eradication of sugarcane crops, there was a stunting of plant growth, with treated plants retaining the same size up to 45 days after herbicide application [51]. The negative effect on growth was evaluated by measuring the plant height, with the treated plants maintaining similar average height values throughout the evaluation period in contrast to controls, which were able to maintain vigorous vegetative growth. The growth stunting was due to the indirect influence of glyphosate on the regulators of plant growth, such as indole-3-acetic acid. This hormone is fundamental for cellular elongation, apical dominance and stem and root growth and is dependent on the shikimate pathway, being inhibited when there is a disruption of EPSPS [36].

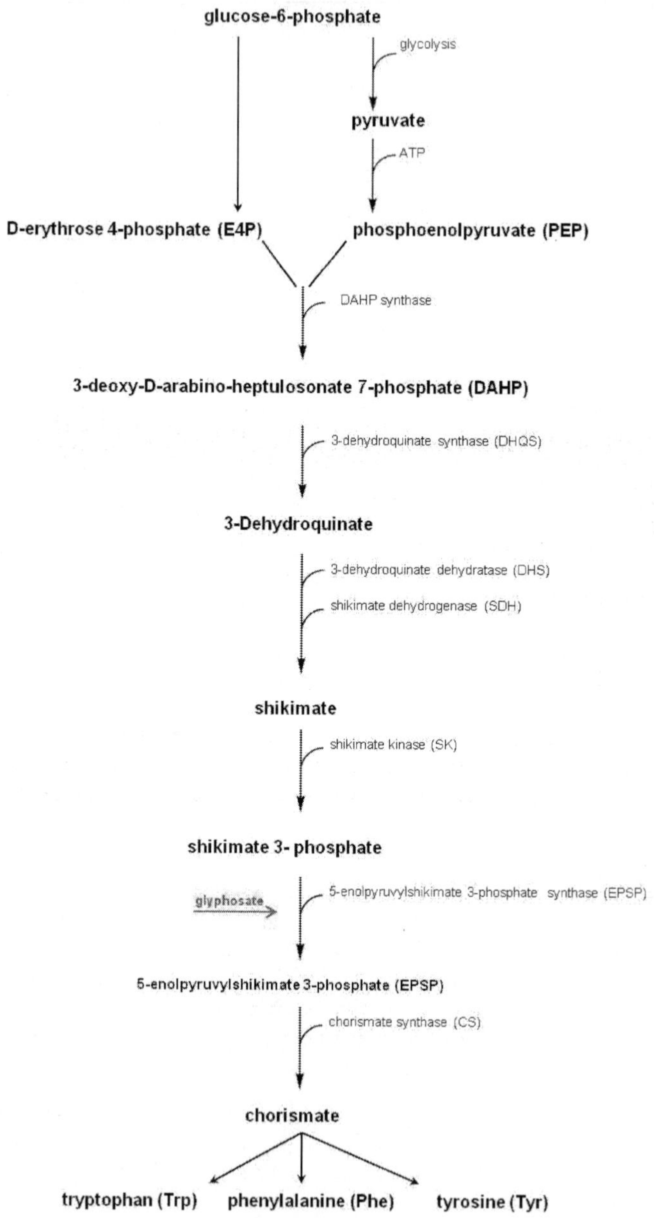

Figure 1. The shikimate pathway and the action of glyphosate on plants.

http://dx.doi.org/10.5772/54958

Figure 2. Sugarcane plants ten days after the application of glyphosate (2880 g a.e. ha^{-1}).

The varying tolerance of sugarcane cultivars to glyphosate was studied [52], who found different sensitivities among cultivars. The authors also classified the genotypes IAC86-2210, IAC83-1313, IAC82-2045, PO83-698 and IAC83-4157 as susceptible to glyphosate, IAC86-3154, IAC87-3184, RB72454 and SP80-1842 as of intermediate susceptibility, and IAC82-3092, IAC87-3396 and RB806043 as tolerant. Nevertheless, complete death, even in the less susceptible cultivars, occurred after 45 days following application. A plant's inherent tolerance is related to the plant's capability for absorption, translocation, metabolism and/or elimination of a herbicide [53]. In [54] also noted that differences in absorption depend primarily on morpho-anatomical characteristics of the species and that in the aerial parts of the plant, absorption is highly influenced by the presence or absence of cuticles. The physicochemical content of the leaf surface is another form of plant resistance to glyphosate [55]. According to these authors, leaves with flat cuticle surfaces and without large quantities of wax can better retain applied droplets. After penetration, the herbicide can then be metabolised into secondary compounds without herbicidal activity, or its potency might be enhanced [56].

The plant's development stage is another factor that should be considered in the eradication of cultivars because plants must be 40 to 80 cm tall at the time of glyphosate application [2], and the total leaf area must be sufficient to intercept the herbicide. The inherent resistance to glyphosate is greater in taller plants [57]. After the formation of culms, plants become more tolerant to the herbicide [58].

A relationship between plant size and glyphosate efficacy was also observed [59] while studying *Conyza bonariensis*. The authors observed that herbicidal efficacy was greater when the plants presented up to two pairs of leaves. Nevertheless, in more advanced stages of development, it was necessary to increase herbicide dosage by up to fivefold.

Glyphosate applied 40 days after the last harvest caused the highest percentage of dead tillers, and also, genotypes IAC87-3184, RB835489 and SP87-344 displayed high to intermediate sensitivity, while IAC91-5155 was considered tolerant to the herbicide [49]. The most effective application time for eradication was 65 days after cane harvesting [60]. The same authors reported that a dose of 960 g a. e. ha^{-1} eradicated the majority of cultivars, except for Co997, which needed 1920 g a. e. ha^{-1} of the herbicide.

Figure 3. Intoxication symptoms caused by glyphosate rates in sugarcane cultivars (IACSP94-2094-4004 and IACSP94-4004) to 27 days after application. Instituto Agronômico de Campinas -IAC, 2012.

The development of transgenic sugarcane plants, particularly those with tolerance to glyphosate, will most certainly change the way in which sugarcane is eradicated. The use of glyphosate will reduce the costs associated with the control of weeds across cycles; however, during the period of cane eradication, the herbicide will now have a limited impact due to the tolerance introduced to the cultivars. In this case, eradication may have to be performed mechanically, which will have a negative impact on soil conservation and might stimulate weed germination from soil propagule banks. From this perspective, it is important to emphasise that research aimed at sugarcane plants tolerant to glyphosate should also consider the use of herbicides in the eradication of future cultivars.

In the eradication of ratoon crops, glyphosate is used to eradicate the crop and also to control emerged weeds. However, when a sugarcane field also possess weeds that are hard to control, such as *Cynodon dactylon* and *Cyperus rotundus*, the use of higher doses of residual herbicides after the application of glyphosate is adopted in a process known as "disinfestation".

Figure 4. Intoxication symptoms caused by glyphosate rates in sugarcane cultivars (IACSP94-2094-4004 and IACSP94-2191) to 45 days after application. Instituto Agronômico de Campinas - IAC, 2012.

The aggressive biological nature of hard to control weed species requires that handling start with the desiccation of the plants to optimise the use of glyphosate in the eradication of ratoon while also being able to introduce residual herbicides at higher doses. In these instances, the use of glyphosate in crop eradication serves the dual role desiccating the ratoons and controlling problematic weeds.

In the time period following eradication but before planting sugarcane, the producer should formulate a strategy and opt for techniques that ensure higher sustainability of the system. These methods include such techniques as crop rotation or planting green-manure crops, although these may still compromise the techniques' sustainability if installed in fields that have been infested with "difficult-to-control weeds" or previously treated with residual herbicides for "disinfestation". The producer should carefully plan to use techniques that generate the most effective soil preparation and handling of weeds while simultaneously ensuring that after the treatments, the soil remains prepared for a new sugarcane plantation.

5. The shikimate pathway and isoenzymatic markers

The shikimate pathway is found only in plants and microorganisms and is completely absent in mammals, birds, reptiles, fish and insects. These organisms extract the aromatic compounds necessary for survival and reproduction from their diet, while plants must produce such compounds because they do not have alternative means to obtain the compounds [23].The shikimate pathway is initiated with the reaction of PEP and erythrose 4-phosphate, a reaction catalysed by the enzyme DAHP synthase (3-deoxy-D-arabino-heptulosonate-7-phosphate synthase [61]. The resulting product is the seven-carbon acyclic intermediate 3-deoxy-D-arabino-heptulosonate-7-phosphate (DAHP).

DAHP is converted to cyclic form through catalysis by 3-dehydroquinate synthase in the presence of NAD^+ as a coenzyme. In this process, 3-dehydroshikimate dehydratase dehydrates the cyclic form of DAHP. Next, shikimate dehydrogenase, in the presence of $NADP^+$ (oxidised NADPH), reduces the cyclic and dehydrated DAHP to shikimate. The molecule is later phosphorylated by the SP3 kinase, which converts a molecule of ATP to ADP. The phosphorylated shikimate subsequently reacts with one molecule of PEP in the presence of 5-enolpyruvylshikimate-3-phosphate (EPSP) synthase, resulting in the production of EPSP. The shikimate pathway terminates with the production of chorismate (chorismic acid) through the dephosphorylation of EPSP by chorismate synthase.

Other metabolites essential to plant life may be produced from chorismate, including the amino acids tryptophan, phenylalanine and tyrosine, as well as vitamin K, ubiquinone and tetrahydrofolate [62, 23]. The amino acid phenylalanine is a precursor not only of proteins but also of other secondary products, such as phenolic compounds, anthocyanins, lignin and promoters and growth inhibitors. Tryptophan is also a precursor to indole-3-acetic acid, which is responsible for apical dominance and is vital for cellular growth and several other regulatory processes. Therefore, inhibitors of the shikimate metabolic pathway represent a strategic alternative in the development of herbicides with low environmental impact, such as glyphosate [23]. In this context, it could be interesting to use protein electrophoresis as a tool to study the eradication of plants by glyphosate using the isoenzymes involved in the metabolic pathway of shikimic acid.

Isoenzymes are the multiple molecular forms of enzymes that perform the same or similar catalytic activities. These enzymes are coded by one or more genes and may play an important role in survival across diverse environments. Isoenzymes are directly affected by both biotic and abiotic stressors [63]. The band intensity and isoenzymatic profile are plant-, tissue- and development stage-specific [64]. Some factors that affect plant metabolism, such as mineral nutrition, low temperature and diseases, among others, influence the activity of isoenzymes, specifically, esterases, peroxidases, phosphatases and phenolases, which in turn generate different expression patterns and levels of activity.Isoenzymatic patterns were used as tools by [65] who concluded that the enzymatic system of malate dehydrogenase is an efficient marker for aerobic respiration in pepper seeds during the maturation period. In Serbia [66] (the Vojvodina region), observed that the shikimate dehydrogenase system is also an efficient isoenzymatic marker for the study of genetic variability and polymorphisms in different almond genotypes.

In [67] was used the same technique in soy cultivars and registered a difference in electrophoretograms in terms of peroxidase activity. In [68] used isoenzymatic markers to identify species of lettuce nematodes, and [69] used the method "in vitro" in sugarcane to observe varietal differences among doses of glyphosate.

6. Esterase isoenzymes in abiotic stress

Enzymatic activity is influenced by stress factors, such as non-optimal temperature or nutrient levels and infection by pathogens. These stress responses subsequently lead to gene activation and, as a consequence, to the emergence of several molecular forms [70]. Because of the involvement of isoenzymes in changes to metabolism and defence mechanisms in plants, studies involving isoenzymes can be used in cases of both biotic and abiotic stress [63]. The authors report that polymorphisms displayed by isoenzymes are intermediate products of gene expression and are closer to the final phenotypical expression than those of DNA polymorphisms.

Esterases are isoenzymes comprising a group of genetically distinct enzymes that are found across a large spectrum of living organisms and that play a large variety of roles; nevertheless, esterases display a common trait of catalysing the hydrolysis of esters, peptides, amides and halide bonds [71]. Esterases can be found as both monomers and dimers [72, 73]. Esterases are significantly linked to lipid metabolism, such as that of membrane phospholipids, due to catalysis of ester hydrolysis [74].

In polyacrylamide gel electrophoresis assays, the α-esterase isoenzyme is detected using naphthyl ester substrates and histochemical stains. Enzyme isoforms with an affinity to hydrolyse α-naphthyl acetate are identified on the gel as black bands derived from the precipitation of α-naphthol, which results from the hydrolysis of α-naphthyl with the Fast blue RR salt stain [75].

Esterase isoenzymes have also been extensively explored in studies of genetic diversity due to their high rate of polymorphisms [76, 77]. In [78] the authors reported that sugarcane cultivars could be identified using esterase isoenzymes, and [79] in studying the parameters for sugarcane differentiation, observed that the electrophoretic profile of esterases is maintained in plants of varying physiological ages, as long as the growth environment is controlled.

Esterase isoenzymes are among the most widely used enzymes in the evaluation of enzymatic alteration in plants that are affected by parasitic nematodes across various pathosystems [80]. In [75] was studied esterase polymorphisms in 16 cultivars of soy that underwent or were spared treatment with glyphosate. The authors observed variation in the sensitivity to α-esterase isoenzymes of the different cultivars and also found that sensitivity did not seem to be connected with the homozygous RR status of the genetically modified plants.

7. Practical results of the isoenzymatic profiles of shikimate dehydrogenase and α-esterase in sugarcane

The tolerance of sugarcane cultivars to chemical eradication using varying doses of glyphosate was investigated [51] using phytotechnical parameters and isoenzymatic markers. The author hypothesised that the study of isoenzymatic profiles of shikimate dehydrogenase and α-esterase could optimise phytotechnical fieldwork observations regarding herbicidal tolerance.

Shikimate dehydrogenase was selected because it is involved in the shikimic acid pathway, which is affected by glyphosate action, and α-esterase was chosen because it is associated with oxidative stress. The isoenzymatic profiles of shikimate dehydrogenase and α-esterase were studied in sugarcane cultivars IACSP94-2094, IACSP94-2101, IACSP93-3046, IACSP94-4004, IAC86-2480 and RB72454 at 8, 24, 48, 72 and 144 hours after the application of glyphosate at doses of 0, 1440, 2160, 2880, 3600 and 4320 g a. e. ha^{-1}. The results showed that the bands for shikimate dehydrogenase tended to position near the cathode (at the top of the gel), while α-esterases were positioned closer to the anode, due to a greater migration during the gel run. The enzymatic system of shikimate dehydrogenase presented bands that were less sharp and also had a lower number of bands (three). There were no observed polymorphisms among cultivars, regardless of whether the data were analysed according to herbicide dosages or in relation to controls. Therefore, the isoforms remained constant among the different cultivars and treatments. In [81] was studied 20 enzymatic systems in the identification of sugarcane cultivars and also did not obtain any promising results using shikimate dehydrogenase. In [82] was studied populations of *Stryphnodendron adstringens*, known in Brazil as *barbatimão,* and also did not find polymorphisms for shikimate dehydrogenase.

The enzymatic system of α-esterase was specific for each studied cultivar, allowing for cultivar identification based on this biochemical marker [51]. This observation corroborates the findings [83], who created an analytical key for sugarcane cultivars and found a different pattern of α-esterase in each of the ten cultivars studied. A large number of bands of the α-esterase enzymatic complex were found with varied band intensity and thickness. The characteristics of this complex can be related to the degree of ploidy of the plant species; sugarcane is polyploid [70, 84].

Cultivars of variety IACSP93-3046 and RB72454 did not present differences in terms of bands owing to the application of glyphosate. These cultivars were considered susceptible to glyphosate based on field studies reporting a percentage of tiller death of 93.16% and 94.25% respectively. Moreover, marked toxic effects were rated as high as 94% for IACSP93-3046 and 95.5 % for RB72454 [51].

Sugarcane cultivar IACSP94-4004 was the only cultivar to show an alteration in its band pattern due to the application of glyphosate. Across all of the treatments in which the herbicide was applied, there were two additional bands that were not present in the controls and that were present from the first assessment 8 hours after the application (HAA) of herbicide to the last assessment at 144 HAA. The appearance of additional bands may be due to the expression of genes from this enzymatic system in response to stress caused by glyphosate treatment, thereby demonstrating that the cultivar response to the herbicide is directly linked to the genotype of each cultivar variety. In fact, cultivar IACSP94-4004 was relatively tolerant to the field experiments, as it was the cultivar to show the lowest average (percentage) of tiller death at 45 HAA (80.15 %), the symptoms of toxicity in this cultivar were less pronounced with an average of 82.5 %, and only glyphosate doses of 3600 g a. e. ha^{-1} caused symptoms similar to those caused by the highest dose of 4320 g a. e. ha^{-1}.

Evaluation of the isoenzymatic system of esterases has been used in other studies to characterise cultivar tolerance. Nevertheless, in the evaluation of sugarcane eradication,

http://dx.doi.org/10.5772/54958

the isoenzymatic profiles of shikimate dehydrogenase and α-esterase did not constitute a reliable tool [51].

8. The use of glyphosate in sugarcane as a ripener

Maturation is one of the most important aspects of sugarcane crops because maturation is directly related to the optimal time-point for harvest/industrial transformation.For the plant to enter the maturation process naturally, one or more sources of stress are necessary, with a gradual reduction in photoperiod, temperature or precipitation being the most effective stimulants [85]. In Brazil, more specifically in the south-central region, the maturation process is initiated in April/May when the climate becomes colder and drier. However, even under favourable conditions, sugarcane maturation may also be induced in responsive cultivars as a strategy to produce high-quality raw material across all of the different phases of harvesting.

To guarantee that maturation is complete, uniform, early and programmed and to avoid undesirable flowering, the sugarcane-ethanol industry has been adopting the use of ripeners (growth regulators) in sugarcane. Ripeners are chemical compounds that induce the translocation and storage of sugars, mainly sucrose, in the culm. Therefore, the goal of the ripeners is both to advance and maintain natural maturation and to provide high-quality raw materials for early industrial transformation, as well as to aid in the handling of cultivars [85].

The same authors state that to artificially induce maturation, growth regulators are applied by aircraft eight to ten months after the last harvest, that is, during the plants' vegetative state. In practice, the months of February and March or October are the periods during which farmers aim to apply the enhancers because they can anticipate the beginning or the end of the harvest.

There are two basic types of ripeners for sugarcane fields, "non-stressors" and "stressors". Non-stressor ripeners do not diminish the plants' growth rate, and their action induces the release of ethylene, the compound responsible for maturation that helps in the accumulation of sucrose in sugarcane culms. Stressor compounds, such as glyphosate, are growth inhibitors that markedly decrease the sugarcane growth rate, making the plants accumulate sucrose instead of expending it as an energy source for growth. This reduction in growth rate forces the plant to mature [85]. In [86] sugarcane plants with stagnated growth stop sprouting new leaves, and as a consequence, the reduced number of phytochromes in chloroplasts becomes insufficient to detect the photoperiod and thus stimulate the transition of the apical bud from vegetative to reproductive.

The effect of glyphosate, after it is applied to sugarcane, has a rapid onset, allowing for an increase of sucrose accumulation in 30 to 40 days after application. The glyphosate dose used is normally 0.3 to 0.4 l ha^{-1} and may reach a maximum of 0.6 – 0.8 l ha^{-1}, leading to differences in maturation speed as a function of dosage. Harvest should be performed when the highest levels of sucrose are reached [12].

9. Conclusions

Glyphosate is a high-efficiency molecule in the sugar-ethanol industrial sector in Brazil. Glyphosate can be used as an herbicide in the control of weeds and the chemical eradication of sugarcane crops, and when applied in low dosages, glyphosate can be used as a ripener.

Independently of where the product is used in the production system, the key factor is the adequacy of the dosage. If cultivar tolerance is known, it is possible to adjust the dose of glyphosate to eradicate sugarcane ratoon crops. Research has shown that a dose of 2 kg ha^{-1} (1440 g a. e. ha^{-1}) of the commercial product Roundup WG eradicates 92.76% of the tillers of the IACSP94-2094 cultivar but only 40.3% of the IACSP94-4004 cultivar. For the most effective eradication of the IACSP94-4004 cultivar, 4 kg ha^{1} (2880 g a. e. ha^{-1}) of chemical is needed. This finding demonstrates that knowledge of plant tolerance can be a valuable tool to adjust glyphosate doses to the appropriate concentrations for ratoon-crop eradication.

Reducing the quantity of applied glyphosate is possible, as long as the crop is sensitive. Environmental and economic benefits can also be obtained by applying lower quantities of the herbicide. Information on cultivar tolerance also allows producers to know when to use higher herbicide concentrations than those recommended in the literature. However, if the producer applies a lower dose than is necessary, the ratoon crop will not be completely eradicated. As a consequence, there will be additional expenditure on a additional herbicide application or on mechanical eradication, which is highly problematic because it causes the greatest disruption to the soil and later leads to the presence of crop stubble in the new sugarcane field.

Acknowledgements

The authors are grateful to FAPESP (Fundação de Amparo à Pesquisa do Estado de São Paulo) for the financial support (Grant # 2010/09016-9).

Author details

Carlos Alberto Mathias Azania[1*], Luciana Rossini Pinto[1], Rodrigo Cabral Adriano[1], Dilermando Perecin[2] and Andréa Padua Azania[1]

*Address all correspondence to: azania@iac.sp.gov.br

1 Instituto Agronômico de Campinas - IAC, Centro de Cana, Ribeirão Preto, Brazil

2 Universidade Estadual Paulista – Unesp, Jaboticabal, Brazil

References

[1] Bacchi OOS, Rolim JC. Dose letal de glifosato para fins de eliminação química de soqueiras. Proceedings of the 2nd Congresso Nacional da Sociedade dos Técnicos Açucareiros e Alcooleiros do Brasil; 1981, Rio de Janeiro. Rio de Janeiro: STAB, 1981. p. 13-20.

[2] Lorenzi H. Pragas da cultura da cana-de-açúcar. Proceedings of 2nd Reunião Técnica Agronômica, 1983, Piracicaba. São Paulo: Copersucar, p. 59-82.

[3] Souza Jr, Perecin D, Azania CAM, Schiavetto AR, Pizzo IV, Candido LS. Tolerância de cultivares de cana-de-açúcar a herbicidas aplicados em pós-emergência. Bragantia 2009; (68): 941-951.

[4] Zera FS, Azania CAM, Schiavetto AR, Lorenzato CM, Freitas GB, Azania AAPM. Tolerância de mamona (*Ricinus communis*) a herbicidas utilizados na cultura da cana-de-açúcar. Nucleus 2011 (8): 453-462.

[5] Machado FBP. Brasil, a doce terra. http://www.canaweb.com.br/conteudo/Historiadosetor.htm/ (accessed 9 February 2001).

[6] Kuva MA. Efeitos de períodos de controle e de convivência das plantas daninhas na cultura da cana-de-açúcar (*Saccharum* sp) no Estado de São Paulo [dissertation]. Universidade de São Paulo, Piracicaba; 1999.

[7] Vidal RA, Trezzi MM. Origem da cultura e sua importância. In: Teoria e prática do manejo de infestantes na cultura da cana-de-açúcar no Brasil. Porto Alegre: 2011.

[8] Casagrande AA, Vasconcelos ACM. Fisiologia da parte aérea. In: Dinardo-Miranda LL, Vasconcelos ACM, Landell MGA (ed.). Cana-de-Açúcar. Campinas: Instituto Agronômico; 2008. p. 57-78.

[9] Scarpari MS, Beauclair EGF. Anatomia e botânica. In: Dinardo-Miranda LL, Vasconcelos, ACM, Landell MGA (ed.). Cana-de-Açúcar. Campinas: Instituto Agronômico; 2008, p. 45-56.

[10] Casagrande AA. Tópicos de morfologia e fisiologia da cana-de-açúcar. Jaboticabal: FUNEP; 1991.

[11] Alexander AG. Sugarcane physiology: a comprehensive study of the Saccharum source-to-sink system. Amsterdam: Elsevier; 1973.

[12] Rodrigues JD. Fisiologia da cana-de-açúcar. Botucatu: Fepaf; 1995.

[13] Agrianual 2007: Anuário da agricultura brasileira. São Paulo: Instituto FNP, 2007. p. 23-28.

[14] Anjos IA, Figueiredo PAM. Aspectos fitotécnicos do plantio. In: Dinardo-Miranda LL, Vasconcelos ACM, Landell MGA (ed). Cana-de-Açúcar. Campinas: Instituto Agronômico; 2008. p. 585 - 598.

[15] FAO 2009 - Food and Agriculture Organization of the United Nations. http://faostat.fao.org/site/339/default.aspx/ (accessed 28 August 2011).

[16] CONAB - Companhia Nacional de Abastecimento. Acompanhamento de safra brasileira: cana-de-açúcar, primeiro levantamento, maio 2011 – Brasília: Conab, 2011. http://www.conab.gov.br/ (accessed 28 de August 2011).

[17] UNICA - Portal da União da Agroindústria Canavieira. Relatório de Sustentabilidade 2010. São Paulo, 2010. http://www.unica.com.br/ (accessed 10 August 2010).

[18] Gonçalves DB. Dilemas do desenvolvimento sustentável na produção canavieira paulista. PhD thesis. Universidade Federal de São Carlos; 2005.

[19] Halter S. História do herbicida agrícola glyphosate. In: Velini ED, Meschede D, Carbonari CA, Trindade MLB. Glyphosate (ed). Botucatu: Fepaf; 2009. p.11-16.

[20] Jones DK, Hammond JI, Relyea RA. Competitive stress can make the herbicide Roundup® more deadly to larval amphibians. Environl Toxicol Chem 2011; (30):446–454.

[21] AGROFIT – Sistema de agrotóxicos fitossanitários. Base de Dados. http://extranet.agricultura.gov.br/agrofit_cons/principal_agrofit_cons. 2011/ (accessed 20 February 2012).

[22] Rodrigues BN, Almeida FS. Guia de herbicidas. 6th ed. Londrina: IAPAR; 2011.

[23] Gruys KJ, Sikorski JA. Inhibitors of tryptophan, phenylalanineand tyrosine biosynthesis as herbicides. In: Singh BK (ed.). Plant amino acids: biochemistry and biotechnology. New York: Marcel Dekker, 1999. p. 357-384.

[24] Arantes SACM, Lavorenti A, Tornisielo, VL. Efeito da calagem na mineralização de 14C-glifosato em solos. Ciência e Agrotecnologia 2011; 35 (2) 234-241.

[25] Franz JE, Mao, MK, Sikorski JA. Glyphosate's molecular mode of action. In: Glyphosate. A unique global herbicide. American Chemical Society. Monograph 189, p. 521-615, 1997.

[26] Galli AJB, Montezuma MC. Alguns aspectos da utilização do herbicida glifosato na agricultura. ACADCOM Editora; 2005. 67p.

[27] Monquero PA, Christoffoleti PJ, Osuna MD, De Prado RA. Absorção, translocação e metabolismo do glyphosate por plantas tolerantes e suscetíveis a este herbicida. Planta Daninha 2004; (22): 445-451.

[28] Duke SO, Powles SB. Glyphosate-resistant crops and weeds: now and in the future. AgBioForum 2009; (12): 346–357.

[29] Dill GM, Sammons RD, Feng PCC, Kohn F, Kretzmer K, Mehrsheikh A, et al. Glyphosate: discovery, development, applications, and properties. In: Nandula, V. K. (ed.) Glyphosate resistance in crops and weeds: history, development, and management. Hoboken: Wiley; 2010. 321p.

[30] Williams GM, Kroes R, Munro IC. Safety evaluation and risk assessment of the herbicide roundup and its active ingredient, glyphosate, for humans. Regul Toxicol Pharmacol 2000; 31 (2) 117-165.

[31] Bernal MH, Solomon KR, Carrasquilla G. Toxicity of formulated glyphosate (Glyphos) and Cosmo-Flux to larval Colombian frogs 1. Laboratory acute toxicity. J Toxicol Environ Health A 2009; (72) 961–965.

[32] Reis MR. Impacto do glyphosate associado ao endossulfan e tebuconazole sobre microbiota do solo na cultura da soja. PhD thesis. Universidade Federal de Viçosa; 2009.

[33] Duke SO, Powles SB. Glyphosate: a once-in-a-century herbicide. Pest Manag Sci 2008; (64) 319–325.

[34] Kruse ND, Trezzi M, VIDAL RA. Herbicidas Inibidores da EPSPs: Revisão de literatura. Revista Brasileira de Herbicidas 2000; 1(2): 139-46.

[35] Silva MD, Peralba MCR, Mattos MLT. Determinação de glifosato e ácido aminometilfosfônico em águas superficiais do Arroio Passo do Pilão. Revista Ecotoxicologia e Meio Ambiente, 2003 (13): 19-28.

[36] Velini ED. Modo de ação do glyphosate. In: Velini ED, Meschede D, Carbonari CA, Trindade MLB (ed.). Glyphosate. Botucatu: Fepaf, 2009, 439p.

[37] Tzin V, Galili G. New insights into the shikimate and aromatic amino acids biosynthesis pathways in plants. Mol Plant 2010; 3(6): 956–972.

[38] Silva MA, Jeronimo EM, Dal'col LA. Perfilhamento e produtividade de cana-de-açúcar com diferentes alturas de corte e épocas de colheita. Pesqui Agropecu Bras 2008; 43(8): 979-986.

[39] Sugawara LM, Rudorff BFT, Freitas CC, Picoli MCA, Adami M. Estimativa de produtividade de cana-de-açúcar (Saccharum officinarum L.) por meio de técnicas de análise de regressão linear múltipla. In: Proceedings of 13rd Simpósio Brasileiro de Sensoriamento Remoto; 2007; Florianópolis: INPE, p.435-442.

[40] Yoshinaga EMS. As políticas de exploração da cana-de-açúcar no Brasil: da ocupação colonial a produção sucroalcooleira moderna [dissertation]. Universidade São Marcos, São Paulo; 2006.

[41] Borba MMZ, Bazzo AM. Estudo econômico do ciclo produtivo da cana-de-açúcar para reforma de canavial, em área de fornecedor do estado de São Paulo. In: 47th Congresso Sober - Sociedade Brasileira de Economia Administração e Sociologia Rural, 2009, Porto Alegre.

[42] King NJ, Mungomery RW, Hugues CG. Manual of cane growing. New York: Elsevier, 1965. 375 p.

[43] Pacheco EP, Cantalice JRB. Compressibilidade, resistência a penetração e intervalo hídrico ótimo de um Argissolo Amarelo cultivado com cana-de-açúcar nos Tabuleiros Costeiros de Alagoas. Rev. Bras. Ciên Solo 2011; 35 (2): 403-415.

[44] Bassinelo AI, Abrahão IS, Valadão MB, Barcellos JET, Piccolo CR. Primeiros resultados de estudos de novas variedades de cana-de-açúcar em solos de cerrado. In: Proceedings of 3rd Congresso Nacional da Stab and 5th Convenção da Actalac; 1984, São Paulo; 1984. p. 206-214.

[45] Kuva MA, Gravena R, Pitelli RA, Christoffoleti PJ, Alves PLCA. Períodos de interferência das plantas daninhas na cultura da cana-de-açúcar. III - Capim-braquiária (Brachiaria decumbens) e capim-colonião (Brachiaria decumbens). Planta Daninha 2003; 21 (1): 37-44.

[46] Dinardo-Miranda LL. Manejo de nematoides e pragas de solo em cana-de-açúcar. In: Campos AP, Vale DW, Araújo ES, Corradi MM, Yamauti MS, Fernandes OA, Freitas S (ed.). Manejo Integrado de Pragas. Jaboticabal: FUNEP, 2006. p. 59-80.

[47] Salvi JV, Matos MA, Milan M. Avaliação do desempenho de dispositivo de corte de base de colhedora de cana-de-açúcar. Engenharia Agrícola 2007; 27 (1): 201-209.

[48] Coleti JT. O preparo de solo sob a ótica conservacionista. In: Dinardo-Miranda LL, Vasconcelos ACM, Landell MGA (ed.). Cana-de-açúcar. Campinas: Instituto Agronômico, 2008. p. 573-584.

[49] Silva MA, Carlin SD, Caputo MM. Tipos de colheita e épocas de aplicação de glifosato na erradicação de soqueiras de cana-de-açúcar. Pesqui Agropecu Bras, 2006; 41 (1): 43-49.

[50] Galli AJB. A molécula glyphosate e a agricultura brasileira. In: Velini ED, Meschede DK, Carbonari CA, Trindade MLB (ed.). Glyphosate. Botucatu: Fepaf, 2009.p. 439.

[51] Adriano RC. Características fitotécnicas e isoenzimáticas em cana-de-açúcar após a aplicação de glyphosate. [dissertation]. Instituto Agronômico, Campinas; 2012.

[52] Silva MA, Rossetto R. Diferenças varietais na eliminação química de soqueiras de cana-de-açúcar. STAB, Açúcar, Álcool e Subprodutos 2002; (20): 24-27.

[53] Christoffoleti PJ. Aspectos da resistência de Plantas Daninhas a Herbicidas. 3rd ed. Campinas: Associação Brasileira de Ação a Resistência de Plantas Daninhas aos Herbicidas (HRAC-BR), 2008. 120p.

[54] Ferreira EA. Manejo de plantas daninhas tolerantes ou resistentes ao glyphosate no Brasil. In: Vargas L. (Ed.). Glyphosate: passado, presente e futuro. São Paulo: SBCPD, 2008.

[55] Heredia A, Casado CG, Laguna L, Reina LL, Serrano JM, Domínguez E. La cutícula vegetal: estrutura y funciones. Ecología 1998; (12) 293-305.

[56] Roman ES. Como funcionam os herbicidas: da biologia a aplicação. Passo Fundo: Gráfica Editora Berthier; 2007. 160p.

[57] Ferreira MO. A eficiência do glifosato na destruição química das socarias de cana-de-açúcar das variedades NA56-79 e CB45-3. STAB, Açúcar, Álcool e Subprodutos 1986; (4): 46-48.

[58] Procópio SO, Silva AA, Vargas L, Ferreira FA. Manejo de plantas daninhas na cultura da cana-de-açúcar. Viçosa: Editora Viçosa, 2003. 150p.

[59] Dinelli G, Marotti I, Bonetti A, Minelli M, Catizone P, Barnes J. Physiological and molecular insights on the mechanisms of resistence in Conyza Canadensis (L.) Cronq biotypes. Pestic Biochem Physiol 2006; (86): 30-41.

[60] Santos AJR, Graciano SP, Bacchi OOS, Kashiwakura Y. Doses e épocas de aplicação de glifosato na erradicação de soqueiras de diferentes variedades de cana-de-açúcar. In: Proceedings of the 3rd Congresso Nacional da Sociedade dos Técnicos Açucareiros e Alcooleiros do Brasil, 1984, São Paulo: STAB; p.276-281.

[61] Lehninger AL, Nelson DL, COX MM. 2nd ed. São Paulo; Sarvier; 1995.

[62] Devine M, Duke SO, Fedtke C. Inhibition of amino acid biosynthesis. In: Physiology of herbicide action. 1993. p. 251-294.

[63] Torggler MGF, Contel EPB, Torggler SP. Isoenzimas – variabilidade genética em plantas. Ribeirão Preto. Sociedade Brasileira de Genética. 1995.

[64] Peirce LC, Brewbaker JL. Applications of isozyme analysis in horticultural science. HortScience 1973; (8): 17-22.

[65] Vidigal DS de, Dias DCFS, Von-Pinho EVR de, Dias LAS dos. Alterações fisiológicas e enzimáticas durante a maturação de sementes de pimenta (Capsicum annum L.). Rev Bras Sementes 2009; 31(2): 129-136.

[66] Colich S, Milatovich D, Nikolich D, Zec G. Dehydrogenase isoenzyme polymorphism in selected almond genotypes (Prusus Amygdalus Batsch.). Bulg J Agric Sci 2009; (15): 552–556.

[67] Menezes SM de, Tillmann MAA, Dode LB, Villela FA. Detecção de soja geneticamente modificada tolerante ao glifosato por métodos baseados na atividade de enzimas. Rev Bras Sementes 2004; 26 (2): 150-155.

[68] Rabello LKC. Quantificação de danos e perdas causados por Meloidogyne spp. em alface (*Lactuca sativa* L.) [dissertation]. Universidade Federal do Espírito Santo; 2010.

[69] Zambrano AY, Demey JR, González V. Selección in vitro de líneas celulares de caña de azúcar resistentes a glifosato®. Agronomía Tropical 2002; 52 (2): 139-160.

[70] Gottlieb LD. Conservation and duplications of isozymes in plants. Science 1982; (216): 73-380.

[71] Walker CH, Mackness MI. Estererases: problems of identification and classification. Biochemical Pharmacology 1983; (32): 3265-3269.

[72] Brune W, Alfenas AC, Junghans TG. Identificações específicas de enzimas em géis. In: Alfenas AC. (ed). Eletroforese e marcadores bioquímicos em plantas e microrganismos. 2nd ed. Viçosa: Editora UFV; 2006. p. 202-328.

[73] Weeden NF, Wendel JF. Genetics of plant isozymes. In: Soltis, D.E. & Soltis, P.S. (eds.). Isozymes in plant biology. London: Champman and Hall. 1990. p. 46-72.

[74] Santos CMR, Menezes NL, Vilella FA. Modificações fisiológicas e bioquímicas em sementes de feijão no armazenamento. Rev Bras Sementes 2005; (27): 104-114.

[75] Valentini LC. Caracterização genética e funcional de α- e β- esterases em cultivares de soja (Glycine max l. merrill) no estado do Paraná. [dissertation]. Universidade Estadual de Maringá; 2007.

[76] Lopes RC, Casali VWD, Barbosa LCA, Cecon PR. Caracterização isoenzimática de oito acessos de erva-de-bicho. Horticultura Brasileira 2003; 21 (3): 433-437.

[77] Martins CC, Bovi MLA, Mori ES, Nakagawa J. Isoenzimas na diferenciação de sementes de três espécies do gênero Euterpe. Revista Árvore 2007; 31 (1): 51-57.

[78] Thom M, Maretzki A. Peroxidase and esterase isozymes in Hawaiian sugarcane. Hawaiian Planter's Record 1970; 58 (6): 81-94.

[79] Gonçalves CHRP, Cury JA, Crocomo OJ. Isoenzimas na identificação e seleção de variedades de cana-de-açúcar (Saccharum spp.). In: Proceedings of 2nd Congresso da Sociedade de Técnicos Açucareiros do Brasil; 1981. p. 329-340.

[80] Medeiros JE. Seleção de bactérias para controle da meloidoginose e atividade isoenzimática de meloeiro parasitado por Meloidogyne incógnita. [dissertation]. Universidade Federal Rural de Pernambuco; 2007.

[81] Gallacher DJ, Lee DJ, Berding N. Use of isozyme phenotypes for rapid discrimination among sugarcane clones. Aust J Agric Res 1995; (46): 601–609.

[82] Glasenapp JS. Estrutura genética e fenóis totais de populações naturais de barbatimão (Stryphnodendron adstringens) [dissertation] Universidade Federal de Viçosa; 2007.

[83] Almeida M, Crócomo OJ. Caracterização bioquímica de cultivares de cana-de-açúcar (*Saccharum* spp.): isoenzimas, proteínas solúveis e valor brix. Sci Agríc 1994; 51 (3): 422-429.

[84] Kajihara D. Caracterização dos genes mustang em gramíneas com ênfase no estudo funcional em cana-de-açúcar [dissertation]. Universidade de São Paulo; 2008.

[85] Bueno ACR, Machado DFSP, Azania AAPM, Azania CAM. Maturação estimulada. Revista Cultivar 2011; 12 (139): 30-32.

[86] Araldi R, Silva FML, Ono EO, Rodrigues JD. Florescimento em cana-de-açúcar. Ciência Rural 2010; 40 (3): 694-702.

Chapter 8

Herbicides Used in Tobacco

William A. Bailey

Additional information is available at the end of the chapter

http://dx.doi.org/10.5772/56008

1. Introduction

Tobacco is a major and economically important crop in many countries worldwide, with 6.91 million tons produced annually, mainly in China, India, Brazil, Zimbabwe, Turkey, Indonesia, Russia, Malawi, nations of the European Union, and the United States [1]. In the United States alone, approximately 360,000 tons are produced annually at a value of more than $USD 1.25 billion [2]. Although there are at least 14 different types of tobacco grown around the world, all are affected by pests. Disease and insect pests are of primary importance in tobacco production, but weeds are also a major focus of pest control in tobacco. Although weeds may not cause as much direct damage to tobacco as diseases and insects, weeds present in tobacco can influence tobacco yield and quality, cause harvest interference, and serve as hosts for disease and insects. Although tobacco is considered to be very competitive with weeds relative to other crops, use of herbicides, usually supplemented with cultivation, is still a primary component of weed control. The objective of the research presented here is to provide a more thorough understanding of the effects of weeds in tobacco and the characteristics of major herbicides available to control these weeds in tobacco production in the United States.

1.1. Competitive effects of weeds on tobacco yield and quality

Weeds directly compete with tobacco for light, water, nutrients, carbon dioxide, and space and can negatively impact tobacco yield and quality. In addition, the quality of the final product may be further affected due to the presence of foreign plant material, referred to in the tobacco industry as Non-Tobacco Related Material (NTRM).

The most direct impact of weed competition in tobacco is reduced leaf yield. Leaf quality can also be negatively affected if weeds physically damage tobacco before or during harvest. Contamination of the harvested tobacco crop by green weed vegetation or reproductive parts of weeds has the largest effect on tobacco quality [3, 4]. Chemical exudates from weedy species that contaminate tobacco leaves and remain until the tobacco is processed can also impact leaf chemical balance and resulting flavor of manufactured tobacco products.

The critical weed-free period is a phrase that is used to describe the period during crop production in which weeds are most likely to reduce crop growth and yield. This is the time period during which weed control efforts must be maintained to prevent crop yield loss. The significance of the critical weed-free period is that, if the crop is maintained weed-free for this period, it will be able to effectively compete with late-emerging weeds without sustaining yield loss. Critical weed-free periods are influenced by the competitiveness of the individual crop species and weed species. For most crops, the critical weed-free period for most weeds is 4 to 6 weeks after crop emergence. Since tobacco is transplanted in the field rather than seeded, it is inherently more competitive with weeds than direct-seeded crops. For this reason, the critical weed-free period for tobacco may be 1 to 2 weeks shorter than for direct seeded crops. In addition, the large leaves which most types of tobacco produce makes it more competitive than many other crops by having a greater ability to reduce photosynthetic ability of weeds growing under the tobacco canopy. Flue-cured tobacco maintained free of common ragweed (*Ambrosia artemesiifolia* L.) for two weeks following transplanting did not sustain significant yield losses from common ragweed that emerged later [4]. For most weed species, maintaining weed-free or near weed-free conditions for 6 weeks after transplanting allows tobacco to shade out weeds that emerge later in the season [5]. In Greece, yield of burley and oriental tobacco increased significantly with weed-free periods of 3 or 4 weeks and decreased when weeds were allowed to compete with tobacco for more than 3 to 4 weeks after transplanting. When yield was reduced due to weed competition, there were also differences in chemical composition of the tobacco [6]. Natural populations of weeds that were allowed to compete with dark tobacco for the entire season resulted in a 28% to 40% reduction in total yield compared to tobacco plots treated with herbicides [7, 8].

If weeds are allowed to compete with tobacco for the entire season, the level of competition that weeds impose is also influenced by the density of the weeds that are present in the crop. In general, crop yield decreases as weed density increases. Different weed species also have different competitive ability with tobacco and thus can effectively compete at lower densities than other species. In general, dicots (broadleaf weeds) are more competitive with tobacco than monocots (grass weeds). Within broadleaf and grass weeds, individual species can be more competitive with tobacco than others. For example, among broadleaf species, Eastern black nightshade (*Solanum ptycanthum* L.) has a more rapid growth, higher photosynthetic ability, and a more erect growth habit than black nightshade (*Solanum nigrum* L.), and is more competitive with tobacco. Among grass species, giant foxtail (*Setaria glauca* L.) is more competitive than either green (*Setaria viridis* L.) or yellow foxtail (*Setaria faberii* L.). Much of these differences in competitiveness can be attributed to differences in plant size among species. Perennial weed species are also generally more competitive and difficult to control in tobacco than annual weed species. Perennial species generally have a more extensive root system and extensive energy reserves than annual species.

Differences in root elongation rate also influence differences in competitiveness by affecting water and nutrient absorption potential. Among weedy broadleaf species, common cocklebur

http://dx.doi.org/10.5772/56008

(*Xanthium strumarium* L.) has the greatest root elongation rate and extracts the greatest amount of moisture per unit area of soil [9]. Under field conditions, the water requirements for various weed species vary from 150 to 1900 kg water per kg dry matter produced. Of the nutrients that weeds and tobacco compete for, nitrogen is often the first nutrient to come into short supply as a result of competition. Weeds are commonly better assimilators of nutrients than crop plants, normally possessing 50 to 100% more nitrogen than the crop plant based on a whole plant dry weight basis [10].

Where water and nutrients are adequate, low light intensity that occurs from shading plays a major role in limiting plant growth. Plants compete for light by positioning their leaves to intercept available light more favorably than neighboring plants. Plants that exhibit more rapid early-season growth and have upright growth to grow taller than neighboring plants will be most successful in competition for light. Broadleaved crops such as tobacco have a distinct competitive advantage over grass plants or sedges that have narrow leaves. Tall, dense crops like tobacco successfully compete with shorter plants for light, particularly when weed emergence occurs later in the season after tobacco is well established and tobacco can easily impose a shading effect on newly emerged weed seedlings.

Aside from directly competing with tobacco to reduce marketable yield and quality, many weed species are troublesome with tobacco due to their ability to interfere with harvest operations. Tobacco crops that are heavily infested with weeds, even relatively non-competitive weeds, can have reduced yield through competition before harvest and even more during harvest. Weed species with twining or climbing growth habits such as morningglory species (*Ipomoea* spp.), honeyvine milkweed (*Ampelamus albidus* [Nutt.] Britt.), or common bindweed (*Convolvulus arvensis* L.) may not be very competitive with tobacco during the growing season, but can cause dramatic losses at harvest, even when weed densities are relatively low. A single climbing weed in a tobacco crop may become entangled in several tobacco plants and cause leaf damage and loss both prior to and during harvest. Infestations from weeds that become entwined around tobacco stalks are troublesome during hand harvest operations but even more troublesome for mechanical harvesting systems. Presence of morningglory at an average density of 1 plant per 10 m^2 has caused a 5% reduction in harvested yield of dark tobacco in Kentucky USA due to damage and leaf loss during hand harvest (W.A. Bailey, unpublished data). Mechanical harvesters that encounter morningglory entwined in tobacco at similar densities would likely incur greater leaf losses as well as sustain extensive damage to the harvester itself. Parts of weedy plants that remain in the tobacco crop through curing are more likely to become NTRM, causing extensive reduction in price and likely reduction in marketing opportunities for future crops.

1.2. Weeds as alternate hosts to other pests in tobacco

Weeds can act as a major host site for other tobacco pests such as diseases, insects, and nematodes. Many weeds that commonly occur around tobacco fields can harbor other pests and result in increased infection on tobacco crops. Generally, weed species that have the closest botanical relationship to tobacco, such as solanaceous weed species, are most likely to harbor pests that can infest tobacco. However, many plant species with little botanical relationship to tobacco can also serve as hosts. For example, *Datura* species such as Jimsonweed are common alternate hosts to at least 12 tobacco diseases, at least one nematodes species, and at least 3 major insect pests of tobacco. *Nicandra* species such as Apple-of-Peru are common alternate hosts to at least 4 major tobacco diseases including blue mold, brown spot, bushy top virus, and vein banding virus.

1.3. Diseases

Table 1 lists weed species that commonly act as alternate hosts for tobacco diseases. Many diseases have an extremely wide host range and so only the number of species, families, genera, or most common host species are listed. Reference materials [11-14] were used to construct Tables 1, 2. and 3.

Disease	Causal Agent	Hosts Species	Plant Families	Common Weedy Hosts
Bacterial Wilt	*Pseudomonas solanacearum*	197	33	Common ragweed (*Ambrosia artemisiifolia* L.) Pennsylvania smartweed (*Polygonum pennsylvanicum* L.)
Hollow stalk	*Erwinia* sp.	120	Solanaceae Brassicaceae Cucurbitaceae	*Solanum* sp.
Wildfire / Angular leafspot	*Pseudomonas syringae* pv. tabaci	Many	Most common: *Solanaceae*	Jimsonweed (*Datura stramonium* L.) Smartweed species (*Polygonum* sp.) Shepards-purse (*Capsella bursa-pastoris* L.) Black nightshade (*Solanum nigrum* L.) Barnyardgrass (*Echinochloa crus-galli* L.) Dandelion (*Taraxacum officinale* Weber)
Tobacco Mosaic Virus (TMV)	Various	350	29 Most common: Solanaceae Compositae Hydrophyllaceae Scrophulariaceae	Horsenettle (*Solanum carolinense* L.) Ground cherry (*Physalis angulata* L.) Jimsonweed (*Datura stramonium* L.)
Vein Banding Virus	Various	many	Most common: *Solanaceae*	(*Solanum* sp.) *Chenopodium* sp.

http://dx.doi.org/10.5772/56008

Disease	Causal Agent	Hosts Species	Plant Families	Common Weedy Hosts
				Groundcherry species (*Physalis* sp.) Apple of Peru Nicandra physaloides (L.) Pers.
Stolbur	Mycoplasma	65	24	Field Bindweed (*Convolvulus arvensis* L.)
Aster yellows	Mycoplasma	175	52	Dodder (*Cuscuta* sp.)
Tomato Spotted Wilt Virus (TSWV)	Various	166	34	Dandelion (*Taraxacum officinale* L.) Spiny amaranth (*Amaranthus spinosus* L.) Jimsonweed (*Datura stramonium* L.) Clasping coneflower (Rudbeckia amplexicaulis Vahl.) Brazilian vervain (Verbena brasiliensis Velloso) Mouseear chickweed (Cerastium vulgatum) Prickly lettuce (Lactuca scariola) Carpetweed (Mollugo verticillata) Blackseed plantain (Plantago rugelii) Hairy buttercup (Ranunculus sardous) Spiny sowthistle (Sonchus asper) Common chickweed (Stellaria media) Hairy bittercress (Cardamine hirsuta) Dogfennel (Eupatorium capillifolium) Carolina geranium (Geranium carolinianum) Purple cudweed (Gnaphalium purpureum)

Disease	Causal Agent	Hosts Species	Plant Families	Common Weedy Hosts
				Blue toadflax (Linaria canadensis) Carolina desert-chicory (Pyrrhopappus carolinianus) Wild radish (Raphanus raphanistrum) Venus' looking-glass (Triodanis perfoliata)
Cucumber Mosaic Virus (CMV)	Various	many	36 dicot families 4 monocot families	Carolina geranium (*Geranium carolineanum* L.) Cutleaf groundcherry (*Physalis angulata* L.) Dayflower (*Commelina nudiflora* L.) American pokeweed (*Phytolacca americana* [L.] *var. rigida* [Small] Caulkins & Wyatt Common Chickweed (*Stellaria media* L.) Jimsonweed (*Datura stramonium* L.) Chenopodium sp.
Tobacco Etch Virus (TEV)	Various	69	11	*Solanum* sp. Jimsonweed (*Datura stramonium* L.)
Tobacco Vein Mottle Virus (TVMV)	Various		*Solanaceae*	Horsenettle (*Solanum carolinense* L.) Cutleaf groundcherry (*Physalis angulata* L.)
Bushy Top Virus	Various		*Solanaceae*	Jimsonweed (*Datura stramonium* L.) Apple of Peru (*Nicandra physaloides* [L.] Scop.)
Peanut Stunt Virus (PSV)	Various		Fabaceae Solanaceae	Kudzu (*Pueraria thumbergiana* [Sieb. & Succ.] Benth.) Jimsonweed (*Datura strumonium* L.)
Alfalfa Mosaic Virus (AMV)	Various	305	47	Jimsonweed (*Datura stramonium* L.)

http://dx.doi.org/10.5772/56008

Disease	Causal Agent	Hosts Species	Plant Families	Common Weedy Hosts
Tobacco Leaf Curl Virus (TLCV)	Various	Many	14 Most common: Malvaceae Euphorbiaceae Fabaceae Solanaceae	*Datura* sp. *Physalis* sp. *Solanum* sp. *Sida* sp. Least snoutbean (*Rhynchosia minima* [L.] DC)
Beet Curly Top Virus (BCTV)	Various	244		Bristly starbur (*Acanthospermum hispidum* L.)
Tobacco Rattle Virus (TRV)	Various	380	Many	Shepards-purse (*Capsella bursa-pastoris* L.) Black nightshade (*Solanum nigrum* L.) Common chickweed (*Stellaria media* L.) Henbit (*Lamium amplexicaule* L.) Redroot pigweed (*Amaranthus retroflexus* L.) Spiny sowthistle (*Sonchus asper* [L.] All.) Flixweed (*Descurainia sophia* L.) Redstem filaree (*Erodium cicutarium* L.)
Tobacco Ringspot Virus (TRSV)	Various	Many	Many Most common: Solanaceae Compositae Cucurbitaceae Scrophulariaceae	Common ragweed (*Ambrosia artemisiifolia* L.) Wild carrot (*Daucus carota* L.) Dandelion (*Taraxacum officinale* L.) Horsenettle (*Solanum carolinense* L.) Groundcherry (*Physalis* sp.) Common pokeweed (*Phytolacca americana* L.) Jimsonweed (*Datura stramonium* L.)

Disease	Causal Agent	Hosts Species	Plant Families	Common Weedy Hosts
Tobacco Streak Virus (TSV)	Various	Many	31	Common burdock (*Arctium minus* [Hill] Bernh.) Field bindweed (*Convolvulus arvensis* L.) Plantain (*Plantago* sp.) White clover (*Trifolium repens* L.) *Crotalaria* sp. Jimsonweed (*Datura stramonium* L.)
Tobacco necrosis virus (TNV)	*Olpidium brassicae* (Wor.) Dang	88	37	
Tobacco stunt virus (TSV)	*Olpidium brassicae* (Wor.) Dang			*Chenopodium* sp.
Potato Virus Y (PVY)	Various		*Solanaceae* (most common), also *Amaranthaceae, Chenopodiaceae, Compositae, Fabaceae*	
Damping off Stem/root rot	*Pythium* sp.		At least 270 genera	
Sore shin	*Rhizoctonia solani* Kuhn	230	66	
Southern Stem/ Root Rot	*Sclerotium rolfsii* Sacc.	189	*Compositae*	
Fusarium wilt	*Fusarium oxysporum* (Schlecht) Wr. f. *nicotianae* Johnson	Many		
Verticillium wilt	*Verticillium alboatrum* Reinke and Berth	250	Dicots	
Olpidium seedling blight	*Olpidium brassicae* (Wor.) Dang	Many	Most common: Cruciferae Graminae Brassicacae	Shepards-purse (*Capsella bursa-pastoris* [L.]Medik) Common lambsquarters (*Chenopodium album* L.)

http://dx.doi.org/10.5772/56008

Disease	Causal Agent	Hosts Species	Plant Families	Common Weedy Hosts
				White poplar (*Populus alba* L.)
Black Root Rot	*Thielaviopsis basicola* (Berk. And Br.) Ferraris	137	33 Most common: Fabacae Solanaceae Cucurbitaceae	
Charcoal Rot	*Macrophominapha seoli* (Maubl.)	>300		
Blue Mold	Peronospora tabacina D. B. Adam	Mainly *Nicotiana*	*Solanaceae*	Poorman's orchid (*Schizanthus pinnatus* Ruiz & Pav. Egyptian henbane) (*Hyoscyamus muticus* L.) Lanceleaf groiundcherry (*Physalis lancifolia* L.) Belladonna (*Atropa belladonna* L.) Apple of Peru *Nicandra physalodes* (L.) Scop.)
Brown Spot	*Alternaria alternata*	56	19 Most common: *Solanaceae*	Jimsonweed (*Datura stramonium* L.) Apple of Peru (*Nicandra physalodes* [L.] Scop.)
Powdery mildew	*Erysiphe cichoracearum* DC	Many	115 genera Main families: Cucurbitaceae Compositae	
Frogeye leafspot	*Cercospora nicotianae* Ellis. & Everhart	28	16	
Anthracnose	Colletotrichum nicotianae Boning	Many	Many	Some grasses Common pokeweed (*Phytolacca americana* L.) Geranium (*Geranium* sp.) Lettuce (*Lactuca* sp.)

Table 1. Common weeds that serve as alternate hosts for tobacco diseases.

Nematodes

Table 2 lists weed species that act as alternate hosts to nematodes that may infect tobacco.

Nematode species	Genus	Number of Hosts Species	Number of Plant Families	Common weedy hosts
Root knot nematode	*Meloidogyne* sp.	>3,000	Most major plant families. Dicots and monocots.	
Tobacco cyst nematode	*Globodera* sp.	At least 45	Most common: *Solanaceae*	
Brown root rot (Lesion nematodes)	*Pratylenchus* sp.	>500	Most common: Graminae Fabaceae Solanaceae Compositae	Large crabgrass (*Digitaria sanguinalis* L.) Horsenettle (*Solanum carolinense* L.)
Stem-break (Stem and Bulb nematode)	*Ditylenchus dipsaci* [Kuhn] Filipjev	>400	44	
Stunt nematode	*Tylenchorhynchus* sp.	Many	Many Common families: Graminae Solanaceae	
Stubby root nematode	*Trichodorus* sp.	At least 51	15 Most common: Fabaceae Graminae Euphorbiaceae	Fescue (*Festuca* sp.) Lettuce (*Lactuca* sp.) Vetch (*Vicia* sp.) Wild onion (*Allium canadense* L.) Lespedeza (*Lespedeza* sp.) Showy crotalaria (*Crotalaria spectabilis* L.) Jimsonweed (*Datura stramonium* L.)

Table 2. Common weeds that serve as hosts for nematodes.

Insects

Table 3 lists weeds species that serve as alternate hosts for insects that may attack tobacco.

Insect	Genus	Number of host species	Number of host families	Common Weedy Hosts
Green peach aphid Red tobacco aphid	Myzus persicae Myzus nicotianae	Many	Many Most common: Solanaceae Amaranthaceae, Chenopodaceae, Compositae, Fabaceae Brassicacae	*Solanum* sp. *Chenopodium* sp. Groundcherry (*Physalis virginiana* Mill.) Virginia pepperweed (*Lepidium virginicum* L.) Tansymustard (*Descurainia pinnata* L.) Curly dock (*Rumex crispus* L.) Jimsonweed (*Datura stramonium* L.) Common chickweed (*Stellaria media* L.) Dayflower (*Commelina* sp.) Kudzu (*Pueraria lobata* L.) Common ragweed (*Ambrosia artemisiifolia* L.)
Western Flower Thrips	*Frankliniella* sp.	Many	Many	Dandelion (*Taraxacum officinale* L.) Spiny amaranth (*Amaranthus spinosus* L.) Jimsonweed (*Datura stramonium* L.)
Flea beetle	*Epitrix* sp.	Many	Many	Horsenettle (*Solanum carolinense* L.) Morningglory sp. (*Ipomoea* sp.)
Cabbage looper	*Trichoplusia ni*	Many	Solanaceae Brassicaceae	Black nightshade (*Solanum nigrum* L.) Wild mustard (*Brassica napus* L.) Peppergrass (*Lepidium* spp.)

Insect	Genus	Number of host species	Number of host families	Common Weedy Hosts
Cutworms	*Lepidoptera* sp.	Many	Many	Field bindweed (*Convolvulus arvensis* L.) Canada thistle (*Cirsium arvense* L.)
Hornworm	*Manduca sexta*		*Solanaceae* only	Horsenettle (*Solanum carolinense* L.) Jimsonweed (*Datura stramonium* L.) Nightshade species
Budworm	*Heliothis virescens* F.	Many	Many	Beardstongue (*Penstemon laevigatus* Aiton) Beggarweed (*Desmodium* spp.) Bicolor lespedeza (*Lespedeza bicolor* Turcz.) Black medic (*Medicago lupulina* L.) Cranesbill (*Geranium dissectum* L.) Deergrass (*Rhexia* spp.) Dock (*Rumex* spp.) Groundcherry (*Physalis* spp.) Japanese honeysuckle (*Lonicera japonica* Thunb.) Lupine (*Lupinus* spp.) Morningglory (*Ipomoea* spp.) Passionflower (*Passiflora* spp.) Prickly sida (*Sida spinosa* L.) Sunflower (*Helianthus* spp.) Toadflax

http://dx.doi.org/10.5772/56008

Insect	Genus	Number of host species	Number of host families	Common Weedy Hosts
				(*Nuttallanthus canadensis* [L.] D.A. Sutton) Velvetleaf (*Abutilon theophrasti* Medik.)

Table 3. Common weeds that serve as hosts for insects.

1.4. Most common and troublesome weeds in tobacco

It is not the intention here to list every possible weed problem that exists in tobacco. Some species can be found in numerous tobacco growing regions while others are region specific. However, several plant families do have species that are common and problematic in many tobacco production regions. According to a weed survey conducted across several tobacco-growing regions of the world in 2006 (W. A. Bailey, unpublished data), the five most common and troublesome weed genera in tobacco are: *Amaranthus, Cyperus, Digitaria, Chenopodium,* and *Ipomoea.* Descriptions of each genera are adapted from references [15, 16, 17]. Table 4 lists the most common and troublesome weeds in the most prevalent tobacco growing regions around the world based on the 2006 survey of tobacco growing regions.

Weed Species	Plant Family	Scientific Name
Broadleaf seed species:		
Redroot pigweed	*Amaranthaceae* (pigweed family)	*Amaranthus retroflexus*
Yellow nutsedge	*Cyperaceae* (sedge family)	*Cyperus esculentus*
Ivyleaf morningglory	*Convolvulaceae* (morningglory family)	*Ipomoea hederacea*
Common lambsquarters	*Chenopodiaceae* (Goosefoot family)	*Chenopodium album*
Common ragweed	*Asteraceae* (sunflower family)	*Ambrosia artemisiifolia*
Horsenettle	*Solanaceae* (nightshade family)	*Solanum carolinense*
Grass weed species:		
Large crabgrass	*Poaceae* (grass family)	*Digitaria sanguinalis*
Goosegrass	*Poaceae* (grass family)	*Eleusine indica*
Fall panicum	*Poaceae* (grass family)	*Panicum dichotomiflorum*
Giant foxtail	*Poaceae* (grass family)	*Setaria faberi*
Johnsongrass	*Poaceae* (grass family)	*Sorghum halepense*

Table 4. Most common and troublesome weeds in tobacco worldwide.

2. Cultural practices for weed control in tobacco

2.1. Site selection, rotation, and scouting

Integrated weed management involves using practices that reduce weed infestations but does not necessarily eliminate all weeds. Weed control can range from poor to excellent, depending on the characteristics of the weed species involved and the effectiveness of the control practices used. A small number of weeds with relatively lower competitive ability than tobacco can be allowed to remain in the crop without negatively influencing yield, quality, or harvest efficiency. Weed control practices available for tobacco can be placed into four general groups: 1) preventative; 2) cultural; 3) mechanical or physical; and 4) chemical.

Preventative weed control involves taking measures to prevent the introduction, establishment, or spread of weed species into areas that are not currently infested with these species. Preventative weed control practices for tobacco can include measures such as using weed-free seed and weed-free transplants, weed-free animal manures if manures are used as a nutrient source, weed-free transplanting and tillage equipment, and elimination of weed infestations in areas bordering tobacco fields. Preventative weed control can also include manually eradicating weeds in and around fields before they can mature and produce seed to proliferate their infestation.

Choosing sites for tobacco production that have low weed populations is also a major means of preventative weed control. Many sites may have good production characteristics, such as well-drained, fertile soil, with minimal potential for erosion or loss from disease, but may contain heavy populations of highly competitive weeds that can limit tobacco production. Some fields may become so infested with heavy populations of troublesome weeds that it is no longer feasible to grow tobacco in those fields, even when the most appropriate herbicides are used correctly. Sites chosen for tobacco production should have relatively low weed populations and, ideally, should not contain weed species that cannot be controlled by herbicides registered for use in tobacco.

Proper site selection for tobacco involves planning, observation, and knowledge of weed populations in fields several seasons prior to growing tobacco in those fields. Entire fields or portions of fields that contain particularly noxious or troublesome weeds should be avoided. Fields being considered for tobacco production should be observed while they are fallow and while they are in production of other crops for at least 2 seasons in order to get an idea of the weed species that are present. Having knowledge of the weed species that will occur in a field and where the heaviest infestations occur will help the grower plan the best choice of herbicide system, application rate and method, and total weed management system.

Once a site is chosen and tobacco is transplanted, scouting during the production season is also an important means of cultural weed control. Scouting involves intensively observing the crop on a weekly basis in at least four random areas of each hectare in the field. Weekly scouting is important to reveal the status of emerging weed problems in the field, but also to observe any potential insect and disease problems that may be developing. Knowing the status of weeds in the field allows for planning of any needed control measures of herbicide applica-

http://dx.doi.org/10.5772/56008

tions, cultivation, or hand weeding. Scouting allows for timely operations that will be more effective than attempting to control weeds after they become more mature.

2.2. Field preparation and cultivation

Where conservation tillage (no-tillage or strip-tillage) practices are not imposed, primary tillage with moldboard plowing, chisel plowing, and disking are the major methods used in field preparation for tobacco in the United States. Primary tillage is the major method of destroying weeds and preparing the ground for tobacco transplanting. Moldboard plowing is the primary means of turning under residue to allow decomposition and is most necessary with grass crops or annual grass weeds, while chisel plowing and disking are secondary tillage practices that aid in destruction of residue and help level the ground in preparation for tobacco transplanting. Field cultivators or mechanical rotary tillers are also used as a finishing tool just prior to transplanting.

Mechanical cultivation is still a necessary supplemental weed control practice in conventional tillage tobacco production because herbicides generally do not control all weeds that occur in tobacco production. Cultivation can also aid in soil aeration when soil crusting occurs, but also contributes to soil erosion and soil drying near the surface. No more than two cultivations are necessary for tobacco. Excessive or late cultivation can injure tobacco root systems, causing problems with water and nutrient uptake while also potentially increasing problems with tobacco mosaic virus, black shank (*Phytophthora nicotianae* Breda de Haan), and Granville wilt (*Pseudomonas solanacearum* E. F. Smith). Cultivation should be made shallow in the top 5 cm of soil so that tobacco roots are not injured and weed seed present below the herbicide treated area are not disturbed and allowed to germinate.

3. Herbicides used for weed control in tobacco

Herbicides play an important role in weed control, particularly in commercial tobacco production in more developed countries. Of all the pesticides used in tobacco production, herbicides make up the smallest percentage, approximately 10.4% [18]. The number of herbicides registered for use in tobacco has remained constant for several years and exhibits little signs of growth. There are approximately 50 different chemicals registered for use as herbicides for tobacco worldwide and they take on many different trade names and formulations depending on which regions they are used in. Recently, the presence of generic manufacturers has played an increasing role with many of these products having varying compositions and labels that may differ significantly from the original manufacturer's specifications. Although several herbicides are registered for control of weed species in tobacco, certain herbicides are not registered in all countries or regions. Readers should refer to herbicide registrations for the specific country or region of interest, and follow use instructions given on all product labels.

Similar to common names of weeds, trade names of herbicides vary around the world depending on the company marketing the product, local regulations, and regulatory parameters. With any pesticide application, it is essential that the correct product be selected for the identified target weed species and that the product has a legal registration for use on tobacco in a given country. There may also be cases where a product has a legal registration for use on tobacco in that country but the tobacco manufacturers do not want the product applied to the crop due to leaf residue issues or other concerns. Over the past two decades, analytical techniques have allowed manufacturing companies to accurately evaluate residue levels of tobacco pesticides on cured leaves. In some cases, these residue levels have prompted companies to discourage the use of some products.

Herbicides may be applied in many different ways, but most herbicides for use in tobacco are applied to the soil prior to weed emergence, and many must be applied prior to tobacco transplanting. Some of these herbicides are applied as pretransplant surface (PRETR) applications and others are applied as pretransplant incorporated (PTI) applications where the herbicide is mechanically incorporated into top 2.5 to 5 cm of soil. Seed of most annual weed species occur in this depth of soil and therefore it is advantageous to keep herbicides at this depth. All soil-applied herbicides need adequate soil moisture in order to be effective, and incorporation increases the availability of moisture for herbicide activation and prevents loss of the herbicide through volatilization into the atmosphere. Only a limited number of herbicides are registered for use in tobacco and none control all weeds that may occur. Therefore, much attention should be given to planning weed control strategies [19, 4].

Spray applicators should always remember to follow application instructions given on the label and also insure that the herbicide is registered for use in tobacco in the area where it is to be applied. The following is a listing and description of herbicides currently used in tobacco in various parts of the world for control of grasses, sedges, and broadleaf weeds. The general application guidelines described and weed spectrum of control are based on the use of these herbicides in tobacco within the United States. Consult the product labels of these herbicides for additional information.

3.1. Herbicides commonly used in tobacco

On a worldwide basis, the most commonly used herbicides for tobacco include alachlor, clomazone, metolachlor, napropamide, pebulate, pendimethalin, sethoxydim, and sulfentrazone. The following are descriptions of the weed control properties and basic use patterns.

3.1.1. Alachlor

Alachlor is a chloroacetamide herbicide that inhibits lipid biosynthesis and the synthesis of proteins, gibberellins, lignin, and anthocyanin production in susceptible plants. Alachlor controls many common annual grasses such as crabgrass (*Digitaria* sp.), foxtail (*Setaria* sp.), goosegrass (*Eleusine indica* [L.] Gaertn.), fall panicum (*Panicum dichotomiflorum* Michx.), and barnyardgrass (*Echinochloa crus-galli* [L.] P. Beauv.); as well as yellow nutsedge (*Cyperus esculentus* L.), but is of limited value for control of broadleaf weeds. Alachlor applications for tobacco are normally applied prior to transplanting and shallowly incorporated in the top 2.5 to 5 cm of soil, but may also be applied pretransplant without incorporation. Alachlor is a liquid formulation and the normal use rate is approximately 2.2 to 3.4 kg ai/ha. NOTE: Alachlor is a restricted use herbicide due to oncogenicity (tumor causing potential in laboratory animals) and alachlor has also been identified as having the potential to leach through the soil into ground water, particularly where soils are coarse and groundwater is near the surface [19, 20].

3.1.2. Clomazone

Clomazone is a carotenoid and chlorophyll inhibitor that causing bleaching/whitening in susceptible plants. Clomazone controls several common annual grasses species such as crabgrass (*Digitaria* spp.), *Panicum* spp., and foxtails (*Setaria* spp.). In addition to grass control, clomazone also controls jimsonweed (*Datura stramonium* L.), common lambsquarters (*Chenopodium album* L.), hairy galinsoga (*Galinsoga quadriradiata* Cav.), common ragweed (*Ambrosia artemisiifolia* L.), and velvetleaf (*Abutilon theophrasti* Medik.). Clomazone is normally applied as a soil surface PRETR application, but can also be applied over-the-top of tobacco within 7 days of transplanting as tobacco shows good tolerance to this herbicide. Although clomazone is usually applied to the soil surface with no incorporation, it can be incorporated into the soil surface provided that caution is taken not to incorporate deeper than 5 cm. Clomazone is available in liquid formulations and the normal use rate is approximately 0.84 to 1.1 kg ai/ha [19, 20].

3.1.3. Metolachlor

Metolachlor is a chloroacetamide herbicide similar to alachlor that has the same mode of action and same basic spectrum of weed activity, controlling numerous annual grass weeds and yellow nutsedge (*Cyperus esculentus* L.), but has limited activity against broadleaf weeds. Metolachlor applications for tobacco are normally applied prior to transplanting and shallowly incorporated in the top 2.5 to 5 cm of soil, but may also be applied pretransplant without incorporation. Metolachlor is normally a liquid formulation and the use rate is approximately 1.1 to 2.1 kg ai/ha [19, 20].

3.1.4. Napropamide

Napropamide is an acid amide herbicide that inhibits several metabolic processes including lipid biosynthesis and the synthesis of proteins and gibberellins. Napropamide is used primarily for the control of annual grasses such as crabgrass (*Digitaria* spp.), *Panicum* spp., and foxtails (*Setaria* spp.). Napropamide also provides some control of small-seeded broadleaf weeds such as pigweed (*Amaranthus* spp.) and common lambsquarters (*Chenopodium album* L.). Napropamide is highly volatile and should be mechanically incorporated immediately after application, and preferably in the same operation as the application. Application of napropamide is normally made prior to transplanting. Napropamide is available in dry and liquid formulations and the normal use rate is approximately 1.1 kg ai/ha [19, 20].

3.1.5. Pebulate

Pebulate is a thiocarbamate herbicide that inhibits lipid formation in sensitive plants. Pebulate controls annual grasses such as crabgrass (*Digitaria* spp.) and foxtails (*Setaria* spp.) as well as suppression of certain small-seeded broadleaf weeds such as pigweeds (*Amaranthus* spp.) and common lambsquarters (*Chenopodium album* L.). In addition, pebulate is one of the few herbicides available for use in tobacco that provides good suppression of nutsedge sp. (*Cyperus* spp.). Similar to napropamide, pebulate is highly volatile and should be incorporated immediately after application, preferably in the same operation. Pebulate is applied prior to tobacco transplanting at a use rate of approximately 4.5 kg ai/ha [19, 20].

3.1.6. *Pendimethalin*

Pendimethalin is a dinitroanaline herbicide that inhibits mitosis in susceptible plants. Pendimethalin provides excellent control of annual grasses and certain small-seeded broadleaf weeds. Pendimethalin provides excellent control of crabgrass species (*Digitaria* spp.), foxtail species (*Setaria* spp.), *Panicum* species, and goosegrass (*Eleusine indica* [L.] Gaertn.), and also provides some control of broadleaf species such as pigweed (*Amaranthus* spp.) and common lambsquarters (*Chenopodium album* L.). Pendimethalin is normally applied as a PTI application to a well-prepared soil surface up to 60 days prior to transplanting tobacco. Pendimethalin should be incorporated into the top 2.5 to 5 cm of soil within 7 days after application. Pendimethalin is available as liquid formulations and normal use rate is approximately 1.4 to 1.7 kg ai/ha [19, 20].

3.1.7. *Sethoxydim*

Sethoxydim is a cyclohexanedione herbicide that inhibits lipid biosynthesis in susceptible grass species. Sethoxydim only controls grasses, so it is totally safe to broadleaf crops such as tobacco. Sethoxydim has no soil residual activity and is the only true postemergence herbicide that can be applied over-the-top of tobacco later than 7 days after transplanting. Sethoxydim may be applied up to 42 days prior to tobacco harvest. Sethoxydim is effective on annual grass species such as crabgrass (*Digitaria* spp.), *Panicum* species, and foxtails (*Setaria* spp.), and also controls perennial grasses such as shattercane (*Sorghum bicolor* L.) and Johnsongrass (*Sorghum halepense* L.). Application must be made to emerged, actively growing grasses to be effective. For perennial shattercane and Johnsongrass, sethoxydim is most effective if grass plants are allowed to get 45 to 60 cm tall before application. Do not cultivate within 5 days before application or 7 days after application. Crop oil concentrate at 1% of the spray volume per hectare is recommended with sethoxydim application. Recommended rates of sethoxydim are approximately 0.3 kg ai/ha. For spot treatment by hand, prepare 1 to 1.5% sethoxydim solution with 1% crop oil concentrate and spray grass plants until wetted [19, 20].

3.1.8. *Sulfentrazone*

Sulfentrazone is an aryl triazolinone herbicide that inhibits photosynthesis by inhibiting the enzyme protoporphyrinogen oxidase. Sulfentrazone provides partial control and suppression of annual grasses such as crabgrass (*Digitaria* spp.), *Panicum* sp., foxtails (*Setaria* spp.), and goosegrass (*Eleusine indica* L.). However, its main attribute is control of nutsedge species (*Cyperus* spp.) and troublesome broadleaf weed species such as nightshade species (*Solanum* spp.), groundcherry species (*Physalis* spp.), morningglory species (*Ipomoea* spp.), smartweed

http://dx.doi.org/10.5772/56008

species (*Polygonum* spp.), pigweed species (*Amaranthus* spp.), and common lambsquarters (*Chenopodium album* L.). Sulfentrazone must be applied prior to transplanting tobacco and should be applied to the soil surface without incorporation. If incorporation is used, it must not be deeper than 5 cm from the soil surface. Currently, sulfentrazone is also marketed in the United States in a prepackaged combination with carfentrazone. Carfentrazone is a postemergence burn down herbicide designed for broadleaf weed control prior to transplanting. Sulfentrazone is available as a liquid formulation and normal use rate is approximately 0.28 to 0.42 kg ai/ha [19, 20].

3.1.9. *Burndown of weeds or cover crops in conservation tillage production systems*

No-tillage and strip-tillage tobacco production requires that any existing vegetation, whether it be weed growth or cover crop, be killed prior to transplanting tobacco without using extensive tillage as in conventional tillage tobacco production. Paraquat is a common herbicide that is used as a burndown prior to tobacco transplanting in no-tillage tobacco in the United States. Paraquat should be applied as a broadcast application to actively growing weeds or cover crops no larger than approximately 15 cm in height. Use rates for paraquat for burndown prior to tobacco transplanting are approximately 0.7 to 1.1 kg ai/ha. Glyphosate may also be used to burndown existing vegetation prior to tobacco transplanting as a broadcast application at approximately 0.28 kg ai/ha. Glyphosate should be applied 30 days or more prior to tobacco transplanting and paraquat should be applied several days prior to tobacco transplanting. Carfentrazone may also be used in conservation tillage tobacco prior to transplanting at use rates up to 0.027 kg ai/ha. Carfentrazone has generally not been as effective as paraquat or glyphosate for pretransplant burndown in conservation tillage tobacco [19].

3.2. Weed control expected from herbicides used in tobacco

Although there are a limited number of herbicides registered for tobacco relative to other crops that occupy more total area, the herbicides available for use in tobacco generally provide adequate weed control, particularly when supplemented with cultivation in conventional tillage production systems.

The following are results from herbicide experiments conducted in dark tobacco in western Kentucky USA from 2005 to 2007. Treatments included all residual herbicides that were currently registered for use in tobacco. Soil type was a Grenada silt loam (fine-silty, mixed, thermic Oxyaquic Fraglossudalf) with 1.8% organic matter and pH of 6.4. Tobacco plots were prepared by conventional tillage with moldboard plowing and disking. Final field preparation and incorporation of herbicide treatments that required incorporation was done with a field cultivator. Fertilization and other crop production practices were according to standard recommendations [21]. Experiments were arranged in a randomized complete block design with 4 replications and plots were 4 rows, 4.1 m wide by 12.2 m long. Herbicide treatments were applied one day prior to transplanting as broadcast applications using CO_2-pressurized sprayers with flat fan nozzles calibrated to deliver 187 L/ha at 120 kPa. 'Narrowleaf Madole' dark tobacco was then transplanted on 1-m row spacing and 81-cm plant spacing within rows. Crop injury and weed control was evaluated using a 0 to 100% scale where 0 = no plant injury and 100 = plant death [22]. Tobacco injury data shown in Table 5 is from 2 weeks following transplanting while weed control data shown in Table 6 is from one week prior to harvest. Dark tobacco was fire-cured using standard practices [21] and yield and quality data are shown in Table 7.

Herbicide treatments evaluated included sulfentrazone, clomazone, sulfentrazone plus clomazone, pendimethalin, pendimethalin followed by sulfentrazone, pebulate, napropamide, and pebulate plus napropamide. All herbicide treatments were applied using maximum use rates allowed on U.S. labels. Sulfentrazone and clomazone treatments were applied as pretransplant applications to the soil surface while pendimethalin, pebulate, and napropamide treatments were incorporated immediately after application. Tobacco was cultivated twice early in the season following transplanting as is the standard practice.

As these data illustrate, there is potential to observe mild crop injury under some conditions following application of these tobacco herbicides (Table 5). Greatest potential for injury occurred following sulfentrazone and pendimethalin applications, although injury was never greater than 11% in any year and tobacco recovered quickly.

These data also illustrate that combinations of two tobacco herbicides provide more effective control of a broader spectrum of weeds than any one tobacco herbicide (Table 6). Sulfentrazone applied alone effectively controlled yellow nutsedge and ivyleaf morningglory, but was not as effective on large crabgrass and common ragweed. Conversely, clomazone was effective on large crabgrass and common ragweed but not as effective on yellow nutsedge and ivyleaf morningglory. The most effective herbicide treatment evaluated across these four weed species was sulfentrazone and clomazone applied together. Pendimethalin followed by sulfentrazone was also a very effective treatment, but did not control common ragweed as well as sulfentrazone plus clomazone. Pebulate plus napropamide also provided better weed control than either herbicide applied alone, but this combination was still not as effective as sulfentrazone plus clomazone or pendimethalin followed by sulfentrazone on the weed species evaluated here.

Although obvious differences in weed control were seen, these differences did not always translate to yield, quality, or gross revenue differences (Table 7). Total yield of dark tobacco treated with herbicides ranged from 2,765 kg/ha with pendimethalin alone to 3,051 kg/ha with pendimethalin followed by sulfentrazone with minimal differences in total yield between treatments. Herbicide treatments increased total yield by at least 359 kg/ha compared to tobacco that was only cultivated without herbicide treatment. Differences in quality grade index were also few, ranging from 61.9 to 70.1 across all treatments. There were no differences is gross revenue between herbicide treatments, with gross revenue ranging from 11,163 to 12,911 $USD/ha with herbicide treated tobacco, compared to 9,377 $USD/ha with tobacco that was only cultivated with no herbicide treatment.

4. Conclusion

Although tobacco is considered a very competitive crop, weeds can directly impact tobacco by limiting yield and quality, and causing interference of harvest and other field operations.

http://dx.doi.org/10.5772/56008

In addition, weeds can more indirectly affect tobacco by harboring several major tobacco diseases, insects, and nematodes. Weed control practices for tobacco include field site selection, rotation, scouting, and many fields receive intensive tillage prior to transplanting and cultivation following transplanting. In many areas of the world, weed control for tobacco is almost exclusively a manual task using hand weeding and animal-drawn cultivation implements. Although tobacco is not a food crop, the high value of tobacco relative to other crops makes manual weed management practices economically feasible in some regions.

In more developed regions, however, the use of herbicides is the main component of weed control practices in tobacco. Mechanical cultivation is still used to supplement herbicides in most fields, as no-tillage or reduced tillage production systems have not been adopted as readily in tobacco as in other crops like corn, soybean, and small grains. Although only a limited number of herbicides are available for use in tobacco compared to grain crops, the herbicides that are available have generally provided adequate weed control, particularly when supplemented with cultivation. Of the herbicides that are available, combinations of two herbicides are generally more effective than a single herbicide and some herbicide combinations are more effective than others. Data presented here indicate that sulfentrazone plus clomazone or pendimethalin followed by sulfentrazone were the most effective herbicide programs for weed control in dark tobacco.

			Tobacco Injury[c]		
Herbicide Treatment	**Application Timing**	**Application Rate**	**2005**	**2006**	**2007**
		--- kg ai/ha ---	--------- 0 to 100% ---------		
Sulfentrazone	PRETR[b]	0.42	2 bc	3 bc	0 b
Clomazone	PRETR	1.12	1 bc	0 c	0 b
Sulfentrazone + Clomazone	PRETR	0.42 + 1.12	3 bc	4 b	0 b
Pendimethalin	PTI[a]	1.66	5 b	11 a	2 a
Pendimethalin fb[a] Sulfentrazone	PTI fb PRETR[b]	1.66 + 0.42	10 a	5 b	2 a
Pebulate	PTI	4.48	2 bc	3 bc	0 b
Napropamide	PTI	2.24	1 bc	2 bc	0 b
Pebulate + Napropamide	PTI	4.48 + 2.24	2 bc	5 b	0 b
Untreated Control	-	-	0 bc	0 c	0 b

[a]Data collected from herbicide trials conducted near Murray, KY USA in 2005, 2006, and 2007. Injury data presented by year.

[b] Abbreviations: fb = followed by; PRETR = pretransplant; PTI = pretransplant incorporated.

[c]Means within a column followed by the same letter are not significantly different according to Fisher's Protected LSD at P=0.05.

Table 5. Early-season tobacco injury observed from herbicide treatments.

Herbicide Treatment	Application Timing	Application Rate	Weed Control[c]			
			Large crabgrass	Yellow nutsedge	Common ragweed	Ivyleaf morningglory
		--- kg ai/ha ---	--------- 0 to 100% ---------			
Sulfentrazone	PRETR[b]	0.42	61 c	91 a	31 e	90 b
Clomazone	PRETR	1.12	86 a	17 c	83 a	62 c
Sulfentrazone + Clomazone	PRETR	0.42 + 1.12	89 a	96 a	85 a	97 a
Pendimethalin	PTI[a]	1.66	89 a	23 c	42 d	73 b
Pendimethalin fb[a] Sulfentrazone	PTI fb PRETR[b]	1.66 + 0.42	96 a	93 a	54 c	94 ab
Pebulate	PTI	4.48	54 c	77 b	53 c	35 de
Napropamide	PTI	2.24	72 b	22 c	68 b	31 e
Pebulate + Napropamide	PTI	4.48 + 2.24	75 b	78 b	71 b	39 d
Untreated Control	-	-	0 d	0 d	0 f	0 f

[a]Data collected from herbicide trials conducted near Murray, KY USA in 2005, 2006, and 2007, weed control data pooled over years.

[b]Abbreviations: fb = followed by; PRETR = pretransplant surface application; PTI = pretransplant incorporated application.

[c]Means within a column followed by the same letter are not significantly different according to Fisher's Protected LSD at P=0.05.

Table 6. Late-season weed control from herbicides and herbicide systems currently used in dark tobacco production in the U.S.[a]

Herbicide Treatment	Application Timing	Application Rate	Stalk Position[ab]				Quality Grade Index[c]	Gross Revenue[d]
			Lug	Second	Leaf	Total		
		kg ai/ha	-------------kg/ha-------------				0-100	$/ha
Sulfentrazone	PRETR	0.42	405 a	580 ab	1992 a	2977 ab	64.9 ab	12,497 a
Clomazone	PRETR	1.12	355 ab	579 ab	2010 a	2943 ab	70.1 a	12,911 a
Sulfentrazone + Clomazone	PRETR	0.42 + 1.12	394 a	595 a	2028 a	3017 ab	64.4 ab	12,598 a
Pendimethalin	PTI	1.66	351 ab	565 ab	1843 a	2765 b	61.9 ab	11,163 ab
Pendimethalin fb[e] Sulfentrazone	PTI fb PRETR	1.66 + 0.42	375 ab	617 a	2059 a	3051 a	63.4 ab	11,883 a

Herbicide Treatment	Application Timing	Application Rate	Stalk Position[ab] Lug	Second	Leaf	Total	Quality Grade Index[c]	Gross Revenue[d]
		kg ai/ha	-------------kg/ha--------------				0-100	$/ha
Pebulate	PTI	4.48	351 ab	569 ab	1958 a	2877 ab	63.6 ab	11,779 a
Napropamide	PTI	2.24	355 ab	594 a	1879 a	2828 ab	66.7 ab	12,067 a
Pebulate + Napropamide	PTI	4.48 + 2.24	370 ab	603 a	2031 a	3004 ab	65.9 ab	12,430 a
Untreated Control	-	-	314 b	499 b	1592 a	2406 c	66.2 ab	9,377 b

[a]Data collected from herbicide trials conducted near Murray, KY USA in 2005, 2006, and 2007. Tobacco yield data pooled over years.

[a]Means within a column followed by the same letter are not significantly different according to Fisher's Protected LSD at P=0.05.

[b]Tobacco leaves removed by stalk position following fire-curing. Lug corresponds to lower stalk leaves, second from midstalk, and leaf from upper stalk.

[c]Quality grade index is a numerical representation of Federal quality grade received for tobacco and is a weighted average of grade index for all stalk positions.

[d]Gross revenue is the total gross value of tobacco (in $USD) based on Federal grade and price support values.

[e]Abbreviations: fb = followed by; PRETR = pretransplant; PTI = pretransplant incorporated.

Table 7. Effect of herbicide treatment on dark-fired tobacco yield, quality grade index, and gross revenue[a].

Author details

William A. Bailey

Address all correspondence to: abailey@uky.edu

Department of Plant & Soil Sciences, University of Kentucky, Research and Education Center, Princeton, KY, USA

References

[1] United Nations Food and Agriculture OrganizationFood and Agriculture Organization Statistical Yearbook 2010. Food and Agriculture Organization Statistical Divison.

http://www.fao.org/docrep/015/am081m/PDF/am081m00b.pdfaccessed 29 September (2012).

[2] National Agriculture Statistics ServiceAgricultural Statistics Annual Report. http://www.nass.usda.gov/Publications/Ag_Statistics/2011/Chapter02.pdfAccessed 29 September (2012).

[3] Collins, W. K, & Hawks, S. N. Jr. Cultivation and weed management. *In* W. K. Collins and S. N. Hawks (eds.) Principles of Flue-Cured Tobacco Production. Raleigh: North Carolina State University. (1993).

[4] Parker, R. G, Fisher, L. R, & Whitley, D. S. Weed management in conventional and no-till burley tobacco. *In* 2007 Burley Tobacco Information. Raleigh: North Carolina Cooperative Extension Service. (2007).

[5] Palmer, G. K, & Pearce, R. C. Light air-cured tobacco. *In* D. L. Davis and M. T. Nielsen (eds.) Tobacco: Production, Chemistry, and Technology. Oxford: Blackwell Science. (1999).

[6] Lolas, P. C. Weed community interference in burley oriental tobacco (*Nicotiana tabacum*). Weed Res. (1986). , 26(1), 1-8.

[7] Bailey, W. A. Comparison of herbicide systems for dark tobacco. Proc. South. Weed Sci. Soc. 60:18. (2007).

[8] Bailey, W. A. Dark tobacco (*Nicotiana tabacum*) tolerance to trifloxysulfuron and halosulfuron. Weed Technol. (2007). , 21, 1016-1022.

[9] Davis, R. G, Weise, A. F, & Pafford, J. L. (1965). Root moisture extraction profiles of various weeds. Weeds 1965;, 13, 98-102.

[10] Greer, H. A. L. Weeds: costly competitors for nutrients. Plant Food Rev. (1966).

[11] Daub, M. E, Echandi, E, Gooding, G. V, Jr, K. J, Jones, G. B, Lucas, C. E, Main, N. T, Powell, S. M, Schneider, H. D, Shew, P. B, & Shoemaker, H. W. Spurr, Jr. H. D. Shew and G. B. Lucas (eds.) Compendium of Tobacco Diseases. St. Paul: American Phytopathological Society. (1991).

[12] Groves, R. L, Walgenbach, J. F, Moyer, J. W, & Kennedy, G. G. The role of weed hosts and tobacco thrips, *Frankliniella fusca*, in the epidemiology of *tomato spotted wilt virus*. Plant Dis. 2002;(2002). , 86(6), 573-582.

[13] Wisler, G. C, & Norris, R. F. Interactions between weeds and cultivated plants as related to management of plant pathogens. Weed Sci. 2005;(2005). , 53, 914-917.

[14] Lucas, G. B. Diseases of Tobacco. Raleigh: Biological Consulting Associates. (1975).

[15] Flora of North America Editorial Committeeeds. Flora of North America North of Mexico. 12 vols. New York and Oxford. (1993).

http://dx.doi.org/10.5772/56008

[16] Elmore, C. D. editor. Southern Weed Science Society Weed Identification Guide. Champaign, IL: Southern Weed Science Society. (1999).

[17] Radford, A. E, Ahles, H. E, & Bell, C. R. Manual of the Vascular Flora of the Carolinas. Chapel Hill: University of North Carolina Press. (1968).

[18] Sigma Database Agrochemical Use on Tobacco. Product Studies Research, Newbury Berks, UK. (1999).

[19] Seebold, K, Green, J. D, & Townsend, L. Pest Management. *In* 2007 Kentucky Tobacco Production Guide. Lexington, KY: Kentucky Cooperative Extension Service. (2007).

[20] Anderson, W. P. Weed Science: Principles and Applications. New York: West Publishing Company. (1996).

[21] Seebold, K. editor. Kentucky & Tennessee Tobacco Production Guide. Lexington, KY: Kentucky Cooperative Extension Service. (2011). , 2011-2012.

[22] Frans, R, Talbert, R, Marx, D, & Crowley, H. Experimental design and techniques for measuring and analyzing plant responses to weed control practices. *in* N. D. Camper, ed. Research Methods in Weed Science. 3rd ed. Champaign, IL: Southern Weed Science Society. (1986). , 29-46.

Section 2

Natural Areas, Aquatic, and Turf Case Studies

Chapter 9

Herbicides for Natural Area Weed Management

Gregory E. MacDonald, Lyn A. Gettys,
Jason A. Ferrell and Brent A. Sellers

Additional information is available at the end of the chapter

http://dx.doi.org/10.5772/56183

1. Introduction

Natural areas represent a significant resource for many countries. In the U.S. natural areas can be defined as conservation lands set aside for preservation or restoration, such as city or county park, private woods, state or national park, Bureau of Land Management (BLM) lands, or other areas [1,2]. In many cases these areas are utilized for recreation, ecosystem services or other non-agricultural purposes [3,4]. Given this broad definition, natural areas encompass a huge portion of the land mass of the United States and represent incredible biological diversity. According to the U.S. National Vegetation Classification in 2012 there are 8 major classifications in the U.S. with 430 groupings and over 6100 associations [5]. Some of the more common ecological communities include deciduous temperate forests, temperate coniferous forests, grasslands, and wetlands such as swamps, tidal marshes, and riparian zones.

Many natural areas are managed to some degree for a variety of uses, but due to the complexity of many natural area systems, the management techniques developed for, and utilized in these areas is diverse. Some areas are managed exclusively for recreation and include water attractions, hiking and biking trails, horse trails, or camping. In these cases, user satisfaction, human health and safety are the primary goals, with ecological community diversity being a secondary, but often equally important, goal [6]. Other areas that are managed for conservation (including hunting), preservation or restoration may not require as intense or frequent management [7].

Vegetation management in natural areas is performed for a variety of purposes but falls broadly into two primary categories: 1) maintaining the existing vegetation at desirable levels and species composition or 2) restoring the ecosystem to a desirable state. With the latter category, restoration can include reintroduction of naturally occurring species, reintroduction of a natural ecological process such as fire or water fluctuations, and/or providing an environment that allows for natural reintroduction/colonization of native species [8].

Within the past 2 decades, vegetation management in natural areas has intensified due to issues with invasive species. Invasive weedy species represent one of the biggest threats to the diversity and utility of many natural areas [9]. Moreover, invasive species are considered to be a major threat to endangered species, second only to habitat loss [10]. Currently there are over 400 invasive non-native plants impacting approximately 133 million acres in the U.S. alone and it is estimated that invasive species are spreading at the rate of 1.7 million acres annually [11]. In 1999, a mandated executive order specifically addressed invasive species and their impacts, leading to the formation of the National Invasive Species Council (NISC) and the Invasive Species Advisory Committee (ISAC) [12]. These organizations and many more at the regional, state and local level dedicated to invasive species management has greatly influenced natural area vegetation management.

This chapter will provide an overview of the types of management practices used in a range of natural area systems and detail those herbicides used in natural areas. Weed management in aquatic systems will not be discussed in this chapter.

2. Herbicide registration and regulation for use in natural areas

Herbicides are labeled for use on a specific crop or site as defined by the U.S. Environmental Protection Agency [13]. Many herbicides can be used in natural areas, but labeling may be restricted to only specific uses within the broader context of a 'natural area'. In addition, many states, agencies, and/or local governments may prohibit or restrict usage of a particular product or compound. It is not the intent of this chapter to list those specific sites where a particular herbicide could be used, but rather provide details of how the herbicide is applied, its mode of action, its spectrum of activity and environmental considerations associated with use.

3. Overview of natural area herbicides and their mechanisms/modes-of-action

This section will provide background of those herbicides used in natural areas and will include information on chemistry, formulations, mode-of-action and selectivity. Specific details to each herbicide are listed in Table 1.

http://dx.doi.org/10.5772/56183

Herbicide (common name)	Mechanism of Action[2]	Rate range kg/ha[3]	Application Methods[4]	Environmental Dissipation[5]	*Common Application Methods and General Spectrum of Control[6]
2,4-D	O (4)	0.56-3.8	F, B, I, CS	Microbial (7-10)	POST – annual, perennial broadleaves
Diclorprop	O (4)	4.1-10.4	F, S, I, CS	(10)	PRE, POST – annual, perennial BL's, brush
Dicamba	O (4)	0.28-2.2	F, S, B, CS	Microbial (4-14)	PRE, POST – annual, perennial BL's, brush
Picloram	O (4)	0.14-1.12	F,S, B, I, CS	Microbial (90-300)	PRE, POST – perennial BL's, brush, trees
Triclopyr	O (4)	0.56-9.0	F, B, I, CS	Microbial (30)	POST – perennial BL's, brush, trees
Fluroxypyr	O (4)	0.14-0.56	F	Microbial (38)	POST – annual, perennial BL's, brush
Clopyralid	O (4)	0.14-0.56	F, S	Microbial (40)	PRE, POST – annual, perennial BL's, brush
Aminopyralid	O (4)	0.09-0.25	F, S, I	Microbial (35)	PRE, POST – annual, perennial BL's, brush
Aminocyclopyrachlor	O (4)	0.06-0.28	F, S, B, I	Microbial (60)	PRE, POST – annual, perennial BL's, brush
Simazine	C_1 (5)	2.2-8.9	S	Microbial (70-90)	PRE – annuals, perennials
Diuron	C_2 (7)	4.5-18	S	Microbial (90)	PRE – annuals, perennials
Tebuthiuron	C_2 (7)	0.84-4.48	S	Microbial (400)	PRE – perennial herbs, brush, trees
Hexazinone	C_1 (5)	2.5-7.5	S	Microbial (90)	PRE – perennial grass, brush, trees
Bromacil	C_1 (5)	1.8-13.4	S	Microbial (60)	PRE – annual, perennial, brush
Prometon	C_1 (5)	8.9-36	S	Microbial (450)	PRE – perennial grass, brush, trees
Glyphosate	G (9)	1.1-5.6	F	Irreversibly bound	POST – annuals, perennials, brush
Fosamine	Z (27)	2.24-26.9	F	Microbial (8)	POST – woody brush, trees
Glufosinate	H (10)	0.32-1.56	F	Irreversibly bound	POST – annuals, limited perennials
Paraquat	D (22)	0.71-1.14	F	Irreversibly bound	POST – annual species, no soil activity
Sethoxydim	A(1)	0.31 - 0.53	F	Microbial (4-11)	POST - annual grasses only
Clethodim	A(1)	0.11 – 0.28	F	Microbial (3)	POST - annual and perennial grasses only
Fluazifop-p-butyl	A(1)	0.13 – 0.42	F	(7-21)	POST - annual and perennial grasses only
Imazapyr	B (2)	0.56 – 1.70	F, S, B, I, CS	Microbial (25-140)	PRE, POST – perennial grass, brush, trees
Imazapic	B (2)	0.05 – 0.21	F, S	Microbial (60-120)	PRE, POST – annuals, perennial grasses
Imazamox	B (2)	0.14 – 0.56	F	Microbial (20-30)	POST – annuals, brush, trees
Chlorsulfuron	B (2)	0.018-0.15	F, S	Hydrolysis (40)	PRE, POST - rangeland annual/perennials
Metsulfuron-methyl	B (2)	0.012-0.17	F, S	Hydrolysis (30)	PRE, POST – annuals, perennials, brush
Sulfometuron-methyl	B (2)	0.065-0.4	F,S	Hydrolysis (20-28)	PRE, POST – annual, perennials, brush
Flumioxazin	E (14)	0.28-0.42	S	Microbial (12-18)	PRE – annual species
Oxyfluorfen	E (14)	0.56-2.24	S	Photolysis (35)	PRE- annual species
Isoxaben	L (21)	0.56-1.12	S	Microbial (50-120)	PRE – seedling annual species
Pendimethalin	K_1 (3)	0.84-3.36	S	Photolysis (44)	PRE – seedling annual species

Herbicide (common name)	Mechanism of Action[2]	Rate range kg/ha[3]	Application Methods[4]	Environmental Dissipation[5]	*Common Application Methods and General Spectrum of Control[6]
Oryzalin	K_1 (3)	2.24-6.72	S	Photolysis (20-90)	PRE – seedling annual species
Diclobenil	L (20)	4.5-22.4	S	Microbial (60)	PRE – seedling annual species, nutsedge
S-Metolachlor	K_3 (15)	1.4-2.8	S	Microbial (67)	PRE – seedling annual species, nutsedge

[1] Information presented derived from sources including, but not limited to: 2007 Herbicide Handbook, Weed Science Society of America, Lawrence, KS. 458p; ExToxNet - The EXtension TOXicology NETwork, http://extoxnet.orst.edu/; Crop Data Management Systems, Inc., http://www.cdms.net/.

[2] Mode of action classification based on Herbicide Resistance Action Committee (HRAC) – [letters and subscript numbers] and the Weed Science Society of America (parentheses). HRAC http://www.hracglobal.com/ WSSA http://www.wssa.net/Weeds/Resistance/WSSA-Mechanism-of-Action.pdf

[3] Rate range based on current label guidelines for control in natural areas or non-cropland sites. Rate expressed in kilograms of active ingredient per hectare.

[4] Application methods include: F - foliar, S – soil, B – basal bark, I – stem injection, CS – cut stump

[5] Environmental dissipation includes the major means of breakdown and half-life range in days in soil. In some cases, the mechanism of breakdown is not available.

[6] abbreviations: POST – postemergence activity/application; PRE – preemergence soil activity; BL's – broadleaf species.

*General application information only – refer to product label and local/state recommendations for specifics on use rates, application methods and timing, species controlled and restrictions for use.

Table 1. Properties and application methods of commonly used herbicides used in natural areas[1].

3.1. Synthetic auxins or growth regulators

The growth regulator herbicides represent the oldest and possibly the most widely used of the herbicides used in natural areas. These materials are mechanistically classified as synthetic auxins [14] and include herbicides in the phenoxycarboxylic acids, benzoic acid and pyridine carboxylic acid (picolinic acid) chemical groups.

2,4-D is the principle herbicide in the chemical group phenoxycarboxylic acids and has been used for broadleaf weed control since the late 1940's. This compound was first noted to have growth regulator properties in 1942, and registered as an herbicide after World War II [15]. There have been 28 different chemical formulations registered for 2,4-D, including the parent acid, amine salts and esters [14]. Salt formulations are characterized by fairly high water solubility and low volatility, while esters are more prone to volatility and more soluble in liquid fertilizers [16]. Ester formulations show greater phytotoxicity per acid equivalent basis due to greater cuticle penetration and foliar uptake. Short chain esters are highly prone to volatilization, and no longer registered for use. As of 2005, there were 9 formulations of 2,4-D supported for reregistration by the United States Environmental Protection Agency [17]. These include the parent acid, the sodium, diethanolamine, dimethylamine, isopropylamine, and triisopropanolamine salts, and the 2-butoxyethyl, 2-ethylhexyl, and isopropyl esters. In general salts are formulated as wettable powders, granules or soluble concentrates, while the water-insoluble esters are formulated as emulsifiable concentrates or mixed with oils or liquid fertilizers.

In addition to 2,4-D, there have been several other phenoxycarboxylic acid herbicides developed that are used in natural areas. These include MCPA, diclorprop, and mecoprop. Once again several formulations of each have been developed, including salt and ester forms. Of the three, diclorprop is used most extensively in natural areas [18], while MCPA and mecoprop are mainly used in grass crop and turf situations for annual and perennial broadleaf weed control [19]. As of 2007, the parent acid and the dimethylamine salt and ethylhexyl ester formulations of dichlorprop are registered for use by the US EPA. This herbicide has better activity on woody brush and trees, compared to 2,4-D. The ester formulation is often used alone or in oil-based carriers for spot specific plant treatments such as fencerows and rights of ways.

For many years, the phenoxy herbicide 2,4,5-T was the standard treatment for woody brush and tree control in pastures and rangeland [20]. This herbicide was highly active on several species and possessed considerable soil persistence, which contributed to its effectiveness. 2,4,5-T was cancelled for use by the U.S. EPA in the early 1980's due to concerns from the contaminant dioxin during certain manufacturing processes. Dioxin has been demonstrated to be a known carcinogen and was present in considerable quantities of 2,4,5-T used during the Vietnam war [21]. The herbicide known as 'Agent Orange' was actually a combination of 2,4,5-T and 2,4-D used for widespread aerial-applied jungle defoliation [22]. However, the levels of dioxin in commercially produced 2,4,5-T after the war were very low, but continuing concerns and public outcry lead to the cancellation of this herbicide [23].

The benzoic acid herbicide chemical family contains only one currently available herbicide for use in natural areas, dicamba. Dicamba is formulated only as a salt, with the following salts registered for use by the US EPA: dimethylamine (DMA) salt, sodium (NA) salt, isopropylamine (IPA) salt, diglycolamine (DGA) salt, and potassium (K) salt [14]. Interestingly, this herbicide can volatilize and move off target, despite being formulated as a salt. Dicamba is highly effective on many weeds in crops and is widely used in pasture/rangeland situations for perennial weed management [24]. It is considered to have superior perennial broadleaf weed control compared to many of the phenoxy herbicides, while still providing selectivity towards crops (primarily corn and sorghum). Dicamba also possesses greater soil persistence than phenoxys, which also contributes to its control [25].

The pyridine or picolinic acid herbicide chemical family comprises several herbicides that are widely used for natural area weed control. In general these herbicides are more potent compared to equivalent rates of phenoxy herbicides, and many possess considerable soil residual activity. The first picolinic acid herbicide developed was picloram in 1963 by Dow Chemical [14]. Similar to 2,4-D, picloram is formulated as salts (triisopropanolamine and potassium) and ester (ethylhexyl/isooctyl). Picloram is used in a wide range of natural areas, particularly open rangeland, for woody brush control [18]. Several formulations are also used in permanent pasture situations for perennial broadleaf weed control. The use of picloram is limited in certain areas over potential groundwater contamination concerns due to high water solubility and relatively long soil half-life. Moreover, many crops are highly sensitive to picloram at very low rates (<ppb), which also limits use in tolerant crops due to rotational concerns [26].

Triclopyr is probably the most widely used picolinic acid herbicide in natural areas, especially for woody brush species [27]. This herbicide is formulated as the triethylamine salt and the butoxyethyl ester, both of which are used across a wide range of natural, forest and pasture/rangeland situations. It possesses good activity on many annual and perennial broadleaf weeds and brush, but at rates slightly higher when compared to picloram [20]. However, unlike picloram, triclopyr has limited soil activity and is generally considered to be non-soil active [14].

The picolinic acid herbicide fluroxypyr also has limited soil activity, and is used primarily for broadleaf weed control in cereals, fallow cropland and pastures. It is formulated as a meptyl and butometyl ester and is often combined with other growth regulator herbicides to broaden weed control spectrum [28]. The use of fluroxypyr in natural areas is limited, primarily rights of ways, mainly due to superior weed control spectrum from other picolinic acid herbicides and labeling restrictions.

Clopyralid is another picolinic herbicide with moderate utility in natural areas. This herbicide was discovered in 1961 by Dow Chemical Company but was not registered for herbicidal use in the U.S. until 1987 [14]. It is mainly formulated as the monoethanolamine salt, but ester formations are also available. Clopyralid has moderate soil persistence and may cause problems with sensitive crops planted after clopyralid use in the previous crop [28]. This herbicide has broadleaf weed activity, similar to the picolinic acid herbicides as a whole, but has greater specificity and therefore selectivity towards many legume, solanaceous and composite type weeds [29,30,31].

Aminopyralid is a relatively new picolinic acid herbicide registered for use in pastures/ rangeland, forestry and natural areas [14]. Aminopyralid is only formulated as the potassium salt. It has moderate soil persistence, and like clopyralid, has specificity towards legume, composite and solanaceous weeds [32]. In fact, one of the primary registrations for this herbicide is for the control of tropical soda apple (*Solanum viarum*) in southeastern U.S. pastures [33]. In other areas of the U.S. the primary target species is composites such as thistles (*Cirsium spp.*) and species of knapweeds (*Centaurea spp.*) [34]. It is formulated as a salt and often combined with other herbicides to increase weed spectrum.

Aminocyclopyrachlor is the most recent herbicide to be registered for use in natural areas [35]. This herbicide possesses the typical growth regulator mode of action, but does not fit within the chemical classifications listed above. The uses of this compound are still being developed, but like aminopyralid and clopyralid, it has remarkable specificity at low use rates [36]. Aminocyclopyrachlor is primarily formulated as a salt, but ester formulations have been tested for basal bark applications in oil carriers. This herbicide is very active on a range of broadleaf species, but also possesses considerable activity on certain grasses, including many perennial grasses [37].

The mode of action of the synthetic auxin herbicides is not well understood, but appears to disrupt the normal cellular and tissue response to auxin. Auxin is present in plants at very small concentrations (nanomolar) and acts as a signaling molecular for a wide range of cellular functions and responses [38]. Auxin levels must be precisely controlled within the plant for normal regulation of plant responses to growth, development and environmental stimuli [39]. Auxin is regulated through two processes; metabolism via biosynthesis, conjugation, de-conjugation and degradation or transport and distribution within and between cells. The distribution of auxin, including directional flow, is regulated by the presence and activity of auxin transporters in the plasma membrane. Because auxins are weak acids, they are dissociated in the presence of neutral cellular pH (7.0) and trapped as anions within the cell. Thus transport out of the cell can be mediated through plasma membrane located facilitators specific for auxin.

Herbicides within this classification are considered auxin mimics, and are thought to act like auxin within plant tissues. Earlier research suggested that these herbicides acted to acidify the cell wall by activating a membrane bound ATPase proton pump and this acidification induced cell elongation [40]. Other work showed an increase in RNA polymerase, leading to increases in cell division and uncontrolled growth. Ethylene generation has also been reported, likely to counteract the stimulatory effect of auxin [41]. However, recent work has shown 2,4-D to be transported by influx carriers into the cell [42] and also through efflux carriers [43]. Due to limited metabolism, the auxin-effect of these herbicides presumably causes rapid cell division in some cells and a complete cessation of growth in other cells. This unregulated growth results in stem twisting, leaf strapping, puckering, and a plethora of other symptoms associated with growth regulator herbicides.

Synthetic auxin herbicides are chemically weak acids, and although some possess soil activity, these herbicides are applied to the foliage of plants. Once applied these herbicides are rapidly absorbed by leaf tissue and remobilized, similar to carbohydrate movement, to areas of meristematic growth via the phloem [14]. They possess the similar anion trapping mechanism as natural auxins, and this likely contributes to their effectiveness in herbicidal activity. Soil uptake of these herbicides occurs through the xylem where upward movement to shoots and leaves takes place. However, once diffusing from the xylem into leaf tissues, the herbicide is transported, in a similar manner to carbohydrates, to regions of meristematic growth.

The ability to metabolize is the primary selectivity mechanism for tolerant plant species. In most cases, grasses are moderately to highly tolerant to growth regulator herbicides through the ability to conjugate these herbicides with amino acids or sugars [25]. Most of these herbicides are slowly degraded regardless of plant species, but grasses appear to have the ability to shunt the herbicide conjugate to the vacuole, where it is either sequestered from sites of action, and/or slowly degraded. Many picolinic acid herbicides such as picloram, aminopyralid and clopryralid are sequestered in the vacuole of tolerant plants, but the compound remains intact and thus herbicidally active [44]. This has lead to many issues with off-target damage due to removal of the herbicide sequestering plant tissue and subsequent release of the herbicide in the environment.

This phenomenon was first observed with picloram, and later with clopyralid and aminopyralid. In the case of picloram, animals grazing on treated forage grasses were observed to have the ability to transfer the herbicide through urination or defecation. Concentrating of the herbicide, coupled with soil persistence lead to problems with sensitive crops planted in fields after grazing. Dried hay, either degraded as plant biomass or via manure, transferred from treated fields to other areas has also been shown to cause problems [45]. Manure from animals fed on treated forage that is used for compost and fertilizer is another source of contamination. More recently, grass clippings from treated turf, primarily clopyralid, can also be a problem [46]. The sequestration rather than degradation, coupled with high sensitivity at very low rates (parts per billion) for many broadleaf crop species is the reason for this major problem. This issue has lead to the cancellation of this herbicide in many areas, due to contamination in municipal compost for use by the general public [47]. Product labels containing these herbicides explicitly restrict the movement of treated plant biomass, and manure from livestock fed with treated forage in an effort to minimize off-target injury.

Recently genes for the metabolism of dicamba and 2,4-D have been inserted from bacteria into soybeans, cotton and corn, affording the ability to utilize these herbicides for weed control [48,49]. However, there are many concerns over the use of this technology, including the accelerated development of resistance by weeds as observed with the widespread use of glyphosate in glyphosate tolerant crops. Several weeds have developed resistance to growth regulator herbicides including kochia (*Kochia scoparia*) and lambsquarters (*Chenopodium album*) resistance to dicamba, yellow starthistle (*Centaurea solstitialis*) resistance to clopyralid and picloram and 2,4-D resistance in common chickweed (*Stellaria media*) and most recently common waterhemp (*Amaranthus tuberculatus*) [50,51]. The mechanism of resistance in most of these cases is not known.

3.2. Acetolactate (ALS) inhibitors

Herbicides within this classification are broadly represented by two major chemical families; the sulfonylureas and the imidazolinones. These herbicides are used in a wide range of cropping systems but many are also used in natural areas [14]. Both chemistries are highlighted by low use rates, low mammalian toxicity, and extreme specificity [52,53]. Interestingly, both classes of herbicide target the same plant enzyme, and were simultaneous discoveries by 2 separate agrochemical companies in the 1980's, DuPont for the sulfonylureas and American Cyanamid for the imidazolinones [54].

The first herbicide registered for use from this class was chlorsulfuron by DuPont in 1982 [52]. Chlorsulfuron is predominantly used in the western United States for broadleaf weed control in cereal grains and pasture/rangelands, but more recently for invasive species control by the Bureau of Land Management [55]. Other sulfonylurea herbicides developed by DuPont include sulfometuron and metsulfuron, which were initially labeled for use in forestry and industrial sites, but later labeling included uses for metsulfuron in pastures and natural areas and uses for sulfometuron for invasive species management [55,56].

Like the synthetic auxin herbicides, sulfonylurea herbicides have activity on a wide range of natural area broadleaf weeds but their activity also includes some grasses [57]. In general, and at rates labeled for use, chlorsulfuron is used for annual and short-lived perennial weed control in open rangeland and natural areas, while sulfometuron and metsulfuron have more control of woody brush and trees [58]. Both of these latter herbicides are used for hardwood control in commercial conifer forests and also for broad spectrum weed control in industrial sites such as railroads, rail yards, highway rights-of-way and electrical substations. However, all three

of herbicides also contain labeling specific to natural areas. Metsulfuron has a special local needs (SLN) label for the control old world climbing fern (*Lygodium microphyllum*) in south Florida natural areas [59].

Extremely low use rates and remarkable specificity set these herbicides apart from the traditional phenoxy herbicides [60]. It is difficult to make broad generalizations regarding the activity of the sulfonylureas because some species are controlled while other species, even within the same genus, are not. Therefore, uses for these products are regional or even local, depending on the species to be controlled and not controlled. These herbicides also have considerable soil activity, and this contributes to their long-lasting control in perennial systems [61]. However, this high level of activity can also cause problems with rotational crops, but this is not a common situation in areas where sulfometuron and metsulfuron are applied [60].

The imidazolinone herbicides used in natural areas include imazapyr, imazamox and imazapic. Imazapyr was first registered in 1985 for use in forestry and industrial sites such as railroads, rail yards, and powerline and highway rights-of-way [53]. At typical use rates, this herbicide has very broad spectrum activity that includes annual and perennial broadleaves, and several brush, vine and hardwood tree species. This herbicide also has tremendous activity on perennial grasses, both rhizomatous and bunch type grasses [62,63]. While initially developed for the industrial market, imazapyr is widely used in many natural areas for invasive species management. Imazapyr does have a registration for use in imidazolinone resistant crops, but its usage as such is limited [64].

Imazapic is registered for use in peanuts and certain forages, but is widely utilized for grass and broadleaf weed management in native perennial grass prairies [65]. Many perennial grasses such as eastern gamma grass, big bluestem grass (*Andropogon gerardii*), indiangrass (*Sorghastrum spp.*), switchgrass (*Panicum virgatum*) and buffalograss (*Bouteloua dactyloides*) have good tolerance to imazapic, although some injury is observed at seedling stages or during spring regrowth. Imazapic is also labeled for wildflower planting and for seedhead suppression of bahiagrass in turf settings. Imazapic is also used for the control of several invasive species in natural areas. These include Dalmatian toadflax (*Linaria vulgaris*), yellow starthistle (*Centaurea solstitialis*), leafy spurge (*Euphorbia esula*), Russian knapweed (*Acroptilon repens*), and tall fescue (*Schedonorus phoenix*) [66,67,68,69].

Imazamox is the most recent registration from the imidazolinone herbicide group in natural areas for the control of submersed and emergent vegetation [70]. It is particularly effective on Chinese tallow tree (*Triadica sebifera*), which is a major invasive species throughout much of the southeastern United States. Imazamox is also effective for several emergent and ditchbank species, and preliminary research indicates good control of cattail (*Typha spp.*). This herbicide has limited grass activity, and is most effective on broadleaf species.

The sulfonylurea herbicides chlorsulfuron, sulfometuron and metsulfuron are formulated as dry flowable granules that readily mix with water. Sulfonylureas are weak acid compounds with very high water solubility [14]. These herbicides are readily absorbed by roots from soil applications and transported via the xylem to shoots and leaves of plants. Once in the leaves, these herbicides are often remobilized in the phloem to growing regions - tracking a similar pattern of flow as carbohydrates. Sulfonylureas are also absorbed from applications to plant foliage, entering the leaves and stems, and translocated to areas of high meristematic activity in manner similar to root uptake [60].

The imidazolinone herbicides are also highly water soluble but formulated as salts. They are generally marketed as aqueous solutions, but some older formulations were dry flowable granules. Imidiazolinone herbicides are variable in soil activity but if present can be readily absorbed by plant roots [71]. They are transported to leaves and stem tissues via the xylem and can be remobilized to meristematic tissues. This pattern of reallocation occurs in the phloem, similar to carbohydrate movement. Imidazolinones are also absorbed from applications to plant foliage, entering the leaves and stems, and translocated to areas of high meristematic activity in manner similar to root uptake [72].

Mechanistically, the imidazolinones and sulfonyl-ureas act in the same manner by inhibiting the activity of the enzyme acetolactate synthase (ALS), which is also referred to as acetohydroxy acid synthase (AHAS, EC 2.2.1.6) [73]. This enzyme catalyzes the conversion of 2-ketobutyrate to 2-acetohydroxybutyrate through the addition of a 2 carbon unit using hydroxymethyl thiamine pyrophosphate (TPP). This is the initial step in the formation of the amino acid isoleucine. The ALS enzyme also catalyzes the conversion of pyruvate to form 2-aectolactate, once again utilizing TPP to add a 2-carbon unit [74]. This reaction is the initial step in the formation of valine and leucine. Thus by inhibiting acetolactate synthase, the formation of three essential branched chain amino acids cannot occur and inhibition occurs through a binding of the herbicide across the channel leading to the active site [75]. Herbicides in both groups bind at entrance of this channel, effectively blocking entrance to substrates and co-factors needed for the reaction to occur.

The inability of the plant to produce these essential amino acids leads to a cessation of protein/ enzyme synthesis and plant growth. Since these compounds accumulate in areas of new growth, meristematic activity is stopped. The plant cannot continue to make new cells and eventually dies [60]. Symptoms from these herbicides are generally manifested as discoloration in the growing regions, especially newly emerging leaves and shoot tips. Internode length is markedly decreased, and leaves may be malformed or misshaped [76]. Generalized chlorosis is a common symptom, although imidazolinones may show purple discoloration, especially in effected grasses. In annual species, a characteristic symptom of sulfonylurea injury is a reddening of the abaxial leaf veins.

Selectivity of these herbicides in plants is primarily metabolism based, and is often mediated through mixed-function oxidases (MFO's) [77]. These compounds catalyze several reactions in plants, including the breakdown of harmful xenobiotics such as herbicides. Tolerant plants generally are able to metabolize suflonyl-ureas and/or imidazolinones through this mechanism, thus imparting selectivity [60]. In cropping systems, crop selectivity is compromised if certain insecticides, such organo-phosphates, are used that disrupt MFO activity, allowing the herbicide to affect the target enzyme [78].

Interestingly, resistance development by weedy species occurs through amino acid substitutions of the target enzyme at the binding site [79]. In most cases, only a single amino acid change will confer resistance, and several substitutions (single amino acid changes) will cause resistance in sulfonyl-ureas. Conversely, very few impart resistance in imidazolinones and only one confers resistance across both herbicide families. The substitutions that confer resistance also appear to have little to no effect on enzyme efficiency, and thus growth of herbicide resistant biotypes varies little from non-resistant biotypes [80].

3.3. Photosynthetic inhibitors

Those herbicides that directly inhibit photosynthesis have been used for several years and were developed in the 1950's and 1960's [81]. While several chemical families are represented within this broad mode of action classification, the substituted ureas and triazines are those used most widely for natural area weed control. These products were originally developed for use in pasture/rangelands and forestry situations, but like several other herbicides, have been adopted for use in natural areas.

The triazine herbicides used in natural areas include hexazinone, simazine, and prometon. Simazine was originally developed for broadleaf and grass weed control in corn and sorghum, but later uses included grass and broadleaf control in established fruit and nut crops, albeit much higher rates of application per acre [82,83]. It was also used in aquatic situations for algae control, sold under the trade name "Aquazine", but this was cancelled in the 1990's [83]. Its use in natural areas currently is limited, primarily because simazine lacks broadspectrum control of perennial plants, particularly brush, vines and trees.

Prometon has been used for many years in industrial settings for broad-spectrum annual and perennial grass and broadleaf weed control [85]. This herbicide has considerable activity on many hardwood tree species, and is often marketed as a soil sterilant. This tremendous activity limits its use in many situations that require selectivity, and that includes forestry and most natural areas. Therefore, labeling as such is confined to areas where little to no vegetation is desired such as powerline substations, under asphalt paving, sidewalks, railyards and similar industrial sites [86]. Consequently prometon use in natural areas is very limited.

Hexazinone is an asymmetrical triazine that was originally developed for use in the conifer forest industry for hardwood control, and often used in a manner called pine release [87]. This situation occurs 2-4 years after pine seedling establishment, where hexazinone is broadcast applied to provide control of regenerating hardwood species, allowing the pines to be 'released' from the competing hardwood saplings. Hexazinone also has a label for use in bahiagrass (*Pasapalum notatum*) and bermudagrass (*Cynodon dactylon*) pastures for the control of broadleaf species, but most often targeting smutgrass (*Sporobulus indicus*) [88]. It can be used in many natural area settings where hardwood tree, brush/shrubs or possibly vines are the target, but many native forbs and some native grasses may also be injured. Hexazinone works wells in areas where pines are the primary species, possibly where undesirable species are dominant under pines, and understory selectivity is not paramount. Once these species have been removed, revegetation can then be accomplished.

Diuron and tebuthiuron comprise those herbicides in the substituted urea chemical family that are used in natural areas. Diuron is similar to simazine in that it was first developed for use in crops – corn and cotton, with later registrations including broadleaf and grassy weeds in established fruit and nut crops [89,90]. Diuron has good activity on a number of annual species, but lacks control of perennial plants. It is often a component in combination herbicides for broad-spectrum weed control in industrial sites such as railroads, railyards, powerline rights of way and substations [86]. The goal of these applications is to provide a vegetation free zone for extended periods of time. The use of diuron in natural areas is limited due to spectrum of activity; too much injury on desirable annual grasses and forbs and limited control of larger, more woody shrubs, vines and trees.

Tebuthiuron however, has tremendous activity on a wide range of woody species, particularly hardwood trees such as as oaks (*Quercus spp.*), maple (*Acer spp.*), poplar (*Populus spp.*), and sweet gum (*Liquidambar styraciflua*). [91]. This species is also very effective on shrubs, vines and herbaceous perennials [92]. It is often used in non-crop land and industrial settings for broadleaf vegetation control, including vines and hardwoods. Tebuthiuron is utilized in powerline corridors and around utility poles to promote healthy grass stands to maintain cover for grazing for livestock and wildlife and also erosion control [93]. This herbicide also is labeled for use in certain forestry situations, primarily for non-desirable vegetation control in conifers [94].

Bromacil is another photosynthetic inhibitor that belongs to the uracil chemical family that has limited uses in natural areas. It has similar use patterns as diuron and simazine, including vegetation management in industrial sites such as powerline substations, railroads, railyards, and rights-of-way [86]. Bromacil can also be used in certain fruit crops such as citrus for broadleaf and grass weed control [14,95]. While this herbicide has tremendous activity on annual species, it has less than adequate control of perennial vines, trees and shrubs compared to other herbicides; therefore wide spread utility in natural areas is limited [96].

As a group, photosynthetic inhibitors have low water solubility and limited foliar uptake [97]. Most are formulated dry as wettable powders or pellets, or liquid as clay-suspended flowables. Hexazinone is the only exception with a liquid formulation. These herbicides are soil applied; even applications over the top of existing foliage are active only when reaching the soil [14]. Photosynthetic inhibitors are readily taken up by plant roots and translocated to leaves and shoots through the water stream facilitated by xylem tissue [98]. Once reaching leaves, these herbicides partition into individual cells. As the plant continues to transpire, more herbicide is moved to the leaves, with older leaves and leaf tips transpiring the most water. These areas tend to demonstrate chlorosis first and most strongly simply because these tissues have transpired more water, and thus taken up more herbicide, compared to newer tissues. This causes the characteristic pattern of chlorosis often observed with these herbicides. Subtle differences in water solubility between herbicides and subsequent partitioning into leaf tissue of various species produce variations in chlorotic patterns, such as veinal chlorosis and/or interveinal chlorosis [99].

Differences in water solubility and to a lesser extent degradation, dictate the uses and selectivity of these products. Diuron, simazine, prometon, and bromacil are very non-water soluble and tend to remain in the upper soil profile [14]. This maintains the herbicides in the zone of germinating annual weeds, thus providing extended weed control. Perennial fruit and nut crops avoid herbicide injury primarily through limited uptake, since the roots of most trees are below the concentrated herbicide zone [100]. Conversely hexazinone and tebuthiuron are more water soluble and move deeper into the soil profile, which limits their utility for long-term vegetation management because annual weeds begin to infest the zone above the herbicide [101]. However, this places these herbicides into the root zone of many perennial forbs, vines, shrubs and trees where it is absorbed and translocated, causing injury and often mortality. Even large trees, especially oaks, can be killed if sufficient herbicide is placed in the root zone. Typically the leaves become chlorotic, necrotic and abscise. New leaves emerge, and follow the same chronological pattern, but generally do not expand to more than half normal size. After 2 to 3 cycles of leaf emergence and abscission, the trees succumb to death due to the lack of carbohydrate reserves needed for growth [102]. Depending on species, rate applied, and geographic location, death can take 1-2 years. Unfortunately, these herbicides are sometimes used in malicious attacks to destroy trees or shrubs; and in some cases trees of historic value, such as the Toomer Oaks on the campus of Auburn University, Auburn, Alabama (tebuthiuron) in 2010 or the Treaty Oak of Austin, Texas (hexazinone) in 1989 [103].

Photosynthetic inhibitors, regardless of chemical family, work in the same manner to interrupt the light reactions of photosynthesis. These reactions serve to capture the light energy from sunlight through excitation of chlorophyll molecules and the subsequent removal of an electron from a molecule of water; producing free oxygen and hydrogen [104]. Electrochemical energy is passed through a series of reactions (mainly photosystem II, cytochrome B, plastocyanin and photosystem I) to form NADPH+H. During this transfer, a proton gradient is formed across the chloroplast membrane, sufficient to generate ATP. These herbicides bind to a protein (specifically the D1 protein) within the photosystem II complex that does not allow electron transfer to occur [81]. This blockage of electron flow inhibits the formation of NADPH +H, and indirectly inhibits ATP formation as well. Energy continues to be absorbed by the chlorophyll molecules and transferred to the reaction centers associated with photosystem II, but cannot be dissipated [105]. This excess, or non-transferable, energy is then passed on to free oxygen, creating radical oxygen. Oxygen is a highly toxic radical that quickly reacts within the chloroplast to form hydroxyl radicals, peroxide, and/or lipoxides. Ultimately chloroplast and other cellular membranes become damaged and leaky, chlorophyll molecules are destroyed, and the tissue degrades.

While many photosynthetic inhibitors can be considered total vegetation control herbicides, certain species have the ability to tolerate these herbicides through metabolism. Metabolism is achieved primarily by glutathione and/or carbohydrate conjugation, whereby the herbicide molecules are bound with these compounds and shuttled to the vacuole for further breakdown [106]. However, in natural area systems - especially at rates typically used, placement and differential uptake is the primary mechanism of selectivity. Many conifers, pines (*Pinus spp.*) in particular, have the ability to tolerate hexazinone presumably through metabolism, but the mechanism is not known.

3.4. Glyphosate

Glyphosate is one of the most widely used herbicides in the world, and has been extensively used in natural areas for nearly 4 decades [107]. It is non-selective and provides control of a wide range of species, including annual and perennial grasses, annual forbs, short lived perennials, vines and many tree species [108,109,110]. It has limited activity on conifers, but time of year dictates use during periods of no or slow growth. This is generally the fall months prior to winter, termed hardening-off [111]. While active on many species of larger perennials, it is often mixed with other herbicides for greater control.

Glyphosate is chemically a weak acid, and is readily translocated in phloem tissues to areas of new growth. It is absorbed through foliar tissues such as leaves, shoot tips and green stems, but uptake is limited by woody tissues. Root uptake is possible, but rarely occurs due to irreversible binding of glyphosate to soil particles once the herbicide comes in contact with the soil. As in the case of other weak acid herbicides, glyphosate accumulates in meristematic regions, following a similar movement to that of carbohydrates [112]. Glyphosate affects the ability of plants to produce essential aromatic amino acids by blocking an initial step in the shikimic acid pathway. More specifically, this herbicide inhibits the activity of 5-enolpyruvyl-shikimate-3phosphate synthase (EPSP synthase) which catalyzes the conversion of EPSP from shikimate-3-phosphate and phosphoenolpyruvate [14]. This enzyme is a key enzyme in the shikimate acid pathway, which produces the aromatic amino acids tryptophan, phenylalanine, and tyrosine, along with a multitude of other secondary compounds including phenolics, flavonoids and coumarins [113]. Glyphosate also greatly influences carbon allocation and flow within the cell, as uncontrolled shikimate accumulation occurs as a result of this inhibition.

The typical symptoms of glyphosate injury include an initial cessation of growth followed by chlorosis in the meristematic regions of growth [14]. Chlorosis is often lighter in color compared to the photosynthetic inhibitors, and in some species may almost appear white or cream colored. Necrosis occurs several days after initial symptoms and complete plant death results in 21 to 35 days depending on species and maturity/size of treated plants. Glyphosate is extremely difficult to metabolize by plants and is readily translocated to areas of new growth [114]. This stability within plant tissues is the reason it has excellent activity on many perennial plants, allowing glyphosate to be 'stored' in overwintering tissues such rhizomes and rootstocks [115]. When plants begin to reallocate carbohydrates for spring regrowth, glyphosate is remobilized to these areas. Another unique symptom of glyphosate, particularly in regrowing perennial species, is the phenomenon of bud fasciation [116]. Bud fasciation is where several buds/shoot tips arise from a single meristematic region, forming a cluster of tightly packed shoots and leaves. The exact mechanism is not well understood, but appears to be related to a loss of apical dominance and deregulation of auxin activity.

Resistance to glyphosate has increased in annual cropping systems (Roundup-Ready technology) but resistance has not been documented in natural areas systems [117]. Several plants have the ability to tolerate and outgrow applications of glyphosate, especially trees, shrubs and woody vines. In these cases, limited uptake and/or dilution within non-metabolically active tissues is the likely reason for poor activity.

3.5. Fosamine

Fosamine has been used in industrial right of way situations for many years and more recently used for invasive species control in natural areas such as natural savannahs and prairies. Brush control is the target for this herbicide, but it can be used for the control of herbaceous weeds such as leafy spurge (*Euphorbia esula*). Fosamine is tolerated by certain species of conifers, but hardwoods and other deciduous trees are often damaged. Fosamine is applied to the foliage of target plants where it is slowly absorbed by leaf tissues [118]. This herbicide has little to no soil activity and is rapidly degraded by soil microbes, limiting its environmental persistence [119]. This herbicide is recommended for late summer/autumn applications – typically one to two months prior to leaf drop. Fosamine appears to have limited translocation out of treated foliage and does not exhibit symptoms on treated tissue [120,121]. The effect of fosamine is not apparent until the following spring where leaves often fail to emerge or if emerged will be small and spindly in appearance. The mechanism of fosamine is not clear, but some evidence suggests an inhibition of mitosis or the inability of new developing cells to effectively transport calcium [14]. The limited translocation within plant tissues allows the use of this herbicide as a 'side-trim' treatment, where a portion of tree can be controlled without affecting the entire tree. This type of application is used in powerline and railroad situations to chemically trim a tree to remove unwanted limbs and foliage [86].

3.6. Inhibitors of Acetyl CoA Carboxylase (ACCase inhibitors)

Herbicides within this group fall into two broad chemical families – the cyclohexanediones or the aryl oxy-phenoxy propionates [14]. There are several herbicides within these families labeled for use in non-crop/natural areas, but the most widely utilized include sethoxydim, clethodim and fluazifop-butyl [86]. These herbicides are characterized by their selectivity towards annual and perennial grasses, with minimal to no activity on other monocots or dicot species [122]. They are primarily applied to the foliage due to a lack of appreciable soil activity through binding to soil particles and rapid microbial degradation

ACCase inhibiting herbicides are applied to the foliage of grasses, where they are readily absorbed. Similar to other weak acid herbicides, they are translocated to areas of meristematic growth following the pattern of carbohydrate flow [112]. Cyclohexanediones and aryl-oxy-phenoxy propionate herbicides inhibit the activity of acetyl CoA carboxylase [123]. This enzyme is the initial step in the formation of fatty acids, which are the primary building blocks of cell membranes and other cellular components necessary for normal growth. New growth is stopped and grasses often become chlorotic or purple in color. Another characteristic symptom is the water soaked browning of stems when pulled from the whorl.

The utility of these herbicides is limited to annual and perennial grass control. Clethodim and fluazifop have superior activity on perennial grasses, and are often used for the control/ suppression of reed canarygrass (*Phalaris arundinacea*), cogongrass (*Imperata cylindrica*), Japanese stiltgrass (*Microstegium vimineum*) to name a few [124,125,126]. However, complete control of well established grass stands is often not achieved with a single application and multiple treatments are usually required.

3.7. Glufosinate and paraquat

Glufosinate and paraquat are contact type herbicides that can be used in a wide range of non-cropland, industrial, rights-of-way areas and natural areas [86]. Both of these herbicides are contact in activity, requiring complete coverage of the target foliage to attain good control [14]. In addition, both paraquat and glufosinate do not possess soil activity due to immediate and irreversible binding to soil particles [127]. These herbicides are very effective on annual broadleaf and grassy weeds, but only marginally effective on well established perennial plants. Since these herbicides do not translocate out of treated foliage, perennial plants can usually regrow following treatment [112].

Glufosinate is rapidly absorbed by leaf tissue and is active in the chloroplast of cells. Specifically, glufosinate inhibits the enzyme glutamine synthase, which catalyzes the incorporation of free ammonia into the amino acid glutamate to form glutamine [128]. This reaction is the primary mechanism by which plants incorporate nitrogen for use in cellular products such as amino acids, nucleotides, enzymes and storage proteins. The lack of nitrogen incorporation, however, is not the primary means by which the plant dies. Free ammonia levels increase in the chloroplast where this molecule begins to uncouple membranes. Uncoupling is the action where membranes can no longer maintain a gradient that drives energy formation in photosynthesis [129]. Damage becomes visible generally after 4 to 5 days and appears as chlorotic lesions followed by rapid necrosis of treated leaves.

Paraquat herbicide was developed in the early 1960's for broad spectrum weed control in non-crop land and other vegetation free sites. Paraquat is rapidly absorbed by leaf tissues and is active primarily in the chloroplast, although it may also impede mitochondrial function [112]. Paraquat affects the light reactions of photosynthesis in the photosystem I complex, more specifically at the site of electron transfer from ferrodoxin to NADPH+H reductase [130,131]. Paraquat does not bind or disrupt enzyme activity, but rather steals/diverts the electron to become a reduced paraquat molecule. Paraquat in this reduced form quickly passes the electron energy to oxygen, creating oxidized paraquat and radical oxygen (O^{2-}). Paraquat becomes reduced again by another electron, oxidized through transfer to oxygen and the cycle continues. Subsequently, the ability of the plant to make NADPH+H is compromised, but more importantly radical oxygen reacts with water and lipids to produce hydrogen peroxide, hydroxyl radicals and lipoxides. These radicals interact with the lipid fraction of membranes, destroying the chloroplast and eventually the plasma membrane [132]. Symptoms from paraquat can be evident within 12 to 24 hours after application. Leaves first appear water soaked, followed quickly by necrotic lesions that coalesce to encompass the entire leaf. High light levels promote faster necrosis, and complete damage is generally achieved within 2-4 days.

Glufosinate and paraquat require good coverage and therefore must be applied in higher carrier volumes compared to systemic herbicides. There appears to be some movement with glufosinate out of treated tissues, but translocation to perennial structures such as rhizomes or tubers does not occur to an appreciable extent [133]. The utility of paraquat and glufosinate for natural area plant management is limited for several reasons. First most weeds in natural areas are perennials, so applications of these herbicides will only provide temporary control.

Secondly, these herbicides are non-selective - causing damage to any plant that is contacted, desirable and undesirable vegetation [14]. Thirdly, these herbicides lack soil activity so long term control cannot be realized.

3.8. Protox inhibitors

There are several herbicides and herbicide families that encompass this mode of action category [14]. Protox inhibitors are primarily used in annual cropping systems for broadleaf, grasses and nutsedge (*Cyperus spp.*) control [134,135,136]. These herbicides possess good soil activity with moderate to long soil persistence and many also have tremendous foliar activity [137].

Protox inhibitors are readily taken up by plant roots and translocated to leaves and shoots through the water stream facilitated by xylem tissue [138]. Once reaching leaves, these herbicides partition into individual cells. As plant continues to transpire, more herbicide is moved to the leaves, with older leaves and leaf tips transpiring the most water. These areas tend to demonstrate damage initially and most strongly simply because these tissues have transpired more water, and thus taken up more herbicide, compared to newer tissues. Damage appears as bronzing or necrotic lesions in leaf tissue. These lesions generally lack pattern, but eventually coalesce into more wide-spread damage and eventual leaf drop. Stem tissues may also exhibit similar necrotic injury. In grasses and sedges, a browning of leaf tissue along the midvein is often observed.

Foliar activity shows a similar pattern, with necrotic lesions developing in random areas on leaf tissues, with complete necrosis occurring in 3-5 days [139]. Even tolerant plants will show some damage from foliar applications on treated tissue, but to a much lesser extent and quickly outgrow the injury. There is no translocation from foliar applications of protox inhibiting herbicides, only those areas contacted will be damaged [139]. However, subsequent damage may occur from root uptake, if an appreciable amount of herbicide reaches the soil and remains active. This is highly dependent on whether the herbicide has soil activity, application rate, and foliar coverage at the time of application.

Protox inhibiting herbicides have a very unique mode of action that was not clearly understood for many years [140]. Mechanistically, these herbicides inhibit the enzyme protoporphyrinogen oxidase which catalyzes the conversion of protoporphyrinogen IX to protoprophyrin IX in the chloroplast [141]. This step is an intermediate process in the production of chlorophyll molecules. Excess protoporphyrinogen IX leaks out of the chloroplast envelope into the cytoplasm where is it converted by a cytoplasmic (insensitive) version of protoporphyrinogen oxidase to protoprophyrin IX [142]. This molecule has the ability to absorb light energy, but can only dissipate this energy to oxygen. This forms singlet oxygen, a highly reactive form of oxygen that quickly interacts to form other highly toxic radicals that destroy cell membranes. Cells become leaky, rupture, die and eventual tissue degradation follows.

Utility of the protox inhibiting herbicides in natural areas is limited. Foliar activity is contact only, therefore perennial plants quickly regrow. In addition, at the rates needed to garner control, selectivity is lost or severely compromised. Appreciable control can be achieved from soil uptake and activity, but those rates of herbicide application necessary may also reduce selectivity, and in some cases may not be within label guidelines. Flumioxazin and oxyfluorfen are protox inhibiting herbicides that may be used in non-crop areas, but applicability to natural areas has not been widely studied. These herbicides may have some use in restoration situations, providing control of undesirable vegetation prior to or immediately after an augmented restoration planting. However, there has been limited research to determine which herbicide product is most effective as a function of selectivity and desirable persistence.

3.9. Growth inhibitors

Herbicides that are categorized as growth inhibitors fall into three major mechanisms of action, but produce the common effect of inhibiting seedling emergence. The three mechanisms include: 1) interruption of mitosis through a blockage of spindle fiber formation, 2) interruption of cell wall formation through an inhibition of cellulose biosynthesis, and 3) interruption of cell membrane formation through a blockage of very long chain fatty acid synthesis. In nearly all situations, these herbicides are applied to the soil where they are absorbed by germinating seedlings, preventing seedling growth [14].

These compounds are characterized by extremely low water solubility, maintaining the herbicides in the upper soil profile [127]. As seeds germinate, the roots and emerging shoots come in contact with the herbicide, where it is rapidly absorbed, inhibiting growth and killing seedlings before they emerge from the soil. These herbicides do not translocate within plant tissues, so the growing regions of the plant must come in contact to be effective. Foliar applications are ineffective because the herbicides remain in the cuticle or epidermal cells, and cannot come in contact with meristematic tissues which are generally shielded within the bud structure. Selectivity is achieved through placement, whereby the shoots of tolerant germinating seedlings can emerge with minimal herbicide uptake in meristematic regions *and* the roots can grow below the treated layer. In cropping systems, this is most often achieved with broadleaf crops possessing hypogeal germination patterns. Perennial crops also exhibit good tolerance because the roots are well below the treated soil layer and foliar uptake is minimal. Examples of growth inhibiting herbicides used in non-crop areas include diclobenil, pendimethalin and metolachlor, but applicability to natural areas has not been widely studied.

4. Herbicide application methods in natural area weed management

This section will detail the various methods used for applying herbicides for management of weedy species in natural areas. The complexity of natural areas dictates a unique and often non-conventional approach to herbicide application to 1) maintain selectivity, 2) provide control of large specimens, and 3) minimize off-target damage to the natural environment. Selectivity is much more difficult to achieve and maintain in natural areas. Herbicides are generally developed for weed control in cropping systems, and then secondarily labeled for use in non-cropland areas. In crops only selectivity towards the crop plant is desired, and damage to all other plants is beneficial, advantageous or inconsequential. However, in natural

http://dx.doi.org/10.5772/56183

Figure 1. Postemergence application of herbicide to woody brush. Photo Courtesy James Miller, U.S. Department of Agriculture, Forest Service, http://www.forestryimages.org

area weed management, only one species is the target and damage to other species is not desirable – especially injury to rare or endangered plants.

4.1. Post-emergence foliar applications

This is the most common method of application, whereby the herbicide in diluted solution is applied as a spray over the top of targeted species (Figure 1). For larger areas, treatments are made to both target and non-target species utilizing an aerial (propeller plane or helicopter), tractor or all-terrain vehicle (ATV) mounted broadcast spray boom. Herbicides are applied as the amount of active ingredient per unit land area, and calibrated to deliver this amount based on carrier volume output. Smaller, more isolated or higher selectivity required sites will utilize a backpack sprayer with a hand-held spray wand or boom. Backpack applications cannot be calibrated in the same manner; herbicides are applied as a percentage of undiluted herbicide in a variable carrier output [18,143].

Aerial applications are highly restricted and only certain herbicides can be applied aerially, and in some cases only during certain times of the year to minimize off-target injury. For example, the state of Florida restricts the use of organo-auxin from aerial applications from January 1 until May 1 of each year [144]. Aerial treatments often utilize very low gallon spray volumes (3-10 gallons per acre) to maximize efficiency with weight and spray volume [145]. This restricts aerial applications to systemic herbicides that are not dependent on high carrier volume for effectiveness.

Tractor or ATV boom-mounted sprayer applications can utilize a range of carrier volumes and thus not restricted to systemic herbicides only. These types of application equipment generally utilize a rear mounted boom with flat fan nozzles. The size or width of the boom varies, but ATV mounted booms are generally less than 15 feet while tractor booms may reach 30 feet or greater. Regardless, boom width is restricted compared to traditional agricultural applications due to unevenness of terrain to be covered, obstacles such as trees, shrubs, etc. and limitations on pump and tank capacity on smaller tractors and ATVs. Boom applications, especially those utilizing boom widths greater than 15-20 feet, require relatively flat ground, uniform height and high density of target species. As such, many land managers cannot utilize this type of equipment in many natural area systems.

Boom-less nozzles are often used in industrial applications and have some merit for use in natural area weed control. These nozzles are specifically designed to produce a multi-stream pattern across a 12-15 foot-wide spray swath. When mounted on an ATV or truck, these nozzles can produce a sizable sprayed area, without the issues associated with a fix boom to avoid obstacles and uneven terrain. However, coverage with these types of nozzles is not uniform and generally high volume output is required to maintain proper spray pattern. In addition, the actual nozzle is very expensive compared to a standard fan flan system. Due to difficulties with application uniformity and issues with achieving selectivity, most natural area weed managers will rely heavily on small backpack sprayers. This type of sprayer consists of a 5 gallon/20 liter (on average) tank, a hand-held, single nozzle spray wand, and a small diaphragm pump with an attached lever. The operator uses the pump to pressurize the tank, forcing the liquid spray mixture through the spray wand. Pressurization is under the control of the operator, and is generally maintained to provide a proper pattern from the adjustable orifice on the spray wand. As the name suggests, the apparatus is worn on the back of the applicator using shoulder straps and often a waist strap to stabilize weight distribution. In most cases, the user operates the wand with one hand and pressurizes the tank with the other.

Backpack applications utilize diluted herbicide solution and mixed as a percent solution; in most cases between 0.5 and 3% solution. Applications are made to target species on a visual 'spray to wetness' observation. To achieve some degree of uniformity among applicators, the basis for adequate spray delivery is when spray droplets begin to drip from the leaf surfaces. This 'spray to runoff' technique is common regardless of target species or herbicide. While it is difficult to accurately measure volume output on a per acre basis, most researchers estimate these types of applications to range from 30 to 50 gallons per acre. In many cases, post-emergence foliar applications contain herbicides with soil residual activity, either from an herbicide that possesses both foliar and soil activity or soil active herbicides that are tank-mixed to provide extended control. Regardless, the application technique is the same for most boom-mounted sprayers. For soil applications using a backpack sprayer, the applicator self-calibrates by placing a known amount of liquid in the sprayer and sprays a defined area. Once the area has been completely sprayed, the amount of liquid used by the applicator is calculated to determine individual spray output per area (in most cases ft^2).

4.2. Soil basal applications

Soil basal applications are used for 2 primary purposes - 1) provide control of an existing plant or group of plants, or 2) provide preventative control of potential plant problems around stationary objects such as power poles. In either scenario, the herbicide is placed in often high concentrations around the base of the treated plant or object. The herbicide may be applied in liquid or granular form, and in a variety of placement patterns to achieve maximum root uptake of the intended target(s). Some herbicides, especially soil active photosynthetic herbicides, are formulated as pellets, which are essentially larger, more concentrated granules. Dry formulated granules or pellets are often easier and more accurate to apply as basal soil treatments. In these situations a certain number of pellets or dry volume of granule is placed as a function of targeted plant circumference. The pattern of placement varies considerably among applicators and may include circular, piles of pellets, or even gridlines in the case of larger infestations [146]. Soil basal herbicides include many of the photosynthetic inhibitors and several of the ALS and growth regulating herbicides. While the growth inhibiting and protox herbicides possess good soil activity, their effectiveness on established and larger plants is limited due to lack of root uptake and translocation or short-term control. Uses are generally restricted to those situations where preventative control is the primary objective.

4.3. Basal-bark applications

Basal-bark applications are utilized to provide control of larger specimens, where over-the-top foliar applications are not feasible for logistical or selectivity reasons. As the name suggests, basal-bark treatments are made near the ground to the trunks of small trees or shrubs [143]. Treatments are applied using a hand-held spray bottle or backpack sprayer to provide a tight stream of liquid onto the bark (Figure 2). Techniques for basal-bark applications vary widely among practitioners and weed specialists, but most agree that complete coverage around the trunk base is necessary for control. The width of the spray band around the tree varies as a function of species, size and herbicide being used, but most common is a 12 inch (30 cm) width band. Applications are generally made to the point of visual dripping or running of the liquid down the bark surface.

Basal-bark treatments utilize an oil carrier (often referred to as basal oil) in which the herbicide is diluted at a high concentration or undiluted [147]. Diesel fuel or kerosene was used as carriers for many years, but environmental and economic restrictions limit current usages in many areas. In some cases, depending on herbicide formulation, the herbicide may be applied in undiluted form. Regardless of carrier, the herbicide must be in an oil soluble/lipophilic form to allow for penetration into the bark tissues. The objective is to maximize herbicide penetration through the outer epidermal layers (periderm) and reach the secondary phloem and cambium [143]. Once reaching these layers, the herbicide may be remobilized in the phloem, penetrate and affect the dividing cambium cells, or possibly enter the water stream via the xylem

Figure 2. Basal bark application to small tree. Photo Courtesy BASF.

sapwood. There is little research as to the actual mechanism of mortality but is surmised that the herbicide is translocated slowly throughout the plant, accumulating in regions of active growth and killing meristematic tissues. The resiliency of many large woody trees and shrubs requires that the herbicide remain available within the plant, and presumably in translocatable form, for a period of time that allows the specimen to exhaust food reserves and/or meristems to provide complete control.

Basal bark herbicides are limited to ester formulations of triclopyr, picloram, 2,4-D, 2,4-DP. Dicamba and oil-soluble formulation of imazapyr have also been used, often in combination with other herbicides [148]. To be effective as a basal treatment, the herbicide must be able to solubilize in oil, which is needed to penetrate the bark layers. The herbicide must also be systemic to allow translocation once reaching the vascular tissues. For these reasons, basal bark treatments are exclusively weak acid herbicides, but only those chemistries that can be formulated to be oil soluble such as esters. Several weak acid herbicides, including the sulfonylureas, are not effective as basal treatments because of low oil solubility.

4.4. Stem injection applications

Stem injection applications are generally made to trees or shrubs with larger than 4 inch (20 cm) diameter trunk bases, which is the upper limit for effective basal treatments. In this type of application – also called hack and squirt, the herbicide is placed into a cut or frill made into the bark of the specimen (Figure 3). A hatchet, axe, machete, or other hand-held cutting device is used to make a downward cut/incision that penetrates the bark to the cambium layer, creating a cavity to contain a small amount of herbicide solution [147]. Although highly dependent on herbicide and species, incisions are made evenly around the trunk, or in the case of larger trees a complete girdle might be necessary. One rule of thumb is one incision per inch of trunk diameter [149]; another is no incisions more than 3 inches (10 cm) apart [150]. Herbicide activity on a given species is generally what dictates the number of cuts that is required. Additionally, it is useful to place these cuts near the base of the stem. Making the application higher on the stem will often increase the likelihood of stem-sprouting below the application site.

Figure 3. Hack and squirt application to larger diameter tree. Photo courtesy James Miller, U.S. Department of Agriculture, Forest Service, http://www.forestryimages.org

Unlike basal bark applications, this type of application can utilize water and oil soluble formulations, providing greater flexibility in herbicide options. In addition to those herbicides mentioned for basal bark, glyphosate, triclopyr amine salt, and hexazinone can be effectively used. Typical concentrations for injections range from 33 to 50% solution in water. In some cases, undiluted herbicide is used. Only a small amount of liquid is placed per cut (< 5 ml) and applied using a single nozzle backpack sprayer, or a hand-held spray bottle. A marker dye is often used to help applicators visualize and keep track of treatment applications. There have been several pieces of equipment developed to 'inject' herbicide into woody plant tissues, combining the mechanical cutting operation with liquid dispensing operation [151]. The 'hypo-hatchet' delivers a pre-measured amount of liquid through a pore in the hatchet blade when inserted into the trunk tissue [152]. Injector bars (Figure 4) contain the herbicide mixture within the bar which is

Figure 4. Stem injection of herbicide into trunk of target tree. Phot credit James Miller, U.S. Department of Agriculture, Forest Service, http://www.forestryimages.org

jabbed into a tree, and a lever is pulled allowing a pre-measured amount of liquid to flow through the end of the bar [147]. Some bar type devices will insert a granular pellet during each injection. Other injection tools include a hand-held gun, with a large diameter needle that can be inserted into softer perennial tissues, once again with a premeasured amount that is injected.

4.5. Cut-stump applications

Cut-stump applications occur, as the name implies, to the cut portion of a felled tree or shrub. The purpose of the application is to prevent regrowth of the plant from shoots arising from the cambium layer of the cut stump. Herbicide is applied to the cut surface, making sure to cover the entire outer cambium layer [86,147]. Placement of the herbicide across the entire stump is not necessary, since the majority of the inner tissues consist of non-living heartwood (Figure 5). Applications should occur within 30 minutes of cutting to avoid the layer becoming scabbed over, reducing herbicide uptake and penetration.

Triclopyr amine or ester, picloram, 2,4-D, dicamba, imazapyr or glyphosate can be used for cut stump applications. Ester formulations can be applied as 25% solution in basal oil, while amine/salt formulations are applied as 50% solution in water. Sometimes undiluted herbicide can be used, but care must be taken to avoid 'flashback'. Flashback is a phenomenon where the herbicide is absorbed by the trunk and roots of the felled specimen, translocated through the root system, and passed through root grafting to the roots of neighboring plants [149]. Neighboring plant roots can also absorb the herbicide from soil around the treated stump, where herbicide is washed

Figure 5. Herbicide application with marker dye made to cut stump, targeting only outer cambium region. Photo credit James Miller, U.S. Department of Agriculture, Forest Service, http://www.forestryimages.org

off the stump or root crown. Regardless, applicators are encouraged to use only the amount necessary to provide control of resprouting, and limit excessive herbicide use.

4.6. Ballistic herbicide application

This unique approach to applying herbicides has been developed by Dr. James Leary with the University of Hawaii [153]. In this system, herbicides are encapsulated in paint ball pellets and distributed to the target species via a commercially available paint ball gun. Each 'ball' contains a known amount of herbicide and rate is calculated by the number of balls fired at each specimen. The applicator targets the apical regions of the plant, or the larger stems to increase the 'splatter' effect that helps distribute the herbicide within the plant architecture.

Dr. Leary has performed nearly all initial testing with imazapyr and triclopyr, which readily translocates within plant tissues. Imazapyr also possesses good soil residual activity, which aids in effectiveness. This technology is still in the evaluation phase, but holds good promise for treating invasive species in remote and inaccessible areas. Most of the treatment evaluations have been performed on the slopes of tropical mountains in Hawaii, where the only means of treatment has previously been a single nozzle suspended from a helicopter. The nozzle is embedded in a heavy ball that helps reduce swaying and the pilot attempts to direct the nozzle over the crown of the targeted specimen. This approach is time consuming, precarious and expensive. With the ballistic approach, the applicator fires a number of balls into the crown to deliver the herbicide. This allows for more specimens to be treated per helicopter flying time, eliminates the need for unwieldy spray equipment and provides for more precise herbicide application [154]. As mentioned, this technology is still under intense evaluation, and commercialization of the process has not been undertaken.

5. Integrated approaches to natural area weed management

Regardless of herbicide or application method used, chemical weed control must be used in an integrated approach for controlling weeds. Other methods of weed control such as prevention, biological, cultural and mechanical techniques are often utilized to complement chemical control programs. For example, mechanical felling of large trees, followed by chemical treatment of regrowth, is a common operational strategy for forestry. Chemical control to provide initial kill of vegetation, followed by the introduction of a biocontrol agent is very effective for management of *Melaleuca quinquenervia* in south Florida. A critical aspect of management in many systems is the use of fire to reduce ground litter, promote seed germination and flowering, and provide control of undesirable species. Fire can also be used to reduce biomass and promote regrowth, which often results in more efficacious herbicide treatments. Conversely, intense fires by excessive fuel generated from invasive species can cause severe damage, especially to desirable over-story trees.

Restoration is another very important component of natural area management. This aspect involves: 1) promoting the existing desirable vegetation through regrowth or regeneration from a seed bank, or 2) intentional planting of desirable species through physical transplanting or sowing of seed. Previous control methods can have a profound effect on restoration. Mechanical tillage can disrupt the seedbank through exposure seed on the surface or bury beyond the point of emergence. Residual activity from herbicides used to control invasive plants can also be deleterious to recolonizing desirable species. Studies to determine herbicide longevity and sensitivity of species is important when developing both control strategies and subsequent restoration plans as components of an overall management plan.

6. Conclusions

Herbicides are a critical component to managing undesirable species in natural areas. However, several considerations must be addressed for effective and environmentally safe usage. Proper herbicide selection, timing of application, type of application methodology and application rate must be adhered according to the product label. Actual site of usage must also fall within product label guidelines. Herbicides should never be used as a stand-alone approach but rather as a component of an integrated long-term management strategy for invasive species control and natural area restoration.

Acknowledgements

This publication is a contribution of the University of Florida Institute for Food and Agricultural Sciences and the Florida Agricultural Experiment Station. The authors also wish to thank

http://dx.doi.org/10.5772/56183

the Center for Aquatic and Invasive Plants at the University of Florida for support in publication of this document.

Author details

Gregory E. MacDonald[1*], Lyn A. Gettys[2], Jason A. Ferrell[1] and Brent A. Sellers[3]

*Address all correspondence to: pineacre@ufl.edu

1 University of Florida Institute of Food and Agricultural Sciences, Department of Agronomy, Gainesville, FL, USA

2 University of Florida Institute of Food and Agricultural Sciences, Department of Agronomy, Fort Lauderdale Research and Education Center, Davie, FL, USA

3 University of Florida Institute of Food and Agricultural Sciences, Department of Agronomy, Range Cattle Research and Education Center, 3401 Experiment Station Rd., Ona, FL, USA

References

[1] Shrader-Frechette, K.S. and E.D. McCoy. 1995. Natural landscapes, natural communities, and natural ecosystems. Forest and Conservation History 39:138-142.

[2] Maser, C. 1990. On the "naturalness" of natural areas: A perspective for the future. Natural Areas Journal 10:129-133

[3] Daily, G.C., S. Alexander, P.R. Ehrlich, L. Goulder, J. Lubchenco, P.A. Matson, H.A. Mooney, S. Postel, S.H. Schneider, D. Tilman, and G.M. Woodwell. 1997. Ecosystem services: benefits supplied to human societies by natural ecosystems. Issues in Ecology Number 2, Spring 1997 18p.

[4] Walls, M. 2009. Parks and recreation in the United States: The National Park System. Resources for the Future (RFF) Backgrounder, January 2009, 14p.

[5] U.S. National Vegetation Classification. 2012. http://usnvc.org/explore-classification/ (accessed 9 February 2013).

[6] Anonymous. 2006. Management Policies 2006. National Park Service. http://www.nps.gov/policy/mp2006.pdf. 180p. (accessed 9 February 2013).

[7] Anonymous. 2012. Conserving the Future: Wildlife Refuges and the Next Generation. U.S. Fish and Wildlife Service. 49 p.

[8] Mitsch, W.J. and S.E. Jørgensen. 2004. Ecological Engineering and Ecosystem Restoration. John Wiley & Sons, Inc., New York. 411 pp.

[9] Olson, L.J. 2006. The Economics of Terrestrial Invasive Species: A Review of the Literature. Agricultural and Resource Economics Review 35(1):178-194

[10] Pimental, D., M. Pimental and A. Wilson. 2007. Plant, Animal, and Microbe Invasive Species in the United States and World. Biol. Invasions Ecol. Studies 193:315-330.

[11] U.S. Forest Service. 2006. Four Threats – Quick Facts Invasive Species. Online access: http://www.fs.fed.us/projects/four-threats/facts/invasive-species.shtml (accessed 9 February 2013).

[12] Federal Register. 1999. Executive Order 13112 of February 3, 1999 - Invasive Species. Federal Register/Vol. 64, No. 25/Monday, February 8, 1999/Presidential Documents pp. 6183-6186. Online access: http://www.gpo.gov/fdsys/pkg/FR-1999-02-08/pdf/99-3184.pdf (accessed 9 February 2013).

[13] Anonymous. 2012. Pesticide Product Labels. U.S. Environmental Protection Agency. http://www.epa.gov/pesticides/regulating/labels/product-labels.htm (accessed 9 February 2013).

[14] Senseman, S. 2007. Herbicide Handbook. Weed Science Society of America, Lawrence, KS. 458p. ISBN 1-89276-56-5.

[15] Zimmerman, P.W. and A.E. Hitchcock. 1942. Substituted phenoxy and benzoic acid growth substances and the relation of structure to physiological activity. Contr. Boyce Thompson Inst. 12:321-343.

[16] Grover, R., J. Maybank and K. Yoshida. 1972. Droplet and Vapor Drift from Butyl Ester and Dimethylamine Salt of 2,4-D. Weed Science 20:320-324.

[17] Anonymous. 2005. Reregistration Eligibility Decision for 2,4-D. U.S. Environmental Protection Agency. http://www.epa.gov/oppsrrd1/REDs/24d_red.pdf. EPA 738-R-05-002. 304 p. (accessed 9 February 2013).

[18] Bovey, R.W. 1976. Response of selected woody plants in the United States to herbicides. Agriculture Handbook No. 493, United States Department of Agriculture. 100 pp.

[19] Johnson, B.J. and R.E. Burns. 1985. Effect of timing of spring herbicides on quality of bermudagrass (Cynodon dactylon). Weed Sci. 33:238-243.

[20] Bovey, R.W. 2001. Woody plants and woody plant management: ecology, safety and environmental impact. 2001 pp. vii + 564 pp.

[21] Akhtar, F.Z., D.H. Garabrant, Ketchum, N.S. and J.E. Michalek. 2004. Cancer in US Air Force Veterans of the Vietnam War. J. of Occupational & Environ. Medicine: 46(2): 123-136.

[22] Mortelmans, K., S. Haworth, W. Speck, and E. Zeiger. 1984. Mutagenicity testing of agent orange components and related chemicals. Toxicology and Applied Pharmacology. 75(1): 137–146.

[23] Gangstad, E.O. 1983. Benefit/Risk analysis of Silvex cancellation. J. Aquatic Plant Manage. 21:65-69.

[24] Scifres, C.J. and G.O. Hoffman. 1972. Comparative susceptibility of honey mesquite to dicamba and 2,4,5-T. J. Range Mgt. 25:143-146.

[25] Morton, H.L., E. D. Robison and R. E. Meyer. 1967. Persistence of 2,4-D, 2,4,5-T, and Dicamba in Range Forage Grasses. Weeds 15:268-271.

[26] Ragab, M. T. H. 1975. Residues of picloram in soil and their effects on crops. Can. J. Soil Sci. 55:55-59.

[27] Harrington, T.B. and J.H. Miller. 2005. Effects of application rate, timing, and formulation of glyphosate and triclopyr on control of Chinese privet (Ligustrum sinense). Weed Techn. 19(1):47-54.

[28] Koger, C.H, J.F. Stritzke, and D.C. Cummings. 2002. Control of Sericea Lespedeza (Lespedeza cuneata) with triclopyr, fluroxypyr, and metsulfuron. Weed Techn. 16(4): 893-900.

[29] Boydston, R.A. and M.D. Seymour. 2002. Volunteer potato (Solanum tuberosum) control with herbicides and cultivation in onion (Allium cepa). Weed Techn. 16(3):620-626.

[30] Enloe, S.F, J.M. DiTomaso, S.B. Orloff, and D.J. Drake. 2005. Perennial grass establishment integrated with clopyralid treatment for yellow starthistle management on annual range. Weed Techn. 19(1):94-101.

[31] Scifres, C.F., R.A. Crane, B.H. Koerth and RC. Flinn. 1988. Roller application of clopyralid for huisache, (Acacia farnesiana) control. Weed Technol. 2:317-322.

[32] Bukun, B., D.L. Shaner, S.J. Nissen, P. Westra, and G. Brunk. 2010. Comparison of the interactions of aminopyralid vs. clopyralid with soil. Weed Sci. 58(4):473-477.

[33] Ferrell, J.A., J.J. Mullahey, K.A. Langeland, and W.N. Kline. 2006. Control of tropical soda apple (Solanum viarum) with aminopyralid. Weed Techn. 20(2):453-457.

[34] Enloe, S.F., G.B. Kyser, S.A. Dewey, V. Peterson, and J.M. DiTomaso. 2008. Russian knapweed (Acroptilon repens) control with low rates of aminopyralid on range and pasture. Inv. Plant Sci. and Manage. 1(4):385-389.

[35] Claus, J.S., R.G. Turner, J.H. Meredith, C.S. Williams and M.J. Holiday. 2010. Aminocyclopyrachlor development and registration update. Proc. South Weed Sci. Soc. 63:178.

[36] Bell, J.L., I.C. Burke and T.S Prather. 2011. Uptake, translocation and metabolism of aminocyclopyrachlor in prickly lettuce, rush skeletonweed and yellow starthistle. Pest Manage. Sci. 67(10):1338-1348.

[37] Flessner, M.L., R.R. Dute, and J.S. McElroy. 2011. Anatomical response of St. Augustinegrass to aminocyclopyrachlor treatment. Weed Sci. 59(2):263-269.

[38] Davies PJ, ed. 2010. The plant hormones: biosynthesis, signal transduction, action!, 3rd edn. Dordrecht, The Netherlands: Springer.

[39] Normanly J, Slovin JP, Cohen JD. 2010. Hormone biosynthesis, metabolism and its regulation. In: Davies PJ, ed. Plant hormones:biosynthesis, signal transduction, action! 3rd edn. Dordrecht, The Netherlands: Springer, pp.36–62.

[40] McQueen-Mason, S.J. and D. J. Cosgrove. 1995. Expansin Mode of Action on Cell Walls (Analysis of Wall Hydrolysis, Stress Relaxation, and Binding). Plant Phys. 133:87-100.

[41] Abeles, F.B. 1966. Auxin stimulation of ethylene evolution. Plant Phys. 41:585-588.

[42] Delbarre A, Muller P, Imhoff V, Guern J. 1996. Comparison of mechanisms controlling uptake and accumulation of 2,4-dichlorophenoxy acetic acid, naphthalene-1-acetic acid, and indole-3-acetic acid in suspension-cultured tobacco cells. Planta 198:532–541.

[43] Hošek, P., M. Kubeš, M. Laňková, P.I. Dobrev, P. Klíma, M. Kohoutová, J. Petrášek, K. Hoyerová, M. Jiřina and Eva Zažímalová. 2012. Auxin transport at cellular level: new insights supported by mathematical modeling. J. Exp. Bot. 63(10):3815-3827.

[44] Scifres, C.J., R.R. Hahn and M.G. Merkle. 1971. Dissipation of picloram from vegetation of semiarid rangelands. Weed Sci. 19:329-332.

[45] Boydston, R.A. 1994. Clopyralid persistence in spearmint (Mentha cardiaca) hay injures potato (Solanum tuberosum). Weed Technol. 8:296-298.

[46] Branham, B.E. and D.W. Lickfeldt. 1997. Effect of pesticide-treated grass clippings used as mulch on ornamental plants. HortScience 32:1216–1219.

[47] Aminopyralid Pesticide Fact Sheet. [Online]. United States Office of Prevention, Pesticides, Environmental Protection, and Toxic Substances. (USOPPEPTS) (2005). Available: http://www.epa.gov/opprd001/factsheets/aminopyralid.pdf [8 May 2010]. Accessed 9 February 2013.

[48] Behrens, M.R., N. Mutlu, S. Chakraborty, R. Dumitru, W.Z. Jiang, B.J. LaVallee, P.L. Herman, T.E. Clemente, D.P. Weeks. 2007. Dicamba Resistance: Enlarging and Preserving Biotechnology-Based Weed Management Strategies. Science 25 vol. 316, no. 5828 pp. 1185-1188.

[49] Wright, T.R., S. Guomin, T.A. Walsh, J.M. Lira, C. Cui, P. Song, M. Zhuang, N.L. Arnold, G. Lin, K. Yau, S.M. Russell, R.M. Cicchillo, M.A. Peterson, D.M. Simpson, N. Zhou, J. Ponsamuel, and Z. Zhang. 2010. Robust crop resistance to broadleaf and grass herbicides provided by aryloxyalkanoate dioxygenase transgenes. Proc. Nat Acad. Sci. 107(47):20240-20245.

[50] Debreuil, D.J., L.F. Friesen, J.N. Morrison. 1996. Growth and seed return of auxin-type herbicide resistant wild mustard (Brassica kaber) in wheat. Weed Sci. 44:871-878.

[51] Bernards, M.L., R.J. Crespo, G.R. Kruger, R.E. Gaussoin, P.J. Tranel. 2012. A waterhemp (Amaranthus tuberculatus) population resistant to 2,4-D. Weed Sci. 60:379-384.

[52] Hay, J.V. 1990. Chemistry of the sulfonylurea herbicides. Pest Sci. 29(3):247-261.

[53] Wepplo, P. 1990. Imidazolinone herbicides: Synthesis and novel chemistry. Pest Sci. 29(3):293-315.

[54] Moberg, W.K. and B. Cross. 1990. Herbicides inhibiting branched-chain amino acid biosynthesis. Pest Sci. 29(3):241-246.

[55] DiTomaso, J.M. 2000. Invasive weeds in rangelands: Species, impacts, and management. Weed Sci. 48(2):255-265.

[56] Hutchinson, J.T and K. A. Langeland. 2008. Response of Selected Nontarget Native Florida Wetland Plant Species to Metsulfuron Methyl. J. Aquat. Plant Manage. 46: 72-76.

[57] Blair, A. M. and T. C. Martin. 1988. A review of the activity, fate, and mode of action of sulfonylurea herbicides. Pestic. Sci. 22:195-219.

[58] Boutin, C., H. Lee, E. T. Peart, P. S. Batchelor and R. J. Maguire. 2000. Effects of the sulfonylurea herbicide metsulfuron methyl on growth and reproduction of five wetland and terrestrial plant species. Environ. Toxicol. Chem. 19:2532-2541.

[59] DuPont. 2003. Escort XP Herbicide Special Local Need 24 (c) Labeling for control of Old World climbing fern in Florida. Wilmington, DE. 2 pp. http://www.dupont.com/ag/us/prodinfo/prodsearch/information/H64613.pdf (accessed 9 February 2013).

[60] Brown, H. M. 1990. Mode of action, crop selectivity, and soil relations of the sulfonylurea herbicides. Pestic. Sci 29:263-281.

[61] Walker, A. and S.J. Welch. 1989. The relative movement and persistence in soil of chlorsulfuron, metsulfuron-methyl and triasulfuron. Weed Res. 29(5):375-383.

[62] Holzmueller, E. and S. Jose. 2010. Response of cogongrass to imazapyr herbicides on a reclaimed phosphate-mine site in central Florida, USA. Ecological Rest. 28(3):300-303.

[63] Mozdzer, T.J., C.J. Hutto, P.A. Clarke, and D.P. Field. 2008. Efficacy of Imazapyr and Glyphosate in the Control of Non-Native Phragmites australis. Restoration Ecology 16(2):221-224.

[64] Zuver, K.A., M.L. Bernards, J.J. Kells, C.L. Sprague, C.R. Medlin, and M.M. Loux. 2006. Evaluation of postemergence weed control strategies in herbicide-resistant isolines of corn (Zea mays). Weed Technol. 20(1):172-178.

[65] Barnes, T.G. 2004. Strategies to convert exotic grass pastures to tall grass prairie communities. Weed Techn. 18(1):1364-1370.

[66] Markle, D.M. and R.G. Lym. 2001. Leafy spurge (Euphorbia esula) control and herbage production with imazapic. Weed Techn. 15(3):474-480.

[67] Ruffner, M.E. and T.G. Barnes. 2010. Natural grassland response to herbicides and application timing for selective control of tall fescue, an invasive cool-season grass. Inv. Plant Sci. and Manage. 3(3):219-228.

[68] Shinn, A.I. and D.C. Thill. 2002. The response of yellow starthistle (Centaurea solstitialis), annual grasses, and smooth brome (Bromus inermis) to imazapic and picloram. Weed Techn. 16(2):366-370.

[69] Shinn, A.I. and D.C. Thill. 2003. The response of yellow starthistle (Centaurea solstitialis), spotted knapweed (Centaurea maculosa), and meadow hawkweed (Hieracium caespitosum) to imazapic. Weed Techn. 17(1):94-101.

[70] SePro 2012. http://www.sepro.com/documents/Clearcast_Label.pdf (accessed 9 February 2013).

[71] Loux, M.M. and K.D. Resse. 1993. Effect of soil type and pH on persistence and carryover of imidazolinone herbicides. Weed Technol. 7:452-458.

[72] Pester, T.A., S.J. Nissen, and P. Westra. 2001. Absorption, translocation, and metabolism of imazamox in jointed goatgrass and feral rye. Weed Sci. 49(5):607-612.

[73] Stidham, M.A. 1991. Herbicides that inhibit acetohydroxyacid synthase. Weed Sci. 39:428-434.

[74] Barak, Z., N. Kogan, N. Gollop, and D.M. Chipman. Importance of AHAS isoenzymes in branched chain amino acid biosynthesis. Z. Barak, D.M. Chipman, and J.V. Schloss (Eds.), Biosynthesis of Branched Chain Amino Acids, VCH Publishers, New York (1990), pp. 91–109.

[75] McCourt, J.A., S.S. Pang, J. King-Scott, L.W. Guddat and R.G. Duggleby. 2006. Herbicide-binding sites revealed in the structure of plant acetohydroxyacid synthase. Proc. Nat Acad. Sci. 103(3):569-573.

[76] Lee, J.M. and M.D.K. Owen. 2000. Comparison of acetolactate synthase enzyme inhibition among resistant and susceptible Xanthium strumarium biotypes. Weed Sci. 48(3): 286-290.

[77] Siminszky, B. 2006. Plant cytochrome P450-mediated herbicide metabolism. Phytochem. Rev. 5:445-458.

[78] Kreuz, K., R. Fonne-Pfister and P.J. Porpiglia. 1991. Organophosphorus insecticides as inhibitors of cytochrome P450-dependent sulfonylurea herbicide metabolism. Abstr. Am. Soc. Plant Physiol. 96:27.

[79] Tranel, P.J. and T.R. Wright. 2002. Resistance of weeds to ALS-inhibiting herbicides: what have we learned? Weed Sci. 50(6):700-712.

[80] Sibony, M. and B. Rubin. 2003. The ecological fitness of ALS-resistant Amaranthus retroflexus and multiple-resistant Amaranthus blitoides. Wed Res. 43(1):40-47.

[81] Draber, W., K. Tietjen, J.F. Kluth and A. Trebst. 1991. Herbicides in Photosynthesis Research. Angew. Chem. Int. Engl. 30:1621-1633.

[82] Buchholtz, K.P. and R.E. Doersch. 1968. Cultivation and herbicides for weed control in corn. Weeds. 16:232-234.

[83] Larsen, R.P. and S.K. Reis. 1960. Simazine for controlling weeds in fruit tree and grape plantings. Weeds. 8:671-677.

[84] Walker, C.R. 1964. Simazine and other s-triazine compounds as aquatic herbicides in fish habitats. Weeds 12(2):134-139.

[85] Muller, G. 2008. History of the discovery and development of triazine herbicides. In: McFarland, J.E. and O.C. Burnside, eds. The Triazine Herbicides. 1st edn. Oxford, England: Elsevier, pp. 13-30.

[86] Ferrell, J.A. 2012. Weed Management in Rights-of-Way and Non-Cropped Areas. Agronomy Department, Florida Cooperative Extension Service, Institute of Food and Agricultural Sciences, University of Florida. SS-AGR-111. http://edis.ifas.ufl.edu/pdffiles/WG/WG06800.pdf (accessed 9 February 2013).

[87] Minogue, P.J., B.R. Zutter and D.H. Gjerstad. 1988. Soil factors and efficacy of hexazinone formulations for loblolly pine (Pinus taeda) release. Weed Sci. 36(3):399-405.

[88] Wilder, B., J.A. Ferrell, B.A. Sellers, and G.E. MacDonald. 2008. Influence of hexazinone on 'Tifton 85' bermudagrass growth and forage quality. Weed Technol. 22(3): 499-501.

[89] Eshel, Y. 1969. Tolerance of cotton to diuron, fluometuron, norea, and prometryne. Weed Sci. 17(4):492-496.

[90] Norton, J.A. and J.B. Storey. 1970. Effect of herbicides on weed control and growth of pecan trees. Weed Sci. 18(4):522-524.

[91] Pettit, R.D. 1979. Effects of picloram and tebuthiuron pellets on snad shinnery oak communities. J. Range Manage. 32(3):196-200.

[92] Scifres, C.J., J.L. Mutz and W.T. Hamilton. 1979. Control of mixed brush with tebuthiuron. J. Range Manage. 32(2):155-158.

[93] Johnson, K.H., R.A. Olson and T.D. Whitson. 1996. Composition and Diversity of Plant and Small Mammal Communities in Tebuthiuron-Treated Big Sagebrush (Artemisia tridentata). Weed Technol. 10(2):404-416.

[94] Waldrop, T.A. and L.F. Thomas. 1988. Precommercial Thinning a Sapling-Sized Loblolly Pine Stand with Fire. South. J. of Applied Forestry. 12(3):203-207.

[95] Takahashi K., Y. Sakai, Y.Harada, and K. Hirose. 1978. Effects of ten years application with bromacil in citrus (satsuma mandarin) orchard. 1. Effects of bromacil on annual variations of species and population. Weed Res. Japan 22(4):198-202.

[96] Bovey, R.W., R.E. Meyer, and J.R. Baur. 1981. Potential herbicides for brush control. J. Range Mange. 34(2):144-148.

[97] Buchel, K.H. 1972. Mechanisms of action and structure activity relations of herbicides that inhibit photosynthesis. Pest. Sci. 3(1):89-110.

[98] McNeil, W.K., J.F. Stritzke, and E. Balser. 1984. Absorption, translocation, and degradation of tebuthiuron and hexazinone in woody species. Weed Sci. 32:739-743.

[99] Yanase, D. and A. Andoh. 1992. Translocation of photosynthesis-inhibiting herbicides in wheat leaves measured by phytofluorography, the chlorophyll fluorescence imaging. Pest. Biochem. and Physiol. 44(1):60-67.

[100] Gardiner, J.A., R C. Rhodes, J.B. Adams, Jr., and E.J. Soboczenski. 1969. Synthesis and studies with 2 –C14 -labeled bromacil and terbacil. J. Agric. Food Chem. 17:980-986.

[101] Chang, S.S. and J.F. Stritzke. 1977. Sorption, movement and dissipation of tebuthiuron in soils. Weed Sci. 25(2):184-187.

[102] Scifres, C.J., J.W. Stuth and R.W. Bovey. 1981. Control of oaks (Quercus spp.) and associated woody species on rangeland with tebuthiuron. Weed Sci. 29(3):270-275.

[103] Laband, D.N., W.C. Morse, S.A. Enebak, A.H. Chappelka. 2012. The Toomer's Oaks tragedy and the importance of cultural environmental services. South. J. Applied For. 36(4):220-222.

[104] Lawlor, D.W. 1987. Photosyntheis: metabolism, control and physiology. Longman Scientific and Technical, Longman Group Limited Essex CM20 2JE, England. 262 p.

[105] Dodge, A.D. 1982. The Role of Light and Oxygen in the Action of Photosynthetic Inhibitor Herbicides. In: D.E. Moreland, J.B. St. John and F.D. Hess, eds. Biochemical Responses Induced by Herbicides. Amercian Chemical Society, pp.57-77.

[106] Shimabukuro, R.H., D.S. Frear, H.R. Swanson and W.C. Walsh. 1971. Glutathione Conjugation - an enzymatic basis for atrazine resistance in corn. Plant Physiol. 47:10-14.

[107] Franz, J.E., M.K. Mao, and J.A. Sikorski. 1997. Glyphosate: a unique global herbicide. American Chemical Society. 653pp. ISBN 0-8412-3458-2.

[108] Johnson, B.J. 1976. Glyphosate for weed control in dormant bermudagrass. Weed Sci. 24(1):140-143.

[109] Stougaard, R.N., G. Kapusta and G. Roskamp. 1984. Early preplant herbicide applications for no-till soybean (Glycine max) weed control. Weed Sci. 32(4):293-298.

[110] Younce, M.H. and W.A. Skroch. 1989. Control of selected perennial weeds with glyphosate. Weed Sci. 37(3):360-364.

[111] Zutter, B.R., P.J. Minogue, D.H. Gjerstad. 1988. Response Following Aerial Applications of Glyphosate for Release of Loblolly Pine in the Virginia Piedmont. South. J. Applied For. 12(1):54-58.

[112] Devine, M., S.O. Duke, and C. Fedtke. 1993. Physiology of Herbicide Action. PTR Prentice-Hall Inc. Englewood Cliffs, NJ. 441 p.

[113] Hollander, H. and N. Amrhein. 1980. The site of the inhibition of the shikimate pathway by glyphosate - I. inhibition by glyphosate of phenylpropanoid synthesis in buckwheat (Fagopyrum esculentum Moench). Plant Physiol. 66:823-829.

[114] Zandstra, B.H. and R.K. Nishimoto. 1977. Movement and activity of glyphosate in purple nutsedge. Weed Sci. 25(3):268-274.

[115] Bradshaw, L.D., S.R. Padgette, S.L. Kimball and B.H. Wells. 1997. Perspectives on glyphosate resistance. Weed Technol. 11(1):189-198.

[116] Putnam, A.R. 1976. Fate of glyphosate in deciduous fruit trees. Weed Sci. 24(4):425-430.

[117] Heap, I. The International Survey of Herbicide Resistant Weeds. www.weedscience.org (accessed 9 February 2013).

[118] Kitchen, L.M., C.E. Rieck, W.W. Witt. 1980. Absorption and translocation of 14C fosamine by three woody plant species. Weed Res. 20(5):285-289.

[119] Han, J.C-Y. 1979. Stability of 14C fosamine ammonium in water and soils. J. Agric. Food Chem. 27(3):564-571.

[120] Mann, R.K., W.W. Witt and C.E. Rieck. 1986. Fosamine Absorption and Translocation in Multiflora Rose (Rosa multiflora). Weed Sci. 34(6):830-833.

[121] Morey, P.R. and B.E. Dahl. 1980. Inhibition of mesquite (Prosopis juliflora var. glandulosa) growth by fosamine. Weed Sci. 28(3):251-255.

[122] Rendina, A.R. and J.M. Felts. 1988. Cyclohexanedione herbicides are selective and potent inhibitors of acetyl-CoA carboxylase from grasses. Plant Physiol. 86(4):983-986.

[123] Rendina, A.R., A.C. Craig-Kennard, J.D. Beaudoin, M.K. Breen. 1990. Inhibition of acetyl-coenzyme A carboxylase by two classes of grass-selective herbicides. J. Agric. Food Chem. 38(5):1282-1287.

[124] Hovich, S.M. and J.A. Reinartz. 2007. Restoring forest in wetlands dominated by reed canarygrass: The effects of pre-planting treatments on early survival of planted stock. Wetlands 27(1):24-39.

[125] Judge, C.A., J.C. Neal, and J.F. Derr. 2005. Preemergence and postemergence control of Japanese stiltgrass (Microstegium vimineum). Weed Technol. 19(1):183-189.

[126] MacDonald, G.E. 2004. Cogongrass (Imperata cylindrica) – Biology, ecology and management. Crit. Rev. in Plant Sci. 23(5):367-380.

[127] Weber, J.B., G.G. Wilkerson, H.M. Linker, J.W. Wilcut, R.B. Leidy, S. Senseman, W.W. Witt, M. Barrett, W.K. Vencill, D.R. Shaw, T.C. Mueller, D.K. Miller, B.J. Brecke, R.E. Talbert, and T.F. Peeper. 2000. A proposal to standardize soil/solution herbicide distribution coefficients. Weed Sci. 48(1):75-88.

[128] Shelp, B.J., C.J. Swanton, B.G. Mersey, J.C. Hall. 1992. Glufosinate (phosphinothricin) inhibition of nitrogen metabolism in barley and green foxtail plants. J. Plant Physiol. 139:605–610.

[129] Wild, A., H. Sauer, and W. Ruhle. 1987. The effect of phosphinothricin (glufosinate) on photosynthesis. I. Inhibition of photosynthesis and accumulation of ammonia. Z. Naturforsch., 42 (1987), pp. 263–269.

[130] Funderburk, Jr., H.H. and J. M. Lawrence. 1964. Mode of Action and Metabolism of Diquat and Paraquat. Weeds 12(4):259-264.

[131] Lehoczki, E., G. Laskay, I. Gaal, and Z. Szigeti. 1992. Mode of action of paraquat in leaves of paraquat-resistant Conyza canadensis (L.) Cronq. Plant, Cell & Environ. 15:531–539.

[132] Vaughn, K. C. and S.O. Duke. 1983. In situ localization of the sites of paraquat action. Plant, Cell & Environ. 6:13–20.

[133] Sellers, B.A., R.J. Smeda, and J. Li. 2004. Glutamine synthetase activity and ammonium accumulation is influenced by time of glufosinate application. Pest. Biochem. and Physiol. 78(1):9-20.

[134] Clewis, S.B., W.J. Everman, D.L. Jordan, and J.W. Wilcut. 2007. Weed management in North Carolina peanuts (Arachis hypogaea) with S-metolachlor, diclosulam, flumioxazin, and sulfentrazone systems. Weed Technol. 21(3):629-635.

[135] Fausey, J.C. and K.A. Renner. 2001. Broadleaf weed control in corn (Zea mays) and soybean (Glycine max) with CGA-248757 and flumiclorac alone and in tank mixtures. Weed Technol. 15(3):399-407.

[136] Peachy, E., D. Doohan and T. Koch. 2012. Selectivity of fomesafen based systems for preemergence weed control in cucurbit crops. Crop Prot. 40:91-97.

[137] Kumert, K.J., G. Sandmann and P. Boger. 1987. Modes of action of diphenylether herbicides. Rev. Weed Sci. 3:35-55.

[138] Duke, S.O., J. Lydon, J.M. Becerril, T.D. Sherman, L.P. Lehnen, Jr. and H. Matsumoto. 1991. Protoporphyrinogen Oxidase-Inhibiting Herbicides. Weed Sci. 39(3):465-473.

[139] Dayan, F.E., H.M. Green, J.D. Weete and H.G. Hancock. 1996. Postemergence Activity of Sulfentrazone: Effects of Surfactants and Leaf Surfaces. Weed Sci. 44(4):797-803.

[140] Hess, F.D. 2000. Light-dependent herbicides: an overview. Weed Sci. 48(2):160-170.

[141] Wright, T.R., E. P. Fuerst, A.G. Ogg, Jr., U.B. Nandihalli and H.J. Lee. 1995. Herbicidal activity of UCC-C4243 and acifluorfen is due to inhibition of protoporphyrinogen oxidase. Weed Sci. 43(1):47-54.

[142] Jacobs, J. M. and N.J. Jacobs. 1993. Porphyrin accumulation and export by isolated barley (Hordeum vulgare) plastids. Plant Physiol. 101:1181–1187.

[143] Ferrell, J.A., K.A. Langeland, and B.A. Sellers. 2012. Herbicide Application Techniques for Woody Plant Control. Agronomy Department, Florida Cooperative Extension Service, Institute of Food and Agricultural Sciences, University of Florida. SS-AGR-260. http://edis.ifas.ufl.edu/ag245 (accessed 9 February 2013).

[144] Fishel, F.M., J.A. Ferrell, G.E. MacDonald, and B.J. Brecke. Florida's Organo-Auxin Herbicide Rule – 2012. Department, Florida Cooperative Extension Service, Institute of Food and Agricultural Sciences, University of Florida. SS-AGR-12. http://edis.ifas.ufl.edu/wg051 (accessed 9 February 2013).

[145] Spillman, J. 1982. Atomizers for the aerial application of herbicides—ideal and available. Crop Prot. 1(4):473-482.

[146] vanPelt, N.S. and N.E. West. 1990. Effects of manual application method on application time, thoroughness and efficacy of tebuthiuron. J. Range Manage. 43(1):39-42.

[147] Miller, J.H., S.T. Manning, and S.F. Enloe. 2010. A management guide for invasive plants in southern forests. Gen. Tech. Rep. SRS-131. Asheville, NC: U.S. Department of Agriculture Forest Service, Southern Research Station. 120 p.

[148] Nelson, L.R., A.W. Ezell, and J.L. Yeiser. 2006. Imazapyr and triclopyr tanks mixes for basal bark control of woody brush in the southeastern United States. New For. 31:173-183.

[149] Kochenderfer, J.D., J.N. Kochenderfer, and G.W. Miller. 2012. Manual herbicide application methods for managing vegetation in Appalachian hardwood forests. Gen. Tech. Rep. NRS 96. Newtown Square, PA: U.S. Department of Agriculture, Forest Service, Northern Research Station. 59 p. (http://www.nrs.fs.fed.us/pubs/gtr/gtr_nrs96.pdf (accessed 9 February 2013).

[150] Langeland, K.A., J.A. Ferrell, B.A. Sellers, G.E. MacDonald, and R.K. Stocker. 2011. Integrated Management of Nonnative Plants in Natural Areas of Florida. Department, Florida Cooperative Extension Service, Institute of Food and Agricultural Sciences, University of Florida. SP 242. http://edis.ifas.ufl.edu/wg209 (accessed 9 February 2013).

[151] Kossuth, S. V., J.F. Young, J.E. Voeller, H.A. Holt. 1982. Year-Round Hardwood Control Using the Hypo-Hatchet Injector. South. J. Applied For. 4(2):73-76.

[152] Bovey, R.W., T. O. Flynt, R. E. Meyer, J. R. Baur and T. E. Riley. 1976. Subsurface Herbicide Applicator for Brush Control. J. Range Manage. 29(4):338-341.

[153] Leary, J.J.K. 2011. Standing Operating Procedures (SOP) for Herbicide Ballistic Technology Operations: Ground and Aerial Herbicide Application. University of Hawaii at Manoa. http://manoa.hawaii.edu/hpicesu/SAFETY/SOP-32_HBT.pdf (accessed 9 February 2013).

[154] Leary, J.J.K., J. Gooding, J. Chapman, A. Radford, B. Mahnken, and L. J. Cox. 2013. Calibration of an Herbicide Ballistic Technology (HBT) helicopter platform targeting Miconia calvescens DC in Hawaii. Invasive Plant Science and Management In-Press.

Chapter 10

Integrated Weed Management Practices for Adoption in the Tropics

Wendy-Ann P. Isaac, Puran Bridgemohan and
Wayne G. Ganpat

Additional information is available at the end of the chapter

http://dx.doi.org/10.5772/55950

1. Introduction

The earth is undergoing a number of irreversible changes as a result of the activities of man, many of which are adversely affecting the environment. Inappropriate methods of agricultural production, especially those stimulated by efforts in pursuit of short-term gains, have been identified as prime contributors to this environmental degradation.

In earlier times, traditional farming in the tropics involved the use of natural resources in adequate quantity for the sustenance of its population, without diminishing the natural resource base. Key elements of that system included multiple cropping and mixed farming, minimum tillage and water conservation techniques, the use of simple hand tools and other low input technologies.

These sustainable farming methods have been described in pejorative terms as drudgery, laborious, and inefficient. Many have been rejected and new technologies and other high energy based inputs have been embraced. These technologies are costly and heavily foreign-exchange dependent. They also disturb the delicate ecological balance resulting in increased occurrence of pests and diseases, shift in noxious weed populations, soil erosion and pollution of the air and water resources.

The situation in the tropical world is exacerbated as many tropical countries are characterised by conditions that are ideal for the prolific growth and development of a range of plant species. Many of these species are generally non-harmful. However, when inappropriate methods of weed control and/or poor crop management strategies are employed, weeds assume noxious potentials. Ready examples are corn grass (*Rottboellia cochinchinensis*), white-top (*Parthenium hysterophorus*) and nutgrass (*Cyperus rotundus*) [33].

Present farming involves substantial reliance on a range of manufactured inputs. The high dependence on herbicides for weed control in the cultivation of rice, maize, bananas, citrus, sugarcane, onions, white potatoes and vegetable crops is not unnoticed. The competition among suppliers of herbicides has resulted in lower costs of these products which has fuelled their use and abuse in the region.

Low-input, sustainable agriculture addresses multiple objectives from increasing profits to maintaining the environment, and builds on multiple systems as integrated pest management (IPM), integrated weed management (IWM), and crop rotation. Integrated weed management involves the combination of a number of weed control practices that reduces the dependence on any one type of control method and also lowers the input of herbicides. This approach is important for the control of perennial weeds that are inadequately controlled by any single method [8]. The application of IWM also includes the knowledge of past annual and perennial weed populations in fields and weed seed bank [7], competitive crop cultivars, improved crop and soil management practices, and appropriate selection of herbicides [52]. In the context of sustainable agriculture, the concept of IWM seems enlightening and applicable.

The objective of this paper is to discuss the various weed management practices for the control of noxious weeds in major cereal, root and vegetable crops in tropical sustainable agriculture and the strategies used over time to promote their adoption by small farmers.

1.1. Common Integrated Weed Management (IWM) strategies in the tropics

Integrated weed management systems are based on an agro-ecosystem approach for the management and control of weeds at economic threshold levels [12]. Many farmers in the tropics today practise the same weed control measures as was practised before the introduction of herbicides [35]. The IWM systems approach includes any or a combination of the following practices that give a crop a comparative advantage in competing with weeds.

1.1.1. Prevention strategies

Prevention strategies include field sanitation and harvesting methods that do not spread weed seeds and vegetative propagules at every step of production (such as seed selection, field preparation, planting, fertilization, irrigation, weed control, harvest and transport) [19]. Such strategies can significantly reduce the infestation of noxious weeds such as nutgrass and white-top [7]. The use of clean crop seed, especially those direct seeded, e.g., maize and legumes, is critical in the prevention of weed problems in new and existing fields. Prevention should be a daily activity, incorporated into the routine of all workers involved in agricultural production, at farm, state and national levels [19]. It is recommended that managers make simple, cost effective modifications to their farm practices to mitigate the risk of introducing new weed seeds to the field. Some of the key considerations as outlined in [19] include:

- diligent monitoring for sources of new weed introductions to the agro-ecosystem;
- proactive government laws and regulations controlling the introduction and movement of plant materials or soil from one location to another;

- destroying vegetative propagules of perennial weeds;
- whenever possible, depleting the soil weed seed bank;
- propagating seeds and seedling transplants in media free of weed propagules;
- preventing weeds from going to seed in crop fields;
- cleaning farm machinery before movement into fields;
- minimizing the presence of weed seed in livestock feed, manures and composts;
- preventing weed seed introduction into rivers and irrigation canals.

For preventive strategies to be fully adopted in an IWM approach, there must be an attitudinal change by farmers and agricultural educators in the tropics. Prevention, although complex, is a very efficient technique for any property size at all crop production stages, from the acquisition of machinery, seed, water and fertilizers, to crop harvest and processing.

1.1.2. Competitive crops and/or smother crops

Crops differ in their competitiveness with weeds based on their emergence, leaf-area expansion, light interception, canopy architecture, leaf-angle, shape and competitiveness. Within a crop species, cultivars may vary in their competitiveness. While the improved varieties may be high yielding, the traditional varieties exhibit multiple adaptations, competitive ability against weeds and require less agricultural input. The use of competitive crops to discourage weeds is an important IWM strategy. To maximise crop production by minimising the impact of weeds, replacement series and addition series designs have been recommended for intercrop, cover crop and green manure selection [41].

Plant height and leaf area index correlate with competitive ability in row crops. These characters allow the crop to outgrow and cover the weeds. Indeterminate varieties of bean, cowpea, squash and cucumber appear to be better competitive than determinate varieties [38, 39]. The indeterminate varieties of these crops have a vining or spreading habit which allows rapid canopy closure, thus suppressing emerging weeds.

Some plants are able to exude chemical substances which suppress the growth of other neighbouring plants. Research in plants with allelopathic potential is ongoing and has revealed a clearer understanding into the genetics of allelopathic activity in certain crops [29].

Smother crops are quickly established and usurp the resources that weeds would otherwise use. The suppression of weeds may be through both competition (resources) and allelopathy [11]. Smother crops include cowpea (*Vigna unguiculata),* forage soya beans (*Glycine max* L.Merril), Sudan grass (*Sorghum bicolor* subsp. *drummondii*), kudzu (*Pueraria phaseoloides*) and pumpkins (*Cucurbita maxima*), which are very effective in suppressing nutgrass (*Cyperus rotundus*) and small broadleaved weeds.

1.1.3. Optimum plant population

Row spacing and seeding rate may influence the ability of the crop to compete with weeds for resources and, therefore, may affect weed management [24, 45]. The rapid closure of the crop canopy can be obtained with a reduction in row spacing [1], an increase in seeding rate [42], and selection of varieties with traits that favour rapid canopy development [13]. It has been reported that rows of 38 cm or less could increase yields and reduce tillage and herbicide requirements because of faster canopy closure [1]. Cereal and vegetable crops can compete with weed growth if they are established at the optimum plant population that allows them to more effectively usurp resources. If crops can reduce incident light by 50 % or more, weeds will seldom become a problem [6, 15]. This approach requires closer intra-and interrow spacings and higher crop densities than normally used.

1.1.4. Cover crops and mulches

Cover crops have long been used extensively in the tropics for soil and water conservation, to maintain soil structure and enhance soil fertility, especially on steep or difficult terrain. They are often referred to as living mulches. The use of leguminous cover plants to suppress weeds in plantation crops in the tropical world dates back many decades, but the integration of the legumes into arable cropping systems has not been developed to a level acceptable to farmers. Cover crops also contribute to pest management and help to suppress unwanted weeds. Its use has been mainly in plantation crops. The introduction of inexpensive nitrogen fertilizers and herbicides encouraged many farmers to discontinue this practice. Cover crops can be intercropped or interplanted with a crop of economic significance. They work by excluding light and limiting weed emergence. Examples of cover crops in the tropics include: *Mucuna pruriens* (L.) DC. (velvet beans), *Desmodium heterocarpon* var *ovalifolium* and *Arachis pintoi* Crap. & Greg. (wild or perennial peanut).

Mulches, on the other hand, may be in the natural form of plant or crop residues or in synthetic form as plastic films or woven synthetic fibres. Other non-living mulches can be either natural materials (plant leaves, stalks, straw, compost and dry soil) or synthetic materials, such as polyethylene, which are used widely in pineapple production. The major disadvantages of plastic films are material costs and difficulty in removal after cropping season. Organic mulches or living mulches are considered cover crops, e.g., mungbean (*Vigna radiata* (L.) Wilczek cv. Local) and have been shown to be an economical alternative to synthetic mulches [36]. Watermelon and tomato farmers in Dominica, West Indies use Guinea grass (*Panicum fasiculatum*) as a mulch and cover crop. The grass is killed using a weed killer such as paraquat, and when it re-grows, it is brush-cut before crop emergence or otherwise left as a residue. The crop is planted directly into the cover crop residue which enhances soil and water conservation and protection from wind.

In root crops, for example cassava, live green legumes e.g. *Desmodium heterophyllum* (Willd) DC [20] with bean *(Phaseolus* sp.) have been used successfully. Both legumes gave better weed control and crop yields than the herbicide and mulch treatments and *Desmodium heterocarpon* var *ovalifolium* in banana [25]. *Stylosanthes guianensis* (Aubl.) SW, too, has been used as a cover crop to suppress weeds in cassava [43, 44]. Legume and dry mulch covers are beneficial because they improve soil organic matter and nutrient status, prevent erosion and suppress weeds [30]. The use of legume covers is, however, expensive because of the cost of seeds and labour for their establishment [53, 56]. It is important to use legume and other crop covers which will not compete with the crop for resources. Moreover, any crop cover used must directly benefit the farmer if adoption of the practice is to be sustained.

Some of the weed species that are easily smothered by live legume covers include: *Ageratum conyzoides* L., *Alternenthera sessilis* L. R. Br. ex Roth, *Mimosa invisa* Mart, *Digitaria orizontalis* Willd, and *Panicum maximum* Jacq. However, some sedges and grasses like *Cyperus rotundus* L (purple nutsedge), *Rottboellia cochinchinensis* (Lour) Clayton (Raoul grass), *Sorghum halepense* L. Pers (Johnson grass) and *Ipomoea* sp. (morning glory) are noxious weeds and difficult to control in root and cereal crops [20, 46].

Both cover crops and mulches offer great agro-ecological potential. They serve as a physical barrier against weed emergence, both conserve the soil and improve the ecological balance of the soil, enhance crop yield and provide several environmental services. These new technologies, however, are not easily accepted by small farmers in the tropics. Notwithstanding, they offer a complex combination of interrelated practices which include: (i) necessary practices so as to ensure the production and retention of sufficient mulch and (ii) complementary practices in order to be able to grow a crop and/or maintain yield levels. This typically implies several adaptations to the entire farm production system. Whether mulching actually is a viable component for smallholder conservation farming in developing countries depends on a number of factors, including bio-physical, technological, farm level and institutional factors. The combination of these factors determines the feasibility of and the economic returns to mulching practices—and thereby farmer acceptance.

The development and dissemination of cover crops and mulches for small farmers in tropical developing countries highlights a number of promising experiences, particularly among banana growers i in St. Vincent, in the Caribbean [25]. The technology offers significant savings through reduced tillage and alleviation of some major crop production constraints such as water conservation, timeliness of land preparation and crop establishment.

1.1.5. Improved husbandry

The basic principles of IWM which include: suppression of weed growth, prevention or suppression of weed seed production, reduction in weed seed bank and prevention or reduction in weed spread, are key elements of all improved husbandry practices. All crop husbandry practices, particularly precision placement and timing of fertiliser application, enhance maximum stimulation of the crop and minimum stimulation of the weed population. Additionally, the use of clean certified seeds, clean farm implements, effective seedbed preparation and seeding methods that improve crop growth, all reduce weed competition [7, 36]. Other management practices including: cultural weed control (intercropping, early planting, optimum plant crop density, and tillage), chemical (minimum herbicide) weed control, mechanical weed control and hoe weeding, have been shown to reduce the competitive effects of weeds on vegetable and cereal crops growth, development and yield [36].

Weeds have a life cycle synchronised to that of the crop such that more weeds emerge with the crop with the onset of rains [33]. Intermittent wetting and drying of weed seeds brought about by early rains preceded by dry spells break seed dormancy [8]. Tillage operations bring buried weed seeds to the surface where they germinate. However, early planting of the crop gives the crop a competitive advantage over the weeds [33, 35].

1.1.6. Irrigation practices

Judicious irrigation practices such as the use of clean water, channels and canals, can reduce the spread of weed seeds to uninfested fields [3, 38]. Flooding is an important component of weed management in rice in the tropical world. In irrigated and flooded systems, the environment in which weed seeds have to germinate is characterized by the existence of low oxygen concentrations. Differential responses between rice and weeds to flooding could be an important component of weed management for the direct-seeded rice crop, since rice is tolerant to flooding, but many weeds, e.g., *Cyperus iria, Fimbristylis miliacea, Leptochloa chinensis, Ludwigia hyssopifolia* and similar weed species are not. However, the timing, duration, and depth of flooding and intensity and frequency of irrigation are critical if germination and growth of a number of weed species are to be effectively suppressed.

Irrigated and upland rice and cereal crops are typically grown with few agricultural inputs. A wide range of weeds infest upland rice, many of which are pan-tropical, including the grass weeds: *Digitaria* spp., *Echinochloa colona, Eleusine indica, Paspalum* spp., and *Rottboellia cochinchinensis,* and the broadleaf weeds: *Commelina* spp., *Ageratum conyzoides, Portulaca oleracea, Amaranthus* spp. and *Euphorbia* spp. The variability of weed species composition in upland rice tends to be greater than in the other production systems, and is dependent upon the ecology the cropping system and the management practices used.

Once weed seedlings have emerged and passed the seedling stage, their growth will not be reduced by flooding. In an irrigated environment, there was no emergence of *Leptochloa chinensis* when rice was flooded 5 days after seeding, but its emergence increased to more than 70 plants m^{-2} when flooding was delayed until 20 days after seeding. In such situations where water is not readily available, early flooding would make the best use of water to control weeds. Introducing flooding after herbicide application or weeding or hoeing could help reduce future weed growth and the need for additional interventions [17, 27].

1.1.7. Inter-row cultivation and minimum tillage

Inter-row cultivation is practical in widely spaced row crops, such as maize, vegetables, sugarcane and banana [8, 19], which have interrow distances of 60 cm or more. Interrow cultivations are done by tractor drawn implements or hand operated rotary tillers. The efficiency of this method is higher than manual methods. Minimum tillage, on the other hand, involves the use of the minimum amount of tillage required for crop production for meeting the tillage requirement under existing soil and climatic conditions [56]. It refers to eliminating excess tillage, e.g., reducing four secondary tillage steps to two [8]. Both operations complement and enhance the efficiency of minimum herbicide input through soil incorporation of pre-plant herbicides.

Successive inter-row cultivation has been effective in reducing weed growth and density. Many weed species exhibit morphological plasticity in response to environmental variation and density. Weeds can compensate for density changes so that total biomass per unit area is held relatively constant. Inter-row cultivation improved crop yield by 33% and 78% [20]. However, herbicide application resulted in yield increases of 57 to 300% [27]. Benefits from inter-row cultivation can be limited by in-row weed growth. Most uncontrolled weed growth occurs in the uncultivated area adjacent to and within the crop row. Therefore, the integration of other mechanical or cultural methods often improves with inter-row cultivation. Inter-row cultivation also has potential as a means of controlling late flushes of weeds, but it should not be considered a stand-alone weed management technique since significant in-row weed growth may limit benefits [7].

Tillage operations have a major impact on distribution of weeds in the soil, weed survival and persistency [8, 10], weed species diversity in a given cropping system [15] and the selection pressure on the weed population. Although not much research has gone into the effect that tillage has on tropical weeds, studies have shown that grass weeds, *Setaria* spp and *Corchorus tridens* were higher under the ripper and basins compared to conventional tillage. Also, broadleaf weeds were less in minimum tillage compared to conventional tillage. Rotation with conventional tillage systems controls the grasses and perennials but other weeds or weed groups may assume numerical dominance. To balance the pressure of tillage, there may be need to consider rotational tillage where appropriate [35].

Tillage affects vertical weed seed distribution in a soil profile and this seed distribution affects weed seed germination by influencing the soil environment surrounding the seeds [12, 13]. There is less soil disturbance with minimum or zero-till systems and, as such, most of the weed seeds are on or near the soil surface after crop planting. In systems with high soil disturbance using conventional tillage, mixing weed seeds uniformly in the tilled soil depth has been found to be beneficial. It was also found that on direct-seeded rice, 77% of the weed seeds were retained in the top 2 cm soil layer under a zero-till system, whereas soil disturbance under a conventional tillage system resulted in 62% of the seeds being buried to a depth of 2-5 cm. The seeds were not present in the 5-10 cm soil layer in the zero-till system [22, 23].

The conditions for seed germination are conducive near the soil surface and therefore there is high germination of the weed seeds that are close to the soil surface under zero-till systems, for example, *Ageratum conyzoides, Eclipta prostrata, Echinochloa colona, Digitaria ciliaris* and *Portulaca oleracea*. The weed seed populations on the top that are not dormant are easily destroyed by the stale seedbed practice. In this practice, weed seeds are allowed to germinate after a light irrigation or shower and are then killed by using a non-selective herbicide or shallow tillage. This practice helps to reduce the size of the weed seed bank in the soil [8]. Conservation agriculture, or zero tillage farming, is an effective solution to stopping agricultural land degradation, for rehabilitation, and sustainable crop production intensification in the tropics [21, 22].

1.1.8. Minimum herbicides

Herbicide use continues to be one of the most important tools in weed management. However, an IWM approach creates an opportunity to reduce herbicide rates and in some instances, just forgo the use of herbicides altogether.

Given the high cost of herbicides in the tropics, smallholders sometimes either reduce the herbicide rate or mix with other herbicides with differing modes of action. These practices are not without risk. Oftentimes, smallholders realise that these practices are inconsequential and there is no recourse with pesticide retail outlets regarding poor herbicide performance if label rates have not been followed. Yet, farmers often cut rates as a cost saving strategy.

The effectiveness of a reduced rate usually depends on the type of herbicide, weed species present, weed pressure, environmental conditions and, of course, the competitiveness of the crop stand. If the weed pressure is high or the weeds are under stress, it is probably advisable to use an integrated approach. However, reduced rates of herbicide may lead to some level of herbicide resistance and thus the approach to be taken must be carefully considered.

The extent of herbicide use in the tropics is closely related to the cost and availability of labour. Large scale rice and banana production in the tropics receive more than two herbicide applications. However, in the smaller farms, only about 50% of the rice area is treated, particularly where rural labour is available. Herbicides replace hand weeding and enable direct seeding which is less labour demanding, compared to transplanting. Herbicides are also used in the transplanted systems, though to a much lesser extent, and in systems particularly where crop rotation is practised.

There is a need to reduce herbicide input in crop production which can complement cultural practices. With proper timing and selected application methods, good control may be achieved with one-fourth to one-half rates of application [7]. Herbicides are becoming more expensive, and by reducing the pesticide load into the environment, the risk of pollution is reduced. This can be achieved by:

i. banded application of herbicides

ii. the use of low volumes to improve glyphosate performance

iii. proper timing of post emergence herbicides

iv. the use of herbicide combinations at low rates

v. the use of newer, more active and more rapidly degradable herbicides, and

vi. monitoring fields to achieve spray decisions

Using lower herbicide dosages would reduce expenditure on herbicides to a fraction of the cost of full label herbicide rates while maintaining efficacy and other benefits derived from herbicide use. Research has revealed that half the recommended dosages of atrazine and nicosulfuron resulted in the lowest weed biomass. Mixing a third of the recommended herbicides of Atrazine and Nicosulfuron resulted in equivalent weed control to the atrazine label recommended dosages. Weed seed production was reduced. Reduced herbicide dosages may fit into the economics of the small farmer and hence has the potential to be a 'small hammer' in the IWM programme [51]. However, as mentioned before because of the risk of herbicide resistance, this decision must be taken carefully.

1.1.9. Crop and herbicide rotation

Crop and herbicide rotation reduce selection pressure on weeds and this allows for the development of resistant ecotypes and biotypes [35]. Crop rotation should include crops with either different cultural practices or morphology that will upset the life cycle of weeds as in white-top (*Partheniuim hysterophorous*) and corn grass (*R. cochininensis*) [7, 33]. Crop rotations and crop diversification are useful tools for weed management, as they encourage operational diversity that in turn can facilitate improved weed management [31]. Manipulating different planting and harvesting dates among crops provides more opportunities for producers to prevent either plant establishment or seed production by weeds. If sufficient differences exist in the germination requirements of crops and weeds, then seed date can be manipulated to the benefit of the crop for example. Weeds then germinate after canopy closure and they become non competitive [35].

However, in the small farm production systems, crop diversification in rotation and even crop succession are limited. The effectiveness of crop rotation in weed suppression may be enhanced by crop sequences that create varying patterns of resource competition, allelopathy, soil disturbance and mechanical damage to certain species. Diversified crop rotations are likely to provide best opportunities for exploiting diverse sets of tactics and ecological processes to suppress weeds [57]

There are only eight modes of action in available herbicides, and as a consequence rotating herbicides is as important as alternating crops, as overuse will increase the risk of single-, cross-, and multiple resistance [29].There is also the potential for a "species shift," as new weed species take over when the population of another diminishes, as a result of an effective herbicide or other control practice. Resistance, however, poses a more serious problem, as it depends on the weed species, the efficacy of the herbicide, and the frequency of herbicide use. Continuous use of a particular herbicide will contribute to resistance, and farmers should rotate two or three herbicides [49]. Additionally, using herbicides with the same mode of action will create an environment for resistance development. To reduce the risk of resistance the following guidelines should be considered:

- Alternate non-chemical with chemical control methods.
- Rotate herbicides, including mode of action of herbicides with the same site of action. Example, Maverick is a sulfonylurea herbicide and Pursuit is an imidazolinone herbicide, but both are group 2 herbicides.
- Tank mix different modes of action to apply different types of materials.
- Rotate crops which differ in their competitiveness against weeds based on life cycle, growth habit, maturity length, etc., so rotating to different crops can help prevent some weed species

from becoming dominant in a given field and control "suspect" herbicide-resistant weeds as if they were an invasive weed species.

Multiple management practices can be used in an integrated plan to prevent or delay the development of herbicide-resistant weed populations. In addition, avoid using herbicides with the same site of action in both fallow years and in the succession crops. Herbicide diversification is the key to preventing resistance, since using one system will create resistant weeds. Herbicide rotation is critical to maintaining grade and delaying resistance. Rotating herbicides with multiple modes of action is critical to delaying the spread of resistance and preventing weeds and volunteers [27].

Currently, there is an increase in the number of resistant weed biotypes, including those resistant to glyphosate, PPO, ALS, dicamba and triazine chemistries. The rapid growth of Respect the Rotation™ is a testament to the urgency with which thousands of growers treat the issue of weed resistance [36]. Glyphosate-resistant weeds are spreading at alarming rates from rampant infestations; 358 biotypes have developed resistance to one or more herbicide groups, including glyphosate, PPO, ALS, dicamba and triazine chemistries.

1.1.10. Intercropping or relay cropping

Intercropping or relay cropping systems are based on the principle that space should be occupied by crops and not weeds [57]. Relay cropping can be practised by market gardeners who harvest their crops by hand. These crops should be planted in such a way that the intercrop provides an effective canopy to shade weeds, or that previous crop residue can be used as a mulch to prevent weed growth in successional crops, e.g., pigeon pea (*Cajanus cajan*) interplanted with maize (*Zea mays*). Occasionally, the second crop in some intercropping systems is for the purpose of weed management. Crops such as velvetbean (*Mucuna pruriens*), lablab (*Lablab purpureus*), *Desmodium heterocarpon* and tropical kudzu (*Pueraria phaseoloides*) have been used successfully as intercrops in banana (*Musa* sp.), cassava (*Manihot esculenta*) and maize for the management of weeds such as watergrass (*Commelina* sp.) and cogongrass (*Imperata cylindrica*) [18, 25, 26] across tropical environments. It was found that intercrops may inhibit weeds by limiting resource capture by weeds or through allelopathic interactions [31], and that weed biomass was reduced in 90 % of the cases when a main crop was intercropped with a "smother" crop. It has also been reported that self-regenerating intercrops reduce establishment costs and can provide weed suppression over years [37].

1.1.11. Biological agents

The use of biological agents such as mycoherbicides, insects and pathogens to control weeds in the tropics is not common. However, the potential for its application to control noxious weeds using monophagous/oligophagous natural enemies must not be overlooked [29]. Table 1.0 shows some of the most successful achievements using this method of control which include: water hyacinth (*Eichhornia crassipes* (Mart.) Solms) using specific insects, white-top (*Parthenium hysterophorus* L.) using a fungus, Christmas bush (*Chromolaena odorata* (L.) King & Robins) using an insect and nutgrass (*Cyperus* spp.) using a fungus.

http://dx.doi.org/10.5772/55950

Classical biological control is the best among the viable options available for sustainable management of invasive weeds, especially where other technologies such as chemical and mechanical control are unacceptable due to cost and adverse impact on the environment [40].

Some of the techniques described for biological control of weeds in developed countries can be safely and efficiently transferred to developing countries with minimal expense for the initial institutional and human-capacity building. It is essential to know the organism to be used as well as the methods for rearing and release and its host range in order to avoid problems with crops. *The Code of Conduct for the Import and Release of Exotic Biological Control Agents* (FAO, 1996), gives good guidance on how to proceed in order to introduce new exotic organisms for biological control.

Weeds	Biological control agents
Eichhornia crassipes (Mart.) Solms Water hyacinth	Weevils: *Neochetina eichhorniae, N. Bruchi* Moth: *Sameodes albigutalis*
Parthenium hysterophorus L. White-top	Fungus: *Puccinia abrupta* var. Partheniicola Zygogramma bicolorata Epiblema strenuana
Chromolaena odorata (L.) King & Robins. Christmas bush	Moth: *Parauchaetes pseudoinsulata*
Lantana camara L. Black sage, Lantana	Lacebug: *Teleonemia scrupulosa*
Cyperus rotundus L. Nutgrass	Fungus: *Puccina canaliculata* Dactylaria higginsii Moth: Bactra spp.
Amaranthus spp.	Fungus: *Phomopsis amaranthicola*
Rottboellia cochinchinensis (Lour.) Corn grass	Fungus: *Sporisorium ophiuri*
Adapted from: [29]	

Table 1. Some organisms used for the biological control of selected weeds

2. Adoption strategies

The traditional top-down approaches, participatory approaches and discovery based teaching methods have all been used to promote integrated weed management.

The top-down method has been by far the most predominant method and widely used in training on weeds and their control. The focus of these sessions was to train farmers how to apply, mostly synthetic pesticides, and emphasised the need for continuous application.

Farmers responded well to these instructional approaches given the severe losses they sustain because of the extent and vigour of weed growth in the tropics and the quick, highly visible effect of synthetic herbicide applications. These class and field sessions have been historically conducted either as stand-alone modules in training courses or as part of the general agronomic practices for field crops. Over the years, extension agents conducted these courses in communities or at centralised farmer training centres. The concept of integrated weed management was not part of the landscape at this time.

In the 1990s, the emergence of farmer participatory approaches to educating farmers gained momentum. Although the focus was on Integrated Pest Management (IPM), weed management was incorporated into learning activities. Farmers, for the first time, were presented with the option of applying a mix of weed management strategies instead of a single chemical option. The aim of farmer participatory approaches is to strengthen farmers' decision-making skills through an understanding of the agro-ecology of their fields. The approach is widely recognised as an integral part of more sustainable and environmentally friendly crop production practices. The flagship method, Farmer Field Schools (FFS), continues to be used as the preferred approach to integrated management mostly of pests and diseases but increasingly included is the management of noxious weeds.

The Farmer Field School approach involved farmers in activities mostly in the field to understand weed dynamics and to involve farmers in decisions to manage weeds using more sustainable approaches. These activities, done on farmers' fields, have been conducted across the Caribbean as part of the FFS approach to integrated pest management. Farmers have been exposed to different weed management strategies which stressed integrated approaches. FFS have been conducted in St Lucia, Suriname, Trinidad and Dominica [4].

The FFS model is flexible, and, in recent times, one component has been singled out for increased use because of the enhanced learning it provides. Discovery-based learning is based on the principles of experiential learning; farmers are guided by a trained facilitator who draws out their knowledge and helps them construct meanings based on their rich field experiences. This has been done in several countries of the Caribbean. Discovery-based learning activities have been used in St Vincent in a Farmer Participatory Research (FPR) process to manage weeds in bananas [25, 26]. Farmers were encouraged to plant several cover crops on their farms to evaluate the efficacy of these crops on weed control in bananas. As farmers carried out these activities, they took the weekly measurements and did simple statistical analysis. They were able to discover for themselves the benefits of alternative approaches to the pesticide approach both for their health and that of consumers in foreign countries who purchase their bananas.

Farmers in Trinidad have also conducted community experiments using paper, used cartons, grass much, plastic, precision irrigation all in an attempt to evaluate alternative weed management strategies. Farmers have discovered for themselves the effects of the various treatments and some of these have been adopted by farmers who are tending to move to the low pesticide/ organic farming methods.

A mix of adoption strategies has been used over the years in an effort to get the right approach to IWM. No silver bullet has been found. It is a work in progress. Given the diverse weed flora, farming experiences and farmer circumstances in the tropical world, scientists, educators and farmers will have to dedicate increased energies towards finding an approach that is economical, culturally acceptable and environment friendly.

3. Research needs for integrated weed management systems for the tropics

There is a need to encompass weed management into improved/integrated crop management systems and to develop research and development programmes that will facilitate a more comprehensive understanding of ecology, physiology, biochemistry, competitiveness/ allelopathic potential and threshold of weeds.

4. Conclusion

The key to a successful weed management programme is the effective insertion into crop management programmes of those control techniques that will minimise the impacts of weeds not controlled by the competing crop. The dependence on overly generalized and increasingly expensive chemical input packages, developed elsewhere under a different set of conditions, and aggressively promoted by Researchers, Extension agents and Agro-chemical companies, must be broken.

The IWM systems approach fits into the work habit of many farmers and gives more effective control than when only chemical methods are used. In addition, yield improvements in the order of 40 to 100 % are realized. While IWM systems are considered technologically sound, the social and environmental advantages, as well as the economic costs associated with the practice, need to be ascertained. If farmers are not convinced of the economic viability of the system, then the technology no matter how sound will not be adopted.

Author details

Wendy-Ann P. Isaac[1], Puran Bridgemohan[3] and Wayne G. Ganpat[2]

1 Department of Food Production, Faculty of Science and Agriculture, The University of the West Indies, St. Augustine, Trinidad

2 Department of Agricultural Economics and Extension, Faculty of Science and Agriculture, The University of the West Indies, St. Augustine, Trinidad

3 The University of Trinidad and Tobago, Trinidad

References

[1] Arce, G. D, Pedersen, P, & Hartzler, R. G. Soyabean seeding rate effects on weed management. Weed Technology (2009). , 23, 17-22.

[2] Bayer CropScienceGlyTol Product Bulletin. Research Triangle Park: Bayer CropScience. n.d. Print. http://www.bayercropscience.us/learning-center/articles?storyId=F5B005DE-7AD2-8DF3-FE6D02522531accessed on 25 September (2012).

[3] Benech-arnold, R. L, Sanchez, R. A, Forcella, F, Kruk, B. C, & Ghersa, C. M. Environmental control of dormancy in weed seed banks in soil. Field Crops Research (2000). , 67, 105-122.

[4] Bekele, I, & Ganpat, W. G. Participatory On-Farm Trials- A Useful Tool". In Sustainable Food Production Practices in the Caribbean Ganpat, W.G. and Isaac, W.P. Ian Randle Publishers, Jamaica (2012). , 409-420.

[5] Brathwaite, R. A. I, & Isaac, W. P. Ecological weed management practices in the Caribbean, In: Sustainable Food Production Practices in the Caribbean, Ganpat, W.G. and Isaac, W.P. (eds.). Ian Randle Publishers, Jamaica, W.I. (2012). , 167-188.

[6] Bridgemohan, P, & Brathwaite, R. A. I. Paper presented at IFS-COSTED Seminar on Vegetable Cropping and Development in Latin America and the Caribbean. October, (1988). Trinidad., 3-8.

[7] Bridgemohan, P, & Brathwaite, R. A. I. Weed management strategies for the control of Rottboellia cochinchinensis in maize in Trinidad. Weed Research (1988). , 29, 433-440.

[8] Bridgemohan, P, Brathwaite, R. A. I, & Mcdavid, C. R. Seed survival and patterns of seedling emergence studies of Rottboellia cochinchinensis (Lour.) W.D. Clayton in cultivated soils. Weed Research (1991). , 31, 265-272.

[9] Bridgemohan, P, & Daisley, L. E. A. D. Paper presented at the 2nd West Indies Agricultural Economics Conference, Belize, Central America, July (1992). , 4-21.

[10] Bridgemohan, P, Brathwaite, R. A. I, & Mc David, C. R. Seed survival and patterns of seedling emergence studies of Rottboellia cochinchinensis in maize in Trinidad. Weed Research (1991). , 31, 265-272.

[11] Bridgemohan, P, & Mc David, C. R. A Model of the competitive relationships between R. cochinchinensis and Zea mays. Annals of Applied Biology. (1993). , 123, 649-656.

[12] Bridgemohan, P, Mc David, C. R, Bekele, I, & Brathwaite, R. A. I. The effects of Rottboellia cochinchinensis on the growth, development and yield of maize. Tropical Pest Management, (1992).

[13] Bussan, A. J, Burnside, O. C, & Puettmann, K. J. Field evaluation of soybean (Glycine max) genotypes for weed competitiveness. Weed Science (1997). , 45, 31-37.

[14] Carter, M. R, & Ivany, J. A. Weed seed bank composition under long term tillage regimes on fine sandy soils in the fine sane sandy soils loam in Atlanta, Canada, Soil and Tillage Research. (2006). and (2), 29- 38.

[15] Centro International De Argricultura Tropical (CIAT) (1979). Annual report 1978, Cali, Colombia, CIAT. , 52-59.

[16] Chauhan, B. S, & Johnson, D. E. Influence of tillage systems on weed seedling emergence pattern in rainfed rice, Soil Tillage Research (2009). , 109, 15-21.

[17] Chauhan, B. S, & Johnson, D. E. The role of seed ecology in improving weed management strategies in the tropics, Advances in Agronomy (2010). , 105, 221-262.

[18] Chikoye, D, Ekeleme, F, & Udensi, U. E. Cogongrass suppression by intercropping cover crops in corn/cassava systems. Weed Science. (2001). , 47, 674-667.

[19] Christoffoleti, P. J, Carvalho, S. J. P, Nicolai, M, & Doohan, D. And VanGessel, M. Prevention strategies in weed management, In: Non-chemical weed management: Principles, concepts and technology, Upadhyaya M.K. and Blackshaw, R.E. (eds) CAB International (2007).

[20] Doll, J. D, & Piededrahita, W. C. Proceedings 3rd International Symposium International Society for Tropical Roots Crop. Ibadan, Nigeria. 2-9 December (1973). Leakey, C.L.A. (ed.) IITA, Ibadan, 399-405.

[21] Friedrich, T, Kassam, A. H, & Shaxson, F. Conservation Agriculture. In: Agriculture for Developing Countries. Science and Technology Options Assessment (STOA) Project. European Parliament. European Technology Assessment Group, Karlsruhe, Germany. (2009).

[22] Friedrich, T, & Kassam, A. H. Adoption of Conservation Agriculture Technologies: Constraints and Opportunities. Invited paper, IV World Congress on Conservation Agriculture, 4-7 February (2009). New Delhi, India. , 257-264.

[23] Glaze, N. Cultural and mechanical manipulation of Cyperus spp. Weed Technology. (1987). , 1, 82-83.

[24] Grichar, W. J, Bessler, B. A, & Brewer, K. D. Effect of row spacing and herbicide dose on weed control and grain sorghum yield. Crop Protection (2004). , 23, 263-267.

[25] Isaac, W. P, Brathwaite, R. A. I, Cohen, J. E, & Bekele, I. Effects of alternative weed management strategies on Commelina diffusa Burm. infestations in Fairtrade banana (Musa spp.) in St. Vincent and the Grenadines. Crop Protection (2007). , 26, 1219-1225.

[26] Isaac, W. P, Brathwaite, R. A. I, Ganpat, W. G, & Bekele, I. The Impact of Selected Cover Crops on Soil Fertility and weed and nematode Suppression through Farmer

Participatory Research by Fairtrade Banana Growers in St. Vincent and the Grenadines: Technical Report. World Journal of Agricultural Science (2007). , 3(3), 371-379.

[27] Johnson, E, & Frick, B. Inter-row cultivation- effective weed control in field pea? Organic Agriculture Centre of Canada (OACC) http://www.agriculture.gov.sk.ca/Default.aspx?DN=88be1afa4-40c7-b6ea-9b0af05b0117Accessed 5 October (2012).

[28] Kasasian, N. Weed Control in the Tropics. Leonard Hill, London. (1971).

[29] Labrada, R. Weed management: A basic component of modern crop production. In Sustainable weed management. Singh, H.P., Batish, D.R., Kumar Kohli, R. (eds). (2005). Food Products Press, and imprint of the Haworth Press Inc., NY.

[30] Lal, R. ed.), (1979). Soil Tillage and Crop Production. International Institute of Tropical Agriculture Proceedings Series 2, Ibadan, Nigeria.

[31] Liebman, M, & Dyck, E. Crop rotation and intercropping strategies for weed management. Ecological Applications (1993). , 3, 92-122.

[32] Liebman, M, & Gallandt, E. R. Many little hammers: ecological management of crop weeds interactions, In L. Jackson, Ed. Ecology in Agriculture. New York Academic. (1997). doiI:10.1016/b978-012378260-1/http://dx.doi.org/10.1016/B978-012378260-1/50010-5, 500100-5.

[33] Mabasa, S, & Rambakudzigba, A. M. Periodicity of weed seedling emergence and the effect of weeding frequency on maize and sorghum, Zimbabwe Journal of Agricultural Research (1993). , 31, 27-41.

[34] Mandumbu, R, Peter, J, Charles, K, & Tibugari, H. Integrated weed management in Zimbabwe's smallholder sector, Where are we?: A Review. Modern Applied Science. (2011). doi:10.5539/mas., 5n5p111, 111.

[35] Mashingaidze, A. B, & Chivinge, O. A. Weed control using reduced herbicide dosages: a Theoretical Framework, Transactions of Zimbabwe Scientific Association (1995). , 63, 12-19.

[36] Mashingaidze, A. B. Improving weed management and crop productivity in maize systems in Zimbabwe, Tropical Resource Paper 57, (2004).

[37] Martin, C. C. Weed control in tropical ley farming systems: a review. Australian Journal of Experimental Agriculture(1996). , 36, 1013-1023.

[38] Mohler, C. L. A model of the effects of tillage on emergence of weed seedlings. Ecological Applications, (1993). , 3, 53-73.

[39] Mohler, C. L. Enhancing the competitive ability of crops. In ecological management of agricultural weeds (2007). Cambridge University Pess

[40] Muniappan, R, Reddy, G. V. P, & Raman, A. A. Biological control of weeds in the tropics and sustainability. In Biological Control of Tropical Weeds using Arthropods,

R. Muniappan, G. V. P. Reddy, and A. Raman. (eds.) Cambridge University Press, (2009).

[41] Maxwell, B. D, & Donovan, O. J.T. Understanding weed-crop interactions to manage weed problems. In Non-Chemical Weed Management, Principles, Concepts and Technology, Upadhyaya, M.K and Blackshaw, R.E. (2007). CABI International

[42] Nice, G. R, Buehring, N. W, & Shaw, D. R. Sicklepod (Senna obtusifolia) response to shading, soybean (Glycine max) row spacing, and population in three management systems. Weed Technology (2001). , 15, 155-162.

[43] Nitis, I. M. Stylosanthes as companion crop to cassava (Manihot esculenta). Faculty of Veterinary Science and Animal Husbandry, Uayana University, Bali, Danpasar. (1977).

[44] Nitis, I. M, & Suarma, M. Undergrowing cassava with Stylosanthes grown under coconut. In Proceedings of the 4th Symposium International Society for Tropical Root Crops. Cock, J. et al. (eds.) International Development Research Centre, Ottawa, Canada. (1977). , 98-103.

[45] Donovan, O, Harker, J. T, Clayton, K. N, Newman, G. W, Robinson, J. C, & Hall, D. L.M. Barley seeding rate influence the effects of variable herbicide rates on wild oat. Weed. Science (2001). , 49, 746-754.

[46] Onochie, B. E. (1975). Critical periods for weed control in cassava in Nigeria, PANS. 1975; , 24, 292-299.

[47] Overland, L. The role of allelopathic substances in the "smother crop" barley. American Journal of Botany (1966). , 53, 423-427.

[48] Putnam, A. R. Vegetable weed control with minimal herbicide inputs. Horticultural Science (1990).

[49] Regnier, E. E, & Janke, R. R. Sustainable Agricultural Systems. Edwards, Lai, Madden and Miller (eds). Soil and Water Conservation Society, Ankeny, Iowa (1990).

[50] Ross, M. A, & Lembi, C. A. (1985). Applied Weed Science. Burgess Publishing Company, Minneapolis, Mn, 340 pp.

[51] Sanyal, D, Bhowmik, P. C, Anderson, R. L, & Shrestha, A. Revisiting the Perspective and Progress of Integrated Weed Management. Weed Science. (2008).

[52] Schweizer, E. E. New technological developments to reduce groundwater contamination by herbicides. Weed Technology (1988). , 2, 223-227.

[53] Sharma, B. M, & Dairo, F. M. Ecophysiological studies on two common weeds associated with cassava crop, Journal Root Crops (1981). , 17, 85-91.

[54] Shaw, W. C. Integrated weed management systems technology for pest management. Weed Science (1982). , 30, 2-12.

[55] Swanton, C. J, Maloney, K. J, Chandler, K, & Gulden, R. (2008). Integrated weed management: Knowledge based weed management systems, Weed Science 2008;, 56, 168-172.

[56] Unamma, R. P. A, Ene, L. S. O, Odurukwe, S. O, & Enyinnia, T. Integrated weed management for cassava intercropped with maize. Weed Research (1986)., 26, 9-17.

[57] Westerman, P. R, Hofman, A, Vet, L. E. M, & Van Der Werf, W. Relative importance of vertebrates and invertebrates in epigaeic weed seed predation in organic cereal fields. Agriculture, Ecosystem and Environment. (2003)., 95, 417-425.

Chapter 11

Integrated Plant Invasion and Bush Encroachment Management on Southern African Rangelands

M. S. Lesoli, M. Gxasheka, T. B. Solomon and
B. Moyo

Additional information is available at the end of the chapter

http://dx.doi.org/10.5772/56182

1. Introduction

1.1. Background

Rangeland could be defined as the land on which indigenous vegetation (climax or natural potential) is predominantly grass, grass-like plants, forbs, or shrubs that are grazed or have the potential to be grazed, and which is managed as a natural ecosystem for grazing livestock and wildlife habitat [1]. Rangeland productivity is threatened by land degradation mostly characterised by soil erosion and invasion by alien plant species. Plant invasion is considered a threat to rangelands because of the suppression of productivity of herbaceous plant species due to the increase of bush cover [2]. In an endeavour to understand the concepts of plant invasion in rangelands, it is important to acknowledge that the terms invasion and encroachment are normally used loosely and commonly interchangeably. However, it is crucial to understand their distinction so that the approaches in addressing their different characteristics and effects on rangelands are informed by clear comprehension. Bush encroachment refers to the spread of plant species into an area where previously it did not occur [18]. Invasion on the other hand, refers to the introduction and spread of an exotic plant species into an area where previously did not occur. Thus, bush encroachment could occur even with indigenous species and it is more defined by plant density than species themselves. Whilst invasion on the other hand, although it includes plant density, focuses on the exoticism of species in question and it is, therefore, more species specific. Furthermore, while encroachment focuses on the woodiness of the species, invasion is not limited to woody species but includes the alien herbaceous species; thus, there are grasses that are classified as invaders. However, in this chapter bush encroachment and invasion are used interchangeably and treated as synonyms.

Other than the suppression of herbaceous by encroaching species, the higher bush density in rangelands reduces land accessibility by livestock, and that subsequently negatively affects the utilisation of rangelands. Furthermore, due to competition for light, water, and nutrients between native and invading species, the grazing capacity of rangelands declines [2, 4] and plant biodiversity becomes compromised [3]. Therefore, invasions are considered one of the largest threats to the ecosystems of the earth [5- 6], and the services that they provide to humanity [5]. These species are characterised by rapid spread and they displace native vegetation and disrupt important ecosystem processes, and that leads to serious environmental impacts [5- 7]. There are a number of sources for invading species, however, in natural ecosystems such as rangelands some alien tree species used in commercial forestry and agroforestry cause major problems as invaders [8]. The effects of bush encroachment, such as an increase in woody vegetation density and cover, and reduction of biomass production in rangelands [9], have been widely reported in Southern Africa [10 – 11]. Invader species can be found in different ecosystems, however, in South Africa, they are a significant environmental problem in terrestrial and freshwater ecosystems [12]. Bush encroachment and invasion on rangelands, therefore, have negative effects on rangeland biological and economic value. Thus, bush encroachment and invasion results in rangeland degradation, which leads to declination of rangeland functional capacity and subsequently on the increased food insecurity and poverty. Hence, introduction of woody plant cover in grasslands and their increase in savanna ecosystems is an indication of rangeland degradation [13]. The foregoing assertion is aligned with the definition of rangeland degradation, which states the reduction or loss of biological and economic productivity arising from inappropriate land use practices [13]. Therefore, if bush encroachment in rangeland is left unchecked, it progresses within grassland ecosystems until a closed canopy woodland thicket occurs [15], which influences vegetation species composition and in turn threatens the sustainability of livestock production as well as wildlife habitat [16] and grassland birds [17]. Thus, the increase in vegetation cover of encroaching species can significantly reduce grass productivity through competition, shading and allelopathic effects.

Invasion phenomenon is becoming an increasing concern to land managers who are seeking cost-effective ways of combating the spread of invasive species [6]. It is important to acknowledge that factors causing invasion are complex [10, 19]. This is because of a large number of predisposing factors and that species behave differently at various environments. Therefore, any ecological and/or economic intervention in managing bush encroachment in rangelands should be anteceded by the comprehensive understanding of the drivers for this phenomenon. Nevertheless, bush encroachment is often associated with overgrazing [20]. This is because of a positive relationship between grazing pressure and woody vegetation cover [13]. There are other reported drivers of bush encroachment such as increased rainfall [21], fire suppression [22], and soil characteristics [23]. It is acknowledged, therefore, that bush encroachment threatens livestock production particularly, grazers [24] and in turn livelihoods of pastoral communities hence researchers, policy makers and practitioners need to understand bush encroachment dynamics and characteristics in order to adapt to live with or control it. Invasive plants in rangelands in the long-term affect livestock industry by lowering forage yield and quality, interfering with grazing accessibility and poisoning animals and subsequently increasing costs of management and production of livestock, and eventually reducing land value. In the wildlife ecosystems, these species affect the wildlife habitat and forage production, deplete soil and water resources, and reduce plant and animal diversity [25]. In general, woody and succulent species invasion in rangelands result in a decline in biodiversity [26], reduction in ecosystem resilience [27] and a greater likelihood of irreversible changes in plant species composition [28].

Grazing is one of the economic ways of utilising rangelands especially in communal and/or pastoral areas. The provisions of biodiversity conservation and ecosystem stability within rangelands maintain the ecological value of these ecosystems. Therefore, maintaining or restoring rangeland ecosystem health and resilience is a critical social imperative to ensure the future supply of the ecosystem services, which are vital for the future well-being of human societies [29]. Such services include provision of stable soils, reliable and clean supply of water, and the natural occurrence of plants, animals and other organisms to meet the aesthetic and cultural values, and to enhance the livelihoods of people living around rangelands [30]. This review chapter explores the phenomenon of plant invasion and bush encroachment in the southern African region; however, reference is made to invasion and encroachment reported beyond the southern African boundaries. Furthermore, although this chapter emphasises bush encroachment and invasion in rangelands or natural ecosystems, the reference is further made from other ecosystems such as cultivated, riparian, and marine areas. This chapter explores plant invasion and encroachment phenomenon in terms of its identified causes, its ecological and economic impact. Furthermore, bush encroachment control practices in rangeland ecosystems and their significance in restoring invaded ecosystems were evaluated. Finally, different methods and approaches used in management of invasion in rangeland are synthesised into an integrated rangeland management approach.

2. Bush encroachment and invasion in rangelands

2.1. The concept of bush encroachment and invasion

Bush encroachment could be defined as an increase in woody plant abundance in grassland and savanna regions accompanied by changes in the herbaceous cover and composition of the natural vegetation [31 - 33]. This section addresses the question of whether bush encroachment and/or invasion are the problem in rangelands and if the phenomenon poses a challenge to natural ecosystems and human livelihoods. South Africa's natural ecosystems such as rangelands are under threat from invasive alien plants [12, 34], the scale of the problem facing mangers of invasive alien plants in South Africa is huge, and thus, about 10 million ha has been invaded [35]. There is some sort of cosmopolitan concern about the effects of bush encroachment and invasion on rangeland ecosystem productivity and sustainability. Thus, human communities and natural ecosystems worldwide are under siege from a growing number of destructive invasive alien species [36]. These species erode natural capital, compromise ecosystem stability, and threaten economic productivity of rangeland ecosystems. Besides the effects of invasion in agriculture, forestry, and human health, biological invasions are also widely recognised as the second-largest global threat to biodiversity. The problem of invasion in rangelands is growing in severity and geographic extent as global trade and travel accelerate, and as human mediated disturbance and increased dissemination of propagules makes ecosystems more susceptible to invasion by alien species [36]. One of the remarkable characters of invasive alien plants is that few, if any, of them are invasive in their countries of origin. Thus, their ability to grow vigorously and produce copious amounts of seeds is kept in check by a host of co-evolved invertebrates and pathogens [6]. Some of these plant species, when transported to a new continent without the attendant enemies, they exhibit "ecological release." This phenomenon allows the introduced species to multiply rapidly in the absence of a host of attendant invertebrates and diseases, with associated tendencies to spread rapidly and to out-compete native species [6].

Mostly livestock and wildlife production depend on rangelands for sustenance as a source of feed and habitat. Rangelands are represented by a variety of ecosystems including desert and rich alluvial valleys, coastal and inland foothills, high mountain meadows and arid inland plains [25]. In the southern African context, the larger space of rangelands is represented by savanna and grassland ecosystems. Savannas are extensive, socioeconomically important ecosystems with a mixture of two life forms, thus, trees and grasses [37, 38, 39]. Whilst in Africa, savannas are the most important ecosystems for raising livestock [40]. Thus, domestic livestock, particularly *Bos* (cattle), *Ovis* (Sheep) and *Equus* (Horses) have grazed many of these areas for many years. As a result, the plant composition has changed greatly from the original ecosystems [41].

Factors and mechanisms regulating bush encroachment by invasive woody plants in rangeland ecosystems are not fully apprehended [2, 42]. However, the dynamics and modalities of bush encroachment are mostly widespread in African [13, 20], Australian [43], and North American and Latin American rangelands [39]. The increase in the tree-grass ration in the savannas has been attributed to the replacement of indigenous herbivores by domestic grazing animals and the intense utilisation of the natural vegetation by domestic livestock [33, 44]. Furthermore, heavy grazing results in reduced fuel loads leading to less frequent and low intensity fire, which reduces the effectiveness of fire in the control of woody vegetation. This heavy grazing further leads to altered competitive interactions between the woody and herbaceous layers due to the removal of grasses [32]. However, a number of times, these phenomenon have been linked to climate change [45] or land use patterns [24] or combination of number of factors [13], both biotic and abiotic in nature. Thus, local climate and long-term climate change in conjunction with grazing effects and fire limitation have been identified as possible causes of bush encroachment [46, 47, 48]. Long-term prohibition of range fire, cultivation of bottomlands and continuous grazing on the remaining portion of the communal rangelands have been reported to have induced the invasion of bush encroachment to a level of more than 60%. This has resulted in reduced grass cover, poor range condition, and subsequently poor livestock productivity [13, 49, 50].

Although there are a myriad of explanations about bush encroachment and invasion in rangelands, the first attempt at a general explanation for bush encroachment was a two-layer hypothesis for tree-grass coexistence [2, 51, 52, 53]. In this model, water is assumed be the major limiting factor for both grassy and woody plants growth. Based on this analogy, it is hypothesized that grasses use only topsoil moisture, and woody plants mostly use subsoil moisture [54]. Therefore, reduction of grass plant density and vigour through practices such as severe grazing, allows more water to percolate into the subsoil, where it is made available for woody plant growth. Subsequently, reduction of grassy vegetation has demonstrated an increase in shrub and tree abundance under heavy grazing [55, 56]. The two-layer model is still widely accepted to explain bush encroachment phenomenon, however, field data and other theoretical models have indicated the contravening evidence [20]. Thus, the release of trees from competition with grass is not required for mass tree recruitment to occur; for example, encroachment of certain species such as *Prosopis glandulosa* is unrelated to herbaceous biomass or density [57]. Furthermore, a spatially explicit simulation model indicates that rooting niche separation might not be sufficient to warrant coexistence under a range of climatic situations [58].

This indicates that the concepts of bush encroachment and invasion in rangelands are by far still complex in terms of causation and/or predisposition factors. There were great differences reported in a number of studies in the degree of niche separation. These variations depend on various abiotic factors, and plant species involved [59, 60, 61]. Therefore, the two mechanisms, heavy grazing and rooting niche separation, do not suffice to serve as the one-dimensionally exclusive explanations for bush encroachment. This is justified by the fact that at initiation of bush encroachment young trees use the same subsurface soil layer as grasses in the sensitive early stages of growth. In addressing the relationship between bush encroachment and grazing, bush encroachment has been reported in areas where grazing was not severe. Therefore, overgrazing in combination with rooting niche separation are not the solitary predisposing factors for bush encroachment; bush encroachment sometimes also occurs on soils too shallow to allow for root separation [62]. This further shows the complexity of comprehending the causes of bush encroachment in grasslands and savannas and that further translates to the complexity of controlling the problem. This, therefore, suggests that there is no panacea in addressing the bush encroachment; therefore, integration of bush encroachment control measures and practices could lead to a sustainable solution than accrediting one method over others.

There are a number of disturbances that have been mooted to be the major determinants of savanna vegetation structure, and savannas have been portrayed as inherently unstable ecosystems. Thus, they are considered to be oscillating in an intermediate state between those of stable grasslands and forests. This is because they are pushed back into the savanna state by frequent disturbances related to human impact, herbivory, fire [61], or drought, and spatial heterogeneities in water, nutrient, and seed distribution [58]. The disturbance hypotheses suggest that bush encroachment occurs as disturbances shift savannas from the open grassland towards the forest extreme of the environmental spectrum. Although disturbance theories may be valid for specific situations, however, they may lack generality [2].

Bush encroachment and invasion by alien plant species may further be, to a certain degree, attributed to climate change. Climate change causes a number of variations in the atmosphere, and such changes could positively or negatively affect vegetation growth performance. One of the effects of climate change is an accumulation of carbon dioxide (CO_2) concentrations in the atmosphere. These increased CO_2 concentrations are likely to have an effect on tree-grass dynamics in savannas. This is because savanna trees and grasses have different photosynthetic pathways, which will respond differently to changes in atmospheric CO_2 accumulations. It is predicted that atmospheric CO_2 is exponentially increasing and will likely double to 700 parts per million (ppm) within the next century [62]. This has a further potential beneficial effect on plant life; the benefit is attributed to the fact that plants take up CO_2 via photosynthesis and use it in photosynthesis to produce carbohydrates. Thus, the higher CO_2 concentration could significantly increase the capacity of plants to absorb and temporarily store excess carbon. The efficiency of plants in the savanna to utilise the high CO_2 concentrations will be influenced to a larger extend by the photosynthetic pathways of different plant species and, therefore, that will influence plant species composition and ecosystem structure. For example, *Acacia* trees have the C3 photosynthetic pathway, which is less efficient, hence, they have a lower net photosynthetic rate at current atmospheric CO_2 levels than the C4 pathway used by most of savanna grasses [62]. However, at the higher atmospheric CO_2 levels than currently experienced, C3 plants will have a higher net photosynthetic rate than C4 plants. Thus, C3 plants should show increases in yield of 20 – 35% with a doubling of atmospheric CO_2, while C4 plants such as grasses should only experience a 10% increase in yield. Furthermore, the increased CO_2 concentrations will improve the competitive ability of trees against grasses. Thus, *Acacia* trees will have more carbon to invest in carbon-based defences against herbivory such as condensed tannins [63, 64].

In an attempt to further explain bush encroachment phenomenon in semi arid and arid environments, it is hypothesised that it is a natural phenomenon occurring in ecological systems governed by patch-dynamic processes [65]. This hypothesis has been based on field observations gained on the spatial distribution of *Acacia reficiens* trees in arid central Namibia. It is argued that encroachment of *A. reficiens* along rainfall gradient increases with increasing rainfall in spite of a relatively constant level of grazing [65]. However, any form of vegetation disturbance in rangelands (grazing, fire, etc.) can create space, and thus, making water and nutrients available for tree establishment due to reduced competition. However, under low soil nitrogen conditions, the nitrogen-fixing trees have a competitive advantage over other plants and, given enough rainfall, may germinate as a group in the bare patches created by the disturbances. The mechanism underlying this hypothesis, which demonstrates how it may be used to explain this phenomenon are such that both tree-grass coexistence and bush encroachment occur in a patch-dynamic system with stochastic rainfall patterns [2]. Nevertheless, it was suggested that in arid and semi-arid savanna ecosystems, woody vegetation needs above-average precipitation for germination and subsequent establishment [66]. To keep the soil moist for a period sufficient for germination and survival through the sensitive early stages of seedling development, several rain events close in succession are necessary [67]. However, in a savanna ecosystem, rainfall is often patchily distributed, in terms of both time and space [46, 68, 69]. Therefore, the spatial overlap of several rainfall events of high frequency in a single year is a rare occurrence in semi-arid and arid ecosystems. In addition to local seed availability, this rainfall frequency is a necessary condition for the creation of a bush encroachment patch.

http://dx.doi.org/10.5772/56182

2.2. Spatial distribution of encroaching and invasive plant species in rangelands

Several estimates have been made of the spatial extent of alien plant invasions in South Africa [36]. The rapid reconnaissance in 1996/97 [35] suggested that about 10 million hectares of South Africa has been invaded by the approximately 180 species that were mapped. In South Africa, there are a number of invading species; however, the principal invaders are trees and shrubs in the genera *Acacia, Hakea* and *Pinus*. However, the majority of invasive and/or encroaching species in rangelands are in the Fabaceae family, which are normally nitrogen-fixing legumes [70]. Localization of invading species distribution is influenced by the landscape formation gradient, thus, there are dense invasions in the mountains and lowlands and along the major river systems [12].The susceptibility of rangelands to bush encroachment and/or invasion varies between the vegetation types. Thus, vegetation types such as grassland and savanna biomes are extensively invaded mostly by species such as Australian wattles (*Acacia* species), other tree species, and a variety of woody scramblers (notably, triffid weed, *Chromolaena odorata*, and brambles, *Rubus* species). Invading trees such as jacaranda (*Jacaranda mimosifolia*) and syringe (*Melia azedarach*) have spread into semi-arid savanna by spreading along perennial rivers. In the Nama Karoo, woody invaders, notably mesquite (*Prosopis* species), have invaded large areas of alluvial plains and seasonal and ephemeral watercourses. Several cacti (*Opuntia* species) and saltbushes (*Atriplex* species) have invaded large areas of the Nama Karoo and Succulent Karoo [71] and the thicket biome in the Eastern Cape [12].

There are a number of species introduced from other continents and can cause significant problems on rangelands. The temporal and spatial spread of an invading organism including plants generally follows a sigmoid curve [72, 73]. Thus, the initial expansion is slow as the founder colony expands and starts new colonies, decreasing again as the potential habitat (invadable area) becomes fully occupied. The increase of invasive species on the given space and time leads to significant changes on the ecosystem integrity. Thus, invasive plants in the new region lead to profound changes in ecosystem processes, community structure, and displacing native species [74]. Therefore, it is fundamental to determine the spread of invading species in terms of time and space prior to development of a plan to control them. Several attempts have been made to prioritize alien species according to their invasive potential in different parts of the world. However, most attention has been given to screening species for their invasive potential prior to their introduction to a region [75, 76].

The ranking of Weeds of National Significance was developed for Australia based on expert scoring of four criteria [77]. These are grounded on their invasiveness, impacts, potential for spread, and socio-economic and environmental values. In South Africa, invasive species were prioritized based on their potential invasiveness, spatial characteristics, potential impacts, and conflicts of interest [78]. The Southern African Plant Invaders Atlas (SAPIA) database contains records for over 500 species of invasive alien plants in South Africa, Lesotho, and Swaziland, with information on their distribution, abundance, and habitat types [79]. There are two lists of invasive alien plants, classified into group species based on similarities in their distribution, abundance, and/or biological traits [80]. The first list contains those species that have already had a substantial impact on natural and semi-natural ecosystems such as rangeland in South Africa. Species demonstrating high value for any of the three components was considered to have high impact and species with high values for all three components have the highest impact. These species are perceived to constitute the prime concern for managers and, therefore, are referred to as the major invaders. Therefore, the presence and abundance of this species could be regarded to be above the economic threshold and warrant economic and ecological attention. Thus, the projects aimed at the prevention and/or control of these species should receive the largest proportion of available funding over the next few decades.

The second list contains those species that currently have a lower impact on natural or semi-natural ecosystems in South Africa. Thus, these species exhibit a lower product of range, abundance, and effect, but appear to have the capacity to exercise greater influence in the future. They are, therefore, termed "emerging invaders," and are currently afforded lower priority in management. However, some of these species are likely to become more important in the future, and could become targets for pre-emptive action such as biocontrol. These species should be carefully monitored to ensure that they do not become major problems. There are 117 major invaders identified in South Africa, and black wattle (*Acacia mearnsii*), white and grey poplars (*Populus alba/canescens*) and mesquite (*Prosopis glandulosa* var. *Torreyana/velutina*) are the three species/species-groups falling within the 'very wide spread-abundant' category [80]. The distribution pattern of these 'very widespread/widespread-abundant' species corresponds to the areas where high overall numbers of invasive alien plants were recorded. Most of the major invaders are found within the 'widespread common' and localised abundant categories. The highest numbers of species in the 'localized-abundant' category are restricted to Western Cape and Natal coasts, and northeastern Mpumalanga and Gauteng (Table 1). A list of 84 emerging invaders identified in South Africa was also presented; a majority (60%) of these species have been listed by the regulations under the Conservation of Agricultural Resources Act (CARA). Emerging invaders account for approximately 2500 records, or 5%, of the SAPIA database, and those species added from other sources [81, 82] and expert knowledge. Almost 20% of the emerging species are classified as riparian species according to expert opinion. A further 17% of these species are estimated to have the potential of expanding over a large part of the country if unmanaged (categories 'large habitat–large propagule pool', 'large habitat–moderate propagule pool' and 'large habitat–small propagule pool'), and almost 80% of species falling in these categories have been afforded legal status [80]. These species are distributed along the eastern coast and northeastern interior, but have not yet been recorded in the Northern Cape and Western Cape.

Most of the emerging invaders (61%) are estimated to have a moderate amount of invasible habitat available within South Africa (categories 'moderate habitat–large propagule pool' and 'moderate habitat– moderate propagule pool'). These categories show a slight difference in species distribution; distribution patterns of the 'moderate habitat–large propagule pool' category are similar to the 'localized–abundant' category of major weeds, whilst distribution patterns for the 'moderate habitat-moderate propagule pool' category show a lower incidence of fynbos invaders. The emerging invaders that are estimated to have a small amount of invasible habitat available but a large current propagule pool size (Table 2) show a very similar distribution pattern to the species which fall into the 'moderate habitat–large propagule pool' category.

http://dx.doi.org/10.5772/56182

Range-abundance	Scientific name	Common name	No of grids-cells	%Grid-cells abundant	Riparian or landscape	CARA category
Very widespread-abundant	*Acacia mearnsii*	Black wattle	432	28	Both	2
	Poplars alba/canescens	White and grey poplars	557	20	Riparian	2
	Prosopis glandulosa var. *Torreyana/velutina*	Honey mesquite/ prosopis	453	15	Both	2
Very widespread-common	*Agave americana*	American agave	433	1	Landscape	Proposed
	Arundo donax	Giant reed	377	14	Riparian	proposed
	Eucalyptus spp.	Gum trees	506	4	Both	1
	Melia azedarach	Seringa	558	7	Both	
	Nicotiana glauca	Wild tobacco	396	3	Both	3
	Opuntia ficus-indica	Sweet prickly pear	863	4	Landscape	1
	Ricinus communis	Castor-oil plant	471	7	Riparian	2
	Salix babylonica	Weeping willow	475	12	Riparian	2
Widespread-abundant	*Acacia cyclops*	Red eye	167	29	Both	2
	Acacia dealbata	Silver wattle	256	24	Riparian	1/2
	Acacia longifolia	Long-leaved wattle	95	24	Both	1
	Acacia saligna	Port Jackson willow	160	28	Both	2
	Ageratina adenophora	Crofton weed	11	19	Riparian	1
	Ageratum colyzoides/ houstonianum	Invading ageratum	74	26	Riparian	1
	Argemone mexicana	Yello–flowered Mexican poppy	29	18	Riparian	1
	Atriplex lindleyi spp. inflata	Sponge-fruit saltbush	164	13	Landscape	3
	Azolla filiculoides	Red water fern	206	36	Riparian	1
	Caesalpina decapetala	Mauritius thorn	128	19	Both	1
	Campuloclinium macrocephalum	Pompom weed	17	25	Both	1
	Cardiospermum grandiflorum/ halicacabum	Balloon vines	63	22	Both	1
	Cestrum aurantiacum/ laevigatum	Inkberry	80	24	Both	1
	Chromolaena odorata	Triffid weed	96	36	Both	1
	Eichlomia crassipes	Water hyacinth	95	22	Riparian	1
	Lantana camara	Lantana	261	27	Both	1
	Pinus pinaster	Cluster pine	86	26	Landscape	2

	Psidium guajava	Guava	167	17	Both	2
	Rubus cuneifolius	American bramble	75	34	Both	1
	Rubus fruticosus	Europian blackberry	89	20	Both	2
	Salix fragilis	Crack willow	75	22	Riparian	2
	Solanum mauritianum	Bugweed	268	21	Both	1
Widespread-common	*Acacia decurrens*	Green wattle	101	21	Both	2
	Acacia melanoxylon	Australian blackwood	138	15	Both	2
	Achyranthes aspera	Burweed	77	4	Both	1
	Ailanthus altissima	Tree-of-heaven	32	5	Both	3
	Anredera cordifolia	Bridal wreath	24	8	Both	1
	Araujia sericifera	Moth catcher	36	2	Both	1
	Atriplex nummularia spp. nummularia	Old-man saltbush	173	7	Both	2
	Bidens formosa	Cosmos	48	11	Riparian	
	Cardiospermum halicacaburn	Heart pea	30	0	Riparian	
	Casuarina equisetifolia	Horsetail tree	24	3	Both	2
	Cereus jamacaru	Queen of the night	127	9	Landscape	1
	Conyza bonariensis	Flax-leaf fleabane	5	0	Riparian	
	Crotalaria agatiflora subsp. imperialis	Bird flower	18	0	Both	Proposed
	Cuscuta campestris	Common dodder	82	1	Both	1
	Datura spp (D. Ferox/ D. Inoxia/D. Stramonium)	Thorn apples	84	1	Riparian	1
	Echium plantagineum/vulgare	Patterson's curse/blue echium	44	14	Both	1
	Eucalytus camaldulensis	Red river gum	123	15	Riparian	2
	Hakea sericea	Silky hakea	78	12	Landscape	1
	Ipomoea alba	Moonflower	23	3	Riparian	1
	Ipomoea indica/purpurea	Morning glories	98	8	Both	1
	Jacaranda mimosifolia	Jacaranda	201	6	Both	3
	Mirabilis jalapa	Four-o'clock	7	0	Landscape	Proposed
Widespread-common	*Morus alba*	White or common mulberry	130	4	Riparian	3
	Opuntia aurantiaca	Jointed cactus	61	5	Landscape	1
	Opuntia imbricata	Imbricate cactus	131	10	Landscape	1
	Opuntia monacantha	Cochineal pricky pear	48	1	Both	1
	Opuntia robusta	Blue-leaf cactus	225	1	Landscape	

	Opuntia stricta	Australian pest pear	108	10	Landscape	1
	Pinus halepensis	Aleppo pine	85	3	Landscape	2
	Pinus patula	Patula pine	90	12	Both	2
	Pinus radiata	Radiata pine	71	12	Landscape	2
	Pinus spp.	Pine trees	126	9	Landscape	
	Pyracantha angustifolia	Yellow fire thorn	143	1	Both	3
	Robinia pseudoacacia	Black locus	110	9	Both	2
	Schinus molle	Pepper tree	232	1	Both	Proposed
	Senna didymobotrya	Peanut butter cassia	142	13	Both	3
	Senna occidentalis	Wild coffee	56	8	Both	
	Sesbania punicea	Red sesbania	325	13	Riparian	1
	Solanum seaforthianum	Potato creeper	33	7	Both	1
	Solanum sisymbriifolium	Dense-thorned bitter apple	40	6	Both	1
	Sorghum halepense	Johnson grass	44	4	Riparian	2
	Tamarix spp. (T. chinensis/T. ramosissima)	Tamarisk	92	4	Riparian	1/3
	Verbena bonariensis	Purple top	58	5	Riparian	
	Verbena tenuisecta	Fine-leaved verbena	14	4	riparian	
	Xanthium strumarium	Large cocklebur	151	12	Both	1
	Zinnia peruviana	Redstar Zinnia	4	0	Both	
Widespread-scarce	*Acacia baileyana*	Bailey's wattle	87	0	Both	3
	Populus nigra var. italica	Lombardy poplar	90	0	Riparian	Proposed
Localized-abundant	*Acacia pycnantha*	Golden wattle	35	25	Landscape	1
	Albizia lebbeck	Lebbeck tree	5	33	No data	1
	Azolla pinnata var. imbricata	Mosquito fern	3	25	Riparian	
	Colocasia esculenta	Elephant's ear	10	21	Riparian	
	Echinopsis spachiana	Torch cactus	57	3	Landscape	1
	Eucalyptus lehmannii	Spider gum	41	13	Landscape	1/2
	Flaveria bidentis	Smelter's bush	19	26	Riparian	
	Hakea drupacea	Sweet hakea	28	7	Landscape	1
	Hakea gibbosa	Rock hakea	18	11	Landscape	1
	Harrisia martinii	Moon cactus	21	43	Both	1
	Hedychium coccineum	Red ginger lily	3	20	Riparian	1
	Hedychium flavescens	Yellow ginger lily	5	40	Both	1
	Hedychium spp.	Ginger lilies	7	25	Riparian	1

Helianthus annuus	Sunflower	5	17	No data	
Leptospermum laevigatum	Australian mrytle	38	30	Landscape	1
Ligustrum vulgare	Common privet	3	20	Riparian	3
Lilium formosanum	Formosa lily	16	21	Landscape	3
Litsea glutinosa	Indian laurel	8	44	Both	1
Macfadyena unguis-cati	Cat's claw creeper	27	27	Both	1
Melilotus alba	White sweet clover	15	40	Riparian	
Metrosideros excelsa	New Zealand bottlebrush	2	25	Riparian	3
Myriophyllum aquaticum	Parrot's feather	48	19	Riparian	1
Nassella trichotoma	Nassella tussock	12	21	Landscape	1
Nerium oleander	Oleander	24	6	Riparian	1
Opuntia fulgida	Chainfruit-cholla/rosea cactus	11	17	Landscape	1
Opuntia lindheimeri/Opunia engelmannii var. linderheimeri	Small round-leaved prickly pear	11	21	Landscape	1
Paraserianthes lophantha	Stinkbean	54	10	Both	1
Parthenium hysterophorus	Parthenium weed	24	37	Riparian	1
Paspalum dilatatum	Common Paspalum	6	33	Riparian	
Pennisetum villosum	Feathertop	22	21	Landscape	1
Pinus elliottii	Slash pine	34	15	Landscape	2
Pistia stratiotes	Water lettuce	27	17	Riparian	1
Pittosporum undulatum	Australian cheesewood	3	0	Both	1
Rumex usambarensis	Rumex	4	20	Landscape	
Salvinia molesta	Salvinia	33	20	Riparian	1
Schinus terebinthifolius	Brazilian pepper tree	32	16	Both	1

N.B: Major invaders grouped according to categories. 'No. grid-cells' is the number of grid-cells where the species has been recorded in the Southern African Plant Invaders Atlas (SAPIA) database; '% grid-cells abundant' is the percentage of grid-cells in South Africa where the species is recorded as very abundant or abundant in the SAPIA database (note: where more than one record with the same species and abundance code occurred within a grid-cell, it was counted as one record); 'Riparian or landscape' is the classification given to a species if more than 75% of its records in the SAPIA database fell into the respective category (if neither the landscape nor riparian records exceeded 75% then the species was classified as 'both'); and 'CARA category' lists the species regulated by the Conservation of Agricultural Resources Act (Act 43 of 1983), where 1 refers to Category 1 prohibited weeds that must be controlled in all situations; 2 includes Category 2 plants with commercial value that may be planted in demarcated areas subject to a permit, provided that steps are taken to control spread; 3 includes Category 3 ornamental plants that may no longer be planted or traded, but may remain in place provided a permit is obtained and steps taken to control their spread; and 'proposed' includes those species that were proposed for listing under the Conservation of Agricultural Resources Act, but require further investigation before they can be included.

Table 1. Major invaders plants species in South Africa according to their categories (Source: [80])

http://dx.doi.org/10.5772/56182

3. Effects of bush encroachment and invasion on rangelands

3.1. Ecological impact

It is important to establish an understanding of ecological effects of bush encroachment on rangeland ecosystems prior to embarking on any bush encroachment intervention. Thus, the degree of invasion should be quantified to help justify the need for, and determine the type of intervention. It is fundamental to characterise invasion and these could be in terms of identification of invading species (morphology, phenology, anatomy, physiology, mode of spread), plant population density, spatial localization (along the landscape, vegetation types, soil type, water distribution), seasonal distribution, their impact on the ecosystem stability (soil cover and biodiversity) and productivity (primary and secondary). The global reviews of plant invasions suggest that the most damaging species transform ecosystems by using excessive amounts of resources, notably, water, light, and oxygen. Invading species achieve these by adding resources such as nitrogen, promoting or suppressing fire, stabilising sand movement, and/or promoting erosion, accumulating litter and accumulating or redistributing salt [82]. Such changes potentially alter the flow, availability, or quality of nutrient resources in biogeochemical cycles. They further modify tropic resources within the food web and alter physical resources such as living space or habitat, sediment, light and water. In addition, invaders are most likely to have substantial effects on ecosystems by rapidly changing the disturbance regime [36]. Thus, dense stands of alien trees and shrubs in rangelands can rapidly reduce abundance and diversity of native plants [83].

Different invading species have similar or specific effects on rangeland ecosystem dynamics. Thus, invasion of black wattle (*Acacia mearnsii*) in South African rangeland ecosystems has negative ecological impacts [8]. These impacts include reduction of surface stream flow, loss of biodiversity, increase in fire hazard, and increases in soil erosion, destabilisation of riverbanks, and loss of recreational opportunities, aesthetic costs, and nitrogen pollution and subsequently loss of grazing potential. An increase in the height and biomass of vegetation increase rainfall interception and transpiration, and decreases stream flow [8]. Alien trees and shrubs increase above ground biomass and evapotranspiration and thereby decrease both surface water runoff and ground water recharge [84]. The reduction of surface water runoff as a result of current invasions was estimated to be 3 300 mm^3, which is about 7% of the national total [35], most of which is coming from the fynbos and grassland biomes [85]. The increased biomass and evapotranspiration rates associated with invasive alien plants arise because of their greater height, root depth, and senescence, compared to the native species that they replace [86]. Invasive plants may influence native ecosystems by exerting resource competition on native plants to altering fire dynamics [87]. Thus, the increased biomass that accompanies plant invasions also result in more intense fires [8, 36, 70] due to an accumulation of fuel loads. On the other hand, the dense stands of invasive trees hamper access for fire management purposes [36], which makes it difficult for fire control in rangelands. The increase in fire intensity due to accumulation of sufficient fuel load subsequently damages vegetation and soil [70], which in turn leads to excessive soil erosion due to soil water repellency caused by fire [36].

Therefore, it suffices to indicate that the alien invasive plants reduce the functional capacity of rangeland ecosystems such as support for livestock and wildlife [36, 70]. This is among others due to competition between invasive plants and grasses that are important for grazing. This competition leads to reduction on performance of a number of ecosystem functions such as grass cover, which subsequently contributes to loss of grazing potential [36]. There is also a significant loss of biodiversity due to competition [70], resulting from the displacement of species-rich indigenous plant communities by single-species stands, and disruption of important ecosystem processes [8]. On the other hand, invasion of riverbanks causes deep channelling followed by slumping during floods and that result in destabilized riverbanks. Subsequently, the invasion along the riverbanks leads to loss of recreational opportunities due to reduction of access for anglers, canoeists, white-water rafters, and swimmers. Invasive plants further detract from the wilderness character of many rural landscapes and conservation areas and that imposes reduction of the aesthetic value of ecosystems. An increase in soil nitrogen levels in nutrient-poor environments can make habitats unsuitable for indigenous plants and more susceptible to invasion by other species, and, in turn, reducing biodiversity.

In order to develop the effective invasion control in rangelands, it is significant to understand the mechanisms that are employed by the invader species to survive and colonise the new ecosystems. There are a number of ways through which invasive plants survive and outcompete the indigenous species in rangelands; one of the mechanisms is their ability to grow rapidly compared to indigenous plants. Thus, invasive alien plants typically grow more rapidly, often increasing the proportion of biomass contributed by alien plants. The large biomass contributed by invasive plants is composed of leaves, bark, seed, flowers, and twigs that become 'terrestrial litter' after abscission [88]. Such litter enters and is retained in water bodies where its rate of breakdown by invertebrate feeding as well as decomposition through fungal and bacterial activity differs from that of inputs from indigenous plants [89]. The often large differences in litter inputs from invasive alien plants relative to indigenous species leads to reduced decomposition rate and dramatically alters the nutrient cycle in rangeland ecosystem [90]. Additions in the biomass contributed by alien plants can increase the amount of metabolised nutrients, which in turn escalates natural eutrophication processes [91] as well as free-floating and rooted aquatic macrophyte invasions [92]. Thus, eutrophication leads to gradual changes in the plant and animal populations and the development of potentially toxic algal blooms and, therefore, a slow decline in water and habitat quality [91]. The level of impact that litter from invasive alien plants has on nutrient cycles is determined by vegetative spread, plant structure, phenology, plant water and nutrient uptake efficiency, photosynthesis type, presence of symbionts and nitrogen fixation, phosphorus content and tissue chemistry such as allelopathy [93].

http://dx.doi.org/10.5772/56182

Habitat–propagule pool size	Scientific name	Common name	Impact	Weediness	Biocontrol	% Weedy relatives	Combined Score	CARA category
Large–large	*Bromus diandrus*	Ripgut brome	0	2	10	5	53	
	Pinus taeda	Loblolly pine	10	1	10	4	87	2
	Tecoma stans	Yellow bells	5	1	10	3	69	1
	Tipuana tipu	Tipu tree	5	1	10	10	73	3
Large–moderate	*Celtis sinensis/*	Chinese nettle tree/						
	Celtis occidentalis/	Common hackberry/						
	Celtis australis	European hackberry	0	1	10	1	45	Proposed
	Cytisus scoparius	Scotch broom	5	5	10	4	86	1
	Pennisetum purpureum	Elephant grass	10	3	10	2	95	Proposed
	Pereskia aculeata	Pereskia	10	1	10	2	87	1
	Rosa rubiginosa	Eglantine	10	3	10	3	96	1
	Toona ciliata	Toon tree	5	1	10	2	64	3
	Ulex europaeus	European gorse	5	5	10	1	80	1
Large–small	*Acacia paradoxa*	Kangaroo thorn	5	2	10	3	69	1
	Pueraria lobata	Kudzu vine	5	3	10	5	76	1
	Triplaris americana	Triplaris	5	0	10	1	62	1
Moderate–large	*Acacia elata*	Pepper tree wattle	5	2	10	3	69	3
	Acacia podalyriifolia	Pearl acacia	5	1	10	3	67	3
	Ardisia crenata	Coralberry tree	5	1	10	0	66	1
	Cinnamomum camphora	Camphor tree	10	2	10	0	90	1/3
	Cotoneaster franchetii	Orange cotoneaster	5	2	10	1	69	3
	Cotoneaster pannosus	Silver-leaf cotoneaster	5	2	10	1	69	3
	Eucalyptus cladocalyx	Sugar gum	5	1	10	2	68	2
	Eucalyptus saligna	Saligna gum	5	1	10	2	66	
	Eugenia uniflora	Surinam cherry	5	2	10	0	68	1

Habitat–propagule pool size	Scientific name	Common name	Impact	Weediness	Biocontrol	% Weedy relatives	Combined Score	CARA category
	Hedychium coronarium	White ginger lily	10	2	10	1	87	1
	Hedychium gardnerianum	Kahili ginger lily	10	3	10	1	92	1
	Ligustrum japonicum	Japanese wax-leaved privet	5	1	10	3	66	3
	Ligustrum lucidum	Chinese wax-leaved privet	5	4	10	3	78	3
	Ligustrum ovalifolium	Californian privet	5	1	10	3	68	3
	Ligustrum sinense	Chinese privet	5	4	10	3	80	3
	Lonicera japonica	Japanese honeysuckle	5	6	10	1	83	Proposed
	Myoporum serratum	Manatoka	5	0	10	2	84	3
	Myoporum tenuifolium ssp. *montanum*	Manatoka	5	0	10	2	69	
	Nephrolepis exaltata	Sword fern	10	0	10	3	82	1
	Pyracantha coccinea	Red firethorn	5	0	10	8	61	
	Spartium junceum	Spanish broom	5	3	10	10	82	1
	Syzygium paniculatum	Australian water pear	5	0	10	0	61	
Moderate–moderate	*Albizia procera*	False lebbeck	5	1	10	2	64	1
	Alhagi maurorum	Camelthorn bush	5	2	10	10	79	11
	Anacardium occidentale	Cashew nut	5	1	10	1	63	
	Callistemon rigidus	Sitt-leavedbottlebrush	0	1	10	1	45	Proposed
	Catharanthus roseus	Madagascar periwinkle	0	2	10	3	51	
	Cestrum parqui	Chilean cestrum	10	3	10	1	91	1
	Cynodon nlemfuensis	East African couch	5	2	10	10	76	

Habitat–propagule pool size	Scientific name	Common name	Impact	Weediness	Biocontrol	% Weedy relatives	Combined Score	CARA category
	Cytisus monspessulanus	Montpellier broom	5	0	10	4	66	1
	Duranta erecta	Forget-me-not	0	1	10	1	44	Proposed
	Eriobotrya japonica	Loquat	0	2	10	0	50	3
	Ficus carica	Fig	0	2	10	0	50	
	Gleditsia triacanthos	Honey locust	5	2	10	1	68	2
	Leucaena leucocephala	Leucaena	5	3	4	3	52	1
	Mangifera indica	Mango	0	1	10	0	46	1
	Montanoa hibiscifolia	Tree daisy	0	1	10	1	44	
	Passiflora edulis	Passion fruit	0	2	10	1	50	1
	Passiflora subpeltata	Granadina	0	1	10	1	46	
	Physalis peruviana	Cape gooseberry	0	2	10	5	54	
	Phytolacca octandra	Forest inkberry	0	2	10	6	55	
	Pyracantha crenulata	Himalayan firethorn	5	1	10	8	73	3
	Senna bicapsularis	Rambling cassia	5	0	10	1	62	3
	Senna pendula var. *glabrata*	Rambling cassia	5	2	10	1	68	3
	Sesbania bispinosa var. *bispinosa*	Spiny sesbania	0	0	10	4	45	
	Sophora japonica	Japanese pagoda tree	0	0	10	2	42	
	Syzygium cumini	Jambolan	5	1	10	0	66	3
	Syzygium jambos	Rose apple	5	1	10	0	66	3
	Tithonia diversifolia	Mexican sunflower	0	1	10	3	48	1
	Ulmus parvifolia	Chinese elm	0	0	10	5	46	
	Verbena brasiliensis	Slender wild verbena	0	1	10	2	45	
Riparian–large	*Canna indica*	Indian shot	5	2	10	10	79	1
	Canna x *generalis*	Garden canna	5	1	10	10	72	

Habitat–propagule pool size	Scientific name	Common name	Impact	Weediness	Biocontrol	% Weedy relatives	Combined Score	CARA category
	Casuarina cunninghamiana	Beefwood	5	1	10	4	69	2
	Cortaderia jubata	Purple Pampas	5	3	10	2	75	1
	Cortaderia selloana	Pampas grass	5	5	10	2	81	1
	Oenothera biennis	Evening primrose	5	1	10	4	67	
	Populus deltoides	Match poplar						Proposed
	Eucalyptus microtheca	Coolabah	0	0	10	2	42	
	Mimosa pigra	Giant sensitive plant	5	4	10	1	76	3
	Myriophyllum spicatum	Spiked water-milfoil	5	4	10	3	80	1
	Oenothera glazioviana	Evening primrose	5	2	10	4	72	
	Oenothera indecora	Evening primrose	5	1	10	4	68	
	Oenothera jamesii	Giant evening primrose	5	0	10	4	64	
	Oenothera laciniata	Cutleaf evening primrose	5	1	10	4	67	
	Oenothera tetraptera	White evening primrose	5	0	10	4	66	
	Parkinsonia aculeata	Jerusalem thorn	5	1	10	0	66	
Small–large	*Alpinia zerumbet*	Shell ginger	5	0	10	0	62	
	Grevillea robusta	Australian silky oak	5	2	10	0	67	3
	Quercus robur	English oak	5	1	10	1	67	

N. B: Scores for 'Impact', 'Weediness', Biocontrol' and 'Weedy relatives' are standardized by dividing the maximum score for that criterion and multiplying by 10. Scores for these four criteria were weighted, with 'Impact', 'Weediness' and Biocontrol' receiving an equal weighting of four, and 'Weedy relatives' receiving a lower weighting of one. The weighted criteria were summed to obtain the 'Combined score' for each species. 'CARA category' lists the species regulated by the Conservation of Agricultural Resources Act (Act 43 of 1983), where 1 refers to Category 1 prohibited weeds that must be controlled in all situations; 2 includes Category 2 plants with commercial value that may be planted in demarcated areas subject to a permit, provided that steps are taken to control spread; 3 includes Category 3 ornamental plants that may no longer be planted or traded, but may remain in place provided a permit is obtained and steps taken to control their spread; and 'proposed' includes those species that were proposed for listing under the Conservation of Agricultural Resources Act, but require further investigation before they can be included.

Table 2. Emerging invaders grouped according to categories (Source: [80])

http://dx.doi.org/10.5772/56182

The majority of invasive and/or encroaching species in rangelands is dominated by the genus *Acacia,* which is the second largest with over 900 species [70]. Australian acacias are important invaders of South African rangeland areas [94]. In the fynbos ecosystems where soil nutrients are generally poor, the invasion by nitrogen-fixing acacias increases nitrogen inputs, and subsequently leads to an increase in soil fertility. Therefore, the massive increase in soil fertility permits acacia species to propagate and outcompete indigenous species [90]. There are a number of *acacia* species found in rangelands and their ability to fix nitrogen has been widely reported; these include *Acacia cyclops, A. dealbata, A. mearnsii* and *A. saligna* [90, 95]. The groundwater on places that were invaded by *A. saligna* has shown elevated NO_3^- and NO_2^- concentrations compared to groundwater in natural ecosystems [94]. The presence of *A. saligna,* as well as the nutrient leaching that occurred after its removal, result in seasonal nitrogen concentrations that are higher than the water quality targets for domestic use (NO_x < 6 mg/l) [94, 96]. Therefore, the removal of alien plants would be beneficial from both a water quantity as well as water quality perspective [94].

In natural communities, plants compete in different ways; one of these ways is chemical interactions in the form of allelopathy [87, 97]. Invasive plants interfere with other plants by releasing allelochemicals into the environment and that negatively affects surrounding plants, thus giving the producer a competitive advantage. Invasive plants possess physiological traits that enable them to exploit ecological opportunities. The word allelopathy comes from the Latin words *allelon,* which means of each other and *pathos,* which means to suffer, which is commonly associated with the chemical inhibition of one species of plants by another [98]. Allelopathy is the process through which invasive plants such as *eucalyptus, Pinus, Chromolaena* and *Lantana* produce biochemicals that influence the growth, survival, and reproduction of indigenous species. However, it is important to note that most of the plant species naturally produce number of allelopathic substances such as monoterpenes and phenols [97]. Phenolics and volatile compounds can be released from eucalyptus foliage. These biochemicals can act as antibiotics in certain soils, possibly affecting nitrogen cycles.

Although it has not been evaluated, the impacts of allelochemicals may subsequently influence water quality through soil erosion or surface runoff processes [70]. Allelochemicals are believed to be present in almost all plant tissues such as leaves, flowers, fruits, stems, roots, rhizomes, seeds, and pollen where they may be released from plants into the environment by means of volatilization, leaching, root exudation, and decomposition of plant residues [99, 100]. Invasive plants use the mechanism of allelopathy to outcompete other plants [87]. Allelochemicals can be found present in litter and on the soil surface where plants grow. Rain assists with the leaching of allelopathic substances into the soil, where they may affect the germination and growth of other plants [97, 101]. Allelopathic substances might play a role in shaping plant community structure in semi-arid and arid environments [97]. Thus, allelopathic substances inhibit plant growth depending on the concentration, leachability, season, and age of the plants [101]. Phytotoxins can persist in the soil and litter layer for long after allelopathic plants senesce, thereby reducing the establishment potential of an area. Allelopathic substances can be present in the soil and often determined by a number of important factors [97]. These factors include the density at which the leaves fall, the rate at which this material decomposes, the distance from other plants and, finally, rainfall [101, 102, 103]. Phenolics signify the main allelopathic compounds that inhibit seed germination, plant growth and other physiological processes that result in changes of floristic composition within a plant community.

Competition between plants can lead to the allelopathic inhibition of germination or growth via phytotoxic chemical releases, which are caused by competing species. However, allelopathy can be extremely difficult to demonstrate in the field due to difficulties in differentiating allelopathic effects from resource competition [87, 99]. Allelochemical compounds are in fact released into the soil and accumulate to levels of toxicity, which leads to inhibition of germination [100]. Allelochemicals released by invasive plants may affect native plant survival and production in a number of ways. These include the modification of the soil microbiota [74, 104], and enhancement of growth of beneficial microbes in their rhizosphere leading to an establishment of positive feedbacks that can contribute to the decrease of native biodiversity [74]. Allelochemicals are further known to inhibit absorption of ions [105]. Other than allelopathic effects, invasive plants exert competition of resource especially through light [87]. Therefore, allelopathy and resource competition operate simultaneously influencing each other and, in the meantime, they are influencing plant community structure [106].

Allelochemicals, as soon as released into the soil, may inhibit germination, shoot, and root growth of other plants, which will affect nutrient uptake thereby destroying the plant's usable source of nutrients [107]. Allelopathy of invasive plants delays the germination and growth of seedlings of other species and eventually hinders their growth completely. Therefore, degree of inhibition due to allelopathy is largely dependent on the concentration of the extracts and, to a lesser extent, on the species from which they were derived [101, 108]. The effects of allelopathy on germination and growth of plants occur through a variety of mechanisms including reduced mitotic activity in roots and hypocotyls, suppressed hormone activity, reduced rate of ion uptake, inhibited photosynthesis, and respiration, inhibit protein formation, decreased permeability of cell membranes and/or inhibition of enzyme action [97]. Plants that germinate at slower rates are often smaller; thereby, this may seriously influence their chances of competing with neighbouring plants for resources such as water [109]. Indirectly, allelopathic effects of invasive species on germination and growth of native species determine their competitive ability against them [97]. The roots of *Aloe ferox* have allelopathic inhibition on tomato seed germination [97]. Accumulation of allelochemicals in the rhizosphere because of root and microbial exudates and/or metabolism may affect the germination. However, under arid conditions germination will be less affected since microbial activities are very low due to low availability of soil moisture [101]. The effects of allelochemicals on the root growth are due to cell division destruction [105]. *L. maackii* also exudes allelopathic compounds from its leaves or roots that inhibit germination and growth of species that grow on the same site [87]. Allelochemicals could be found on any part of the plant; however, the concentration varies with plant parts. The leaf extracts of *L. maackii* appeared to have a more negative effect on seed germination than root extracts [87]. Generally, leaf extract concentrations have a stronger effect on germination of seeds of other plants [87]. However, it is important to note that allelopathic chemicals from one plant can hinder germination of seeds of the same plant. For example, chenopod seed germination can also be inhibited by extracts generated from its leaves [97].

http://dx.doi.org/10.5772/56182

However, all extracts, except the one obtained from the leaves of *E. tomentosa* significantly inhibited the germination of lettuce seed and appeared to stunt the growth of roots and shoots of germinants [97].

There are different allelochemicals exuded by invasive plants; these may have direct and indirect effects on germination and establishment of native species. However, phenolics are widely recognized for their allelopathic potential in plants, and can be found in a variety of tissues. Phytotoxic activity of allelochemicals in soil has been considered as plant-to-plant interaction, which is mediated by chemicals released from the plants [99]. Indirect effects of allelochemicals include its influence on the availability of nutrients in the soil, which may cause changes in soil chemical characteristics [110]. Allelochemicals might inhibit the growth of nitrifying bacteria, which would decrease N-availability at the plant level [111]. Additionally, chemical compounds produced in the process of litter decomposition are inhibitory for both heterotrophic and autotrophic bacteria and fungi [110,111] and, thus, rates of mineralization may be reduced. Allelochemicals such as phenolic acids are considered to have an important influence on nutrient cycling in terrestrial ecosystems [110]. The allelochemicals can produce some changes in the resource exploitation competition in such way that allelochemicals affect the mycorrhizae that allow the plant to absorb the nutrients, which leads to decrease in the soil productivity [106, 112]. Soil microorganisms are affected by root exudates that eventually affect other plant roots. Some chaparral species produce substances, which accumulate on the soil surface and make the soil less wettable [111]. The allelochemicals affect availability and accumulation of inorganic ions, although their activities are influenced by ecological factors such as nutrient limitation, light regime and soil moisture deficiency [106].

Allelochemicals, such as phenolics and terpenoids, play an important role in the inhibition of nitrification and, thus, influence soil productivity of a plant community [113]. Thus, any influence on nutrient dynamics may ultimately affect the growth of plants in the community, which will lead to the increase of invasive plants. Reduced soil fertility may enhance the production of allelochemicals from invasive plants [106]. The addition of plant litter to soil may influence nutrient mobilization and soil pH, which can further influence nutrient immobilization and microbial activity [114]. Therefore, litter can alter the chemistry of the soil in such a way that it inhibits germination of other plants [106]. Chemicals released into the environment by a plant may not necessarily have direct effects on community structure but abiotic soil factors can influence these chemicals. Many phenolic acids have potential to influence microbial population, cause a shift in the microbial community, and eventually affect soil productivity of the area [106]. The soil microflora is directly responsible for decomposition and mineralisation processes and soil fauna is of considerable importance in regulating these processes through influencing the growth and activity of soil microbes [115]. Allelochemicals exuded from roots of invasive plants and residue decomposition play an important role in inhibiting plant pathogens particularly those borne in soil [116]. However, amended soils with allelopathic residues tend to be rich in organic matter [117]. Electrical conductivity (EC) of the amended soils increased as compared to the control and all nutrients were significantly more [117]. Although, earlier reports show that inclusion of plant litter, in addition to releasing putative phytotoxins into the soil medium, alters the soil nutrient dynamics and, thus, affects the plant growth [106, 112, 116]. A similar increase in electrical conductivity of the soil incorporated with residues of allelopathic plants was reported [118]. In fact, the behaviour of the allelopathic compounds present in soil remains unclear [119].

The modes of release of the allelopathic compounds are not specific because they vary from plant to plant [120]. Thus, allelochemicals are released into the environment by root exudation, leaching from aboveground parts, volatilisation, and decomposition of plant material and ultimately enter into the soil [99, 110, 121]. Therefore, allelochemicals may reach other plants through transport such as root exudates into the soil and may induce the inhibitory activity on the other plants. The behaviour of allelochemicals in soil is run by the physicochemical properties including soil organic matter and organisms [99]. The model that has assumptions such as "allelochemicals are released into the soil from living plants and degraded into non-allelopathic substances was developed. Therefore, rate of the release is proportional to the amount of allelochemicals in living plants and rate of allelochemicals degradation is proportional to the amount of allelochemicals released [121]. However, the soil microorganisms were also reported to produce and release allelochemicals [112]. The release of allelochemicals by mature shrubs may inhibit plant germination, survival or growth [111]. Allelopathic content of a plant varies according to its maturity [122]. Allelopathic compounds released from different plant parts can be either released continuously within specific periods such as specific developmental stages or influenced by external factors such as precipitation [123]. The synthesis and exudation of allelochemicals via roots is usually enhanced by stress conditions that the plant encounters such as extreme temperature, drought, and ultraviolet exposure [124].

The visible effects of allelopathy frequently observed are inhibited or delayed seed germination or reduced seedling growth. The diversity of structure among allelochemicals suggests that they have no common mode of action [110]. Plant exudates can also have an indirect effect on the surrounding environment and reduce neighbouring plant germination or growth, independent of toxicity [111]. Allelopathic activities are more pronounced when allelopathic potential species grow under water stress [125]. Phenolic acids that were tested had a similar mode of action such as inhibition of nutrient uptake by roots of plants [126]. In most cases, various allelochemicals take action as growth regulators by inhibiting growth and changing development [112]. The common mode of action of allelochemicals is quite related to the membrane destruction [126]. It was discovered that allelochemicals affect plants on cell division, cell elongation, cell structure, cell wall, ultrastructure of the cell [112, 127]. Phenolic allelochemicals can also lead to increased cell membrane permeability; cell contents spill which lead to the increase of lipid peroxidation, and eventually, slow growth or death of plant tissue occurs [112, 126, 127]. Furthermore, nutrient uptake can be affected negatively by allelochemicals. This occurs when these allelochemicals inhibit nutrient absorption of the plant [127]. The mode of action of benzoic acid involved the inhibition of nutrient uptake by plant roots, which resulted in growth inhibition [126]. The radicle elongation was significantly reduced by the extract of leaves, and leaves and stem at the three concentrations of *Acacia mearnsii*, which signifies that *A. mearnsii* has allelopathic potential [128]. The impact of allelochemicals also have been observed on the respiration of the plants which affect oxygen absorption capacity [127], eventually inhibit photosynthesis by reducing the chlorophyll content which affect photosynthesis rate [98, 112, 126]. There is an inhibition of the activity of hydroxyphenyl-pyruvate dioxygenase (HPPD) enzyme due to isoxaflutole, which results in the inhibition of meristmatic tissue, which leads to inhibition of shoot growth [126]. Therefore, the modes of action of most allelochemicals and phytotoxins are complex and are not clearly understood [126].

The active compound or compounds must be isolated in an amount adequate for identification and for further characterisation in bioassays [110]. Screening of fractions of plant extracts or leachates for their effects on seed germination of various plant species are frequently used to identify phytotoxic compounds [110]. The identification of an active phytotoxic compound from a suspected allelopathic plant does not establish that this is the only compound involved in allelopathy. The release of allelochemicals of different chemical classes from allelopathic plant species has been documented including tannins, cyanogenic glycosides, several flavonoids and phenolic acids [129]. The most clearly identified compounds can be divided into four groups: phenolic acids, hydroxamic acids, alkaloids, and quinones. In the study of allelopathy, plants are identified based on the allelochemical release [120]. Most studies utilized some parts of the plants such as roots, leaves and leaves plus stem to establish the existence of allelochemicals on the identified plants [107, 128].

3.2. Economic impacts

Rangelands contribute to the economy of Southern Africa in a number of ways. They provide agricultural commodities that can be valued in the market such as wool, meat, milk etc. These are the major source of forage for grazing animal which in turn influence animal production. Rangelands further provide benefits that, are not directly related to the agricultural sector, such as wildlife habitat, however, have an impact on the economy through activities that make use of them [130]. Increases in the density of woody plants worldwide are a major threat to livestock production [13, 131], and rangeland biodiversity. Invasive species pose problems for managers of rangelands because they reduce the land's usefulness for grazing activities. In addition, they interfere with other non-agricultural functions that rangelands provide, such as acreage of wildlife habitat and watershed quality. Therefore, in order to realise the impact of invasion on rangelands, it is important to understand the total economic loss that invasive plant infestations create on the economy in relation to both its agricultural and non-agricultural products of the ecosystems [130].

Economic impact of invasive species could be defined as the product of a species' range, abundance and per capita [36, 80, 132]. Although the invasive plants have an ecological implication they also have some economic implications; these could be either positive or negative. Species such as *Acacia mearnsii* (Black wattle) are highly invasive and have spread over an area of almost 2.5 million ha in South Africa [133]. It has significant negative impacts on water resources, biodiversity, and the stability and integrity of riparian ecosystems [8]. These two features, a commercial value on the one hand, and an invasive, damaging ability on the other, give rise to a classic conflict of interests, where the benefits accrue to a number

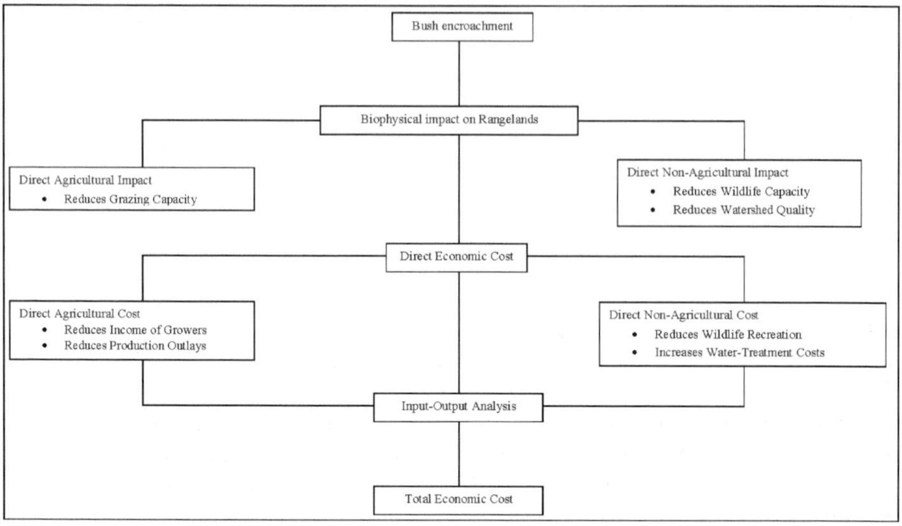

Figure 1. Hypothetical flow chart indicating economic impact of bush encroachment in rangelands (Source [80]).

of people, while the society at large bears the external costs. Furthermore, there are larger reductions of water resulting from the presence and densities of invasive plants. Thus, the potential water reductions in South Africa would be more than 8 times greater if invasive alien plants were to occupy the full extent of their potential range [85]. These invasions come at a significant cost to the economy, estimated at about R6.5 billion per annum, which is about 0.3% of South Africa's GDP of around R 2 000 billion, and with potential to rise to > 5% of GDP if invasive plants were to be allowed to invade all of the suitable habitat [134]. Economic of bush encroachment in rangelands can be divided into agricultural and non-agricultural, direct and indirect impacts, and, further, into primary and secondary impacts (Figure 1). Economic impacts of plant invasions may be related to a decline in cattle carrying capacity (agricultural impact), wildlife carrying capacity, and watershed quality (non-agricultural impacts). Reductions in cattle grazing outlays may account for the direct agricultural costs. In addition, economic impacts may be estimated as reductions in wildland-associated recreation expenditures and increases in expenditures to mitigate damages from runoff and soil erosion to account for the non-agricultural losses. These estimated losses are incorporated into an input–output model of economy to compute total (direct plus secondary) economic costs incurred due to the invasion of noxious weeds [130]. Secondary economic effects of bush encroachment include indirect and induced losses on the economy. Indirect losses are linked to economic sectors not necessarily directly affected by the infestations, but these sectors supply inputs needed by directly affected industries. Induced effects represent changes in household spending patterns, caused by changes in employment that the direct and indirect effects generate.

http://dx.doi.org/10.5772/56182

4. Management of rangelands for bush encroachment and invasion

4.1. Bush encroachment control

Bush encroachment forms dense infestations that rapidly deplete soil moisture, preventing the establishment of other species. As it displaces native vegetation, it reduces wildlife habitat and ecosystem diversity, and suppresses production of nutritious, palatable forage for wildlife and livestock, which leads to a reduction in grazing and wildlife carrying capacity. Soil and water conservation benefits of the regions rangelands also decline; watershed quality declines in areas where the weeds have advanced [135].

Bush encroachment is considered a threat to forage production, which is the feed for the grazing livestock [42]. The threat to the pastoral economy by bush encroachment and invasion is often the main reason for the control of bush encroachment [136]. Bush encroachment control is a disturbance that reduces the threat of bush encroachment by disrupting the invasive woody plant community structure through transformations of biotic environments and habitat conditions in which colonization of the disturbed microhabitat takes place. Bush control methods shift the rangeland vegetation from dominance by woody vegetation to dominance by herbaceous vegetation. This control of the bush is aimed at creating suitable habitat for grazers [137, 138]. Thus, forage production of herbaceous vegetation increases with reduction of woody species. The principle of bush encroachment control is based on the ability of the control method to shift the competition between desired and undesired species. Encroaching species have the higher competitive ability over the native species, which is why they colonise. They build up this competitive advantage by modifying the environment in such a way that growing conditions will suit their needs through a number of ways. These include release of chemical substances that suppresses germination and growth of their competitors (Allelopathy) and modification of soil fertility in the case of acacias through higher nitrogen inputs, which in turn favours their growth. Encroaching species also impose competition for light and through shading and subsequently growth for native species becomes negatively affected. There is also a competition for soil moisture and soil nutrient; in this manner, most of the invasive plants win because of their deeper root systems. Other invasive species produce large numbers of seeds, which normally are dispersed faster, have a shorter dormant time before germination, and colonise. Invasive plants use one or a combination of these mechanisms for survival. Therefore, bush encroachment control reduces the ability of invasive plants to exhibit these survival mechanisms. The use of selective herbicides is aimed at reducing the competitive ability of invasive species through killing them and, in that, species that are not affected by this herbicide gain an advantage. Mechanical methods such as hand clearing targets unwanted plants and create a competitive space for desired plants, thus, without this clearing the invasive species are more competitive. Use of fire to control invasive woody plants is justified by the fact that when woody plants are burned they do not recover or they take a longer time to recover which gives the herbaceous species time to grow with minimal or no competition. In the biological control method, use of herbivores such as goats to selectively-browse on the encroaching species or use of invertebrates that feed on the seed of invading species also reduces the competition against native plants.

It is important to mention that the shift towards herbaceous species dominance, in turn, may induce shifts in herbaceous species that tolerate bush cover and such species might decline in numbers [139]. The changes could cause partial or total reduction of plant biomass [140] by shifting vegetation structure and composition [141]. Furthermore, disturbance can produce changes in the life history strategies of individual species in response to intensities of disturbance forces [140] and the created micro-environmental conditions [142]. Although livestock-forage production of rangelands may support removal of encroaching species to enhance forage production, it is important to note that bush encroachment control methods are management systems [137] that might have varied policy implications for bush control [143]. Therefore, understanding the potential role of different bush encroachment control methods for promoting herbaceous species composition requires recognition of the objectives of resource users and policymakers [144]. Thus, the intended ecosystem status is dependant of the functional characteristics of such an ecosystem.

4.2. Bush encroachment management methods

4.2.1. Rangeland management practices

Grazing management entails management of livestock and vegetation resources. The main livestock decisions made by farmers both in the commercial and communal areas are concerned with livestock type, number and seasonal pattern of movement [145]. Commercial and communal livestock farming are generally regarded as the rangeland management systems and they are distinct in grazing management practices. Thus, communal grazing areas are generally characterised by continuous grazing, which is perceived by most of the scientists to be the root cause of the often-reported land degradation in this system. On the other hand, commercial livestock farming is characterised by structured and objective grazing management practices such as assigning the correct livestock units in proportion to the carrying capacity of the land. These would be done in rotation to give vegetation in grazed areas time to recover such that the rested areas can be grazed again. Understanding the dynamics of bush encroachment in relation to rangeland management systems over a broad range of environments is essential for sustainable management of rangeland ecosystems [146]. Although rangelands are complex ecosystems varying at multiple scales in time and space [147, 148], most management usually intends to maintain or enhance livestock production by reducing plant community variability in space and time [149, 150]. This is usually accomplished by promoting spatially uniform dominance of a few productive forage species. Although it is generally believed that improper grazing practices leading to overgrazing are responsible for bush encroachment, it is not attributed to heavy grazing alone, but is strongly influenced by seasonality, which is a characteristic of arid and semi-arid environments [42]. In combination with seasonality, the ban on fire and exclusion of browsing animals such as goats and camels may also contribute to the invasion of bush encroachment.

Rangeland management practices, particularly fire suppression and overgrazing, have been reported to increase the proportion of some native species [70]. These natives can reduce overall forage quality or quantity (e.g. *Juniperus* spp., *Artemisia tridentata*, and *Gutierrezia* spp.)

or poison livestock (e.g. *Delphinium* spp., *Astragalus* spp., and *Amsinckia menziesii* var. *intermedia*). One of the challenges of managing invasive species is that there is no particular life cycle typical to noxious weeds of rangelands reported [151]. Thus, noxious rangeland weeds can be annuals (e.g. *Centaurea soltitialis, Crupina vulgaris, Bromus tectorum*), biennials (e.g. *Carduus nutans, Conium maculatum, Onopordum acanthium*), long-lived herbaceous perennials (e.g. *Convolvulus arvensis, Centaurea maculosa, Cirsium arvense*), shrubs (e.g. *Gutierrezia* spp., *Artemisia tridentata*), or trees (e.g. *Juniperus* spp., *Prosopis glandulosa*). Although several plant families represent these species, the largest number of noxious species belongs to the Astereceae (sunflower) family.

Effective rangeland management requires sound ecological data about the land being managed; however, obtaining such data is not sufficient to ensure the implementation of restoration practices by land users. Thus, rational decisions at the farm or community, regional and national levels, depend on researchers providing not only ecologically sound but also economical, effective alternatives for land use [152]. In addition, because natural resource depletion and recovery compound over time, it is necessary to assess the sustainability of management alternatives over decadal periods [153]. Furthermore, to determine the true advantage of restoration management, it is necessary to compare the benefits of changing management practices with the cost of not changing current practices, which, rather than maintaining productivity, may lead to loss of production through shifts in plant species composition, accelerated soil erosion, and loss of biodiversity.

4.2.2. Chemical — Herbicides

Chemical control methods are usually expensive to apply and should be considered only under specific circumstances. Thus, their nature are suited primarily to the initial thinning of bush at high density, where there is poor fuel load to support fire, where trees are above the browse line, where the bush is unacceptable to animals and where the herbicide is intended to selectively kill a specific plant [154]. However, herbicides can sometimes be used in follow up operations such as after fire where there is a need for pre-emergence herbicide application intended to kill the seedlings of a target plant in soil. Herbicides have been applied extensively on rangelands to reduce forbs that were considered undesirable, which have been assumed to lead to an increase in grass production and ultimately to an improvement in livestock performance [155]. Herbicides are the primary method of weed control in most rangeland systems [151]. In South Africa, there is a considerable effort taken by the government to address the negative impact of alien invading species on the natural and environmental resources of the country [8].

Herbicides vary in their chemical properties, that make them vary more with their mode of action under different climatic and soil conditions, and they further vary in their methods of application and their effect on the ecosystems. There are two broad groups of herbicides used in rangelands. The first type is composed of the herbicides that are applied on the soil surface and are absorbed by the roots; these are the herbicides that are based on tebuthiuron, ethidimuron or bromacil as their active ingredient [154]. The second group of herbicides is sprayed onto the plant and absorbed directly by the foliage and other above ground parts of the plants;

these herbicides have picloram as the active ingredient. The second group may also have ingredients such as 2, 4-D and 2, 4, 5-T. Soil applied formulations are marketed as granules, wettable powders or as liquid with active ingredients ranging in concentration between 20% and 70%. Granular products can be applied by hand, with some suited to aerial application. Wettable and liquid products are mixed with water and applied on the soil surface adjacent to the stem of the plant. The application rates of soil formulations vary according to clay content, organic matter and pH of the soil. These herbicides remain in the soil inactive until it rains such that the active ingredient can dissolve in water so that the roots can absorb it. Herbicides applied directly to the plant normally have an oil or water base and are applied to either the stem or the leaves of the plant.

In South Africa, of particular note are herbicides containing bromacil (5 – Bromo -3- sec – butyl – 6 – methyluracil) as the active ingredient (a. i.) which are used to control encroaching species. These herbicides include Bushwacker SC (Enviro Weed Comtrol Systems (Pty – Ltd), Bushwacker GG (Enviro Weed Control Systems (Pty Ltd) and Rinkals 400 PA (Dow AgroSciences LLC) e.t.c [156]. These herbicides vary primarily in their bromacil concentration, thus, Bushwacker SC contains 500 g of bromacil per litre, Bushwacker GG contains 200 g of bromacil per kilogram and Rinkals 400 PA contains 400 g of bromacil per kilogram. These herbicides are usually selective within certain application rates, environmental conditions, and methods of application. Bromacil works by interfering with the photosynthetic pathway of plants [157]. Its application is usually done just before the active growth stage of plants, thus, before the wet season stabilizes. It quickly dissolves in soil water and may stay in the soil for several years [157]. Bromacil is readily absorbed through the root system [158] and is a specific powerful mobile inhibitor of photosynthesis [159]. The target plant must be undergoing active photosynthesis for the herbicide to be effective. It inhibits photosynthesis by blocking the photosystem II reaction, thereby, preventing the conversion of sunlight into chemical energy, thus, it blocks the photosynthetic electron transport [159]. Bromacil blocks electron transport from QA to QB in the chloroplast thylakoid membranes by binding to the D-1 protein at the QB binding niche. The electrons that are blocked from passing through photosystem II are transferred through a series of reactions to other reactive toxic compounds. These compounds disrupt cell membranes and cause chloroplast swelling, membrane leakage, and ultimately cellular destruction [160]. Inhibition of photosynthesis thus results in slow starvation of the target plant and eventual death. It is translocated upward via the xylem to foliage and interferes with light-harvesting complexes [159]. In the soil, there is little adsorption of bromacil to soil colloids, therefore, it moves (leaches) through the soil and it can contaminate groundwater [157]; however, it is highly susceptible to microbial degradation [161]. When used as a selective herbicide, it can persist in the soil for one year; however, if it is applied at high concentrations, it can persist for more than one year [161].

The herbicide 2,4-D [(2, 4-dichlorophenoxy) acetic acid] is also a commonly used herbicide in the rangeland vegetation management [162]. Combined estimates of 2,4-D use annually on cropland, pastureland, and rangeland could range from 12.7 to 14.9 million kg [163]. Native and exotic dicots are primary targets of many rangeland herbicide applications [162, 164]. However, these plants also contribute key structural, vegetation, and nutritional elements to

http://dx.doi.org/10.5772/56182

wildlife habitat [165] and livestock diets [166]. Some forbs are foraged by animals especially during the seasons when forage is scarce. Therefore, reducing forbs with herbicide might influence ecosystems across trophic levels and potentially alter ecosystem function. Furthermore, biodiversity has been proposed as a source of stability in managed ecosystems [167, 168]. Therefore, decreasing forb diversity with the use of phenoxy herbicides like 2,4-D alters arthropod habitat and reduces arthropod diversity, which influences higher trophic levels [149,169]. The decrease in forb abundance and diversity beyond normal temporal dynamics could be detrimental to wildlife because forbs also comprise key structural, vegetative, and nutritional elements [165].

Although herbicides are considered effective in controlling weeds, they are often facing the challenge with evolution of resistant weed populations [170, 171]. Thus, depending on both the population's genetic background and ecological scenario, apart from expressing herbicide resistance, weed species adapt to herbicides by phenological changes [172, 173]. Comparisons of herbicide-resistant and susceptible biotypes have shown that populations can vary not only in morphological traits but also in developmental responses, such as relative growth rate, photosynthetic rate or germination rate [174, 175]. Adjusting seed germination time and rate has been considered as one of the potential mechanisms by which annual weeds can improve their competitive ability in agricultural scenarios [173, 176]. Hence, success of annual weed species in cropping systems may be assessed through the degree of synchronization of germination (determined by factors controlling exit from dormancy), ability to germinate at high rates (determined by seed response to environmental factors, mainly temperature), and seed longevity (determined by genotype and seed response to environmental factors promoting ageing).

On the other hand, herbicides have some effects on the environment, thus, some plants and animals, which are not targeted are also exposed. The environmental fate of herbicides is related to chemical and physical properties of the products, amount, and frequency of use, methods of application, abiotic and biotic characteristics of the environment, and meteorological conditions [177]. At the recommended rates of use in agriculture, the half-life of herbicides ranges from up to 1 month (e.g. 2, 4-D), to 3-12 months (e.g. atrazine, trifluralin, metsulphuron methyl), to more than 1 year for picloram, tebuthiuron, pendimethalin, chlorsulphuron, and ethametsulphuron methyl [178, 179]. Persistence can be extended under certain use conditions, for example, high pH soils, and low soil moisture [179]. Residues can accumulate to toxic concentrations with consecutive treatments, and products and their metabolites such as atrazine and chlorsulphuron can exhibit persistent and toxic properties [179].

4.2.3. Mechanical

Mechanical control options include the physical felling or uprooting of plants, often in combination with burning [180]. Mechanical control is labour-intensive and thus expensive to use in extensive and dense infestations, or in remote or rugged areas.

4.2.3.1. Rangeland burning

Fire is regarded as the natural factor of the southern African environment; it is thought to have occurred from time immemorial, and therefore, it is part of ecosystems. Rangeland burning is an important ecological management tool in the maintenance and productivity of grasslands in Southern Africa region [181]. The burning in rangelands is practiced for a number of reasons; one of these reasons is to control bush encroachment. To use fire effectively in rangelands, it is important to understand how it behaves and to develop an insight into the way in which various factors influence such behaviour. Fire intensity is one of the important components of the fire regime [182]. Fire regime can be defined as season and frequency of burning together with type and intensity of fire [18]. The effect of fire on natural ecosystems arises from a response of living organisms to the release of heat energy generated by the combustion of plant material. Thus, it is an oxidation process involving a chain reaction during which the solar energy originally converted into carbon compounds by photosynthesis is released as heat during fire [183]. The effect of fire on vegetation, therefore, depends upon the amount of heat energy, and upon the rate and vertical level at which it is released [184]. The rate of fire is measured in terms of time taken to burn a given unit area, it is affected by a number of factors including fuel load and moisture. The vertical level at which heat energy is released during fire determines the height at which plants will be burned. The plant (tree) height is one of the important factors determining the effect of fire on bushes, thus, as the bushes become taller, the fire intensity required to cause a topkill of the stems and braches become critical. Thus, as the plant height increases, the bushes become resistant to fire [182].

Since the effectiveness of fire in rangeland to control bush encroachment depends largely on the fire intensity, which, in turn, depends on fuel characteristics such as fuel load. It is important to note that fire cannot be applied at all times, thus, there should be considerations on the suitability of the ecosystem to support fire. The high intensity fire is required to control bush encroachment at all phases, thus, controlling coppice growth and bush seedlings or maintaining bush at an available height and in an acceptable state for browsing animals [184]. Use of fire as a control method for bush encroachment, therefore, has higher potential in higher rainfall areas where the soil moisture available is reliable and sufficient to produce fuel load that can support regular fires. The use of fire has to be sustained in order to get good results; this is because the bush can recover through coppice regrowth and seedling recruitment after burning, therefore, there should periodic follow up burn. In moist areas, the frequency of burning required to control bush encroachment depends on the rate at which the bush recovers. The recommended type of fire used in controlling bush encroachment is generally head fire (burning towards the direction of wind); this will mostly occur in the form of surface fire except in extreme conditions where it can develop into crown fire in more densely wooded areas with more flammable foliage. The season of burning should be during the early spring, after the first spring rain. This will ensure the intense fire but with minimal undue deleterious effects on the grass sward. Fire should be applied close to the commencement of the growing season as possible to minimise the length of soil exposure to potential soil erosion.

Reduction of bush encroachment with fire has positive results on herbaceous vegetation biomass production, thus, biomass production is enhanced, and therefore, forage production

http://dx.doi.org/10.5772/56182

increased, which is positive to livestock production. Where fire is used as a regular management tool, it changes species composition, thus, species that are adapted to fire tend to dominate while species that are not favouring fire do not persist. Thus, in South Africa, frequent burning in the False Thornveld of Eastern Cape, favours species such as *Themeda triandra* and has a negative effect on the abundance of *Cymbopogon plurinodes* [185]. Similar results have been observed at the Tall Grassveld of Kwazulu Natal, where *Tristachya leucothrix, Cymbopogon excuvatus* and *cymbopogon validus* became dominant with burning frequency [183]. Furthermore, where higher frequency of fire is used, for example where burning is annual, the bush will be controlled but that has an effect on the basal cover of herbaceous plants, thus, the basal cover becomes poor due to effects of fire on plant vigour. That, in turn, renders the soil susceptible to soil erosion, which is another environmental disaster. Fire remains the cheapest form of management available to conserve and perpetuate natural plant communities. However, its effectiveness is based on clear and objective application of a fire regime, thus frequency, season and intensity may be used effectively to retain the natural element and control the invasive elements in the flora of natural ecosystems [186].

4.2.3.2. Manual/Physical cutting/clearing

Manual and mechanical techniques such as pulling, cutting, and otherwise damaging plants, are used to control some invasive plants, particularly if the population is relatively small. These techniques can be extremely specific, and therefore, minimizing damage to desirable plants. However, manual techniques are generally labour and time intensive. These techniques are effective if the treatments are administered several times to prevent the weed from re-establishing. In the process, labourers and machines may severely trample vegetation and disturb the soil, thus, providing prime conditions for re-invasion by the same or other invasive species.

Bush encroachment reduces grass growth in rangeland as discussed in the previous sections and that results in reduced biomass production, which subsequently affects forage production. The approach that has been used to address the negative impacts of invading species in South Africa has been predominantly physical by clearing alien plants [187]. Clearing of the bush in encroached areas results in an increased dry matter yield and basal cover of herbaceous vegetation [184], which are good indicators for rangeland health if the functional characteristic of such an ecosystem is forage production. Furthermore, species richness of herbaceous plants and relative abundance of few of the species among the initial population that is intolerant of bush cover increase with tree cutting [142]. As a result, the reduction of bush cover can restore herbaceous plant productivity and biodiversity in rangelands [188]. However, there are herbaceous species that have a positive relationship with certain trees, and removal of such trees negatively leads to reduction of these herbaceous species. This decline indicates the shifts in the microenvironment due to the removal of ecologically important trees, thus exposing sensitive herbaceous species to increased light intensity.

It is important, however, to note that although bush cutting has positive results on forage productivity, it has high costs involved [142]. Therefore, it is more applicable on the smaller scale. On the larger scale, where bush clearing is done with heavy implements such as a bulldozer blade, the trees are removed with their roots, which minimises resprouting of encroaching species. However, the soil disturbance generally severely affects the grass layer, but the grasses will often re-establish themselves [154]. The re-establishment of grasses will be following the secondary succession trend, thus the first colonisers are likely to be annual pioneers, which have little forage value. Furthermore, severe soil disturbance may encourage the establishment of a large number of seedlings of some woody plants. This may lead to establishment of a woody community that is denser than the original community.

4.2.4. Biological control of encroaching and invasive species

Biological control has been defined as the use of living organisms to reduce the vigour, reproductive capacity, or effects of weeds [189]. Biological control (biocontrol) involves the deliberate introduction of invertebrates or diseases, and is aimed at reducing the effects of ecological release. Biocontrol is aimed at arriving at a situation where the plant is returned to the status of a non-invasive naturalized alien, that is an alien plant that is able to survive, and even reproduce, but does not invade aggressively in its new habitat [6]. Biological control could be regarded as the only sustainable mechanism to prevent the spread of invasive alien species in the long term [190]. Biocontrol is potentially very cost-effective, and environmentally benign. Despite concerns to the contrary [191], the modern practice of using carefully screened and host-specific biocontrol agents is safe, and "host shifts" have not occurred in the over 350 recorded cases where weed biocontrol agents have been used worldwide [192].

Although there are some inconsistencies in terms of when biocontrol practices were established in South Africa, at least there is an agreement in that biocontrol agents have been released against 47 weed species. The disagreement in literature is such that Olckers and Hill (1999) indicated that in South Africa, biocontrol has been practiced since 1910, and that to date, 103 biocontrol agents (including invertebrates and pathogens) have been released against 47 weed species. Whilst on the other side, it has been suggested that the biological control of weeds has been practiced since 1913 and since then some 47 weed species have been subjected to the effects of approximately 85 species of biocontrol agents [190]. Therefore, based on the cited literature, there is an uncertainty about the years of establishment of biocontrol in South Africa and for this chapter the assumption will be that the biocontrol was adopted for use between 1910 and 1913. Although in South Africa physical methods of controlling the alien species are mostly used, biological control using species-specific invertebrates and pathogens from the plant's country of origin is also a control option; however, there has been a considerable resistance to its use [180]. The seed-feeding weevil is one of the agents that have been released against *Acacia mearnsii* in areas where the wattle is not grown commercially [8]. Nevertheless, plant-attacking agents could potentially be used; however, these compared with seed-attacking agents such as weevils could kill the target plant and therefore, impact severely on commercial prospects. The impact of biological control agents on controlling invasive species vary with species controlled, biological agents introduced, mode of operation of agents and many other factors. The use of biological control measures on invasive plants have been reported in South Africa with varying rates of success. The elaborate example where the invasive plants were controlled with biological control agents was at Kruger National Park

http://dx.doi.org/10.5772/56182

(KNP). The impact of *S. rufinasus* on *A. filiculoides* within the Kruger National Park (KNP) has been exceptionally good. Thus, 100% clearing of the weed was achieved in a few months after release and the infestation has been maintained at that level. The insects are able to survive for long periods in the vicinity and re-establish themselves should the area become re-infested.

Biological control agents such as *Neochetina eichhorniae, Cercospora rodmanii, Orthogalumna terebrantis, Niphograpta albiguttalis, Neochetina bruchi, Eccritotarsus catarinensi* have been used to control invasion by *Eichhornia crassipes* (Water hyacinth). Although proven in many other instances elsewhere to be effective, the agents released within the KNP have had little impact in terms of bringing the infestation under control. This little impact has been ascribed to frequent low level flooding as well as major floods that have repeatedly washed the infestation away, and therefore, preventing large numbers of insects to build up [194]. *Lantana camara* has been cited to be one of the invasive plants at KNP and other areas of South Africa. Two biological control agents viz. *Octotoma scabripennis* (leaf-mining hispine beetle) and *Falconia intermedia* (Lantana sapsucker) have been introduced at KNP. However, *O. scabripennis* failed to establish and the initial trial site for *Falconia intermedia* was reported to have been destroyed by the floods and therefore, both agents have provided insignificant impact on *L. camara* [194].

Opuntia stricta (Sour prickly pear) has been identified as one of the invasive species at KNP and therefore, it was one of the species that were controlled. In an attempt to control this species, two agents have been introduced against it, the first of which being *Cactoblastis cactorum* (phycitid moth) in 1988 [195] and subsequently *Dactylopius opuntiae* (cochineal) in 1996 [196]. The structure of infestations of *O. stricta* changed after the introduction of *C. cactorum* where large plants were replaced by high densities of smaller plants. However, fruit production did not decline and therefore *C. cactorum* failed to provide the degree of control that was expected [195]. Predation and parasitism, especially ant predation of eggs, has a definite impact on the distribution and abundance of *C. cactorum*. *Dactylopius opuntiae*, which had been instrumental in the control against *O. ficusindica*, was released on at least three occasions between 1990 and 1995 yet failed to establish due to the biotype that was used. The Plant Protection Research Institute (PPRI) sourced a different biotype of *D. opuntiae* from Australia, which established well and is reported to be currently destroying large stands in the Skukuza region in South Africa [196].

Pistia stratiotes (Water lettuce) was determined to be one of the invasive species within the Kruger National Park. The snout weevil (*Neohydronomus affinis*) was introduced to control the weeds. The impact of *N. Affinis* on *P. Stratiotes* varied at different infestations throughout the KNP. The other biocontrol agent *Cyrtobagous salviniae* (snout beetle) was released to control *Salvinia molesta* (Kariba weed). The infestations of *S. molesta* at the three areas where the agent was released and established were brought under complete control and have been maintained at that level. *Trichapion lativentre, Rhyssomatus marginatus* and *Neodiplogrammus quadrivattatus* were used to control *Sesbania punicea* (Red Sesbania) at Kruger National Park. The impact of the three agents on plants has been reported to be exceptionally good [194]. The three weevil species have reduced the problem to such an extent that *S. punicea* is under complete control in the area, thereby requiring no further action to be taken. The biological control of *S. punicea* remains the best example of an invasive tree species control.

The use of mammals such as goats in agricultural areas to control bush encroachment has been reported in South Africa [197]. Apart from tree seedlings, which can be affected by smaller browsers, the use of browsers to execute control on woody plants largely excludes wild game [154]. However, elephants have also been reported to be effective in controlling bush encroachment [198, 199]. Nevertheless, their use is confined to large game reserves or game farms where their population should be large enough to make an appreciable impact on the woody vegetation, which could, in turn, lead to serious management problems. The control of bush encroachment by use of mammals such as goats is dependent firstly on the acceptability of plant species that are controlled to these mammals for use as browse, and secondly availability of the browse material. The acceptability relates to the palatability and nutritional value of a browse material to the browser. Browse availability relates to the height at which browse material can be accessed by browsing animals, the browse line for goats is approximately 1.5 m. Boer goats are well suited to controlling woody plants because the intensity and frequency with which they utilise the browse can be controlled. Furthermore, the Boer goats are relatively insensitive to chemical deterrents, such as high tannin levels present in many woody species [154]. Boer goats cannot be used to control dense stands of woody plants whose canopies extend above the browse line of approximately 1.5 m.

4.2.5. Integrated bush encroachment and invasion management

Integrated weed control usually involves a combination of at least three of the primary elements of control - mechanical, chemical and biological [180]. Integrated weed management (IWM) could be defined as a system for the planning and implementation of programs, using an interdisciplinary approach, to select a method for controlling undesirable plant species or group of species using all available methods. These methods generally vary between preventative and restorative domains. The success of preventative encroachment measures mostly depends on the understanding of the causes of encroachment and identification of barriers for natural recovery. Restorative measures depend on the rangeland ecosystem structure and functional characteristics to be restored. Integrated bush encroachment control is a multidisciplinary, ecological approach to managing unwanted plant species in rangeland ecosystems.

However, it is important to note that the decision to use a certain method to control the bush encroachment is informed by the cost of using that method against the benefit. Bush encroachment control methods are management systems [137] that might have varied policy implications for bush control [193]. Therefore, understanding the potential role of different bush encroachment control methods for promoting herbaceous species composition requires recognition of the objectives of resource users and policymakers [142]. The failure to recognise the long-term intended ecosystem status could lead to a subsequent failure to achieve bush encroachment control objectives and that could further lead to land use practice and policy controversy. Thus, the resource users are interested in livestock production through increased plant productivity, while the goal of policymakers is environmental preservation. Therefore, the land use practice imperatives and policy directives should be harmonised to permit both forage production and biodiversity conservation functional characteristics of the ecosystem to thrive.

The increasing invasion of non-indigenous species has become one of the top causes of global biodiversity loss and environmental change [200, 201]. Therefore, there is a need for development of intensive mechanisms to control these invasions in the ecosystems before the natural value of ecosystems is lost permanently. As part of a comprehensive remedial effort to control invasions, assessment and characterisation of invader species will serve as a foundation towards integration of efforts to control invaders species. There is urgency for more rigorous and comprehensive assessments of the impacts and risks associated with plant invasions [202]. Thus, prevention and control strategies can be targeted appropriately if sufficient assessments are conducted [203]. In the approaches toward the control of invasive alien weeds, any intervention needs to be aligned with the different stages of spread and characteristics of a desired ecosystem. The stages of spread can be divided into four broad phases: (i) arrival or entry phase; (ii) adaptation and establishment phase; (iii) an exponential growth phase; and (iv) dominance phase. It is in the exponential growth stage of weed spread that integrated control programmes find a logical relevance. Prevention, and early detection and eradication, are more appropriate for the first two stages, while options may be severely limited once weed populations reach the final stage of total ecosystem domination.

Plant invasions are interdisciplinary both by their impacts and by utility and therefore, assessments should recognize the interdisciplinary nature of the problem of species invasions. Thus, the ecosystem characteristics determine whether the appropriate conditions allow for the establishment of the invasive species, and on the other hand, economic systems affect the state of the ecosystem through its use, and through the prevention and control measures implemented to stop the invasions. Hence, accounting for the economic and ecological links and feedbacks is critical in invasion assessments [204]. It is fundamental to have a clear understanding on different functions of ecosystems, thus, an assessment of rangeland area in terms of its ability to achieve its ecosystem functions. Natural resource managers and farmers at all levels require full knowledge of ecosystem functions. This could be achieved through collating results from experiments in different fields or locations within the context of a more encompassing systems management framework that treats the rangeland ecosystem as a complete bio-economic unit. Therefore, in order to improve decision making, farmers need answers to questions at the systems level, including the biological and economic elements of the rangeland production entities they are attempting to manage.

Most often, a single method is not always effective to achieve sustainable control of the rangeland weeds. This is because of among other reasons some methods can only control bush encroachment at a certain stage and some could leave areas that are treated vulnerable to other forms of landscape hazards. For example, use of fire in rangelands depending on the intensity will burn shoots of woody plants; however, the seeds in the soil could be left to germinate and furthermore, some seeds may be stimulated to germinate by fire. It is also difficult to ascertain a complete kill of unwanted species with fire because normally the basal buds of certain trees remain unburned and therefore resprout. It is for these reasons that the introduction of biological control agents becomes important especially where complete removal of the invading species is anticipated. Use of herbivores works effectively where the intention is to maintain the current stand of encroaching species especially in the savanna where there is coexistence between grasses and trees. There are species which are not preferred by animals for foraging, and use of biological control through herbivores would not be effective; therefore, introduction of invertebrates could be used. However, most of the invertebrates are not readily available in Southern Africa for use at the farm or landscape level. It is impractical to burn certain areas that are encroached; this is sometimes due to poor fuel load that can support high fire intensity needed to burn the woody species. Encroachment in some of these areas cannot be controlled with the use of herbivores (goats) and herbicides could be useful.

All this, therefore, suggests that there are areas and bush encroachment situations where a single method can be used; however, a combination of different methods could be used simultaneously or alternatively in subsequent approaches. Nevertheless, it is important that prior to the implementation of any selected method or any combination or any sequence to develop post encroachment treatment management plan. This is because removal of bush with any technique can leave the land vulnerable to soil erosion or further encroachment of the same or new species. Therefore, a successful long-term management program should be designed to include combinations of mechanical, biological, and chemical control techniques. Numerous mechanical and cultural options have been developed to manage noxious rangeland weeds, including mowing, prescribed burning, timely grazing, and perennial grass reseeding or inter-seeding. Furthermore, several herbicides are registered for use on rangelands and most biological control programs focus on noxious rangeland weed control. Successful management of noxious weeds on rangeland will require the development of a long-term strategic plan incorporating prevention programs, education materials and activities, economical and sustainable multi-year integrated approaches that improve degraded rangeland communities, enhance the utility of the ecosystem, and prevent reinvasion or encroachment by other noxious weed species [151].

There are a number of factors to consider in selection of the bush control method; however, the dominant consideration is the cost of the method. However, there are furthermore considerations beyond the cost of the method. The use of fire in controlling bush encroachment in rangelands is determined by a threshold amount of flammable fine fuel needed to carry fire that is sufficiently intensive to reduce woody plants. Furthermore, to effectively control woody plants with burning, fire must be applied regularly. Many rangelands occur in semi-arid environments in which forage-based livestock production is the primary agricultural activity and intermittent droughts are inevitable [205]). Therefore, accumulating sufficient fine fuel to carry fires in such environments requires the reduction in livestock numbers compared to areas where fire is not used. Hence, sustainable utilisation of semi-arid rangelands depends on complex management of animal species, stocking rates, and the vegetation composition, structure, phenology and quality [129].

The integration of bio-control agents and herbicides in a scientifically sound and rigorous management plan is the first step in a long-term approach to weed management. Such management plans should aim to maximise the benefits of all the respective control options and thereby ensure the infestation is contained and the density reduced to acceptable thresholds. Biological control is used as an important, long-term management solution to numerous weeds worldwide. When carefully integrated into management plans the combination of biocontrol and other control measures may provide effective solutions to the problem, and various methods therefore, should not be used in contradiction to one another. All available knowledge surrounding a particular invasive plant problem needs to be considered when developing such integrated programmes. No single method is likely to prevent either distribution or densification of the plant from or in its current range. Combination of the biological control and herbicides can bring remarkable results; while herbicides are used to contain the infestation to its present range, biological control (invertebrates) is being released into dense stands where it is proving destructively effective in controlling the plants. Goats used in the system that allows coppice growth to be used frequently and severely strongly influence woody plants, that is, provided that their canopies be below the browse line. Where the plants are above browse line, fire can be used to reduce plant height where fuel load is sufficient; however, where fuel load is not sufficient chemical or physical control can be used and, in both cases, goats can be used as follow up control.

In this chapter, integration in the control of invasive species is not limited to control methods themselves in isolation but in all the processes relating to bush encroachment management. Primarily, it is important as the initial stage of integration to identify and characterise invasion/encroachment of species. This should include establishment of their origin, mode of establishment and spread (seeds, cuttings etc), their phenological and morphological characteristics and assessing their favourable growth conditions. It is further important to determine the degree of invasion/encroachment, which will help setting economic and ecological thresholds of invasion. The analysis of the ecological and economic impact of invasion/encroachment in the environment should be carried out prior to any intervention. That will help in determining whether there is a need for intervention and magnitude of such intervention. The need for intervention should be assessed against the set thresholds for invasion. Setting objectives for invasion/encroachment management is very fundamental because the objectives will be used as the yardstick for the control.

A number of factors will guide selection of the approach to control bush encroachment. These factors include species to be controlled, the stage of invasion and landscape of an area. The approach to be selected would be chemical, mechanical and biological depending on the approach suited to the species to be controlled, the major landscape on which the invasion has occurred and the stage of invasion. The method that is ecologically and economically sound and practical should be selected. Integrated bush encroachment approaches may be practiced in combinations that could either be used simultaneously, alternatively or sequentially. In simultaneous integration of bush control methods, more than one method that could complement each other under the prospects of chemical, biological or physical methods used together. Some methods cannot be used simultaneously because of the danger that they can cause on other organisms and environment. For example, the methods that can be integrated simultaneously could be manual clearance and use of goats as browsers. The alternative integration could be executed through turns, thus, one method first and then the other. The alternate integration can be practiced in rotation if planned properly, for example, use fire with a given period in between goat treatment. Thus, burning can be applied every three years while goat use is continued. Sequential integration is executed in succession of methods where one method can be used to prepare for the next method in a sequence. In this integration there should be short-term objectives relating to each method and long-term objectives, which are based on the integrated approach. Thus, mechanical control in the form of fire or physical cutting can be used to reduce plant height to facilitate the use by goats as the maintenance stage of control. Where there is high density of bush, which impairs the movement of animals, or where the bush is above the browse line of goats or where the bush is unacceptable to browsing animals yet the fuel load is poor, mechanical cutting would be useful. This would reduce the bush density, which will open up for goats to be able to browse, that will further open up for grass to grow then fire can be used as a follow up. Where the bush has higher density but there is sufficient fuel load, fire will be the most applicable method. Fire will clear up the bush faster and relatively cost effectively, therefore, where there is enough fuel load fire is recommended as the first on the integration followed by use of goats. Biological control would always be the last in the sequence and it is the approach that helps in achieving long-term bush encroachment control objectives. The use of invertebrates (Weevils) could be integrated with the use of herbivore (goats) since the weevils take care of the seeds and the goats can take of the foliage to maintain the stands.

A post treatment management plan should be part of integration in bush encroachment control, thus, there should be a clear plan on what rangeland management system will be practiced that will ensure that the control objectives are achieved. Thus, some invasion control methods such as the use of fire can leave the soil bare and susceptible to soil erosion and, therefore, there should be a clear objective plan on what practices will be taken immediately after treatment. Furthermore, on the areas that are severely encroached and grass biomass and basal cover are affected, use of herbicides will also leave the soil bare and grazing can worsen the situation and lead to soil erosion. Therefore, as part of integration, exclusion of treated areas to minimise grazing should be considered. This exclusion could be coupled with introduction of plant propagules, thus, revegetation through seeds or seedlings of the grass on the bare patches.

There is a need for periodic monitoring and evaluation as part of integration of the encroachment control. This will help in determining whether the treatment is achieving expected results within the given timeframes. That will help in realising if there is a need for the adjustment of the plan. Effective bush control monitoring and evaluation should be done according to the pre set objectives; it will help in the establishment of whether the objectives are achieved. Performance measures, monitoring, and adaptive management are necessary. Using these methods, status and trends can be tracked, analysis and accountability facilitated, and decisions adapted so that the intended balance among social, economic, and ecological concerns is achieved. Ecosystems' performance appraisal will be important at the end of the integration, this should be a pronouncement of whether the target ecosystem has been reached and should be coupled with sustainability management programme that will eliminate factors that could have lead to encroachment. Ecosystem performance measures can provide a quantitative basis for evaluating how well actions under the integrated bush control approach are meeting stated objectives. Performance measures allow for continuous learning, which broadens understanding about how ecosystems function. There are many approaches to

http://dx.doi.org/10.5772/56182

evaluate performance; however, performance measures should specifically address management goals and objectives and should be quantifiable, expressing status and trends of specific resource values of concern, such as unique ecosystem type.

5. Conclusions and recommendations

Bush encroachment and invasion could be attributed to a number of factors, which by their nature vary with species and locality. These factors cannot easily be ranked according to the strength of causation and/or according to the intensity of their effect on rangeland ecosystems. Factors that are blamed for bush encroachment include improper grazing practices, suppression of fire, drought, rainfall intensity and distribution and climate change. The temporal and spatial distribution of bush encroachment follows a sigmoid distribution curve. Although some invasive species are abundant, they are localised in certain areas whilst, on the other hand, certain species are widely distributed but low in copiousness. There are three major methodological guidelines; these fall under chemical, mechanical/physical and biological and depend on a number of factors within economic and ecological impressions. Bush encroachment occurrences are generally caused by different factors, at different landscapes, by different plant species and with different effects. Therefore, the invasion control methods should consider this variation for success in treatments. Thus, there are areas and invasion situations where a single method can be used; however, a combination of different methods could be used in simultaneous or alternative or subsequence approaches.

Integrated plant invasion management should have four major stages of execution; these are comprised of diagnostic, preventative, control and management. The diagnostic stage should include identification and characterisation of invasion, determination of the degree of invasion, analysis of the ecological and economic impact of invasion, determination of the need for intervention, and setting objectives for intervention. The control stage should include selection of invasion control approach or combinations. Management stage includes post-treatment management, monitoring, evaluation, and ecosystems' performance appraisal. Preventative stage is more practical on the areas that are not yet invaded; at this stage management of areas that are not yet encroached is central. Assessment and characterisation of vulnerable areas for invasion will be important in developing an encroachment prevention plan. It is also important to assess plant invasion predisposing factors; however, these may vary with species and localities. In the diagnostic stage, determination of the level of spread is very fundamental and will serve as the background for selection of the bush encroachment control and management methods. The stage of bush encroachment spread can be divided into four broad phases viz, entry phase, adaptation and establishment phase, an exponential growth phase and dominance phase. It is in the exponential growth stage of weeds spread that integrated control programmes find a logical relevance. Prevention, and early detection and eradication, are more appropriate for the first two stages, while options may be severely limited once weed populations reach the final stage of total ecosystem domination. Although there is massive literature on the plant invasion and bush encroachment, there is still a significant need for further research in establishing fundamental characteristics of bush encroachment phenomenon in rangelands. This will lead in systematic characterisation of bush encroachment and subsequently that will lead to development of more practical and radical yet scientific bush encroachment control and management practices in rangelands.

Author details

M. S. Lesoli[1,2*], M. Gxasheka[2], T. B. Solomon[2] and B. Moyo[1]

*Address all correspondence to: lesolistar@gmail.com

1 Fort Cox College of Agriculture and Forestry, King Williams Town, South Africa

2 Department of Livestock and Pasture Science, University of Fort Hare, Alice, South Africa

References

[1] Allen, V G, Batello, C, Berretta, E J, Hodgson, J, Kothmann, M, Li, X, Mclvor, J, Milne, J, Morris, C, Peeters, A, & Sanderson, M. (2011). An international terminology for grazing lands and grazing animals. Grass and Forage Science , 66, 2-28.

[2] Ward, D. (2005). Do we understand the causes of bush encroachment in Afriucan savannas? African Journal of Range and Forage Science , 22, 101-105.

[3] Wigley, B. J, Bond, W. J, & Hoffman, M. T. (2010). Thicket expansion in a South African savanna under divergent land use: local vs. global drivers? Global Change Biology, , 16, 964-976.

[4] Wiegand, K, Saltz, D, & Ward, D. approach to savanna dynamics and woody plant encroachment- Insights from an arid savanna. Perspectives in Plant Ecology, Evolution and Systematics, , 7, 229-242.

[5] Kaiser, J. (1999). Stemming the tide of invasive species. Science , 285, 1836-1841.

[6] Van Wilgen, B W, De Wit, M P, & Anderson, H J. Le Maitre D C, Kotze I M, Ndala S, Brown B and Rapholo M B (2004). Costs and benefits of biological control of invasive alien plants: case studies from South Africa. South African Journal of Science , 100, 113-122.

[7] Bright, C. (1998). Life Out of Bounds- Bioinvasions in a borderless world. Norton, New York.

[8] De Wit, M P, Crookes, D J, & Van Wilgen, B W. (2001). Conflicts of interest in environmental management: estimating the cost and benefits of a tree invasion. Biological Invasions , 3, 167-178.

http://dx.doi.org/10.5772/56182

[9] Van Auken, O. W. (2009). Causes and consequences of woody plant encroachment into western North American grasslands. Journal of Environental Management, , 90, 2931-2942.

[10] Moleele, N. M, Ringrose, S, Matheson, W, & Vanderpost, C. (2002). More woody plants? The status of bush encroachment in Botswana's grazing areas. Journal of Environmental Management , 64, 3-11.

[11] Wigley, B. J, Bond, W. J, & Hoffman, M. T. (2009). Bush encroachment under three contrasting land-use practices in a mesic South African savanna. African Journal of Ecology , 47, 62-70.

[12] Richardson, D M, Macdonald, I, Hoffman, W, & Henderson, J H. L (1997). Alien plant invasions. In: Cowling R M, Richardson D M and Pierce S M (eds). Vegetation of Southern Africa. Cambridge University Press, Cambridge. , 535-570.

[13] Oba, G, Post, E, Syvertsen, P O, & Stenseth, N C. (2000). Bush cover and range condition assessments in relation to landscape and grazing in southern Ethiopia. Landscape Ecology , 15, 535-546.

[14] Hahn, B D, Richardson, F D, Hoffman, M. T, Roberts, R, Todd, S W, & Carrick, P J. (2005). A simulation model of long-term climate, livestock, and vegetation interactions on communal rangelands in the semi-arid Succulent Karoo, Namaqualand, South Africa. Ecological modeling 183 (1-3): 211- 230.

[15] Ansley, R J, Pinchak, W E, Teague, W R, Kramp, B A, Jones, D L, & Jacoby, P W. (2004). Long-term grass yields following chemical control of honey mesquite. Journal of Range Management , 57, 49-57.

[16] Rollins, D, & Cearley, K. (2004). Integrating wildlife concerns into brush management. In: Hamilton WT, McGinty A, Ueckert D N, Hanselka C W, Lee M R (Eds.), Brush management: past, present and future. Texas A&M University Press, College Station, Texas, , 239-258.

[17] Knopf, F L. (1994). Avian assemblages on altered grasslands. Studies in Avian Biology , 15, 247-257.

[18] Trollope, W, Trollope, W, Bosch, L A, & Veld, O J H. and pasture management terminology in southern Africa. Journal of the Grassland Society of Southern Africa, 7:1, 52-61.

[19] Wiegand, K, Ward, D, & Saltz, D. (2005). Multi-scale patterns and bush encroachment in an arid savanna with a shallow soil layer. Journal of Vegetation Science, , 16, 311-320.

[20] Skarpe, C. (1990). Shrub layer dynamics under different herbivore densities in an arid savanna. Vegetation , 87, 11-18.

[21] Joubert, D. F, Rothauge, A, & Smit, G. N. (2008). A conceptual model of vegetation dynamics in the semiarid Highland savanna of Namibia, with particular reference to bush thickening by Acacia mellifera. Journal of Arid Environments, , 72, 2201-2210.

[22] Van Langevelde, F. Van de vijver C, Kumar L, Van de koppel J, De ridder N, Van andel J, Skidmore AK, Hearne JW, Stroosnijder L, Bond WJ, Prins HHT and Rietkerk M (2003). Effects of fire and herbivory on the stability of savanna ecosystems. Ecology, , 84, 337-350.

[23] Sankaran, M, Ratnam, J, & Hanan, N. (2008). Woody cover in African savannas: the role of resources, fire and herbivory. Global. Ecological Biogeography, , 17, 236-245.

[24] Angassa, A, & Oba, G. (2008). Effects of management and time on mechanisms of bush encroachment in southern Ethiopia. African Journal of Ecology, , 46, 186-196.

[25] DiTomaso J M (2000). Invasive weeds in rangelands: species, impacts, and management. Weed Science , 48, 255-265.

[26] West, N E. (1993). Biodiversity on rangelands. Journal of Range Management , 46, 2-13.

[27] Peterson, G, Allen, G R, & Holling, C S. (1998). Ecological resilience, biodiversity and scale. Ecosystems , 1, 6-18.

[28] Westoby, M, Walker, B H, & Noy-meir, I. (1989). Opportunistic management for rangelands not at equilibrium. Journal of Range Management , 42, 266-274.

[29] Teague, W R, Grant, W E, Kreuter, U P, Diaz-solis, H, Dube, S, Kothmann, M M, Pinchak, W E, & Ansley, R J. (2008). An ecological economic simulation model for assessing fire and grazing management effects on mesquite rangelands in Texas. Ecological Economics , 64, 611-624.

[30] Grice, A C, & Hodgkinson, K C. (2002). Challenges for Rangeland People. In: Grice, A.C., Hodgkinson, K.C. (Eds.), Global Rangelands: Progress and Prospects. CABI Publishing, New York, , 1-11.

[31] Archer, S, Scifres, C, Bassham, C R, & Maggio, R. (1988). Autogenic succession in a subtropical savanna: conversion of grassland to thorn woodland. Ecological Monograph, , 58, 111-127.

[32] Hoffman, T, & Ashwell, A. (2001). Nature divided: Land degradation in South Africa. University of Cape Town Press, Cape Town, South Africa.

[33] Britz, M L, & Ward, D. (2007). Dynamics of woody vegetation in a semi-arid savanna, with a focus on bush encroachment. African Journal of Range and Forage Science, , 24(3), 131-140.

[34] Macdonald, I, Kruger, W, & Ferrer, F J. A A (1986). The ecology and management of biological invasions in Southern Africa. Oxford University Press, Cape Town.

[35] Le Maitre D CVersfeld D B and Chapman R A (2000). The impact of invading alien plants on surface water resources in South Africa: a preliminary assessment. Water S A , 26, 397-408.

[36] Richardson, D M, & Van Wilgen, B W. (2004). Invasive alien plants in South Africa: How well do we understand the ecological impacts? South African Journal of Science , 100(1), 45-52.

[37] Belskey, A J. (1990). Tree/grass ratios in East African savannas: a comparison of existing models. Journal of Biogeography , 17, 483-489.

[38] Scholes, R J, & Walker, B H. (1993). An African Savanna- Synthesis of the Nylsvley Study. Cambridge University Press, Cambridge.

[39] Scholes, R J, & Archer, S R. (1997). Tree-grass interactions in savannas. Annual Review of Ecological Systems , 28, 517-544.

[40] Lumprey, H F. (1983). Pastoralism yesterday and today: the overgrazing problem, In: Bouliere F (ed). Tropical Savannas: Ecosystem of the World. Elsevier, Amsterdam. The Netherlands. , 643-666.

[41] Murphy, A H. (1986). Significance of rangeland weeds for livestock management strategies. Proceedings of the California Weed Conference., 114-116.

[42] Angassa, A, & Oba, G. (2010). Effects of grazing pressure, age of enclosures and seasonality on bush cover dynamics and vegetation composition in southern Ethiopia. Journal of Arid Environments , 74, 111-120.

[43] Burrows, W H, Henry, B K, Back, P V, Hoffman, M B, Tait, L J, Anderson, E R, Menke, N, Danahar, T, Carter, J O, & Mckeon, G M. (2002). Growth and carbon stock change in eucalypt woodlands in northeast Australia: ecological and greenhouse sink implications. Global Change Biology , 8, 769-784.

[44] Britz, M L, & Ward, D. (2007). The effects of soil condition and grazing strategy on plant species composition in a semi-arid savanna. African Journal of Range and Forage Science , 24(2), 51-61.

[45] Archer, S. (1995). Tree-grass dynamics in a Prosopis-thornscrub savanna parkland: reconstructing the past and predicting the future. Ecoscience Aarino T and Martikainen P J 1994. Mineralization of carbon and nitrogen in acid forest soil treated with forest and slow-release nutrients. Plant and Soil, 164:187-193., 2, 83-99.

[46] Ward, D, Ngairorue, B T, Karamata, J, Kapofi, I, Samuels, R, & Ofran, Y. (2000). Effects of communal pastoralism on vegetation and soil in a semi-arid and in an arid region of Namibia. In: White P S, Mucina L, Leps J and Van Der Maarel E (Eds) Vegetation Science in Retrospect and Perspective. Opulus Press, Uppsala, Sweden. , 344-347.

[47] Neilson, R P. (1986). High-resolution climatic ananysis and southwest biogeography. Science , 232, 27-34.

[48] Angassa, A, Oba, G, & Tolera, A. (2012). Bush encroachment control demonstrations and management implications on herbaceous species in savannas of Southern Ethiopia.

[49] Gifford, R M, & Howden, M. (2001). Vegetation thickening in an ecological perspective: significance to national greenhouse gas inventories. Environmental Science and Policy , 4, 59-72.

[50] Oba, G. (1998). Assessment of indigenous range management knowledge of the Borana pastoralists of southern Ethiopia. Commissioned by GTZ-Borana Lowland Pastoral Development Program in collaboration with the Oromiya Regional Bureau for Agricultural Development, Negelle/Borana Ethiopia.

[51] Oba, G, & Kotile, D G. (2001). Assessments of landscape level degradation in southern Ethiopia: pastoralists versus ecologists. Land Degradation and Development , 12, 461-475.

[52] Walter, H. (1939). Grassland, Savanne und Busch der arideren Teile Afrikas in ihrer Ökologischen Bedingtheit. Jahrbuchr für Wissenschaftliche Botanik , 87, 750-860.

[53] Walter, H. (1954). Die Verbuschung, eine Erscheinung der subtropischen Savannengebiete,und ihre Ökologischen Ursachen. Vegetatio 5/, 6, 6-10.

[54] Noy-meir, I. (1982). Stability of plant-herbivore models and possible application to savanna, In: Huntley B J and Walker B H (eds). Ecology of Tropical savannas. Ecological Studies. Springer Verlag, Berlin and Heidelberg, Germany. , 591-609.

[55] Walter, H. (1971). Ecology of Tropical and Subtropical Vegetation. Oliver and Boyd, Edinburgh, UK.

[56] Van Vegten, J A. (1983). Thornbush invasion in eastern Botswana. Vegetation , 56, 3-7.

[57] Perkins, J S, & Thomas, D. S G (1993). Spreading desert or spatially confined environmental impacts? Land degradation and cattle ranching in the Kalahari Desert of Botswana. Land Degradation and Rehabilitation , 4, 179-194.

[58] Brown, J R, & Archer, S. (1999). Shrub invasion of grassland: recruitment is continuous and not regulated by herbaceous biomass or density. Ecology , 80, 2385-2396.

[59] Jeltsch, F, Milton, S J, Dean, W, & Van Rooyen, J. N (1996). Tree spacing and coexistence in semiarid savannas. Journal of Ecology , 84, 583-595.

[60] Helsa, B I, Tieszen, L L, & Boutton, T W. (1985). Seasonal water relations of savanna shrubs and grasses in Kenya. Journal of Arid Environments , 8, 15-31.

[61] Knoop, W T, & Walker, B H. (1985). Interactions of woody and herbaceous vegetation in a southern African savanna. Journal of Ecology , 73, 235-253.

[62] Higgins, S I, Bond, W J, Trollope, W, & Fire, W. resprouting and variability: A recipe for grass-tree coexistence in savanna. Journal of Ecology , 88, 213-229.

[63] Wolfe, D W, & Erickson, J D. (1993). Carbon dioxide effects on plants: uncertainties and implications for modeling crop response to climate change. In: KAISER, H.M. & DRENNEN, T.E. (eds.). Agricultural Dimensions of Global Climate Change. St. Lucie Press, Delray Beach, Florida. , 153-178.

[64] Rohner, C, & Ward, D. (1997). Chemical and mechanical defence against herbivory in two sympatric species of desert Acacia. Journal of Vegetation Science , 8, 717-726.

[65] Ward, D, & Young, T P. (2002). Effects of large mammalian herbivores and ant symbionts on condensed tannins of Acacia drepanolobium in Kenya. Journal of Chemical Ecology , 28, 913-929.

[66] Wiegand, K, Ward, D, & Saltz, D. (2002). Multi-scale patterns in an arid savanna with a single soil layer. Journal of Vegetation Science, submitted.

[67] Connor, O. T.G. (1995). Acacia karoo invasion of grassland: environmental and biotic effects influencing seedling emergence and establishment. Oecologia , 103, 214-223.

[68] Wilson, T B, & Witkowski, E. T F (1998). Water requirements for germination and early seedling establishment in four African savanna woody plant species. Journal of Arid Environments , 38, 541-550.

[69] Sharon, D. (1981). The distribution in space of local rainfall in the Namib Desert. International Journal of Climatology , 1, 69-75.

[70] Prins, H, & Loth, T. P E (1988). Rainfall patterns as background to plant phenology in northern Tanzania. Journal of Biogeography , 15, 451-463.

[71] Chamier, J, & Schachtschneider, K. Le Maitre D C, Ashton P J and van Wilgen B W (2012). Impacts of invasive alien plants on water quality, with particular emphasis on South Africa. Water SA , 38(2), 345-356.

[72] Milton, S J. (1995). Spatial and temporal patterns in the emergence and survival of seedlings in arid Karoo shrubland. Journal of Applied Ecology , 32, 145-156.

[73] Harper, J L. (1977). Population Biology of plants: Academic Press. London.

[74] Birks, H. J B (1989). Holocene isochrone maps and patterns of tree spreading in the British Isles. Journal of Biogeography , 16, 503-540.

[75] Lorenzo, P. Rodri guez-Echeverria S, Gonzalez L and Freitas H (2010). Effect of invasive Acacia dealbata Link on soil microorganisms as determined by PCR-DGGE. Applied Soil Ecology, , 44, 245-251.

[76] Navarantham, S J, & Catley, A. (1986). Quarantine measures to exclude plant pests. In Ecology of Biological Invasions, eds R H Groves and J J Burdon, Cambridge University Press, Cambridge. , 106-112.

[77] Smallwood, K S, & Salmon, T P. (1992). A rating system for potential exotic bird and mammal pests. Biological Conservation , 62, 149-159.

[78] Thorp, J R, & Lynch, R. (2000). The determination of weeds of national significance. National Weeds Strategy Executive Committee, Launceston, Cornwall. Http://www.weeds.org.au/docs/WONS/

[79] Robertson, M P, Villet, M H, Fairbanks, D, Henderson, K, Higgins, L, Hoffman, S I, Le, J H, Maitre, D C, Palmer, A R, Riggs, I, Shackleton, C M, & Zimmermann, H G. (2003). A proposed prioritization system for the management of invasive alien plants in South Africa. South African Journal of Science , 99, 37-43.

[80] Henderson, L. (1998). Southern African Plant Invaders Atlas (SAPIA). Applied Plant Science , 12, 31-32.

[81] Nel, J L, Richerdson, D M, Rouget, M, Mgidi, T N, & Mdzeke, N. Le MAtre D C, B W van Wilgen, Schonegevel L, Henderson L and

[82] Wells, M J, Balsinhas, A A, Joffe, H, Engelbrecht, V M, Harding, G, & Stirton, C H. (1986). Acatalogue of problem plants in southern Africa. Mem. Bot. Surv. S. Afr. 53.

[83] Richardson, D M, Pysek, P, Rejmánek, M, Barbour, M G, Panetta, D F, & West, C J. (2000). Naturalization and invasion of alien plants: concepts and definitions. Diversity Distribution. , 6, 93-107.

[84] Van Der Berckt, T. (2002). The ecological effects of Acacia saligna in a Sand Plain Fynbos ecosystem of the Western Cape, South Africa. M.Sc. thesis, University of Stellenbosch.

[85] Görgens, A, & Van Wilgen, M. B W (2004). Invasive alien plants and water resources: an assessment of current understanding, predictive ability and research challenges. South African Journal of Science 100 (1-2): 27- 34.

[86] Van Wilgen, B W, & Reyers, B. Le Maitre D C, Richardson D M and Schonegevel L (2008). A biome scale assessment of the impact of invasive alien plants on ecosystem services in South Africa. Journal of Environmental Management , 89(4), 336-349.

[87] Calder, I, & Dye, P. (2001). Hydrological impacts of invasive alien plants. Land use Water Research , 1(7), 1-8.

[88] Dorning, M, & Cipollini, D. (2006). Leaf and root extracts of the invasive shrub, Lonicera maackii, inhibit seed germination of three herbs with no autotoxic effects. Plant Ecology, , 184, 287-296.

[89] Aerts, R. (1997). Climate, leaf litter chemistry and leaf litter decomposition in terrestrial ecosystems: a triangular relationship. Oikos , 79-439.

[90] Stewart, B. A, & Davies, B. R. (1990). Allochthonous input and retention in a small mountain stream, South Africa. Hydrobiologia , 202-135.

[91] Stock, WD, & Wienan, . 1995. Impacts of invading N2-fixing Acacia species on patterns of nutrient cycling in two Cape ecosystems: evidence from soil incubation studies and 15N natural abundance values. Oecologia 101 (3) 375-382.

[92] Kalff, J. (2002). Limnology: Inland Water Systems. Prentice Hall, Upper Saddle River, NJ.

[93] Lee, G. F. (1973). Role of phosphorus in eutrophication and diffuse source control. Water Research. 7 (1-2) 111-128.

[94] Ehrenfeld, J. G. (2003). Effects of exotic plant invasions on soil nutrient cycling processes. Ecosystems , 6(6), 503-523.

[95] Jovanovic, N. Z, Israel, S, Tredoux, G, & Soltau, L. Le Maitre D, Rusinga F, Rozanov A and Van Der Merwe N (2009). Nitrogen dynamics in land cleared of alien vegetation (Acacia saligna) and impacts on groundwater at Riverlands Nature Reserve (Western Cape, South Africa). Water SA , 35(1), 37-44.

[96] Virtue JG And Melland RL (2003). The Environmental Weed Risk of Revegetation and Forestry Plants. South Australia. Department of Water, Land and Biodiversity Conservation. Report, DWLBC 2003/02.

[97] Jefferson, L V, & Pennacchi, M. (2003). Allelopathic effects of foliage extracts from four Chenopodiaceae species on seed germination. Journal of Arid Environments, , 55, 275-285.

[98] Weir, T L, Park, S, & Vivanco, J M. (2004). Biochemical and physiological mechanisms mediated by allelochemicals. Current Opinion in Plant Biology 2004, 7:472-479.

[99] Kobayashi, K. (2004). Factors affecting phytotoxic activity of allelochemicals in soil. Weed Biology and Management, , 4, 1-7.

[100] Pour, A P, & Farahbakhsh, H. (2012). Allelopathic Effect of Lemon Balm on Germination and Growth of Pea, Safflower and Wheat. International Research Journal of Applied and Basic Sciences, , 3(2), 309-318.

[101] Saxena, A, Singh, D V, & Joshi, N L. (1996). Autotoxic effects of pearl millet aqueous extracts on seed germination and seedling growth. Journal of Arid Environments , 33, 255-260.

[102] Escudero, A, Albert, M J, Pitta, J M, & Perez-garcia, F. (2000). Inhibitory effects of Artemesia herba-alba on the germination of the gypsophyte Helianthemum squamatum. Plant Ecology , 148, 71-80.

[103] Nilsson, M C, Zackrosson, O, Sterner, O, & Wallstedt, A. (2000). Characterisation of the differential interference effects of two arboreal dwarf shrub species. Oecologia , 123, 122-128.

[104] Callaway, R M, Ridenour, W M, Laboski, T, Weir, T, & Vivanco, J M. (2005). Natural Selection for resistance to the allelopathic effects of invasive plants. Journal of Ecology, , 93, 576-583.

[105] Singh, N B, Pandey, B N, & Singh, A. (2009). Allelopathic effects of Cyperus rotundus extract in vitro and ex vitro on banana. Acta Physiol Plant: , 31, 633-638.

[106] Inderjit and Weiner J (2001). Plant allelochemical interference or soil chemical ecology? Perspectives in Plant Ecology, Evolution and Systematics, , 4, 3-12.

[107] Arowosegbe, S, Wintola, O A, & Afolayan, A J. (2012). Phytochemical constituents and allelopathic effect of Aloe ferox Mill. root extract on tomato. Journal of Medicinal Plants Research, , 6(11), 2094-2099.

[108] Wu, L, Guo, X, & Harivandi, M L. (1998). Allelopathic effects of phenolic acids detected in buffalograss (Buchloe dactyloides) clippings on growth of annual bluegrass (Poa annua) and buffalograss seedlings. Environmental and Experimental Botany, , 39, 159-167.

[109] Witkowski, E, & Growth, F. and competition between seedlings of Protea repens (L.) L. and the alien invasive, Acacia saligna (Labill) Wendl. in relation to nutrient availability. Functional Ecology , 5, 101-110.

[110] Kruse, M, Strandberg, M, & Strandberg, B. (2000). Ecological Effects of Allelopathic Plants- a Review. National Environmental Research Institute, Silkeborg, Denmark. NERI Technical Report (315), 66.

[111] Vila-aiub, M M, Neve, P, Steadman, K J, & Powles, S B. (2005). Ecologicalfitness of a multiple herbicide-resistant Lolium rigidum population: dynamics of seed germination and seedling emergence of resistant and susceptible phenotypes. Journal of Applied Ecology , 42, 288-298.

[112] Reigosa, M J, Sanchez-moreiras, A, & Gonzalez, L. (1999). Ecophysiological Approach in Allelopathy. Critical Reviews in Plant Sciences, , 18(5), 577-608.

[113] White, C S. (1994). Monoterpenes: their effects on ecosystem nutrient cycling. Journal of Chemical Ecology, , 20, 1381-1406.

[114] Inderjit (1996). Plant Phenolics in Allelopathy. Botanical Review, , 62, 186-202.

[115] Wardle, D A, Nilsson, M C, Gallet, C, & Zackrisson, O. (1998). An ecosystem level perspective of allelopathy. Biological Reviews, , 73, 305-319.

[116] Khanh, T D, Chung, M I, Xuan, T D, & Tawata, S. (2005). The Exploitation of Crop Allelopathy in Sustainable Agricultural Production. Journal of Agronomy and Crop Science, , 191, 172-184.

[117] Batish, D R, Lavanya, K, Singh, H P, & Kohli, R K. (2007). Root-mediated Allelopathic Interference of Nettle-leaved Goosefoot (Chenopodium murale) on Wheat (Triticum aestivum). Journal of Agronomy and Crop Science, , 193, 37-44.

http://dx.doi.org/10.5772/56182

[118] Xuan, T. D, Tawata, S, Khanh, T. D, & Chung, I. M. (2005). Decomposition of allelopathic plants in soil. Journal of Agronomy Crop Science. , 191, 162-171.

[119] Chang-hung, C. (1999). Roles of Allelopathy in Plant Biodiversity and Sustainable Agriculture. Critical Reviews in Plant Sciences, , 18(5), 609-636.

[120] Sanchez-moreiras, A M, Weiss, O A, & Reigosa-roger, M J. (2003). Allelopathic Evidence in the Poaceae. Botanical Review, , 69, 300-319.

[121] An, M, Liu, D L, Johnson, I R, & Lovett, J V. (2003). Mathematical modelling of allelopathy: II. The dynamics of allelochemicals from living plants in the environment. Ecological Modelling, , 161, 53-66.

[122] Einhelling, F A, & Leather, G R. (1988). Potentials for exploiting allelopathy to enhance crop production. Journal of Chemical Ecology, , 14, 1829-1844.

[123] Zackrisson, O, & Nilsson, M C. (1992). Allelopathic effects by Empetrum hermaphroditum on seed germination of two boreal tree species. Canadian Journal of Forest Research , 22, 1310-1319.

[124] Bertin, C, Yang, X, & Weston, L A. (2003). The role of root exudates and allelochemicals in the rhizosphere. Plant and Soil, , 256, 67-83.

[125] Inderjit and Moral R D (1997). Is Separating Resource Competition from Allelopathy Realistic? The botanical Review, , 63(3), 221-230.

[126] InderjitWeston L A and Duke S O (2005). Challenges, achievements and opportunities in allelopathy research. Journal of Plant Interactions, , 1(2), 69-8.

[127] Zhao-hui, L, Wang, Q, Ruan, X, Cun-de, P, & An, J. (2010). Phenolics and Plant Allelopathy. Molecules, , 15, 8933-8952.

[128] Fatunbi, A. O, Dube, S, Yakubu, M T, & Tshabalala, T. (2009). Allelopathic Potential of Acacia mearnsii De Wild. World Applied Sciences Journal , 7(12), 1488-1493.

[129] Einhelling, F A. (1996). Interactions involving allelopathy in cropping systems. Agronomy Journal, , 88, 886-893.

[130] Julia, R, Holland, D W, & Guenthner, J. (2007). Assessing the economic impact of invasive species: The case of yellow starthistle (Centaurea solsitialis L.) in the rangelands of Idaho, USA. Journal of Environmental Management , 85, 876-882.

[131] Gemedo-DalleMaass B L and Isselstein, J (2006). Encroachment of woody plants and its impact on pastoral livestock production in the Borana lowlands, southern Oromia, Ethiopia. African Journal of Ecology , 44, 113-299.

[132] Parker, I M, Simberloff, D, Lonsdale, W M, Goodell, K, Wonham, M, Kareiva, P M, Williamson, M, Von Holle, B, Moyle, P B, Byers, J E, & Goldwasser, L. (1999). Impact: towards a framework for understanding the ecological effects of invaders. Biological invasions , 1, 3-19.

[133] Versfeld, D B. Le Maitre D C and Chapman R A (1998). Alien Invading Plants and Water Resources in South Africa. A Preliminary Assessment. Report to the Water Research Commission. CSIR No. ENV/S-C 97154. WRC Report No TT 99/98 September.

[134] De Lange, W J, & Van Wilgen, B W. (2010). An economic assessment of the contribution of weed biological control to the management of invasive alien plants and to the protection of ecosystem services in South Africa. Biological Invasions , 12(12), 4113-4124.

[135] Callihan, R H, Lass, L W, Mccaffrey, J, & Michalson, E. (1996). Yellow starthistle management for small acreages. Current Information Series University of Idaho, Moscow.(1025)

[136] Olson, R A, & Whitson, T D. (2002). Restoring structure in late-successional sagebrush communities by thinning with tebuthiuron. Restoration Ecology , 10, 146-155.

[137] Fulbright, T E. (1996). Viewpoint: A theoretical basis for planning woody plant control to maintain species diversity. Journal of Range Management , 49, 554-559.

[138] Angassa, A, & Oba, G. (2008). Effects of management and time on mechanisms of bush encroachment in southern Ethiopia. African Journal of Ecology , 46, 186-196.

[139] Clarke, P J, Latz, P K, & Albrecht, D E. (2005). Long-term changes in semi-arid vegetation: Invasion of an exotic perennial grass has larger effects than rainfall variability. Journal of Vegetation Science. , 16, 237-248.

[140] Grime, J P. (1977). Evidence for existence of three primary strategies in plants and its relevance to ecological and evolutionary theory. American Naturalist. , 111, 1169-1194.

[141] Laska, G. (2001). The disturbance and vegetation dynamics: a review and an alternative framework. Plant Ecology. , 157, 77-99.

[142] Vetaas, O. R. (1992). Micro-site effects of trees and shrubs in dry savannas. Journal of Vegetation Science. , 3, 337-344.

[143] Olson, R. A, & Whitson, T. D. (2002). Restoring structure in late-successional sagebrush communities by thinning with tebuthiuron. Restoration Ecology. , 10, 146-155.

[144] Angassa, A, Oba, G, & Tolera, A. (2012). Bush encroachment control demonstrations and management implications on herbaceous species in savannas of southern Ethiopia. Tropical and Subtropical Agroecosystems, , 15, 173-185.

[145] Connor, O, Kuyler, T G, Kirkman, P, & Corcoran, K P. B (2010). Which grazing management practices are most appropriate for maintaining biodiversity in South African grassland. African Journal of Range and Forage Science, , 27(2), 67-76.

[146] Kgosikoma, O E, Harvie, B A, & Mojeremane, W. (2012). Bush encroachment in relation to rangeland management systems and environmental conditions in Kalahari ecosystem of Botswana. African Journal of Agricultural Research, , 7(15), 2312-2319.

[147] Fuhlendorf, S D, & Smeins, F E. (1998). Influence of soil depth on plant species response to grazing within a semi-arid savanna. Plant Ecology , 138, 89-96.

[148] Briske, D D, Fuhlendorf, S D, & Smeins, F E. (2005). State-and-transition models, thresholds, and rangeland health: a synthesis of ecological concepts and perspectives. Rangeland Ecology and Management , 58, 1-10.

[149] Fuhlendorf, S D, Engle, D M, Arnold, D C, & Bidwell, T G. (2002). Influence of herbicide application on forb and arthropod communities of North American tallgrass prairies. Agriculture, Ecosystems and Environment , 92, 251-259.

[150] Holechek, J L, Pieper, R D, & Herbel, C H. (2004). Rangeland Management Principles and Practices, fifth ed. Prentice Hall, Upper Saddle River.

[151] DiTomaso J M (2000). Invasive weeds in rangeland: Species, impacts, and management. Weed Science, , 48(2), 255-265.

[152] Teague, W R, Grant, W E, Kreuter, U P, Diaz-solis, H, Dube, S, Kothmann, M M, Pinchak, W E, & Ansley, R J. (2008). An ecological economic simulation model for assessing fire and grazing management effects on mesquite rangelands in Texas. Ecological economics , 64, 611-624.

[153] Teague, W. R. (1996). A research framework to achieve sustainable use of rangeland. Agriculture, Ecosystems and Environment , 57, 91-102.

[154] Smit, G N, Richter, C, & Aucamp, F. A J (1999). Bush encroachment: An approach to understanding and managing the problem. In: Tainton N M (Ed). Veld Management in South Africa. University of Natal Press, Pietermaritzburg.

[155] Fuhlendorf, S D, Engle, D M, Meilia, O, Weir, C M, & Cummings, J R, D. D C (2009). Does herbicide weed control increase livestock production on non-equilibrium rangeland? Agriculture, Ecosystems and Environment , 132, 1-6.

[156] Dube, S, Lesoli, M S, & Fatunbi, A O. (2009). The efficacy and safety of bromacil based herbicide for the control of the invasive bush species in South African rangelands. African Journal of Biotechnology , 8(9), 1776-1781.

[157] EXTONET (1993). Extension Toxicology Network. A Pesticide Information Project of Cooperative Extension Offices of Cornell University, Michigan State University, Oregon State University, and University of California at Davis. (URL:http://pmep.cce.cornell.edu/profiles/extoxnet)

[158] Gangstad, E. O. (1989). Woody brush control. CRC Press, Boca Ralon, Florida. , 103.

[159] Prostko, E. P. (2001). Herbicide mode of action. Extension Weed Specialist. University of Georgia Tifton, GA. http://www.cropsoil.uga.edu/weedsci/slides/newmode/slide21.html#menuX000.

[160] Tu, M, Hurd, C, & Randall, J. M. (2001). Weed Control Methods Handbook: Tools and Techniques for Use in Natural Areas. The Nature Conservancy, http://tncweeds.ucdavis.edu, version: April 2001.

[161] Arteca, R. N. (1995). Plant growth substances; principles and substances. Chapman and Hall. http://books.google.com/books?id=m5yI97kkeMkC&pg=PA289&dq=bromacil&sig=hX9BwLUsOTsa_Q3oCY0vjFDEOhg#PPP1,M1.

[162] Rice, C K, & Stritzke, J F. (1989). Effects of 2, 4-D and atrazine on degraded Oklahoma grasslands. Journal of Range Management , 42, 217-222.

[163] Donaldson, D, Kiely, T, & Grube, A. (2002). Pesticide Industry Sales and Usage: 1998 and 1999 Market Estimates. U.S. Environmental Protection Agency, Washington, DC.

[164] Gillen, R L, Rollins, R D, & Stritzke, J F. (1987). Atrazine, spring burning, and nitrogen for improvement of tallgrass prairie. Journal of Range Management , 40, 444-447.

[165] Koerth, B H. (1996). Chemical manipulation of plants. In: Krausman, P.R. (Ed.), Rangeland Wildlife. Society for Range Management, Denver, , 321-337.

[166] Heitschmidt, R K. and Taylor Jr. C A (1991). Livestock production. In: Heitschmidt, R.K., Stuth, J.W. (Eds.), Grazing Management: An Ecological Perspective. Timber Press, Inc., Portland, , 161-177.

[167] Tilman, D, & Haddi, A E. (1992). Drought and biodiversity in grasslands. Oecologia , 89, 257-264.

[168] Tilman, D, & Downing, J A. (1994). Biodiversity and stability in grasslands. Nature , 367, 363-365.

[169] Taylor, R L, Maxwell, B D, & Boik, R J. (2007). Indirect effects of herbicides on bird food resources and beneficial arthropods. Agriculture Ecosystems and Environment , 116, 157-164.

[170] Powles, S B, & Holtum, J A. (1994). Herbicide Resistance in Plants. Biology and Biochemistry. Lewis Publishers, Boca Raton.

[171] Ghersa, C M, Benech-arnold, R L, & Satorre, E H. and Mart´ınez-Ghersa M A (2000). Advances in weed management strategies. Field Crop Res. , 60, 95-104.

[172] Mortimer, A M. (1997). Phenological adaptation in weeds-an evolutionary response to the use of herbicides? Pestic. Sci. , 51, 299-304.

[173] Mart´ınez-Ghersa M AGhersa C M, Benech-Arnold R L, Mac Donough R, S´anchez R A (2000). Adaptive traits regulating dormancy and germination of invasive species. Plant Spec. Biol. , 15, 127-137.

[174] Holt, J S. (1990). Fitness and ecological adaptability of herbicide-resistant biotypes. In: Green M B, Le Baron H M, Moberg W K (Eds.), Managing Resistance to Agro-

chemicals. From Fundamental Research to Practical Strategies. ACS Symposium Series ACS Washington DC, (421), 419-429.

[175] Vila, M, & Sardans, J. (1999). Plant competition in Mediterranean-type vegetation. Journal of Vegetation Science, , 10, 281-294.

[176] Roush, M L, & Radosevich, S R. (1985). Relation between growth and competitiveness of four annual weeds. Journal of Applied Ecology , 22, 895-905.

[177] Klingman, G C, Ashton, F M, & Noordhoff, L J. (1982). Weed Science: Principles and Practices, 2nd edn. John Wiley, New York, 449 pp.

[178] Emmerich, W E, Helmer, J D, Renard, K G, & Lane, L J. (1984). Fate and effectiveness of tebuthiuron applied to a rangeland watershed. Journal of Environmental Quality, , 13, 382-386.

[179] Anderson, R L, & Barrett, M R. (1985). Residual phytotoxicity of chlorsulfuron in two soils. Journal of Environmental Quality 14:l 1 l-114.

[180] Van Wilgen, B, Richardson, D, & Higgins, S. (2001). Integrated control of invasive alien plants in terrestrial ecosystems. Land use and Water Resources Research , 1(5), 1-6.

[181] Sabitii, E +N, Wamara, J B, Ogen-odoi, A A, & Wein, R W. (1992). The role of fire in pasture and rangeland management. Nomadic Peoples , 31, 107-110.

[182] Trollope, W, & Tainton, W. N M (1986). Effect of fire intensity on the grass and bush components of the Eastern Cape thornveld. Journal of the Grassland Society of Southern Africa , 3(2), 37-42.

[183] Trollope, W. S W (1999). Fire Behaviour. In: N M Tainton (ed) Veld Management in South Africa. University of Natal Press, Pietermaritzburg.

[184] Trollope, W. S. W. (1980). Controlling bush encroachment with fire in the savanna areas of South Africa, Proceedings of the Annual Congresses of the Grassland Society of Southern Africa, , 15(1), 173-177.

[185] Forbes, R G, & Trollope, W. S W (1990). Effect of burning on sweet grassveld in Ciskei and the Eastern Cape. Unpublished paper. University of Fort Hare, Alice.

[186] Groves, R H. (1989). Ecological control of invasive terrestrial plants. In: Drake J A, Mooney H A, di Castri F, Grives R A, Kruger F J, Rejmanek M and Williamson M (eds). Biological Invasions: A Global perspective, Scope 37. John Wiley and Sons Ltd.

[187] Van Wilgen, B W. Le Maitre D C and Cowling R M (1998). Ecosystem services, efficiency, sustainability and equity: South Africa's Working for Water programme. Trends in Ecology and evolution 13: 378.

[188] Ansley, R J, & Castellano, M J. (2006). Strategies for savanna restoration in the southern Great Plains: Effects of fire and herbicides. Restoration Ecology , 14, 420-428.

[189] Reinhardt, C F. (2000). Weed management practices in natural ecosystems: a critical overview. Koedoe , 43(1), 67-74.

[190] Olckers, T. (1999). Introduction: Biological control of weeds in South Africa (1990-1998). African Entomology, Memoir (1)

[191] Louda, S M, Kendall, D, Conner, J, & Simberloff, D. (1997). Ecological effects of an insect introduced for the biological control of weeds. Science , 277, 1088-1090.

[192] Waterhouse, D F, & Norris, K R. (1987). Biological Control: Pacific prospects. Inkata Press, Melbourne.

[193] Olckers, T, & Hill, M P. (1999). Biological control of weeds in South Africa (1990-1998). African Entomology Memoir 1.

[194] Martin, B. W, & Foxcroft, L. C. (2001). CATALOGUE OF BIOLOGICAL CONTROL INTERVENTIONS ON INVASIVE ALIEN PLANTS: KRUGER NATIONAL PARK. Scientific Services Section, Kruger National Park Private Bag X402, SKUKUZA 1350.

[195] Hoffmann, J H, Moran, V C, & Zeller, D A. (1997). Evaluation of Cactoblastis cactorum (Lepidoptera: Phycitidae) as a biological control agent of Opuntia stricta (Cactaceae) in the Kruger National Park, South Africa.

[196] Foxcroft, L C, & Hoffmann, J H. (2000). Dispersal of Dactylopius opuntiae Cockerell (Homoptera: Dactylopidae), a biological control agent of Opuntia stricta (Haworth.) Haworth. (Cactaceae) in the Kruger National Park. Koedoe , 43(2), 1-5.

[197] Belsky, A J. (1984). Role of small browsing mammals in preventing woodland regeneration in the Serengeti National Park, Tanzania. African Journal of Ecology, , 22, 271-279.

[198] Kalamera, M C. (1989). Observations of feeding preference of elephants in the Acacia tortolis woodland of Lake Manyara national Park, Tanzania. African Journal of Ecology, , 27, 325-333.

[199] Lewis, D M. (1991). Observations of tree growth, woodland structure and elephant damage on Colophosprmum mopane in Luangwa Valley, Zambia. African Journal of Ecology, , 29, 207-221.

[200] Mack, R N, Simberloff, D, Lonsdale, W M, Evans, H, Clout, M, & Bazzaz, F A. (2000). Biotic invasions: causes, epidemiology, global consequences and control. Ecological Applications , 10, 689-710.

[201] Sala, O. E. and 18 others), (2000). Biodiersity scenarios for the year 2100. Science , 287, 1779-1774.

[202] Mc Neely, J A, Mooney, H A, Neille, L E, & Schei, P. and Waage J K (Eds.), (2001). Global strategy on invasive alien species. IUCN on behalf of the Global Invasive Species Programme, Gland, Switzerland.

[203] National Invasive Species Council(2001). Meeting the invasive species challenge: national invasive species management plan. National Invasive Species Council, Washington DC.

[204] Perrings, C, Williamson, M, Barbier, E B, Delfino, D, Dalmazzone, S, Shogren, J, Simmons, P, & Watkinson, A. (2002). Biological invasions risks and the public good: an economic perspective. Conservation Ecology 6 (1): 1.

[205] Thurow, T L, & Taylor, C A. Jr. (1999). Viewpoint: The role of drought in range management. Journal of Range Management , 52, 413-419.

Chapter 12

New Natural Herbicide Candidate for *Sicyon angulatus* Control

Jung-Sup Choi and In-Taek Hwang

Additional information is available at the end of the chapter

http://dx.doi.org/10.5772/54964

1. Introduction

Most synthetic herbicides are used for controlling troublesome weed species in modern agriculture all over the world. However, consecutive use of the same herbicide brings about resistant weed problems and many countries are restricting repeated treatment in agricultural lands [1]. For these and environmental reasons, new herbicide discovery and subsequent registration is very challenging. Recently, evaluating natural products of animals, plants, microorganisms and minerals for developing environmental friendly herbicides has increased [2]. Several compounds have been developed or in development as natural herbicides such as bialaphos [3], methoxyhygromycin (MHM) [4], and pelargonic acid [5]. Essential oils such as clove oil and cinnamon oil also contain allelochemicals that control a broad spectrum of weeds and can be used as natural herbicide source [6,7]. Plumbagin isolated from *Drosophyllum lusitanicum* and *Plumbago auriculata* inhibited the seed germination of lettuce and wheat [8,9]. Several classes of natural compounds including triketones, benzoquinones, naphthoquinones and anthraquinones have been reported as hydroxyphenylpyruvate dioxygenase (HPPD) inhibitors and hence the novel classes of HPPD inhibitors could be developed based on their structural backbones [10].

Agricultural research for herbicide discovery with new target site is increasing due to the demand from farmers and multinational companies. Even so, new mode of action have not been succesfully introduced in the past 10 years [2,3]. We have recently reported : 7-keto-8-aminopelargonic acid synthase (EC 2.3.1.47, KAPAS, also known as 8-amino-7-oxononanoate synthase, AONS) and have suggested the potential KAPAS inhibitor triphenyltin [11]. KAPAS is a pyridoxal 5′-hophate dependent enzyme which catalyzes the decarboxylative condensation of L-alanine with pimeloyl-CoA in a stereospecific manner to form7-keto-8-aminopelargonic acid, Coenzyme A, and carbon dioxide in the first committed step of biotin biosynthesis. Perhaps the most important role of biotin is in the carboxylation of acetyl-CoA to give malonyl-CoA, which is the first step in fatty-acid biosynthesis. Since fatty-acid synthesis is essential for the growth and development of most organisims, biotin is thus an essential nutrient for plants and animals. Plants, microorganisms, and some fungi biosynthesize their own biotin, while other organisms require trace amounts of the vitamin in their diet. Therefore, inhibition of the enzymes involved in the biotin biosynthesis pathway can cause irreparable damage to plants but be non-toxic to non-plant organisims, and for this reason, such enzymes can be useful targets for the rational design of inhibitors in the hopes of finding new herbicides [12,13].

Also, we attempted to search for KAPAS inhibitors from plant-derived natural compounds. Several naturally occurring quinones including chrysophanic acid, tanshinones, 5,8-dihydroxy-1,4-naphthoquinone, and plumbagin was selected as potent inhibitors against KAPAS. We evaluated the plumbagin showing most effective KAPAS inhibition, as a natural herbicide under greenhouse and field tests. Field tests were focused on the annual noxious weed species of *Sicyos angulatus* (burcucumber or star-cucumber) which have migrated from eastern North America and have been designated as one of the ecological disturbance plants listed by the Ministry of Environment in Korea. The alien plant *S. angulatus* was first observed in 1989 and rapidly emerged in the marginal of agricultural fields close to riparian zone where it has been rapidly spreading along rivers in Korea over the past two decades [14,15]. Invasion into the natural ecosystems by exotic species is a major global threat to biodiversity. *S. angulatus* was also listed in Federal and State Noxious Weeds, USA and its geographical distribution was published in the OEPP/EPPO Bulletin [16]. It is adapted to wet habitats: deciduous swamps, woodland floodplains, and river floodplains. It also colonizes open habitats along fencerows, roadsides, and woodland borders. *S. angulatus* is found in every state east of the Rocky Mountains and also found in Canada's eastern provinces, Mexico, the Caribbean, and Eastern Asia. It was first introduced to Europe as an ornamental plant, but has since escaped cultivation and become a weedy invasive species. Asaeda et al. [17] reported the most dominant liana species in the floodplain is *S. angulatus* and it was first sighted in Japan in 1952. Ceschin et al. [18] reported exotic species of *S. angulatus* as a new arrival alien in the Tiber River in Rome. Many reports of its invasiveness have been published in the United Kingdom [19], Norway [20], Japan [21], Korea [14], and Spain [22] etc.

In this chapter, we briefly describe the KAPAS inhibitory activity of plumbagin, which showed the most potent inhibition during the preliminary survey of many natural products. Also the herbicidal activity of plumbagin was evaluated under greenhouse conditions and field trials. Physiological responses caused by the plumbagin treatment with respect to cellular leakage, chlorophyll loss and the rescue effect with biotin supplement through tissue section experiments or seed germination are reported. Plumbagin is under examination as a LOHAS (Lifestyles of Health and Sustainability) [23] herbicide against an invasive alien vine plant species.

http://dx.doi.org/10.5772/54964

2. Development for *Sicyon angulatus* control

2.1. Plumbagin preparation

The specimens of *P. auriculata* grown in the greenhouse were collected, and the air-dried root (180 g) was soaked in 2 L of acetone at room temperature for 7 days. The extract was filtered and evaporated to dryness under negative pressure. The concentrated extract (1.5 g) was suspended in 100 ml of water and re-extracted with an equal volume of dichloromethane, which afforded 1.2 g of dichloromethane soluble fraction. The dichloromethane soluble fraction was subjected to silica gel column chromatography eluted with a mixture of hexane and ethyl acetate (20:1) to give 120 mg of plumbagin as a dark yellow crystal. The spectral data of isolated plumbagin (purity > 99%), such as UV, MS and 1H NMR and ^{13}C-NMR were well accorded with the result of Bhattacharyya and Carvalho [24]. For field trial, plumbagin (5-hydroxy-2-methyl-1,4-naphthoquinone) was purchased from Sigma–Aldrich, which was originally isolated from Plumbago indica (Plumbaginaceae). The purity of commercially available plumbagin was estimated over 90% by HPLC.

2.2. Plumbagin as a KAPAS inhibitor

The full-length of AtKAPAS cDNA was amplified and isolated from *Arabidopsis thaliana* cDNA and cloned into MBP fusion vector to generate the *Escherichia coli* expression construct pEMBPek-KAPAS [11]. SDS–PAGE analysis revealed that *E. coli* transformed with MBP fusion vector showed the expression of a very strongly induced fusion protein of ca. 98.2 kDa, which consisted of the AtKAPAS protein of 51.3 kDa and the maltose binding peptide MBP affinity tag of 46.9 kDa [11,32]. Pimeloyl-CoA was synthesized according to the method described previously [25]. KAPAS activity was determined according to the method described previously [12] using a linked assay by monitoring the increase in absorption of NADH at 340 nm using a Microplate Spectrophotometer (Benchmark Plus, Bio-rad, USA), thermostatically controlled at 30°C. At KAPAS protein was expressed in *E. coli* at a very high level, and a significant portion of these proteins was soluble, and their affinity-purified preparations contained a single major polypeptide. The lysates from IPTG-induced *E. coli* containing pEMBPek-KAPAS as well as from *E. coli* harboring control vector MBP fusion vector were loaded onto maltose affinity column (1.1 cm x 30 cm, Millipore, USA). The AtKAPAS protein bound to MBP resin was eluted with 10 mM maltose solution. A typical assay contained 20 mM potassium phosphate (pH 7.5), 1 mM α-ketoglutarate, 0.25 mM thiamine pyrophosphate, 1 mM NAD^+, 3 mM $MgCl_2$, 0.1 unit of α-ketoglutarate dehydrogenase, and 2–10 μg of KAPAS (3 mg protein/ml) in a total volume of 200 μL. L-Alanine and pimeloyl-Co A were added to give the desirable final concentrations. Prior to analysis, enzyme samples were dialyzed for 2 h at 4°C against 20 mM potassium phosphate (pH 7.5) containing 100 μM PLP. The KAPAS concentration in all analyses was 10 μM in 20 mM potassium phosphate (pH 7.5). 96-well microplates containing each 528 natural compounds prepared from various medicinal plants and exotic herbs were evaluated on KAPAS inhibition assay at the concentration of 1 mM. Through the consecutive experiment at lower concentration against samples showing 90% inhibition of KAPAS activity, plumbagin were selected as the most effective KAPAS inhibitor.

IC_{50} value of KAPAS inhibition by plumbagin was calculated from the regression curve prepared with the extensive assay performed with the plumbagin ranged from 0.1 to 250 μM with five replications. A reference was prepared with all components except plumbagin.

Enzyme activity was tested with the partially purified AtKAPAS protein extracted from transgenic *E. coli*. AtKAPAS protein was expressed *in E. coli* at a very high level, and a significant portion of these proteins was soluble, and their affinity-purified preparations contained a single major polypeptide. The inhibitory effect of 528 plant-derived natural compounds collected in Korea Chemical Bank, KRICT on KAPAS was evaluated using the partially purified AtKAPAS protein, *in vitro*. Less than 2% of tested compounds exhibited significant inhibitory effect on KAPAS at the concentration lower than 20 μM. Interestingly, several naturally occurring quinones including chrysophanic acid, tanshinones, 5,8-dihydroxy-1,4-naphthoquinone, and plumbagin were observed to give a potent inhibitory effect on KAPAS. Plumbagin, a natural naphthoquinone demonstrated the most effective inhibitory effect on KAPAS with an IC_{50} of 2.1 μM (Fig. 1).

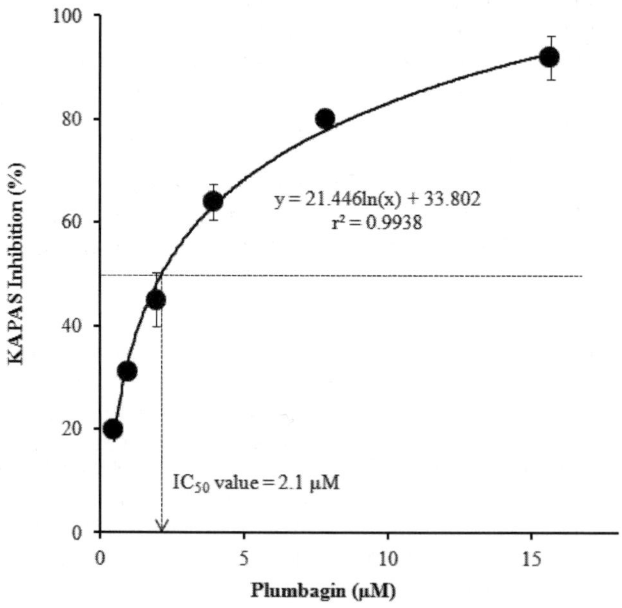

Figure 1. KAPAS inhibition by plumbagin *in vitro* assay. Vertical bars represent standard deviation. In some cases the vertical bar is obscured by the datum symbol.

2.3. Herbicidal activity of plumbagin

2.3.1. Materials and methods

Herbicidal activity and spectrum of plumbagin were investigated against eight weed species, consisting of three grass species of *Sorghum bicolor* (sorghum), *Echinochloa crus-galli* (barnyard grass), *Digitaria sanguinalis* (large crabgrass) and five broad leaf species of *Solanum nigrum* (black nightshade), *Aeschynomene indica* (Indian joint vetch), *Abutilon avicennae* (velvetleaf), *Xanthium strumarium* (common cocklebur), *Calystegia japonica* (Japanese bindweed). Seeds of weeds for foliar application were germinated in a commercial greenhouse substrate (Boo-Nong Soil, Seoul, Korea) and watered with tap water. About five plants were grown at 30/20 ± 3°C, day/night temperature with an about 14 h photoperiod for 12 days under greenhouse. Foliar application was conducted at 12 days after sowing, the test solution was sprayed into the test pot grown with 10 ~ 15 seedlings of sorghum, barnyard grass, large crabgrass, black nightshade, Indian joint vetch and velvetleaf, and two seedling of common cocklebur and Japanese bindweed. Various concentrations of the purified plumbagin from *P. auriculata* prepared with 50% acetone solution containing 0.1% Tween-20 were sprayed onto plants with a laboratory spray gun delivering spray volume of 5 ml per pot. The control treatment recieved the same volume of spray without plumbagin. After treatment, the plants were placed in a vented cabinet to dry and returned to the same greenhouse without replication. At 5 days after treatments, visual injury of plants assessed on a scale from 0 (no injury) to 100 (complete death). A field trial was performed against 10 ~ 15 leaf-stage and 2 ~ 3 m vine length of natural *S. angulatus* habitats around riparian zones in Nam-Han River. Foliar applications were conducted with 1,000 and 2,000 μg/mL of plumbagin in 50% acetone solution containing 0.1% Tween-20 using a laboratory sprayer delivering spray volume of 300 ml/m^2 with a control treatment of the same preparation solution without plumbagin. The field trial was performed from 22th September to 6th October, 2011, and the trial contained three replicates of 1 m^2 plot size. The control value was evaluated visually at 5, 8, and 14 days after treatments. Test plots were situated directly adjacent to each other.

2.3.2. Results

Under greenhouse conditions, all eight weed species were completely controlled by the foliar application of 1,000 and 2,000 μg/mL plumbagin, while 500 μg/mL applications also showed 100% herbicidal efficacy against seven weed species with the exception of *A. avicennae* (Fig. 2). 250 μg/mL applications against eight weeds showed 60 ~ 100% control (Table 1), and especially a concentration as low as 32 μg/mL had a herbicidal efficacy of 70% on *D. sanguinalis* (data not shown). With a plumbagin treatment of eight weed species, the main herbicidal symptoms were desiccation or extensive necrosis within 2 h. The difference of symptoms caused by the plumbagin between grass species and broad leaf species was insignificant after foliar application. Field test results revealed that the natural compound plumbagin controlled alien weed *S. angulatus* completely at 2,000 μg/mL under foliar application. Visual symptoms of plant injury after plumbagin foliar application against natural *S. angulatus* were desiccation or burn down within 2 h after treatment. Control values were evaluated as 95–100% by a visual rating scale of 0–100 at 5, 8, and 14 days after treatment with 1,000 or 2,000 μg/mL. The residual activity lasted for 2 weeks without any regrowth.

Conc.	Herbicidal efficacy (%)[1)]							
(μg/mL)	SORBI	ECHCG	DIGSA	SOLNI	AESIN	ABUTH	XANSI	CAGEH
2000	100	100	100	100	100	100	100	100
1000	100	100	100	100	100	100	100	100
500	100	100	100	100	100	70	100	100
250	90	60	100	100	100	60	90	100

[1)]Herbicida1 activity was determined 7 days after treatment by visual injury. SORBI, Sorghum bicolor (sorghum); ECHCG, Echinochloa crus-galli (barnyard grass); DIGSA, Digitaria sanguinalis (large crabgrass); SOLNI, Solanum nigrum (black nightshade); AESIN, Aeschynomene indica (Indian jointvetch); ABUTH, Abutilon avicennae (velvetleai); XANSI, Xanthium strumarium (common cocklebur); CAGEH, Calystegia japonica (Japanese bindweed). '2,000 pg/ml. can change to 4 kg/ha.

Table 1. Herbicidal efficacy of plumbagin post-emergence foliar application against several weeds in a greenhouse condition

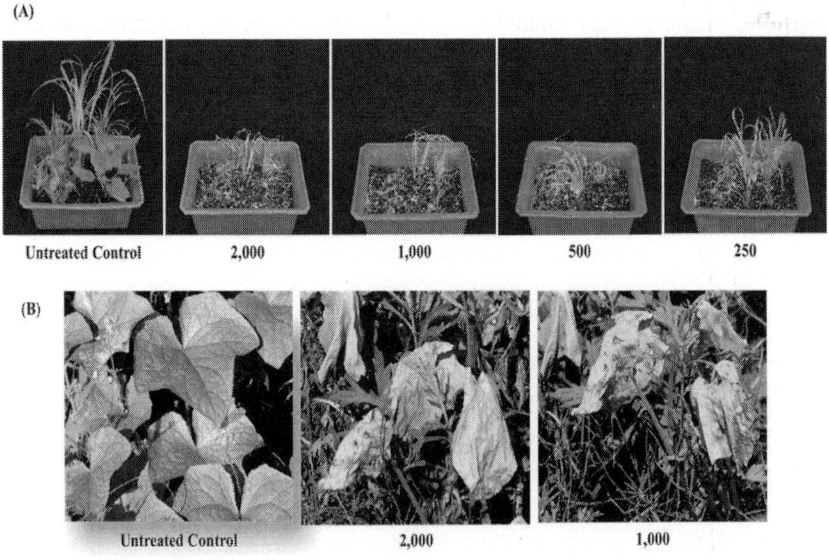

Figure 2. Herbicidal symptoms of post-emergence foliar application of plumbagin (μg/mL). (A) Pot test in a greenhouse condition against 8 weed species. (B) Field trials for *Sicyos angulatus* control. * 2,000 μg/mL can change to 4 kg/ha.

http://dx.doi.org/10.5772/54964

2.4. Reversal study

2.4.1. Materials and methods

Seeds of *A. thaliana* were germinated on a 55 mm Polystyrene Petri-dish lined with one-layer filter paper (Advantec No. 2). One milliliter of each plumbagin solution dissolved in absolute acetone with various concentrations of 0, 25, 50 and 100 μM was dropped evenly onto the filter paper and placed in a vented cabinet to dry. After complete drying, 1 ml of distilled water with or without supplement of 0, 0.25, 0.5 and 1 mM biotin (Sigma, USA) was added, and 30 seeds were placed onto the filter paper in Petri-dish. Each Petri-dish was sealed with laboratory film and incubated in a growth chamber at 25°C, 14/10 h (Light/Dark). Germination inhibition percentages were calculated with the number of germinated *A. thaliana* seeds at 7 days after application. All treatments for each measurement were triplicates.

2.4.2. Results

The inhibited germination of *A. thaliana* seeds treated with plumbagin was significantly rescued in a dose dependent manner by biotin supplement. Germination rate of *A. thaliana* seeds at plumbagin levels of 25, 50, and 100 μM was 33.3%, 23.3%, and 16.7%, respectively. However, the inhibited germination by plumbagin was negated up to 93.3%, 86.7%, and 83.3% with the supplement of 1 mM biotin, and also it was negated up to 66.7%, 63.3%, and 60.0% with the supplement of 0.5 mM biotin, respectively (Table 2, Fig. 3). Biotin supplement apparently rescued the inhibited germination *A. thaliana* seeds caused by the treatment of plumbagin.

Plumbagin	+Biotin (mM)			
(μM)	0	0.25	0.5	1
0	100[1)]	100	100	96.7
25	33.3	60.0	66.7	93.3
50	23.3	53.3	63.3	86.7
100	16.7	46.7	60.0	83.3

[1)]Germination rate of *A. thaliana* seed at 7 days after application.

Table 2. Reversal effect of *Arabidopsis thaliana* seed germination with biotin supplement

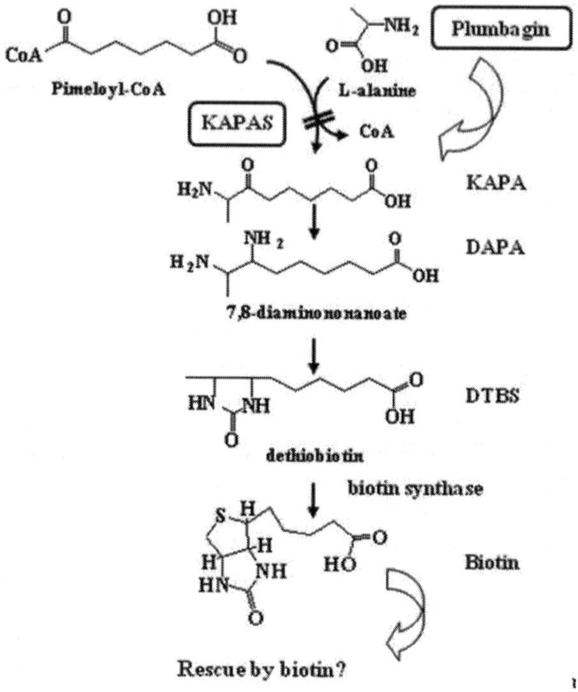

Figure 3. Proposed target site of plumbagin on KAPAS and biotin synthesis pathway in plant.

5. Summary

A new herbicide developed the lifestyle of health and sustainability (LOHAS) initiative is required to satisfy environmental and regulatory pressures. LOHAS describes an estimated $290 billion US marketplace for goods and services focused on health, the environment, social justice, personal development and sustainable living. Approximately 13–19% of the adults in the U.S. are currently considered LOHAS consumers. This is based on surveys of the U.S. adult population estimated at 215 million [23]. Also world-wide consumers demand these types of compounds as potential natural-product based herbicides. In this chapter, we attempted to develop a new herbicide from natural compounds having the new target KAPAS, and we applied this to annual noxious weed species of *S. angulatus* (burcucumber or star-cucumber). Our laboratory has performed molecular genetics dissection using anti-sense approach to identify new target AtKAPAS on the pathway of biotin biosynthesis and to characterize the phenotypic consequences of loss-of-function mutations [11]. The 528 plant-derived natural compounds stored in KRICT Chemical Bank were assessed on the inhibitory effect on KAPAS using the partially purified AtKAPAS protein, *in vitro*. Less than 2% of 528 compounds exhibited inhibitory effect under a concentration of 20 μM. Interestingly, several naturally occurring quinone compounds including chrysophanic acid, tanshinones, 5,8-dihydroxy-1,4-naphthoquinone, and plumbagin were observed to give a potent inhibitory effect on KAPAS. Plumbagin, a natural naphthoquinone demonstrated the most effective inhibition on KAPAS in a concentration-dependent manner, and the IC_{50} was calculated as 2.1 μM. Webster et al. [12] reported that biotin is an essential enzyme cofactor for carboxylase and transcarboxylase reactions. Abell [28] and Pillmoor et al. [29] suggested that if an enzyme is a potential target, a 60–80% inhibition of its activity leads to a severe growth. However, this requires the confirmation of potential target. For the purpose of target validation, a rescue study was carried out. Plumbagin inhibited germination of *A. thaliana* seeds but this effect was rescued by a biotin supplement. From this point of view, our results suggest that strong inhibition of KAPAS by plumbagin leads to restriction on the biotin biosynthesis in plants, ultimately the stems or leaves of plant treated with plumbagin die. Hwang et al. [11] argued that knowledge of biochemical pathways in plants is incomplete, and the next major herbicide target may lie in an unexpected area of plant metabolism; knowledge in detail how plants actually die as a result of inhibition of some known targets is still ambiguous. Also, we should note that the complete inhibition of enzyme activity at some known targets is not necessary for plant death [30]. However, it can be predicted that the herbicidal activity is somewhat connected between the reduced level of target enzyme activity and plant death. The enzyme inhibition results and rescue effect by biotin strongly suggested that the herbicidal activity by foliar treatment was due to the inhibition of KAPAS caused by the plumbagin. The natural chemical plumbagin has been shown by our research to effectively control eight weed species of *S. bicolor, E. crus-galli, D. sanguinalis, S. nigrum, A. indica, A. avicennae, X. strumarium, C. japonica* under non-replicated greenhouse conditions. Also, the foliar application of the natural compound plumbagin at 2,000 Mg/mL has completely controlled 10 ~ 15 leaf-stage and 2 ~ 3 m vine length natural *S. Angulatus,* with sustantial residual activity under field conditions. The residual activity lasted for 2 weeks because regrowth was not observed until then. Visual symptoms of browning and necrosis of leaf tissue after plumbagin foliar applications appear to be introduced by cellular leakage rather than the inhibition of photosynthesis since cellular leakage occurred under light and dark conditions without chlorophyll loss. It seems closely related to the membrane lipid peroxidation as a result of the biotinyl carboxylase and transcarboxylase inhibition attributable to the biotin deficiency by KAPAS inhibition. Biotin is an essential enzyme cofactor for carboxylase and transcarboxylase reactions in plant leaf, and KAPAS inhibition resulted in biotin depletion. As reviewed by Delye et al. [31] and Hwang et al. [32], these pathways in plant have been well established by acetyl-CoA carboxylase (ACCase) inhibiting herbicides, like as aryloxyphenoxypropionates and cyclohexanediones. ACCase is involved in the first step of lipid synthesis. The target site of acetyl-CoA carboxylase is a biotin-dependent enzyme that catalyzes the irreversible carboxylation of acetyl-CoA to produce malonyl-CoA. The inhibition of KAPAS by plumbagin might result in the deficiency of substrate biotin to the biotinyl carboxylase in plants. However, the mechanism of action should be studied for better understanding of whole plant-compound interactions confirmative for this speculation.

In a competing mechanism, proton abstraction is involved with the attack of acetyl-CoA. When the biotin is deficient, the product, malonyl-CoA is not produced. Malonyl-CoA is a building block for new fatty acids and can inhibit the transfer of the fatty acyl group from acyl-CoA to carnitine with carnitine acyltransferase, which inhibits the beta-oxidation of fatty acids in the mitochondria. *S. angulatus* have been designated as one of the ecological disturbance plants by the Ministry of Environment in Korea. *S. angulatus* has spread across the marginal of agricultural field close to riparian zones along the rivers in Korea within the 15 years since its first appearance in 1989 (An Dong), covering more than 110 ha in 2005 [14,15]. The social and agricultural impact, risk assessment, invasion plants identification, and control management methods for alien vine plant such as *Humulus japonica* and *S. angulatus* have become a great problem in Korea. In conclusion, our results show that the herbicidal effect of plumbagin, a naturally occurring naphthoquionone, is closely associated with its inhibitory effect on KAPAS, a new target site of herbicide. Plumbagin and related 1,4-naphthoquinone compounds could be employed as a good chemical lead for an *S. angulatus* herbicide with a new mode of action (Fig. 4).

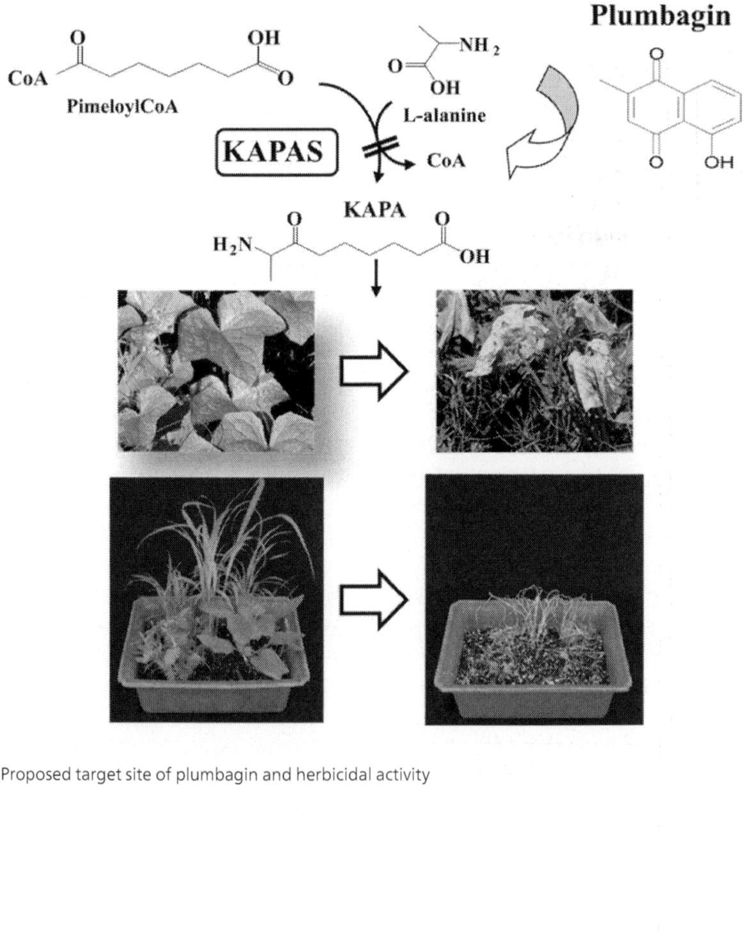

Figure 4. Proposed target site of plumbagin and herbicidal activity

http://dx.doi.org/10.5772/54964

Acknowledgements

This work was supported by the R&D Program of MKE/KEIT [10035386, Biochemical Crop Protecting Agents for LOHAS] and by the KRICT's own project [KK-1104-B0].

Author details

Jung-Sup Choi[1] and In-Taek Hwang[1,2]

1 Korea Research Institute of Chemical Technology, Yusung, Daejon, Republic of Korea

2 Department of Green Chemistry and Environmental Biotechnology, University of Science & Technology, Gajungro, Yuseong-gu, Daejon, Republic of Korea

References

[1] P.J. Tranel, T.R. Wright, Resistance of weeds to ALS-inhibiting herbicides: what have we learned?, Weed Sci 2002;50:700–712.

[2] S.O. Duke, H.K. Abbas, T. Amagasa, T. Tanaka, Phytotoxins of microbial origin with potential for use as herbicides, in: L.G. Copping (Ed.), Crop Protection Agents from Nature: Natural Production and Analogues, Critical Reviews on Applied Chemistry, vol. 35, Society for Chemical Industries, Cambridge, UK, 1996, pp. 82-113.

[3] E. Bayer, K.H. Gugel, K. Hagele, H. Hagenmaier, S. Jessipow, W.A. Konig, H. Zahner, Stoffwechselprodukte von Mikroorganismen. Mitteilung (1) Phosphinothricin and phosphinothrithyl-alanyl-alanin, Helvetica Chimica Acta 1972;55:224–239.

[4] H.B. Lee, C.J. Kim, J.S. Kim, K.S. Hong, K.Y. Cho, A bleaching herbicidal activity of methoxyhygromycin (MHM) produced by an actinomycetes strain Streptomyces sp 8E-12, Lett. Appl. Microbiol. 2003;36:387–391.

[5] M. Fukuda, Y. Tsujino, T. Fujimori, K. Wakabayashi, P. Böger, Phytotoxicity activity of middle-chain fatty acids I: effect on cell constituents, Pest Biochem. Physiol. 2004;80:143–150.

[6] T. Tworkoski, Herbicide effects of essential oils, Weed Sci. 2002;50:425–431.

[7] L.D. Bainard, M.B. Isman, Phyto-toxicity of clove oil and its primary constituent eugenol and the role of leaf epicuticular wax in the susceptibility to these essential oils, Weed Sci. 2006;54:833–837.

[8] S. Goncalves, M. Ferraz, A. Romano, Phytotoxic properties of *Drosophyllum lusitanicum* leaf extracts and its main compound plumbagin, Sci. Hortic. 2009;122:96–101.

[9] J.J.M. Meyer, F. Van der Kooy, A. Joubert, Identification of plumbagin epoxide as a germination inhibitory compound through a rapid bioassay on TLC, S. Afr. J. Bot. 2007;73:654–656.

[10] G. Meazza, B.E. Scheffler, M.R. Tellez, A.M. Rimando, J.G. Romagni, S.O. Duke, D. Nanayakkara, I.K. Khan, E.A. Abourashed, F.E. Dayan, The inhibitory activity of natural products on plant p-hydroxyphenylpyruvate dioxygenase, Phytochemistry 2002; 60:281–288.

[11] I.T. Hwang, J.S. Choi, H.Y. Song, S.J. Cho, H.K. Lim, N.J. Park, D.H. Lee, Validation of 7-keto-8-aminopelargonic acid synthase as a potential herbicide target with lead compound triphenyltin acetate, Pest Biochem. Physiol. 2010;97:24–31.

[12] S.P. Webster, D. Alexeev, D.J. Campopiano, R.M. Watt, M. Alexeeva, L. Sawyer, R.L. Baxter, Mechanism of 8-Amino-7-oxononanoate synthase: spectroscopic, kinetic, and crystallographic studies, Biochemistry 2000;39:516–528.

[13] A. Nudelman, D. Marcovici-Mizrahi, A. Nudelman, D. Flintd, V. Wittenbache, Inhibitors of biotin biosynthesis as potential herbicides, Tetrahedron 2004;60:1731–1748.

[14] J.H. Kil, H.Y. Kong, K.S. Koh & J.M. Kim, Management of Sicyos angulata spread in Korea. In: Neobiota, From Ecology to Conservation, 4th European Conference on Biological Invasions, Vienna (AT), 2006, BfN-Skripten 184, pp. 170.

[15] J.H. Kang, B.S. Jeon, S.W. Lee, Z.R. Choe, S.I. Shim, Enhancement of seed germination by aging, cold stratification, and light quality during desiccation in bur cucumber (*Sicyos angulatus* L.), Korean J. Crop Sci. 2003;48:13–16.

[16] European and Mediterranean Plant Protection Organization, EPPO data sheet on Invasive Alien Plants Sicyos angulatus, OEPP/EPPO Bulletin 2010;40:401–406.

[17] T. Asaeda, MD.H. Rashid, S. Kotagiri, T. Uchida, The role of soil characteristics in the succession of two herbaceous lianas in a modified river flood plain, River Res. Appl. 2011;27:591–601.

[18] S. Ceschin, G. Salerno, S. Bisceglie, A. Kumbaric, Temporal floristic variations as indicator of environmental changes in the Tiber River in Rome, Aquat. Ecol. 2010;44:93–100.

[19] C.G. Hanson and J.L. Mason, Bird seed aliens in Britain. Watsonia 1985;15:237–252. <http://www.watsonia.org.uk/html/watsonia_15.html>.

[20] T. Ouren, Soybean adventitious weeds in Norway, Blyttia 1987;45:175–185.

[21] S. Kurokawa, Invasion of exotic weed seeds into Japan, mixed in imported feed grains, 2001;<http://www.agnet.org/library/eb/497/>.

[22] J.F. Larché, Sicyos angulatus, nouvelle adventice du maïs dans le Sud-Ouest de la France. Phytoma – La Défense des Végétaux, 2004; 571:19–22.

[23] LOHAS on line. <http://www.lohas.com/>.

[24] J. Bhattacharyya, V.R. De Carvalho, Epi-isoshinanolone from *Plumbago scandens*, Phytochemistry 1986;25:764–765.

[25] O. Ploux, A. Marquet, The 8-amino-7-oxopelargonate synthase from *Bacillus sphaericus*. Purification and preliminary characterization of the cloned enzyme overproduced in *Escherichia coli*, Biochem. J. 1992;283:327–331.

[26] W.H. Kenyon, S.O. Duke, K.C. Vaughn, Sequence of effects of acifluorfen on physiological and ultrastructural parameters in cucumber cotyledon discs, Pest Biochem. Physiol. 1985;24:240–250.

[27] J.D. Hiscox, G.F. Israelstam, A method for the extraction of chlorophyll from leaf tissues without maceration, Can. J. Bot. 1979;57:1332–1334.

[28] L.M. Abell, Biochemica lapproaches to herbicide discovery: advances in enzyme target identification and inhibitor design, Weed Sci. 1996;44:734–742.

[29] J.B. Pillmoor, S.D. Lindell, G.G. Briggs, K. Wright, The influences of molecular mechanisms of action on herbicide design, in: N.N. Ragsdale, P.C. Kearney, J.R. Plimmer (Eds.), Processing of the Eighth of the English International Congress of Pesticide Chemistry, America Chemical Society, Washington, DC, 1995, pp. 292–303.

[30] C. Alban, D. Job, R. Douce, Biotin metabolism in plants, Annu. Rev. Plant Physiol. Plant Mol. Biol. 2000;51:17–47.

[31] C. Délye, A. Matéjicek, J. Gasquez, PCR-based detection of resistance to acetyl-CoA carboxylase-inhibiting herbicides in black-grass (*Alopecurus myosuroides* Huds) and ryegrass (*Lolium rigidum* Gaud), Pest Manag. Sci. 2002;58:474–478.

[32] I.T. Hwang, D.H. Lee, And N-J Park, 7-Keto-8-aminopelargonic acid synthase as a potential herbicide target, in: S. Soloneski, M. Larramendy (Eds.), Herbicides, INTECHWEB.ORG, Theory and applications, 2011. p471–494.

Chapter 13

Herbicides in Aquatic Systems

Lyn A. Gettys, William T. Haller and
Gregory E. MacDonald

Additional information is available at the end of the chapter

http://dx.doi.org/10.5772/56015

1. Introduction

Water plays a crucial role in maintaining the health of our global ecosystem. We rely on this valuable resource to provide drinking water, irrigation, and recreation; in addition, appropriate management of our waters is critical for flood control efforts. A diversity of native aquatic plants constitutes an integral part of the aquatic environment. These mixed populations of hydrophytes provide structure, habitat and food for fish, waterfowl, and other wildlife and act as nutrient sinks by removing phosphorus, nitrogen, and other elements from the water column. Many regions of the world – but especially those with mild climates – provide an ideal habitat for many organisms, including aquatic plants. Non-native aquatic plants are frequently introduced to aquatic systems through a number of pathways, including transport by animals, currents, or wind, but the majority of problematic plants are brought in as a result of anthropogenic activities. Human introduction of non-native aquatic plants may be accidental (e.g., via ballast water or as contaminants in desirable flora) or intentional.

2. Aquatic weeds

Many of the worst aquatic weed problems in the United States are the result of intentional introduction. For example, waterhyacinth [*Eichhornia crassipes* (Mart.) Solms] (Fig. 1) was reportedly introduced to the United States at the Southern States Cotton Expo in New Orleans in 1884. Visitors to the Expo were given waterhyacinth plants as souvenirs and many of these plants found their way into the waters of Louisiana, Texas, and Florida [1]. Local legend states that a Florida resident was entranced by the beautiful, showy flowers of this Amazonian native and brought plants back to his water garden near the St. Johns River. The plants grew abundantly and the backyard water gardener decided to share his "bounty of beauties" with others by tossing his extra plants into the St. Johns River [2]. Within a decade, the St. Johns was so clogged with waterhyacinths that navigation had become impossible [1-3].

Figure 1. Waterhyacinth. Photo courtesy Lyn Gettys.

The St. Johns constituted a major shipping passage through Florida; in order to mitigate this important resource and make it available to commercial concerns for the transport of goods, the US Army Corps of Engineers was authorized by the US Congress to use "any means necessary" to clear the system of these noxious weeds [1]. Attempts to control floating waterhyacinths utilized applications of a wide variety of substances, including arsenic, sulfuric acid, and other toxic chemicals [1]. Some of these substances effectively controlled waterhyacinths, but proved toxic to cattle that grazed on treated plants [2]. Feeding deterrents such as rotten eggs and manure were added to chemical applications to discourage grazing, but were ultimately ineffective or too expensive to use under operational conditions [1]. After these disappointing results, resource managers were forced to resort to mechanical control – manually removing plants from the surface of the water and offloading them to shore – in their attempts to clear Florida's waterways (Fig. 2). This method proved expensive and ineffective, as plants grew faster than they could be harvested from the system, but was the only management tool available until the discovery of synthetic herbicides in the 1940s [3]. Waterhyacinth is now controlled in many regions via chemical means (e.g., application of herbicides), but this Brazilian native is still considered one of the world's worst weeds [4, 5] and is intensively managed in virtually all areas the species has managed to invade.

http://dx.doi.org/10.5772/56015

Figure 2. Mechanical harvesting of waterhyacinth. Photo courtesy UF/IFAS Center for Aquatic and Invasive Plants.

Floating weeds such as waterhyacinth are readily visible and many stakeholders understand the need to control these types of noxious species. Submersed invasive species, however, are often hidden from view and the problems associated with them are not readily apparent. Submersed weeds often go unnoticed until they form surface mats; by this point, plants have been growing unchecked, often for months, and the water column is filled with plant material. This is often the case with hydrilla [*Hydrilla verticillata*) (L.f.) Royle] (Fig. 3), a noxious invader with multiple centers of origin that has been called the world's worst weed [6]. Hydrilla was introduced to the United States intentionally via the aquarium industry [7], and historical accounts suggest that some aquarium plant dealers cultivated hydrilla in canals and waters near their nurseries to have a ready supply of plant material for their customers [8]. However, the species has undoubtedly been introduced to the country's waterways repeatedly, as hobbyists regularly dispose of extra aquarium plants by tossing them in the nearest body of water. Because hydrilla is able to root from extremely small fragments [9], other pathways of introduction include waterfowl, other fauna and recreational equipment such as boats and trailers. Hydrilla causes a host of problems in its regions of invasion and greatly reduces ecosystem services and anthropogenic uses of aquatic resources.

Hydrilla can reportedly grow 1 inch (2.5 cm) per day [6], but most researchers agree that this is a gross underestimate of the plant's actual productivity [10]. This noxious weed wreaks havoc on the ecosystem by forming monocultures [11], which serve as poor habitat for resident fauna. Dense plant growth traps heat, raises the temperature of surface water and depletes dissolved oxygen, resulting in conditions that negatively impact fish survival [12, 13]. Hydrilla also obstructs water flow, which can have catastrophic consequences if resource managers need to quickly move water to prevent flooding during tropical storms, hurricanes, and other severe weather events. Recreational uses of hydrilla-infested waters are limited as well; boats motors quickly become clogged and strangled with weeds (Fig. 4), fishing lines are snagged within moments of being cast, and swimmers have reportedly drowned after becoming entangled in hydrilla [14]. Hydrilla is intensively managed in its regions of invasion. Populations of this submersed weed are reduced by a number of means, including mechanical harvesting, hand-pulling, benthic barriers, and biological control organisms such as Asian or

Figure 3. Hydrilla. Photo courtesy William Haller.

Chinese grass carp (*Ctenopharyngodon idella* Val.) [15, 16], but the vast majority of resource managers rely on chemical control to keep the growth of hydrilla in check.

Figure 4. Boat motor clogged with hydrilla. Photo courtesy UF/IFAS Center for Aquatic and Invasive Plants.

Waterhyacinth and hydrilla quickly establish and become invasive in virtually all areas where they have been introduced, but these species are not the only aquatic plants that cause problems in natural systems, reservoirs, and canals through the world. For example, resource managers charged with protecting the waters of the Pacific Northwest and many other parts of the US struggle with invasions of Eurasian watermilfoil (*Myriophyllum spicatum* L.), flowering rush (*Butomus umbellatus* L.), and curlyleaf pondweed (*Potamogeton crispus* L.) (Fig. 5). It is thought that these species were initially introduced through the aquarium and nursery trade, but have since spread throughout the country's waters as a result of improper or inadequate cleaning of contaminated equipment that has been moved from infested sites to pristine waters.

Figure 5. Other common aquatic invaders in the US. Left: curlyleaf pondweed. Right: flowering rush (emerged) and Eurasian watermilfoil (submersed). Photos courtesy Lyn Gettys.

It is clear that aquatic weeds can severely reduce ecosystem functions and limit the use of infested waters for anthropogenic activities such as recreation and flood control. However, invasive aquatic plants can pose serious risks to human health as well. For example, a number of floating species provide ideal conditions for mosquito breeding activities. Even in fast-flowing water, the stagnant water needed for mosquito reproduction is often present in the rosettes of floating weeds such as waterhyacinth and waterlettuce (*Pistia stratiotes* L.) [17-19] (Fig. 6).

3. Weed control methods in aquatic systems

A number of techniques can be employed to control or reduce populations of aquatic weeds. Clearly, the most effective way to avoid the problems associated with invasive plants is through exclusion, or preventing them from entering uninfested aquatic systems. Public education programs that emphasize proper disposal of cultivated introduced plants and animals can be helpful, but target audiences (i.e., pet and aquarium owners) often remain unaware of the ecological consequences associated with the release of these organisms into

Figure 6. Waterlettuce. Photo courtesy Lyn Gettys.

public waters. Although this sort of intentional release is certainly a vector for the introduction of new invaders (see Section 2 of this chapter describing the introduction of waterhyacinth), accidental transfer of aquatic weeds frequently occurs when boats, trailers, and other equipment is moved from an invaded site to a pristine body of water (Fig. 7). The likelihood of introduction via this route can be reduced by requiring careful inspection of any object before movement from one body of water to another. This is especially important when boats and other equipment are being relocated from a body of water that is suspected of or known to harbor invasive species to one that is pristine. These inspections can identify seeds, vegetative fragments, larvae, veligers, and other propagules of invasive aquatic species and ensure their removal before launching at a new site, thus preventing the introduction of exotic organisms into an uninfested body of water. This method has been employed with some success in the northern US, where rigorous boat inspection programs have kept invasive aquatic plants and animals such as zebra and quagga mussels (*Dreissena polymorpha* and *D. rostriformis bugensis*, respectively) from spreading to new sites [20, 21].

When exclusion programs fail and an exotic plant species colonizes a new system, managers often attempt to manually remove the invader as a first line of defense. The methods employed for removal efforts vary and are often dependent on available resources. For example, hand-pulling of target weeds may be effective, especially if the infestation is small and localized, and may be cost-effective if a pool of engaged stakeholders and volunteers can be mobilized to accomplish the task. If the new invader has colonized a relatively large area or has established in water deeper than 1 meter, the use of specialized equipment such as mechanical harvesters (Fig. 8) may be employed. Mechanical removal of aquatic weeds is often viewed by the public as the most "environmentally friendly" control method, especially among clientele that dislike the use of pesticides, and the technique certainly has utility under some circum-

Figure 7. Aquatic weeds on a boat trailer. Photo courtesy Lyn Gettys.

stances. However, a number of factors must be taken into consideration before starting mechanical control efforts, regardless of whether volunteer labor or mechanical harvesters are employed. For example, it may be logistically difficult or prohibitively expensive to dispose of harvested plant material. Resource managers sometimes have access to a nearby "high and dry" site where collected weeds can be stockpiled and allowed to desiccate and decay, but harvested material must often be transported off-site for disposal. This process can add significantly to the cost of the project, especially if the weeds must be disposed of in a landfill that charges tipping fees. As much as 95% of the fresh weight of aquatic weeds is water; a single acre of hydrilla can weigh as much as 24,000 pounds (10,886 kg), but only 1,200 pounds (544 kg) of that weight is plant material and the remaining 22,800 pounds (10,342 kg) is water [22]. Also, removal of weeds by volunteers or mechanical harvesters typically causes fragmentation of plant material and fails to capture root crowns, tubers, seeds, and other propagules in the sediment. Many aquatic weeds – including hydrilla, curlyleaf pondweed, and Eurasian watermilfoil – easily root from fragments and quickly regrow from sediment- borne propagules. As a result, initial observations at many sites that are managed using hand or mechanical removal of aquatic weeds may suggest that these methods have successfully addressed the problem, but control of the new invader is often ephemeral and weed populations regenerate in as little as a few weeks. A third factor to consider when hand-pulling or using mechanical harvesters to remove aquatic weeds is water depth. Volunteers are unlikely to be able to remove plants growing in water that is deeper than 3 feet (1 m) without diving gear and most traditional mechanical harvesters can only remove plant material in the upper 5 feet (1.5 m) of the water column, although newer equipment can harvest weeds in the upper 10 feet (3 m) of water. These factors should be considered before launching a weed removal program, regardless of whether weeds are taken out of the system by hand or by utilizing mechanical harvesters, but there are additional challenges inherent to each method. For example, volunteers tasked with hand-pulling invaders must be adequately trained to ensure that they will be able to successfully identify the target weed, especially when the invader is similar in appearance to desirable native plants that should be allowed to remain in the system. In contrast, mechanical harvesters are "non-selective" – they indiscriminately remove all plant material in the harvesting zone and are unable to distinguish between weeds and native species. Also, mechanical harvesters often result in bycatch, or the removal of fish and other aquatic fauna along with plant material. This problem is most pronounced when shallow-water (upper 5 feet; 1.5 m) harvesters are employed and can result in the removal of up to 28,000 fish per acre [23], but bycatch can be reduced by greater than 99% (removal of around 120 fish per acre) when deep-water (upper 10 feet; 3 m) harvesting is utilized [24].

Figure 8. Mechanical harvesting of hydrilla. Photo courtesy William Haller.

Another method that can provide some control of unwanted aquatic species is biological control, or the use of organisms to reduce weed populations. This technique, often referred to as biocontrol, is based on the concept that most species that become weedy after introduction to a new region are not problematic in their native range due to the presence of endemic predators that keep their growth in check. Identifying and evaluating potential biocontrol agents is an arduous, time-consuming, expensive process. The process typically begins with researchers travelling to the invader's center of origin and collecting insects, pathogens, or other organisms that are found in association with the target weed species. These biological agents are quarantined and subjected to a battery of tests to determine whether they fit the criteria and requirements of successful biocontrol agents. A hallmark of a biocontrol agent is host specificity; in other words, they must cause damage exclusively to the target weed species while leaving other plants untouched [25, 26]. Biocontrol agents should also be able to survive, grow, and reproduce in the invaded range of the weed and ideally, they should be able to form self-sustaining populations without augmentation. Some success has been realized using biocontrol organisms for aquatic weed control; for example, the Asian or Chinese grass carp (*Ctenopharyngodon idella* Val.) (Fig. 9) is well-known as a voracious consumer of hydrilla [27]. Unfortunately, grass carp are somewhat non-selective; although they are most frequently employed to control hydrilla, they will consume and eliminate virtually all submersed

http://dx.doi.org/10.5772/56015

vegetation in an aquatic system. Also, because the grass carp is a non-native introduced species, special precautions must be taken to reduce the likelihood of these biocontrol agents becoming invasive themselves. In Florida and many other states in the US, a permit must be issued by state resource managers before the introduction of grass carp into an aquatic system (although some states prohibit the use of grass carp as biocontrol agents altogether) [28]. In most cases, permit holders must ensure that stocked waters are secured (i.e., water intakes and outflows must be screened) to prevent the fish from escaping into other waters and all released grass carp must be triploid. Triploidy is the presence of an additional set of chromosomes, a condition that is induced by subjecting fish eggs to cold, heat, or pressure shock treatments immediately after artificial fertilization, and renders the grass carp unable to reproduce [29].

Figure 9. Asian grass carp. Photo courtesy William Haller.

Other organisms have also been employed as biocontrol agents. For example, a number of insects and pathogens have been evaluated for control of various aquatic weeds, including the noxious aquatic invader alligatorweed [*Alternanthera philoxeroides* (Mart.) Griseb.]. The most promising of these agents, the alligatorweed flea beetle (*Agasicles hygrophila* Selman and Vogt) (Fig. 10), can reduce populations to the point that more aggressive weed control methods can be reduced or even eliminated, provided winter temperatures in the region are mild enough to allow overwintering of the beetles [30]. Although these and other biocontrol agents have some utility in aquatic weed control, they cannot be relied on to completely eliminate infestations of invasive weeds. True biocontrol agents are host-specific; therefore, populations of the target weed must always be present in order to serve as a host or food source for the agent. As a result, weedy species cannot be eradicated through the actions of a biocontrol agent. When more complete control of aquatic weeds is necessary, resource managers rely heavily on chemical control, or the use of herbicides.

Figure 10. Alligatorweed flea beetle. Photo courtesy Lyn Gettys.

4. Water use and its influence on herbicide selection

A number of factors must be taken into consideration when selecting a herbicide for chemical control. Clearly, the most important criterion is efficacy of the product on the target weed. However, resource managers must also take into account how treated waters will be used. Although some aquatic systems are used for fisheries or crop production (e.g., rice cultivation), most are not used to grow food. Non-production waters targeted for aquatic weed control efforts can be categorized in a number of different ways, but the most common broad groupings include agricultural waters, flood control canals, recreational waters, retention ponds, and "development" waters (man-made lakes and ponds created for aesthetic reasons). Many waters are multi-use and span several of these categories, but this discussion will focus on the primary purpose of each grouping.

Agricultural waters are typically used for crop irrigation and for watering of livestock. A number of herbicides labeled for use in aquatic systems have irrigation and/or livestock watering restrictions. These restrictions preclude the use of treated water for a specific period of time or until the concentration of the herbicide is below a specified level. These restrictions vary among products and may also vary among products with the same active ingredient. Irrigation and livestock watering restrictions are clearly listed on the product label; compliance may be as simple as not using treated water for the appropriate length of time or may require laboratory tests to determine the concentration of herbicide in the water. Intentional or accidental failure to adhere to irrigation restrictions may result in a number of consequences, including – but not limited to – damage to livestock and non-target crop plants, herbicide residues in crops that exceed the allowed tolerance established by the United States Environ-

http://dx.doi.org/10.5772/56015

mental Protection Agency (USEPA), and prosecution by the USEPA for failure to follow label guidelines.

Flood control canals should be able to quickly move large volumes of water. These systems may be used only rarely for their true purpose; however, their ability to function as intended is critical when residential or developed areas are threatened by tropical storms, hurricanes or other extreme weather events. As such, it is critical that these canals be kept clear of aquatic vegetation that may impede the flow of water. A "scorched earth" philosophy and the use of a non-selective herbicide is sometimes employed to ensure that flood control canals remain free of aquatic weeds, and native plants are not exempt from weed control efforts in these systems. This is because even a small population of submersed plants – be it a weed such as hydrilla or a native plant such as eelgrass (*Vallisneria americana* Michx.) (Fig. 11) – can severely restrict water flow and increase the likelihood of flooding. Although the goal of weed control efforts in flood control canals is often to eliminate as much vegetation in the water column and surface as possible, canal banks should remain vegetated (ideally with a well-rooted, non-invasive native species) to prevent erosion during periods of rapid flow.

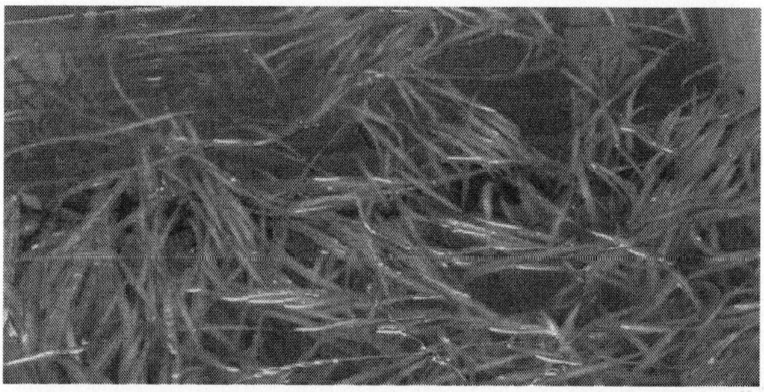

Figure 11. Eelgrass. Photo courtesy Lyn Gettys.

Recreational waters are typically managed to facilitate anthropogenic activities such as fishing, duck hunting, boating, and swimming. As a result, stakeholders – along with expectations and concerns – are many and varied. For example, most research has shown that sport fish populations in natural areas are greatest when submersed plants inhabit 30-40% or less of the water column [31, 32], but many sportfishers believe that dense weeds are necessary to provide good habitat for sportfish such as largemouth bass [33-35]. Also, some aquatic plants – including native species such as pondweed (*Potamogeton* spp.) and invasive weeds such as hydrilla – are eaten by ducks and waterfowl (Fig. 12). In fact, many duck hunters (and some waterfowl scientists) are less than supportive of aquatic vegetation control operations because they say these efforts deplete duck and waterfowl feeding habitat [36, 37]. These and other stakeholders often protest when weed control efforts are undertaken because they suspect reductions in weed coverage will negatively impact their hunting and fishing activities. Although some sportsmen recognize that it is rarely possible to maintain low coverage rates of aquatic weeds, many others fail to appreciate that the unchecked growth characteristic of submersed weeds necessitates weed control efforts that focus on eliminating as much vegetation as possible.

Figure 12. Ducks consuming seeds and vegetation on a pond bank. Photo courtesy Lyn Gettys.

Other recreational activities – such as boating and the use of personal watercrafts such as jet skis – are also directly impacted by aquatic weeds. Access to boat ramps can be restricted by overabundant growth of macrophytes in and around the littoral zone, while dense submersed vegetation can wrap around the propellers of outboard motors and hinder or halt boat operation. In addition, dense submersed vegetation can make swimming and waterskiing difficult, dangerous, or nearly impossible, and can increase the risk of drowning if individuals become entangled in dense weeds.

Retention ponds are by definition designed to be ephemeral; their ultimate purpose is to retain storm water, capture runoff, filter nutrients, and lessen or prevent flooding. Nevertheless, many stakeholders consider retention ponds to be long-term "water features" that enhance the aesthetics of urban and suburban areas. Retention ponds may be used on a limited basis for recreational purposes (e.g., fishing and swimming), but these activities are often restricted by the resource owner to limit liability. Aquatic weed control efforts in retention ponds must take into account stakeholder expectations; for example, if the goal is to reduce or eliminate unsightly algae or submersed weeds while leaving a fringe of ornamental flowering plants in the littoral zone, care must be taken to choose a selective herbicide that will control the target species without causing unacceptable levels of damage to desirable vegetation. Weed control efforts in retention ponds may also be challenging for resource managers due to the high visibility of these sites. Many stakeholders become alarmed at the sight of herbicide applicators wearing "moon suits" (Fig. 13) – a common name for personal protective equipment specified on the herbicide label – and assume that the water is being poisoned. Therefore, it can be useful to ensure that applicators are able to communicate with the public and to assuage fears regarding the toxicity of herbicides labeled for use in aquatic systems.

Figure 13. Herbicide applicator wearing personal protective equipment. Photo courtesy Lyn Gettys.

"Development" waters are man-made lakes and ponds that are created with the primary goal of increased aesthetics. These artificial bodies of water provide residential developers with a source of fill dirt, after which they are able to market adjacent homesites as desirable waterfront property, which are often sold at a premium. They also increase the value of the entire development, which can now be advertised as including ponds and water features. Some development waters are maintained in a pristine, plant-free state and rely on fountains or other hardscape features to provide an attractive visage. Others are planted or aquascaped, either to simulate natural bodies of water or to mimic large-scale water gardens with showy ornamental plants (Fig. 14). Because development waters are rarely connected to public waters, weed problems in these systems are typically the result of introduction by humans, or less often, by waterfowl and wildlife that have visited the development waters after spending time in nearby weed-infested aquatic systems. Anthropogenic introduction of aquatic weeds is frequently intentional, as when property owners dump unwanted aquarium or water garden plants into the development waters. However, the introduction of aquatic weeds can occur inadvertently when invasive species are misidentified and sold as desirable native plants or when propagules of invasive species "hitchhike" as contaminants on the desirable plants that are used for aquascaping [38-40]. Because development waters are considered valuable components of the landscape, they are often intensively managed to ensure that their aesthetic qualities are optimized.

5. Herbicide usage and labeling in aquatic systems — Case studies from Florida (USA)

Herbicides are used extensively to control weeds in crop production and agricultural systems. The terrestrial agrichemical industry in the US is robust; estimated sales in 2007 were $12.454 billion, with 40% of the market ($5.856 billion) attributable to herbicides [41]. In contrast, the

Figure 14. Waterlilies in a development pond. Photo courtesy Lyn Gettys.

market for aquatic weed control products is much smaller; for example, public agencies in Florida spent around $22.5 million in 2005 to manage aquatic invaders in public waters [42]. Any product that is marketed in the US to control pests – including weeds – must first be labeled by the US Environmental Protection Agency (USEPA or the Agency). Obtaining a pesticide label from the USEPA is a time-consuming and expensive undertaking; the Agency requires registrants (the manufacturer or group seeking a pesticide label) to submit data from more than 100 tests before a product can be evaluated for possible labeling, and the testing process typically requires the investment of tens of millions of dollars [43]. These tests are conducted to determine the effects of the experimental pesticide on the organism targeted for control, but also to assess its impact on non-target organisms, human health, and the environment as well. USEPA regulation of pesticides began with the adoption of the Federal Insecticide, Fungicide, and Rodenticide Act (FIFRA) in 1947; FIFRA has since been amended multiple times, most notably by the Federal Environmental Pesticide Control Act of 1972, and continues to serve as the primary process to ensure that human health and the environment are not negatively impacted by the use of pesticides [43].

Because obtaining a pesticide label from the USEPA requires significant financial resources, registrants only request Agency evaluation of products that are likely to capture a market large enough to offset the costs associated with obtaining a pesticide label. As outlined above, aquatic herbicides constitute a small niche market, with limited potential to allow registrants to recoup the funds required for initial labeling of a pesticide. Therefore, most herbicides that are labeled by the USEPA for use in aquatic systems have already been approved by the Agency for terrestrial use. Small-scale testing – such as greenhouse studies evaluating the efficacy of a product on aquatic weeds – may be conducted on a limited basis under specific conditions

(Fig. 15). If these preliminary experiments suggest that a herbicide shows promise as an aquatic weed control agent, the registrant may pursue an aquatic label for the product. Registrants seeking an aquatic label must submit additional data to the Agency, including how the product affects target and non-target aquatic flora and fauna, its persistence in aquatic sediments and water, and the nature and impacts of its decomposition components. These tests are conducted under Experimental Use Permits (EUPs) issued by the USEPA and by state regulatory agencies. For example, the testing of pesticides in Florida waters is conducted under EUPs issued by the USEPA and by the Florida Department of Agriculture and Consumer Services (FDACS). There are a number of restrictions on waters that are treated with experimental products; for example, treated waters may not be used for fishing, swimming, irrigation, drinking, or watering of livestock. Evaluation of an EUP herbicide typically continues for several years until the registrant has sufficient data to submit to the USEPA, along with a proposed aquatic label [43]. The aquatic label includes all of the information found on terrestrial labels, such as the personal protective equipment that is required to handle and apply the herbicide. In addition, aquatic herbicide labels include water use restrictions to prevent harm to human health and the environment. Some products have no limitations on the use of treated waters; however, others may specify that water from the system may be not be used for various purposes until either a certain period of time has elapsed or until the concentration of the herbicide falls below a specified set point.

Figure 15. Efficacy testing in the greenhouse. Photo courtesy William Haller.

It is important to note that all herbicides labeled for aquatic weed control by the USEPA in the US are "general use" pesticides that can be purchased and applied by anyone, including homeowners and unlicensed applicators. However, the USEPA allows states to apply additional restrictions to pesticides; in fact, a number of states classify aquatic herbicides as "restricted use" products that can only be purchased and applied by individuals that have received a state-issued license. Any and all individuals using a pesticide must comply with all of the requirements outlined on the pesticide label. The label is a legally binding document and misuse of a pesticide can result in serious consequences, up to and including the levy of fines and incarceration [43]. Although licensing is not required by federal law to purchase or apply aquatic herbicides, the vast majority of public agencies and private companies that employ applicators to manage aquatic systems specify that all personnel using these products obtain an aquatic pesticide applicator license from the state in which they are employed. This ensures that applicators have been trained and have shown competency in a number of important areas, including label interpretation, proper application techniques, equipment calibration, use of personal protective equipment, and proper disposal methods. Each state has its own requirements for obtaining and keeping a pesticide license. For example, all certified pesticide applicators in Florida must pass at least two written examinations – one that tests core competency and one that evaluates competence in a specific area or category [44]. A number of categories are offered to individuals seeking certification in Florida, and applicators may become licensed in as many categories as desired after the core competency examination has been successfully completed. Most licensees that are charged with applying pesticides in aquatic systems have multiple category certifications, the most common being aquatics, natural areas, and right-of-way. Florida pesticide applicator licenses are valid for four years from the date of issuance, and a license can be renewed in one of two ways. Applicators may submit proof that they have attended training sessions and earned a specified number of continuing education units (CEUs) in core and category areas during the four-year period since the license was issued or last renewed. Alternatively, applicators may re-take core competency and category examinations every four years [44].

6. How environmental factors influence herbicide applications

Herbicide applications to the aquatic environment share some of the challenges associated with treatment of agricultural lands, including drift (the unintended aerial dispersal of herbicides from the treatment area) and damage to desirable non-target plants. However, aquatic herbicide applications are further complicated by a number of factors unique to aquatic systems. For example, herbicides used for weed control in crop production typically reach the target plant at the concentration in which they are applied. In contrast, products employed to control submersed aquatic weeds must travel through the water column to reach their target and thus undergo substantial dilution before coming into contact with the plant. In addition, flow and currents result in the movement of the herbicide out of the treated area, which reduces contact exposure time (the period in which the product maintains contact with the target weed) and further limits efficacy of the treatment [45]. Another factor that complicates herbicide application in aquatic systems is the stratification of waters (Fig. 16), especially within systems in temperate regions. Most bodies of water have three distinct zones or layers, with little mixing among the layers. The upper and lower portions of a body of water are referred to as the epilimnion and hypolimnion, respectively. Water in the epilimnion is directly exposed to ambient air temperatures and therefore tends to be very warm in the summer and cold or frozen during winter. In contrast, water in the hypolimnion maintains a more or less constant temperature all year. The epilimnion and hypolimnion are separated by the thermocline, a layer characterized by drastic temperature changes [45, 46]. The effect of stratification may have little effect on efforts to manage emergent or floating aquatic weeds. However, this phenomenon can have a substantial effect on treatment of submersed invaders, because herbicides applied to the epilimnion are unlikely to penetrate through the thermocline to reach target weeds growing in the hypolimnion.

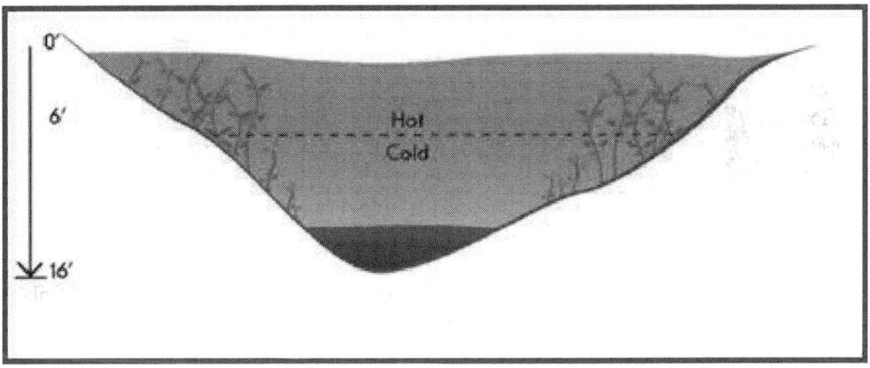

Figure 16. Stratified lake (summer, with warm epilimnion). Illustration courtesy UF/IFAS Center for Aquatic and Invasive Plants.

Another important consideration in the treatment of aquatic systems is the effect of weed control activities on fish that reside in waters targeted for herbicide application. Although the presence of fish does not affect herbicide efficacy, special precautions must be taken to ensure that these and other aquatic denizens are not harmed as a result of weed control efforts. Only a few herbicides labeled for use in bodies of water are inherently dangerous to fish, but fish kills are nonetheless a major concern for applicators working in aquatic systems. The primary reason fish kills occur after weed control activities are undertaken is a reduction in dissolved oxygen (DO), which results from a number of factors [47]. Primary among these factors is the decomposition of vegetative material that has been killed by herbicides and is broken down by aerobic organisms, which deplete DO during the process. Also, photosynthesis by plants that have been killed by herbicides is eliminated and the DO they previously contributed to the water column is no longer produced, further reducing levels of DO. In order to reduce the likelihood of fish kills, most labels for aquatic products specify that herbicides be applied to only a portion of a weed-infested body of water at a time to allow fish to escape from treated areas and to prevent the extreme drop in DO that accompanies the elimination of all vegetation from an aquatic system.

7. Application methods in aquatic systems

Some of the techniques for applying herbicides in aquatic systems are similar to those used for weed control in crop production. This is especially true when the target aquatic invaders are growing along ditchbanks or shorelines or in narrow canals that can be treated using a backpack sprayer or a truck, tractor or other wheeled vehicle. However, herbicide applications to open waters require specialized equipment and tools in order to effectively reach the aquatic weeds that are targeted for control, and the primary vehicle required for aquatic weed control is a boat. The size and disposition of the treatment boat varies and is dependent on the application method to be employed, which is often dictated by the target weed and the form of herbicide being utilized. Aquatic herbicides are typically sold in liquid and granular formulations, and some active ingredients are available in both forms [48]. Granular formulations are most often applied using a boat-mounted spreader (Fig. 17). Most liquid formulations are packaged as concentrates and are applied in dilute form. Dilution is frequently accomplished by adding the concentrate to a boat-mounted tank filled with water. A variety of equipment exists to apply herbicides that have been diluted in an onboard tank; these include handguns (for treating emergent and floating weeds), booms (for treatment of surface water), and trailing weighted hoses (for subsurface treatments) [45, 49]. Regardless of the formulation and application method employed, calibration of application equipment is critically important to ensure that the correct amount of herbicide is introduced to the system. Poorly calibrated equipment may result in the application of too little herbicide, which will likely yield poor weed control and reduced product efficacy. Using an excess amount of herbicide will increase costs associated with the treatment and may result in concentrations above those specified on the product label; as outlined above, this is a violation of federal law and may have serious legal consequences.

Figure 17. Application of granular herbicide using a boat-mounted spreader. Photo courtesy UF/IFAS Center for Aquatic and Invasive Plants.

http://dx.doi.org/10.5772/56015

8. Conclusions

Fresh-water resources are extremely important components of global and local ecosystems. The introduction of exotic invasive species to these systems limits their ability to function as healthy, diverse habitats for native flora and fauna; in addition, anthropogenic uses such as flood control, public safety, and recreation are hindered as well. The most effective method to reduce the impact of aquatic invaders is to prevent their introduction to these valuable and important systems, but invasive species continue to become established in aquatic systems throughout the world. The primary method used to control introduced aquatic weeds in the US is the application of registered aquatic herbicides. Pesticides that are applied to waters in the US are labeled and registered by the USEPA after extensive testing, and most states – including Florida – require that the use of these products be regulated by state agencies as well. Aquatic herbicides represent a small subset of the pesticides labeled by the USEPA and registrants only pursue aquatic labeling of products if there is a market large enough to offset the costs associated with additional registration requirements. A number of unique challenges are associated with weed control in aquatic systems, including the effects of dilution, current, and stratification of water within systems. These challenges can be overcome through the selection of proper herbicides and application methods.

Acknowledgements

This publication is a contribution of the University of Florida Institute for Food and Agricultural Sciences and the Florida Agricultural Experiment Station.

Author details

Lyn A. Gettys[1*], William T. Haller[2*] and Gregory E. MacDonald[3*]

*Address all correspondence to: lgettys@ufl.edu

*Address all correspondence to: whaller@ufl.edu

*Address all correspondence to: pineacre@ufl.edu

1 University of Florida Institute of Food and Agricultural Sciences, Department of Agronomy, Fort Lauderdale Research and Education Center, Davie, FL, USA

2 University of Florida Institute of Food and Agricultural Sciences, Department of Agronomy, Center for Aquatic and Invasive Plants, Gainesville, FL, USA

3 University of Florida Institute of Food and Agricultural Sciences, Department of Agronomy, Gainesville, FL, USA

References

[1] Klorer J. 1909. The water hyacinth problem. Journal of the Association of Engineering Societies 42:33-48.

[2] Webber HJ. 1897. The water hyacinth, and its relation to navigation in Florida. United States Department of Agriculture, Division of Botany. Bulletin No. 18. http://ia600406.us.archive.org/9/items/waterhyacinthits00webb/waterhyacinthits00webb.pdf (accessed 22 August 2012).

[3] Penfound WT, Earle TT. 1948. The biology of the water hyacinth. Ecological Monographs 18(4):447-472. http://www.jstor.org/stable/1948585 (accessed 22 August 2012).

[4] Holm L, Doll J, Holm E, Pancho J, Herberger J. 1997. World weeds: natural histories and distribution. John Wiley, New York, USA.

[5] Lowe S, Browne M, Boudjelas S, De Poorter M. 2000. 100 of the world's worst invasive alien species: a selection from the global invasive species database. The Invasive Species Specialist Group. www.issg.org/booklet.pdf (accessed 22 August 2012).

[6] Langeland KA. 1996. Hydrilla verticillata (L.F.) Royle (Hydrocharitaceae), "The perfect aquatic weed". Castanea 61(3):293-304.

[7] McLane WM. 1969. The aquatic plant business in relation to infestations of exotic aquatic plants in Florida waters. Hyacinth Control Journal 8:48-49.

[8] Schmitz DC, Nelson BV, Nall LE, Schardt JD. 1991. Exotic aquatic plants in Florida: a historical perspective and review of present aquatic plant regulation program. In: Center TD, Doren RF, Hofstetter RL, Myers RL, Whiteaker LD (eds.). Proceedings of a symposium on exotic pest plants, pp. 303-336. November 2-4, 1988, Miami, Florida. United States Department of the Interior, National Park Service, Washington, DC, USA.

[9] Langeland KA, Sutton DL. 1980. Regrowth of hydrilla from axillary buds. Journal of Aquatic Plant Management 18:27-29.

[10] Glomski LM, Netherland MD. 2012. Does hydrilla grow an inch per day? Measuring short-term changes in shoot length to describe invasive potential. Journal of Aquatic Plant Management 50:54-57.

[11] Haller WT, Sutton DL. 1975. Community structure and competition between hydrilla and vallisneria. Hyacinth Control Journal 13:48-50.

[12] Madsen JD. 1997. Methods for management of nonindigenous aquatic plants. In: Luken JO, Thieret JW (eds.). Assessment and management of plant invasions, pp. 145-171. Springer: New York, USA.

[13] Holland LE, Huston ML. 1984. Relationship of young-of-the-year northern pike to aquatic vegetation types in backwaters of the Upper Mississippi River. North American Journal of Fisheries Management 19:18-27.

[14] ENSR. 2005. Rapid response plan for hydrilla (Hydrilla verticillata) in Massachusetts. Massachusetts Department of Conservation and Recreation. Boston, Massachusetts, USA. http://www.mass.gov/dcr/watersupply/lakepond/downloads/rrp/hydrilla.pdf (accessed 23 August 2012).

[15] McGehee JT. 1979. Mechanical hydrilla control in Orange Lake, Florida. Journal of Aquatic Plant Management 17:58-61.

[16] Sutton DL, Vandiver VV. 2006. Grass carp: a fish for biological management of hydrilla and other aquatic weeds in Florida. Bulletin No. 867. Florida Cooperative Extension Service, Institute of Food and Agricultural Sciences, University of Florida. http://edis.ifas.ufl.edu/fa043 (accessed 23 August 2012).

[17] Lounibos LP, Escher RL. 1985. Mosquitoes associated with water lettuce (Pistia stratiotes) in southeastern Florida. Florida Entomologist 68(1):169-178.

[18] Mulrennan JA. 1962. The relationship of mosquito breeding to aquatic plant production. Hyacinth Control Journal 1:6-7.

[19] Seabrook EL. 1962. The correlation of mosquito breeding to hyacinth plants. Hyacinth Control Journal 1:18-19.

[20] Idaho Department of Agriculture, Aquatic Ecosystem Restoration Foundation, and Pacific States Marine Fisheries Commission. 2012. A review of the state of Idaho dreissenid mussel prevention and contingency plans. http://aquatics.org/musselreport.pdf (accessed 14 November 2012).

[21] McNabb TJ. 2012. Aquatic invasive species: spreading prevention activities. Aquatics 34(2):19-20.

[22] Zimba PV, Hopson MS, Smith JP, Colle DE, Shireman JV. 1995. Chemical composition and distribution of submersed aquatic vegetation in Lake Okeechobee, Florida (1989-1991). In: Aumen NG, Wetzel RG (eds.). Ecological studies on the littoral and pelagic systems of Lake Okeechobee, Florida (USA). Advances in Limnology 45:241-246.

[23] Haller WT, Shireman JV, DuRant DF. 1980. Fish harvest resulting from mechanical control of hydrilla. Transactions of the American Fisheries Society 109:517-520.

[24] Haller WT, Jones DK. 2012. Technology and improved efficacy of mechanical control of hydrilla. Aquatics 34(3):17-19.

[25] DeLoach CJ. 1997. Biological control of weeds in the United States and Canada. In: Luken JO, Thieret JW (eds.). Assessment and management of plant invasions, pp. 172-194. Springer: New York, USA.

[26] Cuda JP, Charudattan R, Grodowitz MJ, Newman RM, Shearer JF, Tamayo ML, Villegas B. 2008. Recent advances in biological control of submersed aquatic weeds. Journal of Aquatic Plant Management 46:15-32.

[27] Haller WT. 1994. Probable grass carp stocking scenarios. In: Haller WT (ed.). Proceedings of the Grass Carp Symposium, pp. 236-238, 7-9 March 1994, Gainesville, Florida. US Army Engineer Waterways Experiment Station, Vicksburg, Mississippi, USA.

[28] Colle D. 2009. Grass carp for biocontrol of aquatic weeds. In: Gettys LA, Haller WT, Bellaud M. (eds.). Biology and control of aquatic plants: a best management practices handbook, pp. 61-64. Aquatic Ecosystem Restoration Foundation, Marietta, Georgia, USA.

[29] Cassani JR, Caton WE. 1986. Efficient production of triploid grass carp (Ctenopharyngodon idella) utilizing hydrostatic pressure. Aquaculture 55(1):43-50.

[30] Buckingham GR. 2002. Alligatorweed. In: Van Driesche R, Blossey B, Hoddle M, Lyon S, Reardon R (eds.). Biological control of invasive plants in the eastern United States, pp. 5-15. USDA Forest Service Publication FHTET-2002-04, 413 p.

[31] Canfield DE Jr., Hoyer MV. 1992. Aquatic macrophytes and their relationships to Florida lakes. Final report to the Bureau of Aquatic Plants, Florida Department of Natural Resources. Tallahassee, Florida, USA.

[32] Colle DE, Shireman JV. 1980. Coefficients of condition for largemouth bass, bluegill, and redear sunfish in hydrilla-infested lakes. Transactions of the American Fisheries Society 109:521-531.

[33] Estes JR, Sheaffer WA, Hall EP. 1990. Study I. Fisheries studies of the Orange Lake chain of lakes. Florida Game and Fresh Water Fish Commission. Tallahassee, Florida, USA.

[34] Porak WF, Crawford S, Renfro D, Cailteux RL, Chadwick J. 1990. Study XII. Largemouth bass population responses to aquatic plant management strategies. Florida Game and Fresh Water Fish Commission. Tallahassee, Florida, USA.

[35] Tucker T. 1987. How to fish hydrilla. Bassmaster 20(9):30-34.

[36] Johnson FA, Montalbano F III. 1987. Considering waterfowl habitat in hydrilla control policies. Wildlife Society Bulletin 15(3):466-469.

[37] Anonymous. 2011. Background information for the Fish and Wildlife Conservation Commission's position on hydrilla management. Florida Fish and Wildlife Conservation Commission, Tallahassee, Florida, USA. http://myfwc.com/media/1386747/hydrilla-mgmt-position-background-information.pdf (accessed 31 October 2012).

[38] Les DH. 1996. Hydrilla verticillata threatens New England. Aquatic Exotic News 3(1):1-2.

[39] Maki K, Galatowitsch S. 2004. Movement of invasive aquatic plants into Minnesota (USA) through horticultural trade. Biological Conservation 118(3):389-396.

[40] Keller RP, Lodge DM. 2007. Species invasions from commerce in live aquatic organisms: problems and possible solutions BioScience 57(5):428-436.

[41] US Environmental Protection Agency. 2011. Pesticides industry sales and usage: 2006 and 2007 market estimates. http://www.epa.gov/opp00001/pestsales/07pestsales/market_estimates2007.pdf (accessed 19 November 2012).

[42] Mossler MA, Langeland KA. 2009. Florida crop/pest management profile: aquatic weeds. Document PI-138, Pesticide Information Office, Agronomy Department, Florida Cooperative Extension Service, Institute of Food and Agricultural Sciences, University of Florida http://edis.ifas.ufl.edu/pdffiles/PI/PI17500.pdf (accessed 19 November 2012).

[43] Layne C, Stubbs D. 2009. Requirements for registration of aquatic herbicides. In: Gettys LA, Haller WT, Bellaud M. (eds.). Biology and control of aquatic plants: a best management practices handbook, pp. 145-150. Aquatic Ecosystem Restoration Foundation, Marietta, Georgia, USA.

[44] Florida Department of Agriculture and Consumer Services. 2012. Pesticide applicator licenses. http://www.freshfromflorida.com/onestop/aes/pestapp.html (accessed 19 November 2012).

[45] Haller WT. 2009. Aquatic herbicide application methods. In: Gettys LA, Haller WT, Bellaud M. (eds.). Biology and control of aquatic plants: a best management practices handbook, pp. 151-156. Aquatic Ecosystem Restoration Foundation, Marietta, Georgia, USA.

[46] Mudge CR, Haller WT, Gettys LA. 2011. Thermocline, north versus south: friend or foe. Aquatics 33(1)12-14.

[47] Whitford F, Becovitz J, Robertson B, MacGowan B, Blase G, Avenius B, Donahoe J, Zimmerman D, Blessing A. 2009. What killed the fish? Using observation, sampling, and science to solve the mystery. Purdue Extension Publication PPP-79. http://www.ppp.purdue.edu/Pubs/PPP-79.pdf (accessed online 17 November 2012).

[48] Netherland MD. 2009. Chemical control of aquatic weeds. In: Gettys LA, Haller WT, Bellaud M. (eds.). Biology and control of aquatic plants: a best management practices handbook, pp. 65-77. Aquatic Ecosystem Restoration Foundation, Marietta, Georgia, USA.

[49] Haller B, Gettys L, Glenn M, Reynolds G. 2007. Building weighted trailing hoses. Aquatics 29(4):8-14.

Chapter 14

Herbicide Impact on Seagrass Communities

A. Damien Devault and Hélène Pascaline

Additional information is available at the end of the chapter

http://dx.doi.org/10.5772/55973

1. Introduction

Anthropogenic chemical contamination is of concern due to the continuous decline of ecosystems. Pesticide use impacts whole environmental matrices, especially aquatic ones, because of collecting watershed pollution in streams, rivers, and finally coastal areas. The coastal environment is one of the most vulnerable: global changes (current sea level rising, ocean acidification, global warming) add to land use disruption (soil erosion, chemical uses, urban sprawl) in coastline areas. Moreover, population growth mainly affects this endangered environment because of rural flight and city growth –75% of billions of human beings will live in 100km-large belt around global seas in 2035 [1] imposing urban lifestyle demands. Environmental stress due to such a heterogeneous population repartition will be acute (1) on freshwater, in order to provide it for drinking, industrial and agricultural needs, and (2) on coastal ecosystems because of waste waters and coastline management. Such environmental concerns are critical for tropical countries because of some that are being discovered to be biodiversity hot spots [2].

Human impact is partly due to pesticide use [3-5]. In order to feed a growing population and to manage urban areas, herbicides are often used profusely [6]. These herbicides affect wild fauna and flora through improper use, inefficient (even lack of) wastewater treatment plant effluents or the direct input of herbicide contained by sewage sludge into aquatic environments [7]. For hydrophobic pesticides, contaminated solid phases transfer downstream, in an erosive context due to deforestation and agricultural intensification is involved [8].

Thus, herbicides will contaminate coastal environments [9-10]. In shallow water, such residues will expose remarkable biocoenosis, especially in tropical contexts, because of conserved biodiversity compared to temperate ones, i.e. exposed for decades to aquatic pollution from developed countries' activities.

Considering such shallow waters, herbicides could influence seagrasses. Such tracheophyte flowering plants are the result of terrestrial grass adaptation to marine environments. These monocotyledons colonize shallow water bottoms, especially in silty-sandy substrata because of roots, unlike bryophytes and algae. Thus, in opposition to those non-tracheophyte ones, seagrass meadows could limit sediment erosion and stabilise navigation channels.

2. Seagrasses: Unique origin, physiology and performances

Seagrass is a taxonomic group of about 60 species worldwide likely evolving from a single monocotyledonous flowering plant ancestor (70-100 million years ago), divided into three independent lineages: Hydrocharitaceae, Cymodoceaceae and Zosteraceae [11]. Seagrass species have strong physiological similitude and low interspecies diversity.

As flowering plants, they are anchored in sediments by their roots –what non-tracheophyte marine plant species don't have. Notwithstanding, seagrass only live in submarine environments, even for pollination or other critical steps, unlike other aquatic flowering plants who should use an emerged organ or pass by a terrestrial stage [12]. Seagrasses have some of the highest light requirements (25% of incident radiation when 1% is the average requirement of angiosperm species [13]) even if epidermal chloroplast and internal gas transport systems have been developed, in order to maintain oxidative conditions, despite highly reducing sediment, including toxic sulphide levels, for large amounts of non-photosynthetic tissues [14]. Seagrasses are especially vulnerable to lack of light, mainly due to erosion or eutrophication.

While algae, whose growth is proportioned to the eutrophication level and thus could lead to a dystrophic crisis due to algal necromass decomposition, seagrass growth biomass is sustainable. Seagrass bed increase due to nutrient input makes seagrass meadows, for carbon trapping and storing [15], like corn or sugar cane, among the most efficient trapping plants [16]. Seagrass meadows are a more efficient carbon sink than trees: with an equivalent carbon sequestration per year (about 27 million tons [17]), carbon sequestered in meadows will be buried and therefore partly avoid decomposition in the matte [18]; Pergent et al. [19] estimated this stored amount about a third of the primary production. Living seagrass biomass actually reaches 19.9 billion tons [17].

Moreover, to this biomass should be added suspended matters that seagrass leaves could efficiently sequester because of the blade effect on suspended matter, i.e. acting like a mat, trapping suspended matter and inherent organic matter, and because seagrass decomposition is too long for inducing dead zones [12]. Seagrass blades could drift to the abyss where they are an indispensable carbon contribution for poor-carbon deep sea biocoenosis [20].

Seagrass could be susceptible to exondation because of tides. Such events could be fatal depending on shore temperature. Temperatures of 35°C and greater, not found in the marine environment but possible in pools or during extreme low tide coefficient, could kill seagrass [21, 22] because of photosynthesis interruption; irreparable structural alterations to the PhotoSystem II (PSII) reaction centres induce chloroplast dysfunction, leading the plants to insufficiently jugulate of the reductive conditions in roots.

As marine species, seagrasses are vulnerable to low salinity events and cannot colonize upstream estuaries and freshwater shallows. A 5‰ salinity is the smallest salinity amount compatible with seagrass development (Iversen, 1931 cited by Vermaat et al. [23];[24-26]), whereas seagrass communities can stand waters which are more salty than the global ocean salinity (35‰): Depending on species, seagrasses could stand a salty environment up to 42‰ [27]. Euryhaline seagrass species, i.e. large scale salinity ones, colonize all the climatic areas except polar ones. But, in all of them, these remarkable adaptations are balanced by severe seagrass meadow regression.

3. Seagrass meadows repartition and involved landscape

Seagrasses are present in all the marine ecotones, except polar ones ([28]; Figures 1 and 2).

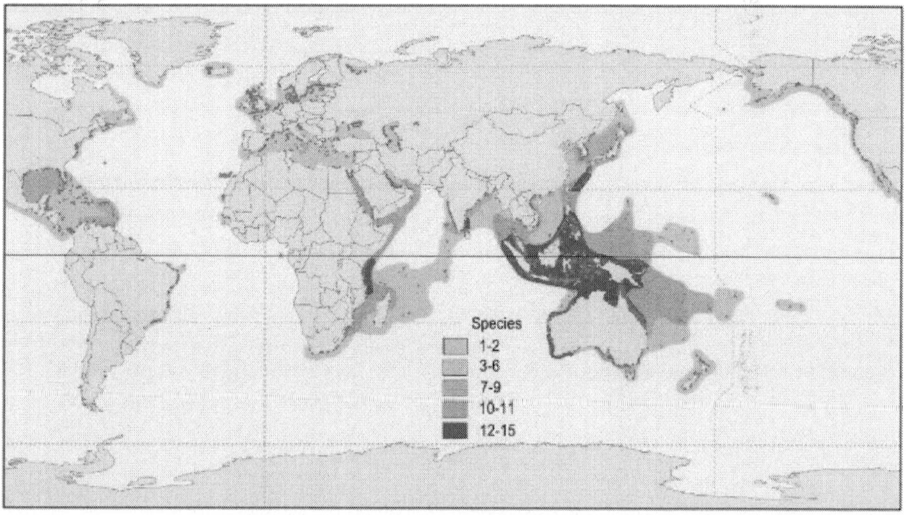

Figure 1. Global seagrass diversity and distribution. Shades of green indicate numbers of species reported for an area; blue points and polygons indicate documented reports of seagrass occurrence (from 2005 UNEP-WCMC).

Temperate areas are marked by seasons with different temperatures, light and precipitation regimes. Land and sea weathers provide extreme wind and flow conditions. Nutrient inputs occur by pulses which seagrass meadows must cope with [29]. Seagrass meadows will consume nutrients in perennial vegetative growth, limiting eutrophication conditions [30].

Ecosystems including seagrasses are listed in Figure 3. In temperate marine water, seagrasses are associated with marshes and kelp beds, and have been providing for centuries ecosystem services to coastal lands [31]. Human use of kelp began as picking fodder, fertilizer –even organic matter- and food on shore. During the late modern time period, dried kelp was used

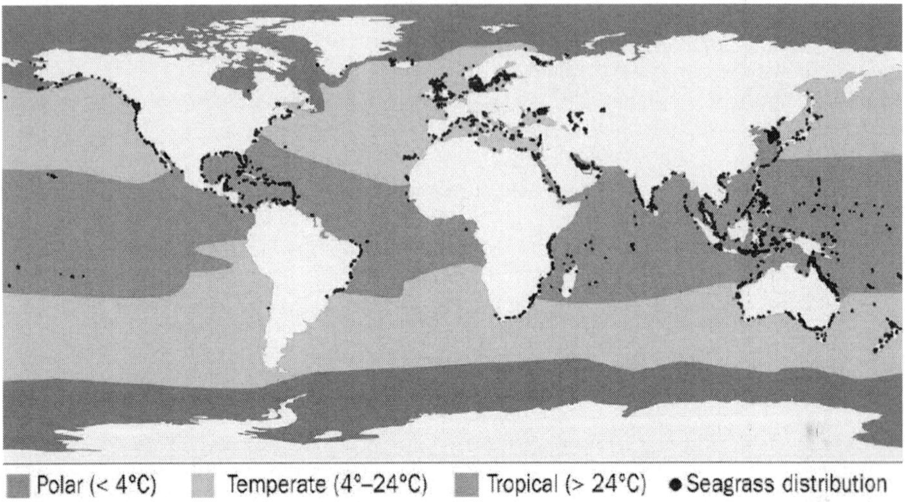

Figure 2. Current global distribution of seagrass in relation to mean ocean temperature. Regional divisions are based on polar (<4 degrees Celsius [°C]), temperate (4°C-24°C), and tropical (>24°C) climate [52].

in order to provide "non-caustic" or "commercial" soda (in fact, sodium carbonate Na_2CO_3) for early industries (glass, photograph, soap, etc.). Seagrass meadows are colonizing near-shore environment mixing or not with seaweeds, i.e. seagrasses mainly settle on movable substrata where their roots could anchor them and seaweeds only settle on hard substrata (rocks or shingles) needed by their basal adhesive organ. Seaweeds are sessile, macroscopic, benthic and multi-cellular algae [32] constituting a polyphyletic community. Seagrass meadows and seaweeds live in adjacent environments and could marginally be in competition. However, the relationship between them does not present the same cooperative side as in tropical areas.

In tropical areas, marine seagrasses are associated with preserved triptych mangrove-meadows-corals [33]: (1) Mangrove stabilises and protects the coastline, limits sediment input in marine environments and holds tidal biodiversity [34]; (2) seagrass meadows, because of blades and roots, limits current movements, enhances suspended particular matter deposition, provides food for endangered species like sea turtles and manatees [35]; (3) coral reefs protect the shore from waves, acting like living breakwater, an especially acute property in tropical areas subjected to typhoons and tsunamis [36]. Triptych partners have a mutual service relationship, i.e. corals are vulnerable to sedimentation limited by mangroves, which are vulnerable to large waves buffered by coral reefs. Moreover, each partner has a nursing role for marine species [37, 38]: for example, considering the eastern area of the Caribbean Sea, 80% of fisheries are located in mangrove, meadow or coral areas for juvenile stages [39]. However, significant landscapes of corresponding environments are limited to some French West Indies bays, exposed to land speculation. Thus, because each triptych partner has its own vulnera-

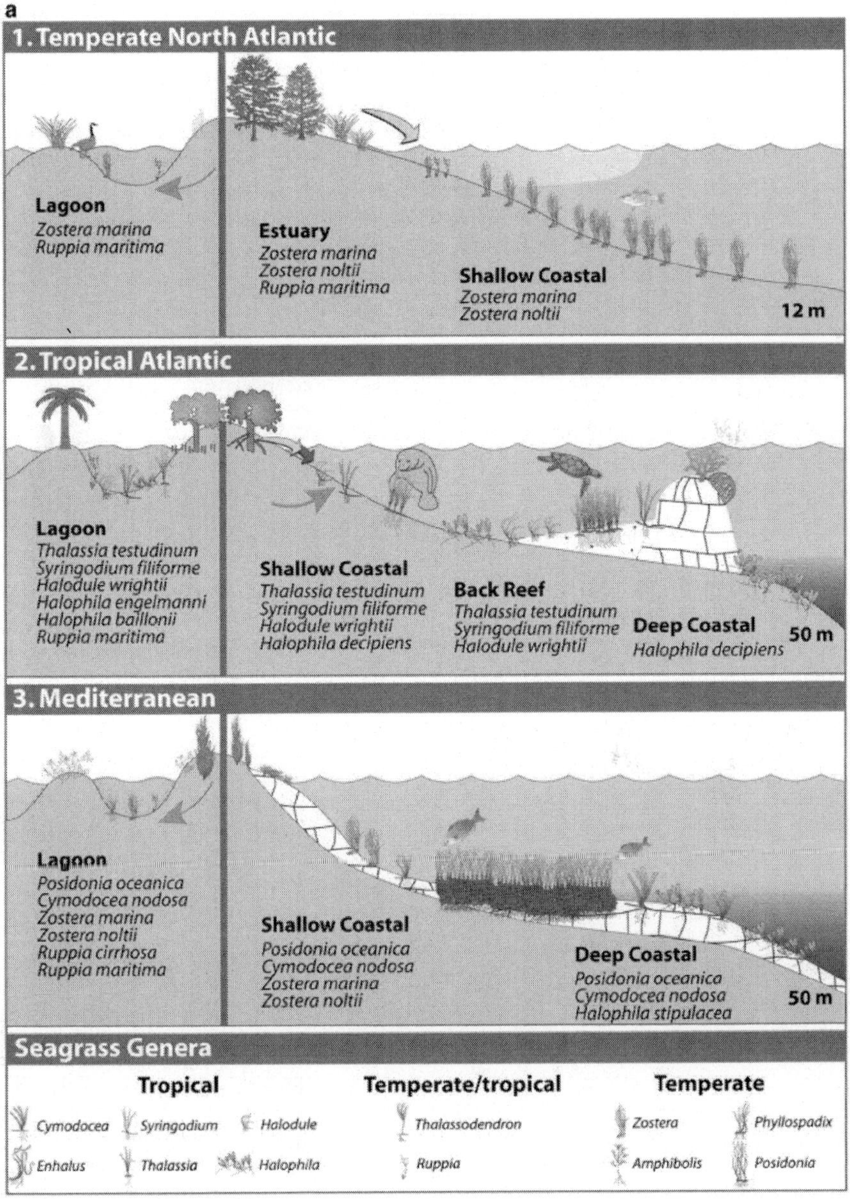

Figure 3. Seagrass habitat diagrams for (a) Bioregions 1–3 and (b) Bioregions 4–6. Major species for each bioregion listed according to dominance within habitats. Maximum reported depths [113].

bilities, this ecosystemic symbiosis is acutely vulnerable; if mangroves are largely unaffected by water quality [40], seagrass meadows are highly susceptible to chemical inputs [12] when corals are mainly sensitive to herbicides because of endosymbiosis [41], i.e. corals include algal symbionts which are highly vulnerable to PSII herbicides, in a few minutes and for ng/L contamination range [42]. The Achilles' heel of each ecological partner endangers the whole ecosystem equilibrium.

Notwithstanding ecological and fishery services, rendered in tropical and temperate zones, seagrass meadows support the detrital food web, perpetuates navigation supports, and have a key position considering carbon and nutrient cycles. Each seagrass annual services have been estimated between $9,000 and $28,000 per acre [43] –globally $1.9 trillion per year by Watson et al. [44] describing seagrass meadows as constituting an endangered capital.

4. Seagrass: The silent fall

Indeed, because of local triptych disruption due to anthropogenic needs or airiness in tropical areas or due to the global environmental decline, seagrass meadows are threatened (Figure 4). All over the world, this unique biocoenosis is regressing; during the last decade, between 20 and 100% in the Gulf of Mexico, depending on the coastal zone, 85% in Florida, 40% in the bay of Arcachon [45], an accelerating loss process [28] leading to an evaluated total loss since 1980 of about 30% of global seagrass meadows, i.e. at the same scale as mangrove regression (-1.8% yr^{-1} [46]). Thus, seagrass meadows are more endangered than the tropical rain forest (-0.5 yr^{-1} [47]) and as precious as it for carbon storage (cf. *supra*). Mangrove and seagrass regression undermine coral reefs, more sensitive to the seagrass meadows' regression and sensitive to another threats (-0.72 to -9% yr^{-1} [48-50]). Each year, about 177,000 km^2 of seagrass meadows, i.e. 1.5% of global seagrass meadows (Ibid.), are lost -about 299 million tonnes of carbon trapped [12].

Worldwide seagrass meadow loss is not balanced by seeding or planting campaigns. First, restoration scales are largely smaller than the seagrass meadows loss; most of them are <1 ha because of costs –even if restoration cost is still less expensive than seagrass loss consequences themselves. Secondly, restoration success rate is low: about 30% [51] or more [52] –but some seagrass species are not transplantable [51] leading them to a more acute endangered situation in lineage, vulnerable because of poor genetic diversity [53]. But restoration initiatives are induced by information about seagrass loss and its consequences. Actually, information lacks in order to know the impact of seagrass meadow fragmentation; such interconnectivity loss is due to human activity because of the declining chemical quality of seawater as well as to building or coastline management [54].

Notwithstanding the alarming situation and issues, the publication rate about seagrass meadow loss remains low; the actual increase of publication numbers and quality about this concern should be proportioned to the global ocean crisis. Mangrove, salt marsh, and coral reefs, in particular, have are three- to one hundred-fold more publications than seagrass

http://dx.doi.org/10.5772/55973

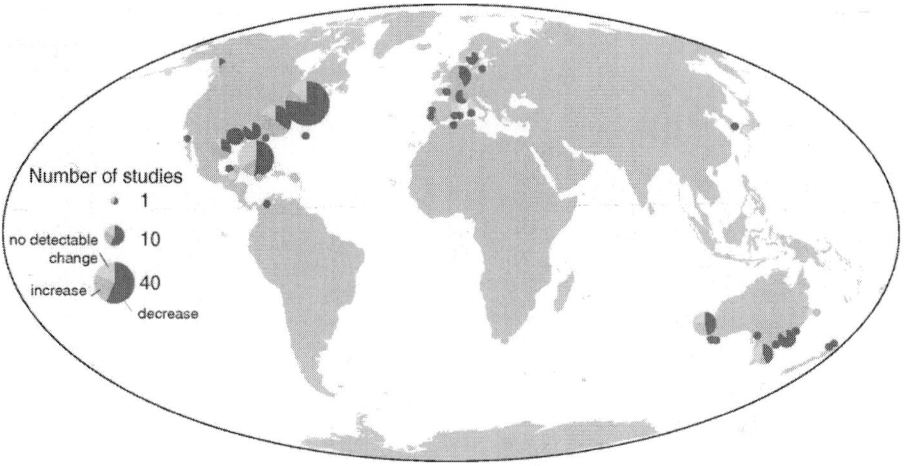

Figure 4. Global map indicating changes in seagrass area plotted by coastline regions. Changes in seagrass areal extent at each site are defined as declining (red) or increasing (green) when areal extent changed by >10%, or no detectable change (yellow) when final area was within ±10% of the initial area. There were 131 sites in North America, 34 sites in Europe, and 40 sites in Australia [28].

meadows [12] even though estimated monetary value of seagrass meadows is more than two-fold more important than mangrove or marshes, and four-fold more than coral reefs [43].

The seagrass loss origin is actually unknown. First of all because, due to perennial growth of seagrass meadows, environmental disease impact is more obviously observed than for other marine plants, and particularly unicellular organisms. Moreover, because of the macroscopic size of seagrass specimens and their eminent services for the environment and for human activity, seagrass meadow loss critically alters coastal activities [51]. Lastly, seagrass meadow loss reveals a long-term impact on the environment when planktonic or fugacious species providing short-term environmental status; such outcomes need high-frequency monitoring for a valuable putting into perspective. Orth et al. [12] defined seagrass as "coastal canaries".

5. Predominant pesticide effects applied to seagrass physiology

Chemical content in seawater is directly, but locally, altered by port uses, i.e. antifouling coatings and urban pesticide uses, and indirectly, but globally, altered by agricultural and urban chemical input due to landscape runoff and subsequent river pollution [55] and when groundwater tables well up through the soil directly at the sea (phenomenon known as Submarine Groundwater Discharge [56]). Their terrestrial impact is well known, as on soils, then on groundwater, surface water [3], sediments [57], biota [4] and human health [58].

Vulnerability of aquatic environments to organic chemicals, and especially pesticides, has been asserted for decades. In many countries, pesticide monitoring is performed for groundwater, often in order to ensure drinking water, as well as for surface water. Incidentally, it is possible to determine the most frequently used herbicides. According to Gilliom [3] for the U.S.A. and to Schäfer et al. [4] in the European Union, 21 herbicides could be identified as being the most frequently used in developed countries, leading them to probably being significantly used in developing ones (Table 1). Haynes et al. [59-61] consider diuron as the most threatening herbicide, even pesticide, for seagrass meadows, partnering the corals of the Great Barrier Reef. However, aquatic plant toxicology is not well-known: BCPC [62] only informs us about two aquatic plant toxicological tests: EC50 (120h) for diuron on *Selenastrum capricornutum* (0.002mg/L) and atrazine EC50 (96h) for *S. capricornutum* (0.01mg/L). Lewis & Devereux [55] provided the first review on non-nutrient anthropogenic chemicals in seagrass ecosystems, summarizing all publications on seagrass –and finding only ten on herbicide impact on seagrass.

	Solubility	Koc	Kow	Application	Effects/metabolic target	Notes
glyphosate	10500		-3.2 (1)	2	block EPSPS-catabolic crossroad for proteins	Very easily complexed
diuron	37.4	400	2.85	6-30 (10-30 total)	PSII, especially on dichotyledons	DT50: 50-180d depending to humidity Hill et al., 1955
atrazine	33		2.5	1,5	PSII, especially on dichotyledons	
simazine	6.2 (1)	103-277: *160*	2,1	1,5 (3 tropical)	PSII	
prometon	750		2.69	10-20	PSII	
amitrole	26000 pH7-c; "/>1384000 d		-0.969 (2)	1-3	triazole	DT90: 15d without anoxia
isoproturon	65		2.5 (1)	1.15	PSII	DT50: 1560d
linuron	63.8 (1)	500-600	3		PSII	DT50: "/>1000d for all pH
metolachlor	488	121-309		1-2,5	PSII	DT50 hydrolysis: "/>200d
S-metolachlor	480	61-369		0.6-1.6	PSII	non hydrolyzable
cyanazine	171			1-3	selectif	
acetochlor	282			3	No definitively known	
metribuzin	1050			0.07-1.45	species-specific; PSII	
bentazone	570 (1)	13.3-176 *142*	0.77 (a); -0.46; -0.55 (b)	1-2,2	species-specific; PSII	winter herbicide; low hydrolysis but high photolysis
EPTC	375		3.2	4.5-6.7	Inhibit lipid synthesis	

	Solubility	Koc	Kow	Application	Effects/metabolic target	Notes
trifluraline	0.221 (calc) 0.395 (field)	4400-40000	483 (1)	0.5-1	microtubule polymerisation inhibition in roots	not hydrolyzable
molinate	1100	121-252	2.86 (pH7.85-7.94 (2))	2.5-5	lipid synthesis disruption: inhibition germination	Photosensible; no degradation after 2 years); not hydrolyzable
norflurazon	34	218-635	2,45 (pH6,5)	0,5-2		photosensible; high volatilizable; shelf-life: 4 years and more
tebuthiuron	2500 (1)		1,82 (1)	0,6-6,87	PSII	
2,4D	311	60 (calc)	-0.75(sp)	0.28-2.3	synthetic auxin	turns crystal in hard water
bromacil	807a; 700; 1287b		1,88a	1.5-8; 5-15 total	PSII	

Table 1. Summary of properties of the predominant herbicides in Europe [4] and in U.S.A.[3]. Solubility is expressed in mg/L, Koc in mL/g, application in kg/ha. Experimental temperature is 25°C without complementary information: (1): 20C; (2): 23°C; (3): Experimental pH is 7 without complementary information: a: pH 5; b: pH 9; c: pH 10; d: pH 4. EPSPS: 5-enolpyruvylshikimate-3-phosphate synthase, an enzyme involved in aromatic amino acids phenylalanine, tyrosine and tryptophan biosynthesis. PSII: inhibition of Hill reaction in photosynthetic electron transport. Triazole: probable interference with carotenoid biosynthesis leading to photooxidation of chlorophyll. Sp: 2,4-D Kow: 2.58-2.83 (pH 1); 0.04-0.33 (pH 5); -0.75 (pH 7) Calc: Value obtained by calculation. Total: total weed killing –in order to obtain a bare soil.

Herbicide effects have been summarized by Jurado et al. [63]. The biochemical target of many herbicides is PhotoSystem II (P680), acting in photosynthesis as a photon-electron converting disruptor. More precisely, the inhibition of Hill reaction (photosynthetic electron transport) is performed in A site by triazines, uraciles pyridazines, in B site by ureas. Acylanilides, diphenyl ethers and nitriles inhibit the Hill reaction too. In the 21 predominant herbicides compilation list [3, 4], 11 herbicides uncouple the biochemical cascade in PSII leading to plastoquinone terminal electron acceptor. Instead of this outcome, formation of unmanaged singlet oxygen provokes lesions proportioned to photosynthesis. Biosynthesis of carotenoids, used to manage singlet oxygen, could be a collateral damage facilitating herbicide effects (pyridazines).

On a plant scale, PSII herbicides lead to a more marked leaf yellowing in new leaves than in old ones, i.e. leafs where photosynthesis has been active, and in places of intense photosynthesis i.e. between leaf veins. Such symptomology should impact on shallow water depth seagrass communities and even save seagrass communities in turbid water. However, light provides food for seagrasses [64] like for the other photosynthetic taxa, but it is a way of detoxifying too. Over time, depending on contamination by such herbicides, light conditions favourable to seagrass communities could be limited by (1) minimal photosynthesis needs, especially high for seagrasses in order to confront sediment anaerobic conditions, and (2) lethal photosynthetic induction, due to poisonous singlet oxygen produced by incident radiation. Indeed, seagrasses reach a 25% requirement of incident radiation [65], due to their submarine adaptation and because of anoxia containment in buried non-photosynthetic tissues [66]. Seagrass incident radiation requirement is to compare with 1% or less of the requirements for other angiosperm species [65]. Seagrasses will physiologically enhance their incident radiation: chloroplast efficiency is modulated for better light capture: adaptation at biochemical level [67] and at organelle scale by conditioning its position in the cell and orientation relatively to the light source [68]. Acting for maximizing photon capture, seagrasses could maximize the risk of lesions due to electron transport alteration by herbicides especially if PSII disruptors mimic a lacks of light because of induced low photosynthetic yield.

The photosynthetic stress hypothesis is strengthened by ten publications (Table 2) observing oxygen production alteration [69-71] and especially oxygen production stimulation for low atrazine concentration (75μg/L) for EC50 at 320μg/L [69]. In the same way (and at same scale, leading to suggest that it is the same phenomenon but with a different descriptor), photosynthesis alteration [72] reaches 120μg/L for immediate (2h) IC50 [42] when pigments, chlorophyll and fluorescence are altered from 10μg/L [73, 74].

Test species	Response parameters	Test duration	Effect concentration (μg/L)	References
Thalassia testudinum	Oxygen production	40h, 88h	320 (EC50)	[71]
Zostera marina	Oxygen production	24h	100i, 1000ti	[70]
Zostera marina	Oxygen production	21-42d	75e, 650i	[69]
Zostera marina	Adenine nucleotides	6h, 21d	10, 100	[76]
	Growth			
	Mortality			
Zostera marina	Growth	10-40d	1900 (first effect, whole plant)	[114]
	Mortality			
	Chlorophyll			
Zostera capricorni	Chlorophyll	10h, 4d (rec)	10, 100	[74]
	Fluorescence			
	Pigments			
Halophila ovalis	Chlorophyll	4d	10	[73]
	Fluorescence			
	Pigments			
Ruppia maritima	Photosynthesis	2h	120 (IC50)	[42]
Ruppia maritima	Growth	35d	2,500, 44,700 (EC50)	[72]
	Photosynthesis			
Halodule whrightii	Growth	22d	10e, 40e, 120e, 420i	[75]

Table 2. Example of toxic effect concentrations reported for atrazine and seagrasses. EC50, IC50 –concentration reducing effect parameter 50% relative to control. rec: recovery; i: inhibition; ti: total inhibition; e: enhancement;

However, some remarks are necessary: (1) like for many environmental monitoring, metrology improvement since decades enhances phenomena perception: concentrations reached 30 years ago could be regarded as overestimated, (2) interspecies heterogeneity could be important: *Halodule wrightii* growth is enhanced at 120µg/L [75] when, at the same concentration *Ruppia maritima* photosynthesis IC50 is reached [42] and *Zostera marina* presents mortalities [76] (3) *Zostera marina* mortalities [76] are noted for concentrations considered as stimulating [69], but in the second case after much more exposition time, (5) the effect is less and recovery is greater *in situ* [74].

Seagrasses are mainly regarded as shadow plants [77]. Indeed, the situation is critical during summer low tides when photoinhibition is a threat. Photoinhibition is defined by Touchette & Burkholder [78] as a reduction in the photosynthetic rate due to other processes such as the toxicological impact of herbicides [79]. Photoinhibition is primarly a photoprotective tool, avoiding PSII excessive photophosphorylation, and dissipating energy as heat [80, 81]. Photoinhibition is obtained by PSII centres rarefaction [81] or inactivation because of D1 protein photolysis surpassing D1 synthesis [80-84] and the increase of xanthophylls cycle's violaxanthin de-epoxidation agent leading to energy dissipation: zeaxanthin. Protein D1 is known to be influenced by ATP in thylakoid lumen; in the case of a lack of ATP production, for example as herbicide impact occurs, seagrass could maintain a high photosynthetic level, initiating a vicious circle leading to the herbicide effect. Such an herbicide trajectory could lead first to photosuppression with excessive UV radiations [85].

The second critical target for seagrass, i.e. especially critical due to aquatic life, is a photorespiration process which is, like for several topics, innovative in the case of seagrasses [78]. Indeed, respiration provides to seagrasses oxidative conditions propitious to life in their partially reductive environment. In order to maintain a redox potential suitable for whole enzymatic activity even in tissues buried into the sediment, seagrasses will actively manage inner gas exchanges. A large part of tissue volume will be due to lacunae or *aerarium*, empty spaces allowing to preserve terrestrial-like conditions for cells, providing oxidative conditions and leaving the leafs erect. Respiration will produce CO_2 for the plant –which will mainly consume its own CO_2 production in order to limit exchanges with external marine environments. Such C3-C4 intermediate plants present, moreover, concentrating carbon systems strengthening ribulose 1,5-bisphosphate carboxylase (Rubisco) in carbon acquisition. 2,4-D photosynthetic pathways are negatively impacted [86], even in micromolar concentrations, by auxin like 2,4-D, leading to up-regulating growth. Because of gas flows to the external environment are more strictly controlled by seagrasses' stomata, which are less numerous than terrestrial plants' ones, auxin-like activity of 2,4-D causes an up - regulation in oxygen production and a subsequent oxygen-inhibition of a key enzyme Rubisco. Lack of carbon-sequestering photosynthetic processes leads to carbon and energy deficits aggravated by messy leaf creation induced by 2,4-D, exhausting the plant, drawing on belowground stocks, and limiting photosynthesis efficiency.

Photorespiration leads to consume O_2 and is considered as protective for photosynthetic electron transport, limiting damage to the photosynthetic apparatus to photo-inactivation during periods of low CO_2 availability and high light intensity [87]. Rates of photorespiration activity are considerably lower in most submersed aquatic plants than in terrestrial ones [78]; if O_2 depletion is too great, anaerobic conditions rule: Krebs cycle's NAD^+ reduction, leading to energy storage in mitochondria by NADH production driving ATP synthesis, is interrupted. NADH accumulates and NAD^+ lacks for critical metabolic processes [88]. Parenthetically, pyruvate is metabolized, leading to fermentation (Davies, 1980) and alcohol content increases, altering whole tissues and thus removing the main obstacle to reductive conditions which are unfit for seagrass life.

Unlike photosynthesis which increases with temperature up to 5–10°C above ambient, respiration rates continue to increase with increasing temperatures in excess of 40°C [78, 89-91]. Light, then depth [92] can also significantly influence respiration; water-column nitrate enrichment tissue NR activity enhances respiration rates in *Z. marina* [93].

6. *In situ*: Chemical cocktail, interaction with metals and temperature increase

If Lewis & Devereux [55] rightly indicated that seagrass are quite non-sensitive to herbicides, based on scientific literature showing the high herbicide concentration reached in order to observe seagrass alteration *in vitro* (Table 2), such results *ex situ* should be weighted by monitoring results, showing everywhere a variegated contamination in space, in time and, moreover, in impact. In the same way, limited impact of herbicides and organic chemicals is mentioned by Waycott et al. [94].

Seagrass meadow contamination by herbicides is well known, as from rivers, as from anti-fouling coatings [55, 95 and therein]. The impact of herbicides on seagrass is more scarcely noted [60, 96, 97], even on a limited scale (3% inhibition of photosynthetic biomaterial assay [98]). But seagrass vulnerability to short but intense contamination has been highlighted [99] and such events could be difficult to monitor. Moreover, short term contamination could be integrated by passive samplers, deployed for weeks, and weighted by the mean concentration in the aquatic environment: depending to monitoring protocol, fugacious pollution could be neglected. Then, seagrass could be resistant to long-term herbicide contamination with severe concentration [61] but vulnerable to toxic pulses [99].

Seagrass physiology is temperature dependant. Seagrass growth is enhanced by temperature increases; the optimal temperature for temperate species is between 11.5°C and 26°C when tropical ones' *preferenda* is between 23°C and 32°C (Lee et al., 2007). Temperature conditions:

–Respiration (see supra); temperature is the predominant factor for respiration control [66, 100],

–Rubisco oxygenase fonction (increased by increasing temperature [88]),

–Sucrose synthase (SS) activity, enhanced in belowground tissues with O_2 decrease and temperature increase [78],

http://dx.doi.org/10.5772/55973

–Sucrose-P synthase (SPS), in the opposite way of terrestrial plants: increasing temperature leads to the increase of SPS activity, which is also influenced by salinity, photosynthesis, CO_2 availability, NH_4^+ and grazing [93, 101, 102].

–C metabolism, i.e. C-sink or C-source depending on temperature [103],

–Stomata function: stomata will be closed in high temperatures, in order to avoid dehydration.

Thus, temperature impacts seagrass growth independently to insulation [104].

Without herbicide impact, 40-45°C is considered as the threshold temperature [22]: for higher ones, irreversible effects are observed, especially at PSII scale. The herbicide presence, even at limited concentrations, could lower such threshold temperature, considering the complex physiologic equilibrium that herbicides could disrupt; such temperature sensibility leads to an enhanced impact of herbicides in warm conditions [105]. Metal accumulation is enhanced by temperature increases [106]. Metals are toxic for seagrasses and especially for PSII [107]. Cu, used in this way as an antifouling alternative, early impacts PSII complex [108-111] in a few days after contamination. The cocktail effect is highlighted [74] for Cu and Irgarol 1051. Gamain [112] shows that herbicide impact is increased in presence of Cu and following temperature: at a temperature for which *Z. noltii* when free of herbicide alteration, even on a biochemical scale, seagrass presents damages in presence of this cocktail. However, these cocktail and summer temperatures are more close to field conditions, especially in tropical waters, than cold conditions and isolate herbicides.

7. Conclusion

Seagrass decline is actually misunderstood. If nutrients increase, it leads to epiphyte proliferation which limits seagrass photosynthesis [104]. Erosion, burying meadows and inducing turbidity limiting photosynthesis, are evoked as the main threat on the seagrass community, chemical interactions could be regarded as underestimated. Even if seagrasses seem to be resilient to herbicide pollution, and even if seagrass recovery has been shown to be better *in situ* than *in vitro*, the cocktail impact seems to be a promising study field. Data concerning seagrass contamination are dramatically scarce despite the precious services that the seagrass community provides, as for economical activity as for environmental concerns like biodiversity preservation and carbon fixation. In the field, seagrass meadows are regressing, and their resilience seem to be altered. Impacts of herbicides on the minimal requirement and on the adaptation to high irradiances are not sufficiently studied leading to observe regression without understanding underlying phenomenology [45]. Seagrass originality involves more largely trans-disciplinarily in order preventing the meadows' decline –but if such a consistent *pièce de résistance* will need appropriate research efforts, the seagrass crisis, taken into account by environmental monitoring like Water Framework, allows, after all, hope for a remediation.

Acknowledgements

Authors wish to thank William and Diana R. Corby for their contributions to their English improvement.

Author details

A. Damien Devault and Hélène Pascaline

Université des Antilles et de la Guyane, EA 929 AIHP-GEODE, groupe Biospheres, Campus de Schoelcher, Schoelcher Cedex, Guiana

References

[1] Haslett, S. K. Coastal Systems, Routledge, (2009). 978-0-41544-060-8

[2] Halpern, B. S. A global map of human impact on marine ecosystems. Science (2008)., 319-948.

[3] Gilliom, R. J. Pesticides in U.S. streams and groundwater. Environmental Science & Technology (2007)., 41(1), 3409-3414.

[4] Schäfer, R. B, Caquet, T, Siimes, K, Mueller, R, Lagadic, L, & Liess, M. Effects of pesticides on community structure and ecosystem functions in agricultural streams of three biogeographical regions in Europe. Science of The Total Environment (2007).

[5] Brodie, J. E, Kroon, F. J, Schaffelke, B, Wolanski, E. C, Lewis, S. E, Devlin, M. J, Bohnet, I. C, Bainbridge, Z. T, Waterhouse, J, & Davis, A. M. Terrestrial pollutant runoff to the Great Barrier Reef: An update of issues, priorities and management responses. Marine Pollution Bulletin (2012).

[6] Ecobichon, D. J. Pesticide use in developing countries. Toxicology (2001).

[7] Janssens, I, Tanghe, T, & Verstraete, W. Micropollutants: A bottleneck in sustainable wastewater treatment. Water Science and Technology (1997)., 35(10), 13-26.

[8] Delpla, I, Jung, A, Baures, V, Clement, E, & Thomas, M. O. Impacts of climate change on surface water quality in relation to drinking water production. Environment International (2009)., 35(8), 1225-1233.

[9] Peters, E. C, Gassman, N. J, Firman, J. C, Richmond, R. H, & Power, E. A. Ecotoxicology of tropical marine ecosystems. Environmental Toxicology and Chemistry (1997)., 16(1), 12-40.

[10] Yamamuro, M. Herbicide-induced macrophyte-to-phytoplankton shifts in Japanese lagoons during the last 50 years: Consequences for ecosystem services and fisheries. Hydrobiologia (2012)., 699(1), 5-19.

[11] Les, D. H, Cleland, M. A, & Waycott, M. Phylogenic studies in the Alismatidea, II: Evolution of the marine angiosperm (seagrasses) and hydrophily. Systematic Botany (1997)., 22-443.

[12] Orth, R. J, Carruthers, T. J. B, Dennison, W. C, Duarte, C. M, Fourqurean, J. W, Heck, K. L, Hughes, A. R, Kendrick, G. A, Kenworthy, W. J, Olyarnik, S, Short, F. T, Waycott, M, & Williams, S. L. A global crisis for seagrass ecosystems. Bioscience (2006)., 56(12), 987-996.

[13] Dennison, W. C, Orth, R. J, Moore, K. A, Stevenson, J. C, Carter, V, Kollar, S, Bergstorm, P. W, & Batuik, R. A. Assessing water quality with submersed aquatic vegetation. BioScience (1993)., 43-86.

[14] Terrados, J, Duarte, C. M, Kamp-nielsen, L, Agawin, N. S. R, Gacia, E, Lacap, D, Fortes, M. D, Borum, J, Lubanski, M, & Greve, T. Are seagrass growth and survival affected by reducing conditions in the sediment? Aquatic Botany (1999)., 65-175.

[15] Duarte, C. M, Middleburg, J, & Caraco, N. Major role of marine vegetation on the oceanic carbon cycle. Biogeosciences (2005)., 2-1.

[16] Mcroy, C. P, & Mcmillan, C. (1977). Production ecology and physiology ofseagrasses. In: P.C. McRoy and C Helfferich (eds.) Seagrass ecosystems: Ascientific Prospective, Marcel Dekker, New York,, 53-87.

[17] Fourqurean, J. W, Duarte, C. M, Kennedy, H, Marbà, N, Holmer, M, Mateo, M. A, Apostolaki, E. T, Kendrick, G. A, Krause-jensen, D, Mcglathery, K. J, & Serrano, O. Seagrass ecosystems as a globally significant carbon stock. Nature Geoscience (2012)., 5-505.

[18] Romero, J, Pergent, G, Pergent-martini, C, Mateo, M. A, & Regnier, C. The detritic compartment in a *Posidonia oceanica* meadow: litter features, decomposition rates and mineral stocks. Marine Ecology (1992)., 13-73.

[19] Pergent, G, Rico-raimondino, V, & Pergent-martini, C. Fate of primary production in Posidonia oceanica meadows of the Mediterranean. Aquatic Botany (1997)., 59-307.

[20] Suchanek, T. H, Williams, S. L, Ogden, J. C, Hubbard, D. K, & Gill, I. P. Utilization of shallow-water seagrass detritus by Carribbean deep-sea macrofauna: δ13C evidence. Deep Sea Research Part A. Oceanographic Research Papers (1985)., 32(2), 201-214.

[21] Marsh, J. A, Dennison, W. C, & Alberte, R. S. Effects of temperature on photosynthesis and respiration in eelgrass (*Zostera marina* L.). Journal of Experimental Marine Biology and Ecology (1986)., 101, 257-267.

[22] Campbell, S. J, Mckenzie, L. J, & Kerville, S. P. Photosynthetic responses of seven tropical seagrasses to elevated seawater temperature. Journal of Experimental Marine Biology and Ecology, , 330(2), 455-468.

[23] Vermaat, J. E, Verhagen, F. C. A, & Lindenburg, D. Contrasting responses in two populations of *Zostera noltii* Hornem. to experimental photoperiod manipulation at two salinities, Aquatic Botany (2000). , 67(3), 179-189.

[24] Den Hartog C(1970). The Seagrasses of the World. North-Holland Publishing Co., Amsterdam. 275 pp.

[25] Giesen, W. B. J. T, & Van Katwijk, M. M. den Hartog C. Eelgrass condition and turbidity in the Dutch Wadden Sea. Aquatic Botany (1990). , 37, 71-85.

[26] Charpentier, A, Grillas, P, Lescuyer, F, Coulet, E, & Auby, I. Spatio-temporal dynamics of a *Zostera noltii* dominated community over a period of fluctuating salinity in a shallow lagoon, Southern France. Estuarine, Coastal and Shelf Science (2005).

[27] Plus, M, & Auby, I. Marc Verlaque, Levavasseur G. Seasonal variations in photosynthetic irradiance response curves of macrophytes from a Mediterranean coastal. lagoon. Aquatic Botany (2005). , 81, 157-173.

[28] Waycott, M, Duarte, C. M, Carruthers, T. J. B, Orth, R. J, Dennison, W. C, Olyarnik, S, Calladine, A, Fourqurean, J. W, & Heck, K. L. Jr, Hughes A. R., Kendricki G. A., Kenworthy W. J., Short F. T., S. L. Williams. Accelerating loss of seagrasses across the globe threatens coastal ecosystems. Proceedings of the National Academy of Sciences of the United States of America. (2009). , 106(30), 12377-12381.

[29] Longstaff, B. J, & Dennison, W. C. Seagrass survival during pulsed turbidity events: The effects of light deprivation on the seagrasses *Halodule pinifiola* and *Halophila ovalis*. Aquatic Botany (1999). , 65-105.

[30] Hemminga, M, & Duarte, C. M. (2000). Seagrass ecology. Cambridge (United Kingdom): Cambridge University Press. 312pp.

[31] Dhargalkar, V. K, & Verlecar, X. N. Southern Ocean seaweeds: A resource for exploration in food and drugs. Aquaculture (2009).

[32] Smith, G. M. (1944). Marine Algae of the Monterey Peninsula, California. Stanford Univ. 2nd Edition. 275pp.

[33] Harborne, A. R, Mumby, P. J, Micheli, F, Perry, C. T, Dahlgren, C. P, Holmes, K. E, & Brumbaugh, D. R. The functional value of Caribbean coral reef, seagrass and mangrove habitats to ecosystem processes. Advances in Marine Biology (2006). , 50-57.

[34] Alongi, D. M. Mangrove forests: Resilience, protection from tsunamis, and responses to global climate change. Estuarine, Coastal and Shelf Science (2008). , 76-1.

[35] Beck, M. W. Heck Jr. K.L., Able K.W., Childers D.L., Eggleston D.B., Gillanders B.M., Halpern B., Hays C.G., Hoshino K., Minello T.J., Orth R.J., Sheridan P.F., Weinstein

M.P.. The identification, conservation and management of estuarine and marine nurseries for fish and invertebrates. Bioscience (2001). , 51, 763-771.

[36] Danielsen, F, Sørensen, M. K, Olwig, M. F, Selvam, V, Parish, F, Burgess, N. D, Hiraishi, T, Karunagaran, V. M, Rasmussen, M. S, Hansen, L. B, Quarto, A, & Suryadiputra, N. The Asian tsunami: a protective role for coastal vegetation. Science (2005).

[37] Nagelkerken, I, Van Der Velde, G, Gorissen, M. W, Meijer, G. J, Hof, t, & Den, T. Hartog C.. Importance of mangroves, seagrass beds and the shallow coral reef as a nursery for important coral reef fishes, using a visual census technique. Estuarine, Coastal and Shelf Science (2000). , 51, 31-44.

[38] Schaffelke, B, Mellors, J, & Duke, N. C. Water quality in the Great Barrier Reef region: responses of mangrove, seagrass and macroalgal communities. Marine Pollution Bulletin (2005).

[39] Vaslet, A, Bouchon-navaro, Y, Louis, M, & Bouchon, C. (2008). Potential effect of mangrove regression for fish species of commercial interest in Guadeloupe. Proceedings of the 61st Gulf and Caribbean Fisheries Institute. November Gosier, Guadeloupe, French West Indies. 7pp., 10-14.

[40] Bell, A. M, & Duke, N. C. Effects of Photosystem II inhibiting herbicides on mangroves-preliminary toxicology trials Marine Pollution Bulletin (2005).

[41] Negri, A, Vollhardt, C, Humphrey, C, Heyward, A, Jones, R, Eaglesham, G, & Fabricius, K. Effects of the herbicide diuron on the early life history stages of coral. Marine Pollution Bulletin (2005).

[42] Jones, T. W, & Winchell, L. Uptake and photosynthetic inhibition by atrazine and its degradation products on four species of submerged vascular plants. Journal of Environmental Quality (1984). , 13-243.

[43] Costanza, R, Arge, d, Degroot, R, Farder, R, Grasso, S, Hannon, M, Limberg, B, Naeem, K, Neill, S. O, Parnelo, R. V, Raskin, J, Sutton, R. G, & Van Den, P. Belt M. The value of the world's ecosystem services and natural capitol. Nature (1997). , 387-253.

[44] Watson, R. A, & Coles, R. G. Lee Long W.J. Simulation estimates of annual yield and landed value for commercial penaeid from a tropical seagrass habitat. Australian journal of Marine and Freshwater Research (1993). , 44-211.

[45] Auby, I, Bost, C, Budzinski, A, Dalloyau, H, Desternes, S, Belles, A, Trut, A, Plus, G, Pere, M, Couzi, C, Feigne, L, & Steinmetz, C. J. ((2011). Seagrass meadow regression in Arcachon Bay : state of the art and cause investigation. Ifremer report. RST/LER/AR/11.007, 155 pp.

[46] Valiela, I, Bowen, J. L, & York, J. K. Mangrove forests: One of the world's threatened major tropical environments. Bioscience (2001). , 51-807.

[47] Achard, F, Eva, H, Stibig, H. J, Mayaux, P, Gallego, J, Richards, T, & Malingreau, J. P. Determination of deforestation rates of the world's humid tropical forests. Science (2002). , 297-999.

[48] Bellwood, D. R, Hughes, T. P, Folke, C, & Nystrom, M. Confronting the coral reef crisis. Nature (2004). , 429-827.

[49] Bruno, J. F, Selig, E. R, Casey, K. S, Page, C. A, Willis, B. L, Harvell, C. D, Sweatman, H, & Melendy, A. M. Thermal stress and coral cover as drivers of coral disease outbreaks. PLoS Biology (2007). , 5-1220.

[50] Gardner, T. A, Cote, I. M, Gill, J. A, Grant, A, & Watkinson, A. R. Long-term region-wide declines in Caribbean corals. Science (2003). , 301-958.

[51] Fonseca, M. S, Kenworthy, W. J, & Thayer, G. W. (1998). Guidelines for the conservation and restoration of seagrasses in the United States and adjacent waters. NOAA Coastal Ocean Program Decision Analysis Series NOAA Coastal Ocean Office, Silver Spring, Maryland.(12)

[52] Green, E. P, & Short, F. T. (2003). World Atlas of Seagrasses. E.P. Green, F.T. Short (Eds.), University of California Press, Berkeley.

[53] Williams, S. L. Reduced genetic diversity in eelgrass transplantations affects both individual and population fitness. Ecological Applications (2001). , 11-1472.

[54] Kenworthy, W. J. The role of sexual reproduction in maintaining populations of Halophila decipiens: Implications for the biodiversity and conservation of tropical seagrass ecosystems. Pacific Conservation Biology (2000). , 5-260.

[55] Lewis, M. A, & Devereux, R. Nonnutrient anthropogenic chemicals in seagrass ecosystems: fate and effects. Environmental Technology and Chemistry (2009). , 28(3), 644-661.

[56] Knee, K. L, & Paytan, A. Submarine Groundwater Discharge: A Source of Nutrients, Metals, and Pollutants to the Coastal Ocean. Treatise on Estuarine and Coastal Science (2011).

[57] Devault, D. A, Merlina, G, Lim, P, Probst, J, Pinelli, L, & Multi-residues, E. analysis of pre-emergence herbicides in fluvial sediments: application to the mid-Garonne River. Journal of Environmental Monitoring (2007). , 9-1009.

[58] Novak, R. J, & Lampman, R. L. (2001). Public Health Pesticides. Handbook of Pesticide Toxicology (Second Edition), , 1(4), 181-201.

[59] Haynes, D, Müller, J, & Carter, S. Pesticide and Herbicide Residues in Sediments and Seagrasses from the Great Barrier Reef World Heritage Area and Queensland Coast. Marine Pollution Bulletin (2000a).

[60] Haynes, D, Müller, J, & Carter, S. Pesticide and Herbicide Residues in Sediments and Seagrasses from the Great Barrier Reef World Heritage Area and Queensland Coast. Marine Pollution Bulletin (2000b).

[61] Haynes, D, Ralph, P, Prange, J, & Dennison, B. The Impact of the Herbicide Diuron on Photosynthesis in Three Species of Tropical Seagrass. Marine Pollution Bulletin (2000c).

[62] BCPC(2007). Pesticide Manual. 15th edition. C.D.S. Tomlin. BCPC Eds., Alton, UK, 1457 pp.

[63] Jurado, A. S, Fernandes, M. A. S, Peixoto, F. P, & Vicente, J. A. F. (2001). Herbicides: the face and the reverse of the coin. An *in vitro* approach to the toxicity of herbicides in non-target organisms. In: Herbicides and Environment, Chapitre 1, A. Kortekamp (Ed.), InTech, Vienna, Austria, 978-9-53307-476-4, 3-43.

[64] Abal, E. G, Loneragan, N, Bowen, P, Perry, C. J, Udy, J. W, & Dennison, W. C. Physiological and morphological responses of the seagrass *Zostera capricorni* Aschers, to light intensity Journal of Experimental Marine Biology and Ecology (1994). , 178(1), 113-129.

[65] Dennison, W. C, Orth, R. J, Moore, K. A, Stevenson, J. C, Carter, V, Kollar, S, Bergstrom, P. W, & Batiuk, R. A. Assessing water quality with submersed aquatic vegetation. BioScience (1993). , 43-86.

[66] Terrados, J, Borum, J, Duarte, C. M, Fortes, M. D, Kamp-nielsen, L, Agawin, N. S. R, & Kenworthy, W. J. Nutrient and mass allocation of South-east Asian seagrasses. Aquatic Botany (1999). , 63-203.

[67] Mazzuca, S, Spadafora, A, Filadoro, D, Vannini, C, Marsoni, M, Cozza, R, Bracale, M, Pangaro, T, & Innocenti, A. M. Seagrass light acclimation: 2-DE protein analysis in Posidonia leaves grown in chronic low light conditions Journal of Experimental Marine Biology and Ecology (2009). , 374(2), 113-122.

[68] Sharon, Y, & Beer, S. Diurnal movements of chloroplasts in *Halophila stipulacea* and their effect on PAM fluorometric measurements of photosynthetic rates Aquatic Botany(2008). , 88(4), 273-276.

[69] Correll, D. L, & Wu, T. L. Atrazine toxicity to submerged vascular plants in simulated estuarine microcosms. Aquatic Botany (1982). , 14-151.

[70] Kemp, W. M, Means, J. C, Jones, T. W, & Stevenson, J. C. (1982). Herbicides in Chesapeake Bay and their effects on submerged aquatic vegetation. In Macalaster EG, Baker DA, Kasper M, eds, Chesapeake Bay Program Technical Studies: A Synthesis. U.S. Environmental Protection Agency, Washington, DC, , 502-567.

[71] Walsh, G. E, Hansen, D. L, & Lawrence, D. A. A flow-through system for exposure of seagrass to pollutants. Marine Environmental Research (1982). , 7-1.

[72] Johnson, J. R, & Bird, K. T. The effects of the herbicide atrazine on *Ruppia maritima* L. growing in autotrophic versus heterotrophic cultures. *Botanica Marina* (1995). , 38-307.

[73] Ralph, P. J. Herbicide toxicity of *Halophila ovalis* assessed by chlorophyll a fluorescence. Aquatic Botany (2000)., 66-141.

[74] Macinnis-ng, C. M. O, & Ralph, P. J. Short-term response and recovery of *Zostera capricorni* photosynthesis after herbicide exposure. Aquatic Botanic (2003)., 76-1.

[75] U.S. Department of Interior. 1989. Atrazine hazards to fish, wildlife, and invertebrates: A synoptic review. Biological Report 85. Contaminant Hazard Reviews Report 78. U.S. Fish and Wildlife Service, Washington, DC.

[76] Delistraty, D. A, & Hershner, C. Effects of the herbicide atrazine on adenine nucleotide levels in *Zostera marina* L. (eelgrass). Aquatic Botany (1983)., 18-353.

[77] Ralph, P. J, & Burchett, M. D. Photosynthetic responses of the seagrass *Halophila ovalis* (RBr.) Hook. F. to high irradiance stress, using chlorophyll a fluorescence. Aquatic Botany (1995)., 51-55.

[78] Touchette, B. W, & Burkholder, J. M. Overview of the physiological ecology of carbon metabolism in seagrasses. Journal of experimental marine biology and ecology. (2000)., 250-169.

[79] Scarlett, A, Donkin, P, Fileman, T. W, Evans, S. V, & Donkin, M. E. Risk posed by the anti-fouling agent Irgarol 1051 to the seagrass, *Zostera marina*.. Aquatic Toxicology (1999a)., 45-159.

[80] Krause, G. H, & Weis, E. Chlorophyll fluorescence and photosynthesis: the basics. Ann. Rev. Plant Physiology, Plant Molecular Biology (1991)., 42-313.

[81] Hanelt, D, Li, J, & Nultsch, W. Tidal dependence of photoinhibition of photosynthesis in marine macrophytes of the South China Sea. *Botanica Acta* (1994)., 107-66.

[82] Ohad, I, Kyle, D. J, & Arntzen, C. J. Membrane protein damage and repair. Removal and replacement of inactivated 32 kilodalton polypeptides in chloroplast membranes. Journal of Cell Biology (1984)., 99-481.

[83] Guenther, J. E, & Melis, A. The physiological significance of photosystem II heterogeneity in chloroplasts. Photosynthesis Research (1990)., 23-105.

[84] Aro, E. M, Virgin, I, & Andersson, B. Photoinhibition of photosystem II. Inactivation, protein damage and turnover. *Biochimica et Biophysica Acta* (1993)., 1143-113.

[85] Dawson, S. P, & Dennison, W. C. Effects of ultraviolet and photosynthetically active radiation on five seagrass species. Marine Biology (1996)., 125-629.

[86] Gareth, L. (2005). The impact of herbicides on biota of the intertidal zone. Master dissertation, Adelaide University, South Australia. 178pp.

[87] Heber, U, Bligny, R, Streb, P, & Douce, R. Photorespiration is essential for the protection of the photosynthrtic apparatus of C3 plants against photoinactivation under sunlight. *Botanica Acta* (1996). , 109-307.

[88] Taiz, L, & Zeiger, E. (1991). Plant physiology, The Benjamin/Cummings Publishing Company, Redwood City, , 1-565.

[89] Drew, E. A. Factors affecting photosynthesis and its seasonal variation in the seaqrasses *Cvmodocea nodosa* (Ucria) Aschers, and ~osldon; oceanic: (L.) Dellle In the Mediterranean. Journal of experimental marine Biology and Ecology (1978). , 31-173.

[90] Bulthuis, D. A. (1983). Effects of temperature on the photosynthesisirradiance curve of the Australian seagrass, Heterozostera tasmanica. Marine Biology Letters , 4, 47-57.

[91] Marsh, J. A, Dennison, W. C, & Alberte, R. S. Effects of temperature on photosynthesis and respiration in eelgrass (Zostera marina L.). Journal of experimental marine Biology and Ecology (1986). , 101-257.

[92] Dennison, W. C, & Alberte, R. S. Photosynthetic response of Zostera manna L. (eelgrass) to in situ manipulations of light intensity. *Oecologia* (Berl.) (1982). , 55-137.

[93] Touchette, B. W. (1999). Physiological and developmental responses of eelgrass (Zostera marina L.) to increases in water-column nitrate and temperature. P.h.D dissertation, North Carolina State University, Raleigh, NC.

[94] Waycott, M, Longstaff, B. J, & Mellors, J. A. Seagrass population dynamics and water quality in the Great Barrier Reef region: A review and future research directions. Marine Pollution Bulletin (2005).

[95] Scarlett, A, Donkin, P, Fileman, T. W, & Morris, R. J. Occurrence of the Antifouling Herbicide, Irgarol 1051, within Coastal-water Seagrasses from Queensland, Australia Marine Pollution Bulletin (1999b). , 38(8), 687-691.

[96] Bester, K. Effects of pesticides on seagrass beds. Helgoland Marine Research (2000).

[97] Mcmahon, K. Bengtson Nash S., Eaglesham G., Müller J. F., Duke N. C., Winderlich S. Herbicide contamination and the potential impact to seagrass meadows in Hervey Bay, Queensland, Australia. Marine Pollution Bulletin (2005).

[98] Bengtson Nash SM., McMahon K., Eaglesham G., Müller J.F. Application of a novel phytotoxicity assay for the detection of herbicides in Hervey Bay and the Great Sandy Straits Marine Pollution Bulletin (2005).

[99] Macinnis-ng, C. M. O, & Ralph, P. J. In situ impact of multiple pulses of metal and herbicide on the seagrass, *Zostera capricorni*. Aquatic Toxicology (2004).

[100] Masini, R, & Manning, C. R. The photosynthetic responses to irradiance and temperature of four meadow-forming seagrass. Aquatic Botany (1997). , 58-21.

[101] Zimmerman, R. C, Kohrs, D. G, & Alberte, R. S. Top-down impact through a bottom-up mechanism: theeffect of limpet grazing on growth, productivity and carbon allocation of Zostera marina L. (eelgrass). *Oecologia* (1996). , 107-560.

[102] Zimmerman, R. C, Kohrs, D. G, Steller, D. L, & Alberte, R. S. Impacts of CO enrichment on productivity and light requirements of eelgrass. Plant Physiology (1997). , 115-599.

[103] Zimmerman, R. C, & Alberte, R. S. (1991). Prediction of the light requirements for eelgrass (*Zostera marina* L.) growth from numerical models. In: Kenworthy, W.J., Haunert, D.E. (Eds.), The Light Requirements of Seagrasses. Proceedings of a workshop to examine the capability of water quality criteria, standards and monitoring programs to protect seagrasses. NOAA Technical Memorandum NMFS-SEFC-287, NOAA, Silver Springs, MD, , 26-37.

[104] Lee, K, Park, S, & Kim, S. R. Y. K.. Effects of irradiance, temperature, and nutrients on growth dynamics of seagrasses: A review. Journal of Experimental Marine Biology and Ecology (2007). , 350-144.

[105] Lucia, M, Anre, J. M, Gonzalez, P, Baudrimont, M, Gontier, K, Maury-brachet, R, & Davail, S. Impact of cadmium on aquatic bird Cairina moschata, Biometals (2009). , 22, 843-853.

[106] Prange, J. A, & Dennison, W. C. Physiological responses of five seagrass species to trace metals, Marine Pollution Bulletin (2000). , 41-327.

[107] Clijsters, H, & Van Assche, F. Inhibition of photosynthesis by heavy metals. Photosynthesis Research (1985). , 7-31.

[108] Cook, C, Kostidou, M, Vardaka, A, & Lanaras, E. T. Effects of copper on the growth, photosynthesis and nutrient concentrations of Phaseolus plants. *Photosynthetica* (1997). , 34-179.

[109] Droppa, M, Terry, N, & Horvath, G. Variation in photosynthetic pigments and plastoquinone contents in sugar beet chloroplasts with changes in leaf copper content. Plant Physiology (1984). , 74-717.

[110] Mohanty, N, Vass, I, & Demeter, S. Copper toxicity affects photosystem II electron transport at the secondary quinone acceptor, QB. Plant Physiology (1989). , 90-175.

[111] Patsikka, E, Kairavuo, M, Sersen, F, Aro, E, & Tyystjarvi, M. E. Excess Copper Predisposes Photosystem II to Photoinhibition in Vivo by Outcompeting Iron and Causing Decrease in Leaf Chlorophyll1, Plant Physiology (2002). , 129-1359.

[112] Gamain, P. (2012). Experimental ecotoxicologic study of the impact of a cocktail of copper and pesticides on *Zostera noltii* seagrass meadow of Arcachon Bay. Master 2 dissertation. Université de Bordeaux 1. 40 pp.

[113] Short, F, Carruthers, T, Dennison, W, & Waycott, M. Global seagrass distribution and diversity: A bioregional model. Journal of Experimental Marine Biology and Ecology (2007). , 350-3.

[114] Schwartzschild, A. C. MacIntyre W.G., Moore K.A., Libelo E.L. *Zostera marina* L. growth response to atrazine in rootrhizome and whole plant exposure experiments. Journal of Experimental Marine Biology and Ecology (1994). , 183-77.

Chapter 15

Transgenic Herbicide-Resistant Turfgrasses

In-Ja Song, Tae-Woong Bae, Markkandan Ganesan,
Jeong-Il Kim, Hyo-Yeon Lee and Pill-Soon Song

Additional information is available at the end of the chapter

http://dx.doi.org/10.5772/56096

1. Introduction

Turfgrasses grow in different habitats for numerous purposes worldwide. They are cultivated for their agronomical, environmental, ornamental, recreational and stock feeding values [1, 2]. Various turfgrasses are used for environmental beautification and for the protection of resources such as land, soil and water. Many varieties of turfgrasses cover home yards, golf courses, parks, soccer fields, and roadsides, etc. To cite a few examples of renewed interest in turfgrasses, they play a significant environmental role in photosynthetically fixing carbon dioxide to evolve oxygen into the atmosphere. In addition to their vast acreage of widespread forage, planting of the grasses in urban areas such as rooftops, parks and, more recently automobile parking lots, contributes to the suppression of urban heat island phenomena [3]. Various causes of soil erosion and losses due to flood washout and landslide can also be circumvented and managed, as the damages are greatly reduced and the conservation of soil moisture and underground water is effectively sustained by the planting of turfgrass varieties. Recreational and sporting activities on the natural turfgrass field, compared to an artificial turf, greatly reduce the risk of personal injuries, thus contributing to the wellbeing of people in general.

Not surprisingly, the worldwide turfgrass market and its associated herbicide sales are substantial; in the United States alone, turfgrass is one of the four major staple crops, second only to corn [4, 5]. In facing the challenge of global warming, turfgrasses are gaining attention of both environmentalists and agronomists for their role in the certified emission reductions. Relatively high production costs of cultivating and maintaining turfgrasses concerns them, however. Healthy swarth growth and well-maintained turf habitats entail herbicide spraying because otherwise dominant weed varieties easily overtake the sward. Annually, their maintenance costs alone run around 4.5 billion dollars in the United States [4, 6]. One of the major costs is certainly herbicidal requirement.

Herbicidal agrochemicals are classified into two categories, selective and non-selective herbicides. The latter kills all plant species, whereas the former is targeted at specific plant(s)/weed(s) for herbicidal action. The biochemical mechanisms of herbicides include the disruptions of (i) the photosynthesis by blocking the photosynthetic reaction centers, electron transport system or photo-oxidative membrane damages, (ii) cell division and root development, (iii) energy transduction and metabolism, (iv) plant growth hormones, (v) biosynthesis of amino acids/proteins and (vi) disruption of other physiologically significant molecules such as chlorophylls and carotenoids, as discussed elsewhere in this volume.

Frequent herbicide applications also pose serious environmental and health concerns, for example, to the authors' residential island of Jeju where there are 30 golf courses open for business. In spite of the current difficulties arising from the public objections, genetically modified turfgrasses with a herbicide-resistant gene provide an effective alternative to the wide applications of agrochemical herbicides. Since the development and ecological impact studies of transgenic herbicide-resistant creeping bentgrass [7, 8] and zoysiagrass [9, 10], several GM varieties of turfgrasses including those of herbicide-resistant cultivars have been developed (see Table 1). Most recently, in reference [11] bentgrass ASR-368 has been patented for its commercial rights. With an increasing number of reports on transgenic herbicide-resistant turfgrasses, it is appropriate to review the subject at this time. Discussion in this chapter focuses on the transgenic herbicide-resistant turfgrasses developed primarily in our laboratory here in Jeju and Gwangju, Korea. For a review of other transgenic grasses with herbicide-resistance traits, see Table 1 and references therein.

Plant species	Cultivar	Method	Marker gene	Target gene	Target trait	References
Agrostis stolonifera (creeping bentgrass)	Crenshaw	*Agrobacterium*	*bar*	*bar*/Rice *tlpd34*	Disease resistance	[16]
	Crenshaw	*Agrobacterium*	*bar*	*bar*/Barley *hva1*	Drought tolerance	[33]
	Crenshaw	*Agrobacterium*	*bar/gus*	*bar/PepEST*	Herbicide resistance/ Disease resistance	[34]
	Crenshaw	*Agrobacterium*	*bar/gus*	*bar*/Maize *Lc+Pl*	Purple-color	[35]
	Crenshaw	*Agrobacterium*	*bar/gus*	*bar/AtBG1*	Herbicide resistance/ Drought tolerance/ dwarf	[36]
	Crenshaw, Penncross	*Agrobacterium*	*bar/gus*	bar	Herbicide resistance	[37]
	Penncross	Electroporation	*bar*	*bar*	Herbicide resistance	[38]
	Penncross	Electroporation	*bar/gus*	*bar*	Herbicide resistance	[39]

http://dx.doi.org/10.5772/56096

Plant species	Cultivar	Method	Marker gene	Target gene	Target trait	References
	Penncross	*Agrobacterium*	*bar*	*bar*/Cowpea *VuNCED*1	Drought/salt tolerance	[40]
	Penncross	*Agrobacterium*	*bar/CP4-EPSPS*	*bar/CP4-EPSPS*	Herbicide resistance	[22]
	Penncross	*Agrobacterium*	*bar*	*bar/ZjLsL*	Herbicide resistance/ dwarf	[41]
	Province Penn-A-4	Biolistics	*bar/gus*	*bar/chitinase +glucanase*	Herbicide resistance/ Disease resistance	[42]
	Penn-A-4	*Agrobacterium*	*hph/gus, bar*	bar	Herbicide resistance	[43]
	Penn-A-4	*Agrobacterium*	*bar*	*bar/Pen4-1*	Herbicide resistance/ Disease resistance	[44]
	Penn-A-4	*Agrobacterium*	*bar*	*bar/AVP1*	Herbicide resistance/ Salt tolerance	[45]
Agrostis palustris (creeping bentgrass)	Suthshore Emerald	Biolistics	*bar/gus*	*bar*	Herbicide resistance	[46]
	Regent Tiger	*Agrobacterium*	*bar/gfp*	*bar*	Herbicide resistance	[47]
	Cobra	Electroporation	*bar*	*bar*	Herbicide resistance	[48]
		Biolistics	*bar*	*bar/hs2*	Herbicide resistance	[49]
Cynodon spp. (bermudagrass)	TifEagle	Biolistics	*bar*	*bar*	Herbicide resistance	[50]
	TifEagle	*Agrobacterium*	*bar/gus*	*bar*	Herbicide resistance	[51]
Dactylis glomerata (orchardgrass)	Embryogen-P	Biolistics	*bar/gus*	*bar*	Herbicide resistance	[52]
	Rapido	Biolistics	*bar/hph/gus*	*bar*	Herbicide resistance	[53]
Festuca arundinacea (tall fescue)		Protoplasts	*bar/hph*	*bar*	Herbicide resistance	[54]
	Alley	Biolistics	*bar*	*bar/Ipt*	Herbicide resistance/ Cole tolerance	[55]
Festuca rubra (red fescue)		Protoplasts	*bar*	*bar*	Herbicide resistance	[56]
Lolium perenne (perennial ryegrass)	Riikka	Biolistics	*bar*	*bar/wft1/wft2*	Herbicide resistance/ Freezing tolerance	[57]
	TopGun	*Agrobacterium*	*bar*	*bar/OsNHX1*	Herbicide resistance/ Salt tolerance	[58]
Panicum virgatum (switchgrass)	Alamo	Biolistics	*bar/gfp*	*bar*	Herbicide resistance	[59]

Plant species	Cultivar	Method	Marker gene	Target gene	Target trait	References
	Alamo	*Agrobacterium*	*bar/gus*	*bar*	Herbicide resistance	[60]
Paspalum notatum (bahiagrass)	Tifton-7	Biolistics	*bar*	*bar*	Herbicide resistance	[61]
	Pensacola	Biolistics	*bar/gus*	*bar*	Herbicide resistance	[62]
Paspalum vaginatum Swartz (Seashore Paspalum)		*Agrobacterium*	*bar/gus*	*bar*	Herbicide resistance	[63]
Zoysia japonica (zoysiagrass)		*Agrobacterium*	*bar/gus*	*bar*	Herbicide resistance	[15]
	Zenith	Biolistics	*bar/hpt*	*bar*	Herbicide resistance	[64]
		Agrobacterium	*bar*	*bar/phyA*	Herbicide resistance/ Shade tolerance	[10]
Zoysia sinica (Chinese lawngrass)		*Agrobacterium*	*bar*	*bar/CBF1*	Herbicide resistance/ Chilling tolerance	[65]

bar: bialaphos resistance gene, gus: β-glucuronidase, hph: hygromycin phosphotransferase. gfp: green fluorescent protein

Table 1. Transgenic herbicide-resistant turfgrasses

2. Turfgrass species

There are some 7,500 turfgrass species of more than 600 genera distributed worldwide. Of these, 30~40 species are cultivated as agronomic plants [1]. Turfgrasses are generally classified into two major species, warm and cold season grasses. The plants are also divided into two groups based on their mechanism of photosynthetic carbon dioxide fixation, C3 and C4 plants. As representative C4 warm season turfgrasses with optimal growth temperatures of 27~35°C, zoysiagrass and Bermuda grass species are widely used for sports fields because of their strong traits such as swarth growth, vegetative propagation and drought tolerance as they are cultivated widely, especially in China, Japan and Korea. However, they tend to grow relatively slowly and particularly with zoysiagrasses prematurely lose their greenness by late autumn. Typical C3 cold season turfgrasses with optimal temperatures in the 15~25°C range include blue grass and bentgrass varieties. The latter is particularly advantageous for the putting greens [1, 4, 5, 12]. In this chapter, the review will be concerned with two main varieties, zoysiagrass (*Zoysia japonica* Steud.) and bentgrass (*Agrostis palustris* L., Crenshaw and Penn-cross varieties), focusing on their herbicide resistant transgenic cultivars.

http://dx.doi.org/10.5772/56096

3. Transgenes and mechanisms of herbicidal action

Turfgrass has been a subject of classical breeding for trait improvement over decades, especially in Japan and United States. However, conventional breeding suffers from such drawbacks as low efficiency, time consuming and labor intensiveness. With an increasing trend in turfgrass cultivation worldwide, excessive applications of herbicides and other agrochemicals over the grass habitats adversely impact the environment, biodiversity and human health [13, 14]. Several attempts to develop GM turfgrass lines with improved traits have been reported; for example, herbicide-resistant turfgrass varieties in references [15], [16], 17] and [10] and insect-resistant turfgrass in reference [18]. A number of laboratories are developing herbicide-resistant and other transgenic turfgrasses with biotic and abiotic stress tolerances (Table 1).

So far, several genes including the two widely adopted ones, *CP4 EPSPS* encoding 5-enolpyruvylshikimate-3-phosphate synthase (EPSPs) and *BAR or PAT* encoding a phosphinothricin acetyl transferase (PAT), have been introduced to generate herbicide-resistant turfgrasses. Other target genes for herbicide resistance include *BXN* (bromoxylnil nitrilase gene), *DHPS* (dihydropteroate synthase gene), *ALS* (acetolactate synthase gene) and others (Table 1). Transgenic bentgrass and zoysiagrass stacked with *BAR* and *PHYA* (phytochrome A) genes conferring herbicide- and shade-resistance traits, respectively, have also been developed [10] and will be reviewed in this chapter.

The widely used herbicide, bialaphos (also phosphinothricin-alanyl-alanine tripeptide, PTT), is an antibiotic produced by certain *Streptomyces* genera and used as an agrochemical, which has been commercialized under the trade name Basta by Bayer Crop Science. It kills plants non-selectively. Bialaphos itself is an inactive compound as a herbicide, but it is cleaved by intracellular peptidases to phosphinothricin (L-PPT), Phosphinothricin (glufosinate) so produced *in situ* binds glutamine synthetase (GS), the key enzyme in the nitrogen fixation in plants, inhibiting its catalytic activity to fix the ammonium with L-glutamate to form glutamine [19] (See Figure 1).

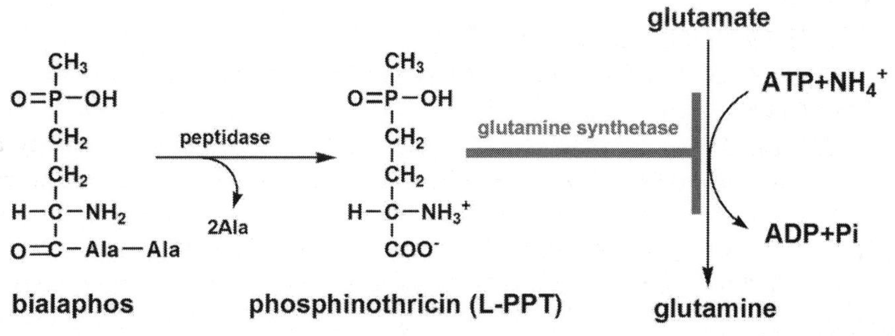

Figure 1. Biochemical mechanism for the herbicidal action of glufosinate through the inhibition of glutamine synthetase by the herbicide.

The glufosinate herbicide causes accumulation of lethal levels of ammonia in both soil bacteria and plant cells. The GS inhibiting activity of glufosinate is lost when its amino group is acetylated by a phosphinothricin acetyl transferase (PAT encoded by *PAT;* also known as *bar or BAR* for bialaphos resistance) (Figure. 2).

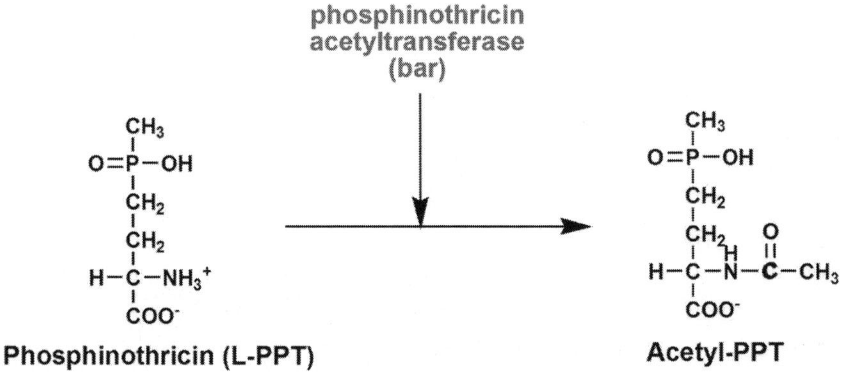

Figure 2. Detoxication of glufosinate by phosphinothricin acetyl transferase (*BAR* or *PAT*).

Thus, a transgenic turfgrass transformed with *BAR* gene becomes resistant to the Basta spray, as glufosinate from the Basta is effectively detoxicated in the plant. The transgenic zoysiagrass and bentgrass developed in our laboratories carry the *BAR* gene isolated from *Streptomyces hygroscopicus* in the soil [10].

Glyphosate is a non-selective herbicidal agent commercialized under the trade name "Round-up" by Monsanto. It exerts its herbicidal action by competitively inhibiting the 5-enolpyruvylshikimate-3-phosphate synthase (EPSPs) centrally involved in the biosynthesis of aromatic amino acids (phenylalanine, tryptophan and tyrosine). Plants treated with glyphosate are killed for the lack of these amino acids in protein biosynthesis. Accumulation of shikimate also leads to cell death, thus contributing to the herbicidal action of glyphosate [20] (Figure. 3).

A transgenic bentgrass carrying the EPSPS gene ("Roundup Ready") then develops resistance to Roundup [7, 21].

Although both *BAR*- and *EPSPS*-.transgenic turfgrasses are yet to be released for agronomic cultivations, second and third generation GM crops including turfgrasses are forthcoming to deal with the intolerance and tolerance being developed to the non-specific herbicides in the transgenic herbicide-resistant turfgrasses and weed plants, respectively. Such next generation crops are also being developed with the hope of leading consumer acceptance. In reference [22] the authors stacked both *BAR* and *CP4 EPSPS* genes in creeping bentgrass to generate dual (glufosinate and glyphosate) herbicide-resistant turfgrasses, hoping that less amounts of two herbicides together are required for weed necrosis than with the greater amount needed with one herbicide alone. The bentgrass species so developed showed an expected degree of

http://dx.doi.org/10.5772/56096

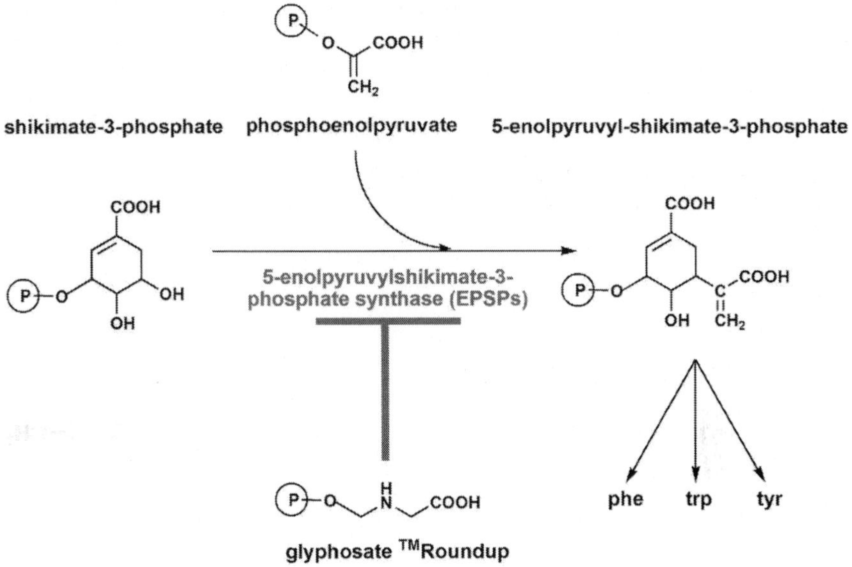

Figure 3. The reaction catalyzed by 5-enolpyruvylshikimate 3-phosphate synthase(EPSPS) (Modified from reference [32])

tolerance to both Basta and Roundup, respectively. While such dual transgene herbicide resistance may counter for a single-transgene plant to lose tolerance to the herbicide and/or for the weeds to develop tolerance to the herbicide, it remains to be seen if this expectation is borne out in natural habitats.

One of the most promising herbicide-resistant traits can be conferred by dicamba monooxygenase gene (*DMO*). Dicamba (3, 6 dichloro-2-methoxybenzoic acid) is an active auxin analog and its presence in the plant cells exaggerate the hormonal effects that lead to the cell and plant death. It is widely used in the Unites States for over four decades. It is a relatively non-toxic and environment-friendly herbicide. Its herbicidal activity is lost in a *DMO*-transgenic crop as dicamba is detoxified to its inactive 3, 6-DCSA (3, 6-dichlorosalicylic acid) [23]. Attempts are being made to generate DMO-transgenic turfgrass plants in several laboratories.

4. Herbicide-resistant zoysiagrass and bentgrass

In a previous report, we discussed the development of the *BAR*-transgenic *Zoysia japonica* Steud., currently undergoing a regulatory approval process under the cultivar name "Jeju Green 21" and compared its phenotypic traits with those of non-transgenic control [9]. Figure 4 (A, B) illustrates the effect of spraying Basta on the test plot containing both control and herbicide-resistant zoysiagrasses. In Figure 4(A), the herbicide-resistant runners were planted in the GMO-spelled area, which continued to grow healthily after Basta spray, showing "Jeju

Green 21" plants growing in "GMO" spell pattern before and after the herbicide treatment at a concentration of 0.1% (w/v) glufosinate. Figure 4(B) shows the mixed turfgrass/weed habitat treated with a 0.5% Basta spray, showing an effective herbicidal killing of the weeds. Non-transgenic grasses are effectively wilted out, whereas the resistant plants remain healthy and indistinguishable from their non-transgenic counterparts physiologically and phenotypically [9]. Figure 5 displays the herbicidal performance of *BAR*-transgenic creeping bentgrass in which a wild type or mutant *PHYA* (*Ser599Ala PHYA*) gene is stacked with the *BAR* gene, *vide infra*. The results show that the gene stacking has not compromised the herbicide-resistance function conferred by the *BAR* gene. Qualitatively, both *BAR*- and *EPSPS*-transgenic bent-grasses effectively tolerate the herbicides, Basta and Roundup, respectively, but quantitative comparisons of the herbicide resistances exhibited by different transgenic zoysiagrass and bentgrass varieties entail further study.

Figure 4. Herbicide resistance assay of putative transgenic zoysiagrass plants. A. 0.8% BASTA® was sprayed onto non-transgenic plants (NT) and bialaphos-resistant zoysiagrass, "GMO" was spelled by removing the plants; GM grass was then planted into the letters, B. 0.5% BASTA® was sprayed onto the weed and bialaphos-resistance zoysiagrass plants.

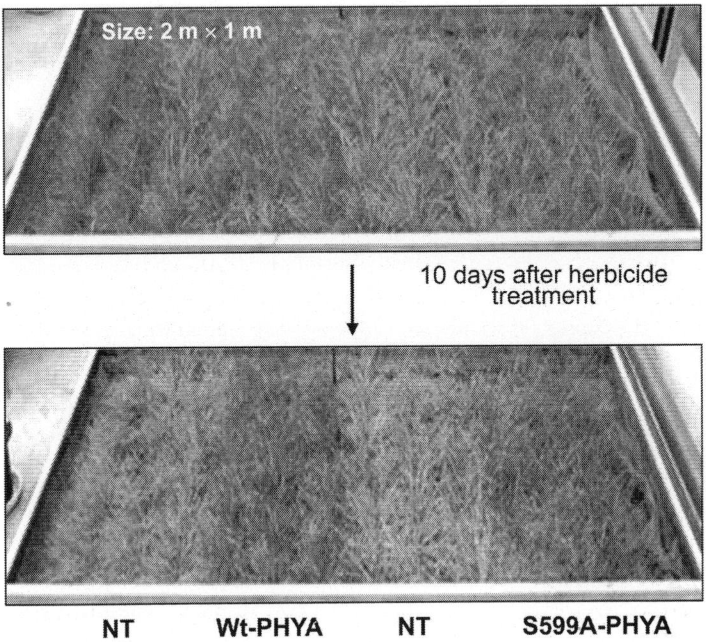

Figure 5. Herbicide resistance assay of putative transgenic creeping bentgrass plants. 0.8% BASTA® was sprayed onto non-transgenic plants (NT) and transgenic plants over-expressing *Wt-PHYA* or *Ser599Ala-PHYA*, and the herbicide resistance of the plants was determined 10 days after the spraying. *Wt-PHYA*, transgenic bentgrass plants with wild-type *PHYA* gene; *Ser599Ala-PHYA*, transgenic bentgrass plants with *Ser599Ala-PHYA* mutant.

When zoysiagrass and possibly other turfgrass species are left unmanaged under natural habitats, their populations and swarth growth are easily overtaken by the dominant weed plants. Figure 6 shows our own observations of herbicide-resistant zoysiagrass plants growing in natural habitats during the four consecutive years (2006~2009). In four years, the ground coverage of zoysiagrass was dominated by the weeds when the grass plot was left unmanaged. On the other hand, the herbicide-resistant plants continued healthy population and swarth growths under managed conditions involving fertilizer applications, herbicide sprays and timely mowings.

Recently, we reported the development and morphological characterization of transgenic *Zoysia japonica* and *Agrostis stolonifera* plants transformed with both *BAR* and *PHYA* genes [1]. The two transgenes confer herbicide resistance and shade tolerance to the grass, respectively. We developed these turfgrass plants by harboring wild-type *Avena PHYA* or *Ser599Ala PHYA* mutant (*S599A-phytochrome A hyperactive mutant* gene [24]) on the *BAR*-decked *pCAMBIA3301* vector in order to confer both herbicide and shade tolerant phenotypes to them. The transgenic plants with *Ser599Ala-PHYA* and *Wt-PHYA* also displayed the shorter phenotypes desired, in addition to their herbicide resistance trait (Figure 7).

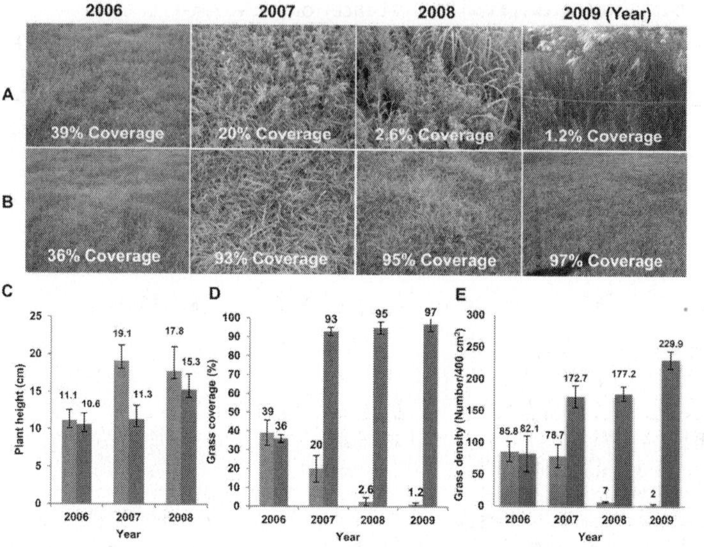

Figure 6. Survival of the transgenic herbicide-resistant zoysiagrass during 4 years (2006-2009) in natural habitats. A. Natural habitats during 4 years, B. Managed field, C. Plant height of zoysiagrass, D. Grass coverage of zoysiagrass, E. Grass density of zoysiagrass. Blue bar, natural habitat; red bar, managed field.

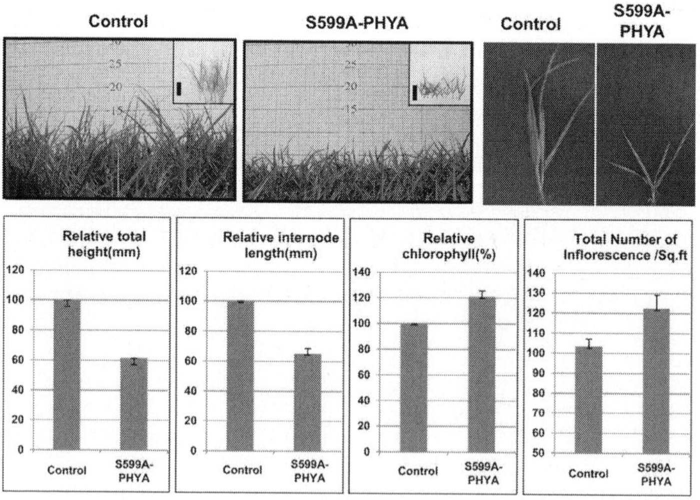

Figure 7. Growth performance of transgenic zoysiagrass plants over-expressing *Ser599Ala-PHYA* showed short phenotypes compared with control plants (*BAR* gene) under field conditions. Bar in insert 1 cm.

We observed a delay in necrosis (senescence) of *Ser599Ala-PHYA* leaves under outdoor conditions in early winter (Figure 8). During the rejuvenation of zoysiagrass after the winter season, various weeds began to dominate over the transgenic turfgrass habitats. However, zoysiagrass plants expressing both *BAR* and *Ser599Ala-PHYA* genes exhibited a significant increase in tiller number and runner length relative to the non-transgenic controls [10]. These traits will be helpful for the zoysiagrass plants to compete effectively with the weeds, especially in disrupting the germination of unwanted weeds.

Figure 8. Photographic view of browning (necrosis) in zoysiagrass transformant lines in early winter. NT, non-transgenic zoysiagrass plants; HR, herbicide-resistant zoysiagrass plants with *BAR* gene; *Wt-PHYA*, transgenic zoysiagrass plants with wild-type *PHYA* gene; *Ser599Ala-PHYA* 2-14 & 2-18 transformant lines, transgenic zoysiagrass plants with *Ser599Ala-PHYA* mutant gene.

5. Environmental risk assessment

To commercialize any of the transgenic turfgrass varieties listed in Table 1, their environmental risks must be assessed under their natural habitats [7, 8, 9, 25]. This chapter briefly reviews our own studies and discusses attempts to block or minimize the risks of gene flow from the transgenic turfgrass habitats to the plants at neighboring and remote sites. For example, in reference [26] and [27] the workers introduced a male-sterility gene into GM crops to block the escape of a transgene from the latter, and this strategy may be applied to turfgrasses. We developed a sterile herbicide-resistant zoysiagrass through γ-radiation mutation, making the latter unbolting and deficient in fertile pollens [28, 29]. The γ-radiation generated herbicide-resistant zoysiagrass can be cultivated in agronomic habitats for eventual commercialization [25].

A preliminary study showed that the transgene (*BAR*) of herbicide-resistant *Zoysia japonica* unintentionally escaped from the test plants to the close neighbored non-transgenic zoysiagrass species [9]. However, the introgression is likely to be suppressed under natural conditions (see Figure. 6) and can be easily terminated by applying non-specific herbicides such as glyphosate and paraquat [25].

According to the "Weed risk assessments for Hawaii and Pacific Islands" database (http://www.botany.hawaii.edu/faculty/daehler/wra/default.htm), transgenic *Zoysia japonica and Zoysia tenuifolia* are classified as being L grade, i.e. not currently recognized as invasive in

Hawaii, and not likely to have major ecological or economic impacts on other Pacific Islands based on the HP-WRA screening process. On the other hand, bentgrass (*Agrostis stolonifera*) belongs to an H grade group of plants, suggesting that transgenic herbicide-resistant bentgrass is a higher risk turfgrass than the zoysiagrass; according to the Hawaii database, *Agrostis stolonifera* is likely to be invasive in Hawaii and on other Pacific Islands as determined by the HP-WRA screening process. In fact, the transgene of the Roundup Ready creeping bentgrass introgressed other recipient plant species 3.8 km away from the test plot [8]. In conclusion, the herbicide-resistant zoysiagrass developed in our laboratory poses substantially less risk of transgene flow than the bentgrass (Figure. 5).

Although the risk of transgene escape and flow from the genetically modified zoysiagrass is low, pollen-induced gene flow cannot be completely discounted. In reference [30] we examined the pollen releases from the defined boundary of *BAR* –transgenic *Zoysia japonica* habitats as a function of physical variables including the boundary, temperature, atmospheric humidity, and lighting condition/duration. Results suggest that zoysiagrass' pollen escape is essentially limited to the close neighborhood, in contrast to bentgrass pollens.

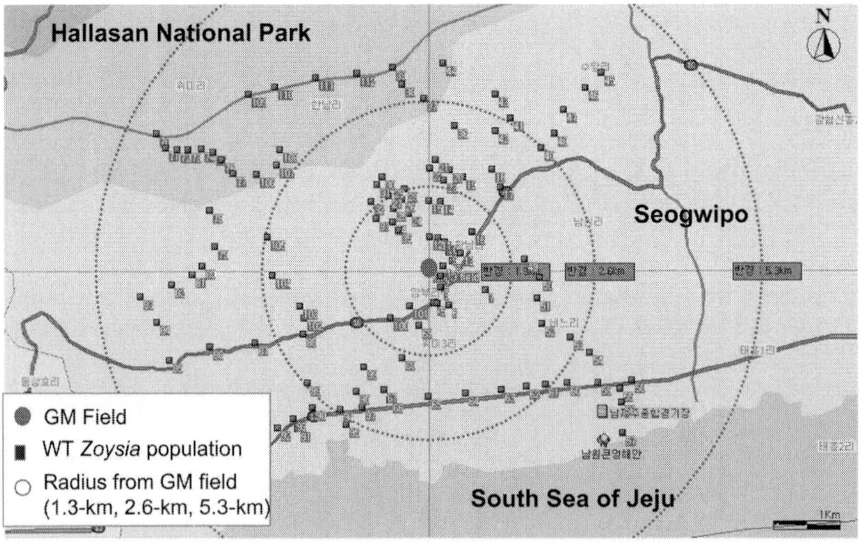

Figure 9. Monitoring for the potential gene flow from the genetically modified zoysiagrass to wild-type zoysiagrass plants within a 5-km radius in natural habitat. Samples were taken from 112 zones (448 sites): *Zoysia japonica* 96 zones (384 sites) and *Zoysia matrella* 16 zones (64 sites).

http://dx.doi.org/10.5772/56096

Figure 9 shows the sites in Jeju Island monitored for the potential gene flow from the herbicide-resistant *Zoysia japonica* to wild-type zoysiagrass within a 5-km radius in natural habitat. No introgression was observed at these sites as of this writing.

6. Commercial potentials and outlook

Turfgrass is a highly value-added crop in terms of commercial profits per land acreage, when compared to other crops. Turfgrasses sward vigorously through vegetative propagation and swarth growth. According to TPI data (Turfgrass Producers International), the turfgrass market size increased by 35% during the five year (2002-2007) period [31]. Based on the data available, transgenic zoysiagrasses pose considerably less risk of transgene escape than does bentgrass. Furthermore, the former can be effectively propagated vegetatively, and sterile herbicide-resistant zoysiagrass (and bentgrass) can be developed through γ-radiation treatment [30]. This will circumvent to a large extent the public's objections to genetically modified plants and their unintended escapes.

7. Conclusion

We compiled a table of transgenic herbicide-resistant turfgrass varieties in various stages of development and eventual agronomic cultivations. As can be seen in Table 1 of this chapter, several transgenes have been introduced into zoysiagrass, bentgrass and other lawn grass species primarily through Agrobacterium-mediated transformation and biolistic transfection. These grasses all exhibit resistance to their intended herbicides such as Basta, Roundup and others, but how well each of the transgenics developed performs in test plots and natural habitats cannot be assessed at this point largely because quantitative data such as the dose-response curves and the outdoor performances are lacking in most cases. In this chapter, we focused our discussion to the *BAR* transgenic *Zoysia japonica* and *Agrostis stolonifera* species. We conclude that these cultivars offer promising potentials as environmentally friendly and economically beneficial turfgrass varieties, especially the former, for Jeju Island and elsewhere.

Acknowledgements

This research was supported by Next-Generation Biogreen 21 Program, Rural Development Administration, Republic of Korea (Grant No. PJ00949901), Basic Science Research Program (NRF Grant No. 2012R1A1A2000706 to PSS, 2012-0004335) and the Priority Research Centers Program (2012048080) through the National Research Foundation of Korea (NRF) funded by the Ministry of Education, Science and Technology.

Author details

In-Ja Song[1], Tae-Woong Bae[1], Markkandan Ganesan[1], Jeong-Il Kim[2], Hyo-Yeon Lee[1*] and Pill-Soon Song[1*]

*Address all correspondence to: pssong@gmail.com, hyoyeon@jejunu.ac.kr

1 Faculty of Biotechnology and Subtropical Horticulture Research Institute, Jeju National University, Jeju, Korea

2 Department of Biotechnology and Kumho Life Science Laboratory, Chonnam National University, Gwangju, Korea

References

[1] Kim K.N. Introductory turfgrass science. (in Korean) Sahmyook University; 2005.

[2] Pessarakli M. Turfgrass Management and Physiology. USA: CRC Press; 2007.

[3] Takebayashi H., Moriyama M. Study on the urban heat island mitigation effect achieved by converting to grass-covered parking. Solar Energy 2009; 83(8) 1211-1223.

[4] Lee L. Turfgrass biotechnology. Plant Science 1996; 115(1) 1-8.

[5] Spangenberg G., Wang Z.Y., Potrykus I. Biotechnology in Forage and Turf Grass Improvement. Berlin: Springer; 1998.

[6] Zilinskas B.A., Wang X. Genetic transformation of turfgrass, In: Liang GH, Skinner DZ, (eds). Genetically Modified Crops: Their Development, Uses, and Risks. New York: Food Product Press; 2004. p309-350.

[7] Watrud L.S., Lee E.H., Fairbrother A., Burdick C., Reichman J.R., Bollman M., Storm M., KIng G., Van de Water P.K. Evidence for landscape-level, pollen-mediated gene flow from genetically modified creeping bentgrass with *CP4 EPSPS* as a marker. Proceedings of the National Academy of Sciences 2004; 101(40) 14533-14538.

[8] Reichman J.R., Watrud L.S., Lee E.H., Burdick C.A., Bollman M.A., Storm M.J., King G.A., Mallory-Smith C. Establishment of transgenic herbicide-resistant creeping bentgrass (*Agrostis stolonifera* L.) in nonagronomic habitats. Molecular Ecology 2006; 15(13) 4243-4255.

[9] Bae T.W., Vanjildorj E., Song S.Y., Nishiguchi S., Yang S.S., Song I.J., Chandrasekhar T., Kang T.W., Kim J.L., Koh Y.J., Park S.Y., Lee J., Lee Y.E., Ryu K.H., Riu K.Z., Song P.S., Lee H.Y. Environmental risk assessment of genetically engineered herbicide-tolerant *Zoysia japonica*. Journal of Environmental Quality 2008; 37(1) 207-218.

[10] Ganesan M., Han Y.J., Bae T.W., Hwang O.J., Chandrasekkhar T., Shin A.Y., Goh C.H., Nishiguchi S., Song I.J., Lee H.Y., Kim J.I., Song P.S. Overexpression of phytochrome A and its hyperactive mutant improves shade tolerance and turf quality in creeping bentgrass and zoysiagrass. Planta 2012; 236(4) 1135-1150.

[11] Guo S.X., Harriman R., Lee L., Nelson E.K. Bentgrass event ASR-368 and compositions and methods for detection thereof, United States Patent Number 7569747B2; 2009.

[12] Fry J., Huang B. Applied Turfgrass Science and Physiology. Hoboken, NJ, USA: John Wiley & Sons; 2004.

[13] Choi J.S., Fermanian T.W., Wehner D.J., Spomer L.A. Effect of temperature, moisture and soil texture on DCPA degradation. Agronomy Journal 1990; 80(1) 108-113.

[14] Schleicher L.C., Shea P.J., Stougaard R.N., Tupy D.R. Efficacy and dissipation of dithiopyr and pendimethalin in perennial ryegrass (*Lolium perenne*) turf. Weed Science 1995; 43(1) 140-148.

[15] Toyama K., Bae C.H., Kang J.G., Lim Y.P., Adachi T., Riu K.Z., Song P.S., Lee H.Y. Production of herbicide-tolerant zoysiagrass by *Agrobacterium*-mediated transformation. Molecules and Cells 2003; 16(1) 19-27.

[16] Fu D., Tisserat N.A., Xiao Y., Settleb D., Muthukrishnan S., Liang G.H. Overexpression of rice TLPD34 enhances dollar-spot resistance in transgenic bentgrass. Plant Science 2005; 168(3) 671-680.

[17] Ge Y, Norton T, Wang ZY. Transgenic Zoysiagrass (*Zoysia japonica*) plants obtained by *Agrobacterium*-mediated transformation. Plant Cell Reports 2006; 25(8) 792-798.

[18] Zhang L., Wu D., Zhang L., Yang C. *Agrobacterium* mediated transformation of Japanese lawn grass (*Zoysia japonica* Steud.) containing a synthetic cryIA(b) gene from *Bacillus thuringiensis*. Plant Breed. 2007; 126(4) 428-432.

[19] Bayer E., Gugel K.H., Hägele K., Hagenmaier H., Jessipow S., König W.A., Zöhner H. Phosphinothricin and Phosphinothritcyl-Alanyl-Alanin. Helvetica Chimica Acta 1972: 55 224-239.

[20] Weed Science Society of America. WSSA: Society, Press Room: Weed Control. http://www.wssa.net/WSSA/PressRoom/index.htm (accessed 19 Dec 2007).

[21] Nelson E., Stone T. Petition for determination of non-regulated status: Roundup Ready Creeping Bent grass Event ASF368. Petition #01-TR-054U [www.aphis.usda.gov/brs/not_reg.html] 2003.

[22] Lee K.W., Kim K.Y., Kim K.H., Lee B.H., Kim J.S., Lee S.H. Development of antibiotic marker-free creeping bentgrass resistance against herbicides. Acta Biochim Biophys Sin 2011; 43(1) 13-18.

[23] Behrens M.R., Mutlu N., Chakraborty S., Dumitru R., Jiang W.Z., LaVallee B.J., Herman P.L., Clemente T.E., Weeks D.P. Dicamba resistance: enlarging and preserving

biotechnology-based weed management strategies. Science 2007; 316(5828) 1185-1188.

[24] Kim J.I., Shen Y., Han Y.J., Park J.E., Kirchenbauer D., Soh M.S., Nagy F., Schäfer E., Song P.S. Phytochrome phosphorylation modulates light signaling by influencing the protein-protein interaction. The Plant Cell 2004; 16(10) 2629-2640.

[25] Bae T.W., Kang H.G., Song I.J., Sun H.J., Ko S.M., Song P.S., Lee H.Y. Environmental risk assessment of genetically modified herbicide-tolerant zoysiagrass (Event: Jeju Green21). (in Korean) Journal of Plant Biotechnology 2011; 38(2) 105-116.

[26] Khan M.S. Plant biology: engineered male sterility. Nature 2005; 436: 783-785.

[27] Ruiz O.N., Daniell H. Engineering cytoplasmic male sterility via the chloroplast genome by expression of beta-ketothiolase. Plant Physiology 2005; 138(3) 1232-1246.

[28] Bae T.W., Kim J., Song I.J., Song S.Y., Lim P.O., Song P.S., Lee H.Y. Production of unbolting lines through gamma-ray irradiation mutagenesis in genetically modified herbicide-tolerant *Zoysia japonica*. Breeding Science 2009; 59(1) 103-105.

[29] Bae T.W., Song I.J., Kang H.G., Jeong O.C., Sun H.J., Ko S.M., Lim P.O., Song P.S., Song S.J., Lee H.Y. Selection of male-sterile and dwarfism genetically modified *Zoysia japonica* through gamma irradiation. (in Korean) Journal of Radiation Industry 2010; 4(3) 239-246.

[30] Kang H.G., Bae T.W., Jeong O.C., Sun H.J., Lim P.O., Lee H.Y. Evaluation of viability, shedding pattern, and longevity of pollen from genetically, modified (GM) herbicide-tolerant and wild-type zoysiagrass (*Zoysia japonica* Steud.). Journal of Plant Biology 2009; 52(6) 630-634.

[31] Turfgrass Producers International. TPI: Professional Resources, TPI products, Surveys: 2007 USDA AG Census report. http://www.turfgrasssod.org/pages/resources/usda-ag census-reports (accessed April 2009).

[32] Priestman M.A., Funke T., Singh I.M., Crupper S.S., Schönbrunn E. 5-enolpyruvylshikimate 3-phosphate synthase from *Staphylococcus aureus* is insensitive to glyphosate. Federation of European Biochemical Societies 2005; 579(3) 728-732.

[33] Fu D., Huang B., Xiao Y, Muthukrishnan S, Liang G.H. Overexpression of barley *hva1* gene in creeping bentgrass for improving drought tolerance. Plant Cell Reports 2007; 26 467-477.

[34] Cho K.C., Han Y.J., Kim S.J., Lee S.S., Hwang O.J., Song P.S., Kim Y.S., Kim J.I. Resistance to *Rhizoctonia solani* AG-2-2 (IIIB) in creeping bentgrass plants transformed with pepper esterase gene *PepEST*. Plant Pathology 2011; 60(4) 631-639.

[35] Han Y.J., Kim Y.M., Lee J.Y, Kim S.J., Cho K.C., Chandrasekhar T., Song P.S., Woo Y.M., Kim J.I. Production of purple-colored creeping bentgrass using maize tran-

scription factor genes *Pl* and *Lc* through *Agrobacterium*-mediated transformation. Plant Cell Reports 2009; 28(3) 397-406.

[36] Han Y.J., Cho K.C., Hwang O.J., Choi Y.S., Shin A.Y., Hwang I., Kim J.I. Overexpression of an *Arabidopsis* b-glucosidase gene enhances drought resistance with dwarf phenotype in creeping bentgrass. Plant Cell Reports 2012; 31(9) 1677-1686.

[37] Kim S.J., Lee J.Y., Kim Y.M., Yang S.S., Hwang O.J., Hong N.J., Kim K.M., Lee H.Y., Song P..S, Kim J.I. *Agrobacterium*-mediated high-efficiency transformation of creeping bentgrass with herbicide resistance. Journal of Plant Biology 2007; 50(5) 577-585.

[38] Asano Y., Ito Y., Fukami M., Morifuji A. Production of herbicide resistant transgenic creeping bent plants. International Turfgrass Society Research Journal 2007; 8 261-267.

[39] Asano Y., Ito Y., Fukami M., Sugiura K., Fujiie A. Herbicide-resistant transgenic creeping bentgrass plants obtained by electroporation using an altered buffer. Plant Cell Reports 1998; 17(12) 963-967.

[40] Aswath C.R., Kim S.H., Mo S.Y., Kim D.W. Transgenic plants of creeping bent grass harboring the stress inducible gene, 9-cis-epoxycarotenoid dioxygenase, are highly tolerant to drought and NaCl stress. Plant Growth Regulation 2005; 47(2/3) 129-139.

[41] Yang D.H, Sun H.J., Goh C.H., Song P.S., Bae T.W., Song I.J., Lim Y.P., Lim P.O., Lee H.Y. Cloning of a *Zoysia ZjLsL* and its overexpression to induce axillary meristem initiation and tiller formation in *Arabidopsis* and bentgrass. Plant Biology 2012; 14(3) 411-419.

[42] Wang Y., Kausch A.P., Chandlee J.M., Luo H., Ruemmele B.A., Browning M., Jackson N., Goldsmith M.R. Co-transfer and expression of chitinase, glucanase, and *bar* genes in creeping bentgrass for conferring fungal disease resistance. Plant Science 2003; 165(3) 497-506.

[43] Luo H., Hu Q., Nelson K., Longo C., Kausch A.P., Chandlee J.M., Wipff J.K., Fricker C.R. *Agrobacterium* tumefaciens-mediated creeping bentgrass (*Agrostis stolonifera* L.) transformation using phosphinothricin selection results in a high frequency of single-copy transgene integration. Plant Cell Reports 2004; 22(9) 645-652.

[44] Zhou M., Hu Q., Li Z., Li D., Chen C.F., Luo H. Expression of a novel antimicrobial peptide Penaeidin4-1 in creeping bentgrass (*Agrostis stolonifera* L.) enhances plant fungal disease resistance. PLoS One 2011; 6(9) 1-12.

[45] Li Z., Baldwin C.M., Hu Q., Liu H., Luo H. Heterologous expression of *Arabidopsis* H +-pyrophosphatase enhances salt tolerance in transgenic creeping bentgrass (*Agrostis stolonifera* L.). Plant, Cell & Environment 2010; 33(2) 272-289.

[46] Hartman C.L., Lee L., Day P.R., Tumer N.E. Herbicide Resistant Turfgrass (*Agrostis palustris* Huds.) by Biolistic Transformation. Nature Biotechnology 1994; 12 919-923.

[47] Chai M.L., Wang B.L., Kim J.Y., Lee J.M., Kim D.H. *Agrobacterium*-mediated transformation of herbicide resistance in creeping bentgrass and colonial bentgrass. Journal of Zhejiang University Science 2003; 4(3) 346-351

[48] Lee L., Laramore C.L., Day P.R., Tumer N.E. Transformation and regeneration of creeping bentgrass (*Agrostis palustris* Huds.) protoplasts. Crop Science 1996; 36(2) 401-406.

[49] Chai B., Maqbool S.B., Hajela R.K., Green D., Vargas Jr J.M., Warkentin D., Sabzikar R., Sticklen M.B. Cloning of a chitinase-like cDNA (*hs2*), its transfer to creeping bentgrass (*Agrostis palustris* Huds.) and development of brown patch (*Rhizoctonia solani*) disease resistant transgenic lines. Plant Science 2002; 163(2) 183-193.

[50] Goldman J.J., Hanna W.W., Fleming G.H., Ozias-Akins P. Ploidy variation among herbicide-resistant bermudagrass plants of cv. TifEagle transformed with the *bar* gene. Plant Cell Reports 2004; 22(8) 553-560.

[51] Hu F., Zhang L., Wang X., Ding J., Wu D.. *Agrobacterium*-mediated transformed transgenic triploid bermudagrass (*Cynodon dactylon* X *C. transvaalensis*) plants are highly resistant to the glufosinate herbicide Liberty. Plant Cell, Tissue and Organ Culture 2005; 83(1) 13-19.

[52] Denchev P.D., Songstad D.D., McDaniel J.K., Conger B.V. Transgenic orchardgrass (*Dactylis glomerata*) plants by direct embryogenesis from microprojectile bombarded leaf cells. Plant Cell Reports 1997; 16(12) 813-819.

[53] Cho M.J., Choi H.W., Lemaux P.G. Transformed T0 orchardgrass (*Dactylis glomerata* L.) plants produced from highly regenerative tissues derived from mature seeds. Plant Cell Reports 2001; 20(4) 318-324.

[54] Wang Z.Y., Takamizo T., Iglesias V.A., Osusky M., Nagel J., Potrykus I., Spangenberg G. Transgenic plants of tall fescue (*Festuca arundinacea* Schreb.) obtained by direct gene transfer to protoplasts. Biotechnology 1992; 10(6) 691-696.

[55] Hu Y., Jia W., Wang J., Zhang Y., Yang L., Lin Z. Transgenic tall fescue containing the *Agrobacterium tumefaciens* ipt gene shows enhanced cold tolerance. Plant Cell Reports 2005; 23(10-11) 705-709.

[56] Spangenberg G., Wang Z.Y., Nagel J., Potrykus I. Protoplast culture and generation of transgenic plants in red fescue (*Festuca rubra* L.). Plant Science 1994; 97(1) 83-94.

[57] Hisano H., Kanazawa A., Kawakami A., Yoshida M., Shimamoto Y., Yamada T. Transgenic perennial ryegrass plants expressing wheat fructosyltransferase genes accumulate increased amounts of fructan and acquire increased tolerance on a cellular level to freezing. Plant Science 204; 167(4) 861-868.

[58] Wu Y.Y., Chen Q.J., Chen M., Chen J., Wang X.C. Salt-tolerant transgenic perennial ryegrass (*Lolium perenne* L.) obtained by *Agrobacterium tumefaciens*-mediated transformation of the vacuolar Na^+/H^+ antiporter gene. Plant Science 2005; 169(1) 65-73.

[59] Richards H.A., Rudas V.A., Sun H., McDaiel J.K., Tomaszewski Z., Conger B.V. Construction of a GFP-BAR plasmid and its use for switchgrass transformation. Plant Cell Reports 2001; 20(1) 48-54.

[60] Somleva M.N., Tomaszewski Z., Conger B.V. *Agrobacterium*-mediated genetic transformation of switchgrass. Crop Science 2002; 42(6) 2080-2087.

[61] Smith R.L., Grando M.F., Li Y.Y., Seib J.C., Shatters R.G. Transformation of bahiagrass (*Paspalum notatum* Flugge). Plant Cell Reports. 2002; 20(11) 1017-1021.

[62] Gondo T., Tsurta S.I., Akashi R., Kawamura O., Hoffmann F. Green, herbicide-resistant plants by particle inflow gun-mediated gene transfer to diploid bahiagrass (*Paspalum notatum*). Journal of Plant Physiology 2005; 162(12) 1367-1375.

[63] Kim K.M., Song I.J., Lee H.Y., Raymer P., Kim B.S., Kim W. Development of seashore paspalum turfgrass with herbicide resistance. 2009; Korean Journal of crop science 54(4) 427-432.

[64] Lim S.H., Kang B.C., Shin H.K.. Herbicide Resistant Turfgrass (*Zoysia japonica* cv. 'Zenith') Plants by Particle bombardment-mediated Transformation. 2004; Korean journal of turfgrass science 18(4) 211 – 219.

[65] Li R.F., Wei J.H., Wang H.Z., He J., Sun Z.Y. Development of highly regenerable callus lines and *Agrobacterium*-mediated transformation of Chinese lawngrass (*Zoysia sinica* Hance) with a cold inducible transcription factor, CBF1. Plant Cell, Tissue and Organ Culture 2006; 85(3): 297-305.

Section 3

Research Reviews

Chapter 16

Toxicity of Herbicides: Impact on Aquatic and Soil Biota and Human Health

Maria Aparecida Marin-Morales,
Bruna de Campos Ventura-Camargo and
Márcia Miyuki Hoshina

Additional information is available at the end of the chapter

http://dx.doi.org/10.5772/55851

1. Introduction

During the last decades, the scientific community, including government and non-government organizations have increased their interest in detecting and controlling the environmental agents responsible for damages to the human health and sustainability of the ecosystems. This interest has been intensified by the frightening increase on the reports of the anthropogenic action on the environment responsible for damages to the ozone layer, accidental release of wastes and radioactive gases, as well as contamination by pesticides used in agriculture. However, the growth of the human population and of the activities associated with agriculture, industrialization and urbanization have contributed to the depredation of the biodiversity and genetic variability, resulting in the compromise of several species, including man [9].

After the industrial revolution, a great number of chemical substances have been released into the terrestrial and aquatic environments and in the atmosphere. These substances can be transported and transformed by different processes, whose transformation by-products can cause adverse effects on man, as well as damages to the terrestrial and aquatic ecosystems. Several studies have shown the presence of residues of several chemical substances in the air, water, soil, food and organisms in general [10].

Environmental pollution by genotoxic and mutagenic products affects the exposed organism and its future generations, this fact is observed both for animals, and in this case man is included, and for the other groups of organisms such as plants and microorganisms. In order to evaluate the consequences of the anthropogenic activities on the ecosystem it is necessary that the scientific community pays a special attention in the search for understanding the modes of action of xenobiotics present in the ecosystem in the biota exposed. For this, extensive, detailed and ordered studies of the contaminants must be developed with the purpose of preventing the biological impairment, such as inductions of alterations in the genetic materials of the organisms [11].

Some studies have been performed in the attempt to evaluate the behaviour, transformations and effects of chemical agents, both in the environment and in the organisms. Toxicology establishes the limits of concentration or quantity of chemical substances acceptable in the environment by studies on the toxic effects of these substances in the organism and ecosystems [12].

Considering that the use of agrochemicals, such as herbicides, have caused a great environmental contamination, due to their widespread use, it has become indispensable to perform the assessment of the toxicity of these compounds.

1.1. The importance of herbicides

Living beings are exposed to the action of numerous agents that are potentially toxic. These agents can be physical, chemical or biological and can provoke in the organisms physiological, biochemical, pathological effects and, in some cases, genetic effects [13]. A great variety of chemical substances with mutagenic potential, both natural and synthetic, have been investigated. Many of these substances are found in food, pharmaceutical drugs, pesticides and in complexes of domestic and industrial effluents. It is known that these compounds can cause detrimental inheritable changes in the genetic material, without these changes being expressed immediately [14]. Thus, several compounds dispersed in the environment can represent danger to human health, since they present a potential to induce mutations [15].

The production of food can occur both by agricultural activities and by livestock. The yield of food production is directly related with the relationship established between the species of interest for production and the other plant, animal, microbial and parasitic biological systems that compete for resources available in the environment [16]. Among the species that jeopardize the agricultural production there are the weeds that, when invade crops, can cause significant loss in the yield and quality of the harvest [17]. Therefore, in order to enhance the productivity and the quality of crops, the removal of weeds from agriculture becomes important.

Before the introduction of selective herbicides as an agricultural practice, the removal of weeds was accomplished manually in an extremely laborious form. Thus, the farmers sought other forms to control weeds, such as, integrating other weed control practices such as crop rotation, tillage and fallow systems [17].

The introduction of selective herbicides in the late 40's and the constant production of new herbicides in the following decades gave farmers a new tool in the control of weeds [17]. Therefore, the process of modernization of agriculture introduced, in the 60's, the use of new biological varieties considered more productive, but dependent on chemical fertilizers and intensive use of pesticides, in order to increase productivity. The use of these chemical agents resulted in the increase of productivity, but, on the other hand, brought adverse consequences, since many are harmful substances for man and the environment. The world practice of using agrochemicals for long periods, often indiscriminate and abusive, has raised concerns among the public authorities and experts of public health and sustainability of natural resources [16].

Many agrochemicals are very toxic substances whose absorption in man are almost exclusively oral and can also occur by inhalation or dermally. As a consequence of the human exposure to pesticides, a series of disturbances can be observed, such as gastric, neurological and muscular [18].

Among the pesticides, the main agents of intoxication are the herbicides and insecticides. According to Vasilescu and Medvedovici [19], herbicides are defined as any substance, individually or in mixtures, whose function is to control, destroy, repel or mitigate the growth of weeds in a crop.

The use of herbicides, despite the fact that they are characterized as a highly effective tool in the control of weeds, has led to a change in the phytosociological composition of weeds and to a selection of biotypes resistant to herbicides, besides also causing impacts in the environment and human health. According to He et al. [20], herbicides are the most used chemical substances throughout the world. During the 90's, the global pesticide sales remained relatively constant, between 270 and 300 billions of US dollars, and 47% of this value corresponded to herbicides and 79% to insecticides. Since 2007, herbicides assumed the first place among the three major categories of pesticides (insecticides, fungicides/ bactericides, herbicides) [21].

The use of herbicides to control weeds has been a common practice in global agriculture, mainly with the objective to increase agricultural production. However, when these chemicals are used in an uncontrolled manner, they can cause impacts on non-target organisms, especially on those that live in aquatic environments [22].

According to Chevreuil et al. [23], Kim and Feagley [24] and Abdel-Ramham et al. [25], most of the toxic effects of the herbicides on animals and plants were insufficiently investigated. As a consequence of the lack of information about the action of herbicides in the biological environment, these chemical agents can also represent a problem to human health [26, 27]. The impact of a pesticide in the environment depends on its dispersion mode and its concentration, as well as its own toxicity [28]. The mutagenic effects of the herbicides can result from several reactions with the organism, as a direct action of the compound on the nuclear DNA; incorporation in the DNA during cell replication; interference in the activity of the mitotic or meiotic division, resulting in incorrect division of the cell [29].

Some herbicides interfere directly in the cell division of plants, elongation and/or cell differentiation, causing disturbances in the functioning of the roots or vascular tissues [30]. In animals, herbicides can act in several tissues or organs and, sometimes, are associated with tumorigenic processes [31].

Jurado et al. [32] listed the general advantages and disadvantages of using herbicides. In this list, the authors cited as advantages: kill unwanted plants; help crops grow since it eliminates weeds that compete with crops for water, nutrients and sunlight; can be safely used in plantations, while the manual or mechanical removal processes of weeds can cause damages to crops; can be used in geographically close crops; in most cases, only one application of the herbicide is sufficient to control the weeds, while the other methods must be constantly used; are easy to use; have fast action; are relatively inexpensive and are economically more viable than manual removal; non-selective herbicides can be used to eliminate vegetation cover in areas intended for the construction of residences and/or roads; to eradicate plants bearing diseases; and since some herbicides are biodegradable, they can become relatively inert after some time. The disadvantages listed by the authors are: some herbicides are not biodegradable and, thus, can persist in the environment for a long period of time; all herbicides are, at least, mildly toxic; can cause diseases and even accidental death (case of paraquat); can be carried into rivers by rainwater or be leached to groundwater polluting these environments; some herbicides can accumulate in the food chain and are toxic for animals, including man.

1.2. Herbicides classification

According to Moreland [33], herbicides are designated by common names approved by the Weed Science Society of America (WSSA) or by the British Standards Institution. Organic herbicides are classified according to their application method, chemical affinity, structural similarity, and by their mode of action [34]. In relation to the application methods, herbicides can be classified into two groups: soil application and foliar application. According to Jurado et al. [32], all the herbicides applied in the pre-planting (surface or incorporation) and pre-emergence (in crops, weeds or both) are classified as herbicides of soil application and those applied in the post-emergence are classified as foliar application.

Moreover, herbicides can be classified according to their mode of action. Following, it will be presented the classes of herbicides, according to their mode of action, based in the classification of Moreland [33] :

1. *chloroplast-associated reactions:* photo-induced electron transport and reaction coupled to phosphorylation occur in the chloroplast, any interference in these reactions inhibit the photosynthetic activity. Herbicides that inhibit the photo-chemically induced reactions are divided into the following classes:

a. electron transport inhibitors: electron transport is inhibited when one or more intermediary electron carriers are removed or inactivated or even when there is interference in the phosphorylation. Example: diuron, atrazine.

b. uncouplers: uncouplers dissociate the electron transport of the ATP formation through the dissipation of the energetic state of the thylakoid membrane, before the energy can be used to perform the high endergonic reaction of ADP phosphorylation. Example: perfluidone.

c. energy transference inhibitors: inhibition of energy transference inhibitors acts directly in the phosphorylation, as well as inhibitors of the electron transport, which inhibit both the electron flow and the formation of ATP in coupled systems. Example: 1,2,3-thiadiazol-phenylurea, nitrofen.

d. inhibitory uncouplers: the term "inhibitory uncouplers" was used by Moreland [33] to indicate that the herbicides interfere in reactions affected by electron transport inhibitors and by uncouplers; These "inhibitory uncouplers" inhibit the basal transport, uncoupled and coupled of electrons. The herbicides classified in this group affect both the electron transport and the gradient of protons. Examples: acylanilides, dinitrophenols, imidazole, bromofenoxim.

e. electron acceptors: the compounds classified in this group are able to compete with some component of electron transport and consequently suffer reduction. Examples: diquat, paraquat.

f. inhibitors of the carotenoid synthesis: this class of herbicides acts to inhibit the synthesis of carotenoids, resulting in accumulation of precursors of carotenoid devoid colour (phytoene and phytofluene). The inhibition of carotenoid synthesis leads to the degradation of chlorophyll in the presence of light; degradation of 70s ribosomes; inhibition of the synthesis of proteins and loss of plastids. Examples: amitrole, dichlormate, SAN6706.

2. *mitochondrial electron transport and phosphorylation:* herbicides that interfere in the mitochondrial system are classified as:

a. electron transport inhibitors: defined as substances that have the ability to interrupt the electron flow in some point of the respiratory chain, acting in one of the complexes. Examples: diphenylether herbicides.

b. uncouplers: in appropriate concentrations, the classic uncouplers, that are weak lipophilic acids or bases, prevent the phosphorylation of ADP without interfering in the electron transport. Generally, any compound that promotes the dissipation of the energy generated by the electron transport, except for the production of ATP, can be considered as uncoupler. Example: isopropyl ester glyphosate.

c. energy transfer inhibitors: compounds of this group inhibit the phosphorylating electron transport, when the apparatus of energy conservation of the mitochondria is intact and the inhibition is circumvented by uncouplers. They combine with an intermediary in the coupling energy chain and, thus, block the phosphorylation sequence that leads to the ATP formation. No herbicide seems to act as an energy transfer inhibitor.

d. inhibitory uncouplers: most of the herbicides that interfere in the oxidative phosphorylation present a great variety of responses and are classified as uncoupling inhibitors. At low molar concentrations, herbicides fulfil almost all, if not all, of the requirements established for uncouplers, but at high concentrations they act as electron transport inhibitors. Herbicides that present this behaviour are the same classified as uncoupler inhibitors of the photoinduced reactions in the chloroplast. Example: perfluidone.

3. *interactions with membrane:* herbicides can affect the structure and function of membranes directly or indirectly. When the herbicides disaggregates a membrane, they can influence directly the transport processes by interacting with the protein compounds, such as, ATPases and by altering the permeability by physicochemical interactions, or indirectly by modulating the supply of ATP needed to energize the membrane. Interactions with the membrane can cause:

a. compositional alterations: can modify or alter the composition of lipids in the membrane and can also act in the metabolism and synthesis of lipids. Examples: dinoben, chloram-bem, perfluidone.

b. effects in the permeability and integrity. Examples: paraquat, diquat, oryfluorfen, oryzalin.

4. *cell division:* herbicides may suppress cell division by interfering in the synthesis or active transport of precursors into the nucleus, which are necessary for the synthesis of DNA during interphase; modify the physical or chemical properties of the DNA or of their complexes; interfere in the formation and function of the spindle; and/or inhibits the formation of the cell wall. Several of the processes mentioned previously need energy and, therefore, interferences in the amount of energy caused by an herbicide could modulate the mitotic activity. The effects of the inhibitors of the cell division are dependent on the concentration and vary according to the species and the type of tissue. There is a relationship between cell division and cellular energy. In higher plants, cell division is prevented or suppressed in conditions in which the glycolysis or the oxidative phosphorylation is inhibited. Another form of the herbicide to alter cell division would be interacting with the microtubules, since these cellular structures are responsible for the orientation and movement of chromosomes during cell division. Examples of herbicides that interfere in cell division: N-phenylcarbamates, ioxynil, trifluralin.

5. *Synthesis ofDNA, RNA and protein:* there are correlations between inhibition of RNA and protein synthesis and low concentration of ATP in tissues and these correlations suggest that interferences in the energy production, necessary to perform biosynthetic reactions, could be the mechanism by which the herbicides could express their effects. Moreover, they can inhibit the synthesis of DNA or RNA by altering the chromatin integrity and, in these cases, the synthesis of proteins is also affected. Examples: glyphosate, trifluralin.

The herbicides can still be classified according to the chemical affinity. Table 1 shows the chemical classes and examples of each class, according to Rao [34].

Class of the herbicide	Examples of herbicides
Acetamides	Acetochlor, alachlor, butachlor, dimethenamid, metolachlor, napropamide, pronamide, propachlor, propanil
Aliphatics	Chlorinated aliphatic acid (TCA), acrolein, dalapon
Arsenicals	Disodium methanearsonate (DSMA), monosodium methanearsonate (MSMA), cacodylic acid
Benzamides	Isoxaben
Benzoics	Dicamba
Benzothiadiazoles	Bentazon
Bipyridiliums	Diquat, paraquat
Carbamates	Asulam, desmedipham, phenmedipham

http://dx.doi.org/10.5772/55851

Class of the herbicide	Examples of herbicides
Cineoles	Cinmethylin
Cyclohexanediones (cyclohexenones)	Clethodin, cycloxidim, sethoxydim, tralkoxydim
Dinitroaniniles	Benefin, ethalfluralin, fluchloralin, pendimethalin, prodiamine, trifluralin
Diphenylethers	Acifluorfen, bifenox, fluoroglycofen, fomesafen, lactofen, oxyfluorfen
Imidazolidinones	Buthidazole
Imidazolinones	Imazapyr, imazaquin, imazethapyr, imazamethabenz
Imines	CGA-248757
Isoxazolidinones	Clomazone
Nitriles	Bromoxynil, dichlobenil, ioxynil
Oxadiazoles	Oxadiazon
Oxadiazolidines	Methazole
Phenols	Dinoseb
Phenoxyalkanoic acids Phenoxyacetics Phenoxybutyrics Arylophenoxy propionics	2,4-D, MCPA, 2,4,5-T 2,4-DB Dichlorprop, diclofop, fenoxaprop, fluazifop-P, quizalofop-P
N-phenylphthalimides	Flumiclorac
Phenylpyridazines	Pyridate
Phenyl Triazinones (Aryl Triazinones)	Sulfentrazone
Phthalamates	Naptalam
Pyrazoliums	Difenzoquat
Pyridazinones	Norflurazon, pyrazon
Pyridinecarboxylic Acids	Clopyralid, picloram, triclopyr
Pyridines	Dithiopyr, thiazopyr
Pyridinones	Fluridone
Pyrimidinythio-benzoates (Benzoates)	Pyrithiobac
Quinolinecaryoxylic acids	Quinclorac
Sulfonylureas	Bensulfuron, chlorimuron, chlorsulfuron, halosulfuron, metsulfuron, nicosulfuron, primisulfuron, prosulfuron, sulfometuron, thifensulfuron, triasulfuron, tribenuron
Tetrahydropyrimidinones	Yet to be commercialized
Thiocarbamates	Butylate, diallate, EPTC, molinate, pebulate, thiobencarb, triallate
Triazines	Ametryn, atrazine, cyanazine, hexazinone, prometryn, simazine
Triazinones	Metribuzin
Triazoles	Amitrole
Triazolopyrimidine Sulfonanilides	Flumetsulam
Uracils	Bromacil, terbacil, UCC-C4243
Ureas	Diuron, fluometuron, linuron, tebuthiuron,
Unclassified herbicides	Bensulide, ethofumesate, fosamine, glufosinate, glyphosate, tridiphane

Table 1. Classification of the herbicides according to the chemical affinity.

1.3. Aquatic and soil contamination due to the presence of herbicides

When a herbicide is used to control weeds, sometimes a majority of the compound ends up in the environment, whether it is in the soil, water, atmosphere or in the products harvested [17]. Due to the widespread use of these chemicals over the years, there has been an accumulation of these residues in the environment, which is causing alarming contaminations in the ecosystems [35] and negative damages to the biota. To Bolognesi and Merlo [3], the widespread use of herbicides has drawn the attention of researchers concerned with the risks that they can promote on the environment and human health, since they are chemicals considered contaminants commonly present in hydric resources and soils. According to the same authors, herbicides represent a high toxicity to target species but it can be also toxic, at different levels, to non-target species, such as human beings. Herbicides can cause deleterious effects on organisms and human health, both by their direct and indirect action [2]. Among the biological effects of these chemicals, it can be cited genetic damages, diverse physiological alterations and even death of the organisms exposed. Some herbicides, when at low concentrations, cannot cause immediate detectable effects in the organisms, but, in long term can reduce their lifespan longevity [4]. Herbicides can affect the organisms in different ways. As with other pesticides, the accumulation rate of these chemicals on biota depends on the type of the associated food chain, besides the physicochemical characteristics (chemical stability, solubility, photodecomposition, sorption in the soil) of the herbicide [5-6]. Thus, despite the existence of several toxicological studies carried out with herbicides, in different organisms, to quantify the impacts of these pollutants and know their mechanisms of action [7, 8, 2], there is a great need to expand even more the knowledge about the effects of different herbicides in aquatic and terrestrial ecosystems. Data obtained from *in situ, ex situ, in vivo* and *in vitro* tests, derived from experiments of simulation, occupational exposure or environmental contaminations, need to enhance so that it is possible to obtain even more consistent information about the action of these compounds.

According to Jurado et al. [32], when herbicides are applied in agricultural areas they can have different destinations, since being degraded by microorganisms or by non- biological means or even be transported by water, to areas distant from the application site. Thus, according to the same authors, the organisms can be then exposed to a great number of these xenobiotics as well as their metabolites.

The fate of the compound in the soil depends on the characteristics of the compound and the soil. The hydrogenionic properties of a compound in the soil determines its sorption characteristics, such as, acid herbicides in soils with normal pH are negatively charged and consequently are movable in most of the soils [17]. Some groups of pesticides are neutral in soils with normal pH but due to electronic dislocations in the molecules, they can bind to soil colloids by several forms [36].

According to Kudsk and Streiberg [17], during the last two decades, several studies have been completed to predict the behaviour of pesticides in the soil. Despite the numerous efforts to assess the effects of herbicides in the soil, there are conflicting data in the literature on the subject, where some studies show that the residues of pesticides can be sources of carbon and energy to microorganisms, and then are degraded and assimilated by them, while other reports affirm that pesticides produce deleterious effects to the organisms and biochemical and enzymatic processes in the soil [37]. According to Hussain et al. [37], in general, the application of pesticides, and here it is also included herbicides, made long term, can cause a disturbance in the biochemical balance of the soil, which can reduce its fertility and productivity.

Once in the soil, herbicides can suffer alteration in their structure and composition, due to the action of physical, chemical and biological processes. This action on the herbicides is the one that will determine their activity and persistence in the soil. Some molecules, when incorporated into the soil, are reduced by volatilization and photo-decomposition. Once in the soil, herbicides can suffer the action of microorganisms, which, added to the high humidity and high temperature, can have their decomposition favoured [38]. If they are not absorbed by plants, they can become strongly adsorbed on the organic matter present in the colloidal fraction of the soil, be carried by rainwater and/or irrigation and even be leachate, thus reaching surface or groundwater [39].

The prediction of the availability of herbicides to plants has two purposes: 1. ensure that the herbicide reaches the roots in concentrations high enough to control weeds, without compromising the agricultural productivity; 2. predict if the compound is mobile in the soil to estimate how much of the herbicide can be leachate from the roots zone to groundwater [17].

The contamination of aquatic environments by herbicides has been characterized as a major world concern. This aquatic contamination is due to the use of these products in the control of aquatic plants, leachate and runoff of agricultural areas [40]. According to He et al. [20], it is a growing public concern about the amount of herbicides that have been introduced into the environment by leachate and runoff, not to mention that the contaminations of the aquatic environments generally occur by a mixture of these compounds and not by isolated substances.

Guzzella et al. [1] did a survey on the presence of herbicides in groundwater in a highly cultivated region of northern Italy. The researchers monitored for two years the presence of 5 active ingredients and 17 metabolites resulting from these compounds. The authors verified that atrazine, although banned in Italy since 1986, was the major contaminant of the groundwater of the sites studied, they also observed that the concentration of at least one of the compounds studied exceeded the maximum allowed concentration in 59% of the samples likely due in both cases to off-label herbicide use. This scenario could be, in long term, a serious problem for the quality of this water, which is used as drinking water.

Toccalino et al. [41] carried out a study to verify the potential of chemical mixtures existing in samples of groundwater used for public supply. In these samples, the most common organic contaminants were herbicides, disinfection by-products and solvents. The authors concluded that the combined concentrations of the contaminants can be a potential concern for more than half of the samples studied and that, even though the water destined to public supply pass through treatments to reduce contaminations and meet the legislations, it can still contain mixtures at worrying concentrations.

Saka [42] evaluated the toxicity of three herbicides (simetryn, mefenacet and thiobencarb) commonly used in rice planting in Japan, on the test organism *Silurana tropicalis* (tadpoles). The authors observed that the three herbicides, particularly thiobencarb, are toxic for tadpoles

(LD50 test), even for concentrations found in waters where the rice is cultivated. In a similar study carried out by Liu et al. [43], it was observed that the effect of the herbicide butachlor (most used herbicide in rice planting in Taiwan and Southeast Asia) on the organism *Fejervarya limnocharis* (alpine cricket frog) exposed to concentrations used in the field. In this study no effect on the growth of tadpoles of *F. limnocharis* was observed, but there was a negative action on survival, development and time of metamorphosis. The authors suggested that the herbicide butachlor can cause serious impacts on anurans that reproduce in rice fields, but this impact varies from species to species.

In a study conducted by Ventura et al. [8], it was observed that the herbicide atrazine has a genotoxic and mutagenic effect on the species *Oreochromis niloticus* (Nile tilapia). In this study, the authors observed that the herbicide can interfere in the genetic material of the organisms exposed, even at doses considered residual, which led the authors to suggest that residual doses of atrazine, resulting from leaching of soils of crops near water bodies, can interfere in a negative form in the stability of aquatic ecosystems.

Bouilly et al. [44] studied the impact of the herbicide diuron on *Crassostrea gigas* (Pacific oyster) and observed that the herbicide can cause irreversible damages to the genetic material of the organism studied. Moreover, the authors affirm that, due to the persistence of diuron in environments adjacent to its application site and that it is preferably used in spring, the pollution caused by its use causes negative impact in the aquatic organisms during the breeding season.

In general, when herbicides contaminate the aquatic ecosystem, they can cause deleterious effects on the organisms of this system. Thus, organisms that live in regions impacted by these substances, whose breeding period coincides with the application period of the herbicides, can suffer serious risks of development and survival of their offspring.

Hladik et al. [45] evaluated the presence of two herbicides (chloroacetamide and triazine), as well as their by-products, in drinking water samples of the Midwest region of the United States. The authors detected the presence of neutral chloroacetamide degradates in median concentrations (1 to 50 ng/L) of the water samples. Furthermore, they found that neither the original chloroacetamide herbicides nor their degradation products were efficiently removed by conventional water treatment processes (coagulation/flocculation, filtration, chlorination). According to Bannink [46], about 40% of the drinking water from Netherlands is derived from surface water. The Dutch water companies are facing problems with the water quality due to contamination by herbicides used to eliminate ruderal plants. These data serve as alerts for the presence of herbicides and their degradation products in drinking water, pointing out the need for the development of new treatment systems that could be more efficient to eliminate this class of contaminants.

According to Ying and Williams [40], organic herbicides, when in aquatic ecosystems, can be distributed in several compartments depending on their solubility in water. These compartments include water, aquatic organisms, suspended sediment and bottom sediment. The more hydrophilic the organic pesticide, the more it is transported to the aqueous phase, and the more hydrophobic a pesticide is, the more it will be associated to the organic carbon of the suspended and bottom sediment [47]. The sorption of the herbicides in sediments in suspension can reduce the degradation rate of the herbicides in water, and the movement of the sediment in suspension can transport the pesticides from one place to another, entering into the tissue of organisms or settling on the bottom [40].

A study conducted by Jacomini et al. [48] evaluated the contamination of three matrices (water, sediment and bivalve molluscs) collected in rivers influenced by crops of sugar cane in São Paulo State-Brazil. In this study, the authors observed that the highest concentrations of residues of the herbicide ametrin were present in the sediment, showing the persistence of this compound in the sediments of rivers and its potential to mobilize between the compartments of the aquatic system, such as water and biota.

When the herbicides are dispersed in the water or sediments in suspension of the rivers, they can end up in other ecosystems such as estuaries. Duke et al. [49], when studying the effect of herbicides on mangroves of the Mackay region, found out that diuron, and even other herbicides, are potentially responsible for the mangrove dieback. According to the authors, the consequences for this death would be the impoverishment of the quality of the coastal water with an increase of the turbidity, nutrients and sediment deposition, problems in the fixation of seedlings and consequent erosion of the estuaries.

In a review conducted by Jones [50], the author highlights the contamination of marine environments by herbicides (such as diuron), discussing that the contamination of these environments can occur by transport of these substances of agricultural or non cultivated areas (roadsides, sports fields, train tracks), runoff by storms and tailwater irrigation release), pulverizations and accidental spills. These contaminations mean that the photochemical efficiency of intracellular symbiotic algae of the coral, in long term, may be compromised, leading to a loss in the symbiotic relationship of the coral with the algae and a consequent bleaching of corals. Still considering the marine ecosystem, Lewis et al. [51] verified that the runoff of pesticides from agricultural areas influence the health of the Great Barrier Reef in Australia and can disturb this sensitive ecosystem.

Considering the prior literature, it is likely possible that the effects of herbicides do not occur only at the places that they are applied but also in places distant from their application. Moreover, herbicides can induce alterations in non-target organisms, altering the survival and the equilibrium of the ecosystems, whether they are aquatic or terrestrial. Thus, much care must be taken when introducing these substances into the environment and more studies should be conducted in order to thoroughly understand the environmental consequences that herbicides can cause.

2. The effects of herbicides using different bioassays and test-organisms

Many studies have evaluated the impact of different chemical classes of herbicides using different doses, organisms and bioassays, focusing on toxic, cytotoxic, genotoxic, mutagenic, embryotoxic, teratogenic, carcinogenic and estrogenic effects.

With respect to the toxicity, some herbicides pose major concerns when applied in regions close to water resources due to their highly toxic potential to many aquatic organisms [52].

Biological tests of toxicity and mutagenicity are, according to Moraes [53], indispensable for the evaluation of the reactions of living organisms to environmental pollution and also for the identification of the potential synergistic effects of several pollutants. The impact that toxic materials can promote in the integrity and function of DNA of several organisms has been investigated [54]. Several biomarkers have been used as tools for the detection of the toxic, genotoxic and mutagenic effects of pollution. Among them we can cite the presence of DNA adducts, chromosome aberrations, breaks in the DNA strands, micronuclei formation and other nuclear abnormalities, besides induction of cell death [55].

Most of the tests used to detect the mutagenic potential of chemical substances are based on the investigation of possible inductions of chromosome damages such as structural alterations, formation of micronuclei, sister chromatid exchanges, assessment of mutant genes or damages in the DNA, using different test organisms, such as bacteria, plants and animals, both *in vitro* and *in vivo* [56].

According to Veiga [57], it is possible to estimate the genotoxic, mutagenic, carcinogenic and teratogenic effects of agrochemicals by relatively simple methods. Several studies have been carried out by several researchers concerned with the harmful effects of pesticides in an attempt to verify their possible physiological [58, 59], mutagenic [7, 8, 60, 61, 62] and carcinogenic effects [63].

The interaction between different methods of evaluating the toxic, genotoxic and mutagenic potential provides a more global and comprehensive view of the effect of a chemical agent. For the monitoring of organisms exposed to chemical agents, the chromosome aberration test, micronucleus test and comet assay have been widely used [64]. A few studies also have shown the toxic effects of chemicals, by cell death processes, both necrotic and apoptotic [65].

According to Kristen [66], the dramatic expansion in the production of xenobiotic compounds by anthropogenic activities has compromised the environment by the introduction of millions of chemicals with toxic potential to biological systems.

Cytogenetic tests are adequate to identify the harmful effects of substances, in their several concentrations and different periods of exposure. These tests, generally performed with test organisms, are commonly applied in biomonitoring to the extent of pollution and in the evaluation of the combined effects of toxic and mutagenic substances on the organisms in the natural environment [53]. Micronuclei assays are efficient to assess the mutagenic activity of herbicides both in laboratorial and field assays [67]. The comet assay can be used to evaluate damages in proliferating cells or not, in *in vitro* or *in vivo* tests and can be applied with the purpose of genotoxicological analyses [68]. According to these same authors, these tests are considered one of the best tools to biomonitor several chemical compounds, including herbicides. According to Ribas et al. [69], the simplicity, reproducibility and rapidity of the comet test, associated to the ability of this assay in evaluating damages in the DNA, makes this technique highly applicable to environmental genotoxicology.

The toxic, cytotoxic, genotoxic, mutagenic, embryotoxic, teratogenic carcinogenic and estrogenic effects caused by herbicides on various organisms could be exemplified by studies as described below.

2.1. Atrazine

Atrazine is a triazinic herbicide, classified as moderately toxic of pre- and post-emergence, used for the control of weeds in crops of asparagus, corn, sorghum, sugarcane and pineapple [70]. According to Eldridge et al. [71], triazinic herbicides are among the most used pesticides in agriculture due to their ability to inhibit the photosynthesis of weeds in crops [16].

Triazine herbicides are extensively used in the United States to control grass, sedge and broadleaf weedsduring the cultivation of maize, wheat, sorghum, sugarcane and conifers [72]. In Brazil, these herbicides are widely used on crops of sugarcane and maize. Due to the widespread use of triazine herbicides in the agriculture and, therefore, its high exposure potential for humans, the United States Environmental Protection Agency (USEPA) has conducted a special review on the published and non published data of several triazine herbicides [73]. According to Nwani et al. [22], the herbicide atrazine is widely used in crops worldwide. The dangers, both toxic and genotoxic of this herbicide have been revised; however, there is an urgent need for more detailed studies on the mode of action of this compound. Atrazine has been tested in several systems, but there are shortcomings in relation to certain tests performed and some evidences of the genotoxic effects, *in vivo*, still need to be confirmed [74].

Several studies using the test system *Aspergillus* have shown that atrazine is not mutagenic to these organisms [75, 76, 77], although it is considered mutagenic for other test systems such as *Drosophila melanogaster* [78, 79]. According to Ribas et al. [74], atrazine was responsible for a significant frequency of aneuploidies in *Neurospora crassa*, given by the chromosomal non-disjunction in *Aspergillus nidulans*, and by the induction of loss of sexual chromosomes in *Drosophila melanogaster*.

Sorghum plants treated with atrazine presented an increase in the number of their chromosomes, multinucleated cells, aneuploidy and polyploidy, and abnormalities in the mother cells of the pollen grain, which suggests that this herbicide interferes in the stability and also in the meiosis [80].

Popa et al. [70] observed that atrazine, when applied in high concentrations in maize seedlings, can induce chromosome breaks, visualized by the presence of single and paired chromosome fragments; a high frequency of chromatids and chromosome bridges; lagging chromosomes and presence of heteropolyploid or polyploid cells. Grant and Owens [81] showed that atrazine induced chromosome breaks (in mitosis and meiosis) in the species *Pisum sativum* and *Allium cepa*.

Hayes et al. [82] investigated the effect of the herbicide atrazine on wild leopard frogs (*Rana pipens*), in different regions of the United States. The authors observed that a great percentage of males exposed to the herbicide presented abnormalities in the gonads, such as development retardation and hermaphroditism. This effect can, in long term, lead to a decline in the amphibian population of the sites contaminated with this herbicide.

According to Gammon et al. [83], some publications have reported a possible feminization of frogs, both in laboratorial assays and field studies. This effect is mainly due to the action of the enzyme aromatase; however, published research not shown the measures of this enzyme. Thus, there are doubts about the feminization theory, except for the studies that presented a great number of frogs with morphological alterations related to very high levels of atrazine.

Nwani et al. [22] evaluated the genotoxic and mutagenic effects of the herbicide Rasayanzine, whose active ingredient is atrazine, using the comet assay and micronucleus test, in erythrocytes and gill cells of the fish *Channa punctatus.* By the data analysis of the two cell types, significant effects for all the concentrations (4.24, 5.30 and 8.48 mg/L) and exposure periods tested (1, 3, 5, 7, 14, 21, 28 and 35 days) were observed. The highest damages were observed for the highest concentrations and exposure times, showing the genotoxic and mutagenic dose-response potential of atrazine for the aquatic organism. Furthermore, it was found that gills were more sensitive to the action of the herbicide, when compared to erythrocytes. From the results obtained, the authors suggested a careful and judicious use of the herbicide atrazine in order to protect the aquatic ecosystems and human population.

A study carried out by Çavas [84] compared the genotoxic effects of the active ingredient atrazine and its commercial formulation Gesaprim, in the concentrations of 5, 10 and 15 μg/L, by the comet assay and micronucleus test, in erythrocytes of the fish gibel carp (*Carassius auratus*). The results showed that there was a significant increase in the frequencies of the micronuclei and DNA strand breaks in the erythrocytes treated with all the concentrations of the commercial formulation of atrazine, showing the genotoxic and mutagenic potential of Gesaprim for this species of fish. While the commercial formulation presented a high genotoxic potential, the assays showed that the active ingredient atrazine is not genotoxic, suggesting that the adjuvants present in Gesaprim must be the responsible for the genotoxic effects observed in this species of fish. Despite the comparative analysis of the genotoxicity between the active ingredient and the commercial product has showed to be a very effective tool for the discovery of genotoxic environmental risks, it is not easy to determine the exact identity of the products used as adjuvants and of the agents of surface action of pesticides due to the existence of the patent protection system.

Atrazine has also been tested to evaluate the ability to induce cytogenetic damages in rodents. Meisner et al. [85] submitted rats to 20 ppm of atrazine (by water ingestion) and did not observe, after exposure to the herbicide, an increase in the number of chromosome aberrations. In a similar study, Roloff et al. [86] reported that there was no significant increase of chromosome aberrations in cells of rat bone marrow, when they were fed with 20 ppm of atrazine.

Wu et al. [87] assessed the embryotoxic and teratogenic effects of atrazine, at the doses of 25, 100 and 200 mg/Kg/day, in Sprague-Dawley rats. Prenatal exposure to the highest dose of the herbicide tested caused hypospadias in 10.23% of male newborn rats, and the lowest dose induced diverse embryotoxic damages in some individuals. According to Modic et al. [88], high doses of atrazine (50 or 200 mg/kg/day), administered daily in male Wistar rats at 60 days of age, promoted alterations in the levels of several hormones in the serum of these individuals, observed by slight increases in the levels of androstenedione testosterone, estradiol, estrone, progesterone and corticosterone, quantified by radioimmunoassay.

To obtain more concise data on the genotoxicity of triazine herbicides, Tennant et al. [89] used the comet assay methodology, which showed to be highly sensitive for the detection of low rates of damages in the DNA. According to these authors, the comet assay showed that atrazine induced a small increase in the damages in the DNA in leukocytes of rats. Moreover, by the comet assay, Clements et al. [90] reported that atrazine induced a significant increase in the frequencies of damages in the DNA of erythrocytes of bullfrog tadpoles, noting the genotoxic potential of this herbicide for this species of amphibian, from the concentration of 4.8 mg/L.

Studies about the cytotoxicity, genotoxicity and mutagenicity of the atrazine herbicide (oral gavage - dose 400 mg/kg/day), carried out by Campos-Pereira et al. [91], have shown the induction of lipid peroxidation and liver damage, death of hepatocytes, and micronucleus formation in exposed Wistar rats. Tests performed by Ventura et al. [8] showed that the same triazine pesticide was able to induce significant DNA fragmentation when using the comet assay, and nuclear alterations and micronuclei using the micronucleus test in *Oreochromis niloticus* (Nile tilapia) erythrocytes exposed to different concentrations of atrazine (6.25, 12.5, 25 μg/L), thus corroborating the studies performed by Campos-Pereira et al. [91].

Ruiz and Marzin [92] assessed the genotoxic and mutagenic effects of the herbicide atrazine by two *in vitro* assays (*Salmonella* assay and SOS Chromotest), one to detect bacterial mutagenicity and the other to verify primary damages in the DNA. The assays were carried out both in the absence and in the presence of S9 fractions from rat liver homogenate (Sprague-Dawley). The authors found that the herbicide atrazine did not present genotoxic potential neither to the *in vitro* test with Salmonella/microsome nor by the SOS Chromotest, both in the absence and in the presence of the S9 fractions, when the strains were exposed to atrazine.

In vitro studies, performed with human lymphocytes, treated with 0.10 ppm of atrazine, detected a slight increase in the chromosome aberrations rates [85]. However, for concentrations below 0.001 ppm of this herbicide, chromosome aberrations were not detected [86] Lioi et al. [93] observed a small increase in the number of sister chromatid exchange but a great increase of chromosome aberrations in human lymphocytes exposed to atrazine. Meisner et al. [94] observed a significant increase in the frequency of chromosome breaks in human blood cells exposed to 1 ppm of the herbicide atrazine.

The genotoxicity of herbicides, such as atrazine, has also been evaluated by the comet assay by the use of human blood lymphocytes. According to Ribas et al. [69], blood cells treated with the herbicide atrazine, at concentrations of 50-200 μg/l, showed an extensive migration of DNA, mainly at concentrations of 100 and 200 μg/l.

In mammalian test systems, submitted to the action of the herbicide atrazine, most of the results seem to be negative, except for the results of Loprieno and Adler [95], who obtained a significant increase in the frequency of chromosome aberrations in bone marrow cells of rats, and the data obtained by Meisner et al. [94], who described an induction of chromosome aberrations in cultured human lymphocytes. While the results from bacteria and mammal test systems are almost all negative, atrazine exhibits clear mutagenic effects in different plant test systems, by inducing chromosome aberrations in *Hordeum vulgare* and *Vicia faba* [96, 97], in *Zea mays* [98], in *Sorghum vulgare* [99] and in *Allium cepa* [62] ; sister chromatid exchanges in maize [100] ; and point mutation in maize [98].

Studies performed by Zeljezic et al. [101] had already reported that atrazine does not present genotoxicity or capacity to induce apoptosis or necrosis in human lymphocytes, while the treatment of these cells with the commercial formulation, Gesaprim, significantly increased the rates of damages in DNA, observed by the comet assay. Srivastava and Mishra [102] observed results that are in agreement with the findings of Zeljezic et al. [101] and Çavas [84], in which the exposure to different concentrations of Gesaprim inhibited the mitotic index and increased the frequencies of micronuclei and chromosome aberrations in somatic cells of *Allium cepa* and *Vicia faba*.

2.2. Atrazine and butachlor

Toxic effects of atrazine, alone or associated with the herbicide butachlor, for the freshwater species such as the green alga *Scenedesmus obliquus* and the cladoceran *Daphnia carinata*, were evaluated, showing values of 96 h-EC50 for *S. obliquus* (atrazine= 0.0147 mg/L and butachlor= 2.31 mg/L, and of 48h-LC50 for *D. carinata* (atrazine= 60.6 mg/L and butachlor= 3.40 mg/L) [20]. These results suggest that atrazine has a highly toxic potential for *S. obliquus* and slightly toxic for *D. carinata*, while butachlor exhibits a moderate toxic potential for both organisms. Now, the analysis of the mixture atrazine-butachlor allowed the authors to verify that the toxic effects were significantly antagonistic for *S. obliquus*, and that there was no significant synergism for *D. carinata* [20].

2.3. Atrazine, simazine ande cyanazine

Simazine and cyanazine, as well as atrazine, are widely used as triazine herbicides of pre- and post-emergence weed control, whose residues have been carried to the source of drinking water of several agricultural communities. These compounds also present a potential risk to humans, mainly due to their presence in food [103]. Studies on the effect of atrazine, simazine and cyanazine performed by Kligerman et al. [104], found that there was not a significant increase in the sister chromatid exchanges and chromosome aberrations in cultured human lymphocytes exposed to these herbicides, up to the solubility limit in aqueous solution using 0.5% of dimethyl sulfoxide. However, Adler [105] observed that doses of 1500 and 2000 mg/Kg of atrazine, administered by oral gavage in rats, induced dominant lethal mutations and chromatin breaks in the bone marrow of these organisms.

Kligerman et al. [103] observed that the association of the herbicides atrazine, simazine and cyanazine did not induce micronuclei in polychromatic erythrocytes of bone marrow of female rats (C57B1/6) exposed by intraperitoneal injection, even when very high doses of these herbicides were administered (125, 250 and 500 mg/Kg of atrazine; 500, 1000 and 2000 mg/Kg of simazine; 100, 200 and 400 mg/kg of cyanazine), showing an absence of genotoxic potential of these compounds for the organism tested.

On the other hand, Hrelia et al. [106] showed that males and females of Sprague-Dawley rats exposed by oral gavage to doses of 56, 112 and 224 mg/kg of cyanazine, did not present significant increases in chromosome aberrations.

Taets et al. [107] evaluated the clastogenic potential of environmental concentrations of the triazine herbicides simazine (0.001 to 0.004 μg/mL), cyanazine (0.003 to 0.012 μg/mL) and atrazine (0.003 to 0.018 μg/mL), in Chinese Hamster Ovary (CHO) cells, using flow cytometry assay. The authors proved the clastogenic action for the herbicides atrazine and cyanazine, proven by the high indices of damages in the cells exposed to atrazine and by the significant frequencies of damages observed in the cells exposed to cyanazine.

2.4. Terbutryn

The herbicide terbutryn is an s-triazine herbicide used pre- and post-emergence and widely used worldwide as an agent to control grass, sedge, and broadleaf weeds in vegetables, cereals and fruit trees. It is an herbicide persistent in the environment, which tends to dislocate by the flow of water and leachate [108].

An *in vitro* study performed by Moretti et al. [108] investigated the genotoxicity of the herbicide terbutryn, by analyzing the relationship between the cytogenetic damage, evaluated by the assays of SCE (sister chromatid exchanges) and MN (micronucleus), and the primary damage in the DNA, assessed by the comet assay, in leukocytes newly-isolated from peripheral human blood. The results showed that terbutryn did not produce significant increases of SCE or MN, both in the absence and in the presence of the metabolic activation system from rat liver (S9 fraction), although terbutryn has induced primary damages in the DNA in a more pronounced form in the absence of S9. The apparent lack of sensitivity of the assays of SCE and MN test for the genotoxicity of terbutryn, in comparison to the comet assay, can be attributed to the generation of specific types of damages, since the SCE and MN are determined in proliferative cells and are sensitive indicators of lesions that survive for, at least, one mitotic cycle, while the comet assay identifies repairable lesions in the DNA of on resting (G0) cells. According to these results, the authors suggest that terbutryn must be considered a genotoxic compound.

2.5. 2,4-D (2,4-dichlorophenoxyacetic acid)

The 2,4-D (2,4-dichlorophenoxyacetic acid) is an herbicide from the group of the polychlorinated aromatic hydrocarbons that has been widely used throughout the world [109] since 1944, to control broadleaf weeds and woody plants [110]. Its action mimics the auxin of plants [111]. According to Martínez-Tabche et al. [112], this herbicide mimics the action of the hormone indole acetic acid, when used in small quantities but it is highly cytotoxic in high concentrations.

According to Ateeq et al. [113], the increase in the frequency of micronuclei and altered cells was significant, when erythrocytes of catfish (*Clarias batrachus*) were analyzed, after exposure to the herbicides 2,4-D and butachlor. There was a positive dose-response relationship in all exposures to the two herbicides and in all exposure periods tested.

Studies carried out by Suwalsky et al. [114] in nerve cells of *Caudiverbera caudiverbera* demonstrated the toxicity of the herbicide 2,4-D. The authors observed a reduction in the dose-dependent response to nerve stimulation in the simpact junction of the frog when they were exposed to this herbicide. This reduction is probably due to a mechanism of lipid perturbation and interference in the properties of the plasma membrane, such as protein conformation and/or interaction with protein receptors, which leads to an inhibition of the glandular chloride channel from the mucosal skin of this test organism.

According to Gómez et al. [115], the main and most common entrance route of 2,4-D in fish is through gills. This herbicide can cause several adverse symptoms to these organisms, such as bleeding, increased damage to the kidneys and renal functions, as well as hepatic degeneration.

Martínez-Tabche et al. [112] evaluated the toxicity of different concentrations of the herbicides 2,4-D and paraquat (0, 5, 75 and 150 mg/L), using several assays (acute lethality test, lipid peroxidation assay by quantification of MDA – Malondialdehyde – and comet assay) in rainbow trout (*Oncorhynchus mykiss*). For the acute lethality tests, it was observed a more evident toxic action for the organisms exposed to the treatment of 24 h with the herbicide paraquat, which presented high indices of mortality, analyzed by the values of LC_{50} (LC_{50} of paraquat = 0.084 mg/L; LC_{50} of 2,4-D = 362.38 mg/L). The authors also showed that individuals exposed to the two higher concentrations of both herbicides had apnoea and white spots in their scales. All concentrations of 2,4-D and paraquat induced a significant increase in the DNA damages and the amount of MDA in the gills exposed.

González et al. [116] proved the genotoxicity of 2,4-D due to a significant increase of SCE in CHO cells treated with the concentrations of 2 to 4 ug/mL of this herbicide. Madrigal-Bujaidar et al. [117] also showed the genotoxic potential of 2,4-D, due to a clastogenc effect of this herbicide at the doses of 100 and 200 mg/Kg, detected by a significant increase of SCE in bone marrow cells and germ cells of rats. Soloneski et al. [118] studied the genotoxic effects of different concentrations (0, 10, 25, 50 and 100 mg/mL) of the herbicide 2,4-D (2,4-dichlorophenoxyacetic) and its commercial derivative 2,4-D DMA (Dimethylamine 2,4-D salt), by the SCE assay and analyses of cell cycle progression and mitotic index human lymphocytes maintained in culture, in the presence (human whole blood - WBC) and absence (plasma leukocyte cultures - PLC) of erythrocytes. These compounds did not induce significant frequencies of SCE and only the concentration of 100 mg/mL of 2,4-D caused alterations in the progression of the cell cycle in PLC, while the different concentrations of 2,4-D and 2,4-D DMA induced a significant increase in the frequency of SCE and a significant delay in the cell proliferation rates in WBC. Moreover, both 2,4-D and 2,4-D DMA presented a dose-response inhibition of the mitotic activity in PLC and WBC. Based on these results, the authors concluded that the herbicide and its commercial derivative presented genotoxic potential, which was higher in the presence of human erythrocytes.

Morgan et al. [119] showed, by embryotoxicity and teratogenicity assays carried out with *Xenopus* (FETAX - frog embryo teratogenic assay – *Xenopus*), that high concentrations of 2,4-D, induce potentially more embryotoxic effects than teratogenic in frog embryos, demonstrated by the values of EC50 and LC50 of 245 mg/L and 254 mg/L, respectively, and by the Teratogenic Index of 1.04. Moreover, the same authors compared the teratogenic action of the herbicide atrazine in relation to 2,4-D, showing that atrazine is potentially more teratogenic than 2,4-D, for frog embryos.

The estrogenic potential of 4 herbicides (triclopyr; 2,4-D; diquat dibromide and glyphosate), was evaluated by the *in vivo* de vitellogenin assay with rainbow trout. A significant estrogenic potential was shown for 2,4-D, since it induced a 93 fold increase in the levels of plasma vitellogenin of the fish treated with this herbicide during 7 days [120].

2.6. Glyphosate

Glyphosate is a non-selective organophosphorus, broad spectrum, post-emergence herbicide, widely used in agriculture, mainly to control grasses, sedges, and broadleaf weeds [121]. Its action occurs by the inhibition of the biosynthesis of aromatic amino acids [122]. Its main mode of action is by the inhibition of the enzyme 5-enolpyruvylshikimate-3-phosphate synthase (EPSPS), which is essential in plants for the synthesis of the referred amino acids. Since this enzyme is absent in animals, this herbicide should be relatively non toxic for these organisms [123]. There are many conflicting data on the toxicity of glyphosate and its commercial formulations.

According to Solomon and Thompson [124], environmental toxicology of glyphosate has been extensively reviewed by a series of international regulatory agencies. According to the authors, as glyphosate binds strongly with organic matter, it is considered immobile in soils and sediments. This binding also removes glyphosate from water, reducing efficiently, the exposure of aquatic organisms. As the acute exposures are most likely to occur, the measures of effect are the most adequate for the purpose of risk assessment. However, in general, the authors affirm that glyphosate presents a low potential of acute toxicity for wild animals, including mammals, birds, fish and aquatic invertebrates.

Williams et al. [125] carried out a critical review on the toxicity of the herbicide RoundUp™ and of its active ingredient glyphosate. The analysis of the toxicity data, carried out by pattern tests and evaluation criteria, indicated that there is no evidence that glyphosate causes direct damages in the DNA, both in assays performed *in vitro* and *in vivo*. The authors concluded that Roundup™ and its components do not represent a risk for the induction of inheritable/somatic mutations in humans. Furthermore, the authors assert that, by the studies performed, glyphosate is not carcinogenic or teratogenic, nor does it cause significant adverse effects in the reproduction, development or in the endocrine system of humans and other mammals and, therefore, does not represent a risk for the health of human beings.

A study on the impact of the herbicide glyphosate and its commercial formulation Roundup™, in three microorganisms of food interest (*Geotrichum candidum, Lactococcus lactis subsp. cremoris* and *Lactobacillus delbrueckii subsp. bulgaricus*), showed that Roundup™ has an inhibitory effect on the microbial growth and a microbiocide effect at concentrations lower than the recommended for agricultural use. It was also observed in this study that glyphosate did not induce significant toxic effects for the three microorganisms studied. These differences between the toxic actions resulted from Roundup™ and glyphosate could be explained by a possible amplified effect of the commercial formulation due to the presence, according to Cox

[126] of adjuvants, such as polyethoxylated tallowamine (POEA), used for a better stability and penetration of the chemical compound [127].

Relyea [128] assessed the toxic potential of environmentally relevant concentrations of glyphosate on three species of tadpoles (wood frog [*Rana sylvatica* or *sylvaticus Lithobates*], leopard frog [*Rana pipiens pipiens* or L.], and American toad [*Bufo americanus* or *Anaxyrus americanus*]), by morphological analysis of individuals, before and after the application of the herbicide, showing that there is a significant induction of morphological alterations in the tadpoles of the three species. Specifically in the case of the wood frog and leopard frog, the exposure to the chemical compound has led to an evident alteration of the size of the tadpole tail, suggesting that the herbicide could be activating physiological mechanisms of development that are normally used as defence responses against predators. These results showed that glyphosate can have widespread and relevant effects on non target species, contradicting other studies, such as the one performed by Solomon and Thompson [124], who affirmed the inexistence or irrelevance of the toxicity of this compound on organisms and the environment.

Studies on the genotoxic potential of the active ingredient glyphosate, present in the commercial formulation Roundup, were performed on the roots of smooth hawksbeard (*Crepis capillaris* L.), in the concentrations of 0.05, 0.1, 0.5 and 1.0% of the active ingredient and for polychromatic erythrocytes (PCEs) of the bone marrow of C57BL rat, at doses inferior to half the LD_{50} (1080 mg/Kg). In these studies the chromosome aberrations assay and micronucleus test were used, which showed that this chemical compound did not induce significant responses for any of the biological systems tested [129].

Martini et al. [123] studied the effects of the commercial formulation of glyphosate in the proliferation, survival and differentiation of the 3T3-L1 fibroblasts (a mammal cell line), by the cell viability test with Trypan, MTT test, enzymatic activity assay of caspase-3 and staining assay with annexin-V and propidium iodide. The results showed that glyphosate inhibits the cell proliferation and induces apoptosis in a dose-dependent way, besides decreasing significantly the ability of the fibroblasts to differentiate to adipocytes. These data suggest the occurrence of important cell damages mediated by the action of this herbicide, indicating that glyphosate presents a potential risk factor for human health and the environment.

Dallegrave et al. [130] evaluated the teratogenicity of the herbicide glyphosate, marketed in Brazil as Roundup (36% of glyphosate and 18% of the surfactant polyoxyethyleneamine), to females of Wistar rats. The females treated orally with three different doses of glyphosate (500, 750, 1000 mg/Kg) from the 6th to the 15th day of gestation. After performing caesarean sections on day 21 of gestation, the number of corpora lutea, implantations, live and dead foetuses and reabsorptions, as well as the external malformations and skeletal malformation were recorded and analyzed. It was observed a mortality rate of 50% of the females treated with the highest concentration of glyphosate; the authors verified that there was a dose-response relationship directly proportional to the increase in the number of skeletal alterations found. These results led the authors to conclude that the commercial formulation of glyphosate (Roundup) is toxic for females of Wistar rats and is able to induce a delay in the fetal skeletal development of this species. It is important to consider that the toxicity and teratogenicity observed can result from both the action of glyphosate as well asthe surfactant present in the commercial formulation.

The oral administration of high doses of glyphosate (3500 mg/Kg) in Charles River COBS CD rats, between the 6th to the 19th day of pregnancy, and in rabbits, between the 6th to the 27th day of pregnancy, showed significant indices of maternal mortality for both species, as well as increase in the number of foetuses with reduced ossification of sternebrae [131], proving the toxicity and teratogenicity of this concentration of the herbicide for the organisms tested.

2.7. 2,4-D and glyphosate

Relyea [132] performed a study to observe the impact of two herbicides (glyphosate and 2,4-D) in the biodiversity of aquatic communities containing algae and more 25 species of animals. In this study the author observed that 2,4-D did not cause great impacts in the community and this is in agreement with previous studies that showed that this substance presents high LC-50 for several species. However, glyphosate had great impact in the community, causing a decrease of 22% of the species richness, while 2,4-D did not cause effects on this diversity. The authors also observed that neither of the two herbicides caused reduction in the periphyton biomass.

2.8. Diquat

Reglone is a bypiridylium herbicide, whose active ingredient is diquat (1,1′-ethylene -2,2′-ipyridyl dibromide), and of foliar application, used to eliminate weeds of different crops [133]. Reglone, in the concentrations tested (0.005, 0.01, 0.05 and 0.1% of the active ingredient for *Crepis capillaris* L.; 34.17 and 8.5 mg/Kg for mouse bone marrow polychromatic erythrocytes - PCEs), did not induce chromosome aberrations in any test system but promoted an increase in the frequency of micronuclei in both plant cells and PCEs [129], and thus is considered a potential mutagenic herbicide for these test organisms.

2.9. Pendimethalin

The herbicide Stomp 330, belongs to the dinitroanilines class, whose active ingredient is pendimethalin [N-(1-ethylpropyl)-2,6-dinitro-3,4-xylidine], it is applied as a systematic selective herbicide of the soil [133]. The responses of the two test systems for Stomp were very different: the concentrations tested (0.005, 0.1, 0.2 and 0.4% of the active ingredient for *Crepis capillaris* L.; 122.2, 244.5 and 489 mg/Kg for rats - PCEs) did not cause significant increases in the frequencies of chromosome aberrations in plant cells, but increased its incidence in rat cells, moreover, it induced an increase in the frequency of micronuclei in both test systems. This could be explained by the proven aneugenic effect of this herbicide, since all the concentrations tested produced C-mitoses in the assays with PCEs [129].

2.10. Paraquat

Paraquat (1,1′-dimethyl-4-4′-bipyridium dichloride) is a non-selective herbicide with fast action, widely used worldwide, mainly in the pre-harvest of cotton and potato and also to control a broad spectrum of weeds [134, 135, 136]. According to Tortorelli et al. [134], paraquat is able to modify the activity of several enzymes of fish, affecting the cardiac contraction and opercular ventilation, effects that can alter the initial development of these organisms. According to Tomita et al. [137], paraquat causes oxidative stress in different species of fish by generating elevated levels of superoxide ion.

A study conducted by D'Souza et al. [138] evaluated the toxicity of the herbicide paraquat for germ cells of male Sprague-Dawley rats by dermal exposure to this chemical. The authors verified that paraquat, even at low doses, significantly reduced the amount of spermatozoa, increased the frequency of spermatozoa bearing abnormalities and the mortality rate of these germ cells, as well as affected the mobility of the spermatozoa of the individuals studied, showing that the herbicide is a cytotoxic and genotoxic agent for the germ cells of this organism.

Hanada [136], analyzing the karyotype of species of *Rana ornativentris*, after exposure for 6 hours to the herbicide paraquat at the concentrations of 10^{-8} to 10^{-6} M, showed that this compound is able to induce genotoxic effects in this organism. The author observed that paraquat promoted, in a dose-dependent manner, a significant increase in the quantity of chromosome breaks in leukocytes of this test organism, suggesting that this species of anuran is highly sensitive to the genotoxic action of the herbicide.

According to Bus et al. [139], the genotoxic action of paraquat may be associated with the transference of a single electron of reduced oxygen to paraquat, forming superoxide ions. The singlet oxygen can be formed from the superoxide ion and subsequently react with lipids to form hydroperoxides and fatty acids. According to Tanaka and Amano [140], lipid peroxidation is responsible for the origin of several chromosome aberrations. Bauer Dial and Dial [141] still affirm that the oxidative stress induced by paraquat may be related to the teratogenic action of this compound to embryos and tadpoles of anurans.

Speit et al. [142] evaluated the genotoxic potential of the herbicide paraquat in Chinese hamster V79 cells, by chromosome aberrations and comet assays. Using a modified protocol of the comet assay with the modified protein FPG (formamidopyrimidine-DNA glycosylase), a repair enzyme that specifically nicks the DNA at sites of 8-oxo-guanines and formamidopyrimidines, it was not possible to detect oxidative damages in the bases of DNA after treatment with paraquat. Now, when the cells were treated directly on the slides, after lysis (i.e., after the cell membrane barrier has been eliminated), a significant increase in the migration of DNA was observed, only after treatment with high concentrations of the herbicide. Thus, the authors verified that the herbicide induced chromosome aberrations but was not able to induce relevant DNA lesions to promote mutations in the gene HPRT in cultured V79 cells.

Ribas et al. [135] assessed the cytotoxic, genotoxic and mutagenic potentials of different concentrations of the herbicide paraquat (0, 1, 5, 25, 50, 100, 250, 500, 1000, 2000 and 4000 μg/mL), by the assays of SCE, chromosome aberrations and micronuclei, in lymphocytes maintained in culture. The results showed that paraquat is an agent that induces cytotoxicity for lymphocytes, since it promoted the reduction in the nuclear division rate in all the concentrations tested and a significant decrease in the cell proliferation rates, when the cells were exposed to the highest concentration of the herbicide. In relation to the genotoxicity, the herbicide induced a significant increase in the frequencies of SCE of the lymphocytes treated, whose damage was not modified by co-treatment with the metabolic activation (S9 fraction of rat liver), but the data on the chromosome aberrations and micronuclei assays were not significant, which led the authors to conclude that paraquat is an inductor of primary damages in the DNA, although they have not shown that it has a clastogenic action.

A study performed by Hoffman and Eastin [143] evaluated the embryotoxic and teratogenic effects of two insecticides (lindane and toxaphene) and two herbicides (paraquat and 2,4,5-T), by external treatment of eggs of mallard duck (*Anas platyrhynchos*), using concentrations of field application. The authors showed that paraquat was the most significantly embryotoxic compound for this organism, independent of the type of vehicle in which the herbicide was associated, besides proving that paraquat impaired the growth of the organisms and was slightly teratogenic. The LC_{50} for this species was 1.5 Kg of the active ingredient/hectare in aqueous emulsion and 1 lb/acre in oil vehicles. When the organisms treated with paraquat were compared to the ones exposed to the herbicide 2,4,5-T, they presented little damages and it was observed few individuals bearing severe defects.

3. Harmful effects of herbicides on human health

The harmful effects of herbicides on human health are determined by several factors, such as the chemical class of those compounds, dose, time, and exposure route. Herbicides can be toxic to humans at high and lower doses [144]. The prolonged exposure can lead to a number of health effects, including the induction of diseases such as cancer and neurodegenerative [145, 146], reproductive and developmental changes [147] and respiratory effects [148].

Doll and Peto [149] estimated that 35% of all cases of cancer in the U.S. population originate from diet, and the herbicides present in foods are responsible. Estrogenicity assays made by Hernández et al. [150] show that organochlorine pesticides may act as endocrine disruption through more than one mechanism, including agonist or antagonist effects of different receptors. Chloro-s-triazize herbicides, pre-emergent pesticides used worldwide, have been generally considered as chemical compounds of low toxic potential for humans; however, there are many controversies on this issue. According to several international agencies, including the Environmental Protection Agency (EPA), Development for Environmental Assessment Center of the United States and IARC Monographs (International Agency for Research on Cancer), the herbicide atrazine, for example, was classified as a chemical agent probably carcinogenic to humans, although the basis for this conclusion is only evidenced in other animals [151, 152]. Due to the fact that atrazine induce mammary tumours in female Sprague-Dawley rats, the Peer Review Committee of the EPA Office of Pesticide Program (OPP) also concluded that atrazine should be considered in the Possibly Carcinogenic to Humans Group [153]. However, EPA [154] has classified this herbicide as a compound probably non carcinogenic to humans.

Some experimental studies have shown that exposure of humans to high doses of atrazine can result in an increased loss of body weight. However, a great number of epidemiological studies carried out with workers occupationally exposed to triazine herbicides indicate that these compounds do not have carcinogenic potential for these individuals. By analyses of different studies, it was observed that, although the chloro-s-triazine herbicides interfere in the endocrine responses of different species of mammals, their potential impact on humans seem to be mainly related to reproduction and development and not with human carcinogenesis [155].

Gammon et al. [83] discussed the extensive list of epidemiological studies with the herbicide atrazine, which describes that the carcinogenic potential of this compound to humans is not conclusive, although some studies have indicated a relationship between a high risk of prostate cancer and exposure to the herbicide.

Mladinic et al. [156] evaluated the genotoxic and mutagenic effects of low concentrations of the herbicides glyphosate and terbuthylazine, considered safe and, therefore, considered possible to occur in occupational and residential exposures (ADI – Acceptable Daily Intake, REL – Residential Exposure Level, OEL – Occupational Exposure Level, and 1/100 and 1/16 LD_{50} – Lethal Dose 50% - oral, rat), in human lymphocytes, with and without the use of metabolic activation (S9 fraction), by the FSH cytome assay, using pan-centromeric DNA probes to assess the content of micronuclei and other chromatinic instabilities. The authors verified that the frequencies of micronuclei, nuclear buds and nucleoplasmic bridges of cells treated with glyphosate slightly increased after the concentration of OEL 3.5 μg/mL, but no concentration induced an increase of the centromeric signals (C+) or DAPI (DAPI+). Now, the treatment with the herbicide terbuthylazine without metabolic activation showed a dose-response increase in the frequency of micronuclei of the lymphocytes exposed, and the significant data were from the concentration of 0.0008 μg/mL (REL) tested. The concentrations ADI (0.00058 μg/mL), REL (0.0008 μg/mL) and OEL (0.008 μg/mL) of terbuthylazine induced a significant occurrence of micronuclei hybridized with the centromeric probe (C+), regardless the presence or absence of S9, and of nuclear buds containing centromeric signals, only in the presence of S9. By the results obtained, it was suggested that the lowest concentrations of glyphosate do not have relevant harmful effects for the DNA molecule, while terbuthylazine presents a predominant aneugenic potential for the genetic material of human lymphocytes.

Terbuthylazine belongs to the chloro-s-triazine herbicides class, which inhibits the photosynthesis of weeds, by reaching the photosystem II. It is a chemical used for a variety of crops, such as maize, sugarcane, olive and pineapple [157]. Since the banishment of atrazine in European countries in 2006, terbuthylazine was recommended as its substitute. Due to the fact that the herbicide terbuthylazine is suspect of causing diseases in humans, such as non-Hodgkin lymphoma and lung cancer, Mladinic et al. [158] evaluated the effects of prolonged exposure (14 days) to low concentrations of this compound (0.58 ng/ml and 8 ng/ml) in human lymphocytes, using the comet assay and the comet-FISH assay (with the c-Myc and TP 53 genes). Treatment with terbuthylazine induced the migration of fragments of DNA in a significant manner, only for the highest concentration treated. The results showed an impairment of the structural integrity of c-Myc and TP 53, due to the prolonged exposure of human lymphocytes to terbuthylazine. The fact that several copies of TP53 were affected by the herbicide can indicate its ability to negatively interfere in the control of the cell cycle. However, the authors concluded that, for a more detailed assessment of the risk of cancer associated with exposure to terbuthylazine, it should be evaluated the impact of this pesticide on other housekeeping genes and markers.

Mladinic et al. [122] evaluated the genotoxic potential, by the comet assay and FISH, and oxidative damages, by the TBARS lipid peroxidation, of different concentrations of glyphosate (three similar to those observed in residential and occupational exposures and two related to LC_{50}) in human lymphocytes. The comet assay showed that the concentration of 580 μg/mL promoted a significant increase in the tail length, while the concentration of 92.8 μg/mL caused an increase in the tail intensity, both in relation to the control test. With the addition of the S9 fraction, the tail length was significantly increased for all the concentrations tested. When the lymphocytes were exposed to the three highest concentrations without S9, there was an increase in the frequency of micronuclei, nuclear buds and nucleoplasmic bridges. The addition of a metabolic activation system only promoted a significant increase of the nuclear instabilities for the highest concentration tested. The values of TBARS significantly increased with the increase of the concentrations tested, regardless the presence or absence of the S9 fraction. Due to the fact that dose-dependent effects for all the assays used were not observed, the authors concluded that these concentrations of glyphosate are not relevant for human exposure, since they did not present a significant risk for human health.

According to Mladinic et al. [122], the increase in the number of crops genetically modified used in assays and diagnosis of resistance to glyphosate, may be related to the fact that these crops tolerate increasingly higher concentrations of the active ingredient necessary for an effective control of weeds, which results from the introduction of increasing amounts of glyphosate into the environment. Thus, some epidemiological studies have shown that human exposure to glyphosate present in the environment is correlated to the development of diseases such as the non-Hodgkin lymphoma [159, 160].

According to He et al. [161], paraquat, the second most widely used herbicide in the world, is able to selectively accumulate in human lungs by causing oxidative injury and fibrosis, leading several individuals to death. Chronic exposure to this herbicide is also associated with hepatic lesions, kidney failure and Parkinson's disease [162, 163].

Studies carried out by He et al. [161] evaluated the toxicity of paraquat on BEAS-2B normal cells (human bronchial epithelial cells), showing that it is dose-dependent and results in mitochondrial damages, oxidative stress, death of lung cells exposed, as well as production of cytokines, pro-fibrogenic growth factors and transformation of myofibroblasts. The authors also proved that administration of resveratrol, a polyphenolic phytoalexin naturally produced by several plants, to control bacteria and fungi, was able to inhibit the production of reactive oxygen species, inflammations and fibrotic reactions induced by paraquat, by the activation of the Nrf2 signaling (Nuclear Factor Erythroid-2), revealing a new molecular mechanism for the intervention against oxidative damages and pulmonary fibrosis resulted from the action of toxic chemical compounds.

The study on the influence of a complex mixture of herbicides (atrazine, 2,4-D, alachlor, ciazine and malathion) in workers occupationally exposed to them, was carried out using cytogenetic methods standardly established (chromosome aberrations and micronucleus assay) and the comet assay technique. This assay showed a significant increase in the DNA migration ($P<0.001$), suggesting that long-term exposure to the pesticides could cause damages in the genome of somatic cells and, therefore, would represent a potential risk to human health [164].

4. Conclusion

The authors present in this manuscript the bioassays and the test-systems most commonly used to evaluate the effects of herbicides and the test-organisms to best suit the assessments of herbicide effects. In these considerations, the authors attempted to present the most sensitive and efficient organisms capable of detecting environmental contamination resulting from the action of these chemical agents. Additionally, we present in this paper the need to carry out research aimed at more effective methods to prevent and/or reduce the deleterious effects of such compounds on the environment, the biota potentially exposed, and especially to human health.

In this study it was addressed several studies that used different methodologies, which evaluated the toxicity and action of herbicides on different non-target organisms, including human species. The table below summarizes the main researches addressed in the text.

Herbicide	Test-organism	Endpoint	Results	Tested concentrations	References
Atrazine	Erythrocytes of Nile tilapia (*Oreochromis niloticus*)	micronucleus test; comet assay	increase in the DNA fragmentation; induction of micronuclei and nuclear abnormalities in all tested concentrations	6.25, 12.5, 25 µg/L	[8] Ventura et al., 2008
Atrazine	Wild leopard frogs (*Rana pipiens*)	toxicity assay	induction of abnormalities in the gonads; developmental delay and hermaphroditism (≥ 0.1ppb)	0.01, 0.1, 0.4, 0.8, 1, 10, 25, 200 ppb	[82] Hayes et al., 2002
Atrazine	*Sorghum vulgare*	chromosome aberration assay	induction of multinucleated, aneuploid and polyploid cells; abnormalities in the mother cells of pollen grains; meiotic instability	2.7 Kg a.i./ha	[80] Liang et al., 1967
Atrazine	Human lymphocytes	chromosome aberration assay	increase in the chromosome aberrations frequency at 0.10 ppm	0.01, 1, 0.10 mg/ml	[85] Meisner et al., 1992
Atrazine	Human lymphocytes	chromosome aberration assay; SCE	increase in the frequency of chromosome aberrations; increase in the frequency of sister chromatid exchange in all tested concentrations	5, 8.5, 17, 51 µM	[93] Lioi et al., 1998
Atrazine	Human blood cells	chromosome aberration assay	Significant increase of chromosome breaks	1 ppm	[94] Meisner et al., 1993
Atrazine	Rat	chromosome aberration assay	there was no significant increase in the frequency of chromosome aberrations at 20 ppm	20 ppm	[85] Meisner et al., 1992
Atrazine	Bone marrow cells of rats	chromosome aberration assay	there was no significant increase in the frequency of chromosome aberrations	20 ppm	[86] Roloff et al., 1992
Atrazine	Human lymphocytes	chromosome aberration assay	induction of chromosome aberrations	0.0001 µg/mL	[94] Meisner et al., 1993

Herbicide	Test-organism	Endpoint	Results	Tested concentrations	References
Atrazine	Rat leukocytes	comet assay	increase in the damages in the DNA for 500 mg/Kg	125, 250, 500 mg/Kg	[89] Tennant et al., 2001
Atrazine	Erythrocytes of bullfrog tadpoles	comet assay	significant increase in the DNA damages, from the concentration of 4.8 mg/L	4.8, 19.75, 77, 308 mg/L	[90] Clements et al., 1997
Atrazine	Human lymphocytes	comet assay	significant increase in the DNA damages, mainly at the concentrations of 100 and 200 µg/L	50, 100, 200 µg/L	[69] Ribas et al., 1995
Atrazine	Hepatocytes of Wistar rats	lipid peroxidation assay; micronucleus test	increase in the rates of lipid peroxidation, hepatic damages, death of hepatocytes and induction of micronuclei.	400 ppm	[91] Campos-Pereira et al., 2012
Atrazine	Erythrocytes and gill cels of the fish *Channa punctatus*	micronucleus test; comet assay	induction of damages in the DNA and micronuclei, in the tested concentrations, in all the exposure periods (from 1 to 35 days), with more significant effects in the highest concentrations and exposure periods; higher sensitivity for gill cells	4.24, 5.30. 8.48 mg/L	[22] Nwani et al., 2011
Atrazine	Erythrocytes of the gibel carp fish (*Carassius auratus*)	micronucleus test; comet assay	significant induction of DNA strand breaks and micronuclei, in all tested concentrations of the commercial product (Gesaprim), but there was not a induction of these genotoxic and mutagenic effects for the active ingredient.	5, 10, 15 µg/L	[84] Çavas, 2011
Atrazine	Human lymphocytes	comet assay	significant increase of damage in the DNA exposed to the commercial product Gesaprim, but there was no induction of genotoxicity for the active ingredient atrazine, for all tested concentrations.	0.047, 0.47, 4.7 ug/L	[101] Zeljezic et al., 2006
Atrazine	Somatic cells of *Allium cepa* and *Vicia faba*	chromosome aberration assay; micronucleus test	significant inhibition of the mitotic index, significant increase in the frequencies of micronuclei and chromosome aberrations of both test organisms, when exposed to the commercial product Gesaprim, but there was no induction of any significant effects when cells were exposed to the active ingredient atrazine, for all tested concentrations.	*A. cepa:* 15, 30, 60 mg/L; *V. faba:* 17,5, 35, 70 mg/L	[102] Srivastava and Mishra, 2009
Atrazine	*Salmonella* and hepatic cells of Sprague-Dawley rats	*Salmonella* assay and SOS Chromotest	there was no significant induction of genotoxic damages nor mutagenic	1 – 1000 µg/ plate	[92] Ruiz and Marzin, 1997

Herbicide	Test-organism	Endpoint	Results	Tested concentrations	References
Atrazine	Sprague-Dawley rats	embryotoxic and teratogenic tests	induction of hypospadias in male newborns at 200 ppm and diverse embryotoxic damages at 25 ppm.	25, 100, 200 mg/kg/d	[87] Wu et al., 2007
Atrazine	Wistar rats	Radioimmunoassay	alterations in the levels of testosterone, androstenedione, estradiol, estrone, progesterone and corticosterone to 50 or 200 ppm for 60 days	50, 200 mg / kg / day	[88] Modic et al., 2004
Atrazine, Simazine and Cyanazine	Human lymphocytes	chromosome aberration assay and SCE	there was no significant increase of chromosome aberrations and sister chromatid exchanges	0.5, 5, 50 ppb	[104] Kligerman et al., 1993
Atrazine, Simazine and Cyanazine	Polychromatic erythrocytes of the bone marrow of female C57B1/6 rats	micronucleus test	there was no significant induction of micronuclei	0, 125, 250, 500 mg/kg	[103] Kligerman et al., 2000
Atrazine, Simazine and Cyanazine	Chinese Hamster Ovary – CHO – cells	flow cytometry assay	significant induction of chromosome damages by atrazine for the tested concentrations, proven clastogenic potential of cyanazine	0.003 μg/mL, 0.018 μg/mL(atrazine); 0.003 μg/mL, 0.012 μg/mL (cyanazine)	[107] Taets et al., 1998
Atrazine and Butachlor	Green alga *Scenedesmus obliquus* and cladoceran *Daphnia carinata*	acute toxicity assay	atrazine is highly toxic for *S. obliquus* and slightly toxic for *D. carinata* and butachlor is moderately toxic for both; the toxic effects of the mixture of the herbicides were significantly antagonistic for *S. obliquus* and there was no significative synergism for *D. carinata*	*S. obliquus*: 0, 0.5, 1, 2, 4, 8 mg/L (butachlor) and 0, 0.008, 0.016, 0.032, 0.064, 0.128 mg/L (atrazine) / *D. carinata*: 0, 1, 1.8, 3, 5, 8 mg/L (butachlor) and 0, 7.5, 15, 30, 60, 120 mg/L (atrazine)	[20] He et al., 2012
Butachlor	Alpine cricket frog (*Fejervarya limnocharis*)	chromosome aberration assay	affected the survival, development and metamorphosis time of tadpoles in different concentrations; DNA damage (0.4-0.8 mg/L)	ranging from 0.025 to 3.2 mg/l	[43] Liu et al., 2011
Terbutryn	Human leukocytes	micronucleus test; comet assay; SCE	there was no significant induction of micronuclei and SCE; significant induction of DNA damages for all tested concentrations	0, 5, 10, 50, 100, 150 μg/mL	[108] Moretti et al., 2002
2,4-D	*Caudiverbera caudiverbera* frog	toxicity assay	dose-dependent reduction in the response of the simpatic junction to	0.01, 0.1, 1 mM	[114] Suwalsky et al., 1999

http://dx.doi.org/10.5772/55851

Herbicide	Test-organism	Endpoint	Results	Tested concentrations	References
			nerve stimulation due to inhibition of the glandular chloride channel in mucosa skin		
2,4-D	Gills of different species of fishes	toxicity assay	bleeding, renal increase, impairment of the renal functions and hepatic degeneration	400 mg/L	[115] Gómez et al., 1998
2,4-D	Chinese Hamster Ovary – CHO – cells	SCE	significant increase in the sister chromatid exchange at 2 and 4 μg/ml	2, 4, 6, 10 μg/mL	[116] González et al., 2005
2,4-D	Bone marrow and germ cells of rats	SCE	significant increase in the sister chromatid exchange at 100 and 200 ppm, for both cell types	50,100, 200 mg/kg	[117] Madrigal-Bujaidar et al., 2001
2,4-D	Frog *Xenopus*	FETAX - frog embryo teratogenic assay	significant induction of embryotoxic and teratogenic effects	245 mg/L	[119] Morgan et al., 1996
2,4-D and Butachlor	Erythrocytes of the catfish (*Clarias batrachus*)	chromosome aberration assay; micronucleus test	significant increase in the frequency of micronuclei and altered cells in a dose-response manner for both herbicides	2,4-D: 25, 50, 75ppm; Butachlor: 1, 2, 2.5ppm	[113] Ateeq et al., 2002
2,4-D and Paraquat	Rainbow trout (*Oncorhynchus mykiss*)	acute lethality test, lipid peroxidation assay by quantification of MDA; comet assay	toxic action more evident for paraquat (high indices of mortality); apnea and white spots in the scales of individuals exposed to the 2 herbicides; increase in the rates of MDA and damages in the DNA after exposure to all concentrations of the tested herbicides	2,4-D: 316, 346, 389, 436, 489 mg/L; Paraquat: 0.055, 0.066, 0.083, 0.116, 0.133 mg/L	[112] Martínez-Tabche et al., 2004
2,4-D and 2,4-D DMA	Humanh lymphocytes and erythrocytes	SCE; analysis of the cell cycle progression and mitotic index	alterations in the cell cycle and induction of SCE for some concentrations only with more significant genotoxic effects for erythrocytes	10, 25, 50, 100 μg/mL	[118] Soloneski et al., 2007
2,4-D; Triclopyr; Diquat dibromide; glyphosate	Rainbow trout (*Oncorhynchus mykiss*)	Vitellogenin estrogenic assay	significant increase in the levels of vitellogenin of the plasma of fishes exposed to 2,4-D	0.11, 1.64, 2.07, 1.25 mg/L	[120] Xie et al., 2005
Glyphosate	*Geotrichum candidum*, *Lactococcus lactis* subsp. *Cremoris; Lactobacillus delbrueckii* subsp. *bulgaricus*	microbial growth assay	inhibition of microbial growth by the commercial product Roundup; microbiocide effect at concentrations lower than the recommended by agricultural use for the commercial product Roundup; non induction of significant toxic effects for the three microorganisms by the active ingredient glyphosate	0.1, 1, 10, 100, 1000, 10000 ppm	[127] Clair et al., 2012
Glyphosate	Tadpoles of wood frog (*Rana sylvatica* or *Sylvaticus lithobates*),	acute toxicity assay	significant induction of morphological alterations of tadpoles of the three species; for the wood frogs and	0, 1, 2, or 3 mg acid equivalents [a.e.] /L of	[128] Relyea, 2012

Herbicide	Test-organism	Endpoint	Results	Tested concentrations	References
	leopard frog (*Rana pipiens pipiens* or L.), and American toad (*Bufo americanus* or *Anaxyrus americanus*)		leopard frogs, exposure to glyphosate affected the size of the tail of tadpoles, for all tested concentrations	Roundup Original MAX	
Glyphosate	Roots from the smooth hawksbeard (*Crepis capillaris* L.); polychromatic erythrocytes of the bone marrow of C57BL rat	chromosome aberration assay; micronucleus assay	there was no induction of genotoxic and/or mutagenic effects for any of the species	*Crepis capillaris*: 0.05, 0.1, 0.5, 1 %; erythrocytes: doses inferior to half the LD_{50} (1080 mg/Kg)	[129] Dimitrov et al., 2006
Glyphosate	Female Wistar rats	acute toxicity assay; teratogenicity assay	high mortality index of females treated with the highest concentration of the commercial product Roundup; increase in the dose-response of fetal skeletal alterations	500, 750, 1000 mg/kg	[130] Dallegrave et al., 2003
Glyphosate	Human lymphocytes	comet assay; FISH; lipid peroxidation assay – TBARS	significant increase in the DNA migration at 580 µg/mL; significant increase of the comet tail intensity at 92.8 µg/mL; greater lesion in the DNA in the presence of S9; increase in the frequency micronuclei, nuclear buds and nucleoplasmic bridges, without S9; significant increase of nuclear instabilities in the highest concentration tested with S9; significant dose-response increase of the levels of TBARS	0.5, 2.91, 3.5, 92.8, 580 µg/mL	[122] Mladinic et al., 2009
Glyphosate adn 2,4-D	Algae and 25 species of aquatic animals	acute toxicity assay	there was no reduction in the biomass of periphyton by the 2 herbicides; there was no great impacts to the aquatic community by 2,4-D; high impact to the aquatic community by glyphosate by the significative decrease in the species richness	0, 1, 2, or 3 mg acid equivalents [a.e.] /L of Roundup Original MAX	[132] Relyea, 2005
Glyphosate and Terbuthylazine	Human lymphocytes	cytome FISH	glyphosate caused an increase in the frequencies of micronuclei, nuclear buds and nucleoplasmic bridges of clells treated (3.5 µg/mL onward), but without induction of centromeric signals; terbuthylazine induced an increase in the frequency of micronuclei hybridized with	0.5, 2.91, 3.50, 92.8, 580 µg/mL (glyphosate); 0,00058, 0,0008, 0,008, 25, 156,5 µg/mL (terbuthylazine)	[156] Mladinic et al., 2009

Herbicide	Test-organism	Endpoint	Results	Tested concentrations	References
			centromeric probe and nuclear buds with centromeric signals in the presence of S9 (0.008 ug/mL onward)		
Terbuthylazine	Human lymphocytes	comet assay; comet assay-FISH	induction of the migration of fragments of DNA, significant only at the highest concentration; impairment of the structural integrity of c-Myc and TP 53 due to prolonged exposure to terbuthylazine	Terbuthylazine: 0.58 ng/ml, 8 ng/ml; carbofuran: 8 ng/ml, 21.6 ng/ml	[158] Mladinic et al., 2012
Paraquat	Several species of fishes	acute toxicity assay; enzyme activity assay	alteration in the activity of different enzymes; negative effects on cardiac contraction and opercular ventilation	0.1-2.0 mg/L	[134] Tortorelli et al., 1990
Paraquat	Several species of fishes	enzyme activity assay	induction of oxidative stress; increase in the levels of SOD	0.2-50 mM	[137] Tomita et al., 2007
Paraquat	Germ cells of Sprague-Dawley rats	cytotoxicity assay	reduction in the quantity of spermatozoa; increase in the mortality rates and abnormalities in spermatozoa for the higher concentrations	0, 6, 15, 30 mg/kg	[138] D'Souza et al., 2006
Paraquat	Leukocytes of *Rana ornativentris*	conventional cytogenetics assay	genotoxic effects, such as chromosome breaks	10^{-6} M	[136] Hanada, 2011
Paraquat	Human lymphocytes	chromosome aberration assay; micronucleus test; SCE	reduction in the cell division index; decrease in the cell proliferation rates; significant increase in the frequencies of SCE (50 μg/mL for 24h treatment; 4000 μg/mL for 2h treatment), significant increase in the MN frequencies (concentrations ≥ 25 μg/mL)	0, 1, 5, 25, 50, 250, 500, 1000, 2000, 4000 μg/mL	[135] Ribas et al., 1998
Paraquat	BEAS 2B normal cells (human bronchial epithelial cells)	cytotoxicity assay, oxidative stress assay	mitochondrial damage; oxidative stress; cell death; production of cytokines, pro-fibrogenic growth facts and transformation of myofibroblast	10 uM	He et al., 2012
Diuron	Pacific oyster (*Crassostrea* gigas)	toxicity assay	irreversible damages to the genetic material, negative impacts in the reproduction of aquatic organisms	300 ng/L, 3 μg/L	[44] Bouilly et al., 2007
Diquat	Roots of smooth hawksbeard (*Crepis capillaris* L.); polychromatic erythrocytes of the bone marrow of C57BL rat	chromosome aberration test; micronucleus test	there was no induction of chromosome aberrations for any test system; significant increase of the frequency of micronuclei for both test systems	*Crepis capillaris*: 0.005, 0.01, 0.05, 0.1%; erythrocytes: 8.5, 34.17 mg/Kg	[129] Dimitrov et al., 2006

Herbicide	Test-organism	Endpoint	Results	Tested concentrations	References
Pendimethalin	Roots of smooth hawksbeard (*Crepis capillaris* L.); polychromatic erythrocytes of the bone marrow of C57BL rat	chromosome aberration test; micronucleus test	there was no significant increase in the frequencies of chromosome aberrations in plant cells, but an increase of their incidence in cells of rats; significant increase in the frequency of micronuclei for both test systems.	*Crepis capillaris*: 0.005, 0.1, 0.2, 0.4%; erythrocytes: 122.2, 244.5, 489 mg/Kg	[129] Dimitrov et al., 2006
Simetryn, mefenacet and thiobencarb	*Silurana tropicalis*	toxicity assay	toxic effects for tadpoles, more significant for thiobencarb	Thiobencarb: 6.85-2.92 mM	[42] Saka, 2010
Complex mixture of pesticides (atrazine, 2,4-D, alachlor, ciazine and malathion)	Workers exposed	chromosome aberration assay; micronucleus test; comet assay	significant increase in the migration of the DNA	Mixture of various concentrations of pesticides	[163] Garaj-Vrhovac and Zeljezic, 2002

Table 2. List o the main researches carried out with several bioindicators to evaluate the toxicity of herbicides.

Author details

Maria Aparecida Marin-Morales*, Bruna de Campos Ventura-Camargo and Márcia Miyuki Hoshina

*Address all correspondence to: mamm@rc.unesp.br

Department of Biology, Institute of Biosciences, São Paulo State University (UNESP), SP, Brazil

References

[1] Guzzella, L.; Pozzoni, F.; Giuliano, G. Herbicide contamination of surficial groundwater in Northern Italy. Environmental Pollution, v. 142, p. 344-353, 2006.

[2] Kortekamp, A. Herbicides and Environment. Kroatia, 2011, 760 p.

[3] Bolognesi, C.; Merlo, F.D. Pesticides: Human Health Effects. Encyclopedia of Environmental Health, p. 438-453, 2011.

[4] Nehls, S.; Segner, H. Detection of DNA damage in two cell lines from rainbow trout, RTG-W1, using the comet assay. Environmental Toxicology, v. 16, p. 321-329, 2001.

[5] Spacie, A.; Hamelink, J.L. Bioaccumulation, in: RAND, G.M.; PETROCELLI, S.R. (Eds.), Fundamentals of Aquatic Toxicology: Methods and Applications, Hemisphere, New York, 1985, pp. 495-525.

[6] Grillo, R.; Santos, N.Z.P.; Maruyama, C.R.; Rosa, A.H.; De Lima, R.; Fraceto, L.F. Poly(Rmvarepsilon-caprolactone)nanocapsules as carrier systems for herbicides: physico-chemical characterization and genotoxicity evaluation. Journal of Hazardous Materials, 2012, doi:10.1016/j.jhazmat.2012.06.019

[7] Fernandes, T.C.C; Mazzeo, D.E.C.; Marin-Morales, M.A. Mechanism of micronuclei formation in polyploidizated cells of *Allium cepa* exposed to trifluralin herbicide. Pesticide Biochemistry and Physiology, v. 88, n. 3, p. 252-259, 2007.

[8] Ventura, B.C.; Angelis, D.F.; Marin-Morales, M.A. Mutagenic and genotoxic effects of the Atrazine herbicide in *Oreochromis niloticus* (Perciformes, Cichlidae) detected by the micronuclei test and the comet assay. Pesticide Biochemistry and Physiology, v. 90, p. 42-51, 2008.

[9] Silva, J.; Fonseca, M.B. Genética Toxicológica. 1 ed. Brasil: Alcance, 2003. 471p.

[10] Bertoletti, E. Companhia de Tecnologia de Saneamento Ambiental: Ensaios biológicos com organismos aquáticos e sua ação no controle da poluição de São Paulo. 1 ed. Brasil: [s.n.] , 1996. 29p.

[11] Ribeiro, L.R.; Salvadori, D.M.F.; Marques, E.K. Mutagênese Ambiental. 1 ed. Brasil: ULBRA, 2003. 355p.

[12] RAND, G.M.; PETROCELLI, S.R. Fundamentals Of Aquatic toxicology: methods and applications. Hemisphere, Washington, v. 42, n. 1, p. 1-28, 1985.

[13] Arnaiz, R.R. Las Toxinas Ambientales y sus Efectos Genéticos. 2 ed. México: [s.n.] , 1995. 267 p

[14] Vogel, E.W. Assessment of chemically induced genotoxic events. In: Prospectives and Limitations, The Netherlanlands: Universitaire Pers Leiden, 1982. p. 24.

[15] Tavares, D.C. Estudos da possível ação genotóxica do alcalóide boldina em sistemas de células de mamífero "in vitro" e "in vivo". 1991. 205 f. Tese (Mestrado em Medicina) - Faculdade de Medicina de Ribeirão Preto, Universidade do Estado de São Paulo, Ribeirão Preto.

[16] Ueta, J.; Pereira, N.L.; Shuhama, I.K.; Cerdeira, A.L. Biodegradação de herbicidas e biorremediação: Microrganismos degradadores do herbicida atrazine. 1 ed. Brasil: [s.n.] , 1997. 545p.

[17] Kudsk, P., Streibig, J.C. Herbicides: a two-edged sword. Weed Res., v. 43, p. 90-102, 2003.

[18] Alves, A. Usos e Abusos. Ciência Hoje, São Paulo, v. 4, n. 22, p. 49-52, 1986.

[19] Vasilescu, M.N. Medvedovici, A.V. Herbicides. Encyclopedia of Analytical Science. 2nd ed. Elsevier, Oxford, p. 243-260, 2005. http://dx.doi.org/10.1016/B0-12-369397-7/00256-9

[20] He, H., Yu, J., Chen, G., Li, W., He, J., Li, H. Acute toxicity of butachlor and atrazine to freshwater green alga *Scenedesmus obliquus* and cladoceran *Daphnia carinata*. Ecotox. Environ. Saf., v. 80, p. 91-96, 2012a.

[21] Zhang, W., Jiang, F., Ou, J. Global pesticide consumption and pollution: with Chinas as a focus. Proceedings of the International Academy of Ecology and Environmental Sciences, v. 1(2), p.125-144, 2011.

[22] Nwani, C.D., Nagpure, N.S., Kumar, R., Kushwaha, B., Kumar, P., Lakra, W.S. Mutagenic and genotoxic assessment of atrazine-based herbicide to freshwater fish *Channa puntatus* (Bloch) using micronucleus test and single cell gel electrophoresis. Environmental Toxicology and Pharmacology, v. 31, p. 314-322, 2011.

[23] Chevreuil, M.; Garmouma, M.; Teil, M.J.; Chesterikoff, A. Occurrence of organochlorines (PCBs, pesticides) and herbicides (triazines, phenylureas) in the atmosphere and in the follout from urban and rural stations of Paris area. Science of the Environment, [S.l.] , v. 182, p. 25-37, 1996.

[24] Kim, J.H.; Feagley, S.E. Adsorption and leaching of trifluralin, metolachlor, and metribuzin in a commerce soil. Journal of Environmental Science and Health-B: Pesticides and Food Contaminants, New York, v. 33, p. 529-546, 1998.

[25] Abdel-Rahmam, A.R.; Wauchope, R.D.; Truman, C.C.; Dowler, C.C. Runoff and leaching of atrazine and alachlor on a sandy soil as affected by application in sprinkler irrigation. Journal of Environmental Science and Health-B: Pesticides and Food Contaminants, v. 34, p. 381-396, 1999.

[26] Munger, R.; Isacson, P.; Hu, S.; Burns, T.; Hanson, J.; Lynch, C.F.; Cherryholmes, K.; Vandorpe, P.; Hausler, Jr. W. J. Intrauterine growth retardation in Iowa communities with herbicides-contaminated drinking water supplies. Environmental Health Perspectives, v. 105, p. 308-314, 1997.

[27] Gorell, J.M.; Jhonson, C.C.; Rybicki, B.A.; Peterson, E.L.; Ricchardson, R.J. The Risk of Parkinson's disease with exposure to pesticides, farming, well water, and rural living. Neurology, Heidelberg, v. 50, p. 1346-1350, 1998.

[28] Vander Werf, H.M.G. Assessing the impact of pesticides on the environment. Agriculture, Ecosystems and Environment, The Netherlands, v. 60, p. 81-96, 1996.

[29] Timbrell, J.A. Introduction to Toxicology. 2. ed. Estados Unidos: Taylor & Francis, 1999. 167p.

[30] Linck, A.J. Effects on the cytology and fine structure of plant cells. Herbicides, [S.l.] , v. 1, p. 83-121, 1979.

[31] Natarajan, A.T. Chromosome Aberrations: past, present and future. Mutation Research, Leiden, v. 504, p. 3-16, 2002.

[32] Jurado, A.S., Fernandes, M.A.S., Videira, R.A., Peixoto, F.P., Vicente, J.A.F. Herbicides: The Face and the Reverse of the Coin. An *in vitro* Approach to the Toxicity of Herbicides in Non-Target Organisms. In: KORTEKAMP, A. (Ed.) Herbicides and Environment. Kroatia, 2011, p. 3- 45 p.

[33] Moreland, D.E. Mechanisms of action of herbicides. Ann. Rev. Plant Physiol., v. 31, p. 597-638, 1980.

[34] Rao, V.S. Principles of Weed Science, New Hampshire, USA, 2nd Ed, 2000, 559 p.

[35] Parsons, B.; Witt, J.M. Pesticides in groundwater in the U.S.A. A report of a 1988 survey of US States. EM8406, Oregon State University Extension Service. Archives of Environmental Contamination and Toxicology, v. 18, p. 734-747, 1989.

[36] Hance, R.J. Some continuing uncertainties in knowledge of herbicide behavior in the soil. Annals of Applied Biology, v.110, p.195-202, 1987.

[37] Hussain, S., Siddique, T., Saleem, M., Arshad, M., Khalid, A. Impact of pesticides on soil microbial diversity, enzymes and biochemical reactions. Advances in Agronomy, v.102, p.159-200, 2009.

[38] Moura, M.A.M., Franco, D.A.S., Matallo, M.B., Impacto de herbicidas sobre os recursos hídricos. Revista Tecnologia & Inovação Agropecuária, v. 1(1), p. 142-151, 2008.

[39] Roman, E.E., Beckie, H., Vargas, L., Hall, L., Rizzardi, M.A., Wolf, T.M. Como funcionam os herbicidas da biologia à aplicação. Passo Fundo, Brasil, 2007, 158 p.

[40] Ying, G.-G., Williams, B. Laboratory study on the interaction between herbicides and sediments in water systems. Environmental Pollution, v.107, p. 399-405, 2000.

[41] Toccalino, P.L., Norman, J.E., Scott, J.C. Chemical mixtures in untreated water from public-supply wells in the U.S. – Occurrence, composition and potential toxicity. Science of Total Environment, v. 431, p. 262-270, 2012.

[42] Saka, M. Acute toxicity of rice paddy herbicides simetryn, mefenacet, and thiobencarb to *Silurana tropicalis* tadpole. Ecotox. Environ. Saf., v. 73, p. 1165-1169, 2010.

[43] Liu, W.Y., Wang, C.Y., Wang, T.S., Fellers, G.M., Lai, B.C., Kam, Y.C. Impacts of the herbicide butachlor on the larvae of a paddy field breeding frog (*Fejervarya limnocharis*) in subtropical Taiwan. Ecotoxicology, v. 20, p. 377-384, 2011.

[44] Bouilly, K., Bonnard, M., Cagnaire, B., Renault, T., Lapègue, S. Impact of Diuron on Aneuploidy and Hemocyte Parameters in Pacific Oyster *Crassostrea gigas*. Arch. Environ. Contam. Toxicol., v.52, p.58-63, 2007.

[45] Hladik, M.L., Bouwer, E.J., Roberts, A.L. Neutral degradates of chloroacetamide herbicides: Occurrence in drinking water and removal during conventional water treatment. Water Research, v.42, p.4905-4914, 2008.

[46] Bannink, A.D. How Dutch drinking water production is affected by the use of herbicides on pavements. Water Sci, Technol., v.49 (3), p.173-181, 2004.

[47] Rostad, C.E. From the 1998 drought to the 1993 flood: transport of halogenated organic compounds with the Mississipi River suspended sediment at Thebes, Illinois. Environ. Sci. Tecnol., v.31, p.1308-1312, 1997.

[48] Jacomini, A.E., Camargo, P.B., Avelar, W.E.P., Bonato, P.S. Assessment of Ametryn Contamination in River Water, River Sediment, and Mollusk Bivalves in São Paulo State, Brazil. Arch. Environ. Contam. Toxicol., v. 60, p. 452-461, 2011.

[49] Duke, N.C., Bell, A.M., Pederson, D.K., Roelfsema, C.M., Nash, S.B. Herbicides implicated as the cause of severe mangrove dieback in the Mackay region, NE Australia: consequences for marine plant habitats of the GBR World Heritage Area. Marine Pollution Bulletin, v. 51, p.308-324, 2005.

[50] Jones, R., The ecotoxicological effects of Photosystem II herbicides on corals. Marine Pollution Bulletin, v.51, p.495-506, 2005.

[51] Lewis, S.E., Brodie, J.E., Bainbridge, Z.T., Rohde, K.W., Davis, A.M., Masters, B.L., Maughan, M., Devlin, D.J., Mueller, J.C., Schaffelke, B. Herbicides: A new threat to the Great Barrier Reef. Environmental Pollution, v. 157, p. 2470-2484, 2009.

[52] Polard, T., Jean, S., Gauthier, L., Laplanche, C., Merlina, G., Sánches-Pérez, J.M., Pinelli, E. Mutagenic impact on fish of runoff events in agricultural areas in south west France. Aquatic Toxicology, v.17, p. 126-134, 2011.

[53] Moraes, D.S.L. Avaliação dos potenciais tóxicos, citotóxicos e genotóxicos de águas ambientais de Corumbá-MS em raízes de *Allium cepa*. 2000. 158 f. Tese (Mestrado em Genética e Melhoramento) – Universidade Estadual de Londrina, Londrina.

[54] Mccarthy, J.F.; Shugart, L.R. Biomarkers of environmental contamination. 1 ed. Estados Unidos: Lewis, 1990. 382p.

[55] Bombail, V.; Dennis, A.W.; Gordon, E.; Batty, J. Application of the comet and micronucleus assays to butterfish (*Pholis gunnellus*) erythrocytes from the Firth of Forth, Scotland. Chemosphere, Oxford, v. 44, p. 383-392, 2001.

[56] Peña, L.F.M. Uso do teste de micronúcleo em eritrócitos circulantes de peixes para monitorização de um local do rio Tibagi e avaliação da genotoxidade de agrotóxicos em bioensaios. 1996. 199 f. Tese (Mestrado em Genética e Melhoramento) – Universidade Estadual de Londrina, Londrina.

[57] Veiga, A.B. O uso do teste de *Allium cepa* para detectar a toxicidade do inseticida Nuvacron. 1995, 58 f. Monografia (Conclusão do Curso de Ciências Biológicas) - Universidade Estadual de Londrina, Londrina.

[58] Almeida, W. F. Acúmulo de inseticidas no homem e sua significação epidemiológica. O biológico, São Paulo, v. 6, p. 171-183, 1974.

[59] Pavanelli, E.A.S. Efeito de biocidas sobre a polinização e germinação de sementes de orquídeas dos gêneros *Cattleya* Lsl. e *Laelia* Lsl. (Orchidaceae). 1995. 179 f. Tese (Doutorado em Botânica) – Instituto de Biociências, Universidade Estadual Paulista, Rio Claro.

[60] Dassenoy, B.; Meyer, J.A. Mutagenic effects of benomyl on *Fucarion oxysporum*. Mutation Research, Amsterdam ,v .21, p. 119-120, 1973.

[61] Sakamoto, E.T.; Takahashi, C.S. Efeitos dos fungicidas Dithane M-45, Benlate e Vitavax 75 PM sobre os índices mitóticos dos meristemas radiculares de *Allium cepa*. In: Anais da SBPC, São Paulo, v. 31, p. 573, 1979.

[62] Ventura, B. C. Avaliação dos efeitos citotóxicos, genotóxicos e mutagênicos do herbicida atrazine, utilizando *Allium cepa* e *Oreochromis niloticus* como sistemas-testes. 2004. 105f. Dissertação (Mestrado em Biologia Celular e Molecular) – Instituto de Biociências, Universidade Estadual Paulista, Rio Claro, 2004.

[63] Terracini, B. Valutazione della carcinogenecita deghi idrocarburi clorutati usati come pesticide. Tumori, Milano, v. 53, p. 601-618, 1977.

[64] HAgMAR, L.; Bonassi, S.; Stromberg, U.; Brogger, a.; Knudsen, L.E.; Norppa, H.; Reuterwall, C. Chromosomal aberrations in lymphocytes predict cancer: a report from the European Study Group on Cytogenetic Biomarkers and Health (ESCTH). Cancer Research, Baltimore, v. 58, p. 4117-4121, 1998.

[65] ZAKERI, Z.; LOCKSHIN, R.A. Cell death during development. Journal of Immunological Methods, [S.l.] , v. 265, p. 3-20, 2002.

[66] KRISTEN, U. Use of higher plants as screens for toxicity assessment. Toxicology in vitro, United Kingdom, v. 11, p. 181-191, 1997.

[67] Grisolia, C.K.; Starling, F.L.R.M. Micronuclei monitoring of fishes from Lake Paranoá, under influence of sewage treatment plant discharges. Mutation Research, Amsterdam, v. 491, p. 39-44, 2001.

[68] Monteith, D.K.; Vanstone, J. Comparison of the microgel electrophoresis assay and other assays for genotoxicity in the detection of the DNA damage. Mutation Research, Amsterdam, v. 345, n. 3-4, p. 97-103, 1995.

[69] Ribas, G.; Frenzili, G.; Barale, R.; Marcos, R. Herbicide-induced DNA damage in human lymphocytes evaluated by the single-gel electrophoresis (SCGE) assay. Mutation Research, Amsterdam, v. 344, p. 41-54, 1995.

[70] Popa, N.E.; Zakrzhevskaya, A.M.; Kozhokaru, R.V.; Enaki, D.K. Cytogenetic effect of some herbicides on maize seedlings. Weed Abstracts, Farnham Royal, v. 35, n. 1, p. 50, 1986.

[71] Eldridge, J.C; Wetzel, L.T.; Stevens, J.T.; Simpkins, J.W. The mammary tumor response in triazine-treated female rats: a threshold-mediated interaction with strain and species-specific reproductive senescence. Steroids, Califórnia, v. 4, p. 672-678, 1999.

[72] Worthing, C.R.; Walker, S.B. The pesticide manual. 7 ed. U.K.: The Lavenham Press, 1983. 589p.

[73] Goldman, L.R. Atrazine, simazine and cyanazine: Notice of initiation of special review in federal register, Estados Unidos, s.n.: 60412-60443. 1994.

[74] Ribas, G.; Surrallés, J.; Carbonell, E.; Creus, A.; Xamena, N.; Marcos, R. Lack of genotoxicity of the herbicide atrazine in cultured human lymphocytes. Mutation Research, Amsterdam, v. 416, p. 93-99, 1998a.

[75] Summer, D.D.; Cassidy, I.M.; Szolics, I.M.; Marco, G.J. Evaluation of the mutagenic potential of corn (*Zea mays* L.) grown in untreated and a atrazine (A Atrex) treated soil in the field. Drug Chemical and Toxicology, [S.l.] , v. 7, p. 243-257, 1984.

[76] Kappas, A. On the mutagenic and the recombinogenic activity of ceratin herbicides in *Salmonella typhimurium* and in *Aspergillus nidulans*. Mutation Research, Amsterdam, v. 204, p. 615-621, 1988.

[77] Butler, M.A.; Hoagland, R.E. Genotoxicity assessment of atrazine and some major metabolities in the Ames test. Bulletin of Environmental Contamination and Toxicology, Florida, v. 43, p. 797-804, 1989.

[78] Murnik, M.R.; Nash, C.L. Mutagenicity of triazine herbicides atrazine, cyanazine, and simazine in *Drosophila melanogaster*. Journal of Toxicology Environmental Health, [S.l.] , v. 3, p. 691-697, 1977.

[79] Torres, C.; Ribas, G.; Xamena, N.; Creus, A.; Marcos, R. Genotoxicity of four herbicides in *Drosophila* wing spot tests. Mutation Research, Amsterdam, v. 280, p. 291-295, 1992.

[80] Liang, G.H.L.; Feltner, K.C.; Liang, Y.T.S.; Morrill, J.L. Cytogenetic effects and responses of agronomic characters in grain sorghum (*Sorghum vulgare* Pers.) following atrazine application. Crop Science, New York, v. 7, n. 3, p. 245-248, 1967.

[81] Grant, W.F.; Owens, E.T. Chromosome aberration assays in *Pisum* for the study of environmental mutagens. Mutation Research, Amsterdam, v. 188, p. 93-118, 2001.

[82] Hayes, T., Haston, K., Tsui, M., Hoang, A., Haeffele, C., Vonk, A. Herbicides: Feminization of male frogs in the wild. Nature, v.419, p.895-896, 2002.

[83] Gammon, D.W.; Aldous, C.N.; Carr Jr, W.C.; Sanborn, J.R.; Pfeifer, K.F. A risk assessment of atrazine use in California: human health and ecological aspects. Pest Manag Sci, v. 61, p. 331-355, 2005.

[84] Çavas, T. In vivo genotoxicity evaluation of atrazine and atrazine–based herbicide on fish *Carassius auratus* using the micronucleus test and the comet assay. Food and Chemical Toxicology, v. 49, p. 1431-1435, 2011.

[85] Meisner, L.F.; Belluck, D.A.; Rolloff, B.D. Cytogenetic effects of alachlor and/or atrazine *in vivo* and *in vitro*. Environmental and Molecular Mutagenesis, New York, v. 19, p. 77-82, 1992.

[86] Roloff, B.D.; Belluck, D.A.; Meisner, L.F. Cytogenetic studies of herbicide interactions *in vitro* and *in vivo* using atrazine and linuron. Environmental Toxicology, New York, v. 22, p. 267-271, 1992.

[87] Wu, Y.G.; Li, S.K.; Xin, Z.C.; Wang, Y.S.; Shou, K.R.; Gao, H.; Li, Y.Q. The establishment of hypospadias rat model and embryoteratogenic test of Atrazine. Zhonghua Zheng Xing Wai Ke Za Zhi, v. 23(4), p. 340-343, 2007.

[88] Modic, W.; Ferrell, J.; Wood, C.; Laskey, J.; Cooper, R.; Laws, S. Atrazine alters steroidogenesis in male Wistar rats. Toxicologist, v. 78, p. 117, 2004.

[89] Tennant, A.H.; Peng, B; Kligerman, A.D. Genotoxicity studies of triazine herbicides: in vivo studies using the alkaline single gel (SCG) assay. Mutation Research, Amsterdam, v. 493, p. 1-10, 2001.

[90] Clements, C.; Ralph, S.; Petras, M. Genotoxicity of selected herbicides in tadpoles *Rana catesbeiana,* using the alkaline single-cell gel DNA electrophoresis (comet) assay. Environmental and Molecular Mutagenesis, New York, v. 29, p. 277-288, 1997.

[91] Campos-Pereira, F.D.; Oliveira, C.A; Pigoso, A.A.; Silva-Zacarin, E.C.M; Barbieri, R.; Spatti, E.F.; Marin-Morales, M.A.; Severi-Aguiar, G.D.C. Early cytotoxic and genotoxic effects of atrazine on Wistar rat liver: A morphological, immuno-histochemical, biochemical, and molecular study. Ecotoxicology and Environmental Safety, v. 78, p. 170-177, 2012.

[92] Ruiz, M.J., Marzin, D. Genotoxicity of six pesticides by *Salmonella* mutagenicity test and SOS chromotest. Mutation Research, v. 390, p. 245-255, 1997.

[93] Lioi, M.B.; Scarfi, M.R.; Santoro, A.; Barbieri, R.; Zeni, O.; Salvemini, F.; Berardino, D.D.; Ursini, M.V. Cytogenetic damage and induction of pro-oxidant state in human lymphocytes exposed in vitro to gliphosate, vinclozolin, atrazine, and DPX-E9636. Environmental and Molecular Mutagenesis, New York, v. 32, p. 39-46, 1998.

[94] Meisner, L.F.; Roloff, B.D.; Belluck, D.A. *In vitro* effects of N-nitrosoatrazine on chromosome breakage. Environmental Toxicology, New York, v. 24, p. 108-112, 1993.

[95] Loprieno, N.; Adler, I.D. Cooperative Programme of the EEC on short-term assays for mutagenicity. In: MONTESANO, R.; BARTSCH, H.; TOMATIS, L. (eds.), Molecu-

lar and Cellular aspects of carcinogen screening tests, France: Science Publisher, 1980. p. 331-341.

[96] Wuu, K.D.; Grant, W.F. Morphological and somatic chromosomal aberrations induced by pesticides in barley (*Hordeum vulgare*). Canadian Journal of Genetic and Cytology, [S.l.] , v. 8, p. 481-501, 1966.

[97] Wuu, K.D.; Grant, W.F. Chromosomal aberrations in somatic cells of *Vicia faba* by pesticides. Nucleus, [S.l.] , v. 10, p. 37-46, 1967.

[98] Plewa, M.J.; Wagner, E.D.; Gentile, G.J.; Gentile, J.M. An evaluation of the genotoxic properties of herbicides following plant and animal activation. Mutation Research, Amsterdam, v. 136, p. 233-245, 1984.

[99] Lee, K.C.; Rao, G.M.; Barnett, F.L.; Liang, G.H. Further evidence of meiotic instability induced by atrazine in grain sorghum. Cytologia, Tokyo, v. 34, p. 697-702, 1974.

[100] Chou, T.S.; Weber, D.F. The effect of the atrazine on sister-chromatid exchanges in maize. Genetics, Califórnia, v. 97, p. 521, 1981.

[101] Zeljezic, D.; Garaj-Vrhovac, V.; Perkovic, P. Evaluation of DNA damage induced by atrazine and atrazine-based herbicide in human lymphocytes in vitro using a comet and DNA diffusion assay. Toxicology in Vitro, v. 20, p. 923-935, 2006.

[102] Srivastava, K.; Mishra, K.K. Cytogenetic effects of commercially formulated atrazine on the somatic cells of *Allium cepa* and *Vicia faba*. Pesticides Biochemistry and Physiology, v. 93, p. 8-12, 2009.

[103] Kligerman, A.D.; Doerr, C.L.; Tennant, A.H.; Peng, B. Cytogenetic studies of three triazine herbicides II. In vivo micronucleus studies in mouse bone marrow. Mutation Research, v. 471, p. 107-112, 2000.

[104] Kligerman, A.D.; Chapin, R.E., Erexson, G.L.; Germolec, D.R.; Kwanyuen, P.; Yang, R.S. Analyses of cytogenetic damage in rodents following exposure to simulated groundwater contaminated with pesticides and a fertilizer. Mutation Research, Amsterdam, v. 300, p. 125-134, 1993.

[105] Adler, I.D. A review of the coordinated research effort on the comparison of the test systems for the detection of mutagenic effects, sponsored by the E.C.C. Mutation Research, Amsterdam, v. 74, p. 77-93, 1980.

[106] Hrelia, P.; Vigagni, F.; Maffei, F.; Morotti, M.; Colacci, A.; Perocco, P.; Grilli, S.; Cantelli-Forti, G. Genetic safety evaluation of pesticides in different short-term tests. Mutation Research, v. 321, p. 219-228, 1994.

[107] Taets, C.; Aref, S.; Rayburn, A.L. The Clastogenic Potential of Triazine Herbicide Combinations Found in Potable Water Supplies. Environmental Health Perspectives, v. 106 (4), 1998.

[108] Moretti, M.; Marcarelli, M.; Villarini, M.; Fatigoni, C.; Scassellati-Sforzolini, G.; Pasquini, R. *In vitro* testing for genotoxicity of the herbicide terbutryn: cytogenetic and primary DNA damage. Toxicology in Vitro, v. 16, p. 81-88, 2002.

[109] Clausen, M.; Leier, G.; White, I. Comparison of the cytotoxicity and DNA-damaging properties of 2,4-D and U 46 D fluid (dimethylammonium salt of 2,4-D). Archives of Toxicology, v. 64, p. 497-501, 1990.

[110] IARC. Some fumigants, the herbicides 2,4-D and 2,4,5-T, chlorinated dibenzodioxins and miscellaneous industrial chemicals. IARC Monogr Eval Carcinog Risk Chem Man 1977;15:111-48.

[111] Osterloh, J.; Lotti, M.; Pond, S.M. Toxicologic studies in a fatal overdose of 2,4-D, MCPP, and chlorpyrifos. J Anal Toxicol, v. 7, p. 125-129, 1983.

[112] Martínez-Tabcge, L.; Madrigal-Bujaidar, E.; Negrete, T. Genotoxicity and lipoperoxidation produced by paraquat and 2,4-Dichlorophenoxyacetic acid in the gills of rainbow trout (*Oncorhynchus mikiss*). Bull Environ Contam Toxicol, v. 73, p.146-152, 2004.

[113] Ateeq, B.; Abdul-Farah, M.; Ali, M.N.; Ahmad, W. Induction of micronuclei and erythrocyte alterations in the catfish *Clarias batrachus* by 2,4-dichlorophenoxyacetic acid and butachlor. Mutation Research, Amsterdam, v. 518, p. 135-144, 2002.

[114] Suwalsky, M.; Quevedo, L.; Norris, B.; Benites, M. Toxic Action of the Herbicide 2,4-D on the Neuroepithelial Synapse and on the Nonstimulated Skin of the Frog *Caudiverbera caudiverbera*. Bull. Environ. Contam. Toxicol., v. 62, p. 570-577, 1999.

[115] Gómez, L.;, Masot, J.; Martínez, S.; Durán, E.; Soler, F.; Romero, V. Acute 2,4-D poisoning in tench (*Tinca tinca* L.): lesions in the hematopoietic portion of the kidney. Arch Environ Contam Toxicol, v. 35, p. 479-483, 1998.

[116] González, M.; Soloneski, S.; Reigosa, M.A.; Larramendy, M.L. Effect of the herbicide 2,4-dichlorophenoxyacetic acid (2,4-D) and its derivative 2,4-D dichlorophenoxyacetic acid dimethylamine salt (2,4-D DMA). I. Genotoxic evaluation on Chinese hamster ovary (CHO) cells. Toxicology in Vitro, v. 19, p. 289-297, 2005.

[117] Madrigal-Bujaidar, E.; Hernandez-Ceruelos, A.; Chamorro, G. Induction of sister chromatid exchanges by 2,4-dichlorophenoxyacetic acid in somatic and germ cells of mice exposed *in vivo*. Food Chem Toxicol, v. 39, p. 941-946, 2001.

[118] Soloneski, S.; González, N.V.; Reigosa, M.A.; Larramendy, M.L. Herbicide 2,4-dichlorophenoxyacetic acid (2,4-D)-induced cytogenetic damage in human lymphocytes *in vitro* in presence of erythrocytes. Cell Biology International, v. 31, p. 1316-1322, 2007.

[119] Morgan, M.K.; Scheuerman, P.R.; Bishop, C.S.; Pyles, R.A. Teratogenic potential of atrazine and 2,4-D using FETAX. J Toxicol Environ Health, v. 48 (2), p.151-168, 1996.

[120] Xie, L; Thrippleton, K.; Irwin, M.A.; Siemering, G.S.; Mekebri, A.; Crane, D.; Berry, K.; Schlenk D. Evaluation of estrogenic activities of aquatic herbicides and surfac-

tants using an rainbow trout vitellogenin assay. Toxicology Science, v. 87(2), p. 391-398, 2005.

[121] Smith, E.A.; Oehme, F.W. The biological activity of glyphosate to plants and animals; a literature review. Vet. Hum. Toxicol., v. 34, p. 531-543, 1992.

[122] Mladinic, M.; Berend, S.; Vrdoljak, A.L.; Kopjar, N.; Radic, B.; Zeljezic, D. Evaluation of genome damage and its relation to oxidative stress induced by glyphosate in human lymphocytes *in vitro*. Environmental and Molecular Mutagenesis, v. 50, p. 800-807, 2009a.

[123] Martini, C.N.; Gabrielli, M.; Vila, M.D.C. A commercial formulation of glyphosate inhibits proliferation and differentiation to adipocytes and induces apoptosis in 3T3-L1 fibroblasts. Toxicology in Vitro, v. 26, p. 1007-1013, 2012.

[124] Solomon, K.; Thompson, D. Ecological Risk Assessment for Aquatic Organisms from Over-Water Uses of Glyphosate. Journal of Toxicology and Environmental Health, Part B: Critical Reviews, v. 6 (3), p. 289-324, 2003.

[125] Williams, G.M.; Kroes, R.; Munro, I.C. Safety Evaluation and Risk Assessment of the Herbicide Roundup and Its Active Ingredient, Glyphosate, for Humans. Regulatory Toxicology and Pharmacology, v. 31, p. 117-165, 2000.

[126] Cox, C. Herbicide factsheet – Glyphosate. J Pest Reform, v. 24, p. 10-15, 2004.

[127] Clair, E.; Linn, L.; Travert, C.; Amiel, C.; Séralini, G.E.; Panoff, J.M. Effects of Roundup(®) and glyphosate on three food microorganisms: *Geotrichum candidum, Lactococcus lactis subsp. cremoris and Lactobacillus delbrueckii subsp. bulgaricus.* Curr Microbiol., v. 64 (5), p. 486-491, 2012.

[128] Relyea, R.A. New effects of Roundup on amphibians: predators reduce herbicide mortality; herbicides induce antipredator morphology. Ecol. Appl., v. 22 (2), p. 634-647, 2012.

[129] Dimitrov, B.D.; Gadeva, P.G.; Benova, D.K.; Bineva, M.V. Comparative genotoxicity of the herbicides Roundup, Stomp and Reglone in plant and mammalian test systems. Mutagenesis, v. 21 (6), p. 375-382, 2006.

[130] Dallegrave, E.; Mantese, F.D.; Coelho, R.S.; Pereira, J.D.; Dalsenter, P.R.; Langeloh, A. The teratogenic potential of the herbicide glyphosate-Roundup in Wistar rats. Toxicology Letters, v. 142, p. 45-52, 2003.

[131] WHO (World Health Organization), 1994. Glyphosate. Environmental Health Criteria. 159, pp. 1-177.

[132] Relyea, R.A., The impact of insecticides and herbicides on the biodiversity and productivity of aquatic communities. Ecological Applications, v. 15 (2), p. 618-627, 2005.

[133] Fetvadjieva, N.; Straka, F.; Michailova, P.; Balinov, I.; Lubenov, I.; Balinova, A.; Pelov, V.; Karsova, V.; Tsvetkov, D. In: Fetvadjieva,N. (ed.), Handbook of Pesticides. 2nd revised edn. Zemizdat Inc., Sofia, pp. 330, 1994.

[134] Tortorelli, M.C.; Hernandez, D.A.; Rey Vazquez, G.; Salibian, A. Effects of paraquat on mortality and cardiorespiratory function of catfish fry *Plecostomus commersoni*. Arch Environ Contam Toxicol, v. 19, p. 523-529, 1990.

[135] Ribas, G.; Surrallés, J.; Carbonell, E.; Xamena, N.; Creus, A.; Marcos, R.. Genotoxic Evaluation of the Herbicide Paraquat in Cultured Human Lymphocytes. Teratogenesis, Carcinogenesis, and Mutagenesis, v. 17, p. 339-347, 1998b.

[136] Hanada, H. Dl-α-tocopherol enhances the herbicide 1,1′-dimetyl-4,4′–bipyridium dichloride (Paraquat, PQ) genotoxicity in cultured anuran leukocytes. Hereditas, v. 148, p. 118-124, 2011.

[137] Tomita, M.; Okuyama, T.; Ishikawa, T.; Idaka, K.; Nohno, T. The role of nitric oxide in paraquat-induced cytotoxicity in the human A549 lung carcinoma cell line. Free Rad Res, v. 34, p. 193-202, 2001.

[138] D'souza, U.J.; Narayana, K.; Zain, A.; Raju, S.; Nizam, H.M.; Noriah, O. Dermal exposure to the herbicide-paraquat results in genotoxic and cytotoxic damage to germ cells in the male rat. Folia Morphol (Warsz), v. 65 (1), p. 6-10, 2006.

[139] Bus, J.S.; Aust, S.D.; Gibson, J.E. Superoxide and singlet oxygen-catalyzed lipid peroxidation as a possible mechanism for paraquat (methyl viologen) toxicity. Biochem. Biophys. Res. Comm., v. 58, p. 749-755, 1974.

[140] Tanaka, R.; Amano, Y. Genotoxic effects of paraquat and diquat evaluated by sister-chromatid exchange,chromosomal aberration and cell-cycle rate. Toxicology in Vitro, v. 3, p. 53-57, 1989.

[141] Bauer Dial, C.A.; Dial, N.A. Lethal effects of consumption of fi eld levels of paraquat-contaminated plants on frog tadpoles. Bull. Environ. Contam. Toxicol., v. 55, p. 870-877, 1995.

[142] Speit, G.; Haupter, S.; Hartmann, A. Evaluation of the genotoxic properties of paraquat in V79 Chinese hamster cells. Mutation Research, v. 412 (2), p. 187-193, 1998.

[143] Hoffman, D.J.; Eastin Jr, W.C. Effects of lindane, paraquat, toxaphene, and 2,4,5-trichlorophenoxyacetic acid on mallard embryo development. Arch Environ Contam Toxicol., v. 11, p. 79-86, 1982.

[144] Zeliger, H.I., Human toxicology of chemical mixtures. In: Toxic Consequences Beyond the Impact of One-component Product and Environmental Exposures. 2nd ed. Elsevier, Oxford, 2011.

[145] Bassil, K.L.; VAKIL, C.; SANBORN, M.; COLE, D.C.; KAUR, J.S.; KERR, K.J. Cancer health 585 effects of pesticides: systematic review. Can. Fam. Physician, v. 53, p. 1704-1711, 2007.

[146] Parrón, T.; Requena, M.; Hernández, A.F.; Alarcón, R. Association between environmental exposure to pesticides and neurodegenerative diseases. Toxicol. Appl. Pharmacol., v. 256, p. 379-385, 2011.

[147] Hanke, W.; Jurewicz, J., The risk of adverse reproductive and developmental disorders due to occupational pesticide exposure: an overview of current epidemiological evidence. Int. J. Occup. Med. Environ. Health, v. 17, p. 223-243, 2004.

[148] Hernández, A.F.; Parrón, T.; Alarcón, R. Pesticides and asthma. Curr. Opin. Allergy Clin. Immunol., v. 11, p. 90-96, 2011.

[149] Doll, R.; Peto, R. The causes of cancer: quantitative estimates of avoidable risks of cancer in the United States today. Journal National Cancer Institute, v. 66, p. 1191-1308, 1981.

[150] Hernández, A.F.; Parrón, T.; Tsatsakis, A.M.; Requena, M.; Alarcón; López-Guarnido, O. Toxic effects of pesticide mixtures at a molecular level: Their relevance to human health. Toxicology, 2012 (*in press*).

[151] Jones, F.; Fawell, J.K. Lessons learnt from the river DEE Pollution Incident. Public Health in Proceedings of the Word conference on chemicals accidents, Harvard, v. 4, p. 223-226, 1987.

[152] Waters, M.D.; Stack, H.F.; Jackson, M.A. Genetic toxicology data in the evaluation of potential human environmental carcinogens. Mutation Research, Amsterdam, v. 437, p. 21-49, 1999.

[153] USEPA. Atrazine: Carcinogenicity characterization and hazard assessment, office of pesticide programs, health effects division. 1999. Http:/ www.epa.gov/scipoly/sap/ #jan

[154] EPA, 2003. US Environmental Protection Agency. October 31, 2003, revised atrazine interim reregistration eligibility decision (IRED). Office of Prevention, Pesticides and Toxic Substances, 2003.

[155] Jowa, L.; HOWD, R. Should atrazine and related chlorotriazines be considered carcinogenic for human health risk assessment? J Environ Sci Health C Environ Carcinog Ecotoxicol Rev., v. 29 (2), p. 91-144, 2011.

[156] Mladinic, M.; Perkovic, P.; Zeljezic, D. Characterization of chromatin instabilities induced by glyphosate, terbuthylazine and carbofuran using cytome FISH assay. Toxicology Letters, v. 189, p. 130-137, 2009b.

[157] Gebel, T.; Kevekordes, S.; Pav, K.; Edenharder, R.; Dunkelberg, H. *In vivo* genotoxicity of selected herbicides in the mouse bone-marrow micronucleus test. Arch. Toxicol., v. 71, 193-197, 1997.

[158] Mladinic, M.; Zeljezic, D.; Shaposhnikov, S.A.; Collins, A.R. The use of FISH-comet to detect c-Myc and TP 53 damage in extended-term lymphocyte cultures treated with terbuthylazine and carbofuran. Toxicology Letters, v. 211, p. 62-69, 2012.

[159] Hardell, L.; Eriksson, M.; Nordstrom, M. Exposure to pesticides as risk factor for non-Hodgkin's lymphoma and hairy cell leukemia: Pooled analysis of two Swedish case-control studies. Leuk Lymp, v. 43, p. 1043-1049, 2002.

[160] De Roos, A.J.; Zahm, S.H.; Cantor, K.P.; Weisenburger, D.D.; Holmes, F.F.; Burmeister, L.F.; Blair, A. Integrative assessment of multiple pesticides as risk factors for non-Hodgkin's lymphoma among men. Occup Environ Med, v. 60, p.11, 2003.

[161] He, X.; Wang, L.; Szklarz, G.; Bi, Y.; Ma, Q. Resveratrol Inhibits Paraquat-Induced Oxidative Stress and Fibrogenic Response by Activating the Nuclear Factor Erythroid 2-Related Factor 2 Pathway. The Journal of Pharmacology and Experimental Therapeutics, v. 342, p. 81-90, 2012b.

[162] Ossowska, K; Smiałowska, M.; Kuter, K.; Wieron′ Ska, J.; Zieba, B.; Wardas, J.; Nowak, P.; DABROWSKA, J.; BORTEL, A.; BIEDKA, I. et al. Degeneration of dopaminergic mesocortical neurons and activation of compensatory processes induced by a long-term paraquat administration in rats: implications for Parkinson's disease. Neuroscience, v. 141, p. 2155-2165, 2006.

[163] Tanner, C.M.; Kamel, F.; Ross, G.W.; Hoppin, J.A.; Goldman, S.M.; Korell, M.; Marras, C.; Bhudhikanok, G.S.; Kasten, M.; Chade, A.R. et al. Rotenone, paraquat, and Parkinson's disease. Environ Health Perspect, v. 119, p. 866-872, 2011.

[164] Garaj-Vrhovac, V.; Zeljezic, D. Assessment of genome damage in a population of Croatian workers employed in pesticide production by chromosomal aberration analysis, micronucleus assay and Comet assay. Journal of Applied Toxicology, Chichester, v. 22, n. 4, p. 249-255, 2002.

Chapter 17

Herbicide Resistant Weeds: The Technology and Weed Management

Jamal R. Qasem

Additional information is available at the end of the chapter

http://dx.doi.org/10.5772/56036

1. Introduction

Pest resistance to control methods in general is not an isolated phenomenon but usually expected and well demonstrated when any method is repeatedly applied over a long period of time without being changed or modified in nature, structure, principals of application or formulation. All pests that growers must control in agricultural land have the capacity to become resistant to whatever tactic is used to control them [11]. It is usually expressed as a gradual adaptation or "fitness" of some individuals or populations of the targeted pest or organism to the frequently applied control methods and available conditions. This adaptation may be physical, morphological or phenological, physiological, anatomical or biochemical or could result from the interaction between any two or more of these. It may also be due to some genetic changes as mutations occur on the key site at which a specific method operates. These mutations are at least partially dominant and inherited. Traits are conferred by modifications to single nuclear genes. This indicates that the rate of resistance evolution will be driven by mutation, the intensity of selection, the dominance and relative fitness of mutations in presence or absence of the herbicide and by dispersal of resistance alleles within and between weed populations [28]. However, no proof that the herbicides cause the mutations leads to resistance [37]. However, most often resistance is controlled by a single, dominant or semi-dominant gene [38] although recessive genes control of herbicide resistant trait in natural weed populations has been also implicated in resistance to dintroanaline, while wild populations exposed to herbicide stresses for the first time may efficiently express herbicide-resistant genes.

Most weed modifications and adaptations, if not all, are advantageous to the pest, since allow its escape on time and/or place and thus avoid external hazard or threat to its existence and genetic line. Resistance therefore should not be confused with natural tolerance or low susceptibility due to a normal physiological or behavioristic property of an unselected population [23].

Organisms are varied in sensitivity, responses and thus adaptability to such conditions and in responses to any treatment or imposed external factors. Tolerance and then gradual resistance of agricultural pests to any control method or environmental stress is thus a strategy through which organisms/ or pests encounter hazards and maintain life and therefore may be applied to any method of pest or weed control including prevention, mechanical, cultural, physical, biological and chemical [30]. For example, weeds resisting soil mulch cover show some morphological and/or physical characteristics that allow penetration of the mulch layer; also, flooding of resistant species possess water impermeable seed coat or generate O_2 and reduce CO_2 penetration. Firing or flaming is resisted through presence of a hard seed coat or deeply buried regenerative propagules; certain weed species show feedback mechanisms or luxury accumulation of mineral nutrients and thus avoid toxicity; high temperature and low soil moisture harmful effects are avoided by adoption of secondary or enforced seed dormancy, while harmful effects of excessive light is avoided by some morpho-physiological alterations. Soil acidity may be encountered in the microhabitat by root exudates or selective mineral absorption and salinity by excretion of salt through different mechanisms and formation of salt glands or vacuoles or shedding salt saturated organs; microbes attack is avoided by production of repellent allelochemicals, and pests through some morpho-chemical adaptations. However, the mechanism behind tolerance or resistance is different and based on the type of target pest or the hazard imposed.

Herbicides represent one of the external factors and form a group of synthetic- plus some biochemicals used to suppress or kill unwanted vegetation and are a major component of pesticides. They assist in management and restoration of areas invaded by invasive species. Herbicides are a major technological tool and responsible, in part, for an agricultural revolution and increase in food production in the last few decades. However, at present this technology faces radical changes in effectiveness under field conditions that lead in different cases to failure of weed control operation due to continued development of weed tolerance/resistance and evolution and limitations in the herbicide industry and development.

2. Agriculture practices and weed evolution

General weed control methods (tillage, hoeing, hand weeding, flooding, cuttings or mowing, flaming, use of general herbicides) are all nonselective and usually applied to a composite weed species or vegetation of inter and intra-specific variations in richness, morphology, growth habit and responses. Each species may adapt, or not, to any of these methods. Since weeds are widely different in mechanisms by which they encounter hazards they are exposed to, they are different in plasticity and responses. With continued use of a single control method for a long period of time, species migrate, flourish or die. Flourishing species gradually became better fit and adapted, and increase in number and population size in absence of others. The only surviving individuals are those possessing rare single gene mutations and evolved resistance will be monogenic, resulting in a large change in the resistance phenotype. However, when doses are lower and selection acts within the range of standing genetic variation, polygenic responses will be possible and resistance will evolve by a gradual change in the mean susceptibility of the population [28]. On the other hand, population of not or less adapted individuals, decline in growth and number until greatly suppressed, limited and may become extinct. Therefore, with continuous dependence on a single method of weed control, a weed population is usually shifting toward better adapted species or individuals that cope well with existing control measures and new conditions. Self-thinning of a weed population is continued toward complete tolerance to employed control measures. Therefore, weeds adapted to mowing tend to grow short, in a rosette form, creeping above the soil surface or show high plasticity and softness of aerial parts and stems and become difficult to mow and also escape hand weeding. Deep rooted weed species are difficult to pull out even by soil tillers. Seasonal dormancy and shifts in the weed population in the growing season is well recognized for certain weed species such as *Senecio vulgaris* [29; 37], while physiological adaptation of *Echinochloa crusi-galli* and *Cyperus rotundus* to flooding conditions and the role of Alcohol dehydroginase enzyme (Adh) in *E. crusi-galli* is well documented [5; 14]. Similar adaptations of *Cirsium arvense* ecotypes to temperature variations [43] and *Typha anguistifolia* and *Typha latifolia* genetic and clonal variations [27; 40] have also been reported. In this regard, it is important to differentiate between tolerance and resistance of weeds to herbicides. Tolerance is the inherited ability of a species to survive and reproduce after herbicide treatment; it refers to the natural variability to herbicides and exists within individuals of a species and quickly evolves. It usually refers to relatively minor or gradual differences in intraspecific variability. Resistance is the inherited ability of a plant or a biotype to survive and reproduce following exposure to a dose of herbicide that is normally lethal to wild type [16; 23; 30; 37]. Therefore, it is a decreased response of a population of weed to herbicides as a result of their application. However, both terms sometimes are misused or used interchangeably.

Tolerant weed species are less harmed by herbicides; they exhibit a certain degree of avoidance or adaptation strategy that allows recovery and thus escape control measures. They may respond by timing stomata closure or having sunken pores or stomata, thick waxy cutical on upper leaf surface, encased growing points or some biochemical, physiological or anatomical properties better developed by time until they become best fit and adapted to applied herbicides and become thereafter resistant. This, however, leads to gradual but radical changes in the weed population composition and distribution spectrum at which resistant individuals or certain weed species increased and dominate and susceptible ones are reduced and replaced. Adaptation or exclusion of the less tolerant species depends on performance of these by time. Generally a weed population becomes rich in individuals and poor in species with the continuous use of the same herbicide or different herbicides of similar mode/mechanism of action. This shift does not however, reflect better competitiveness or higher regenerative ability but most likely due to absence of sensitive highly competing species or forms that allow resistant individuals to utilize more resources [9; 22].

In cultivated fields, associating weeds bear more resemblance to crop plants in morphology, physiology and responses to control measures and other agricultural practices in general. They mimic crops from sowing and germination until harvest. Since herbicides used on crop plants are selective, weeds respond by exhibiting similar morphology, physiology and biochemistry as crop plants to avoid hazards. However, weeds derived from crop plants as hybrids, crop relatives or wild-weedy forms are better fit to such conditions than others. Weed-crop associations also exist between weed species of different taxa from crop plants. In this case, the longer the use of the same herbicide/s, the greater the close association between crops and certain well performed weed species that later transfer into adapted weed races. Crop relative weeds however, are of great potential to intra- and inter- gene exchange and efficient mating system among themselves and with crops, thus become best adapted and more difficult to control.

3. Selection pressure and weed races

With continuous use of the same agricultural practice/s, interspecies selection occurs and plant species are gradually purified (intraspecific selection) by time until they become best adapted. Since all control measures including herbicides aim to eliminate weeds without causing injury to crop plants, weeds respond by developing mechanism/s allowing escape of chemical hazards. Under such conditions, sensitive individuals are first limited or disappear. Tolerant individuals increase in number and accumulate tolerance until they become resistant. Therefore, a resistant population of any weed species is exposed to long-term selection pressure through which it is purified and performs well under prevailing conditions in absence of sensitive weed species. With continuous exposure to herbicide pressure, a population of resistance is usually developed.

Weeds tend to avoid herbicide toxicity by changing normal growth habits, or exhibiting some phenological (such as changes in germination patterns), physical and/or physiological changes through which they adjust emergence time, external appearance or physiology. These however, are inherited traits that allow plants to survive herbicide treatments. One best adaptation is that of weeds similar to crop plants in most or all growth aspects. These form weed races similar to crop plants and well adapted to their habitats. Among reported weed races are *Camelina sativa* to flax crop, *Echinochloa crus-galli var. Oryzicola* that associate with rice and the weedy wild rice or red rice in India and east-south Africa [8; 20]. All are genetically irrelevant to crop plants. However, in some cases weed races are of the same botanical family or belong to the same crop species. This kind of association leads to development of "crop-races" that possess weedy characters very well adapted to cultural practices; they are similar to crop plants in most growth aspects and difficult to control by herbicides or other control methods including hand weeding. They take an advantage from conditions under which crop plants are growing until they become difficult to leave their habitats or even become dependent on crop plants in their growth and environment. These weeds are specialized to certain crop plants or cultivars. Moreover, many genetically related species can exchange genes with crop individuals and mimic crops. It can be concluded that any agricultural practice exerts selection pressure and may become troublesome to farmers when repeatedly applied for a long period. Its positive impact on crop growth and productivity is usually negated with time until it becomes a real trouble. Its residual negative effects may not possible to overcome for a long period after abandonment.

4. Field evidence of weed resistance and herbicide resistance protocol

In the field all growth patterns and distribution of weed species may be observed. Some species grow in colonies, in certain growth patterns, forming an ecological niche, sporadically distributed, or randomly scattered within crop plants. Certain species are dominant while others show moderate growth or are suppressed while some grow vigorous or have limited growth and short stature. This however, depends on the microhabitat and place they occupy in the field and their performance. Under intense cultivation and thick crop stands, individuals of certain weed species express phenotypic plasticity (phenotypes) at which they change/ modify their appearance, reduce or drop lower branches and thus lateral growth, elongate and increase cell divisions, overtopping crop plants and trapping light, although some shade tolerant species perform better under such conditions. Phenotypic plasticity modifying the mode of growth and energy allocation in response to environmental changes is considered to be important adaptive mechanism. These phenological variations can be easily observed among different weed species. Uniform application of herbicides in the field should equally affect all individuals of a single weed species. When herbicides are best timed and properly applied they should yield similar mode of action on species individuals. While differences in influence of a herbicide on different weed species is expected, hence differences in taxonomy, morphology, physiology and biochemistry, but such differences among individuals of a single species should have resulted from some morphogenetic or other variations within the same or different populations of that species. Certain individuals are totally killed, others less injured and some escape control unharmed. When the same herbicide or herbicides of the same mechanism of action are used, it becomes clearer that previously less or unaffected individuals should exhibit similar responses as were first shown. Gradually these individuals increase in number and growth until they dominate the site with continuous use of the same herbicide or its analogues while sensitive individuals are suppressed or removed. This however, takes a relatively long time for the population to shift from susceptible to complete resistant and depends on herbicide, environment and plant factors. These are positive signs on possible herbicide-resistance development in the field. If less affected or unharmed individuals in the first herbicide application are killed or severely injured in repeated treatments then there should be another cause of escape or partial control at first application and herbicide resistance should be then excluded. On the other hand, unharmed individuals may also tolerate higher application rates. Therefore, farmers should keep observing changes in the weed population as long as the herbicides are in use. They must get familiarized with weed species, populations and densities at pre- and post- herbicide treatments, comparing weed growth, performance and densities and recording any changes in populations thereafter. Less or unharmed individuals of any species should be followed up throughout subsequent applications of the same herbicide or herbicides of similar mode of action.

Sometimes partial effect or failure of the applied herbicide to control certain weed species or individual weeds in the first application may be thought as due to wrong calibration, misapplication, incomplete coverage treatment by a general herbicide or unsprayed gaps resulting from low sprayer boom during spray, unfavorable weather conditions, improper timing of herbicide application, and weed flushes after application of a non-resisted herbicide [16]. This could be easily judged in the repeated application to these species or individuals. When the herbicide failed to control these for the second time or at higher rates then resistance may be underway. With continued use of the same herbicide for different times, resistant individuals aggregate forming irregular patches while other weeds are controlled. A patch of uncontrolled weeds starts spreading and healthy weeds are mixed with uncontrolled weeds of the same species (Fig. 1).

Therefore irregularly shaped patches of a single weed species in the field are an indicator of herbicide resistance, especially when:

- There are no other apparent application problems.
- Other weed species on the herbicide label are effectively controlled.
- Field history indicates extensive use of the same herbicide or herbicides of the same mechanism of action.
- No or minimal herbicide symptoms appear on the single uncontrolled weed species.
- There has been a previous failure to control the same species or population in the same field with the same herbicide or with herbicides of the same site of action.

However, the rate at which a resistant weed population is selected depends on the number and frequency of herbicide applications it receives, the size of the population and its genetic diversity, and characteristics of the herbicide target site. Resistance buildup is accelerated when the management of crops does not include different weed control methods that limit herbicide use. In addition, this may be greatly enhanced in conservation or zero tillage because weeds are not killed by mechanical disturbance and general herbicides.

5. Interaction between environment and genetics

Growth and productivity of any plant species are mainly influenced by genetics, ecology and their interactions. Weeds are different from crops in their responses to both factors. They are more flexible and thus better responsive and adapted to extremes in environmental conditions such as high temperature, freezing, excessive light, salinity, drought, etc. Tolerance of weeds and better responses are mainly due to better and rapid interaction between environment and genetics compared to crop plants. In addition, the long term breeding and selection pressure imposed on crop plants has lead to selection of less adapted species or cultivars that are highly sensitive to ecological stresses and deficient in certain characteristics that offer protection or defense mechanisms against unfavorable environment. Weed fitness in natural habitats and their rapid responses to the changing environment allow evolution of weed ecotypes, genotypes, biotypes or phenotypes. Some of the basic differences in the definitions of pest resistance depend on these terms. The basic unit of plant classifications is the "species" that is defined as a group of individuals displaying common characteristics and having the ability to mate and produce fully viable progeny. A species usually consists of several to many populations. A population is a group of organisms within a species that co-exist in time and space [35; 36] and share a distinct range of genetic variations. While a genotype is the sum of the genetic coding or the genome of an individual, a biotype may not be coincident with genotype as an individual has many genes. Certain genes may be expressed or unexpressed and not pertain to the phenotype associated with the biotype. A biotype is a phenotype that consistently expresses or exhibits a specific trait or set of traits; it represents a group of individuals or a population within a species with a distinctive genetic variation of biochemical or morphological traits. Phenotype refers to the physiological and morphological profile of the expressed gene in an individual [42]. A single genotype can produce different phenotypes in response to environmental conditions and the fundamental properties of organisms are known as phenotypic plasticity. The epigenetic change is thus reflecting the alteration of phenotype (morphological or biochemical) without change in either the coding sequence of a gene or the upstream promoter region. Therefore biotypes within the same species may be developed due to this interaction. On the other hand, ecotype is a population within a species that has developed distinctive morphological or physiological characters (herbicide resistance) in response to a specific environment and persists when individuals are moved to a different environment. Ecotypes are of different germination and growth optima for the same environmental factor and phenotypes may be emerged and observed in weed populations. These alter their morphological features in response to certain prevailing environmental conditions which aim at protection of their individuals against unfavorable ecological stresses. Somatic polymorphism of certain weed species is well recognized and expressed as seed polymorphism of different morphological or physiological requirements for germination on different parts of the same weed individual. These however, are somatic rather than genetically based differences.

6. Herbicide resistance and crop relative weeds

Crop relative weeds are usually derived from the same species of crop plants and thus are genetically related. Most crop species have wild relatives and can interact with them under field conditions. Examples are radish, carrots, vetch, celery, lettuce, fennel, eggplants, wheat, barley, oat, etc. In addition, crop plants which are domesticated from wild forms possess a high degree of compatibility with crops. These are referred to as wild and weedy relatives, in spite of the fact that all species are related because their cells can read a common genetic code [15]. Crop weedy relatives are genetically compatible with crop plants and easily exchange genes. The emerged hybrids may become noxious weeds with certain weedy characteristics derived from both crop plants and wild forms. They could exhibit a certain degree of dormancy that is usually weak or absent in its parents and possess other weed traits making them difficult to control. These new generations have the ability to resist environmental hazards much better than parents and can exist and dominate in both productive and unproductive habitats. These

Figure 1. Three resistant weed species (a, b, c) to glyphosate herbicide at different growth stages and spray times. (a). *Conyza canadensis* resistant to glyphosate until harvest stage of wheat. Source http://www.sciencephoto.com/media/ courtesy of the Montana State University (b). A field infested by suspected glyphosate- resistant *Kochia*, after the field was sprayed with three applications of glyphosate. Photo181407/enlarge Southern Agricultural Research Center. By Dillon Tabish, 08-11-12.Available at: http://www.flatheadbeacon.com/articles/article/scientists_discover_possible_herbicide_resistant_weed_in_montana/29184 (c). Palmir Amaranth (*Amaranthus palmeri*) resistance to glyphosate in corn at early growth. Source: E. Larson, April 21st, 2011.Availableat:http:// www.mississippi-crops.com/2011/04/21/how -to-deal-with-glyphosate-resistance-and- weed-issues-in-corn/.

are of a high genetic plasticity allowing their individuals to adapt to extensive herbicide applications and thus resist chemical treatments. Crop-weed crossed forms can easily exchange genes with crop plants as well as with weedy relatives and therefore are becoming troublesome weeds in fields with genetically modified crops.

7. Gene flow potential with wild/weedy relatives of world crops

In nature, genetic information is transferred between different individuals, populations, and generations (to progeny) and across spatial dimensions [2; 15]. This phenomenon, known as gene flow, serves as a mechanism to maintain the biological diversity that helps to ensure long-term survival of populations and species in various environments.

Gene flow is a critical determinant of population genetic structure, playing an important role in both evolutionary and applied plant population genetics [12]. It is also known as 'migration' [13] or admixture [1] and can be defined as the movement of genes between populations of a species and between these populations and inter-fertile relatives [39; 41], conferring new traits, the biophysical characteristics of the organism to individuals of the recipient population [34].

Gene flow could occur through dispersal of pollen (via outcrossing between sexually compatible individuals within or among populations) or seeds (via seed dispersal), or vegetative parts capable of clonal propagation [34; 41]. Pollen dispersal is the typical method for such exchange of genetic information [15] and pollinating visitors or other agents including wind, animal, water current and other factors could play a significant role in this issue. This happens by cross-pollination (hybridization), that is, the pollination of members of one population or genetic pool with that of another [34]. These are natural and ordinary phenomena that occur in conventional as well as genetically modified crops.

Movement of pollen away from its site of production can result in true gene flow only if (1) the pollen first effects fertilization to form seeds, and (2) seeds germinate, produce plants that express the gene (i.e., are not silen8ced), and are able to reproduce [15]. Gene flow can be from crop to crop or landrace, from crop to wild relative, and even from wild relative to crop plant [34]. Spread of this phenomenon would lead to radical changes in vegetation composition and weed ecological distribution and their economic significance.

However, two types of gene flow are known; horizontal and vertical. Stewart [39] showed that 'horizontal' gene flow is the movement of genes between disparate, unrelated species, such as between plants and microbes while horizontal gene flow is more theoretic.

Among the world's 180 most damaging weeds, however, cause 90% of all crop losses, only five groups (related weeds of rice, sorghum, rape seed, sugarcane, and oats) are sexually compatible with the most important crops (Table 1). This fact emphasizes that the number of weed-crop crosses likely to lead to extremely troublesome or unmanageable problems is small.

Weed crosses with herbicide-tolerant biotech crops are likely to be favored in some agricultural fields where the herbicide is used. In areas where little or no herbicide is applied (e.g., native lands), the weed–biotech crop crosses will not be favored [15]. Self-pollinating crops are considered of low risk in terms of gene flow to weeds. Roundup Ready, Clearfield, or Liberty Link canola, in contrast, could pollinate nearby herbicide-susceptible canola as well as weedy canola relatives, resulting in volunteer canola plants and weeds that may be resistant to several herbicide families [38]. However, several pieces of evidence clearly show an escape of weedy transgene from fields via seed flow and this escape occurs via man-mediated long-distance dispersal events [4]. Other results revealed that development of weed resistance via selection pressure from repeated herbicide applications in herbicide resistant crops (in the absence of gene flow), often poses greater risks than that from gene flow to related weed species [15].

Rank	Crop	Scientific Name	Related weeds: sexually compatible with crops
1	Wheat	Triticum aestivum Triticum durum	T.aestivum Aegilops cylindrical A. tauschii A. triumcialis Agropyron spp
2	Rice	Oryza sativa Oryza glaberrima	O. sativa O. glaberrima O. barthii O. longistaminata O. rufipogon O. punctata
3	Maize	*Zea mays*	*Z. mays ssp Mexicana*
4	Soybean	*Glycine max*	*G. soya*
5	Barley	*Hordeum vulgare*	*H. spontaneum*
6	Sorghum	*Sorghum bicolor*	S. bicolor S. almum S. halepense S. propinguum S. sudanense
7	Canola	*Brassica napus, B. rapa, B. juncea*	*B. napus, B. rapa, B. nigra*
8	Sunflower	*Helianthus annus*	*Helianthus annus*

Source: Different references

Table 1. Examples of some important food crops and their sexually compatible weed species

In this regard, biotech crops conferring stress tolerance (e.g., to water deficits, diseases, insects, salt stress, or nutritional deficiencies) may need more scrutiny because their crosses with weedy relatives may impart selective advantages in both agricultural and nonagricultural areas. Thus, some traits obtained from biotech crops could theoretically facilitate development into problematic weedy or wild species [15].

The economic consequences due to gene flow from biotech crops will primarily impact the agricultural fields in which those crops are grown, but potentially could impact natural areas given the proper rare combination of sexually compatible relatives, favorable environment, and reproductive/fitness advantages. As an example, rice grown in tropical countries may be relatively more prone to such processes because of the substantial populations of its wild/weedy relatives that grow naturally in or adjacent to the rice-producing areas [8; 26].

Crop-wild hybridization may also create genotypes with the potential to displace parental taxa in new environments [7]. However, the most important variable affecting gene flow is the degree of relatedness and distance between the crop and the weed, because gene flow is only possible if close relatives are growing near the crop. As a result the possibility of gene flow depends mainly on presence of wild or weedy relatives [11]. Transgene (s) transfer may have unpredictable and out of control ecological impacts under intensive cultivation of biotech crops [25]. While different crops can exchange genes with wild relatives, gene escape to wild or weedy relatives and its ecological impacts are outrated. The ecological consequences of gene flow however, depends on the amount of transgenes moved out to a wild population and the genetically modified traits and whether they have an evolutionary advantage under natural selection pressure or not and if enhanced fitness of wild and weedy relatives then the transgene followed by gene flow would persist and spread rapidly in the population of wild relatives through introgression, invade a new area and outcompete other individuals under natural conditions [24]. Weeds receiving transgenes will continue to evolve when exposed to selection pressure and it becomes nearly impossible to move them out from the environments if they can persist and spread in the populations.

8. Transgenic crops and weed evolution

The development of crops that are resistant to herbicides is a relatively new technology aimed to improve weed control in agricultural land. Herbicide-resistant crops can be created by standard methods of plant breeding, but the use of genetic engineering techniques is more usual. Herbicide-resistant crops are made resistant by either transgene technology or by selection in cell or tissue culture for mutations that confer herbicide resistance [10]. Glyphosate and glufosinate are herbicides most used in this regard. For example, soybean, corn, cotton, sugar beet, and canola are available as glyphosate- resistant cultivars and some are now widely planted in different countries. Importance of genetically engineered crops is to:

- Develop crops more tolerant/resistant to herbicides and thus increase herbicides uses and selectivity.
- Eliminate possible injury effects of soil persistent herbicides to crop plants.
- Increase options for weed control when the number of herbicides is limited, such as in minor crops.
- Effective control of certain difficult weed species and widening of weed control spectrum
- Achieve more effective weed control
- Increase bio-safety and enhance better eco-friendly use of new and less toxic herbicides
- May be more cost- effective weed control method

However, public concern about the impact of genetically modified crops on the natural environment encouraged more studies on this aspect in the last few years. Among the possible impacts, the 'escape' of the transgene, either through dispersal of the crop plant outside the agricultural area or through hybridization with wild relatives and thus increase the possibility of "weediness" [41].

In the majority of instances, there is a very low probability that an approved biotech crop introduction could create an environmental risk different from that of a nonbiotech version of the same crop. This however, does not lessen the serious concerns about possible consequences of the escape of transgenes into the environment [41]. Examples of the risks mentioned in the context of gene flow from genetically modified plants are: i) new emerged weeds resulting from an escape by the crop itself; ii) super weeds resulted by hybridization of a (wild/weedy) species with the transgenic crop; iii) genetic erosion (loss of original diversity of wild relatives). To date, all instances of weeds becoming resistant have resulted from the weed evolving its own biochemical mechanism and not by acquiring genes for resistance from the crop. However, in some cases it would be possible for the herbicide resistance gene to flow from the crop to the weed [11].

Possible consequences of hybridization and introgression depend on the plant, gene, trait, and ecological factors [39]. In the case where transgenes might be introgressed into "weedy wild relatives", there are concerns about exacerbating "weediness" traits or even the disruption of natural ecosystems. Therefore, to assess the risk of gene flow it needs to be examined not only the probability of genes moving between plants, but how possible is it for the new plants to survive [39].

In general, people ideally would like to minimize or prevent gene flow from transgenic organisms to weedy wild relatives or to places where extensive crop breeding takes place [39]. Three approaches to gene flow mitigation are possible [3].The first is by keeping the genetic modification out of the pollen, preventing the formation of pollen, and keeping the pollen inside the flower. It requires transplastomic plants hence the modified DNA is not situated in the cell's nucleus but is present in plastids, which are cellular compartments outside the nucleus. The second approach relies on male sterile plants unable to produce functioning flowers and therefore cannot release viable pollen. Cytoplasmic male sterile plants are known to produce higher yields. The third approach works by preventing the flowers from opening "cleistogamy" that occurs naturally in some plants. Cleistogamous plants produce flowers which either open only partly or not at all.

However, herbicide-resistant genes have no ecological significance in places where the corresponding herbicide is not used. When paired with a gene that might have an effect in a natural ecosystem, there is a potential problem with gene flow. Repeated application of the herbicide (especially general herbicides) would select for and protect crosses and backcrosses, increasing the possibility of successful gene flow to wild, related species [10].

9. Weed control spectrum of selective herbicides and population shifts

Some plants are genetically tolerant to certain herbicides while others have evolved resistance after repeated exposure to an herbicide. Tolerant and resistant plants usually degrade or metabolize the chemical to nonphytotoxic substances. In some cases of resistance, such as with triazine herbicides, the herbicide does not reach the key site in treated plants. Although tolerance and resistance are common, herbicide selectivity among plants is often conditional; thus it depends on plant, herbicide and environment factors.

Some of the factors that influence herbicide selectivity are as follows:

http://dx.doi.org/10.5772/56036

- Physiological or biochemical tolerance to the herbicide
- Herbicide application rate
- Time of application
- Herbicide formulations and surfactants used.
- Growth stage of weed and crop or other plant development
- Weather patterns (temperature, light, wind, rain, etc.)
- Variation in microenvironment or micro- topography
- Variation in resource level
- Soil type and pH

Many of the principles and practices of how herbicides used or applied to attain selective chemical and effective weed control are important. These involve the role of plant morphology and physiology, chemical properties, and environmental factors [31]. Herbicide selectivity in one way or another is in direct link with herbicide resistance. Crops are resistant to herbicides selectively used to kill weeds. Even with repeated treatment, crop plants can resist or tolerate higher rates of selective applied herbicide or repeated treatments. This depends on some level of tolerance/resistance higher in crop plants compared with weeds for that specific herbicide or herbicide group. For example, Syrian marjoram (*Origanum syriacum*) was found to withstand up to 4 times higher rates of oxadiazon and oxyfluorfen herbicides either applied on foliage parts or through the soil [32; 33]. Certainly many factors have an important role in giving a resistant value for crop plants. Some of these are listed below:

9.1. Plant factors and herbicide selectivity

Plant factors that influence the way weeds and crops respond to herbicides are genetic inheritance, age, growth rate, morphology, growth form and anatomy, and physiological and biochemical processes. The most effective use of herbicides results from considering these factors when selecting an herbicide or application method.

9.2. Plant age and growth rate

Weed seedlings or young plants are usually killed more easily than large or mature vegetation. In addition, some preemergence herbicides that suppress seed germination are often not effective when used to control larger, better established plants. Plants that are growing rapidly or in shaded places generally are more susceptible to herbicides than are plants of slow growth or unshaded.

9.3. Morphology

The morphology or growth habit of plants can determine the degree of sensitivity to some herbicides. Morphological differences in root structure, location of growing points, and leaf properties between crops or other desirable plants and weeds can determine the selectivity pattern of some herbicides. Annual weeds in a perennial crop, meadow, or pasture usually can be controlled by herbicides because of their different root distribution and structure compared to those of perennial plants. For example, perennial crops such as alfalfa can recover from moderate contact herbicide injury to foliage whereas annual weeds, because of their small size and shallow root system, will be killed by the same herbicide application.

The meristematic regions of most grasses, such as cereal crops and grassy weeds, are located at the base of the plant or even below the soil surface. The growing points are protected from herbicide exposure by the foliage or soil that surrounds them. Thus, herbicide that contacts only foliage may injure some leaves but will not typically impair the ability of the plant to grow. In contrast, most dicot plants have their meristems exposed at shoot tips and leaf axils. For this reason, these plants are more susceptible than grasses to foliage-applied herbicides, especially of contact action.

Leaf properties of some plants can impart selectivity to certain herbicides, while other plants are effectively controlled. Spray droplets do not adhere well to the surfaces of narrow, upright, waxy leaves that characterize many monocot plants like cereals, onion, and most grasses. Thus, spray droplets do not adequately cover such leaves following herbicide application and the effect of the herbicide is reduced. In contrast, dicot plants have relatively wide leaves that are usually horizontal to the main stem. Leaves of dicot plants, therefore, intercept more spray solution than leaves of grasses and spray droplets spread more evenly over dicot foliage. Herbicide effectiveness is best when spray interception and coverage are greatest and with use of surfactants. However, ecological factors and geographical regions under which weeds are growing have significant influence on herbicide selectivity and rates of applications since they affect or modify weeds morphology and internal anatomy.

9.4. Physiological and biochemical processes

Plant physiology influences herbicide passage after its application. This process is called "absorption". The extent of herbicide movement in a plant- "translocation"- after it has been absorbed is also a physiological process. Both absorption and translocation are important processes governing herbicide activity and vary markedly among plant species. Generally, plant species that readily absorb and translocate herbicides are most easily killed.

Biochemical and biophysical processes are also important plant factors determining herbicide selectivity. Herbicide adsorption can be responsible for differential herbicide susceptibility among plant species. During this process an herbicide is bound so tightly by cellular constituents (usually cell walls) that it cannot be translocated readily and thus is inactivated. Membrane stability is another biochemical/biophysical process that results in herbicide selectivity among plants. In this case, the cell membranes of tolerant plants can withstand the disruptive action of the herbicide. The ability of carrot to withstand the toxicity of certain oils is an example of this form of herbicide selectivity.

9.5. Genetic inheritance

Plant species within a genus usually respond to herbicides in a similar manner, while responses to herbicides by plants in different genera often vary. The reason is that plants with similar taxonomic traits often have similar morphogenetic and enzymatic components. Thus, crops and weeds that belong to the same genera are usually susceptible to the same herbicides and are similarly affected since they have similar biochemistry. This rule is not absolute, however, because varieties of many crops are known to respond differently to the same herbicide and weeds usually adopt different mechanisms of herbicide resistance while crop plants have lost many of their traits in breeding programs that present in wild relatives.

10. Herbicides and edaphic factors

Soil factors affect herbicide performance and their effectiveness. These including soil-organic matter content, microorganism populations, soil water table and moisture content and soil pH. Organic matter acts through adsorption and release of chemical molecules. Certain herbicides are tightly adsorbed on soil particles and thus become unavailable to weeds. These molecules may be totally inactivated upon their release. Therefore weed control may be complete or not based on the amount of the herbicide adsorbed and whether the held amount on soil colloids is compensated or not before applied. The higher the percentage of organic matter and clay particles, the greater the adsorption in amount and time of herbicide molecules and the lower the herbicide activity and *vice versa*. This requires that some operations should be well managed when soil applied herbicides are used including their incorporation or placement in/on the soil.

Activity of soil microorganisms is another factor affecting activity of soil- applied herbicides and persistence. Microorganisms may degrade herbicide molecules and feed on organic herbicides. In general, favorable soil factors to microorganism populations stimulate their activity and thus rapid herbicide degradation. Therefore, soil-microbe population is an important factor in increasing or decreasing herbicide persistence and weed control duration.

Soil water also affects herbicide activity and performance. When high amounts of soil water are available or at high soil water levels, herbicide molecules may by hydrated. On the other hand, moisture is necessary to transfer herbicide molecules into the root system and then translocate these upward to vegetative parts through the xylem.

Soil pH affects cation exchange capacity of soil particles. Salt or mineral forms of certain herbicides may interact with soil particles under these conditions by exchanging cations or anions and thus lead to breakdown of herbicide molecules and inactivation.

All above soil factors and others such as soil- root temperature and soil mechanical properties can affect herbicide activity and performance and their effectiveness in controlling weed species and herbicide selectivity. Weeds may become adapted to certain soil conditions, escape control operations and lead to dominance of well adapted species or populations.

11. Weed resistance and dormancy, avoidance and weed density

Dormancy is the state at which seeds in the soil or buds are not germinating or growing due to external conditions exert influences on physiological and biochemical internal processes including enzymes activities, food transport to embryo and metabolism. This state is keeping seeds or buds safe until the cause of dormancy is over. This behavior is important to maintain genetic line and continuity of the species in changeable environment. Under conditions of herbicide application, some of these chemicals are absorbed by seeds or dormant buds while others are not. These result differences in germination, emergence and growth patterns of different weed species. However, some herbicides may stimulate seed germination while others inhibit this process or even kill seed embryo. Differences also exist in hardness and permeability of seed coat of different weed species at which species of Chenopodiaceae and Fabaceae are good examples. These characters cause differences in germination and growth of seedlings and may confer another cause of herbicide resistance. Avoidance of herbicide toxicity may result from seed interring into dormancy and not further responding to the applied herbicide with no absorption or translocation of the herbicide into the embryo. In addition, herbicide molecules may be deactivated or degraded inside the seed itself by some oxidative enzymes or may bound into certain constituent inside the seed.

On the other hand, stimulation of weed seeds to germinate using certain herbicides also exist and allows higher seedlings emergence and partitioning of herbicide molecules among individuals of weed species. Division of herbicide molecules among high number of emerged seedlings would further diluted herbicide inside weed plants.

All above mentioned factors should be considered when herbicide-resistance is discussed. These may cause great differences in weed growth patterns and distribution in the field.

12. Weed resistance updates and resistance mechanisms

With continued dependence on herbicides for weed control and with the absence of other methods and herbicide rotation, the resistance problem is extenuated and the number of resistant weed species and biotypes is dramatically increased. At present, the reported herbicide resistant weeds are approaching 393 (species and their biotypes). These represent 211 species (124 dicots and 87 monocots) and detected from over 680,000 fields [21; 44] reported from 61 countries from all over the globe. However, the highest number of resistant species was reported from the advanced countries indicating efficient and rapid detection with available technology to diagnose, discover and deal with this issue. However, the highest number of weeds reported resist the main three groups of herbicides based on site of action including; the ALS (127 weeds), Photosystem II (69) and the ACCase (42) inhibitors. The highest number of weed resistant species and biotypes came from the USA (141), Australia (61) and Canada (58). Most numbers of resistant species belong to the families Poaceae, Asteraceae and Amaranthaceae and most frequently mentioned are genera of *Amaranthus* (30 times and 11 species), *Echinochloa* (23 times and 6 species), *Lolium* (20 times and 4 species), *Alopecurus* (12 times and 3 species), *Avena* (11 times and 3 species), *Bromus* (11 times and 5 species), *Conyza* (10 times and 3 species), *Setaria* (9 times and 5 species), *Poa* (8 times and one species), *Ambrosia* (7 times and 2 species), *Digitaria* (6 times and 4 species), *Phalaris* (6 times and 3 species), *Hordeum* (5 times and 2 species) and *Sorghum* (6 times and 3 species). Most are of the grass family usually exhibiting distinct morphological features allowing wide dispersal and escape of herbicide treatment such as encased growing points, vertical leaf arrangement and thick waxy cuticle that reduce herbicide penetration and lead to herbicide droplets bouncing off leaves. Other genera reported are characterized by their prolific seed production and/ or seed polymorphism. All above mentioned genera however, showed multiple resistance to different herbicides groups. Most resisted are herbicides widely and repeatedly used including: glyphosate, paraquat, atrazine and 2,4-D and others used in fields cultivated by genetically modified crops. Some recently developed herbicides are also resisted including chlorsulfuron and sufonylurea group. This phenomenon demonstrates that the herbicide industry and development is far behind weed evolution. On the other hand, weed species and biotypes showing multiple resistance are most common and some are among the world's worst weeds [19] including: *Amaranthus* spp., *Echinochloa* spp., *Avena* spp. and *Chenopodium album* characterized by their polymorphic seed production and phenotypic plasticity. This reflects a great ability to maintain and exhibit high plasticity and possess various mechanisms of herbicide resistance.

The precise molecular mechanism of resistance varies with different plants, but in general plants resist herbicides in one of the following ways:

- Avoiding the herbicide by not absorbing it or, if absorbed, the weed compartmentalizing it away from its target site.
- Reducing the uptake or herbicide uptake is not enough to injure the weed or reach lethal level.
- Changing the structure of the target site of the herbicide so the plant is no longer sensitive
- Reduce herbicide translocation to the key site or binding it into certain plant constituent
- Sequestration by complete physical removal of the herbicide from the key site
- Target site mutation and changes in structure lead to insensitive plants and failure herbicide binding.
- Deactivating the herbicide by chemical alteration or herbicide metabolism before reaching target site

However, resistance mechanisms through which different weed species resist herbicide treatments are many and varied but most are physio-chemically based (Table 2).

Herbicide Group	Site of Action	HRAC Group
ALS inhibitors	Inhibition of acetolactate synthase ALS (acetohydroxyacid synthase AHAS)	B
Photosystem II inhibitors	Inhibition of photosynthesis at photosystem II	C1
ACCase inhibitors	Inhibition of acetyl CoA carboxylase (ACCase)	A
Synthetic Auxins	Synthetic auxins (action like indoleacetic acid)	O
Bipyridiliums	Photosystem-I-electron diversion	D
Glycines	Inhibition of EPSP synthase	G
Ureas and amides	Inhibition of photosynthesis at photosystem II	C2
Dinitroanilines and others	Microtubule assembly inhibition	K1
Thiocarbamates and others	Inhibition of lipid synthesis - not ACCase inhibition	N
PPO inhibitors	Inhibition of protoporphyrinogen oxidase (PPO)	E
Triazoles, ureas, isoxazolidiones	Bleaching: Inhibition of carotenoid biosynthesis (unknown target)	F3
Nitriles and others	Inhibition of photosynthesis at photosystem II	C3
Chloroacetamides and others	Inhibition of cell division (Inhibition of very long chain fatty acids)	K3
Carotenoid biosynthesis inhibitors	Bleaching: Inhibition of carotenoid biosynthesis at the phytoene desaturase step (PDS)	F1
Glutamine synthase inhibitors	Inhibition of glutamine synthetase	H
Arylaminopropionic acids	Unknown	Z
Unknown	Unknown	Z
4-HPPD inhibitors	Bleaching: Inhibition of 4-hydroxyphenyl-pyruvate-dioxygenase (4-HPPD)	F2
Mitosis inhibitors	Inhibition of mitosis / microtubule polymerization inhibitor	K2
Cellulose inhibitors	Inhibition of cell wall (cellulose) synthesis	L

Source: 21; Updated: November, 2012

Table 2. Herbicide resistant weeds summary table (Thursday, November 08, 2012)

13. Factors enhancing herbicide resistance

All natural weed populations, regardless of the application of any herbicide, probably contain biotypes that resist herbicides. Repeated application of an herbicide exposes the weed population to a selection pressure which may lead to an increase in the number of surviving resistant individuals in the population. As a consequence, the resistant weed population may increase to a level that adequate weed control cannot be achieved by the application of that herbicide [18]. Factors enhancing herbicide resistance include: the use of a single herbicide or herbicides of same mechanism of action, same formulation, same method of application, time of application, weather conditions during spraying, weed-density and application rate, surfactants, herbicide family and mechanism of action, crop rotation, and employed control methods.

Because weeds contain a tremendous amount of genetic variation that allows them to survive under a variety of environmental conditions, the development of a resistant species is brought about through selection pressure imposed by the continuous use of an herbicide or herbicides of similar mechanism of action. Long residual pre-emergence herbicides or repeated application of post-emergence herbicides will further increase selection pressure.

Factors in general that can lead to or accelerate the development of herbicide resistance include weed characteristics, chemical properties and cultural practices.

Weed characteristics conducive to rapid development of resistance to a particular herbicide include:

- Weeds having short life cycles (annuals).
- High seed production.
- Level of selection pressure imposed by the herbicide
- Relatively rapid turnover of the seed bank due to high percentage of seed germination each year (i.e., little seed dormancy).
- Several reproductive generations per growing season.
- Extreme susceptibility to a particular herbicide.
- One weed which would normally be controlled but not controlled while others were removed.
- High frequency of resistant gene (s).

Herbicide characteristics which lead to rapid development of herbicide resistance in weed biotypes include:

- A single site of action of the same herbicide continuously is used.
- Broad spectrum of weed control.
- Long residual activity in the soil.

Cultural practices can also increase the selection pressure for the development of herbicide-resistant biotypes. In general, complete reliance on herbicides for weed control can greatly enhance the occurrence of herbicide-resistant weeds. Other factors include:

- Shift from crop rotations towards mono cropping.
- Little cultivation or zero tillage for weed control or no elimination of weeds that escape herbicide control.
- Continuous or repeated use of a single herbicide or several herbicides that have the same mechanism of action.
- High herbicide use rate relative to the amount needed for weed control.
- Complete weed control

- Orchard and vineyard weeds.
- Roadside weeds.

14. Management of herbicide resistance

Herbicide-resistant weed populations can be managed following an integrated weed control program. The following practices are important for an effective management strategy:

- Herbicide rotation. Adopting this method, it should be known that herbicides of different chemical families may have the same site of action.
- Using mixtures of herbicides with different modes of action and overlapping weed spectrums. This would help in managing evolution of weed resistance.
- Crop rotation. Crops differ in their competitiveness against weeds. Plant crops having a different season of growth, different registered herbicides and crops for which there are alternate methods of weed control. Rotation breaks down weed population and prevents the build up of resistance to herbicides. In addition, different crops may require different types of herbicides and thus herbicides may be rotated as well. However, some herbicide groups include different chemicals that can be used in different crops; therefore crop rotation alone may not be enough to avoid resistance development in this case.
- Herbicides with the same site of action should not be applied or used in both fallow years and in the crop(s) planted within 3 years.
- Growers should keep rotating methods of weed control. Non-chemical control techniques including tillage, hand-weeding before flowering, mulching, soil solarization, prevention methods of weed dispersal (certified seed, clean equipments, use a power washer or compressed air to remove seeds).
- Herbicide-resistant weeds should be controlled before flowering and seed setting.
- Farmers should only use non- or short-residual herbicides and avoid using persistent chemicals and not applying them repeatedly within a growing season. This method would reduce the selection of herbicide-resistant weed biotypes. However, repeated applications within a single growing season of certain herbicides (paraquat, glyphosate) also lead to development of resistant weed populations.
- Where possible mechanical weed control such as rotary hoeing and cultivation is recommended to be combined with herbicide treatments.
- Weed escapes of resistant biotypes may be eliminated by cultivation in row crops. Fallow tillage can control herbicide-resistant and susceptible weed populations when they emerge at about the same time.
- Accurate record keeping. Farmers should be familiar with the history of herbicides use in their fields. Also keep tracking the weed species that have been present in a given field and

of how well particular herbicides have controlled them. Farmers should check for weedy patches in patterns consistent with application problems and hand-weeding these patches.

- Always weed free crop seeds should be used that greatly minimize introduction seeds of herbicide-resistant biotypes.
- Implementation of integrated weed management. This is important for effective control of all weeds including herbicide-resistance.
- Monitoring fields for weed escapes for resistant and susceptible biotypes. A resistance problem may not become visible until 30 percent or more of the weed population is no longer controlled. Check to see if the escapes are of one species or a mixture of species. If a mixture, the problem is more likely related to the environment or the herbicide application. If only one species was not controlled, the problem is likely to be resistance, especially if the species was controlled by the herbicide in the past and if the same herbicide has been used repeatedly in the field.
- Implementation of prevention methods of weed control. All measures aimed at prevention of weed introduction to fields and their dispersal should be strictly followed including governmental quarantine regulations.
- Alternating spring and winter crops, thus tillage and herbicides are used at different times in the different crops. Weed biotypes that survive in one crop could be killed in the other.
- Changing herbicide program, if weed resistance occurs, herbicides with other sites of action and other weed management practices must be used in an integrated management strategy. However, weed management strategies that discourage the evolution of herbicide resistance should include the following:
    - Use herbicide only when necessary and where possible herbicide application should be based on economic threshold.
    - Apply herbicides in tank mixed, pre-packed, or sequential mixtures of multiple site of action.
    - Never use unregistered mixtures, follow label recommendation at all times
    - Regularly monitor your crops so that resistant patches can be observed in time to be controlled with, for instance, spot spraying.
    - Apply the herbicide at the correct leaf stage of the weed and the crop.
    - Calibrate sprayer correctly before using herbicides
    - Planting new herbicide-resistant crop varieties should not result in more than two consecutive applications of herbicides with the same site of action against the same weed unless other effective control practices are also included in the management system.
    - Respond quickly to changes in weed populations to restrict spread of weeds that may have been selected for resistance.

  - Encourage railroads, public utilities, highway departments and similar organizations that use total vegetation control programs and vegetation management systems that do not lead to selection of herbicide resistant weeds. Resistant weeds from total vegetation control areas frequently spread to cropland. Chemical companies, governmental agencies, and farm organizations can all help in this effort.

- To keep herbicide-resistant weeds under control, the following strategies should be also incorporated into a weed management plan:

  - Clean tillage and harvest equipments before moved from infested to clean fields from weed resistant species.

  - Total weed control in uncultivated places or sites

  - Close cultivation

  - Monitor hand weeding to insure more than 90% removal of weeds in the crop row.

  - Prevention of weed seed spread through:

– Use of clean equipment.

– Enter the field with resistant plants last.

– Use a power washer or compressed air to remove seeds.

– Recognizing patterns of weed escapes typical of resistant plants

– Watch for small weed patches that appear in the same place in the next crop.

– Watch for weed patches that do not have a regular shape that would indicate an herbicide application problem.

Herbicide resistance however, provides a basic understanding of the genetic basis of weediness, while the development of weed genomics would provide three predictable and useful outcomes. The first is the identification of genes that could improve crop yields. The second is to improve our understanding of the evolution of herbicide resistance and the to aid in the identification of novel herbicide targets. Currently, there is little (if any) solid predictive capability of why some weeds develop resistance and others do not. Third, our understanding of weed biology would be exponentially expanded [6].

Research has recently been performed to assess the ability to cripple the effect of transgenes. The goal here is for the transgenic effect to not be as strong if it went to a wild relative. In one case, the genetic background of the crop weakened the weedy relative. In another case, the weakness was built into the genetic construct, called *transgenic mitigation*, in which an herbicide-resistant gene was paired with a dwarfing gene. In either case, transgenic weeds were less competitive than their non-transgenic parent weeds [39].

http://dx.doi.org/10.5772/56036

15. Conclusion

Weeds either leave (disappear), adapt, tolerate or resist any unfavorable environmental conditions that influence their normal growth and life strategies. Herbicide resistance is a complex phenomenon resulting from altered herbicide target enzyme, enhanced herbicide metabolism or reduced herbicide absorption and/or translocation. It is a survival strategy through which many successful weed species and/or biotypes counteract or escape chemical hazards. Weeds expressing this phenomenon have developed some morpho- (behaviorist), physio-, and/ or biochemical mechanism/s allowing existence. However, two theories are mainly considered: the mutation and the natural selection [17]. Colonizers, as well as some specialist weeds of high seed production and polymorphic characteristics, have rapid responses to prevailing environmental conditions and high ability to express herbicide-resistant genes and exhibit wide ecological variations [28]. This phenomenon is well documented in agricultural as well as other disturbed habitats while the list of weed resistant species gets longer with continued dependence on herbicides for weed control. From the information presented in this chapter, it is clearly demonstrated that herbicide resistance in weeds is far exceeding herbicide technology and industry. Most problematic weed species are genetically related to major food crops including wheat, rice and maize. This may pose another danger for the genetic industry and genetically engineered crops of wild relatives. Away from weed biology and resistance control, methods of weed control must be integrated and continuously rotated for effective weed control and prevention of weed resistance. This however, may not be achieved in absence of information and field data and well managed weed control strategies, considering all the factors that influence weed life and development.

Author details

Jamal R. Qasem

Address all correspondence to: jrqasem@ju.edu.jo

Department of Plant Protection, Faculty of Agriculture, University of Jordan, Amman, Jordan

References

[1] Anonymous-I. (2010) Evolution. Retrieved December, 2010. Available at: *http://www.biologydaily.com/biology/Evolution.*

[2] Anonymous-II. (2010). Gene flow. Retrieved December, 2010. Available at: *http://en.wikipedia.org/wiki/Gene_flow.*

[3] Anonymous-III. (2010) Gene flow mitigation. Retrieved December, 2010. Available at: *http://en.wikipedia.org/wiki/Gene_flow*.

[4] Arnaud, J.F.; Viard, F.; Delescluse, M. and Cuguen, J. (2003) Evidence for gene flow via seed dispersal from crop to wild relatives in *Beta vulgaris* (Chenopodiaceae): consequences for the release of genetically modified crop species with weedy lineages. *The Royal Society, Published online, 19 March, http://www.ncbi.nlm.nih.gov/pmc/articles*.

[5] Barrett, S.C. (1988). Genetics and evolution of agricultural weeds. In: *Weed management in Agroecosystems: Ecological Approaches*. Altieri, M.A. and Liebman, M. (eds.). CRC Press, Inc, Baco Raton, Florida.

[6] Basu, C.; Halfhill, M.D.; Mueller, T.C and Stewart Jr, C. N. (2004). Weed genomics: new tools to understand weed biology. *Tends in Plant Science* 9(8), 391-398.

[7] Campbell, L.G.; Snow, A.A and Ridley, C.E. (2006). Weed evolution after crop gene introgression: greater survival and fecundity of hybrids in a new environment. *Ecology Letters* 9, 1198–1209.

[8] Chen, L.I.J., Lee, D.S., Song, Z.P., Shu, H.S. and Lu, B.R. (2004). Gene flow from cultivated rice (*Oryza sativa*) to its weedy and wild relatives. *Annals of Botany* 93, 67-73.

[9] Conard, S.G. and Radosevich, S.R. (1979). Ecological fitness of *Senecio vulgaris* and *Amaranthus retroflexus* biotypes susceptible or resistant to atrazine. *Journal of Applied Ecology* 16, 171-177.

[10] Duke, S.O.and Cerdeira, A.L. (2005). Transgenic herbicide-resistant crops. *Outlooks of Pest management*. DOI10.1564/16 oct06, pages 2008-211. Research Information Ltd.

[11] DuPont (2008). Herbicide resistant crops and weed management: scientific summary and the DuPont perspective. Available at: http://www2.dupont.com/Biotechnology/en_US/science_knowledge/herbicide_resstance.

[12] Ellstrand, N.C. (1992). Gene flow among seed plant populations. *New Forests Journal* 6(1-4), 241-256.

[13] Ellstrand, N.C. (2003).Current knowledge of gene flow in plants: implications for transgene flow. *The Royal Society, published online 12 May at: http://www.ncbi.nlm.nih.gov/pmc/articles*.

[14] Fuentes, R.G., Baltazar, A.M., Merca, F.E., Ismail, A.M. and Johnson, D.E. (2010). Morphological and physiological responses of lowland purple nutsedge (*Cyperus rotundus* L.) to flooding. *AoB Plants* Vol. 2010.plq010, doi: 10.1093/oobpla/plq010.

[15] Gealy D.R.; Bradford K.J.; Hall L.; Hellmich R.; Raybould A. and Wolt J. (2007). Implications of gene flow in the scale-up and commercial use of biotechnology-derived crops: Economic and policy considerations. *Issue Paper 37. Council for Agricultural Science and Technology (CAST), Ames, Iowa.*

[16] Gunsolus, J.L.(2008). Herbicide resistant weeds. 2002 Regents of the University of Minnesota.. University of Minnesota Extension. Available at:http://www.extension.umn.edu/distribution/cropsystems/DC6077.html.

[17] Hager, A.G. and Refsell, D. (2008). Weed resistance to herbicides. In: *2008 Illinois Agricultural Pest Management Handbook*. Department of Crop Science, University of Illinois.

[18] Herbicide Resistance Action Committee (HRAC). (2009). Guidelines to the management of herbicide resistance. Available at:http://www.hracglobal.com/Publications/ManagementofHerbicideResistance/tabid/225/Default.asps(1-16)1/18/2009.

[19] Holm, L.G., Plucknett, D.L., Pancho, J.V. and Herberger, J.P. (1977). *The World's Worst Weeds: Distribution and Biology*. University of Hawaii, Honolulu. USA.

[20] Holzner, W. and Numata, M. (eds.). (1982). *Biology and Ecology of Weeds*. Dr. W. Junk Publishers. The Hague-Boston-London.

[21] http://www.weedscience.org/summary/UspeciesMOA.asp?1stMOAID=19.

[22] Kending, A. (2009). Herbicide resistance in weeds. University of Missouri Extension. Available at: http://extension.missouri.edu/publications/DisplayPub.aspx?P=G4907.

[23] LeBaron, H.M. and Gressel, J.(eds.). (1982). *Herbicide Resistance in Plants*. John Wiley & Sons, New York.

[24] Lu, B-Rong and Yang, C. (2009). Gene flow from genetically modified rice to its wild relatives: assessing potential ecological consequences. *Biotechnology Advances* 27, 1083-1091.

[25] Lu, B-Rong. (2008). Transgene escape from GM crops and potential biosafety consequences: an environmental perspective. International Center for Genetic Engineering and Biotechnology (ICGEB), vol. 4. Collection of Biosafety Reviews 66-141.

[26] Lu, B-Rong. and Snow, A. (2005). Gene flow from genetically modified rice and its environmental consequences. *BioScience*, 55, 669–678.

[27] Mashburn S. J. Sharitz, R. R. and Smith, M. H. (1978). Genetic variation among *Typha* populations of the southeastern United States. *Evolution* 32: 681-685.

[28] Neve, P. (2007). Challenges for herbicide resistance evolution management: 50 years after Harper. *Weed Research* 47, 365-369.

[29] Putwain, P.D., Scott, K.R. and Holliday, R.J. (1982). The nature of resistance to traizine herbicides: case histories of phenology and population studies. In: *Herbicide Resistance in Plants;* LeBaron, H.M. and Gressel, J. (eds.), John Wiley & Sons, New York pp. 99-115.

[30] Qasem, J. R. (2003). *Weeds and their Control*. University of Jordan Publications. Amman, Jordan. 628 pp.

[31] Qasem, J. R. (2011). Herbicides applications: Problems and considerations, In, *Herbicides and Environment,* Andreas Kortekamp (Ed.), *ISBN*: 978-953-307-476-4, InTech, Available from: http://www.intechopen.com/articles/show/title/herbicides-applications-problems-and-considerations.

[32] Qasem, J. R. and Al-Jebury, I. S. (2001). Weed control in marjoram (*Origanum syriacum* L.) under field conditions. Dirasat 28, 194-207.

[33] Qasem, J.R. and Foy, C.L. (2006). Selective weed control in Syrian marjoram (*Origanum syriacum*L.) with oxadiazon and oxyfluorfen herbicides. *Weed Technology,* 20 (3), 670-676.

[34] Quist D. (2010). Vertical (Trans) gene Flow: Implications for crop diversity and wild relatives published. *Third World Network, Penang, Malaysia Jutaprint, ISBN:* 978-967-5412-26-4.

[35] Radosevich, S. R., Holt, J.S. and Ghersa, C.M. (2007). *Ecology of Weeds and Invasive Plants. Relationship to Agriculture and Natural Resource Management.* Hobokin, N.J.: Willey Interscience.

[36] Radosevich, S., Holt, J. and Ghersa, C. (1997). *Weed Ecology: Implication for Management* . 2nd Edition. John Wiley & Sons Inc. New York.

[37] Smith, C.M. and Namuth, D. (2005). Herbicide resistance: mechanisms, inheritance, and molecular genetics. eLearn & Grow Library, 7pp. available at: http://plantandsoil.unl.edu/croptechnology2005/pageincludes/printModule.jsp?inform.

[38] Smith, C.M.; Hulting, A.; Thill, D.; Morishita, D. and Krenz, J. (2007) Herbicide-resistant weeds and their management, weed control strategies. In: *Pacific Northwest Conservation Tillage Handbook, University of Idaho.*

[39] Stewart N.C. (2008). Gene flow and the risk of transgene spread. University of Tennessee. Retrieved December 2010 Available at: http://agribiotech.info/details/StewartGeneFlow%20Mar%208%20-%2003.pdf.

[40] Tsyusko, O. V., Smith, M. H., Sharitz, R. R. and Glenn, T. C. (2010). Genetic and clonal diversity of two cattail species, *Typha latifolia* and *T. angustifolia* (Typhaceae), from Ukraine. *American Journal of Botany,* 97 (12), 2061-2067.

[41] van de Wiel, C.; Groot, M.; and Den Nijs, H. (2005). Gene flow from crops to wild plants and its population ecological consequences in the context of GM-crop biosafety including some recent experiences from lettuce. *Wageningen UR Frontis Series, Volume 7 Environmental Costs and Benefits of Transgenic Crops.* Retrieved, December 2010 Available at: http://library.wur.nl/ojs/index.php/frontis/article/viewArticle/914.

[42] Vencill, W., Grey, T. and Culpepper, S. (2011). Resistance of weed to herbicides. In: *Herbicides and Environment* (Kortekamp, A. ed.). pp. 585-594. InTech, Available at: http://www.intechopen.com/articles/show/title/herbicides-applications-problems-and-considerations.

[43] White, D.E. (1979). *Physiological Adaptations in Two Ecotypes of Canada Thistle* (*Cirsium arvense* (L.) Scope. MSc. Thesis, University of California, Davis.

[44] WSSA, (2012). Data from the international survey of herbicide resistant weeds obtainable at WeedScience.com.website.

Chapter 18

Pesticide Tank Mixes: An Environmental Point of View

Valdemar Luiz Tornisielo, Rafael Grossi Botelho,
Paulo Alexandre de Toledo Alves,
Eloana Janice Bonfleur and
Sergio Henrique Monteiro

Additional information is available at the end of the chapter

http://dx.doi.org/10.5772/55948

1. Introduction

During the last decades, human activity has affected the aquatic and terrestrial ecosystems' sustainability. None of these activities has damaged the environment as severely as agricultural practices.

Current agricultural practices have negatively affected aquatic and terrestrial ecosystems by destroying habitats, deforesting to increase cropping areas and applying pesticides.

Pesticides are a heterogeneous category of chemical products destined to pest, disease and weed control including several types, such as insecticides, fungicides, herbicides, nematicides and others.

Nowadays, such chemical product applications have been considered the most efficient plant protection procedures and have significantly contributed to the improvement of crop productivity.

Nevertheless, the claimed objective of supplying the population with enough food does not justify damaging the environment, just because small quantities of pesticides are known to efficiently control pests, diseases and weeds. However, most of them are rapidly spread out affecting all living beings (flora and fauna, including humans).

The use of chemical molecules in agriculture increased after the Second World War with the advent of DDT (dichloro-diphenyl-trichloroethane). DDT was discovered in 1939 by Paul Müller (Swiss entomologist) and its worldwide use was rapidly expanded due to its large action range, low cost and efficiency in the control of tropical disease vectors, such as typhoid fever and malaria [1].

After the release of DDT, a large range of molecule groups destined to crop protection were developed and commercialized. In 1962, the book "Silent Spring" was the first act of environment manifest against DDT, describing the bird population decrease (from the top of the food chain) attributed to its indiscriminate use.

After the 1960's, the use of chemical products in agriculture rapidly increased and it was associated with the appearance of environmental and human health problems.

The frequent and incorrect use of pesticides have caused soil, atmosphere, food and water resource (superficial/underwater) contaminations, negatively affecting aquatic and terrestrial organisms as well as frequently causing toxicity to the human population.

Therefore, studies are urgently needed to make environmental monitoring procedures viable in order to detect potential contamination risks and give support to public actions for environmental safety and agriculture sustainability.

Currently, product mixtures (associations between one or more molecules) are applied in agriculture instead of individual molecules; therefore, previous studies that focused on only one molecule should now consider molecule mixtures.

The existence of such a large variety of pests, diseases and weeds affecting yields have led farmers to use product mixtures, aiming at efficiently managing crop protection. Such mixtures, also called product associations, enter the environment in a different way compared to the individual product application. Thus, more studies are required about these mixture-environment interactions and possible interactions between molecules and consequent interferences in the environment.

Although mixtures have been intensively studied concerning their agronomic efficacy, little information is found about their implications on environmental safety.

In this chapter, the tank mixture subject is approached from an environmental point of view, explaining the chemical product mixture interactions and the possible contaminant effects. Studies on the product-environment interactions are presented to provide the main available information as support to future studies and decisions in environmental sustainability and safety.

2. Agronomic characteristics of tank mixtures

Tank mixtures are associations among two or more chemical products (pesticides) or among chemical products and fertilizers in a unique tank for application in crops. This practice is common in Australia, Canada, U.S.A and United Kingdom, where there are recommendations on application procedures, incompatibilities, and safety [2].

Concerning agricultural practices, the tank mixture of two or more chemical products might be a good application strategy, saving fuel and labor-hours, causing less soil compaction, and possibly providing a larger pest control range and efficacy, when compared to the single product application. For these reasons, this technique is preferred by farmers [3].

Nevertheless, the herbicide mixture might induce, for instance, interactions before or after reaching the target-plant, by altering the product action in synergistic, antagonistic or additive ways. One common practice is the simultaneous application of herbicides with and without residual effect in order to increase the weed species control range and/or the control period. Another practice is the addition of adjuvants to improve herbicide performance to control weeds. The simultaneous application of pesticides (concerning the species-target to be controlled) might induce undesirable (antagonistic, synergistic or additive) reactions, depending on the herbicide type and plant species [4]. When the mixture induces an antagonistic reaction, it means that a lower weed control action than expected is observed. When the mixture induces a synergistic reaction, it means that a higher weed control than expected is observed. And, finally, when the mixture induces an additive reaction, it means that no change in weed control is observed.

Several studies have elucidated the questions about synergistic and antagonistic effects of active ingredient mixtures on weed control, for instance, the studies with glyphosate reported by Vidal et al. [4], Shaw and Arnold [3], Selleck and Baird [5].

The application of pesticides plus adjuvants has also been a usual practice. The adjuvant enhances the active ingredient action [6]. In other words, the adjuvant substance induces the herbicide molecule uptake by leaf tissues, by accelerating the product penetration through plant cuticles. The most common types are the biosurfactants, mineral or vegetal oils, synthetic or natural polymers, humectants, organic salts, buffer solutions, and others [7].

The tank mixture practice or different individual pesticide applications at short intervals might result in multiple pesticide residues on foods, as observed by Gebara et al. [8], when monitoring food samples in São Paulo metropolis, Brazil, during the period between 1994 and 2001. The authors found multiple pesticide residues in 5.8% of vegetable samples analyzed and 11.4% of fruit samples.

Gebara et al. [9] alerted for the violation risk of the Theoretical Maximum Dietary Intake (TMDI), which is calculated by the relationship between the Limit of Maximum Residues (LMR, mg kg^{-1}) established for a pesticide in a food and the daily consumption (DC, kg day^{-1}), based on the individual diet. The presence of multiple pesticide residues in foods due to the use of tank mixtures, might lead to the extrapolation of toxicological parameters for the acceptable daily intake (ADI), mainly for children and nursing women.

3. Pesticide tank mixtures environmental effects

3.1. Soil

Weed control with pesticide tank mixtures has been widely studied concerning mixture effectiveness, component antagonism and/or synergism. However, there is little information on environmental issues.

Knowledge on soil-herbicide interactions when herbicide mixtures are applied is extremely relevant. However, few studies on herbicide associations and their soil interactions can be found, because most studies are restricted to the individual molecule behavior.

When a pesticide is released in the environment, it will probably enter the soil by direct application, or indirectly, by crop residue incorporation into the soil and molecule transport by spraying derivation. In the soil, several processes might occur, that is, molecule retention (adsorption, absorption), transformation (decomposition, degradation) and transport (spraying derivation, volatilization, lixiviation, superficial runoff). Such processes will determine the molecule destiny, persistence and agronomic efficiency. The main factors influencing those processes are the climatic conditions, the pesticide physical-chemical properties and the soil physical-chemical attributes. According to Oliveira [10], the complex molecule retention process by soil sorption/desorption directly or indirectly influences other factor activities.

Knowledge on pesticide physical-chemical properties is fundamental to predict soil interactions, potential contamination and transport risks when in the soil solution or associated to sediments. Studies on pesticide mixtures have been restricted to their phytotoxicity effects and few were dedicated to the interactions between two or more associated molecules.

Alves [11] demonstrated that ametryn mineralization half-life is longer when associated to glyphosate than when applied alone; but there was a synergistic effect in the soil, because ametryn half-life was 15 days for the ametryn + glyphosate mixture and 20 days for isolated ametryn in the soil. In the same study, the author observed increased glyphosate mineralization half-life from 55 to 119 days, when comparing single glyphosate and glyphosate + ametryn treatments, respectively; the glyphosate soil half-life could not be determined due to its strong soil sorption during extractions.

Yet in studies of soil microbial activity, Alves [11] observed that glyphosate (at a higher rate) enhanced microbial activity; meanwhile isolated ametryn (at a lower rate) negatively affected microbial activity, but a less negative effect of ametryn + glyphosate mixture (at a lower rate) was observed compared with single ametryn at the same rate. The ametryn + glyphosate mixture (at a higher rate) increased the microbial activity, evidencing a stronger mixture synergistic effect.

Alves [11] also studied the herbicide sorption/desorption in a red Ultisol. High glyphosate and low ametryn sorption were observed when herbicides were applied alone. Higher soil sorption was observed for both herbicides in mixture than for the single molecules. Low glyphosate desorption occurred at all rates in both application procedures (alone or in mixture), but ametryn desorption decreased when applied in mixture.

White et al. [12] studied the effects of chlorothalonil, tebuconazole, flutriafol and cyproconazole fungicides on the metolachlor herbicide dissipation kinetics. Significantly lower metolachlor dissipation was observed with chlorothalonil, when compared with soil treatments without chlorothalonil or with other fungicides. The authors observed significant reduction in metolachlor metabolites probably attributed to the fungicide effect on glutathione S-transferase enzyme activity. Overall, chlorothalonil fungicide induced a two-fold increase in metolachlor persistence.

Ke-Bin et al. [13] observed that atrazine and bentazon herbicides showed longer lag-phase and lower degradation rate when applied in tank mixture in a maize crop. Therefore, the association of atrazine-bentazon had longer soil persistence which means that higher environmental potential contamination risks might be expected.

The effect of glyphosate on atrazine degradation was studied by Krutz et al. [14] in a silt clayey soil (pH 8.3 and 10.6 g kg^{-1} of organic-C) from the Texas region in USA. Atrazine degradation was inversely related to glyphosate rate and microbial activity during an eight-day period, evidencing that glyphosate enhanced microbial activity and inhibited atrazine degradation. The authors discussed that atrazine degradation, when in association, is mainly a microbial mechanism, and the degradation reduction might be explained by a lower enzymatic activity and/or by microbial population suppression by glyphosate.

Similar results were reported by Haney et al. [15] for the same soil type, demonstrating the atrazine and glyphosate effects on soil microbial activity evaluated through the soil carbon (C) and nitrogen (N) mineralization. Soil plots treated with the herbicide mixture showed higher microbial activity than plots treated with single atrazine. The evaluated soil C and N flows allowed understanding of the microbial preference for glyphosate because this herbicide's complete mineralization occurred in 14 days, followed by fast atrazine degradation.

Zablotowicz et al. [16] studied the effects of glufosinate (herbicide), ammonium sulfate (fertilizer) and both products in mixture on atrazine mineralization. The authors observed decreased atrazine mineralization when the product mixture was applied. The authors explained that an alteration in ^{14}C-atrazine molecule partition into its metabolites and residues would occur caused by ammonium sulfate that would restrict the triazine ring cleavage. Such results evidenced that the application of glufosinate combined to a mineral N source might increase soil atrazine persistence, increasing its residual effect.

Lancaster et al. [17] observed that glyphosate increased soil C mineralization and fluometuron microbial degradation. The authors suggested that the increasing C mineralization might be related to the increasing fluometuron degradation or to a priming glyphosate effect.

Concerning the glyphosate and diflufenican association, Tejada [18] observed longer degradation periods for both herbicides in mixture than for the individual molecules. Furthermore, the glyphosate-diflufenican association increased both herbicide toxicities to the soil biological activity (measured by the microbial C biomass and enzyme activities - dehydrogenase, urease, β-glycosidase, phosphatase and arylsulfatase) and the individual herbicide persistence.

Pereira et al. [19] evaluated the application of isolated glyphosate and associated to endosulfan on the soil microbial activity in soybeans and observed reduced microbial activity and biomass, and also, reduced metabolic quotient.

In genetically modified glyphosate-tolerant maize cultivars, it is possible to mix glyphosate and atrazine. In the USA, there are a number of commercially available associations, among them, glufosinate or glyphosate mixed with atrazine [20]. Bonfleur et al. [21] observed that glyphosate mineralization was not affected by atrazine presence in a tropical soil. However, increased atrazine mineralization (measured by the $^{14}CO_2$ release) was observed with increasing glyphosate rates. The authors observed a 100-day variation in the atrazine half-life when associated with a two-fold glyphosate rate. Therefore, the glyphosate-atrazine tank mixture allowed atrazine persistence reduction in the soil. The authors said that a possible explanation is the glyphosate contribution to the microorganisms as source of N, and this N supply might decrease the initial atrazine immobilization when this is the only substrate, and then, increasing its mineralization.

Fogg and Boxall [22] observed inhibitory effects of an isoproturon-chlorothalonil mixture on the isoproturon degradation in soils. Isoproturon half-life (DT50) values varied from 18.5 to 71.5 days when combined with chlorothalonil. This might be explained by the TPN-OH chlorothalonil metabolite inhibition and the reduction in the soil microorganism population involved in isoproturon degradation.

The soil degradation of pendimethalin (herbicide) was significantly reduced when mixed with mancozeb (fungicide) or mancozeb+thiamethoxam (insecticide) [23]. Pendimethalin herbicide half-life increased from 26.9 to 62.2 days when in single and combined (mancozeb + thiamethoxam) applications, respectively, in a sandy soil. On the other hand, the same authors observed that pendimethalin degradation is not affected by the presence of isolated metribuzin or thiamethoxam.

Several studies have pointed out the adjuvant influence on pesticide destiny in the environment, specifically their persistence and bioavailability. Cabrera [24], in laboratory studies, affirmed that metazachlor herbicide added to oil and surfactant showed reduced degradation rates and increased residues in the soil. Similar results to other pesticides were reported by Kucharski and Sadowski [25] and Rodríguez-Cruz et al. [26]. In a field experiment, Kucharski et al. [27] observed a 43% increase in lenacil herbicide residues in the superficial soil layer, with the addition of adjuvants (oil and surfactant).

High mobility pesticides used together with adjuvants present decreased movement along the soil profile. Reddy and Singh [28] evaluated bromacil and diuron herbicides lixiviation in soil columns. In treatments with adjuvant addition, the authors observed significant lower bromacil vertical movement and no effect on diuron movement. These two herbicides present distinct physical-chemical characteristics that explain their differential movement abilities in the soil. Thus, bromacil is an acidic molecule with high water solubility (815 mg L^{-1}); meanwhile diuron is a non-ionic herbicide of low water solubility (42 mg L^{-1}). From the environmental point of view, the adjuvant effect was positive in the case of bromacil, but the agronomic efficacy was restricted.

The results found in the literature have highlighted the interactions existing among several molecules, especially in the soil, but such interactions might be different under other environment compartments. For this reason, studies on environmental pesticide behavior and destination must include all aspects, bringing together laboratory and field approaches.

3.2. Water: An ecotoxicological approach for pesticide mixtures

According to Botelho et al. [29], water resource contamination has currently been considered one of the greatest environmental problems on Earth.

Pesticides applied to field crops are released in the environment mainly through lixiviation (when molecules move into the soil and reach the underground waters), superficial runoff (when molecules move together with soil and water runoff), and spraying derivation (when molecules are carried by wind during pesticide spraying).

The situation is complex once crop diversity allied to the high number and diversity of pesticide products usually applied to field crops, and the short distances between fields and aquatic areas have exposed the water resources not only to individual products but also to all their associations [30].

Several products, mainly herbicides and insecticides, are common superficial water contaminants, due to their large application in agriculture and residential areas. Therefore, there is an increasing concern about superficial and underground water contamination, due to the lack of information on pesticide impacts mainly in aquatic systems.

In Brazil, several studies have been carried out to determine the presence of pesticides in aquatic ecosystems. Armas et al. [31] evaluated the presence of herbicides in the superficial water and sediments of Corumbataí River (State of São Paulo, Brazil). The authors found several herbicides - ametryn, atrazine, simazine, hexazinone, glyphosate and clomazone – and triazines were specifically found in higher levels, above the limits allowed for potable water by Brazilian legislation. Dores et al. [32] found herbicide residues from the triazine group and their metabolites, as well as metribuzin, metolachlor and trifluralin residues. Among the Brazilian literature, the research works of Caldas et al. [33], Lanchote et al. [34], Filizola et al. [35], Laabs et al. [36], Dores et al. [37], Jacomini et al. [38] are pointed out.

Other interesting results can be found in the literature: Benvenuto et al. [39] determined the presence of eleven pesticides in superficial waters of Italy and Spain and observed concentration values varying between 0.002 and 0.087 $\mu g\ L^{-1}$. Yu et al. [40] determined the presence of nine (among eleven pesticides evaluated) herbicides of the triazine group in all water samples analyzed. Similar determinations were made by Ma et al. [41], Palma et al. [42], Balinova and Mondesky [43] and Segura et al. [44].

Understanding of how pesticides affect aquatic environments has been a challenge to researchers, and the science of ecotoxicology has helped to answer many questions on this subject.

The "ecotoxicology" term was first suggested by the French toxicologist René Truhaut, during the *Committee of the International Council of Scientific Unions* (ICSU) meeting, in June 1969, in Stockholm (Sweden) [45]. According to this author, Ecotoxicology is the science that studies the effects of natural or synthetic substances on living beings, populations and communities, animal or vegetal, terrestrial or aquatic, constituting the biosphere, including the substance interaction with the environment where they live in an integrated context [46].

Usually, ecotoxicological experiments follow standardized protocols developed by international organizations, for example, the Environmental Protection Agency (EPA); the Organization for Cooperation and Economical Development (OCDE); and the Brazilian Agency of Technical Norms (ABNT).

The toxicity tests allow evaluating the environmental contamination by different pollutant sources, such as agricultural, industrial and domestic residues, sediments, medicines and chemical products overall, as well as the results of their synergistic and antagonistic effects [47-48]. The ecotoxicological tests can also detect the toxic agent or mixture capacity of causing deleterious effects on living organisms, allowing determination of the harmful concentration ranges, and how and where the effects are expressed [49].

Several parameters have been used to determine the xenobiotic effects in different organisms. Among these variables, the lethality [50-51], immobility [52], gill alterations [53-56], and reproduction [57-59] are pointed out.

The ecotoxicological experiments consist of exposing living organisms to several concentrations of a specific product and evaluating the results that might be expressed according to the test type. For instance, the acute test consists of short-term exposure of organisms to several product concentrations, and then, the species life cycle is evaluated; the toxicity indicative parameters more frequently used are: lethality (expressed by the average lethal concentration - LC_{50}), and immobility (expressed by the observable toxic concentration effect - EC_{50}). It is important to highlight that both parameters take into consideration the effects for 50% of the organisms tested under the specific experiment conditions [60-61]. In the case of a chronic test, the organism is submitted to long-term product exposure and the observable effects are usually focused on organism reproduction, behavior, morphology, and size, among others.

Water quality tests have been important tools aiming to minimize the pollution effects on aquatic ecosystems and to implement remediation and monitoring programs, and for that, the ecotoxicological tests have been used.

In the case of pesticide mixtures, the ecotoxicological tests to determine toxicity effects are difficult to interpret, because toxicity symptoms might depend on interactions occurring among different chemical molecules in solution and their accumulative quantities in organisms [61].

When analyzing mixture toxicity effects, some approaches and definitions must be established. In the aquatic ecotoxicology, two different models have been used to describe the relationships between single compound effects and their mixtures: concentration addition model (CA) and independent action model (IA) [62]. In the CA model, each mixture component toxicity effect is induced through a same mechanism, meanwhile in the IA model, the combined components show different actions, inducing a unique toxicological response, but via distinct reactions within the organisms [63]. Nevertheless, both models are used as references to predict the expected mixture toxicity effect, based on the known toxicity of the individual compounds [62].

For a long time, there has been concern about mixture impacts on aquatic ecosystems, not only from pesticides but also from other compound groups, and several discussions and reviews have been reported. In 1984, Hermens and collaborators investigated organic mixture effects on mortality and reproduction of *Daphnia magna* microcrustacean, after exposure to 14 products with different modes of action. The authors observed more severe toxicity effects on mixture-treated organisms than with individual products, although the chronic test results with the mixture showed less severe symptoms [64]. Strmac and Braunbeck [65] observed several structure and biochemical alterations in rainbow trout hepatocytes submitted to a 20-component mixture, including pesticides. Delorenzo and Serrano [66] evaluated the effects of atrazine (herbicide), chlorpyrifos (insecticide) and chlorothalonil (fungicide) on the *Dunaliella tertiolecta* algae growth; the results of atrazine - chlorpyrifos mixture showed an additive toxicity pattern, meanwhile atrazine - chlorothalonil mixture showed a synergistic effect. Yet, the authors observed a two-fold higher toxicity effect of atrazine – Chlorothalonil mixture than the individual products. Choung et al. [67] observed that relatively high atrazine rates increased the terbufos (insecticide) toxicity to *Ceriodaphnia dubia* microcrustacean.

4. Final remarks

Pesticide tank mixtures are currently and frequently used not only in developed countries with specific regulatory legislation for the practice, but also in all agricultural countries where information on harmful effects do not directly reach farmers.

From the agronomic point of view, an effective pest control with pesticide mixtures will depend on the molecule compatibility and also on specific control tests. When the farmer uses two chemically incompatible substances in tank mixture, high losses in crop yield and equipment problems might occur, for example, sprayer nozzle obstruction due to chemical reaction between molecules and subsequent compound precipitation.

Although the pesticide tank mixture may appear to be an efficient pest control practice with synergistic results, the aspects concerning environmental safety must be considered. Little specific information on associated pesticide residues is available in the literature concerning withholding periods and overall environmental behavior.

When a single pesticide is applied, the expected environmental results should be similar to previous results reported for the pesticide registration and before its commercial release. The environment (mainly aquatic and soil medium) is a large contaminant reservoir, where the chemical compounds used in agriculture can be found together. In spite of that, it is important to reinforce that a single pesticide interacts quite differently with the medium, compared to the mixture interaction, as already discussed in this chapter.

In light of the large global demand for food and the increasing crop productivity in the same cropping area, it is imperative to consider the environmental safety questions concerning tank chemical mixture applications in agriculture.

This is a relatively new science area that demands urgent studies on environmental safety, ecotoxicology and toxicology, in order to make highly prevalent the declaration of the United Nation Organization about the planet environment: "*The man has the fundamental right to liberty, equality and enjoyment of adequate life conditions, under an environment of such quality that allows him living a dignifying life and well-being, and he is carrier of the solemn duty of protecting and improving the environment for the present and future generations*" [68].

Acknowledgements

The authors are grateful to the Research Foundation of the State of São Paulo (FAPESP) and to the National Council for Scientific and Technological Development (CNPQ).

Author details

Valdemar Luiz Tornisielo, Rafael Grossi Botelho, Paulo Alexandre de Toledo Alves, Eloana Janice Bonfleur and Sergio Henrique Monteiro

Laboratory of Ecotoxicology, Center for Nuclear Energy in Agriculture, University of São Paulo, Piracicaba, SP, Brazil

References

[1] Amato, D. C, Torres JPM, Malm O. DDT (dicloro difenil tricloroetano): toxicidade e contaminação ambiental- uma revisão. Química Nova (2002). a) , 995-1002.

[2] CanadáMinistry of Agriculture. Safety precautions- mixing and loading pesticides. Vancouver, (2011). http://www.al.gov.bc.ca/pesticides/d_5.htmaccessed 15 February 2012).

[3] Shaw, D. R, & Arnold, J. C. Weed control from herbicide combinations with glyphosate. Weed Technology (2002). , 16(1), 1-6.

[4] Vidal, R. A, Machry, M, Hernandes, G. C, & Fleck, N. G. Antagonismo na associação de glyphosate e triazinas. Planta Daninha (2003). , 21(2), 301-306.

[5] Selleck, G. W, & Baird, D. D. Antagonism with glyphosate and residual herbicide combinations. Weed Science (1981). , 29(2), 185-190.

[6] International Union of Pure and Applied ChemistryIUPAC: Pesticide Formulations. http://agrochemicals.iupac.org/index.php?option=com_sobi2&sobi2Task=sobi2Details&catid=3&sobi2Id=38&Itemid=19accessed 15 September (2012).

[7] Cronfeld, P, Lader, K, & Baur, P. Classification of Adjuvants and Adjuvant Blends by Effects on Cuticular Penetration. In: Viets AK, Tann RS, Mueninghoff JC. (Eds.) Pesticide Formulations and Application Systems: Twentieth Volume, ASTM STP 1400, American Society for Testing and Materials. West Conshohocken PA; (2001). , 81-94.

[8] Gebara, A. B. Ciscato CHP, Ferreira M da S, Monteiro SH. Pesticide Residues in Vegetables and Fruits Monitored in São Paulo City, Brazil, 1994-2001. Bulletin of Environmental Contamination and Toxicology (2005). , 75(1), 163-169.

[9] Gebara, A. B. Ciscato CHP, Monteiro SH, Souza GS. Pesticide Residues in some Commodities: Dietary Risk for Children. Bulletin of Environmental Contamination and Toxicology (2011). , 86(5), 506-510.

[10] Oliveira, M. F. Comportamento de herbicidas no ambiente. In: Oliveira Jr., Constantin RSJ. (Eds.) Plantas daninhas e seu manejo. Guaíba: Agropecuária; (2001). , 315-362.

[11] Alves PATComportamento dos herbicidas ametrina e glifosato aplicados em associação em solo de cultivo de cana-de-açúcar. PhD Thesis. University of São Paulo; (2012).

[12] White, P. M, Potter, T. L, & Culbreath, A. K. Fungicide dissipation and impact on metolachlor aerobic soil degradation and soil microbial dynamics. Science of the Total Environment (2010). , 408(6), 1393-1402.

[13] Ke-bin, L. I, Cheng, J, Wang, X, Zhou, Y, & Liu, W. Degradation of herbicides atrazine and bentazone applied alone and in combination in soils. Pedosphere (2008). , 18(2), 265-272.

[14] Krutz, L. J, Senseman, S. A, & Haney, R. L. Effect of Roundup Ultra on atrazine degradation in soil. Biology and Fertility of Soils (2003). , 38(2), 115-118.

[15] Haney, R. L, Senseman, S. A, Krutz, L. J, & Hons, F. M. Soil carbon and nitrogen mineralization as affected by atrazine and glyphosate. Biology and Fertility of Soils (2002). , 35(1), 35-40.

[16] Zablotowicz, R. M, Krutz, L. J, Weaver, M. A, Accinelli, C, & Reddy, K. N. Glufosinate and ammonium sulfate inhibit atrazine degradation in adapted soils. Biology and Fertility of Soils (2008). , 45(1), 19-26.

[17] Lancaster, S. H, Haney, R. L, Senseman, S. A, Kenerley, C. M, & Hons, F. M. Microbial degradation of Fluometuron is influenced by Roundup Weather MAX. Journal of Agricultural and Food Chemistry (2008). , 56(18), 8588-8593.

[18] Tejada, M. Evolution of soil biological properties after addition of glyphosate, diflufenican and glyphosate + diflufenican herbicides. Chemosphere (2009). , 76(3), 365-373.

[19] Pereira, J. L, Picanço, M. C, Silva, A. A, & Santos, E. A. Tomé HVV, Olarte JB. Effects of glyphosate and endosulfan on soil microorganisms in soybean crop. Planta Daninha (2008). , 26(4), 825-830.

[20] Owen MDKCurrent use of transgenic herbicide-resistant soybean and corn in the USA. Crop Protection (2000).

[21] Bonfleur, E. J, Lavorenti, A, & Tornisielo, V. L. Mineralization and degradation of glyphosate and atrazine applied in combination in a Brazilian oxisol. Journal of Environmental Science and Health: Part B-Pesticides Food Contaminants and Agricultural Wastes (2011). , 46(1), 69-75.

[22] Fogg, P. Boxall ABA. Degradation of Pesticides in Biobeds: The Effect of Concentration and Pesticide Mixtures. Journal of Agricultural and Food Chemistry (2003). , 51(18), 5344-5349.

[23] Swarcewicz, M. K, & Gregorczyk, A. J. The effects of pesticide mixtures on degradation of pendimethalin in soils. Environmental Monitoring and Assessment (2012). , 184(5), 3077-3084.

[24] Cabrera, D, Lopez-pineiro, A, Albarran, A, & Pena, D. Direct and residual effects on diuron behaviour and persistence following two-phase olive mill waste addition to soil. Geoderma (2010).

[25] Kucharski, M, & Sadowski, J. Influence of adjuvants on behavior of phenmedipham in plant and soil environment. Polish Journal of Agronomy (2009). , 1(1), 32-36.

[26] Rodríguez-cruz, M. S, Sánchez-martín, M. J, Andrades, M. S, & Sánchez-camazano, M. Retention of pesticides in soil columns modified in situ and ex situ with a cationic surfactant. Science of the Total Environment (2007).

[27] [27]Kucharski, M, Sadowski, J, Wujek, B, & Trajdos, J. Influence of adjuvants addition on lenacil residues in plant and soil. Polish Journal of Agronomy (2011). , 5(5), 39-42.

[28] Reddy, K. N, & Singh, M. Effect of Acrylic Polymer Adjuvants on Leaching of Bromacil, Diuron, Norflurazon, and Simazine in Soil Columns. Bulletin of Environmental Contamination and Toxicology (1993). , 50(3), 449-457.

[29] Botelho, R. G, Cury, J. P, Tornisielo, V. L, & Santos, J. B. Herbicides and the Aquatic Environment. In: Mohammed N A El-G H. (Ed.) Herbicides- Properties, Synthesis and Control of Weeds. Rijeka: InTech; (2012). , 149-164.

[30] Gilliom, R. J, Barbash, J. E, Crawford, C. G, Hamilton, P. A, Martin, J. D, Nakagaki, N, Nowell, L. H, Scott, J. C, Stackelberg, P. E, Thelin, G. P, & Wolock, D. M. (2006). The quality of our nation's waters. Pesticides in the nation's streams and ground water, US Geological Survey, Reston, VA, 1992-2001.

[31] Armas, E. D. Monteiro RTR, Antunes PM, Santos MAPF, Camargo PB, Abakerli RB. Diagnóstico espaço-temporal da ocorrência de herbicidas nas águas superficiais e sedimentos do rio corumbataı́ e principais afluentes. Química Nova (2007). , 30(5), 1119-1127.

[32] Dores EFGCCarbo L, Ribeiro ML, De-Lamonica-Freire EM. Pesticide Levels in Ground and Surface Waters of Primavera do Leste Region, Mato Grosso, Brazil. Journal of Chromatographic Science (2008). , 46(7), 585-590.

[33] Caldas, E. D, & Coelho, R. Souza LCKR. Organochlorine pesticides in water, sediment and fish of Paranoá Lake of Brasilia, Brazil. Bulletin of Environmental Contamination and Toxicology (1999). , 62(2), 199-206.

[34] Lanchote, V. L, Bonato, P. S, & Cerdeira, A. L. Santos NAG, Carvalho D, Gomes MA. HPLC screening and GC-MS confirmation of triazine herbicides residues in drinking water from sugar cane area in Brazil. Water, Air, Soil Pollution (2000).

[35] Filizola, F. F, & Ferracini, V. L. Sans LMA, Gomes MAF, Ferreira CJA. Monitoramento e avaliação do risco de contaminação por pesticidas em água superficial e subterrânea na região de Guairá. Pesquisa Agropecuária Brasileira (2002)., 37(5), 659-667.

[36] Laabs, V, Amelung, W, Pinto, A. A, Wantzen, M, Silva, C. J, & Zech, W. Pesticides in Surface Water, Sediment, and Rainfall of the Northeastern Pantanal Basin, Brazil. Journal of Environmental Quality (2002)., 31(5), 1636-1648.

[37] Dores EFGCNavickiene S, Cunha MLF, Carbo L, Ribeiro ML, De-Lamonica-Freire EM. Multiresidue determination of herbicides in environmental waters from "Primavera do Leste" region (Middle West of Brazil) by SPE-GC-NPD. Journal of Brazilian Chemical Society (2006)., 17(5), 866-873.

[38] Jacomini, A. E, Camargo, P. B, & Bonato, P. S. Determination of ametryn in river water, river sediment and bivalve mussels by liquid chromatography-tandem mass spectrometry. Journal Brazilian Chemical Society (2009)., 20(1), 107-116.

[39] Benvenuto, F, Marín, J. M, Sancho, J. V, Canobbio, S, Mezzanotte, V, & Hernández, F. Simultaneous determination of triazines and their main transformation products in surface and urban wastewater by ultra-high-pressure liquid chromatography-tandem mass spectrometry. Analytical and Bioanalytical Chemistry (2010). , 397(7), 2791-2805.

[40] Yu, Z. G, Qin, Z, Ji, H. R, Du, X, Chen, Y. H, Pan, P, Wang, H, & Liu, Y. Y. Application of SPE Using Multi-Walled Carbon Nanotubes as Adsorbent and Rapid Resolution LC-MS-MS for the Simultaneous Determination of 11 Triazine Herbicides Residues in River Water. Chromatographia (2010).

[41] Ma, W. T, Fu, K. K, Cai, Z, & Jiang, G. B. Gas chromatography/mass spectrometry applied for the analysis of triazine herbicides in environmental waters. Chemosphere (2003)., 52(9), 1627-1632.

[42] Palma, G, Sánchez, A, Olave, Y, Encina, F, Palma, R, & Barra, R. Pesticide levels in surface waters in an agricultural-forestry basin in Southern Chile. Chemosphere (2004)., 57(8), 763-70.

[43] Balinova, A. M, & Mondesky, M. Pesticide contamination of ground and surface water in Bulgarian Danube plain. Journal of Environmental Science and Health, Part B. Pesticides, Food Contaminants, and Agricultural Wastes (1999)., 34(1), 33-46.

[44] Segura, P. A, Mcleod, S. L, Loemoine, P, Sauvé, S, & Gagnon, C. Quantification of carbamazepine and atrazine and screening of suspect organic contaminants in surface and drinking waters. Chemosphere (2011 8)., 2011(84), 8-1085.

[45] Truhaut, R. Ecotoxicology: objectives, principles and perspectives. Ecotoxicology and Environmental Safety (1977). , 1(2), 151-173.

[46] Plaa, G. L. Present status: toxic substances in the environment. Canadian Journal of physiology and Pharmacology (1982). , 60(7), 1010-1016.

[47] Lombardi, J. V. Fundamentos de toxicologia aquatica. In: Ranzani-Paiva MJT, Takemota RM, Lizama MAP. (Eds.) Sanidade de organismos aquáticos. São Paulo: Varela; (2004). , 263-272.

[48] Marschner, A. Biologische Bodensanierung und ihre Erfolgskontrolle durch Biomonitoring. In: Oehlmann J., Markert B. (Eds.) Okotoxikologie-Okosystemare Ansatze und Methoden. Ecomed: Landsberg; (1999). , 568-576.

[49] Magalhães, D. P. Ferrão Filho AS. A ecotoxicologia como ferramenta no biomonitoramento de ecossistemas aquáticos. Oecologia Autralis (2008). , 12(3), 355-381.

[50] Botelho, R. G, Santos, J. B, Oliveira, T. A, & Braga, R. R. Byrro ECM. Toxicidade aguda de herbicidas a tilápia (*Oreochromis niloticus*). Planta Daninha (2009). , 27(3), 621-626.

[51] Novelli, A, Vieira, B. H, Cordeiro, D, Cappelini, L. T, Vieira, E. M, & Espíndola, E. L. *Lethal effects of abamectin on the aquatic organisms Daphnia similis, Chironomus xanthus and Danio rerio. Chemosphere* (2012 8). , 2012(86), 1-36.

[52] Botelho, R. G, Inafuku, M. M, & Maranho, L. A. Machado Neto L, Olinda RA, Dias CT, Tornisielo VL. Toxicidade aguda e crônica do extrato de nim (*Azadirachta indica*) para *Ceriodaphnia dubia*. Pesticidas: revista de ecotoxicologia e meio ambiente (2010). , 20(1), 29-34.

[53] Poleksic, V, & Mitrovic-tutundzic, V. Fish gills as a monitor of sublethal and hronic effects of pollution. In: Muller R., Lloyd R. (Eds.) Sublethal and Chronic Effects of Pollutants on Freshwater Fish. Oxford; (1994). , 339-352.

[54] Camargo MMPFernandes MN, Martinez CBR. How aluminium exposure promotes osmoregulatory disturbances in the neotropical freshwater fish *Prochilus lineatus*. Aquatic Toxicology (2009). , 94(1), 40-46.

[55] Botelho, R. G, Santos, J. B, Fernandes, K. M, & Neves, C. A. Effects of atrazine and picloram on grass carp: acute toxicity and histological assessment. Toxicological and Environmental Chemistry (2011). , 94(1), 121-127.

[56] Paulino, M. G. Souza NES, Fernandes MN. Subchronic exposure to atrazine induces biochemical and histopathological changes in the gills of a Neotropical freshwater fish, *Prochilodus lineatus*. Ecotoxicology and Environmental Safety (2012). , 80(1), 6-13.

[57] Brausch, J. M, & Salice, C. J. Effects of an environmentally realistic pesticide mixture on Daphnia magna exposed for two generations. Archives of Environmental Contamination Toxicology (2011). , 2011(61), 2-272.

[58] Botelho, R. G. Machado Neto L, Olinda RA, Dias CTS, Tornisielo VL. Water Quality Assessment in Piracicamirim Creek Upstream and Downstream a Sugar and Ethanol Industry Through Toxicity Tests With Cladocerans. Brazilian Archives of Biology and Technology (2012). , 55(4), 631-636.

[59] Manar, R, Vasseur, P, & Bessi, H. Chronic toxicity of chlordane to *Daphnia magna* and *Ceriodaphnia dubia*: a comparative study. *Environmental Toxicology* (2012 2). , 2012(27), 2-90.

[60] Cooney, J. D. Freshwater tests. In: Rand GM. (ed.) Fundamentals of Aquatic Toxicology: Effects, Environmental Fate And Risk Assessment. New York: CRC; (1995). , 71-102.

[61] Walker, C. H, Hopkin, S. P, Sibly, R. M, & Peakall, D. B. Principles of ecotoxicology. Taylor and Francis; (2001).

[62] Mccarty, L. S, & Borgert, C. J. Review of the toxicity of chemical mixtures containing at least one organochlorine. Regulatory Toxicology and Pharmacology (2006). , 45(2), 104-118.

[63] Bliss, C. I. The toxicity of poisons applied jointly. Annals of Applied Biology (1939). , 26(3), 585-615.

[64] Hermens, J, Canton, H, Steyger, N, & Wegman, R. Joint effects of a mixture of 14 chemicals on mortality and inhibition of reproduction of Daphnia magna. Aquatic Toxicology (1984). , 5(4), 315-322.

[65] Strmac, M, & Braunbeck, T. Cytological and Biochemical Effects of a Mixture of 20 Pollutants on Isolated Rainbow Trout (*Oncorhynchus mykiss*) Hepatocytes. Ecotoxicology and Environmental Safety (2002). , 53(2), 293-304.

[66] Delorenzo, M. E, & Serrano, L. Individual and Mixture Toxicity of Three Pesticides; Atrazine, Chlorpyrifos, and Chlorothalonil to the Marine Phytoplankton Species *Dunaliella tertiolecta*. Journal of Environmental Science and Health Part B- Pesticides, Food Contaminants, and Agricultural Wastes (2003). , 38(5), 529-538.

[67] Choung, C. B, Hyne, R. V, Stevens, M. M, & Hose, G. C. Toxicity of the Insecticide Terbufos, its Oxidation Metabolites, and the Herbicide Atrazine in Binary Mixtures to *Ceriodaphnia dubia*. Archives of Environmental Contamination and Toxicology (2011). , 60(3), 417-425.

[68] Organização Das Nações Unidas- ONUDeclaração sobre o meio ambiente. Estocolmo, (1972).

Chapter 19

Characterization, Modes of Action and Effects of Trifluralin: A Review

Thaís C. C. Fernandes, Marcos A. Pizano and Maria A. Marin-Morales

Additional information is available at the end of the chapter

http://dx.doi.org/10.5772/55169

1. Introduction

The use of chemicals to control human diseases, plagues and weeds in agriculture started in the late 19th century, but only after the Second World War did this practice follow rather scientific criteria [1]. According to targets against which they are designated, the chemicals used in agriculture are called insecticides, fungicides, herbicides, nematicides, among others [2].

All pesticides have the common priority of stopping a metabolic process essential to undesirable organisms, for which they are toxic. These chemicals act directly upon the organisms, eliminating or controlling them, such as interfering in their reproductive process [3].

Among agricultural pesticides, herbicides comprise the most employed group in agriculture. The main function of these chemicals is to control weeds, weed competition reduces productivity, without significantly impacting crop yield. Weeds tend to compete with crops by extracting essential elements from the soil, water, intercepting light and CO_2, interfering in the culture development and affecting agricultural production practices including harvest [4]. Herbicides are also used for eliminating plants from both road, railways, and riversides [3].

The mechanism of action of some herbicides on organisms is not completely understood [5]. Lack of detailed information about the action of herbicides on the biological environment may cause damage to human health [1], [6] and [7].

Herbicides may be classified according to different criteria related to their properties, characteristics, use, efficiency, permanence in the environment and mechanism of action. As for their chemical features, herbicides may be classified as carbamates, amides, diphenyl ethers, amino phosphates, and dinitroanilines, among others [8].

Classification of herbicides based on their mechanism of action has changed over time, both according to the discovery of new herbicides and the elucidation of site of action of the herbicide on plants. The internationally accepted classification is the one proposed by the Herbicide Resistance Action Committee (HRAC). In it, the herbicides are classified in alphabetical order in accordance with their sites of action and chemical classes (Table 1). Herbicides having unknown site of action are grouped under Z until identification. The (numeric) Weed Science Society of America (WSSA) classification system is also listed in Table 1 [5].

HRAC	SITES OF ACTION	CHEMICAL GROUP	WSSA
A	Inhibition of Acetyl-CoA Carboxylase (ACCase)	Aryloxyphenoxypropionates (FOPs)	1
		Ciclohexanodiones (DIMs)	1
		Phenylpyrazolones (DENs)	1
B	Inhibition of Acetolactate Synthase (ALS) (or acetohydrxy acid synthase AHAS)	Sulfonylureas	2
		Imidazolinones	2
		Triazolopyrimidines	2
		Pirimidinil(tio)benzoates	2
		Sulfonylaminocarbonyl-triazolinones	2
C1	Inhibition of Phtosynthesis in photosystem II	Triazines	5
		Triazinones	5
		Triazolinones	5
		Uracils	5
		Pyridazinone	5
		Phenyl Carbamates	5
C2	Inhibition of Phtosynthesis in photosystem II	Ureas	7
		Amides	7
C3	Inhibition of Phtosynthesis in photosystem II	Nitriles	6
		Benzotiadiazinones	6
		Phenyl-pyridazines	6
D	Inhibition of Phtosynthesis in photosystem I	Bipiridiliuns	22
E	Inhibition of Protoporphyrinogen Oxidase (PPO)	Diphenyl ethers	14
		Phenylpyrazoles	14
		N-phenylftalimidas	14
		Thiadiazoles	14
		Oxadiazoles	14
		Triazolinones	14
		Oxazolidinediones	14
		Pyirimidinediones	14
		Others	14
F1	Inhibition of carotenoid biosynthesis in naphytoenedesaturase (PDS)	Pyridazinones	12
		Pyridine Carboxamides	12
		Others	12
F2	Inhibition of carotenoid biosynthesis in 4-hydroxyphenyl-pyruvate-dioxygenase (4HPPD)	Triacetones	27
		Isoxazoles	27
		Pyrazoles	27

http://dx.doi.org/10.5772/55169

HRAC	SITES OF ACTION	CHEMICAL GROUP	WSSA
		Others	27
F3	Inhibition of carotenoid biosynthesis (unknown target)	Triazoles	11
		Isoxazolidinones	13
		Diphenyl ethers	11
G	Inhibition of EPSP synthase	Glycines	9
H	Inhibition of glutamine synthase	Phosphinic acid	10
I	Inhibition of DHP (dihydropteroate synthase)	Carbamates	18
K1	Inhibition of microtubule assembly	Dinitroanilines	3
		Phosphoramidates	53
		Pyridines	3
		Benzamides	3
		Benzoic acid	3
K2	Inhibition of mitosis	Carbamates	23
K3	Inhibition of cell cycle	Chloroacetamides	15
		Acetamides	15
		Tetrazolinones	15
		Others	15
L	Inhibition of cell wall (cellulose) synthesis	Nitriles	20
		Benzamides	21
		Triazolocarboxamides	27
		Quinolinocarboxylic acid	26/27
M	Decouplers (cell membrane disruptors)	Dinitrophenols	24
N	Inhibition of lipid synthesis (different from ACCase inhibitors)	Tiocarbamates	8
		Phosphoroditioates	8
		Benzofurans	16
		Chlorocarbonic acid	26
P	Auxin mimics	Phenoxicarboxylic acid	4
		Benzoic acid	4
		Pyridinecarboxylic acid	4
		Quinolinocarboxylic acid	4
		Others	4
Q	Auxin transport inhibitors	Ftalamates	19
		Semicarbazones	19
R	...	...	
S	...	...	
.	...	...	
Z	Unknown	Arylamino Propionic acid	25
		Pirazoliuns	26
		Organoarsenicals	17
		Others	

WSSA. Weed Science Society of America; **HRAC.** Herbicide Resistance Action Committee.

Table 1. Herbicide Classification in accordance with their mechanism of action.

2. Trifluralin identification and characterisitcs

Trifluralin belongs to the dinitroaniline group which has the aniline structure as a basis, containing NO_2 molecules at 2 and 6 or 3 and 5 positions of the benzene ring. This group has more than ten different herbicides, among which are trifluralin, dinitramine, oryzalin and pendimethalin [8].

Trifluralin has been used in agriculture since 1963 [9]. This herbicide is registered separately or in mixtures, and used in the following crops: *Glycine max*, citrus, *Coffea arábica* under formation, *Gossypium hirsutum, Arachis hypogaea, Phaseolus vulgaris, Allium sativum, Ricinus communis, Manihot esculenta, Helianthus annuus, Solanum melongena, Daucus carota, Abelmoschus esculentus, Brassica oleracea, Brassica oleracea capitata, Brassica oleracea botrytis, Capsicum annuum, Lycopersicon esculentum*, and ornamental plants [10].

Trifluralin is available either in emulsifiable concentrate or in crystalline solid both formulations of the yellow-orange color. It is not quite soluble in water (0.3 to 0.6 mg/L solubility at 25°C) [9], it is mildly volatile (1.1. 10^{-4} mmHg pressure vapor at 25°C), its density is 1.36 g/cm^3 at 22°C, it is considered alkaline and long-lasting in the environment (120-240 days) [8]. Trifluralin has a high affinity to soil [11], is relatively immobile and has a half-life of 3 to 18 weeks, depending on the soil and the geographical location [12].

Trifluralin chemical composition is α,α,α–trifluoro–2–6-dinitro–N–N– dipropyl–p–toluidine [13]. The chemical structure formula is shown in Figure 1.

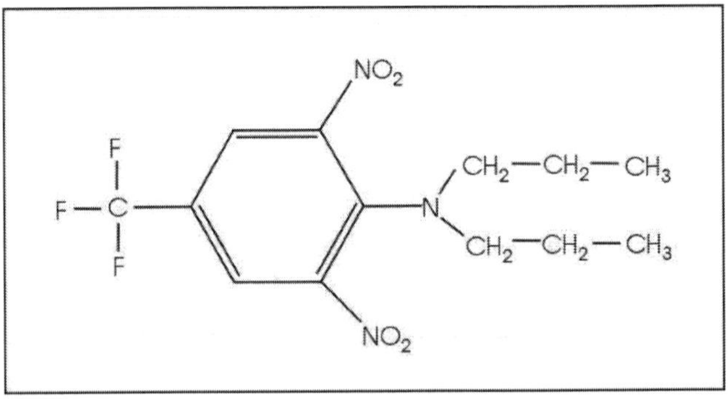

Figure 1. Trifluralin chemical structure formule.

Trifluralin commercial products contain nitrosodipropylamine, a carcinogenic contaminant (NDPA) [14]. This compound reacts with 0^6-guanine DNA and may cause mutation [15]. On account of concerns about this characteristic, the *Environmental Protection Agency* (EPA) demanded that industries make sure products containing trifluralin active principle had nitrosodipropylamine 0.5 ppm concentrations at the most [14].

http://dx.doi.org/10.5772/55169

USEPA (1999) [16] classifies trifluralin as group C: possibly carcinogenic to humans, based on evidences with animals, not with humans.

3. Trifluralin behavior in the environment

3.1. Behavior in soil

Trifluralin is strongly adsorbed by organic matter colloids and not much by clay ones. In organic matter rich soils, adsorption prevents absorption of the product by plant roots. Therefore, the use of this herbicide under such conditions is not advisable [10]. Leaching, as well as soil lateral movement is quite reduced compared to some pesticides [17]. Its main characteristic is soil persistence resulting from low mobility, which can cause damage to crops following its application [12].

Such herbicides as trifluralin, applied in pre-emergence, act better when soil humidity is between high and elevated. Therefore, the herbicide may at least be partially solubilized and distributed in the first layers of the soil surface, which will protect it from losses [8].

This herbicide degradation in soil occurs through chemical, microbial pathways and photolysis. Chemical degradation promotes dealkylation of the amino group, reduction from the nitro to the amino group, partial oxidation from the trifluoromethyl to the carboxyl group and, subsequently, degradation into smaller fragments (Figure 2).

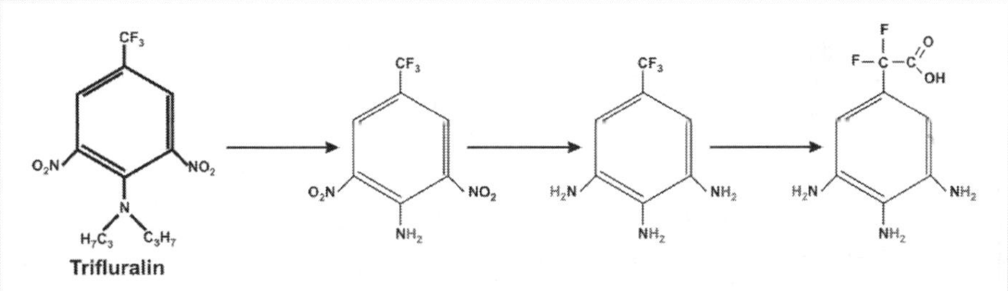

Figure 2. Possible sequence of events that occur during trifluralin chemical degradation.

Microbial degradation may occur under aerobic and anaerobic conditions (Figure 3). However, it is observed that degradation occurs mainly under anaerobic conditions, as the ones observed in poorly drained soils, when there is subsequent rainfall. Under anaerobic conditions, within the same time period, 98% of trifluralin degrades, whereas under aerobic conditions only 25% of the product decomposes. Among the fungi capable of decomposing trifluralin are *Sclerotiumrolfsii*, *Aspergillusniger*, *Fusarium*sp and *Tricoderma*sp [10]. According to Carter and Camper [18], trifluralin may also be degraded by *Pseudomonas* sp.

Trifluralin is also sensitive to degradation by ultraviolet rays, and its volatility is one of the main factors of product loss in the soil as well [19] and [20]. Trifluralin photodecomposition generally involves three processes: propylamine oxidative dealkylation, cyclization and nitro group reduction (Figure 4) [21].

The first product of trifluralin photolysis, according to Dimou et al. [21] and illustrated in Figure 3, seems to be a mono-dealkylate deriving from the main compound, originating compound 1. Dealkylation is attributed to the free radical oxidation. Another intermediate of photodegradation appears to be formed by cyclization reactions. The compounds 4 and 5 are apparently formed by reaction among trifluralinpropylamine α carbon and the NO_2 group of compound 1, ant they are identified as 2– ethyl -7nitro-1-propyl-5 (trifluoromethyl)-1*H*-benzimidazole and 2-ethyl-4 nitro-6- (trifluoromethyl)-1*H*-enzimidazole, respectively. The benzimidazoledealkylate (compound 4) is the most stable photoproduct, which can last in the environment longer, making its detection possible. This product may be formed by the reaction of compound 5 dealkylation.

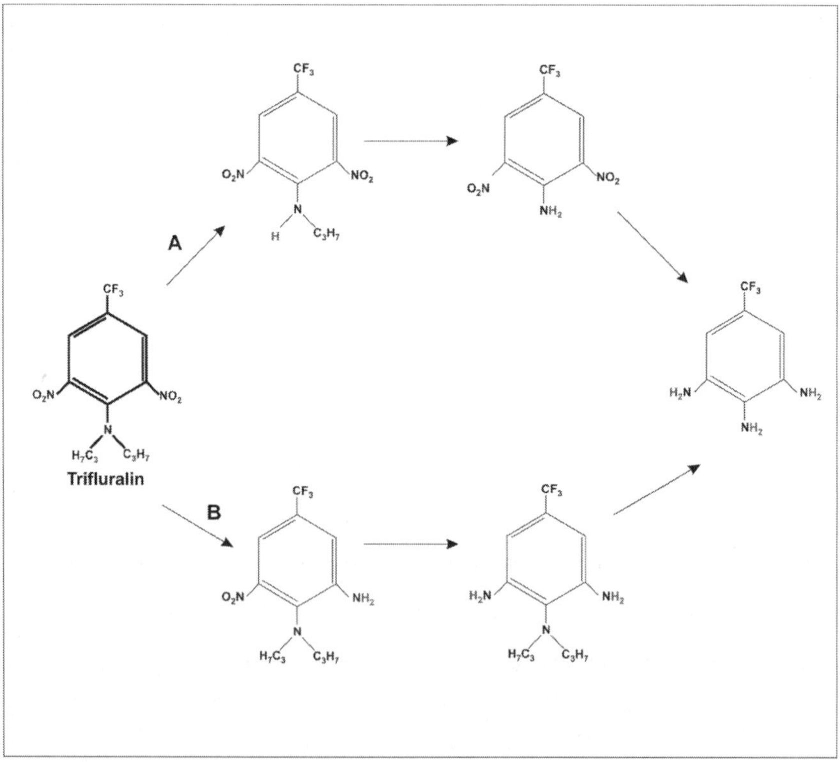

Figure 3. Trifluralin microbial degradation by aerobic (A) and anaerobic (B) pathways. Source: Audus [22].

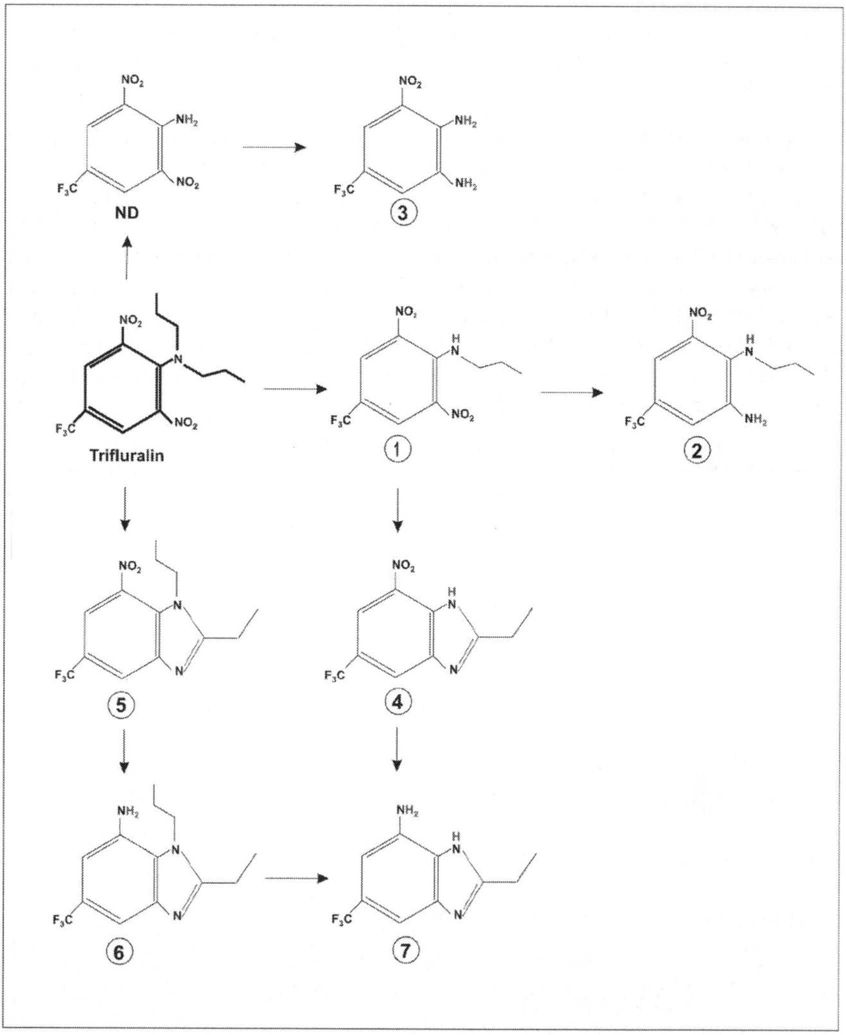

Figure 4. Trifluralin photodegradation. *ND=Substance not detected in the source. Modified scheme by Dimou et al. [21].

Compounds 4 and 5 can be reduced in water by not so clear mechanisms [23], straight from the aryl hydroxylamine formation [24] to form compound 7 and 6, respectively. According to the same author these products have also been formed during trifluralin chemical degradation. Compound 2 and 3 are formed from NO_2 to NH_2 group reduction of compound 1 and 2,6-dinitro-4- (trifluoromethyl) benzenamine (compound ND), respectively. These compounds have also been identified during trifluralin chemical degradation [24], showing that this pathway also happens in other processes, besides photodegradation [21].

Trifluralin average persistence in soil for the recommended doses under field conditions is of 1.8 ppm residue after 180 days following application [25]. However, according to the same author, this persistence may vary in accordance with the kind of soil and climatic conditions.

3.2. Herbicide behavior in water

Water contamination with trifluralin may occur by sediment leaching while equipment is being cleaned, or due to accidental spills. Nevertheless, only 0.5% of the quantity applied to the soil in field conditions is leached and may consequently contaminate water sources. This percentage means a rather low water contamination, representing smaller concentrations than 1.0 $\mu g\ L^{-1}$. As a consequence, trifluralin is not commonly detected in surface water [9] and [26].

While Zimmerman et al. [26], Dayama and Coupe [27], Thurman et al. [28] and were carrying out analyses in the Mississippi River, they detected extremely low levels of trifluralin (lower than 0.1 g/L). Once this herbicide is widely used, the authors ascertain that low concentrations of it detected in surface water may be attributed to its low mobility in soil and low solubility in water (lower than 1 mg/L). USEPA [29] and the European Community legislation [30] established limits of 2µg/ L and 0.1µg/ Ltrifluralin in drinking water, respectively. According to Dimou et al. [21], trifluralin degradation in water is influenced by the presence of nitrate ions, which accelerate photolysis reaction. Products derived from this reaction have either low or no toxicity, when compared to the whole product.

3.3. Herbicide behavior in the air

Grover et al. [31] ascertain that trifluralin is quickly dissipated in the atmosphere. Depending on the season of the year, about 25% of the product applied is volatilized, but only 2-3 $\mu g/m^3$ at the most of trifluralin is found in the air, soon after its application, to less than 100ng/m^3 a few hours later [32]. According to the United States Environment Protection Agency (1993) [33], an average 0.27 ng/m^3 concentration of herbicide, varying from 0 to 3.4 ng/m^3, was found in the Canadian atmosphere between 1988 and 1989.

Mongar and Miller [34] state that low concentrations of this herbicide found in the atmosphere are due to both trifluralin quick reaction with the hydroxyl radical (OH) and the photolysis reaction, which promotes the product degradation.Nonetheless, Waite et al. [32] verified that of the five most used herbicides on the Canadian prairies, trifluralin was the most frequently found in the air (79% of samples).

3.4. Herbicide behavior in plants

Trifluralin is a pre-emergence herbicide which must be incorporated into the soil and applied soon after sowing, when the plant seeds are beginning the germination process [36]. The herbicide absorption occurs mainly by the hypocotyl, then by the seedling radicles, at the beginning of germination [10].

Trifluralin's main mechanism of action is the inhibition of cell mitosis. This herbicide typically acts on the meristems and tissues of underground organs, such as roots, epicotyls, hypocotyls, plumules, rhizomes, bulbs and seeds [8].

The inhibition of radicle development by trifluralin action, both on main root growth and the emission of secondary roots, is quite evident in some dicotyledons. Thickening of the hypocotyls also commonly occurs [8], as well as swollen root tips [36]. According to Almeida [25], trifluralin induces several biochemical changes in higher plants, including alterations of carbohydrate, lipid, nitrogen concentrations and, especially, nucleic acid alterations. Therefore, the product affects cell division in meristematic tissues, thus inhibiting seed germination and the formation of new radicle and hypocotyl cells.

Bayer et al. [37] report that trifluralin promotes a decrease in the zone of meristematic tissues and the interruption of mitosis in the roots of wheat, cotton and onions. The onion cells treated with trifluralin showed to be small, dense and multinucleated, abnormal, weak and aberrant [38].Studies conducted by Fernandes [39] using *Allium cepa* showed that the toxicity of trifluralin residual concentrations might induce changes in that plant. The author observed that the herbicide promoted plant growth inhibition, higher turgidity, weakness and thickness of the roots, in relation to the control treatment.

Plants grown in soils treated with trifluralin exhibited residues on the roots only. No residue was found on the leaves, fruit and seeds [25]. These results indicate that trifluralin is not transported by sap into other plat tissues.

4. Trifluralin mechanisms of action

Plant growth and development depend on mitosis in their meristematic regions. Cell division is a process that requires different cell organelles, structures and the products of many genes to be working correctly. Dinitroanilines, the family to which trifluralin, phosphoride amides and N-phenyl carbonates belong, are microtubule-depolymerizing chemical compounds [5], [40], [41], [42] and[43]. According to Senseman [36], the herbicide-trifluralin complex inhibits microtubule polymerization, leading to physical misconfiguration and loss of function. As a consequence, the mitotic spindle does not form, causing misalignment and chromosome separation during mitosis. In addition to that, the so-called spindle apparatus is not formed.

Microtubules are subcellular structure filaments, basically made up of heterodimeric tubulin protein (Figure 5A) [44]. They have important cellular functions, which are directly related to mitosis and indirectly related to organism development. These structures are involved in several cellular processes such as chromosome migration, cellular structure maintenance, cellulose microfibril orientation and organization, cell wall formation, intracellular movement, as well as cellular differentiation [42] and [45]. Most sets of cell microtubules are labile and their functions depend on this lability.The mitotic spindle is one of the most extraordinary examples, whose formation is brought about after disorganization of cytoplasmic microtubule at the beginning of mitosis. For this reason, the mitotic spindle is targeted by various specific anti-mitotic drugs, which interfere in the exchange of tubulin subunits between the microtubules and the pool of free tubulins [46].

In-vitro analyses of *Chlamydomonas reinhardii* showed that trifluralin specifically binds tubulins, demonstrating that it is the first subcellular target of dinitroaniline action [47]. Trifluralinsub-micromolar concentrations totally blocked cytokinesis and inhibit nuclear division in *Toxoplasma gondii* by interfering in intracellular spindle and in other cytoskeletal components [48].

According to Anthony and Hussey [47], the herbicide-tubulin complex is related to the suppression of microtubule growth. With minus-end specific microtubule depolymerization, the tubules progressively start to get shorter, eventually leading to total loss of microtubule (Figure 5B). The author still states that cortical microtubules are among the most resistant to trifluralin action and microtubule spindles and fragments are among the most sensitive to the herbicide action.

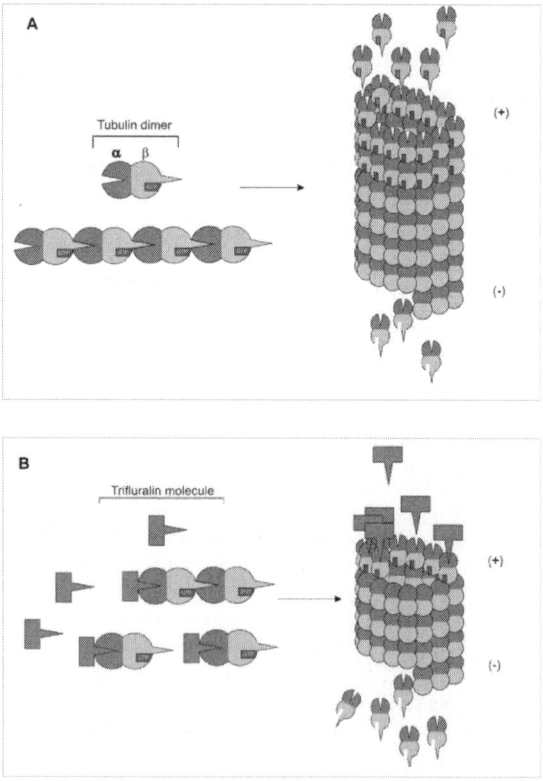

Figure 5. A. tubulin dimers forming the microtubule; **B.** herbicide-tubulin complex preventing microtubule polymerization.

http://dx.doi.org/10.5772/55169

Anthony et al. [49] ascertained that, as a rule, the tubulin sequence is the most preserved among the different organisms; and this preservation is related to the basic functions of microtubules. Mahresh and Larry [50], however, believe that, depending on the organism, dinitroaniline herbicides have different affinities to tubulins, since they do not interact with vertebrate tubulins, although they interact with plant and *Chlamydomonas* tubulins. This situation is reinforced with data from Anthony and Hussey [47], Baird et al. [51], Breviário and Nick [52] and Yemets and Blume [53], who ascertain that dinitroaniline herbicides are compounds with higher specificity for binding plant tubulins than to those of vertebrates.

Studies on plant resistance to dinitroanilines showed that some plant species own a natural mutation which bring about a change in base pairs, and consequently in their genetic code. One of these alterations of base causes a change in the amino acids of the tubulin protein. Threonine, a normal amino acid at position 239, is changed into isoleucine, stopping group NO_2 of the dinitroaniline herbicides from binding the tubulin molecule, thus preventing its mechanism of action (Figure 6) [47].

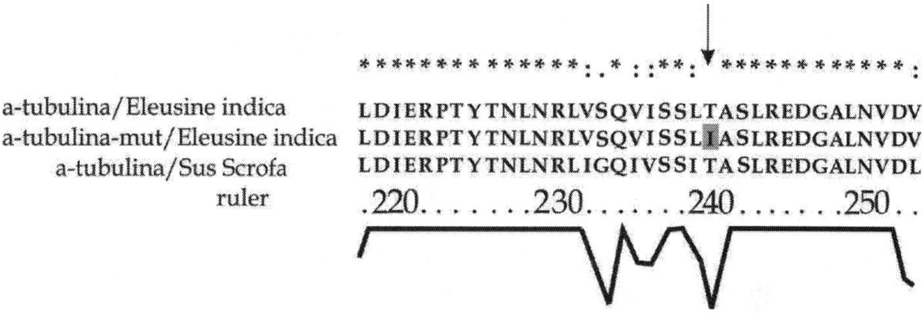

Figure 6. Alignment of amino acid sequence of α-tubulins, evidencing the position of substitution in the mutating tubulin from *Eleusine indica* (Thr 239 into Ile- represented in black and indicated with an arrow). Modified from Blume et al. [54].

From these pieces of information, it would be intuitive to hypothesize the idea that the smallest affinity of trifluralin to vertebrates should be owing to the fact they do not have the amino acid at position 239, seemingly the herbicide target site. Nevertheless, it can be seen in Figure 7 that the threonine amino acid at position 239 of the α-tubulin protein is present in plants, parasites and vertebrates, including man.

However, Hashim et al. [58] found mutations in the α-tubulin gene expression which changed the amino acid synthesis at a different position than that found by Anthony and Hussey [47]. According to Hashim et al. [58], *Alopecurus aequalis* plants that underwent mutations, which altered the amino acid synthesis at positions 202, 136 and 125 of the α-tubulin, also brought about resistance to trifluralin.

Sree et al. [59], Hansen et al. [60] and Vidakovié-Cifrek et al. [61], ascertain that trifluralin can inhibit microtubule polymerization by binding tubulin. However, it can also cause changes in the ion calcium concentration in cytoplasm and influence polymerization and depolymerization regulation of microtubules. According to Hertel et al. [62], changes in the quantity of free Ca^{2+} in cytoplasm, due to trifluralin action, can alter calcium-dependant biochemical and physiological processes, in addition to causing problems to microtubules, either in animals or in plants. Vidakovié-Cifrek et al. [61] report that trifluralin may increase the concentration of Ca^{2+} ions in cytoplasm, influencing onion root mitosis.

Due to trifluralin chemical structure, this herbicide tends to receive two electrons, which significantly increases its toxicity, since the group NH_2 hydrogen of trifluralin tends to bind the polar group of cellular membranes and cause disorganization to its structure, eventually bringing function disorders [63]. This disorganization in the membrane structure seems to interfere mainly in the permeability of plasma and mitochondrial membranes. Trifluralin changes the permeability of membranes because it promotes a collapse in their electric potential, making Ca^{+2} efflux of the mitochondrial inner membranes and Ca^{+2} go from the outer to the inner surface of the cell membrane via uniporters, thus increasing the concentrations of such ions in the inner cytoplasmic membrane.

Since low levels of calcium are needed for polymerization, Hepler [64] ascertains that mitotic spindles may undergo disorders due to the high levels of this ion. Low concentrations of free calcium in the cytoplasm (0.1-0.2 μM) are essential to prevent phosphorus precipitation, compete with Mg^{2+} for binding sites and act as a secondary messenger [65].

According to Alberts et al. [46], Ca^{+2} is important for regulating mitochondrial enzyme activity, and it is imported from the cytosol through an H+ electrostatic gradient. It is also believed that this process is important to remove Ca^{+2} from the cytosol when cytosolic Ca^{+2} levels get dangerously high.

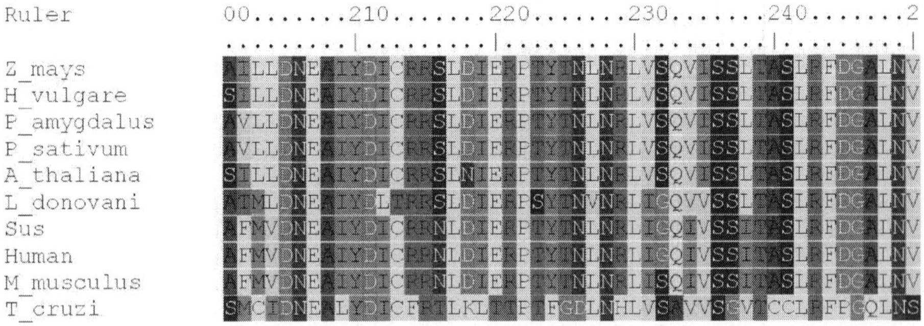

Figure 7. Comparisons among sequences of α-tubulin amino acids of species *Zea mays* (vegetable), *Hordeumvulgare* (vegetable), *Arabidopsis thaliana* (vegetable) *Prunus amygdalus* (vegetable), *Pisum sativum* (vegetable),Leishmania *donovani* (parasite), *Trypanosoma cruzi* (vegetable),*Mus musculus* (vertebrate), *Sus scrofa* (vertebrate) and *Homo sapiens*. The sequences were obtained from the data base at NCBI (National Center of Biotechnology Information) in accordance with the codes P14641, Y08490, P29511, P33629, U12589, U09612, M97956, P05213, P02550 and P04687, respectively [55]. The sequences were aligned by means of the ClustalW program [56], using default parameters. The alignment was then analyzed using the MPALign program [57].

http://dx.doi.org/10.5772/55169

Another important factor to be considered is the derivate generation through pesticide biodegradation [66] and [67]. One of the byproducts of trifluralin biodegradation is an aniline: 2,6dinitroaniline (Figure 8) [68].

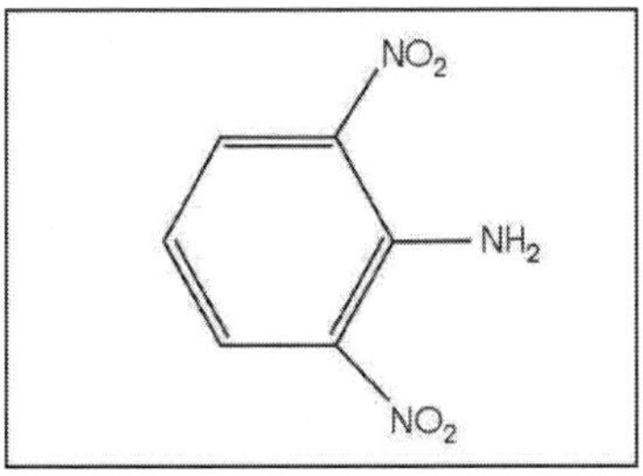

Figure 8. Chemical structure of 2,6 dinitroaniline.

Anilines are compounds that cause a variety of toxic effects depending on the structural changes they undergo. Several studies demonstrate that anilines and halogens can induce metahemoglobin formation and also be toxic to the kidneys and the liver, either treated *in vitro* or *in vivo* [69] and [70]. Aminophenols, the primary products of aniline metabolism, are compounds related to neurotoxicity induction [69].

5. Trifluralin toxic effect

Although many researchers and international governmental agencies have investigated and published trifluralin toxic effects on different fields, whether they are related to either acute or chronic toxicity, cytotoxicity, genotoxicity, mutagenicity and carcinogenicity, the results shown are confusing and often contradictory.

According to the W.H.O (World Health Organization) [70], trifluralin causes hemoglobin oxidation (by forming metahemoglobin), red blood cell destruction, besides being toxic to the kidneys and the liver, and stimulating depression in the central nervous system. It may cause vomiting, diarrhea, weakness, profuse sweating, loss of sight, memory and concentration, and dermatitis as well. This herbicide is considered to be neurotoxic and gastrointestinal irritant. It can lead to death because of ventricular fibrillation [71], although several authors [10], [72], [73], [74], [75] and [76], ascertain that trifluralin is a low toxicity substance.

Trifluralin lethal concentrations and doses for vertebrates and invertebrates are shown in Table 2.

Treatment	Species	Group	Popular Name	Toxicity
Cl50 (48h)	*Lepomis macrochirus*	Fish	Bluegill	19 μg L^{-1}
CL50 (48h)	*Mola mola*	Fish	Ocean sunfish	19μg L^{-1}
CL50 (48h)	*Cyprinus carpio*	Fish	Common carp	1.0mg L^{-1}
CL50 (96h)	*Oncorhynchus mykiss*	Fish	Rainbow trout	0,21mgL^{-1}
CL50 (96h)	*Oncorhynchus mykiss, Lepomis macrochirus, Mola mola*	(Young)fish	Rainbow trout, Bluegill, Ocean sunfish	10-90μg L^{-1}
CL50 (48h)	*Daphnia magma*	Micro-crustacean	-	0,56 mgL^{-1}
CL50 (96h)	*Procambarus clarkia*	Crustacean	Lobster	12mgL^{-1}
DL50 (oral)	*Apis mellifera*	Insect	Honey bee	0,011mg bee-1
DL50 (oral)	*Mus musculus*	Mammal	Laboratory mice	>500 mg kg^{-1}
DL50 (oral)	*Ratus norvegicus*	Mammal	Laboratory mice	> 10.000 mg kg^{-1}
DL50 (oral)	-	Mammal	Dog	> 200 mg kg^{-1}
DL50 (oral)	-	Mammal	Rabbit	> 200 mg kg^{-1}
DL50 (oral)	-	Bird	Hen	> 200 mg kg^{-1}

Data extracted from Gangolli [77].

Table 2. TrifluralinCL50 and DL50 for different organisms

Meister [78] conducted tests with animals and verified that trifluralin does not have any toxic effect on them when they are exposed to the product either through ingestion, inhalation or when in contact with the skin. Nauseas and severe gastrointestinal discomfort may occur after trifluralin ingestion. When placed in the rabbit eyes, it produced a mild irritation, which was reverted within seven days. In humans, it may induce skin allergies and, when inhaled, it may irritate the throat and the lungs.

Table 3 shows some information regarding trifluralin chronic, sub-acute and sub- chronic toxicity to different organisms.

Treatment	Species	Group	Popular Name	Toxicity	Symptoms
LOEC	*Amphiprion percula*	Fish	clownfish	5μg L^{-1}	-
NOEL	*Amphiprion percula*	Fish	clownfish	2μ L^{-1}	-
CE50 (10 days)	*Chlorococcum* sp	Protozoa	-	2,5 mg L^{-1}	-
Sub-acute(dermis -14 days)	Oryctolagus caniculus	Mammal	Rabbit	2mL Kg^{-1}	diarrhea and mild erythema
Sub-chronic(ingestion - 3 months)	*Ratus norvegicus*	Mammal	Mouse	25, 50 e 100 mg kg^{-1} dia^{-1}	no effects produced on either survival or appearance *

*Liver weight of the animals submitted to the 50 and 100mg Kg^{-1} diet somehow showed to be higher, when compared to the control animals. Data extracted from Gangolli [77].

Table 3. Data on trifluralin sub-acute, chronic and sub-chronic toxicity.

According to the Occupational Health Service [79], prolonged skin contact with trifluralin may cause allergic dermatitis. The WSSA [80] states that administering trifluralin to dogs while washing them for two years does not cause toxic effects. However, in trifluralin chronic assays conducted with 60 animals (F344 mice), which received 0.813, 3250 and 6500 ppm dietary does for two years, damage to their liver and kidneys were observed [81].

Worthing [71] states that trifluralin is highly toxic and neurotoxic. The author ascertains that the herbicide is capable of accumulating in the adipose tissue and inhibiting the immunologic function of the thymus. Trifluralin is regarded as possibly teratogenic and fetal toxicity.It has the property of altering the endocrine and reproductive system, and it reduces the quantity of semen, besides increasing the number of abnormal sperm.

In studies conducted by Ovidi et al. [82], they tested trifluralin concentration of 1.53 mg/ml and observed that the herbicide exerts a specific effect on the reproductive system in plants, by direct action on the formation of the pollinic tubes, since it causes complete microtubule depolymerization. The authors even suggest that pollinic microtubule cytoskeleton may be used as bioindicators for studies on toxicity induced by aneugenic agents such as trifluralin.

As a general rule, the effects of pesticides may be diversified, such as the direct reaction with nuclear DNA; incorporation of DNA during cellular replication; interference in mitosis or meiosis, resulting from incorrect cell division [83].

Genotoxic effects may lead to DNA breaks, causing loss of genetic material and mutations which lead to cell death or result in carcinogenesis. Genotoxicity is assessed by different tests, carry out with several organisms and provide safe, precise information regarding their potential to damage the DNA.There are a number of reports evaluating trifluralin for genotoxicity, immunotoxicity, and reproductive toxicity, although the results are not entirely consistent, trifluralin does not appear to be strongly genotoxic [84].

Chromosome aberration tests have shown evidences of trifluralin mutagenicity for different plant species [85], [86], [87], [88], [89] and [90]. Könen and Çavas [91], Peña [92] and Canevari [93] ascertained that the herbicide is capable of inducing significant microtubule rates in *Oreochromis niloticus*. Kaya et al. [94] also ascertained that the herbicide may be considered genotoxic to *Drosophila melanogaster*, since it exhibited positive outcomes for the Somatic Mutation and Recombination Test (SMART). Tests conducted in the bone marrow of mice exposed to trifluralin showed that it is potentially genotoxic [95] and it is also capable of influencing serum concentration of reproductive and metabolic hormones, especially thyroxin [96]. Nonetheless, tests performed on bacteria [14], on *Drosophyla melanogaster* conducted by Bryant and Murnik [97] and Foureman [98], on cells taken from the bone marrow of mice conducted by Nehéz et al. [99], Pilinkaya [100], Gebel et al., [95], and on cell culture conducted by IARC [101] and Ribas et al. [35 and 102] demonstrated contradictory results. According to Chan and Fong [103], Bhattacharya et al. [104] and Esteves et al. [105], due to its characteristics, mechanisms of action and, especially its reduced effects on human cells, trifluralin can be regarded as a promising substance for fighting Leishmaniasis. There is also research that confirms the use of trifluralin as a powerful antiparasitic to treat *Trypanosoma* [106] and [107], *Toxoplasma* [48] and *Plasmodium* [108].

Studies carried out by Peña [92] and Canevari [93] indicate that low trifluralin concentrations may induce mutagenic effects. These authors observed significant presence of micronuclei in erythrocytes of fish submitted to acute treatments with this herbicide. When the micronuclei diameters were measured by Canevari [93], data indicated that they could be derived from losses of whole chromosomes, thus proving the aneugenic effect of the herbicide due to the pesticide interference in the mitotic spindle.

*Allium cepa*meristematic cells treated with trifluralin also presented problems during mitosis, such as polyploidies, C-metaphases, multipolar anaphases, anaphase-telophase chromatin bridges, chromosome delay and loss of genetic material [89]. (Figure 9).

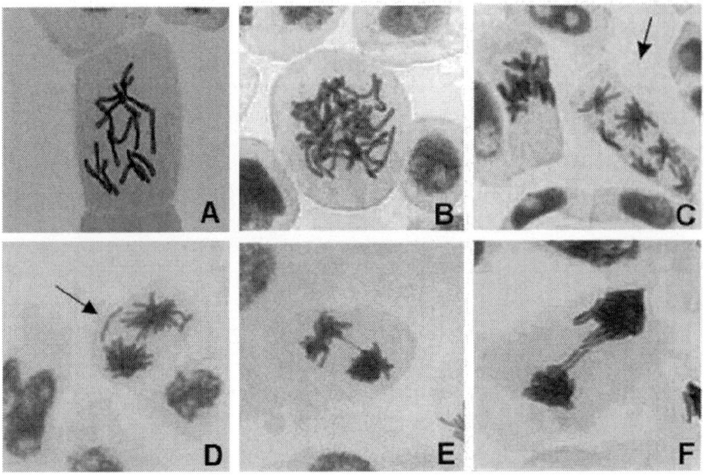

Figure 9. Meristematic cells of *Allium cepa* treated with trifluralin. **A.** C-metaphase; **B.**polyploid cell; **C.** multipolar cell; **D**. loss of genetic material; **E.**chromosome bridge; **F.**telophase with chromosome delay.

According to Fernandes et al. [88], in the bioassays with root meristems of *Allium cepa* treated with trifluralin, a large amount of interphase cells with more than one nucleus and cells with micronuclei and a mini cell were observed (Figure 10).

Lignowski and Scott [85] observed C-metaphases, micronuclei, amoeboid nuclei and polyploidies in root meristems of wheat and onion submitted to trifluralin action. Due to the occurrence of irregular metaphases, they concluded that the mitotic spindle might have been broken owing to the herbicide action on it.

Bioassays performed with trifluralin, using *Pisum sativum* as test material revealed a positive action of the herbicide with the increase in chromosome alterations, C-mitosis and anti-mitosis effects [87].

Fernandes et al. [89] ascertained that, among the root meristems of *Allium cepa* under division, trifluralin promotes a significant increase in the irregular metaphase rate. These data corroborate the statement of Lignowski and Scott [85], Lee et al. [109], Dow et al. [110], Werbovetz et al. [111] and Ovidi et al. [82], who characterized trifluralin as a powerful microtubule inhibitor, which is therefore capable of accumulating a large amount of meristematic cells in metaphase.

Genotoxicity tests using the comet assay in human lymphocyte cultures showed that trifluralin produced a significant increase in the length of the comet's tail. This increase is due to DNA breaks, since there is an induction of nucleotide excision repair, resulting from damage caused by the herbicide action [103]. As for the frequency of comet-bearing cells, the author observed that, after 48 hours of exposure to the herbicide, few tailed nucleoides were found. These results proved to be statistically significant, though.

According to Ribas et al. [35], trifluralin has a genotoxic effect on human cell cultures because it causes a decrease in cell proliferation. The same author ascertains that this herbicide has not revealed carcinogenic effects, since it caused little induction exchange between sister chromatids.The micronucleus test conducted by Ribas et al. [35], used for detecting aneugenic activity, has also produced a negative response, which contradicts studies carried out by several other authors [88], [89], [91], [92], [97], [112], among others) who ascertain that trifluralin brings about chromosome aberrations and nuclear alterations resulting from problems in the mitotic spindle

According to Kang et al. [113], trifluralin is not associated with bladder, kidney, liver, leukemia, colorectal or hematopoietic-lymphatic cancers. The authors only suggest a possible connection between trifluralin exposure and the risk of colon cancer in human beings, but the inconsistency per exposure level and a small number of colon cancers indicate that this could be an incidental finding.

Data from the National Cancer Institute (NCI) [114] report that mice subjected to trifluralin chronic exposure, at low concentrations, had an increase in hepatocellular carcinoma and higher incidence of alveolar bronchial adenomas. An increase in bladder cancer was also verified in mice exposed to low trifluralin concentrations. It was observed that, when male mice were submitted to high doses of trifluralin, they presented higher incidence of follicular cell and thyroid gland tumors [115]. Trifluralin has been reported to cause a significant increase

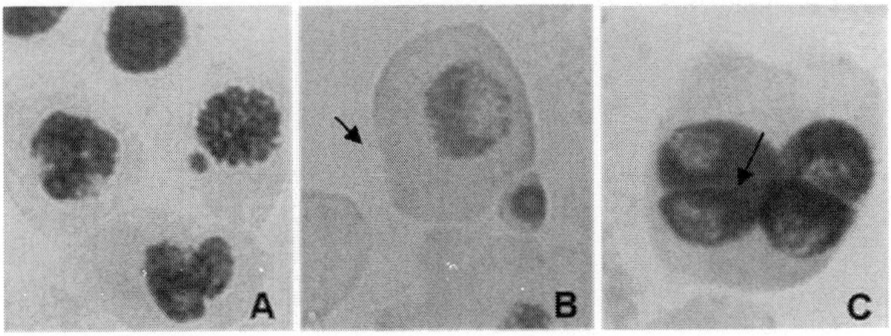

Figure 10. Meristematic cells of *Allium* cepa treated with trifluralin. **A.** cell with micronucleus; **B.** cell with micronucleus and an adjacent mini cell; **C.**polynucleated cell.

in thyroid follicular cell tumors in male Fischer 344 rats only at the highest dietary dose of 6500ppm in a 2-year chronic study [115].

6. Final considerations

The increase in agricultural productivity has occurred thanks to several factors, among which are improvements in genetics, agricultural machinery and the use of substances that allow control of weeds in agriculture.

The use of pesticides has generated discussions and controversy among the scientific community and its users, registering advantageous and disadvantageous recommendations in different ways. Among contrary recommendations to the use of pesticides, we can point out lack of detailed studies on the action of such chemicals on the exposed organisms, making it impossible to associate their action with the emergence of eventual problems. In the soil, trifluralin is moderately persistent, which might jeopardize organisms that are eventually exposed to it. Trifluralin is a substance that has a microtubule-depolymerizing activity, which prevents cell division, a fact that might compromise organism development.

Existing reports characterize trifluralin as a highly acute toxic substance to fish, but there are not enough descriptions of its chronic toxicity and cytotoxic effect. Studies mainly related to its genotoxic, mutagenic and carcinogenic potential are mostly inconclusive or even contradictory. There is little information about the toxicity of products derived from trifluralin degradation and its effects on the organisms.

Author details

Thaís C. C. Fernandes, Marcos A. Pizano and Maria A. Marin-Morales*

*Address all correspondence to: mamm@rc.unesp.br

Universidade Estadual Paulista, IB-Campus de Rio Claro, Rio Claro/SP, Brasil

References

[1] Londres, F. Agrotóxicos no Brasil: um guiaparaaçãoemdefesa da vida. – Rio de Janeiro: AS-PTA – Assessoria e Serviços a ProjetosemAgriculturaAlternativa, 2011. 190 p.

[2] Kotaka, E. T., Zambrone, F.A.D. Contribuições para a construção de diretrizes de avaliação do risco toxicológico de agrotóxicos. Campinas, SP: ILSI Brasil, 2001.

[3] Baird, C. QuímicaAmbiental. Porto Alegre: Bookman, 2002.

[4] Lorenzi, H. Manual de identificação e controle de plantas daninhas. Nova Odessa: EditoraPlantarum. 1990.

[5] Oliveira Jr., R.S. Mecanismo de açãoherbicidas. In: Biologia e Manejo de PlantasDaninhas. Oliveira Jr., R.S., Constantin, J., Inoue, M.H (Eds) Omnipax, 2011.

[6] Munger, R., Isacson, P., Hu, S., Burns, T., Hanson, J., Lynch, C.F., Cherryholmes, K., Vandorpe, P., Hausler, Jr. W. J. Intrauterine growth redardation in Iowa communities with herbicides-contaminated drinking watersupplies. Environ. Health Perspect., Research Triangle Park v. 105, p. 308-314, 1997.

[7] Gorell, J.M., Jhonson, C.C., Rybicki, B.A., Peterson, E.L., Ricchardson, R.J. The risk of Parkinson's disease with exposure to pesticides, farmin, well water, and rural living. Neurology, Madras v.50, p.1346-1350, 1998.

[8] Deuber, R. Botânica das plantasdaninhas. In: DEUBER, R. Ciência das plantasdaninhas. Jaboticabal: FUNEP, 1992.

[9] Grover, R., Wolt, J.D., Cessna, A. J., Schiefer, H.B. Environmental fate of trifluralin. Rev. Environ. Contam. Toxicol. v. 153, p. 1-64, 1997

[10] Rodrigues, B. N., Almeida, F. S.Guia de herbicidas, 5ª ed., Grafmarke: Londrina, 2005.

[11] Sanders Pf, Seiber Jn. A chamber for measuring volatilization of pesticides for model soil and water disposal system. Chemosphere, Oxford v.12, p. 999-1012, 1983.

[12] Calderon, M.J., Hermosín, M.C., Cornejo, J. Y Moreno, F. Movilidad de trifluralina en laboreo tradicional y de conservación. Estudios de la Zona No Saturada del Suelo. Eds. R. Muñoz-Carpena, A. Ritter, C. Tascón: 1999. Tenerife, p.83-88.

[13] Bellinaso M De. L., Henrique L.A., Gaylarde C.C., Greer C.W. Genes similar tonaphthalenedioxygenase genes in trifluralin-degrading bacteria. Pest Manag. Sci., Sussexv. 5, p. 474-478, 2004.

[14] U.S. Environmental Protection Agency. 1987. Trifluralin health advisory. Office of Drinking Water, Washington, DC.

[15] Cooper, M.T., Porter, T.D. Mutagenicity of nitrosamines in methyltransferase-deficient strains of *Salmonella typhimurium*coexpressing human cytochrome P450 2E1 and reductase. Mut. Res., AmsterdamV. 6, P.45-52, 2000.

[16] U.S. Environmental Protection Agency. 1999. Chemicals Evaluated for Carcinogenic Potential Science Information Management Branch Health Effects Division Office of Pesticide Programs.

[17] Laabs, V., Amelung, W., Pinto, A., Altstaedt, A., Zech, W. Leaching and degradation of corn and soybean pesticides in an Oxisol of the Brazilian Cerrados Chemosphere. v. 41, p. 1441-1449, 2000.

[18] Carter, N. D., Camper, N. D. Soil enrichment studies with trifluralin. Weed. Sci, Champaign. v. 23, p. 71-74, 1975.

[19] Selim H.M., Zhu H. Retention and mobility of deltamethrin in soils. Transport. Soil Sci. v. 167, p. 580-589, 2002.

[20] Cooke, C. M., Shaw G., Collins, C. Determination of solid-liquid partition coefficients (K_d) for the herbicides isoproturon and trifluralin in five UK agricultural soils.Environ. Pollut., Barking v. 132, p. 541-552, 2004.

[21] Dimou, A. D., Sakkas, V. A., Albanis, T. A. Trifluralin photolysis in natural waters and under the presence of isolated organic matter and nitrate ions: kinetics and photoproduct analysis. J. of Photochem. Photobiol., A, Chem, Lausanne v. 163, p. 473-480, 2004.

[22] Audus , L.J. Herbicides. London:Academic Press,1980, p. 608.

[23] Crosby, D. G. Fate of organic pesticides in the aquatic environment. Adv. Chem. Ser., Washingtonv. 111, p. 173, 1972.

[24] Klupinski, T. P., Chin, Y. P. Abiotic Degradation of Trifluralin by Fe(II): Kinetics and Transformation Pathways. Environ. Sci. Technol., Easton v. 37, p. 1311-1318, 2003.

[25] Almeida, F.S. Guia de herbicidas; recomendações para o uso adequado em plantio direto e convencional. Londrina, PR. 1985.

[26] Zimmerman, L.R., Thurman, E.M., Bastian, K.C. Detection of persistent organic pollutants in the Mississippi Delta using semipermeable membrane devices. Sci. Total Environ., Amsterdam v. 248, p. 1, 2000.

[27] Dayama, A., Coupe, R.H. Jr. Pesticides in the Yazoo River and BoguePhalia, February through September 1996. In: Daniel JB, editor. Proceedings of the 27th Mississippi Water Resources Conference, Jackson, MS, March 25–26, 1997. Mississippi Water Resources Institute, Starkville MS. p.127–132 , 1997.

[28] Thurman, E.M., Zimmerman, L.R., Scribner EA, Coupe RH Jr. Occurrence of Cotton Pesticides in Surface Water of the Mississippi Embayment. US Geological Survey Fact Sheet. v.4, p. 22-98, 1998.

[29] U.S. Environmental Protection Agency. 2001 (nov). Environmental Law Institute Research Report na Opportunities for Advancing Environmental Justice: Na Analusis of US-EPA, Washington, DC.

[30] E. C. (European Communities). Directive Relating to the Quality of Water Intended for Human Comsumption 1982, 80/778/EEC, oficce for official. Publications of the European Communities, L-2985 Luxemborg.

[31] Grover, R., Cessna, A.J., Waite, D.T. Volatilization losses na transport in air of triazine herbicides. In: Le Baron, H.M., Gianessi, L.P., Mcfarland, J., Burnside, O.C., editors. The triazine herbicides. Amsterdam, the Netherlands: Elservier Science B.V., 2000.

[32] Waite, A.D.T., Bailey, A. P. Sproull, B. J.F., Quiring, A. D.V., Chau, B.D..F J., Bailey, C. J. Cessna, C. Atmospheric concentrations and dry and wet deposits of some herbicides currently used on the Canadian Prairies. Chemosphere, Oxford. v. 58, p.693–703, 2005.

[33] U.S. Environmental Protection Agency. 1993. Health advisories for drinking waters contaminants, Lewis Publishers, Boca laton, FL, USA .

[34] Mongar, K., Miller, G.C. Vapor phase photolysis of trifluralin in an outdoor chamber: Chemosphere, Oxford v. 17, p. 2183–2188, 1988.

[35] Ribas, G. J. S., Carbonell, E. N. X., Creus, A., Marcos, R. Genotoxic evaluation of the herbicide trifluralin on human lymphocytes exposed *in vitro*. Mutat. Res., Amsterdam v. 371, p. 15-21, 1996.

[36] Senseman, S.A. Herbicide Handbook, Ninth Edition. Weed Sci. Soc. Am. Champaign, IL: 458 pp. 2007.

[37] Bayer D.E., Foy C.L., Mallory T.E., Cutter E.G. Morphological & histological effects of trifluralin on root development. Am. J. Bot. v.54, p. 945-952, 1967.

[38] Hacskaylo J., Amato V.A. Effect of trifluralin on roots of corn & cotton. Weed Sci. v. 16, p. 513-515, 1968.

[39] Fernandes, T.C.C. Uso do teste de Allium cepanadetecção da toxicidade e genotoxicidade do herbicidatrifluralina. Monografia (Bacharel), UniversidadeEstadualPaulista, Rio Claro/SP, 2002.

[40] Morejohn, L.C., Bureau, T.E., Molé-Bajer, J., Bajer, A., Fosket, D. E. Oryzalin, a dinitroaniline herbicide, binds to plant tubulin and inhibits microtubule polymerization *in vitro*. Plant.,Berlim v. 172, p.41-147, 1987.

[41] Verhoeven, H.A., Ramulu, K.S., Dijkhuis, P.A. comparison of the effects of various spindle toxins on metaphase arrest and formation of micronuclei in cell-suspension cultures of *Nicotianaplumbaginifolia*. Plant., Berlin v. 182, p. 408-411, 1990.

[42] Morejohn, L.C. The molecular pharmacology of plant tubuline and microtubules: The cytoskeletal basis of plant growth and form In: ed. Lloyd C.W.: 1991. Academic Press, London, p.29-43.

[43] Ramulu, K.S., Verhoeven, H.A., Dijkhuis, P. Mitotic blocking, micronucleation, and chromosome doubling by oryzalin, amiprophos-methyl and colchicine in potato. Protoplasma, New York v. 160, p. 65-71, 1995.

[44] Quader, H. Cytoskeleton: Microtubules. Prog. Bot., Berlin v. 59, p. 375-395, 1997.

[45] Jordan, M.A., Wilson, L. The use and action of drugs in analyzing mitosis. Methods in Cell Biol., New York v. 61, p. 267-295, 1999.

[46] Alberts, B., Johnson, A., Lewis, J., Raff, M., Roberts, K., Walter, P. Molecular Biology of the Cell. Garland Science, New York , 4^{th}ed, 2002.

[47] Anthony, R. G., Hussey, P. J. Dinitroaniline herbicide resistance and the microtubule cytokeleton. Trends Plant Sci., Oxford v. 4, n.3, p. 112-116, 1999.

[48] Stokkermans, T.J.W., Artzman, J.D.S., Keenen, K., Morrissette, N.S., Tilney, L.G., Roos, D.S. inhibition of *Toxoplasma gondii* replication by dinitroaniline herbicides. Exp. Parasitol., San Diego v. 84, p. 355-370, 1996.

[49] Anthony, R. G., Waldin, T. R., Ray J. A., Bright, S. W. J., Hussey, P. J. Herbicide resistance caused by spontaneus mutation of the cytoskeletal protein tubulin. Nature, New York v. 393, p. 260-263, 1998.

[50] Mahresh, K. U., Larry D. N. Mode of dinitroaniline herbicide action. Plant. Physiol., Minneapolis v. 66, p. 1048-1052, 1980.

[51] Baird, W. V., Blume, YaB., WICK, S. Microtubular and cytoskeletal mutants. In: ed. Nick, P. Springer: 2000, p. 159-91.

[52] Breviário, D., Nick, P. Plant tubulins: a melting pot for basic questions and promising applicarions. Transgenic Res., London v. 9, p. 383-93, 2000.

[53] Yemets, A.I., Blume, Y.A.B. Resistence to herbicides with antimicrotubular activity: from natural mutants to transgenic plants. Russ J. Plant. Physiol., New York v. 46, p. 899-907, 1999.

[54] Blume, Ya.B., Nyporko, A. Yu., Yemets, A. I., Baird, W.V. Structural modeling of the interaction of plant α-tubulin with dinitroaniline and phosphoroamidate herbicides. Cell Biol. Int., London v. 27, p. 171-174, 2003.

[55] Yamamoto, E., Zeng, L., Baird, W.V. a-Tubulin missense mutations correlate with antimicrotubule drug resistence in *Eleusineindica*. Plant Cell, Berlin v. 10, p. 297-308, 1998.

[56] Higgins, D., Thompson, J., Higgins D. G., Gibs, T. J. ClustalW: improving the sensitivy of progressive multiple sequeraligment through sequence weighting, position-specific gap penaltie weight matrix choice. Nucleic Acids Res., Oxford, v. 22, p. 4673-4680. 1994.

[57] Arnold, F. C., Debonzi, D. H.MPAlign: Graphical and multiplatform tool for molecular alignments. Proceedings of II International Conference on Bioinformatics and Computational Biology, Angra dos Reis,2004.

[58] Hashim, S., Jan, A., Sunohara, Y., Hachinohe, M., Ohdan, H., Matsumoto, H. Mutation of alpha-tubulin genes in trifluralin-resistant water foxtail (Alopecurusaequalis). Pest. Manag. Sci. v. 68, p. 422–429, 2012.

[59] Sree, K.R., Verhoeven, H.A., Dijkhuis, P. Mitotic dynamics of micronuclei induced by amiprophos-metyl and prospects for chromosome-mediated gene transfer in plants. Tag, Berlin v. 75, p. 575-584, 1988.

[60] Hansen, A. L., Gertz, A., Joersbo, B.,Andesrsen, S.B. Antimicrotubule herbicide for in vitro chromosome doubling in *Beta vulgaris* L. ovule culture. Euphytica, Wageningen v. 101, p. 231-237, 1998.

[61] Vidakovié-Cifrek, M., Pavlica, I., Regula, D.P. Cytogenetic damage in shollot (*Allium cepa*) root meristems induced by oil industry "high-density brines". Arch. Environ. Contam. Toxicol. v. 43, p. 284-291, 2002.

[62] Hertel C., Quader H, Robinson D. G., Roos I., Carafoli E., Marme D. Herbicides and fungicidesstimulate Ca2+ effluxfromratlivermitochondria. FebsLett. Amsterdam v. 127, n.1 p. 37-39, 1981.

[63] Argese, E., Bettiol, C., Fasolo, M., Zambon, A., Agnoli, F. Substituted aniline interation with submitochondrial particles and quantitative struture-activity relationships. Biochim. Biophy. Acta, Amsterdam v. 1558, p. 151-160, 2002.

[64] Hepler, P.K. Calcium and mitosis. IntVerCytol. v. 2, p. 1273-1282, 1992.

[65] Marschner, H. Mineral nutrition of higher plants. London: Academic Press, Harcourt Brace, 1988.

[66] Fishbein, L. The Handbook of Environmental Chemistry. Part C- Anthrop. Comp. Berlin, v. 3, p 1-40, 1984.

[67] Hong, S.K., Anestis, D.K., Henderson, T.T., Rankin, G.O.Haloaniline induced in vitro nephrotoxicity: Effects of 4-haloanilines and 3,5-dihaloanilines. Toxicol. Lett. v. 114, 125-133. 2000.

[68] Wang S., Arnold W.A. Abiotic reduction of dinitroaniline herbicides. Water Res. Supl. 37, v. 17, p. 4191-201. 2003.

[69] Valentovick, M.A., Ball, J.G., Hong, S.K., Rogers, B.A., Meadows, M.K., Harmon, R.C., Rankin, G.O. In vitro toxicity of 2-and 4-chloroaniline: comparisons ith 4-amino-3-chlorophenol, 2-amino-5-chlorophenol e aminophenols. Toxicol. In Vitro, Oxford v. 10, p. 713-720, 1996.

[70] W.H.O. World Health Organization: Public health impact of pesticides in agriculture. Geneva, 1992.

[71] Worthing C.R, ed. *The pesticide manual*, 9th ed. Farnham, British Crop Protection Council, 1991.

[72] Worth, H.M. The toxicological evaluation of benefin and trifluralin. I: Pesticides Simposia: Inter-American Conference on toxicology and Occupational Medicine, Deichmann, W.B., Penalver, R.A., Radomski, J.L., Eds. Halos and Associates, Miami, 1970.

[73] Bem-Dyke, R., Sanderson, D.M., Noakes, D.N. Acute toxicity data for pesticides-1970. Pest Control., London v. 9, p. 119-127, 1970.

[74] Landonin, V. F., Hassan, A., Winteringham, F. P. W. Dinitroaniline pesticides. Chemosphere, Oxford v. 9, p. 67-69, 1980.

[75] Gaines, T. B., And Linder, R. E. Acute toxicity of pesticides in adult and weanling rats. Fundam. Appl. Toxicol., Akron v. 7, p. 299-308, 1986.

[76] Royal Society Of Chemists. Trifluralin. In The Agrochemical Handbook, 2nd ed., Update 5, p. A412. Graham, Cambridge. 1990.

[77] Gangolli, S. The dictionary of substances and their effects. Cambridge: Royal Society of Chemistry, v. 7. 1999, 998p.

[78] Meister, R.T. Farm Chemical Handbook '92. Willoughby: Meister Publishing Company, 1992.

[79] Occupational Health Services.MSDs for Trifluralin. OHS Inc., Secaucus, NJ. 1991.

[80] Wssa Herbicide Handbook Committee. Herbicide Handbook of the Weed Science Society of America, 6th Ed. WSSA, Champaign,1989.

[81] U.S. Environmental Protection Agency. 1989 (jan). Health Advisory Summary: Trifluralin. USEPA, Washington, DC.

[82] Ovidi, E., Gambellini, G., Taddei, A.R., Cai, G., Casino,C.D., Ceci, M., Rondíni, S., Tiezzi, A. Herbicides and themicrotubularapparatus of *Nicotianatabacum*pollentube: immunofluorescence and immunogoldlabellingstudies. Toxicol. in Vitro, Oxford v. 15, p.143-151, 2001.

[83] Timbrell, J.A. Introduction to Toxicology. London: Taylor & Francis, 1999.

[84] Garriott, M.L., Adams, E.R., Probst, G.S., Emmerson, J.L., Oberly, T.J., Kindig, D.E.F., Neal, S.B., Bewsey, B.J., Rexroat, M.A. Genotoxicity studies on the preemergence herbicide Trifluralin. Mutat. Res. v. 260, p. 187–193, 1991.

[85] Lignowski, E.M., Scott, E.G. Effect of trifluralin on mitosis. Weed Sci. v. 20, p. 267-270, 1972.

[86] Wu, T.P. Some cytological effects of treflan and mitomycin C on root tips of Viciafaba. Taiwania, Taipiei v. 17, p. 248-254, 1972.

[87] Grigorento, N.K., Fasilchenko, V.F., Merezhinski, Y.G., Morgun, V.V., Logvinenko, V.F., Sharmankin, S.V. Cytogenetic activity of a herbicide treflan, and its metabolites as applied to maize.Tsiol. Genet.v. 20, p. 294-298, 1986.

[88] Grant, W.F., Owens, E.T. Chromosome aberration assays in Pisum for the study of environmental mutagens. Mutat.Res., Amsterdam. v. 188, p. 93-118, 2001.

[89] Fernandes, T.C.C., Mazzeo, D.E.C., Marin Morales, M.A. Mechanism of micronuclei formation in polyploidizated cells of *Allium cepa* exposed to trifluralin herbicide. Pesticide Biochemistry and Physiology. v. 88, p. 252-259, 2007.

[90] Fernandes, T.C.C., Mazzeo, D.E.C., Marin Morales, M.A.Origin of nuclear and chromosomal alterations derived from the action of an aneugenic agent—Trifluralin herbicide Ecotoxicology and Environmental Safety. v. 72, p. 1680–1686, 2009.

[91] Könen, S., Çavas, T. Genotoxicity testing of the herbicide Trifluralin and its commercial formulation treflan using the piscine micronucleus test. Environmental and Molecular Mutagenesis, v.49, p.434-438, 2008.

[92] Peña, L.F.M. Uso do teste de micronúcleoemeritrócitoscirculantes de peixesparamonitorização de um local do rioTibagi e avaliação da genotoxidade de agrotóxicosembioensaios. Londrina. 1996. [Tese de mestradoemGenética e Melhoramento – UniversidadeEstadual de Londrina].

[93] Canevari, R.A. Avaliação dos efeitosgenotóxicos e diâmetro dos micronúcleosobtidosem*Prochiloduslineatus* (Pisces, *Prochilodontidae*) submetidos a tratamentosagudos com o inseticidaazodrin e o herbicidatrifluralina. Londrina. 1996. [Monografia (Bacharelado) embiologiageralUniversidadeEstadual de Londrina].

[94] Kaya B., Marcos R., Yanikoglu A., Creus A. Evaluation of the genotoxicity of four herbicides in the wing spot test of *Drosophila melanogaster* using two different strains. Mutat. Res., Amsterdam v. 557, p. 53-62, 2004.

[95] Gebel, T., Kevekordes, S., Pav, K., Edenharder, R., Dunkelberg, H. *In vivo*genotoxicity of selected herbicides in the mouse bone-marrow micronucleus test. Arch. Toxicol. v. 71, p. 193–197, 1997.

[96] Rawlings, N.C., Cook, S.J., Waldbillig, D. Effects of the pesticides carbofuran, chlorpyrifos, dimethoate, lindane, triallate, trifluralin, 2,4-D, and pentachlorophenol on the metabolic endocrine and reproductive endocrine system in ewes. J. Toxicol. Environ. Health A. v. 54, p. 21–36, 1998.

[97] Bryant, M.L., Murnik, M.R. Mutagenicity of the herbicide trifluralin inDrosophila melanogaster. Mutat. Res., Amsterdam v. 53 , p. 235, 1977.

[98] Foureman, P.A. Identification of aneuploidy inducing chemicals in *Drosophila*.Environ. Mutagen., New York v. 3, p. 319, 1981.

[99] Nehéz, M.; Páldy, A. Selypes, A. And Berencsi, G. Experiments on the mutagenic effects of two pesticides, DNOC and trifluralin. Mutat Res., Amsterdam v. 74, p. 202-203, 1980.

[100] Pilinskaya, M.S.A Evaluation of the cytogenetic effect of the herbicide treflan and of a number of its metabolites on mammalian somatic cells. Tsitol. Genetic. v. 21, p. 131-135, 1987.

[101] Iarc (1991) Iarc Monographs on the evolution of carcinogenic risks to humans. Occupational Exposures insecticide Application, and Some Pesticides. v. 53. Lyon, France.

[102] Ribas, G., Frenzilli, G., Barale, R., Marcos, R. Herbicide-induced damage in human lymphocytes evalueated by the single-cell gel eletrophoresis (SCGE) assay. Mutat. Res., Amsterdam v. 344, p. 41-54, 1995.

[103] Chan, M.M; Fong, D. Inhibition of Leishmanias but not host macrophages by the antimicrotubulin herbicide trifluralin. Science, v. 249, p. 924–926, 1990.

[104] Bhattacharya, G., Salem, M. M., Werbovetz, K. A. Antileishmanialdinitroaniline sulfonamides with activity against parasite tubulin. Bioorganic and Medicinal Chemistry Letters. v.12, p. 2395–2398, 2002.

[105] Esteves, M.A., Fragiadaki, I., Lopes, R., Scoulica, E., Cruz. M.E.M. Synthesis and biological evaluation of trifluralin analogues as antileishmanial agents Bioorganic & Medicinal Chemistry. v. 18, p. 274–281. 2010.

[106] Bogitsh, B.J., Middleton, O.L., Ribeiro-Rodrigues, R. Effects of the antitubulin drug trifluralin on the proliferation and metacyclogenesis of Trypanosomacruziepimastigotes. Parasitology Research. v. 85, p. 475–480, 1999.

[107] Traub-Cseko, Y.M., Ramalho-Ortigao, J.M., Dantas, A.P., De Castro, S.L., Barbosa, H.S., Downing, K.H. Dinitroanilineherbicidesagainstprotozoan parasites: the case of Trypanosomacruzi. Trends in Parasitology. v. 17, p. 136–141 2001.

[108] Nath, J., Schneider, I., Antimalarial effects of the antitubulin herbicide trifluralin: studies in Plasmodium falciparum. Clinical Research. v. 40, p. A331, 1992.

[109] Lee, J.H., Arumuganathan, K.; Yen, Y., Kaeppler, S., Baenziger, P. S. Root tip cell cycle synchronization and metaphase-chromosome isolation suitable for flow sorting in common wheat (*Triticumaestivum* L.). Genome, Otawa v.40, p.633-638, 1997.

[110] Dow, G., Reynoldson, J., Thompson, A. Comparative efficacy of two tubulin inhibitors, aldabenzole and trifluralin, against *Plasmodium berghei*. Parasitol. Int., Tokio v. 47, p.133-281, 1998.

[111] Werbovetz, K.A., Brendle, J.J., Sackett, D.L. Purification, characterization, and drug susceptibility of tubulin from Leishmania. Mol. Bioch. Parasitol., Amsterdamv. 98, p. 53-65, 1999.

[112] Donna, A., Betta, P.G., Gagliardi, F., Ghiazza, G.F., Gallareto, M. and Gabutto,V. Preliminary experimental contribution to the study of possible carcinogenic activity of two herbicides containing atrazine-simazine and trrifluralin as active principle.Pathologica, Gênova v. 73, p. 707-721, 1981.

[113] Kang, D., Park, S.K., Beane-Freeman, L., Lynch, C.F., Knott, C.E. Sandler,D.P. Hoppin, J.A., Dosemeci, M., Coble, J.,Lubin, J., Blair, A., Alavanja., M. Cancer incidence

among pesticide applicators exposed to trifluralin in the Agricultural Health Study. Environmental Research. v. 107, p. 271–276, 2008.

[114] N.C.I. Institute Nacional Cancer. Biossay of trifluralin for possible carcinogenicity Bethesda, MD, 2000.

[115] Emmerson, J.L., Pierce, E.C., Mcgrath, J.P. The chronic toxicity of compound 36352 (trifluralin) given as a compound of the diet to the fischer 344 rats for two years. Studies r-87 and R-97 (unpublished study received September 18, 1980 under 1471-35; submitted by Elanco Products Co., Division of Eli Lilly and Co., Indianapolis, IN), 1980.

Chapter 20

Allelochemicals as Bioherbicides — Present and Perspectives

Dorota Soltys, Urszula Krasuska,
Renata Bogatek and Agnieszka Gniazdowska

Additional information is available at the end of the chapter

http://dx.doi.org/10.5772/56185

1. Introduction

Since the first implementation of synthetic herbicides in crop protection systems, weeds have continuously developed resistance. As a main reason of such evolution, long-lasting exploitation of herbicides with one target site in plants is considered. This has been the case with the first widely-used triazine herbicides, photosynthesis inhibitors, which have effectively eliminated a wide range of weeds. Unfortunately, inappropriate adjustment of herbicides to weed species occupying fields, application of herbicides at the incorrect developmental stage and in unsuitable weather conditions have contributed to the accumulation of active compounds in the soil, accumulation of weed species and acceleration evolution of resistant biotypes [1]. To date, there have been 211 species and 393 biotypes of herbicide resistant weeds identified [2]. Most of them are resistant to B, C1 and A groups of herbicides, inhibitors of: acetolactate synthase (ALS), photosystem II and acetyl CoA carboxylase, respectively. Ten species pose the biggest threat for crops due to causing yield losses, including the most important herbicide-resistant species which are characterized by multiple resistances: rigid ryegrass (*Lolium rigidum* Gaud.), wild oat (*Avena fatua* L.) and redroot pigweed (*Amaranthus retroflexus* L.).

Evolution of weeds resistant to herbicides demands new solutions to cope with the problem since economic losses generated by weeds can be higher than those caused by other pests. Due to the fact that abandoning chemical weed control is, with current agricultural practices, rather impossible, it is necessary to create new classes of herbicides with new mechanisms of action and target sites not previously exploited. Presently used synthetic herbicides are not approved for use in organic agriculture. Moreover, using crop protection chemicals also need public acceptance. [3]. The number of synthetic chemicals with new target sites are decreasing dramatically. Eco-friendly trends in weed management force scientists to reach for innovative sources and tools. Natural compounds pose a great field for the discovery of new environmentally safe herbicides, so called "bioherbicides", which are based on compounds produced by living organisms. According to the CAS (Chemical Abstracts Service) registry, among the 24 million organic compounds, a large group of secondary plant metabolites is represented. Some of these compounds take part in allelopathic interactions.

2. Allelopathic interactions and allelopathic compounds

Allelopathy is considered a multi-dimensional phenomenon occurring constantly in natural and anthropogenic ecosystems [4]. It is defined as the interaction between plants and microorganisms by a variety of compounds usually referred to as allelopathins, allelochemicals, or allelopathic compounds. This review is focused mainly on compounds taking part in complex allelopathic interactions between higher plants. However, determination of quality, quantity, direct or indirect effects of allelopathins on plant or microorganism communities in the natural environment is very difficult owing to the multi-dimensional character of those interactions. The development of analytical techniques allowing better specification of direct effects of allelopathins, have moved the exploration (or the research on) of this phenomenon from fields into laboratories. The term "allelopathy" refers rather to interactions occurring in the natural environment [5]. For studies with plant extracts, allelopathins isolated from plant tissue, collected from exudates or even synthetic compounds identical to natural ones, it was established the term "phytotoxicity" to distinguish allelopathy (as a phenomenon occurring in natural environment) from studies conducted in laboratory.

Allelopathins are products of the secondary metabolism and are non-nutritional primary metabolites [6,7]. These compounds belong to numerous chemical groups including: triketones, terpenes, benzoquinones, coumarins, flavonoids, terpenoids, strigolactones, phenolic acids, tannins lignin, fatty acids and nonprotein aminoacids. A wide range of these biochemicals are synthesized during the shikimate pathway [8] or, in the case of essential oils, from the soprenoid pathway. Allelochemicals can be classified into 10 categories [9] according to their different structures and properties:

1. water-soluble organic acids, straight-chain alcohols, aliphatic aldehydes, and ketones;
2. simple lactones;
3. long-chain fatty acids and polyacetylenes;
4. quinines (benzoquinone, anthraquinone and complex quinines);
5. phenolics;
6. cinnamic acid and its derivatives;
7. coumarins;
8. flavonoids;

9. tannins;

10. steroids and terpenoids (sesquiterpene lactones, diterpenes, and triterpenoids).

Allelochemicals are released into the environment by plant organs such as roots, rhizomes, leaves, stems, bark, flowers, fruits and seeds (Figure 1a). The huge number of allelopathic interactions is typically negative in character, with positive relations being rare. Allelopathic compounds affect germination and growth of neighboring plants by disruption of various physiological processes including photosynthesis, respiration, water and hormonal balance. The underlying cause of their action is mainly inhibition of enzyme activity. Ability of an allelochemical to inhibit or delay plant growth and/or seed germination is usually defined as its "allelopathic (or phytotoxic) potential". An excellent example of allelopathic interaction is seen in soil exhaustion due to the accumulation of allelopathins that can be prevented by using fertilizers and rotating crops. Plants producing allelopathins are considered as "donor" organisms while the plants which allelopathins are directed to are referred to as "target" plants or "acceptors". The after-effects and strength of allelopathic interactions are diverse due to modifications of the allelopathins taking place in soil (Fig 1b). Most of the allelochemicals penetrate the soil as already plant-active compounds, e.g. phenolic acids, cyanamide, momilactones, heliannuols etc. Some have to be modified into the active form by microorganisms or by specific environmental conditions (pH, moisture, temperature, light, oxygen etc.), e.g. juglone, benzoxazolin-2-one (BOA), 2-amino-3-H-phenoxazin-3-one (APO).

3. Advantages and disadvantages of allelopathins as bioherbicides

Mode of action of some allelochemicals is similar to synthetic herbicides. These features have allowed them to be considered for possible use in weed management as bioherbicides. However, the field of knowledge is poorly studied but it is a very attractive area to explore.

Allelochemicals are highly attractive as new classes of herbicides due to a variety of advantages. However, in the perspective of bioherbicides based on allelopathins, effects caused by these compounds on target plants are also classified as "phytotoxic".

Most of allelopathins are totally or partially water-soluble which makes them easier to apply without additional surfactants [3, 10]. Their chemical structure is more environmentally friendly than synthetic ones. They possess higher oxygen- and nitrogen-rich molecules with relatively few so called 'heavy atoms', a halogen substitute, and are characterized by the absence of 'unnatural' rings. These properties decrease a chemical's environmental half-life, prevent accumulation of the compound in soil and eventual influence on non-target organisms. On the other hand, these properties are an allelochemical's Achille's heel due to less than satisfactory duration of activity. Structure complexity generates more stereocenters making them more reactive and unstable. Therefore, rapid degradation of one of the chemical groups can significantly decrease bioactivity of the whole compound.

The diversity of allelopathins makes them promising tools possessing specific properties in discovering novel, specific target sites in acceptor plants. Even if they inhibit photosynthesis

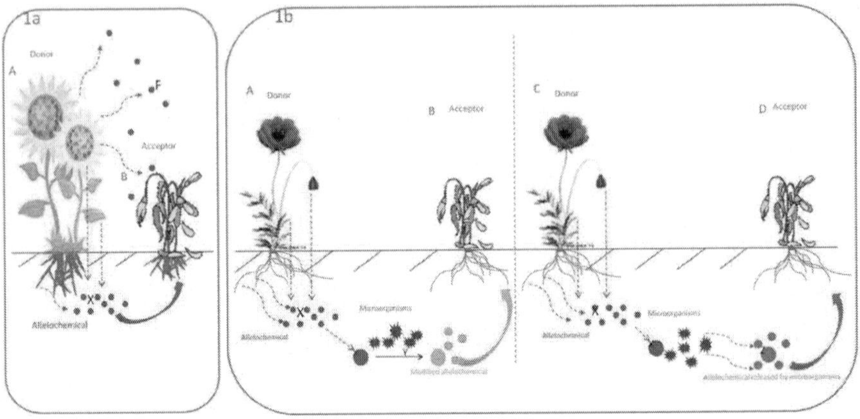

Figure 1. Multi-dimensional nature of allelopathic interacions. (1a) Plant A releases allelochemicals X and F which directly affect growth of plant B. (1b) left side; Plant A releases allelochemical X which is modified or activated by microorganisms to allelochemical Y that affects growth of plant B. (1b) right side; Plant A releases allelochemical X which stimulates microorganisms to produce allelochemical Z that affects growth of plant B.

or respiration, they may also bind to proteins at different sites than synthetic herbicides [11, 12]. This provides the opportunity to eliminate weeds that are already resistant to commercialized herbicides with the same mode of action. Allelochemicals are also characterized by multi-site action in plants without high specificity which is achieved in the case of synthetic herbicides. Therefore, this feature excludes the application of an allelopathic compound as a selective herbicide or totally prohibits its usage in weed management. On the other hand, effects of allelopathins in acceptor plants are highly dose-dependent [13]. This allows the opportunity to search out compounds exhibiting selectivity. Generally, monocotyledonous plants are more resistant to allelochemicals than dicotyledonous ones. Therefore, usage of a compound as a potential herbicide is possible but rather restricted to cultivation of exact crops with a defined weed composition.

The route of discovery is much more complicated with allelopathins. In contrast to synthetic herbicides where synthesis, bioassay, evaluation and quantitative structure-active relationship follow Quantitative Structure-Activity Relationship (QSAR), allelochemicals have to be first isolated from plant extracts [14]. The amount of recovered compounds is usually low in comparison to chemical synthesis. After extraction, purification and selection of the most attractive compound and determination of its mode of action in plants is done. At the end of the process, similar to synthetic herbicides, allelpathins are subjected to QSAR. The long discovery process is usually offset by a shorter, less expensive track of registration [15]. It is worth noting that before an allelochemical can become an herbicide, the following conditions have to be performed: phytotoxic activity at the range between 10^{-5} and 10^{-7} M, identified chemical structure, known mode of action in plants, time of residence in soil, possible influence on microbial ecology and non-target plants, possible toxic properties on human health and profitability of production on a commercial scale [16].

A high number of limitations does not exclude allelochemicals as possible herbicides. In particular, they can be alternatives in weed management strategy. Widely developed bioinformatics and cheminformatics support development of new herbicides [3, 15, 16]. Identified chemical structure of a particular allelochemical is a starting point to design a product with the compound-like properties using computer programs. Thanks to cheminformatics we are able to predict the potential structure of analogues and make several modifications, which make it more or less active, with higher environmental stability, as it was done for leptospermone. We may also predict the target site of compound action in plants due to comparison studies. Similar structure of a compound to a commercialized herbicide or other natural compound whose mode of action is well-known may allow us to predict the target site.

4. Allelopathic plant extracts as bioherbicides

Plant protection is effective but rather costly and problematic due to environmental pollution. Exploration of the allelopathic potential of some species allows the introduction of alternative techniques for weed management, e.g. extracts from allelopathic plants can be applied as foliar sprays. Apart from decreasing the costs of herbicide application, this method also improves crop production.

The best known examples of natural bioherbicides are phytotoxic water extracts from herbage of sorghum (*Sorghum bicolor* (L.) Moench.) (sorgaab) and sunflower (*Helianthus annuus* L.) (sunfaag) which can be effectively used in plant protection without yield losses.

Effects of sorgaab on weeds is time- and dose-depend but is typically used at 5% or 10% (w/v) concentration as double spray 20/30 and 40/60 days after sowing (DAS) or after seedling transplantation (AT) [17-19]. The best results to account for net profits have been elicited with a double spray of 10% extract in cotton (*Gossypim hirsutum* L.), soybean (*Glycine max* L.), wheat (*Triticum aestivum* L.) or rice (*Oryza sativa* L.). The highest efficacy of such extract applications has been verified in rice on reduction of barnyard grass (*Echinochloa cruss-galli* L.) biomass by 40%, without significant changes in weed density and accompanied yield increase by 18%.

Sunfaag has been widely used in wheat. The extract has been usually applied three times at 7-day intervals starting between 3-4 weeks post-emergence. This system of application has reduced biomass of the two most commonly occurring weeds, lambsquarters (*Chenopodium album* L.) and toothed dock (*Rumex dentatus* L.), by 70% and 97% respectively, although it has not eliminated all weed species in field. It has improved wheat biomass by 7-8% in comparison to weed free control without significant changes in number of tillers and total seed biomass. The herbicidal efficiency calculated as the effectiveness of sunfaag in comparison to synthetic herbicides showed a quite high value, 60% efficiency index. Weed management systems require high concentrations of sunfaag ranging up to 80% and can generate economic losses due to the necessity of cultivating higher amounts of sorghum or sunflower that also required an appropriate cultivation system [20, 21]. Therefore, sunfaag can be applied as a pre-emergence herbicide with much lower doses. The most promising application system has considered usage of 10% (w/v) extract at pre-emergence + 25 DAS + 35 DAS. Following the application, there has been noted a remarkably reduced population of wild oat, lesser swinecress (*Coronopus didymus* L.) and littleseed canarygrass (*Phalaris minor* Retz.) without affecting germination of wheat and increased wheat yield in 7% [22]. However, the inhibitory effect on weed growth and crop yield is selective and highly dependent on duration or term of sorgaab and sunfaag application.

Aqueous extracts of sorghum and sunflower are effective on weed growth but unfortunately might not be profitable enough in crop production; however, crop allelopathy can be manipulated for achieving sustainable weed management. Combination of phytotoxic crop water extracts with lower rates of herbicides may provide reduced weed control levels with reduced herbicide usage. The interesting review of allelopathic crop plants in weed management strategy is presented in reference [23]. Two field studies were conducted utilizing water extracts of sorghum, sunflower and rapeseed (*Brassica napus* L.) with reduced glyphosate dosage for controlling purple nutsedge (*Cyperus rotundus* L) in cotton [24]. Sorghum and rapeseed water extracts were tank mixed (at 15 or 18 L ha^{-1}) in different combinations with reduced rates of glyphosate by 767 and 575 g active substance (a.s.) ha^{-1} and sprayed as directed post emergence at 21 DAS. Purple nutsedge density and dry weight were suppressed by 78% to 95% and 83% to 95%, respectively, when different crop water extracts were used in combination with a reduced rate of glyphosate. Seed cotton yield was improved from 15-21% in sorgaab and rape water extract combinations with reduced rates of glyphosate (67-75%). Similar research has been conducted on water extracts of sorghum with sunflower in combination with herbicides in wheat, soybean, rice, and canola (*Brassica* sp.) [25, 26]. Both extracts, in combination with herbicides, have the same or even better effect on inhibition of growth of the following weeds: littleseed canarygrass and lesser swinecress, compared to single synthetic herbicide applications [25, 26]. Spraying of wheat seedlings 30 DAS with sorgaab+sunfaag (18 L each ha^{-1}) with mesosulfuron+idosulfuron (4.32 g a.s. ha^{-1}) has the same effect on total weed density (reduction up to 90% in relation to control) as application of mesosulfuron+idosulfuron used alone, but with higher doses (120 g a.s. ha^{-1}). Herbicidal solution has also improved yield parameters, both in relation to control and in relation to single herbicide application: fertile tillers (10%), spikelets per spike (11%) and grains per spike (10%) [26]. In cotton, application of both extracts at 18 L ha^{-1} each with glyphosate (767 g a.s. ha^{-1}) 21 DAS has been the most effective in density reduction of the highly competitive weed purple nutsedge up to 93% [24]. However, the greatest benefit in wheat is the usage of a sorgaab/sunfaag combination which lowered by 70% doses of metribuzin and phenaxaprop (at 57 g a.s. ha^{-1}), applied at 18 L each ha^{-1}. In turn, in cotton, application of the same rates of extracts per ha with glyphosate (767 g a.s. ha^{-1}) seems to be the most economically reasonable costs of following weed management method [24, 25].

Selectivity of plant extracts on weeds without any negative implications on crop productivity is probably due to differences in the physiological stage of plants and following plant competition. Sunfaag has been applied when wheat seedlings were 3-4 weeks old while lambsquarters and toothed dock 1-week old at the stage of three to four leaf [20, 21].

High allelopathic potential conditioned by glucosinolates and isothiocyanates is present in Brassica sp. [27, 28]. Isothiocyanates have been strong suppressants of germination of spiny sowthistle (*Sonchus asper* L. Hill), scentless mayweed (*Matricaria inodora* L.), smooth pigweed (*Amaranthus hybridus* L.), barnyardgrass, blackgrass (*Alopecurus myosuroides* Huds.) and wheat [28]. Black mustard (*Brassica nigra* L.) extract of different plant parts like leaf, stem, flower and root have inhibited germination and radicle length of wild oat. Inhibitory effects on germination increased with increasing concentration of extract solution of the fresh plant parts [29]. Some experiments were conducted also using garden radish (*Raphanus sativus* L.) extract on germination of 25 weed and 32 crop species [30]. Garden radish extracts totally inhibited germination of 11 weeds such as Johnsongrass (*Sorghum halelense* L. Pers.), *Alhagi* spp., blackgrass (*Alopecurus myosuroides* Huds.), shepherd's-purse (*Capsella bursa-pastoris* L. Medik.), field bindweed (*Convolvulus arvensis* L.), dodder (*Cuscuta* sp.), carrot (*Daucus carota* L.), shortpod mustard (*Hirschfeldia incana* L.), *Ochtodium aegyptiacum* (L.), and shortfruit hedgemustard (*Sisymbrium polyceratium* L.), and 4 crop species namely lettuce (*Lactuca sativa* L.), tobacco (*Nicotiana tabacum* L.), bean (*Phaseolus vulgaris* L.), and clover (*Trifolium* sp.). Garden radish extracts at different rates (100, 66, 50 and 33% of pure extract) did not affect germination of wheat, cotton, and maize (*Zea mays* L.), but affected soybean germination at the 100% extract rate in vitro. Rhizome regeneration of Johnsongrass was inhibited by 54-99% depending on extract concentration. Regeneration of bermudagrass (*Cynodon dactylon* L. Pers.) rhizomes was inhibited to a lower extent at all concentrations; for instance, 54% inhibition occurred at the highest extract concentration. Lower extract rates stimulated redroot pigweed germination, while 66 and 100% extracts inhibited germination by 21 and 42%, respectively. Inhibition reached only 56 and 49% at the highest extract concentration for common purslane (*Portulaca oleracea* L.) and cocklebur (*Xanthium strumarium* L.), respectively. Garden radish residues which were cut into pieces and incorporated into the growing medium decreased weed intensity and increased maize yield [31].

Legumes crops may also be applied as a source of allelochemicals useful in weed suppression. Mulch of dead pea plants could be used to control growth of weeds. Pea cover crop has regulated germination and growth of lady's thumb (*Polygonum persicaria* L.), smooth pigweed, smallflower galinsoga, and common lambsquarters. Similarly, the aqueous leachates (1%) of all four legumes, velvetbean (*Mucuna deeringiana* (Bort.) Merr.), jackbean (*Canavalia ensiformis* (L.) DC.), jumbiebean (*Leucaena leucocephala* (Lam.) de Wit), and wild tamarind (*Lysiloma latisiliquum* (L.) Benth.), have been shown to suppress weeds [32]. These plants exhibited strong phytotoxic effects on the radicle growth of barnyardgrass, alegría (*Amaranthus* ssp.) and amaranth (*Amaranthus hypochondriacus* L.) [33]. Russian knapweed (*Acroptilon repens*) control is difficult in many crops. Allelopathic effects of extracts and plant parts of alfalfa (*Medicago sativa* L.) on Russian knapweed were reported both in Petri dishes and pot experiments [34]. Alfalfa has been recommended in fields with high mugwort (*Artemisia vulgaris* L.) infestation, as it decreased mugwort to 89% under field conditions, while extracts of alfalfa vegetative parts inhibited mugwort germination up to 83% in Petri dish assays.

Application of plant extracts as pre-emergence or as early post emergence herbicides resulted in reduction of doses of synthetic herbicide due to their synergistic or additive action. However, not all phytotoxic extracts are effective enough to inhibit weed growth or germination when applied as spray even when plants show high allelopathic potential as mulch, intercropping system or in rotation. This may be the result of masking the activity of one compound by another in water solution or other factors such as impossibility of extract penetration through the cuticle [12]. A new opportunity to enhance effectiveness of usage of bioherbicides based on natural extracts is associated with extraction of individual allelochemicals and/or its comparison with synthetic herbicides. The extraction of sesquiterpene lactone, dehydrozaluzanin C (DHZ) produced among Compositae family serves as an example [34]. Comparison studies of isolated DHZ (1 mM) and the commercial herbicide Logran® showed high inhibitory activity of DHZ on dicotyledonous plants while the synthetic herbicide showed no activity [34]. Also pure 2-benzoxazolinone (BOA) isolated from several graminaceous crops such as rye (*Secale cereale* L.), maize and wheat was active similarly as herbicide but its stability in the environment was much shorter than the synthetic herbicide [35].

5. Plant allelopathins as sources of bioherbicides

Plant phytotoxic extracts, after evaluation, can be successfully used in integrated weed management. However, as was aforementioned, not all systems of its application under field conditions are suitable and profitable enough. To circumvent masking effects of one allelopathin by another in plant extract, research is now focused on isolation and application of a single, specific compound for the purpose of weed elimination. The list of allelochemicals isolated from various plants that may act as inhibitors of weed seed germination and/or weed growth are summarized in Table 1. A purified allelopathic compound may act on target plants with much higher or much lower strength. Even in situations when an allelopathin is active at unprofitably high doses but has a favorable environmental profile, it still may be a source to explore due to several reasons such as biodegradability. Modifications of chemical structure can make a compound more active on target plants while preserving desire properties.

Herein, examples of purified allelopathins with possible roles as herbicides are described. Some herbicides based on modified allelopathins already launched on the market are also included.

5.1. Sorgoleone

The inhibitory effect of sorghum on various plant species has been known for many years. Accumulation of sorghum phytotoxins in soil affects crop growth and imposes the need for a crop rotation system. Besides crops, weeds are also vulnerable to its allelopathic influence [16, 36]. Sorghum toxicity is mainly determined by both hydrophilic phenols in herbage, as well as hydrophobic sorgoleone and its analogs exuded by the root hairs [37, 38]. Therefore, sorghum herbage reach can be successfully used against weeds as a foliar spray as it is discussed in detail in the previous chapter.

Compounds	Botanical source	Sensitive weeds
Glucosinolates, Isothiocyanates	mustard (*Brassica* sp.) garden radish (*Raphanus sativus*)	spiny sowthistle (*Sonchus asper* L. Hill), scentless mayweed (*Matricaria inodora* L.), smooth pigweed (*Amaranthus hybridus* L.), barnyardgrass (*Echinochloa cruss-galli* L. Beauv.), slender meadow foxtail or blackgrass (*Alopecurus myosuroides* Huds.), *Alhagi* spp., *Cachia maritime*, Shepherd's-purse(*Capsella bursa-pastoris* L.), morning glory (*Convolvulus arvensis* L.), dodders (*Cuscuta* spp.), wild carrot or bird's nest (*Daucus carota* L.), shortpod mustard, buchanweed or hoary mustard(*Hirschfeldia incana* L.), *Ochtodium aegyptiacum* (L.), shortfruit hedgemustard (*Sisymbrium polyceratium* L.)
Sorgoleone	sorghum (*Sorghum bicolor* L. Moench)	littleseed canarygrass (*Phalaris minor* Retz.), lesser swinecress (*Coronopus didymus* L.), purple nutsedge (*Cyperus rotundus* L.), black nightshade (*Solanum nigrum* L.), redroot pigweed (*Amaranthus retroflexus* L.), common ragweed (*Ambrosia atrtemisiflora* L.), sicklepod (*Cassia obtusifolia* L.)
Momilactone	rice (*Oryza sativa L.*), moss (*Hypnum plumaeform*)	barnyardgrass, (*Echinochloa colonum* L.), livid amaranth(*Amaranthus lividus* L.), hairy crabgrass (*Digitaria sanguinalis* L.), annual meadow grass, annual bluegrass or poa (*Poa annua* L.)
Artemisinin	annual wormwood (*Artemisia annua* L.)	redroot pigweed, pitted morning-glory (*Ipomoea lacunose* L.), common purslane (*Portulaca oleracea* L.), annual wormwood, duckweed (*Lemna minor* L.), algae (*Pseudokirchneriella subcapitata*)
Leptospermone	bottle brush (*Callistemon citrinus*), manuka (*Leptospermum scoparium* J.R., G. Forst)	barnyard grass, hairy crabgrass, yellow foxtail (*Setaria glauca* L.), california red oat (*Avena sativa* L.), Indian mustard (*Brassica juncea* L.), curly dock (*Rumex crispus* L.)
Essential oils	eucalyptus (*Eucalyptus* sp.)	barnyard grass, *Cassia occidentalis*, annual ryegrass (*Lolium rigidum*)
Sarmentine	pepper (*Piper* sp.)	barnyard grass, redroot pigweed, crabgrass, Sprangletop (*Leptochloa filiformis* Lam.), dandelion (*Taraxacum* sp.), lambsquarter or wild spinach (*Chenopodium album* L.), annual bluegrass or poa, morning glory or bindweed, wild mustard, curly dock

Table 1. Allelopathic compounds isolated from plants that exhibit inhibitory potential on seed germination and growth of weeds

However, allelochemical sorgoleone has enormous potential as an herbicide due to its high activity against various weed species. Studies conducted under laboratory conditions have shown that low doses of sorgoleone (100 μM) inhibit growth of the following weeds by 80%, black nightshade (*Solanum nigrum* L.), redroot pigweed, common ragweed (*Ambrosia atrtemisiflora* L.), and by 40% of sicklepod (*Cassia obtusifolia* L.), hairy crabgrass (*Digitaria sanguinalis* L.), velvetleaf (*Abutilon theophrasti* Medik.), barnyardgrass and tef (*Eragrostis tef* Zucc., Trotter) [11, 16].

Sorgoleone released into the soil may act as a pre-emergence herbicide. Its persistence in the soil during or after sorghum cultivation inhibits germination and growth of small-seeded weeds, e.g. hairy crabgrass and green bristlegrass (*Setaria viridis* (L.) Beauv.), due to its better absorption and translocation within the small seeds than in large seeds [39]. However, strength and final effect on seeds or seedling physiology is multifactor-dependent. Sorgoleone sorbs strongly to the organic matter. This allows an extended persistence in the soil but unfortunately, significantly reduces its bioavailability. Moreover, the dynamics of decomposition significantly influences sorgoleone bioactivity, e.g. the methoxy- group of the aromatic ring is decomposed by 26% 48 h after exudation; however, some amounts of sorgoleone are also extractable after 6 weeks [40, 41]. Nevertheless, constitutive production of the compound allows a continuous supply and accumulation in the soil around 1.5 cm of root zone [42].

Inhibition of H^+-ATPase in plant roots makes sorgoleone an effective growth inhibitor and potential post-emergence herbicide [43]. Decreased activity of that enzyme affects ion uptake and water balance by decreasing water uptake and affecting plant growth. Redroot pigweed, Jimson weed (*Datura stramonium* L.) and tef grown in hydroponic culture with 10 μM sorgoleone were characterized by lower H^+-ATPase activity in roots. Presence of sorgoleone in nutrient solution significantly suppressed growth and evoked brown coloration and necrosis [43, 44].

Sorgoleone may be taken up by roots but cannot be translocated acropetally by xylem due to high lipophilic properties. Therefore, its application as a post-emergence herbicide may be limited. However, as a spray (0.6 kg ha^{-1}), it has inhibited growth by 12% of green foxtail (*Setaria faberi* Herrm.), by 40-50% purslane, hairy crabgrass and velvetleaf, and up to 80-90% of common ragweed, redroot pigweed, and black nightshade [40].

Due to the structural similarity of sorgoleone to plastoquinon, it acts as a photosystem II (PSII) inhibitor [11, 43]. It binds to the niche of the D1 protein in PSII, gathers electrons and does not allow reoxidation of plastoquinon A by the secondary electron acceptor, plastoquinone B. Competition studies under sorgoleone *versus* synthetic herbicides such as atrazine, diuron, metribuzin and bentazon have shown that sorgoleone is an atrazine competitive inhibitor [11, 12]. Moreover, the I_{50} of sorgoleone is 0.1 μM and similar to other PSII inhibitors. It is worth mentioning that sorgoleone belongs to the His215 family of PSII inhibitors, while atrazine belongs to Ser264. Mutation in Ser264 of the D1 protein is responsible for resistance to triazines as well as other non-triazine herbicides, leading to cross-resistance. However, plants resistant to atrazine, with a QB binding site on PSII mutation (Ser264), are not resistant to sorgoleone. Application of sorgoleone is particularly justified in the case of triazine-resistant biotypes of redroot pigweed, due to the same physiological effects as applications of atrazine in redroot pigweed-susceptible biotypes [11]. These properties make sorgoleone a potential early post-emergence herbicide when applied as a spray with much less environmental implications than atrazine. Therefore, inhibition of photosynthesis is the main target site of sorgoleone action in young seedlings but its mode of action in older plants may be different [12]. Sorgoleone can be a useful inhibitor of *p*-hydroxyphenylpyruvate dioxygenase (HPPD), which takes part in α-tocopherol and plastoquinone synthesis. Inhibition of that enzyme leads to a decreased pool of available plastoquinone and indirectly affects activity of phytoene desaturase, a key enzyme in carotenoid synthesis. Such sequence of events causes declining carotenoid levels and affects photosynthesis [45]. Currently used triketone herbicides (e.g. sulcotrione, isoxaflutole) have the same mechanism of action on HPPD as sorgoleone, irreversible competitive inhibition, with I_{50} = 0.4 μM. Triketone herbicides are considered by the U.S. Environmental Protection Agency (EPA) to be a low environmental risk. They are usually utilized as selective herbicides to eliminate broadleaf weeds in corn [10]. It follows, due to similar action and chemical structure and environmental friendly profile, sorgoleone might also be useful as a selective herbicide; however, such comparison studies have yet to be conducted. Then, its mode of action also cannot explain whether it is more or less active on broadleaf or grass weeds species [44].

5.2. Momilactones

Extracts and residues of rice, the well-known cereal plant, also have allelopathic potential. Among isolated secondary metabolites, phenolic acids, hydroxamic acids, fatty acids, terpenes and indoles were identified [46]. The key role in rice allelopathy plays momilactone A and B isolated from root exudates. High allelopathic rice varieties release up to 2-3 μg of momilactone B per day [3]. These compounds inhibited the growth of typical weeds in rice, e.g. barnyard grass and awnless barnyard grass (*Echinochloa colona* (L.) Link.) at concentrations higher than 1 μM and 10 μM, respectively. Furthermore, phytotoxic abilities of momilacton A and B were also demonstrated on livid pigweed (*Amaranthus lividus* L.), hairy crabgrass and annual bluegrass (*Poa annua* L.) at concentrations higher than 60 μM and 12 μM, respectively [47]. The experiment has shown that momilactone B is secreted by rice roots into the rhizosphere over the entire life cycle [48]. Momilactone A and B belong to the diterpenoid phytoalexins which are known as antimicrobial secondary metabolites generated in response to signal molecules called elicitors (especially biotic elicitors) [49]. Both compounds thought to be unique to rice, recently have been found in the moss (*Hypnum plumaeforme* Wils.), a taxonomically distinct plant [49]. Despite the ability of momilactone A and B to inhibit plant growth, its mode of action in plants is still unknown.

5.3. Artemisinin

Artemisinin is a sesquiterpenoid lactone of annual wormwood (*Artemisia annua* L.). It is synthesized and sequestered in glandular trichomes located on the leaves and flowers [51]. It can also be excreted by the roots or root hairs, but only at the beginning of the growing season; therefore, dead leaves are the major source of artemisinin in soils [52]. Artemisinin is also lost from annual wormwood by rain runoff but to a minor degree (<0.5%),. This allelopathin is well known as a promising anti-malaric agent but also as a phytotoxin selective mainly to broadleaf weeds. Artemisinin (at 33 μM) significantly reduced shoot and root growth of lettuce, redroot pigweed, pitted morning-glory (*Ipomoea lacunose* L.) common purslane and annual wormwood [53]. However, the same treatment had no effect on sorghum or velvetleaf. Several studies have been aimed at identifying the molecular target site of this compound as well as the structural requirements for herbicidal activity [53-55].The effect of artemisinin is most evident on root growth and chlorophyll content. In onion root tips, artemisinin (10 - 100 μM) decreased the mitotic index, provoked abnormal mitotic figures and caused structural modifications of chromosomes [55]. However, no definite target site has yet been identified. The most recent studies on rice sprayed with 1.86 μM artemisinin indicated its inhibiting abilities on photosynthetic electron transport [56]. Artemisinin site of action is probably plastoquinone B in photosystem II. Interestingly, as authors suggest, this effect is caused not directly by artemisinin itself, but rather by an unidentified artemisinin-metabolite occured in the plant after artemisinin application [56].

Other controversies around the phytotoxic potential of artemisinin arose when the dichloromethane extracts of annual wormwood leaves containing artemisinin showed a stronger phytotoxic effect on redroot pigweed seed germination and seedling growth than pure artemisinin [57]. Moreover, aqueous extract with disposed artemisinin had equal inhibitory effects on both physiological processes as allelopathin alone. This experiment suggests a marginal role of artemisinin in plant extract and joint action of other allelochemicals. Although, most studies analyzing allelopathic weed–crop interferences using annual wormwood were conducted under laboratory and greenhouse conditions [58].

Toxic studies on duckweed (*Lemna minor* L.) and the fresh water algae *Pseudokirchneriella subcapitata* (Korshikov) had EC_{50} values 0.24 and 0.19 mg L^{-1} respectively, with growth rate as endpoint corresponding to those of the herbicide atrazine [59]. These profiles questioned environmental safety of artemisinin for the purpose as a bioherbicide. It may be a result of its complex chemical structure, but this compound may be used as the basis for a new herbicide, based on artemisinin chemical structure. Such attempts have already been made using artemisinin's analogues [55]. Four of the tested 12 analogues inhibited germination and root growth of lettuce, *Arabidopsis thaliana* (L.) and duckweed at extremely low concentrations (3 μM).

5.4. Leptospermone

Leptospermone (1-hydroxy-2-isovaloryl-4,4,6,6-tetramethyl cyclohexen-3,5-dione) is a natural triketone produced by the roots of the bottlebrush (*Callistemon citrinus* Curtis) [60]. In its pure form, it was tested both pre- and post-emergence on a range of plant species including: hairy crabgrass, yellow foxtail (*Setaria glauca* (L.) P. Beauv.), barnyard grass, California red oat (*Avena sativa* L.), redroot pigweed, Indian mustard (*Brassica juncea* L.) and curly dock (*Rumex crispus* L.). Leptospermone is a strong p- hydroxyphenylpyruvate dioxygenase (HPPD) inhibitor with I_{50} values 3 μg mL^{-1}[61]. Inhibition of this enzyme leads to disruption in carotenoid biosynthesis and loss of chlorophyll. Unfortunately, a pure compound rate of 9000 g a.s. ha^{-1} was required to give acceptable weed control. Such high doses excluded leptospermone from commercial development. The structure of this allelochemical was used as a basis for development of synthetic analogues including mesotrione (trade name Callisto), an herbicide produced by Syngenta AG. Mesotrione is applied for control of broadleaved weeds in maize. The rates of mesotrione are in the range from 75 to 225 g a.s. ha^{-1} (around 100 times more potent than leptospermone) [60].

However, leptospermone has lately been found as the main herbicidal component of manuka oil (*Leptospermum scoparium* J.R., G. Forst) [61]. Manuka oil (1%) applied as post-emergence spray, significantly decreased growth and dry weight of redroot pigweed, barnyardgrass, velvetleaf and hairy crabgrass. Though, hairy crabgrass seedlings that emerged after manuka oil application were totally blanched. Pre-emergence application of 0.17% manuka oil which corresponds to 0.2 L ha^{-1} of leptospermone inhibited hairy crabgrass growth by 50%. The pre-emergence effects are mainly dependent on its persistence in soil. Average time of leptospermone half-life in soil was calculated at 15 days while applied as a compound of manuka oil time extended by 3 days. This clearly shows that half-life of active compounds may be longer in mixture than applied alone due to additive or synergistic action. This type of leptospermone application poses another possibility of usage for this compound in its natural form without chemical modification of the structure [61].

5.5. Essential oils

Lately, there has been a growing interest for using essential oils as allelopathins with bioherbicide potential. Some of them have already been commercialized and successfully launched in organic agriculture. They disrupt the cuticle and contribute to desiccation or burn down young tissues. Examples of this are the commercially available bioherbicide with the trade name of GreenMatch EX which consists of lemongrass (*Cymbopogon* sp.) oils or Interceptor™ with 10% pine (*Pinus sylvestris* L.) oil [3]. Essential oils are complex mixtures of monoterpenes, sesquiterpenes, and aromatic phenols, oxides, ethers, alcohols, esters, aldehydes and ketones [62]. The main terpenoids of volatile essential oils are monoterpenes (C10) and sesquiterpenes (C15). It has been well documented that essential oils found in foliage of eucalyptus (*Eucalyptus* sp.) show phytotoxic potential. During field experiments it has been reported that common weeds such as coffee senna (*Cassia occidentalis* L.) and barnyardgrass sprayed with different concentrations of eucalyptus oil (from 5 % to 10 % v/v with 0.05 % v/v Tween-80) exhibited dose-dependent and species-dependent levels of injury. Coffee senna plants were more sensitive to the eucalyptus oil than barnyardgrass [62]. Phytotoxicity of eucalyptus oil is due to the components such as 1,8-cineole, citronellal, citronellol, citronellyl acetate, p-cymene, eucamalol, limonene, linalool, α-pinene, γ-terpinene, α-terpineol, alloocimene, and aromadendrene [62]. Pre-emergence herbicidal activity of 1,8-cineole 3, and 1,4-cineole 4 were tested against rigid ryegrass and garden radish var. Long Scarlet in laboratory-based bioassays. 1,8-cineole and its derivatives showed a dose-dependent herbicidal activity against both weed species [64]. Laboratory studies [64, 65] also have shown that soil-applied 1,8-cineole suppressed the growth of several weeds. However, field reports demonstrated that 1,8-cineole alone has poor herbicidal activity [67, 68]. The commercial herbicide cinmethylin is a 2-benzyl ether substituted analog of the monoterpene 1,4-cineole (1-methyl-4-(1-methylethyl)-7-oxabicyclo heptane). This compound was discovered and partially developed by Shell Chemicals as a derivative of the allelopathic natural monoterpene, 1,8-cineole [69]. The benzyl ether substitution appears to decrease the volatility of the cineole ring by several orders of magnitude thereby rendering it more suitable for herbicide use [70]. Cinmethylin is a moderately effective growth inhibitor used for monocot weed control [71]. Despite the fact that it has been used commercially in both Europe and Japan and has been studied experimentally for several decades, the mechanism of action of this herbicide is still unknown [54, 72]. Cinmethylin was commercialized outside the United States in 1982 under the trade names of Cinch and Argold. Cinmethylin is active on several important grasses in rice; Echinochloa sp., Cyperus sp. and heartshape false pickerelweed (*Monochoria viginalis* Burm.f.) at rates from 25 to 100 g a.s. ha^{-1} [73].

5.6. Sarmentine

Sarmentine was first isolated from long pepper (*Piper longum* L.) fruits [74] but is also present in varied organs of other *Piper* species (Huang and Asolkan patent). It has been known as a medicinal plant with many beneficial multidirectional properties on human health. However, methanol extract of long pepper dry fruits has been shown to be suppressive to lettuce [75]. Purification and fractioning of long pepper crude extract allows the dissection of the active compound – sarmentine, a molecule with a long unsaturated fatty acid chain and pyrrolidine. Due to the hydrophobic properties, sarmentine is suspended with surfactants, 0.2 % glycospere O-20, 2% ethanol and 0.1% sodium lauryl sulfate. As a foliar spray, it is active at 2.5 mg mL^{-1}, but its high phytotoxicity is manifested at 5 mg mL^{-1}. Higher concentrations of sarmentine caused almost 100% mortality of redroot pigweed, barnyardgrass, bindweed (*Convonvulus* sp.), hairy crabgrass, sprangletop (*Leptochloa* sp.), annual bluegrass, wild mustard (*Sinapis arvensis* L.), curly dock with impaired effects on horseweed (*Conyza canadensis* (L.) Cronquist) and sedge (*Carex* sp.) growth under laboratory conditions. First phytotoxic symptoms such as bent stems and contact necrosis, have been visible 30 minutes after application; however full-blown implications were seen 7 h after spraying. The most likely mechanism of sarmentine action on plants is disruption of the plant cuticle which leads to disruption of cell membranes and lipid peroxidation followed by formation of radicals [76, 77].

As an herbicide, sarmentine and its derivatives may be both obtained from fruits of long pepper and successfully chemically synthesized [75]. Despite the fact that the compound is active under laboratory conditions, its chemical and biological instability under field conditions may limit its application as an herbicide. However, it has been shown that crucial for sarmentine herbicidal activity is the presence of an amine bond with a secondary amine. Replacement of the acid moiety with structurally similar fatty acids has not changed its phytotoxic potential. Moreover, natural herbicides based on sarmentine may also contain other derivatives with similar modes of action on plants but higher environmental stability [75]. Sarmentine may be successfully applied in combination with synthetic herbicides, e.g. aryloxyphenoxypropionic, benzoic acid, dicarboximide, organophosphorus, triazine, sulfonamide herbicides and with many others. This gives an

http://dx.doi.org/10.5772/56185

opportunity to further the structural modification that makes the compound more stable without any disadvantages on bioherbicide action in plants. It is worth noting that sarmentine has already been patented as an herbicide but not commercialized yet [75].

6. Biotechnology in bioherbicide investigation

A lot of effort has been done to explore the nature of allelopathic interactions. Studies on allelopathic compounds greatly increased thanks to chemical and biochemical techniques, which improved identification and knowledge about its mode of action. Since then, the crucial role of secondary metabolites synthesized and released by plants became better understood. It has been clearly demonstrated that allelopathins may take part in very complex inter- and intra-specific ecological interactions including soil microorganisms. However, despite the extensive research carried out under laboratory conditions, the higher level of such interactions at the ecosystem level has not been sufficiently explored. Structure, chemical properties, and mode of action in plants of multitude allelochemicals are already known but, unfortunately, only a part-per thousand of them have been successfully introduced in agricultural practices. This is mainly due to limitations of compounds as plant protection agents but also due to extended field experiments. A very important aspect that allows the introduction of allelopathy to natural weed management is knowledge about biology of donor and target plants and the exact chemicals responsible for the interaction [78]. All formerly described limitations of natural compounds as bioherbicides decreasing in case of plant extracts as herbicides due to simple and low cost of application. However, separation of one, specific compound that is the most interesting for us among hundreds synthesized by plants often required information about its synthesis *in vivo*.

One of the problems is to obtain adequate amounts of the compound, when its chemical synthesis is impossible or collection of plants, unprofitable. Increased synthesis of an allelopathin gives triple profits. First of all, enhanced allelopathic potential of a plant makes it more competitive against weeds. Second of all, increased concentration of a compound makes plant extract more active. Thirdly, this allows collection of the compound at a sufficient amount and makes it more profitable. However, it is much easier to obtain active compounds from the crop species than wild living ones. Difficulties in introducing plants to cultivation are due to the low ability to grow outside their natural ecosystem [79].

Cells and organ cultures provide opportunities to circumvent these limitations. Abilities of undifferentiated and differentiated cells to produce allelochemicals may be commercialized in bioreactors using cell suspension cultures [79]. Such attempts have been made on Artemisia suspension culture for artemisinin production; however, obtained amounts of that compound were insufficient. The addition of β-cyclodextrins to the growing medium has increased artemisinin synthesis up to 300% [80]. Allelochemicals produced by roots may be obtained from hairy root cultures, both *via* callogenesis or infection with *Agrobacterium tumefaciens.* Transgenic hairy roots are characterized by high genetic stability and facility to accumulate metabolites. The hairy root system already has been applied to increased production of phenolic compounds of nettleleaf goosefoot (*Chenopodium murale* Linn.) [81] and gossypol of cotton [82]. Active growth of roots and rapid colonization of the bioreactor allows rapidly reaching target weight, necessary to obtain an adequate quantity of the compound extracted from plants or growing medium.

The recombinant DNA technology can be useful to improve allelochemical production. Enhancing or suppression of gene expression, metabolic engineering and genetic transformation are promising new tools for allelochemical synthesis [79]. This approach is based on elucidation of the metabolic pathway, enzyme activities and identification of genes encoding crucial enzymes, associated with metabolite (allelochemical) synthesis.

Allelopathy is a quantitative trait. A genetic analysis of quantitative trait loci (QTL) is a promising approach to identify genes underlying this trait. Only a few crops are under genetic screening for its allelopathic properties including: rice, wheat, barley and oat [83, 84]. The first QTL map associated with allelopathic properties was developed in rice. A segregating population derived from a cross of two cultivars varying with allelopathic potential against barnyardgrass. The map contained 140 DNA markers with four main-effects QTL located on chromosome 2, 3 and 8 [85]. Proteomic studies on allelopathy of rice against barnyardgrass confirmed the crucial role of three enzymes: phenylalanine ammonia-lyse (PAL), thioredoxin and 3-hydroxy-3-methilglutarilcoenzyme A reductase 3 (HMGR) is highly involved in phenols biosynthesis [86]. Such a genetic approach may allow the location of the gene in the genome and better understanding of its function in plant allelopathy and create the chance of applying marker assisted selection (MAS) to enhance allelopathic abilities.

Just like breeding programs allow improved crop production, they may also improve production of allelopathic compounds increasing allelopathic potential.

Scopoletin has been known as allelopathic root exudates of oats (*Avena* sp.) that affects growth of neighboring plants. Screening of 3000 of Avena accessions has shown varying ability to scopoletin production. Twenty five of them have exuded higher amounts of scopoletin than control cultivar Garry, of which 4 were threefold more than the control [87]. Variation in allelopathin production was also discovered for sorgoleone of seven sorghum accessions [38] nomilacton A and B of 8 rice accessions [88] DIBOA and DIMBOA of 14 rye cultivars [88], gramine of 43 lines of modern cultivar of barley (*Hordeum vulgare* L.) and wild progenitor *H. spontaneum* (C. Koch) [90]. Enhanced production of active compounds from allelopathic plants can be developed by efficient breeding - selection of individuals with high allelopathic ability. Identification of a single gene, arranged in synthesis of allelopathin already has been performed for sorgoleone. *SOR1* (or compatible *SbDES3*) expression is specific for root hairs of two species of sorghum (*S. bicolor* and *S. halepense*) and associated with sorgoleone synthesis, while it is not expressed in other organs of sorghum *SOR1* encodes novel fatty acid desaturase (FAD), involved in the formation of a specific bond at $16{:}3\Delta^{9,12,15}$ pattern [91, 92]. Comparative studies of FAD derived from sorghum with other desaturases showed high similarity to omega-3 fatty acid desaturases (FAD3) [93]. However, none of the hitherto known desaturases can synthesize double bonding at this unique pattern along the aliphathic chain of the sorgoleone molecule. Characterization of this gene allows an overexpression of *SOR1* and increased sorgoleone synthesis and improved allelopathic potential of sorghum, as well as easier collection of the compound. Moreover, the well-known pathway of sorgoleone synthesis and characteristic of candidate genes may be a promising source of introducing sorgoleone production to grass crops [94].

The situation becomes more complicated when more than one gene encoding special enzymes is required to increase synthesis of a plant compound. Such difficulties have been encountered for DIBOA, synthesized by various grass species [95]. In maize, biosynthesis of this compound is determined by five genes (*Bx1* to *Bx5*) encoding three enzymes: tryptophan synthase α homolog, cytochrome P-450-dependent monooxygenase [95].

Monoterpenes are a large family of compounds produced by a varied family of aromatic plants. Some of the monoterpenes also take part in allelopathic interactions, e.g. linalool, cineole camphene, pinene, limonene, etc. Currently, metabolic engineering allows improved production of specific compounds in heterologous systems [96]. The most interesting are monoterpene synthases which catalyzed geranyl diphosphate (GPP) into output structure of numerous monoterpenes family, e.g. enhanced expression of limonene synthase in transgenic peppermint (*Mentha piperita* L.) has increased yield of monoterpenes. An alternative approach is to change the density of secretory structures by both plant hormone and transcriptional factors manipulation. Such attempts already have been made in annual wormwood and *A. thaliana*. It was recently found that the number of glandular trichomes increased in response to jasmonic acid. Spraying of annual wormwood with this hormone significantly increased density of these structures on leaves what was accompanied with higher artemisinin content [51]. This was an effect of enhanced expression of gene encoding enzymes taking part in artemisinin biosynthesis. On the other hand, in Arabidopsis, co-expression of two positive transcriptional factors (GL1, and R protein of maize) has significantly improved the number of trichomes [96].

However, we have to bear in mind that biosynthesis of natural compounds can be limited to organs, tissues or even cells. Specific locations of compound synthesis, accumulation or secretion often make that compound toxic to other tissues within the same plant organism. Moreover, even successful transformation of a plant does not guarantee successful and sufficient production of a desirable compound. The gene of (S)-linalool synthase (*Lis*) of fairy fans (*Clarkia breweri* Gray), constitutively expressed in transgenic petunia (*Petunia hybrida* Hook.), has produced linalool but in its glycosylated, non-volatile form [96].

All presented techniques provide greater knowledge on allelopathy. However, better understanding of such complex interactions among this phenomenon bring us one step forward to development of new strategies in weed management and finding new herbicides and new herbicidal target sites.

7. Conclusions

The phenomena of allelopathy and phytotoxic interactions between plants are strongly expanding branches of biological science. Allelochemicals, as a group of substances also called biocommunicators, seem to be a fruitful challenge for combining traditional agricultural practices and new approaches in pest management strategies. Allelochemicals have already been used to defend crops against pathogens, insects or nematodes, parallel to some attempts to use them for weed control. Crop rotation, cover crops, dead and living mulches are being employed in agriculture. Both in natural and agricultural ecosystems allelopathic interactions are involved in practically every aspect of plant growth, as they can play the role of stimulants and suppressants. Complex plant-plant and plant-microbe interactions in ecosystems and currently developing studies on molecular, cytological and physiological levels bring us to a better understanding of processes occurring around us. The ancient knowledge of well-known toxic properties of water extracts of a variety of allelopathic plants give us a basis that could be used in the creation of a novel approach in weed control.

Some allelochemicals, mainly these that are mentioned in the text above, may act as a starting point for production of new bioherbicides with novel target sites, not previously exploited, as the understanding of their mode of action is still growing. Creation of bioherbicides based on allelochemicals generates the opportunity to exploit natural compounds in plant protection and shows the possibility to cope with evolved weed resistance to herbicides. Despite the fact that we have extensive knowledge about the chemical nature of natural compounds, we can synthesize its analogues, and we have basically explored its phytotoxic potential, we still have insufficient data. Until recently, most studies on phytotoxicity have been conducted under laboratory conditions due to the ability to eliminate other environmental factors such us temperature, soil texture and its chemical and physical properties. Such approach allows the recognition of only direct effects of allelochemical action. There is still a great need to transfer laboratory data into field conditions. Such experiments are not willing to be taken on due to troublesome field experiments dependent on environmental conditions and a few year repetitions. New tools of molecular genetics, proteomics and metabolomics profiling as well as modern and sophisticated methods of chemistry and biochemistry will lead to the creation of substances, maybe based on the structure of particular compounds occurring in nature, which could be used without any risks as selective and eco-friendly herbicides.

Author details

Dorota Soltys[1*], Urszula Krasuska[1], Renata Bogatek[2] and Agnieszka Gniazdowska[2]

*Address all correspondence to: d.soltys@ihar.edu.pl

1 Laboratory of Biotechnology, Plant Breeding and Acclimatization Institute - National Research Institute, Mlochow, Poland

2 Department of Plant Physiology, Warsaw University of Life Sciences – SGGW, Warsaw, Poland

References

[1] Rola, H, Marczewska, K, & Kucharski, M. Zjawisko odporności chwastów na herbicydy w uprawach rolniczych. Studia i Raporty IUNG-PIB (2007). , 8-29.

[2] International Survey of Herbicide Resistance Weedshttp://www.weedscience.org/In.aspaccessed 31 October (2012).

[3] Dayan, F. E, Cantrell, C. L, & Duke, S. O. Natural products in crop protection. Bioorganic & Medicinal Chemistry (2009). , 17(12), 4022-4034.

[4] Gniazdowska, A, & Bogatek, R. Alleopathic interaction between plants. Multiside action of allelochemicals. Acta Physiologiae Plantarum (2005). B) , 395-407.

[5] Soltys, D, Rudzinska-langwald, A, Gniazdowska, A, Wisniewska, A, & Bogatek, R. Inhibition of tomato (*Solanum lycopersicum* L.) root growth by cyanamide is due to altered cell division, phytohormone balance and expansin gene expression. Planta (2012). , 236(5), 1629-1638.

[6] Weir, T. L, Park, S-W, & Vivanco, J. M. Biochemical and physiological mechanisms mediated by allelochemicals. Current Opinion in Plant Biology (2004). , 7(4), 472-479.

[7] Iqbal, A, & Fry, S. C. Potent endogenous allelopathic compounds in *Lepidium sativum* seed exudate: effects on epidermal cell growth in *Amaranthus caudatus* seedlings. Journal of Experimental Botany (2012). , 63(7), 2595-2604.

[8] Hussain, M. I, & Reigosa, M. J. Allelochemical stress inhibits growth, leaf water relations, PSII photochemistry, non-photochemical fluorescence quenching, and heat energy dissipation in three C3 perennial species. Journal of Experimental Botany (2011). , 62(13), 4533-4545.

[9] Li, Z-H, Wang, Q, Ruan, X, Pan, C-D, & Jiang, D-A. Phenolics and Plant Allelopathy Molecules (2010). doi:10.3390/molecules15128933, 15(12), 8933-8952.

[10] Vyvyan, W. R. Allelochemicals as leads for new herbicides and agrochemicals. Tetrahedron (2002). , 58(9), 1632-1646.

[11] Nimbal, C. I, Yerkes, C. N, Weston, L. A, & Weller, S. C. Herbicidal activity and site of action of the natural product sorgoleone. Pesticide Biochemistry and Physiolology (1996). , 54(2), 73-83.

[12] Dayan, F. E, Howell, J, & Widenhamer, J. D. Dynamic root exudation of sorgoleone and its *in planta* mechanism of action. Journal of Experimental Botany (2009). , 60(7), 2107-2117.

[13] Belz, R. G, Hurle, K, & Duke, S. O. Dose-response- A challenge for allelopathy? Nonlinearity in Biology, Toxicology and Medicine (2005). , 3(2), 173-211.

[14] Duke, S. O. Natural pesticides from plants. In: J. Janick and J.E. Simon (eds.) Advances in new crops. Portland: Timber Press; (1990). , 511-517.

[15] Dayan, F. E, Owens, D. K, & Duke, S. O. Rationale for a natural products approach to herbicide discovery. Pest Management Science (2012). , 68(4), 519-528.

[16] Bhowmik, P. C. Inderjit. Challenges and opportunities in implementing allelopathy for natural weed management. Crop Protection (2003). , 22(4), 661-671.

[17] Irshad, A, & Cheema, Z. A. (2005). Comparative efficacy of sorghum allelopathic potential for controlling barnyardgrass in rice. Proceedings of the 4th World Congress on Allelopathy, Wagga Wagga, New South Wales, Australia. http://www.regional.org.au/au/allelopathy/2005/2/4/2220_irshada.htm

[18] Khaliq, A, Cheema, Z. A, Mukhtar, M. A, & Ahmad, S. M. Evaluation of sorghum (*Sorghum bicolor*) water extract for weed control in soybean. International Journal of Agriculture and Biology (1999). , 1(1), 23-26.

[19] Cheema, Z. A, & Khaliq, A. Use of sorghum allelopathic properties to control weeds in irrigated wheat in semi arid region of Punjab. Agriculture, Ecosystems & Environment. (2000).

[20] Anjum, T, & Bajwa, R. Field appraisal of herbicide potential of sunflower leaf extract against Rumex dentatus. Field Crops Research (2007).

[21] Anjum, T, & Bajwa, R. The effect of sunflower lear extracts on *Chenopodium album* in wheat fields in Pakistan. Crop Protection (2007). , 26(9), 1390-1394.

[22] Naseem, M, Aslam, M, Ansar, M, & Azhar, M. Allelopathic effects of sunflower water extract on weed control and wheat productivity. Pakistan Journal of Weed Science Research (2009). , 15(1), 107-116.

[23] Bhadoria PBSAllelopathy: a natural way towards weed management. American Journal of Experimental Agriculture (2011). , 1(1), 7-20.

[24] Iqbal, J, Cheema, Z. A, & Mushtaq, M. Allelopathic crop water extracts reduce the herbicide dose for weed control in cotton (*Gossypium hirsutum*). International Journal of Agriculture and Biology (2009). , 11(4), 360-366.

[25] Razzaq, Z. A, Cheema, K, Jabran, K, Farooq, M, Khaliq, A, & Haider, G. Basra SMA. Weed management in wheat through combination of allelopathic water extract with reduced doses of herbicides. Pakistan Journal of Weed Science Research (2010). , 16(3), 247-256.

[26] Razzaq, A, Cheema, Z. A, Jabran, K, Hussain, M, Farooq, M, & Zafar, M. Reduced herbicide doses used together with allelopathic sorghum and sunflower water extracts for weed control in wheat. Journal of Plant Protection Research (2012). , 52(2), 281-285.

[27] Fenwick, G. R, Heaney, R. K, & Mullin, W. J. Glucosinolates and their breakdown products in food and food plants. Critical Reviews in Food Science and Nutrition (1983)., 18-123.

[28] Petersen, J, Belz, R, Walker, F, & Hurle, K. (2001). Weed suppression by release of isothiocyanates from turnip-rape mulch. Agronomy Journal 2001;, 93(1), 37-43.

[29] Turk, M. A, & Tawaha, A. M. Allelopathic effect of black mustard (*Brassica nigra* L.) on germination and growth of wild oat (*Avena fatua* L.). Crop Protection (2003)., 22(4), 673-677.

[30] Uygur, F. N, Koseli, F, & Cinar, A. Die allelopathische Wirkung von *Raphanus sativus* L. Journal of Plant Diseases and Protection Zeitschrift für Pflanzenkrankheiten und Pflanzenschutz, Sonderheft (1990). XII, 259-264.

[31] Uludag, A, Uremis, I, Arslam, M, & Gozcu, D. Allelopathy studies in weed science in Turkey- a review Journal of Plant Diseases and Protection Zeitschrift für Pflanzenkrankheiten und Pflanzenschutz Sonderheft (2006). XX, 419-426.

[32] Akemo, M. C, Regnier, E. E, & Bennett, M. A. Weed suppression in spring-sown rye (*Secale cereale*)-pea (*Pisum sativum*) cover crop mixes. Weed Technology (2000)., 14(3), 545-549.

[33] Caamal-maldonado, J. A, Jiménez-osornio, J. J, Torres-barragán, A, & Anaya, A. L. The use of allelopathic legume cover and mulch species for weed control in cropping systems. Agronomy Journal (2001)., 93(1), 27-36.

[34] Macías, F. A. Galino JCG, Molinillo JMG, Castellano D. Dehydrozaluzanin C: a potent plant growth regulator with potential use as a natural herbicide template. Phytochemistry (2000)., 54(2), 165-171.

[35] Reigosa, M. J, Gonzalez, L, Sanches-moreiras, A, Duran, B, Puime, D, Fernandez, D. A, & Bolano, J. C. Comparison of physiological effects of allelochemicals and commercial herbicides. Allelopathy Journal (2001)., 8(2), 211-220.

[36] Alsaadawi, I. S, Al-ekeelie, M. H. S, & Al-hamzawi, M. K. Differential allelopathic potential of grain sorghum genotypes to weeds. Allelopathy Journal (2007)., 19(1), 153-160.

[37] Lehle, F. R, & Putman, A. R. Allelopathic potential of sorghum (Sorghum bicolor): isolation of seed germination inhibitors. Journal of Chemical Ecology (1983)., 9(8), 1223-1234.

[38] Czarnota, M. A, Rimando, A. M, & Weston, L. A. Evaluation of root exudates of seven sorghum accessions. Journal of Chemical Ecology (2003)., 29(9), 2073-2083.

[39] Netzly, D. H, & Butler, L. G. Roots of sorghum exude hydrophobic droplets containing biologically active components. Crop Science (1986)., 26(4), 775-778.

[40] Czarnota, M. A, Paul, R. N, Dayan, F. E, Nimbal, H. I, & Weston, L. A. Mode of action, localization of production, chemical nature, and activity of sorgoleone: a potent

PSII inhibitor in Sorghum spp. root exudates. Weed Technology (2001). , 15(4), 813-825.

[41] Weston, L. A, & Czarnota, M. A. Activity and persistence of sorgoleone, a long-chain hydroquinone produced by Sorghum bicolor. Journal of Crop Production (2001). , 4(2), 363-377.

[42] Trezzi, M. M, Vidal, R. A, & Dick, D. P. Peralba MCR, Kruse ND. Sorptive behavior of sorgoleone in soil in two solvent systems and determination of its lipophilicity. Journal of Environmental Science and Health (2006). , 41(4), 345-356.

[43] Hejl, A. M, & Koster, K. L. The allelochemical sorgoleone inhibits root H^{+}-ATPase and water uptake. Journal of Chemical Ecology (2004). , 30(11), 2181-2191.

[44] Einhellig, F. A, & Souza, I. F. Phytotoxicity of sorgoleone found in grain sorghum root exudates. Journal of Chemical Ecology (1992). , 18(1), 1-11.

[45] Meazza, G, Scheffler, B. E, Tellez, M. R, Rimando, A. M, Romagni, J. G, Duke, S. O, Nanayakkara, D, Khan, I. A, Abourashed, E. A, & Dayan, F. E. The inhibitory activity of natural products on plant p-hydroxyphenylpyruvate dioxygenase. Phytochemistry (2002). , 59(3), 281-288.

[46] Kato-noguchi, H, Hasegawa, M, Ino, T, Ota, K, & Kujime, H. Contribution of momilactone A and B to rice allelopathy. Journal of Plant Physiology (2010). , 167(10), 787-791.

[47] Chung, I-M, Hahn, S-J, & Ahmad, A. Confirmation of potential herbicidal agents in hulls of rice, Oryza sativa. Journal of Chemical Ecology (2005). , 31(6), 1339-52.

[48] Kato-noguchi, H, Ota, K, & Ino, T. Release of momilactone A and B from rice plants into the rhizosphere and its bioactivities. Allelopathy Journal (2008). , 22(2), 321-8.

[49] Okada, A, Okada, K, Miyamoto, K, Koga, J, Shibuya, N, Nojiri, H, & Yamane, H. OsTGAP1, a bZIP transcription factor, coordinately regulates the inductive production of diterpenoid phytoalexins in rice. The Journal of Biological Chemistry (2009). , 284(39), 26510-26518.

[50] Kato-noguchi, H. Convergent or parallel molecular evolution of momilactone A and B: Potent allelochemicals, momilactones have been found only in rice and the moss *Hypnum plumaeforme*. Journal of Plant Physiology (2011). , 168-1511.

[51] Nguyen, K. T, Arseault, P. R, & Wethers, P. J. Trichomes + roots + ROS = artemisinin: regulating artemisinin biosynthesis in *Artemisia annua* L. In Vitro Cellular and Developmental Biology- Plant (2011). , 47(3), 329-338.

[52] Jessing, K. K, Cedergreen, N, Mayer, P, Libous-bailey, L, Strobel, B. W, Rimando, A, & Duke, S. O. Loss of artemisinin produced by *Artemisia annua* L. to the soil environment. Industrial Crops and Products. (2013). , 43-132.

[53] Duke, S. O, Vaughn, K. C, Croom, E. M, & Elsohly, H. N. (1987). Artemisinin, a constituent of annual wormwood (*Artemisia annua*) is a selective phytotoxin. Weed Science 1987;, 35(4), 499-505.

[54] DiTomaso JMDuke SO. Is polyamine biosynthesis a possible site of action of cinmethylin and artemisinin? Pesticide Biochemistry and Physiology (1991). , 39(2), 158-167.

[55] Dayan, F. E, Hernandez, A, Allen, S. N, Moraces, R. M, Vroman, J. A, Avery, M. A, & Duke, S. O. Comparative phytotoxicity of artimisinin and several sesquiterpene analogues. Phytochemistry (1999). , 50(4), 607-614.

[56] Bharati, A, Kar, M, & Sabat, S. C. Artemisinin inhibits chloroplast electron transport activity: mode of action. PLOS ONE (2012). e38942. doi:10.1371/journal.pone.0038942http://www.plosone.org/article/info%3Adoi%2F10.1371%2Fjournal.pone.0038942

[57] Lydon, J, Teasdale, J. R, & Chen, P. K. Allelopathic activity of annual wormwood (*Artemisia annua*) and its role of artemisinin. Weed Science (1997). , 45-807.

[58] InderjitNilsen ET. Bioassays and field studies for allelopathy in terrestrial plants: progress and problems. Critical Reviews in Plant Sciences (2003).

[59] Jessing, K. K. Production of biomedicine under different climatic conditions- Artemisinin as study case IOP Conf. Series: Earth and Environmental Science 6 (2009). doi: 10.1088/http://iopscience.iop.org/1755-1315/6/34/342026/pdf/1755-1315_6_34_342026.pdf

[60] Cornes, D. (2006). Callisto: A very successful maize herbicide inspired by allelochemistry. Maize Association of Australia 6th Triennial Conference. http://www.regional.org.au/au/allelopathy/2005/2/7/2636_cornesd.htm

[61] Dayan, F. E, Howell, J. L, Marais, J. P, Ferreira, D, & Koivunen, M. Manuka oil, a natural herbicide with preemergence activity. Weed Science (2011). , 59(4), 464-469.

[62] Batish, D. R, Singh, H. P, Kohli, R. K, & Kaur, S. Eucalyptus essential oil as a natural pesticide. Forest Ecology and Management (2008). , 256(12), 2166-2174.

[63] Batish, D. R, Setia, N, Singh, H. P, & Kohli, R. K. Phytotoxicity of lemon-scented eucalypt oil and its potential use as a bioherbicide. Crop Protection (2004). , 23(12), 1209-1214.

[64] Barton, A. F, Dell, B, & Knight, A. R. Herbicidal activity of cineole derivatives. Journal of Agricultural and Food Chemistry 210;, 58(18), 10147-55.

[65] Vaughn, S. F, & Spencer, G. F. Volatile monoterpenes as potential parent structures for new herbicides. Weed Science (1993). , 41(1), 114-119.

[66] Romagni, J. G, Allen, S. N, & Dayan, F. E. Allelopathic effects of volatile cineoles on two weedy plant species. Journal of Chemical Ecology (2000). , 26(1), 303-313.

[67] Halligan, J. P. Toxic terpenes from *Artemisia californica*. Ecology (1975). , 56(4), 999-1003.

[68] Heisey, R. M, & Delwiche, C. C. Phytotoxic volatiles from *Trichostema lanceolatum*. American Journal of Botany (1984). , 71(6), 821-828.

[69] Grayson, B. T, Williams, K. S, Freehauf, P. A, Pease, R. R, Ziesel, W. T, Sereno, R. L, & Reinsfelder, R. E. The physical and chemical properties of the herbicide cinmethylin. Pesticide Science (1987). , 21(2), 143-153.

[70] Vaughn, S. F, & Spencer, G. F. Synthesis and herbicidal activity of modified monoterpenes structurally similar to cinmethylin. Weed Science (1996). , 44(1), 7-11.

[71] Russell, S. G, Monaco, T. J, & Weber, J. B. Influence of soil moisture on phytotoxicity of cinmethylin to various crops. Weed Science (1991). , 39(3), 402-407.

[72] Baum, S. F, Karanastasis, L, & Rost, T. L. Morphogenetic effect of the herbicide Cinch on Arabidopsis thaliana root development. Journal of Plant Growth Regulation (1998). , 17(2), 107-114.

[73] Duke, S. O. Allelopathy: Current status of research and the future of the discipline: A commentary. Allelopathy Journal (2010). , 25(1), 17-30.

[74] Huang, H, Morgan, C. M, Asolkar, N. R, Koivunen, M. E, & Marrone, P. G. Phytotoxicity of sarmentine isolated from long pepper (*Piper longum*) fruit. Journal of Agricultural and Food Chemistry (2010). , 58(18), 9994-10000.

[75] Huang, H, & Asolkar, N. R. (2011). Use of sarmentine and its analogs for controlling plant pests. Patent Patentdocs: http://www.faqs.org/patents/accessed 27 January 2011).(20110021358)

[76] Fukuda, M, Tsujino, Y, Fujimori, T, Wakabayashi, K, & Böger, P. Phytotoxic activity of middle-chain fatty acids I: effects on cell constituents. Pesticide Biochemistry and Physiolology (2004). , 80(3), 143-150.

[77] Lederer, B, Fujimori, T, Tsujino, Y, Wakabayashi, K, & Böger, P. Phytotoxic activity of middle-chain fatty acids II: peroxidation and membrane effects. Pesticide Biochemistry and Physiolology (2004). , 80(3), 151-156.

[78] Macías, F. A, Molinillo, J. M, Varela, R. M, & Galindo, J. C. Allelopathy--a natural alternative for weed control. *Pest Management Science* (2007). , 63(4), 327-48.

[79] Bourgaud, F, Gravot, A, Milesi, S, & Gontier, E. Production of plant secondary metabolites: a historical perspective. Plant Science (2001). , 161(5), 839-851.

[80] Durante, M, Caretto, S, Quarta, A, De Paolis, A, Nisi, R, & Mita, G. Cyclodextrins enhance artemisinin production in *Artemisia annua* suspension cell cultures. Applied Microbiology and Biotechnology (2011). , 90(6), 1905-1913.

[81] Mitic, N, Damitrovic, S, Djordjevic, M, Zdravkovic-korac, S, Nikolic, R, Raspora, M, Djordjevic, T, Maksimovic, V, Živkovic, S, Krstic-miloševic, D, Stanišic, M, & Nin-

kovic, S. Use of *Chenopodium murale* L. transgenic hairy root in vitro culture system as a new tool for allelopathic assays. Journal of Plant Physiology (2012). , 169(12), 1203-1211.

[82] Triplett, B. A, Moss, S. C, Bland, J. M, & Dowd, M. K. Induction of hairy root cultures from *Gossypium hirsutum* and *Gossypium barbadense* to produce gossypol and related compounds. In Vitro Cellular & Developmental Biology- Plant (2008). , 44(6), 508-517.

[83] Olofsdotter, M, Jensen, L. B, & Courtois, B. Improving crop competitive ability using allelopathy- an example from rice. Plant Breeding (2002). , 121(1), 1-9.

[84] Belz, R. G. Allelopathy in crop/weed interactions- an update. Pest Management Science (2007). , 63(4), 308-326.

[85] Jensen, L. B, Courtois, B, Shen, L, Li, Z, Olofsdotter, M, & Mauleon, R. P. Locating genes controlling allelopathic effects against *Echinochloa crus-galli* (L.) in upland rice. Agricultural Journal (2001). , 93(1), 21-26.

[86] Lin, W-X, He, H-Q, Shen, L-H, Chen, X-X, Ke, Y, Guo, Y-C, & He, H-B. A proteomic approach to analysing rice allelopathy on barnyard grass (*Echinochloa crus-galli* L.). 4th International Crop Science Congress 26.(2004). Queensland, Australia. http://www.cropscience.org.au/icsc2004/poster/2/4/1/1414_xionglw.htm, 08-1.

[87] Fay, P. K, & Duke, W. B. An assessment of allelopathic potential in *Avena* germplasm. Weed Science (1977). , 25-224.

[88] Kato-noguchi, H. Allelopathic substance in rice root exudates: rediscovery of momilactone B as an allelochemical. Journal of Plant Physiology (2004). , 161(3), 271-276.

[89] Copaja, S. V, Villarroel, E, Bravo, H. R, Pizarro, L, & Argandona, V. H. Hydroxamic acids in *Secale cereale* L. and the relationship with their antifeedant and allelopathic properties. Zeitschrift fuer Naturforschung Section C Journal of Biosciences (2006). , 61-670.

[90] Lovett, J. V. Hoult AHC. (1992). Gramine: the occurrence of a self defence chemical in barley, *Hordeum vulgare* L. In: Hutchinson KJ, Vickery PJ. (eds) Looking Back- Planning Ahead conference proceedings, February Australian Agronomy Conference. "". Edited by Proceedings of the 6th Australian Agronomy Conference, 1992, The University of New England, Armidale, New South Wales. http://www.regional.org.au/au/asa/1992/concurrent/alternative-practices-plant-protection/p.htm#TopOfPage, 10-14.

[91] Pan, Z, Rimando, A. M, Baerson, S. R, Fishbein, M, & Duke, S. O. Functional characterization of desaturases involved in the formation of the terminal double bond of an unusual 16:3Δ 9,12,15 fatty acid isolated from *Sorghum bicolor* root hairs. Journal of Biological Chemistry (2007). , 282(7), 4326-4335.

[92] Yang, X, Scheffler, B. E, & Weston, L. A. SOR1, a gene associated with bioherbicide production in sorghum root hairs. Journal of Experimental Botany (2004). , 55-2251.

[93] Yang, X, Owens, T. G, Scheffler, B. E, & Weston, L. A. Manipulation of root hair development and sorgoleone production in sorghum seedlings. Journal of Chemical Ecology (2004). , 30(1), 199-213.

[94] Weston, L. A, & Duke, S. O. Weed and crop allelopathy. Critical Reviews in Plant Sciences (2003).

[95] Frey, M, Chomet, P, Glawischnig, E, Stettner, C, Grün, S, Winklmair, A, Wolfgang, E, Bacher, A, Meeley, R. B, Briggs, S. P, Simcox, K, & Gierl, A. Analysis of a chemical plant defense mechanism in grasses. Science (1997). , 277(3526), 696-699.

[96] Mahmoud, S. S, & Croteau, R. B. Strategies for transgenic manipulation of monoterpene biosynthesis in plants. Trends in Plant Science (2002). , 7(8), 366-373.

Chapter 21

Managing Commelina Species: Prospects and Limitations

Wendy-Ann Isaac, Zongjun Gao and Mei Li

Additional information is available at the end of the chapter

http://dx.doi.org/10.5772/55842

1. Introduction

Commelina species, notably *C. communis* L, *C. diffusa* Burm, *C. elegans* Kunth. and *C. benghalensis* L. as well as their biotypes, are perennial herbs of Neotropical origin which now have a pantropical distribution. Members of this family (Commelindeae: Commelinaceae) are common throughout the Caribbean, North and Latin America, Africa, Asia, the Middle East and parts of Oceania [18, 27, 28, 63, 64]. There are 500 - 600 species reported in the family Commelinaceae [50]. Recent data indicates that the Commelinaceae family contains 23 genera and at least 225 species native to or naturalized in the New World and 23 genera and about 200 species in the Neotropics [41] and also website reports of 50 genera and 700 species [16, 31]. There are 170 species of Commelina in the warmer regions of the world and 50 species of Murdannia occurring in the tropics and warm temperate regions worldwide with Tropical Asia having the greatest diversity [17].

Wilson [84] presented a comprehensive review on Commelina species and its management with emphasis on chemical weed control in 1981. Since Wilson's review much has been written about the weedy members of this family, notably Commelina species [84]. Indeed, the CAB ABSTRACTS Database contains well over 1200 references on Commelinaceae from 1981 to the present. *Commelina benghalensis* in particular has been the most reported species with several reports of research conducted on its control in southern states of the United States of America (USA) including Alabama, Florida, Georgia, Louisiana and North Carolina [18, 74, 75, 78-81]. Many of these studies should be consulted for basic details of the biology and ecology. The National American Plant Protection Organization (NAPPO) offers a comprehensive global distribution list of this weed species [47].

The current review is an attempt to provide an update on the status of the weedy Commelina species in agricultural production systems. This review is based on world literature over the last 45 years and considers major Commelina species found in the tropics and warm temperate regions in relation to their status, distribution, biology and spread and management.

2. Weed Status

Commelina benghalensis (Tropical spiderwort or Benghal dayflower) has become increasingly important, gaining pest significance in agronomic production systems in the southeastern coastal plain of the United States of America (USA) in crops such as cotton (*Gossypium* spp.) and peanut (*Arachis hypogea*) [70, 71] and in the North China Plain in crops such as potato (*Solanum tuberosum*) and summer corn (*Zea mays*) [37, 71, 72, Li et al. unpublished data 2007). It is commonly associated with wet locations. This weed was in fact listed as a Federal Noxious weed in Florida and Georgia where it is the most troublesome weed in cotton and a pest in peanut, corn (*Zea mays*), soybean (*Glycine max*), nursery stock and orchards [81]. This species which was first observed in USA in 1928 [18] gained noxious weed status in 1983 [81]. Between 1998 to 2001 and then to 2004 this weed which was ranked among the top 39 most troublesome weeds across all crops by Georgia extension agents (in 1998) moved to the 9th most troublesome (in 2001) to the most troublesome cotton weed in Georgia (in 2003) [77] and Florida (2004) and the 3rd most troublesome weed of peanut in several south Georgia counties [54, 80]. In Georgia alone the weed is estimated to infest more than 80,000 ha [80-82] with a confirmed presence in 29 Georgia counties [54]. It is also observed throughout the panhandle and central Florida and listed by the United States Department of Agriculture (USDA) as appearing in more than 12 Florida counties [82].

Commelina communis has become one of the three most troublesome weeds in soybean fields in the Northeast China, and has caused significant reduction in production and quality of soybean [42]. Commelina species, namely *C. diffusa* and *elegans,* were reported as the 3rd most troublesome weed in the Caribbean where they are a serious problem of banana and other crops in the Windward Islands of Dominica, Grenada, St. Lucia and St. Vincent and the Grenadines [24]. Presently, Commelina species, commonly called watergrass, caner grass, pond grass, spiderwort, spreading dayflower, wandering Jew or French weed in these Islands, are by far the most serious in these countries. *Commelina diffusa* was once encouraged as a ground cover to reduce soil erosion [13] and has been identified as the host of the reniformis nematode *Rotylenchulus reniformis* [57], the banana lesion nematode *Pratylenchus goodeyi* [87] and recent data have confirmed its association with the burrowing nematode *Radopholus similis* [55]. These nematodes all contribute to significant reductions in banana production particularly *R. similis,* which may reduce banana production by more than 50 % and decrease the production duration of banana fields [55].

3. Biology and spread

Commelina species are C-3, monocotyledonous plants and therefore have a high efficiency of CO_2 uptake at low irradiance [34]; therefore, they tolerate shade very well and could become persistent. They are both annuals and perennials and therefore dominate the fallow vegetation because they are most competitive due to their growth and regeneration characteristics [72].

The plant is propagated mainly by seeds, stem cuttings and rooting from nodes and pieces [19, 46, 74, 75]. Plants may arise asexually when buds grow into autonomous, adventitiously erect leafy shoots, which later become separated from each other [12]. Occasionally the buds may sprout and grow into erect shoots directly without undergoing a period of inactivity [12]. The plant produces roots readily at the nodes of the creeping stems and will do so especially when broken or cut [27, 28]. Farmers in the Windward Islands report that Commelina species may be intensified when cut with a weed whacker as stolons spread more extensively.

The stems of Commelina species have a high moisture content and once it is well rooted the plant can survive for long periods without moisture [84]. This fact is evident in young banana plantations in the Windward Islands where stems become dried and shrivelled due to the direct contact with solar radiation particularly in the dry season. However, at the onset of rains and when the canopy of the banana closes, stems regain moisture, re-establish and rapidly begin to spread by runners which root at the nodes.

The mature aerial seeds of *C. benghalensis* are produced within 14 to 22 days after flower opening [74] and in some instances, e.g., the rice paddies of the Philippines, can produce in excess of 1,600 seeds/plant [53] or even 12,000 seeds/m^2 [74], whereas seeds grown from underground seeds are capable of producing 8,000 seeds/m^2 [74]. In cultivated areas the plant is spread by irrigation water and waterways. Animals may also spread the seeds.

Commelina species has gained noxious weed status in the Windward Islands because of several factors. Firstly, the fact that the weed was encouraged as a groundcover was compounded by inappropriate agricultural practices, notably irrational herbicide use which farmers have relied on for decades. The non-judicious use of herbicides has created imbalances and disturbances within the ecosystem in these Islands causing resistant biotypes. Secondly, the move within recent years by banana growers to adopt a Fairtrade system which uses no herbicides has catapulted the spread to an all-time high in the Windward Islands. Farmers have been forced to rely on the use of the cutlass or weed whacker as the only alternative strategies which have further intensified the problem by spreading plant propagules [30]. Most importantly these Islands which are characterized by hilly landscapes have ideal moist conditions for the proliferation of Commelina species. Finally, many of the banana plantations have been farmed for several years with virtually no crop rotations or tillage practices and this has further contributed to the stabilization of Commelina species populations.

In the USA, its sudden emergence as a noxious weed is attributed to crop production practices which are well suited for prolific weed growth such as minimum – tillage production (which is undertaken in conjunction with the use of glyphosate – resistant crops) and extreme tolerance to glyphosate [79-81]. The weed appears to be well-suited for high input agricultural production where high levels of fertilizers, irrigation and herbicides are used [79, 80]. The spread of *C. benghalensis* is attributed, in part, to the adoption of weed management programmes that lack the use of residual herbicides along with the adoption of reduced-tillage production practices [54]. Additionally, after introduction, invasive species often go long periods of time (lag period) during which the pest increases in distribution or density without being noticed as an obvious pest [54].

4. Economic impact in crop production

Three species of the Commelinaceae family are considered to be major problem weeds in cropping systems where they have become persistent and difficult to manage [27]. *Commelina benghalensis* is the most important of the three and it occurs as a weed in 25 different crops in 28 countries [27]. This weed has gained high importance in peanut and cotton in the southern United States [78, 79]. *Commelina diffusa* occurs as a weed in 17 crops in 26 countries and *Murdannia nudiflora* occurs as a weed in 16 crops in 23 countries [27].

Commelina diffusa thrives on cultivated soils of cocoa (*Theobroma cacao*), citrus, root crops such as dasheen (*Colocasia esculenta*) that tolerate water, and it is also a major weed in sugarcane (*Saccharum officinarum*), upland rice (*Oryza sativa*), soybean (*Glycine max*), cassava (*Manihot esculenta*), corn (*Zea mays*), banana and plantain (*Musa* spp.) [27]. *Commelina benghalensis* has been reported as a principal weed in upland rice in India and the Philippines, tea (*Camellia sinensis*) in India, coffee (*Coffee arabica*) in Tanzania and Kenya, soybean in the Philippines and cotton and maize in Kenya [27, 47]. This species is common in rice in Sri Lanka, sugarcane in India, the Philippines and Mozambique, cassava in Taiwan and maize in Zimbabwe [9]. *Commelina benghalensis* was reported as a weed of jute (*Corchorus olitorius*), sisal (*Agave sisalana*), beans (*Phaseolus* spp.), pastures, sweet potatoes (*Ipomoea batatas*), vineyards and barley (*Hordeum vulgare*) and other cereals in many countries [7].

Because of Commelina's vigorous growth habit, which allows the plant to form dense pure stands, they may compete easily with low growing crops such as vegetables, pulses and cereals as well as pasture grasses and legumes by smothering them [27]. Because Commelina species is a broadleaved weed it is generally not considered highly competitive for nutrients however this fact is not well researched and its allelopathic potential also needs to be ascertained. Invasive species such as *C. benghalensis* had higher plant growth rate at high nutrient availability and across water availability compared to a related non – invasive, but alien, congener, *C. bracteosa* Hassk. [6]. Interestingly, severe stunting has been reported in *C. diffusa* caused by high nitrogen [59] and altered growth and physiological characteristics for different *C. erecta* clones with increased phosphorus supply [71]. Results from systematic studies on the influence of *C. benghalensis* populations on crop yield are limited [54]. Increased reduction in above-ground and root dry matter as well as a 100% reduction in the number of leaves in lettuce (*Lactuca sativa*) plants were recorded with 1% and 3% hydro – alcoholic extracts of *C.benghalensis* suggesting its allelopathic potential [68].

Studies on the critical periods of interference in Commelina species are limited. Generally crops are affected most severely during the first 2 – 5 weeds of crop growth although mature plants can also be affected [7]. *Commelina benghalensis* in particular may affect crop growth and yield but this varies with environmental conditions [47]. Research aimed at evaluating the periods of interference of *C. benghalensis* in the initial growth of coffee seedlings reported prevention periods of 15 to 88 and 22 to 38 days after coffee seedling sowing under winter and summer conditions, respectively [11]. In cotton it was found that yield loss from *C. benghalensis* can be minimized by planting cotton early in the growing season, prior to substantial emergence of the weed [81].

5. Pests and diseases associated with commelina species

Commelina diffusa is an alternate host plant for the nematodes *Rotylenchulus reniformis*, *Helicotylenchus* spp., *Pratylenchus* spp., *Meloidogyne* sp. and *Radopholus similis* in banana [13, 27, 29, 44, 55, 57, 60, 87] and coffee [58]. The plant is also a collateral host of *Helicotylenchus dihystera* infecting guava fields [35]. *Commelina benghalensis* has also been identified as an alternate host of the southern root-knot nematode (*Meloidogyne incognita*) [55]. The southern root-knot nematode is widely distributed across cotton regions in Georgia [54]. Snails and slugs feed on *C. diffusa* plants and these affect crops such as pineapple and soybean [84].

Five viruses have been found naturally infecting species of Commelinaceae. Aneilema a potyvirus has also been found infecting 15 species of the Commelinaceae family including 4 of Commelina. There have been reports of *Commelina diffusa* potyvirus, which causes a mosaic in *Commelina diffusa* and *C. benghalensis* [2]. The virus is transmitted by two insect vectors, *Aphis gossypi* and *Myzus persicae;* Aphididae. It is transmitted in a non – persistent manner. The virus is transmitted by mechanical inoculation and not by grafting or contact between plants or by seeds. The isolate for cucumber mosaic virus (CMV) is originally from *Commelina elegans* but it is transmitted by Aphis gossypi, and not *Myzus persicae. Commelina diffusa* is susceptible to Commelina X potexvirus, Commelina yellow mottle badnavirus, Spring beauty latent bromovirus, Tradescantia – *Zebrina potyvirus,* spotted wilt and Cherry leaf roll nepovirus [2]. However, *Commelina elegans* is insusceptible to Tradescantia – *Zebrina potyvirus*. U2- tobacco mosaic virus has also been found infecting *C. communis* and *Z. pendula*. Brome mosaic virus isolates have been identified [70] infecting *C. diffusa* and *C. communis* in Fayetteveille, Arkansas, USA.

6. Methods of management in selected crops

Wilson's review on the control of these weed species was directed towards finding suitable chemicals for their control in the early stages of growth, summarizing results of trials from difference parts of the world [84]. However, he suggested that since dense mats of plant material make chemical weed control of older plants difficult, removal by hand is the only effective control at that stage [84].

Currently, chemical control is still generally considered the only practical means of controlling large infestations of Commelina species [78-82]. However, no single method of control seems to be effective for control of Commelina spp. in any crop. The difficulty lies in its ability for regeneration after attempted management even by cultural, mechanical or chemical control.

An Integrated Management Strategy (IWM) is therefore suggested for the best control of this weed species. A multi-component approach including an effective herbicide for successful management has been suggested [80-82].

7. Chemical management

Herbicides are not usually very effective against most Commelina species. The first verified resistance was registered in 1957, when *C. diffusa* biotypes were identified in the United States [26]. *Commelina elegans* has shown resistance to growth – regulator type herbicides [32]. Control using herbicides is, however, variable depending on the herbicide, accuracy of leaf coverage and environmental conditions [7]. Spraying with a selective or non – selective herbicide may work but repeated treatments are required for regrowth. Plants should not be under moisture stress when sprayed. Surfactants will improve penetration into the waxy-coated leaves.

Many standard herbicides have relatively low activity on species of Commelina [84]. These include 2,4-D, propanil, butachlor, trifluralin and pendimethalin. Treatment with 2,4-D or MCPA at the pre-emergent stage has been shown to be ineffective and although a reasonable kill of very young seedlings can be obtained, the plants develop a rapid resistance with age [32]. Particular biotypes are resistant to 2,4-D and they may be cross resistant to other Group O / 4 herbicides [83]. It has been found that one biotype of *C. diffusa* could withstand five times the dosage of a susceptible species [83].

In rice, bentazone, molinate, oxyfluorfen and bifenox are herbicides with good activity [7]. Post-emergent sequential treatments of propanil followed by nitrogen or of molinate followed by KN3 controlled *C. diffusa* in rice [61]. In soybean, bentazone and metribuzin are effective [7]. In corn, combination of bromoxynil and 2,4-D butylate produced a synergistic effect in post-emergent control of 3-4 leaf stage *C. communis* [85]. In plantation crops such as banana, paraquat is not always effective but mixture with diuron is recommended [7]. Dinoseb has been found to kill seedlings as well as dalapon but paraquat is reported to be relatively ineffective [32]. Prodiamine has been reported to be effective in ornamental fern beds [62]. Extreme tolerance to glyphosate has been documented [54]. Glyphosate has been shown to be effective but additives or mixtures may be needed for good results at moderate doses [7]. However, *C. diffusa* has been reported to have larger possibilities of recovery after glyphosate application because of its larger starch reservation [71].

Resistance to residual herbicides has also been reported and relatively high doses of simazine and diuron appear to be necessary to achieve control [32]. Recent studies on use of residual herbicides have identified Dual Magnum® (s-metolachlor) (applied as a preplant incorporated, pre-emergent and post-emergent) as providing excellent residual control (>80%) of *C. benghalensis* in peanut [54]. Atrazine and Dual Magnum®, two commonly used corn herbicides used in the USA, also gave good to excellent residual activity on *C. benghalensis* [3]. The most effective herbicide control strategies for *C. benghalensis* involve combinations of both pre-emergence and postemergence conventional herbicides [54]. These include preemergence herbicides with residual activity such as Axiom® (flufenacet + metribuzin), Dual Magnum®

Canopy SP® (metribuzin + chlorimuron) and Sencor® (metribuzin) and postemergence herbicides with fair to good activity such as Basagran®, Classic® (acetochlor) and Pursuit® (Imazethapyr). Gramoxone Max® and Aim® (acetochlor) can be used post-directed. In evaluating the effectiveness of several pre-emergence herbicides in suppressing *C. benghalensis* emergence, it was reported that s-metolachlor (at 1.07 and 1.60 kg a.i./ha), clomazone (at 0.42 and 1.05 kg a.i./ha) and flumetron (at 1.68 kg a.i./ha) provided ≥ 80% control at 6 weeks after treatment (WAT) in cotton [80]. It was stressed that the application of herbicides with soil residual activity will be crucial for the management of *C. benghalensis* [80].

In the Windward Islands, farmers started using paraquat around 1989 and noticed that it was ineffective. In an interview on August 10, 2002, Paddy Thomas, an experienced banana grower and pesticide salesman in St. Vincent and the Grenadines revealed that farmers started using gramocil (paraquat + diuron) at high doses for example and this too was not effective and resistance in Commelina spp. began to show. He also stated that Reglone, Round – up and Talent (paraquat + asulam) have also been used with little success for the control of Commelina species in the Windward Islands. Glufosinate has since been promoted as an environmentally-friendly option for the control of broad-leaved weeds including Commelina species.

Studies were conducted into the efficacy of glufosinate for weed control in coffee plantations and it was found that it did not effectively control Commelina spp. at a rate of 0.3 – 0.6 kg a.i. / ha, however, paracol and gardoprim suppressed this perennial weed better [50]. Fomasefen and lactofen have shown good potential for control of this broadleaf weed [10]. Glufosinate (240 g a.i./ha) and fomasefen (WIP 276 g a.i./ha) were used in St. Vincent and the Grenadines in Fairtrade banana fields to compare their efficacy in controlling *C. diffusa* [30]. They were both applied at the early post-emergence, 3-5 leaf stage with a backpack sprayer using a TJ-8002 fan-nozzle. Regrowth of *C. diffusa* and other weeds were observed 6 weeks after application with glufosinate, however, no regrowth was observed for up to 3 months with fomasefen. Fomasefen, however, caused damage by burning banana suckers and leaves (about 30%) of established banana plants [30]. Studies were conducted to evaluate the efficacy of several post-emergence herbicides in controlling *C. communis* in soybean, the results showed that imazethapyr (150 g a.i./ha), cloransulam-methyl (31.5 g a.i./ha), fomesafen (375 g a.i./ha) and mixture (756 g a.i./ha) of fomesafen plus imazethapyr with clomazone provided > 80% control of this weed at 30 days after treatment (DAT) [36, 37, 65, 67]. The efficacy of imazethapyr (90 g a.i./ha) in controlling *C. communis* reduced with increased leaf stage, and the control levels at 15 DAT were 100% (at 1 leaf stage), 89.17% (at 2 leaf stage), 56.45% (at 3 leaf stage) and 52.71% (at 4 leaf stage), respectively [41]. Therefore, the optimal application time of imazethapyr was 1-2 leaf stage of *C. communis* [41].

To screen more suitable herbicides for control of *C. benghalensis* and *C. communis* and determine the level of weed control provided by a single application of selected post-emergence herbicides, greenhouse studies on the laboratory toxicity of 23 herbicides to these weeds were conducted in 2010 [21]. The results indicated that, as for *C. benghalensis*, mesotrione, lactofen, oxyfluorfen, clomazone and flumioxazin provide complete control (100%), oxadiazon, fomesafen, metribuzin, acifluorfen, isoproturon, MCPA-sodium, carfentrazone-ethyl, fluroxypyr, fluoroglycofen-ethyl and bentazone are herbicides with excellent activity (90.0 - 100%

control), paraquat, 2,4-D butylate, rimsulfuron and thifensulfuron-methyl are herbicides with good activity (80.0 - 90.0% control), and nicosulfuron, bensulfuron-methyl, dicamba and glyphosate-isopropylammonium are relatively ineffective (< 80.0% control) at their own recommended dose, respectively. As for *C. communis*, mesotrione and thifensulfuron-methyl provide complete control (100%); metribuzin, paraquat, carfentrazone-ethyl, 2,4-D butylate, nicosulfuron, MCPA-sodium, fluroxypyr, flumioxazin and acifluorfen are herbicides with excellent activity (90.0 - 100% control); rimsulfuron, lactofen and fomesafen are herbicides with good activity (80.0 - 90.0% control); and glyphosate-isopropylammonium, bensulfuron-methyl, fluoroglycofen-ethyl, bentazone, clomazone, oxadiazon, oxyfluorfen, isoproturon and dicamba are relatively ineffective (< 80.0% control) at their own recommended dose, respectively. There are 19 and 14 herbicides which provided good to excellent control (> 80%) to *C. benghalensis* and *C. communis* under greenhouse conditions, respectively. However, the performance of those herbicides applied in different crops to control *C. benghalensis* and *C. communis* also needs to be ascertained.

8. Cultural management

This method depends on the crop infested, land size, level of technology available, value of crop, labour availability and costs, availability of draft power and the associated equipment and availability of herbicides [47]. The document further indicates that the methods currently used include proper land preparation, hand hoeing and pulling, removing the plants from the fields and drying, use of ox-drawn and tractor drawn cultivation, slashing and herbicide application. *Commelina diffusa* is very difficult to control manually as the stolons are cut into small pieces which can easily regenerate. Hand weeding and rolling the weed up like a carpet is considered suitable for removal of small infestations [30], if care is taken to remove every last piece. In Uganda, it was reported that heaping of stubborn weeds of Commelina plants is practical during the rainy season to speed up rotting and reduce the frequency of weeding [48]. In the dry season, heaps are then scattered as the dry conditions desiccate Commelina stems rapidly. A small percent of Ugandan farmers (5.9%) dig ditches and bury Commelina species, turning it into manure. Some farmers in St. Vincent have also tried this technique in the field with varying success.

A potential solution to overcoming Commelina weed infestations in banana is by intercropping with a fast, low – growing shade tolerant cover crop. This can be done by intercropping with melons, *Mucuna pruriens* (negra and ceniza), tropical alfalfa, *Cajanus cajan, Vigna radiata* (mung bean), *V. unguiculata* (cowpea), *Crotalaria juncea, Indigofera endecaphylla, Phaseolus trinervius,* and *Ipomea batatas* (sweet potato) which have rapid canopy coverage to suppress the establishment of weeds. Melon (*Colocynthis citrullus* L.) planted at a density of 5,000 plants/ha suppressed weed growth of *Commelina diffusa* for five months, enhancing establishment and yield of melon in Nigeria [49]. Use of vigorous healthy planting material and close spacing of the crop may also be used. It has been shown that spacings of 1.2 x 1.2 m (6,944 plants/ha) and 1.5 x 1.2 m (4,444 plants/ha) gave high yields and "natural" control of these weeds [8, 66].

Field studies conducted in St. Vincent and the Grenadines in 2003/2004 compared several treatments including 3 cover crops in suppressing *Commelina diffusa* weed infestations in banana at 63 days after application (DAA) [30]. The cover crops included *Arachis pintoi* (wild peanuts) which was sown by seed and stem cuttings, 16 cm apart, *Mucuna pruriens* (velvet beans) drilled 30 cm apart and *Desmodium heterocarpon* var *ovalifolium* (CIAT 13651) broadcast at a rate of 5 kg/ha. Best results were obtained from *Desmodium heterocarpon* (86.7%) followed by *Arachis pintoi* (52.1%) and *Mucuna pruriens* (43.3%). *Desmodium heterocarpon* was also found to be competitive to *C. diffusa* significantly suppressing its growth in Farmer Participatory Research trials also conducted in St. Vincent in 2005/2006 [30].

Mulching is another viable option for management of the weed. Mulching with rice straw, cut bush, grass, coffee hulls, water hyacinth or even the dead or senescent banana leaves, pruned suckers and old stems could significantly suppress weed growth. Black plastic mulch also provides good weed control as it stifles weed seed growth and development when light penetration is reduced. There are no reports of work done on the use of these mulches for suppression of Commelina species. In field studies in St. Vincent and the Grenadines in 2003/2004 three dead mulches were compared using senescent banana leaves (traditional practice of farmers) applied to a depth of 3-5 cm, coffee hulls applied to a depth of 3-5 cm and black plastic polyethylene tarp at 1.0 mils thickness [30]. Results indicate a 94.5% and 95.6% suppression of weeds including *C. diffusa* with coffee hulls and banana mulch treatments respectively and 100% suppression with black plastic mulch.

9. Mechanical management

Commelina diffusa is particularly difficult to control by cultivation, partly because broken pieces of the stem readily take root and underground stems with pale, reduced leaves and flowers are often produced [32]. The plant is easy to rake up, roll up or hand pull and very small infestations can be dug out. It can be bagged and well baked in the sun, however, follow – up work is essential as any small fragment of the stem remaining will regrow and needs to be removed and destroyed off - site. Mechanical control using the weed whacker may also contribute the spread of stem cuttings in addition to damaging the banana root system as much of the plant lies within the top 15 cm of the soil [30].

To investigate the effect of cutting and depth on the regeneration potential of *C. diffusa* greenhouse studies were conducted in 2004/2005 (Isaac et al. unpublished data 2005) using three cutting types: tip cuttings (2 nodes, 2 leaves), 2 node pieces only and 1 node, 1 leaf piece buried at depths including 0 (control), 2.5, 5.0 and 7.0 cm to demonstrate emergence patterns. These cuttings were intended to simulate cuttings made from a weed whacker and the practice of burying the weed. Regeneration was observed from all cuttings from 0 – 5.0 cm depths but no growth was observed at 7.0 cm. *C. diffusa* dry matter (DM) was highest at surface level (0cm - control) for all cuttings and reduced with increased depth. Results indicate that for effective management of *C. diffusa* by cutting, nodes must be reduced to less than half with no leaves which may starve the plants' photosynthetic ability and hence suppress regeneration. Burial should be up to 5.0 cm to ensure that there is no emergence of the weed. Similar studies [5] indicated that cuttings buried deeper than 2 cm failed to regenerate.

Research has shown that soil solarization, a hydrothermal process of heating moist soil, can successfully disinfect soil pests and control weeds [1, 4, 15, 56]. Soil solarization by covering with plastic sheeting for 6 weeks in the warmer months will weaken the plant. After removing the plastic any regrowth can be dug out or sprayed, however, this method will not be effective in full shade. Solarization can be used alone or in combination with other chemicals or biological agents as the framework for an IPM programme for soilborne pests in open fields. In field trials in St. Vincent, soil solarization using clear polyethylene plastic at 0.5 mils under Fairtrade banana plants showed variable suppression of *C. diffusa* as the weed emerged under the clear plastic showing chlorotic and suppressed growth symptoms, resuming its full growth potential after removal of the plastic covering 2 months after application (Isaac et al. unpublished data 2005). Seed germination of *C. benghalensis* was found to increase by soil solarization in studies conducted in Brazil [43].

10. Organic management

Attempts have also been made to find organic treatments for control of Commelina species in banana in St. Vincent and the Grenadines [30]. DTE corn weed blocker (corn gluten meal) pre-emergent weed blocker and slow release fertilizer (9-1-0) which controls emerging weeds was applied at a rate of 10 kg/ha. Burnout® (concentrated vinegar and acetic acid) (20%), urea (20%), and fertilizer solution (20%) were also used to evaluate their efficacy on the control of Commelina species and other weed species. All treatments showed varying levels of control for up to 3 weeks. Best results were obtained from Burnout® which caused phytotoxic damage on the leaves of actively growing plants offering 43% control. This was followed by urea (41%), fertilizer solution (34%) and corn weed blocker (20%). Urea, fertilizer and corn weed blocker treatments resulted in the general stunting of plants in addition to the burning of leaves. However, stems and roots remained intact. Similar results using treatments high in nitrogen were obtained in Russia [59] where seed production of *C. benghalensis* and stunted growth under artificial dense competition in cereals resulted. These results indicate that there is no evidence that this Commelina species competes for nitrogen. In fact the species does not pose any threat in competing for nutrients with banana. Repeat applications of these treatments are therefore necessary for the effective management of Commelina species in organic farming systems.

Studies conducted in Brazil in soybean-wheat rotations under no-tillage conditions showed reductions in the seedbank of *C. benghalensis* in areas infested with *Brachiaria plantaginea* [73]. Analysis of the soluble fraction of *B. plantaginea* indicated a predominance of aconitic acid (AA) among the aliphatic acids and ferulic acid (FA) among the phenolic acids. Laboratory bioassays using *C. benghalensis* were carried out to evaluate phytotoxic effects of pure organic acid solutions and dilute extracts of *B. plantaginea* on seeds germination, root development and fungal germination and AA and FA solutions and the extract of *B. plantaginea* extract reduced germination and root length of *C. benghalensis* [73]. Both AA and FA have the potential for use as bio-herbicides.

11. Biological management

There have not been many reports on biological control of Commelina species. *Commelina diffusa* is grazed by small ruminants, pigs and cows. Because this species is very fleshy and has a high moisture content, it is difficult to use it as fodder for domestic stock [27]. However, recent research has indicated that *C. diffusa* compared well with many commonly used fodder crops and could contribute as a protein source for ruminants on smallholder farms [30]. There have also been reports of foraging of this weed by *Gallus domesticus* (chickens) [30].

There are no reports of promising insect candidates for biological control reported on Commelina spp. in the USA [63, 64]. In Korea and China there have been reports of *Lema concinnpennis* and *Lema scutellaris* (Coleoptera: Chrysomelidae) two leaf-feeding species on *C. communis* [86]. *Noelema sexpunctata* (Coleoptera: Chrysomelidae) another leaf-feeding species was also reported on *C. communis* [45].

In Central Virginia, USA, *Pycnodees medius* (Hemiptera: Miridae) was found to cause tissue necrosis on *C. communis* [33]. Various insects were also screened for their potential as biocontrol agents of weeds in rice and it was found that *Necrobis ruficollis* (blue beetle), *Rhaphidopalpa africana* (yellow beetle), *Conocephalus* sp., *Tetragrnathidae* spp. and *Paracinema tricolor* (grasshopper) were promising [45]. Feeding and nymphal development (up to 3rd and 4th instar) of *Cornop aquaticaum* (grasshopper) were reported on *C. africana* L., and *Murdannia africana* (Vahl.) [25]. It was also observed that *Rhaphidopalpa africana* beetles fed more than the others on the weed, *C. benghalensis* L. [25].

There are records of agromyzid leaf miners which may be promising sources of candidate biological control agents [75]. *Liriomyza commelinae* (Diptera: Agromyzidae), a leaf-miner, was however reported on *C. diffusa* in Jamaica [20, 61]. *Commelina diffusa* is the main food plant of *L. commelinae*, however, it is susceptible to predation by the formicid: *Crematogaster brevispinosa* as well as competition and exposure to the sun (high temperatures) which causes high mortality [20].

There are prospects for the management of invasive alien weeds in Latin America using co-evolved fungal pathogens in selected species from the genera Commelina [14]. Pathogens recorded in the native range of Commelina species include: *Cercospora benghalensis* Chidd., *Cylindrosporium kilimandscharium* Allesch. (Hyphomycete), *Kordyana celebensis* Gaum, (Exobasidiales: Brachybasidiaceae), *Phakopsora tecta* H.S. Jacks and Holw (Uredinales: Phakopsoraceae), *Septoria commelinae* Canonaco (Coelomycete), *Uromyces commelinae* Cooke (Uredinales: Pucciniaceae), *Phoma herbarum* [14, 23, 76]. These mycobiota would appear to be good potential agents for classical biological control (CBC) [14]. Although some of the most promising (e.g. the rusts *Phakopsora tecta* and *Uromyces commelinae*) are already present in the New World, they are restricted to certain regions and could be redistributed [14]. The uredinal state of a rust was found widespread on *C. diffusa* in Hawaii [22] sometimes causing death of parts above ground. Studies aimed at identifying mycoherbicidal biocontrol agents have been conducted in Brazil on three endemic pathogens of *C. benghalensis* which were: a bacterium (Erwinia sp.) and two fungi (*Corynespora cassiicola* and *Cercospora* sp.) [38, 39].

12. Conclusion and recommendations

The Commelina species are very persistent, noxious weeds which must be managed using an integrated approach to weed management. Weed management strategies that are narrowly focused will ultimately cause shifts in weed populations to species that no longer respond to the strategy resulting in adapted species, tolerant species or herbicide-resistant biotypes [51], which is the case with Commelina species in cropping systems. The integrated approach should utilize alternative strategies such as those mentioned in this paper including the most practical options, cultural and mechanical not negating the judicious use of herbicides. Such combinations should provide significant management levels of Commelina species for both conventional as well as organic growers using a pesticide free production PFP approach. Utilization of the useful benefits of Commelina species after uprooting will also serve to check the heavy use of herbicides in cropping systems.

The integrated approach must begin very early as once an infestation is really entrenched it presents several difficulties because of the pernicious growth habit of this weed. Successful management of *C. benghalensis* will require a multi-component approach including an effective herbicide that provides soil residual activity [80]. Recent studies on the management of Commelina species have, however, still focused primarily on effective herbicides and herbicide mixtures for their control despite hard evidence of the development of herbicide-resistant biotypes. Additionally, the adoption within recent years of GM crops particularly herbicide – resistant crops presents serious issues involving their negative ecological impact as already there are reports of Commelina species prominence in some agroecosystems due to simple and significant selection pressure brought to bear by these herbicide – resistant crops and the concomitant use of the herbicide [52].

The best way to control Commelina species for small holders in developing countries would be by implementing an integrated approach that embraces a variety of options which should be attuned to the individual farmer's agronomic and socio – economic conditions (soil type, climate, costs, local practices and preferences). For example, in banana growing areas in the Windward Islands, the growth of the weed can be suppressed by a single application of a herbicide or weed whacking very early before extensive spread of the weed followed by planting a competitive cover crop like *Desmodium heterocarpon* that would not only prevent re-invasion but improve soil fertility.

Future research in developing effective management strategies for *Commelina benghalensis* should:

- Develop an accurate predictive model for *C. benghalensis* germination

- Evaluate the seedbank longevity of *C. benghalensis*
- Determine the primary dispersal mechanism(s)
- Characterize the environmental limits of *C. benghalensis* in the U.S.A. [80].

Surely this list can be expanded to include other Commelina species such as *C. diffusa* which is definitely a problematic weed in the cropping systems in the Windward Islands. The research direction should also:

- Determine threshold levels of *C. diffusa* in crops such as banana
- Evaluate the allelopathic potential of Commelina species by extracting hydro alcoholic compounds which could be used as a possible bioherbicide in controlling other problem weeds
- Screen for mycobiota with good potential for CBC such as the rust species Uromyces commilinae which has been identified in several Caribbean Islands.
- Determine the reasons for reduced seed production of *C. diffusa* species found under banana fields in the Windward Islands as compared to higher seed numbers (both aerial and underground) of *C. benghalensis* species in the USA.

Author details

Wendy-Ann Isaac[1], Zongjun Gao[2] and Mei Li[2]

1 Department of Food Production, Faculty of Food and Agriculture, The University of the West Indies, St. Augustine, Trinidad

2 Institute of Plant Protection, Shandong Academy of Agricultural Sciences, Jinan, China

References

[1] Abu-irmaileh, B. E. Weed control in vegetables by soil solarization, In FAO Plant Production and Protection Paper (Amman, Jordon), (1991).

[2] Baker, C. A, & Zettler, F. W. Viruses infecting wild and cultivated species of the Commelinaceae. Plant Disease (1988). , 72(6), 513-518.

[3] Barnes, J. Managing hairy wandering Jew. Queensland Government, Department of Primary Industries Publication #QL03056; (2003).

[4] Benjamin, A, & Rubin, B. Soil solarization as a means of weed control. Phytoparasitica (1982).

[5] Budd, G. D, Thomas, P. E. L, & Allison, J. C. S. Vegetative regeneration depth of germination and seed dormancy in *Commelina benghalensis* L. Rhodesian Journal of Agricultural Research (1979). , 17, 151-153.

[6] Burns, J. H. A comparison of invasive and non-invasive dayflowers (Commelinaceae) across experimental nutrient and water gradients. Diversity and Distributions (2004). , 10, 387-397.

[7] CABICrop Protection Compendium, Global Module, (2002). edition. Wallingford, UK. CAB International. Available from: http://www.cabi/compendia/cpc/index.htmaccessed 15 January 2007)

[8] Chako, E. K, & Reddy, A. Effect of planting distance and intercropping with cowpea on weed growth in banana. In: Proceedings of the 8th Asian-Pacific Weed Science Society Conference. (1981). , 137-141.

[9] Chivinge, O. A, & Kawisi, M. The effect of node number on the regeneration of wandering Jew (*Commelina benghalensis* L.). Zimbabwe Journal of Agricultural Research (1989). , 27(2), 131-138.

[10] Carmona, A. Flex (fomasafen) and cobra (lactofen): Two products with potential for broadleaf weed control for leguminous covers in oil palm plantations. ASD Oil Palm Papers (1991). , 4, 1-5.

[11] Dias, T. C. S, Alves, P. L. C. A, & Lemes, L. N. Interference periods of *Commelina benghalensis* after coffee establishment. Planta daninha (2005). , 23(3), 398-404.

[12] Duke, J. A, & Ayensu, E. S. Medicinal plants of China. Reference Publications, Inc. (1985).

[13] Edmunds, J. E. Association of *Rotylenchulus reniformis* with 'Robusta' banana and Commelina sp. roots in the Windward Islands. Tropical Agriculture, Trinidad (1971). , 1971(48), 1-55.

[14] Ellison, C. A, & Barreto, R. W. Prospects for the management of invasive alien weeds using co-evolved fungal pathogens: a Latin American Perspective, Biological Invasions (2004). , 6(1), 23-45.

[15] Elmore, C, & Heefketh, K. A. Soil solarization an integrated approach to weed control. Proceeding of the 35th Annual California Weed Conference, 143. Department of Botany, California University, Davis CA 95616, USA. (1983).

[16] Explore biodiversity and the wild classroomCommelinaceae: (Spiderwort family). (2002). http://www.explorebiodiversity.com/Plants/Commelinaceae.htmaccessed 10 February 2006)

[17] Fish, L. Commelinaceae. In O.A. Leistner (ed.) Seed Plants of Southern Africa. Strelitza. National Botanical Institute, Pretoria. (2000). , 591-593.

[18] Faden, R. B. The misconstrued and rare species of Commelina (Commelinaceae) in the eastern United States. Annals of Missouri Botanical Gardens (1993). , 80, 208-218.

[19] Fournet, J, & Hammerton, J. L. Weeds of the Lesser Antilles. Institute of National Research Agronomy, Paris, France. (1991).

[20] Freeman, B. E, & Smith, D. C. Variation of density-dependence with spatial scale in the leaf-mining fly *Liromyza commelinae* (Diptera: Agromyzidae). Ecological Entomology (1990). , 15, 265-274.

[21] Gao ZongJunLi Mei and Gao XingXiang. Laboratory toxicity of 20 herbicides against Bengal dayflower (*Commelina bengalensis* L.). (Abstract) VI International Weed Science Congress, Hangzhou, China, 17- 22 June, 2012. Published by the International Weed Science Society, (2012). , 140.

[22] Gardener, D. E. Rust on *Commelina diffusa* in Hawaii. Plant Disease (1981). , 65(8), 690-691.

[23] Gu ZuMin, Ji MingShan, Li XingHai and Qi ZhiQiu. Effects of environmental factors on effectiveness of *Phoma herbarum* strain SYAU-06 against *Commelina communis*. *Chinese* Journal of Biological Control (2009). in Chinese with English abstract)., 25(4), 355-358.

[24] Hammerton, J. L. Weed Problems and Weed Control in the Commonwealth Caribbean, Tropical Pest Management (1981). , 27(3), 379-387.

[25] Hill, M. P, & Oberholzer, I. G. Host specificity of grasshopper, *Cornops aquaticum,* a natural enemy of water hyacinth. In Proceedings of the X International Symposium on Biological Control of Weeds. Spencer, N.R. (Ed.) 4- 14 July 1999, Montana State University, Bozeman, Montana, USA. , 349-356.

[26] Hilton, H. W. Herbicide Tolerant Strains of Weeds. Honolulu, HI: Hawaiian Sugar Planters Association Annual Rep. (1957). , 69-72.

[27] Holm, L. G, Pluknett, D. L, Pancho, J. V, & Herberger, J. P. The World's Worst Weeds: Distribution and Biology. The University Press of Hawaii, Honolulu. (1977).

[28] Holm, L. G, Doll, J, Holm, E, Pancho, J. V, & Herberger, J. P. World Weeds, Natural Histories and Distribution. JohnWiley and Sons, New York, (1997).

[29] Inserra, R. N, Dunn, R. A, Mcsorley, R, Langdon, K. R, & Richmer, A. Y. Weed hosts of *Rotylenchulus reniformis* in ornamental nurseries of Southern Florida. Nematology Circular (Gainesville) (1989).

[30] Isaac, W. A. I, Brathwaite, R. A. I, Cohen, J. E, & Bekele, I. Effects of alternative weed management strategies on *Commelina diffusa* Burm. infestations in Fairtrade banana (*Musa* spp.) in St. Vincent and the Grenadines. Crop Protection (2007a). , 26-1219.

[31] Isaac, W. A. I, & Brathwaite, R. A. I. *Commelina* species- Review of the weed status of the genus and possibilities for alternative weed management in the tropics. AgroThesis(2007b). , 5(1), 3-18.

[32] Ivens, G. W. East African Weeds and their Control. Oxford University Press, Nairobi, Kenya. (1964).

[33] Johnson, S. R. *Commelina communis* (Commelinacea) as host to *Pycnoderes medius* knight (Hemiptera: Miridae) in central Virginia, USA. The Entomologist (1997). , 116, 205-206.

[34] Kennedy, R. A, Eastburn, J. L, & Jensen, K. G. C. C4 evolution of intermediate characteristics. American Journal of Botany (1980). , 67, 1207-17.

[35] Khan, R. M, Kumar, S, & Reddy, P. P. Role of plant parasitic nematodes and fungi in guava wilt. Pest Management in Horticultural Ecosystems (2001). , 7(2), 152-161.

[36] Liu Bo, Li HaiYan, Li Feng and Wang XianFeng.Dayflower control efficacy of a fomesafen plus imazethapyr plus clomazone 18% EC package mixture in soybeans. Agrochemicals (2006). in Chinese with English abstract)., 45(9), 636-638.

[37] Li Wei Dong, Wang Guang Xiang and Zhang Ge Chuan.Control effect of fomesafen 250 g/L SL on broadleaf weeds in soybean field. Journal of Anhui Agricultural Sciences (2011). in Chinese with English abstract)., 39(18), 10934-10935.

[38] Lustosa, D. C, Oliveira, J. R, & Barreto, R. W. Ocorrência de uma bacteriose em *Commelina benghalensis* (L) no Brasil, In 33 Congresso Brasileiro de Fitopatologia, 2000, Belém. Fitopatologia Brasileira (2000). , 25, 324-324.

[39] Lustosa, D. C, & Barreto, R. W. Primeiro relato de Cercospora commelinicola Chupp em Commelina benghalensis L. no Brasil. In: 34 Congresso Brasileiro de Fitopatologia, (2001). São Pedro. Fitopatologia Brasileira 2001; , 26, 364-364.

[40] Lu Xing Tao, Zhang Tian Tian, Zhang Yong, Kong Fan Hua, Ma Wei Yong, Ma Shi Zhong and Zhang Cheng Ling.Weed control efficacy and potao safety of rimsulfuron. Agrochemicals (2011). in Chinese with English abstract)., 50(11), 845-847.

[41] Ma Hong, Guan ChengHong and Tao Bo.The tolerance to imazethapyr in different leaf stages of dayflower (*Commelina communis* L.). Acta Phytophylacica Sinica (2009). in Chinese with English abstract)., 36(5), 450-454.

[42] Ma Hong, Guan ChengHong and Tao Bo.Tolerance to imazethapyr and physiological difference of dayflower (*Commelina communis* L.) at different leaf stages. Chinese journal of oil crop sciences (2010). in Chinese with English abstract)., 32(1), 136-138.

[43] Marenco, R. A, & Castro-lustosa, D. Soil solarization for weed control in carrot. Pesquisa Agropecuária Brasileira, Brasilia (2000). , 35(10), 2025-2032.

[44] Mead, F. W. Bureau of nematology: detections of special interest. Triology Technical Report (1990). , 29(1), 3-6.

[45] Morton, T. C, & Vencl, F. V. Larval beetles from a defence from recycled host-plant chemicals discharged as fecal wastes. Journal of Chemical Ecology (1998). , 24, 765-785.

[46] Myint, A. Common weeds of Guyana. Demerara, Guyana : National Agricultural Research Institute. (1994).

[47] National American Plant Protection Organization (NAPPO)Pest facts sheet- *Commelina benghalensis* L. June (2003). pp.

[48] Nkwiine, C, Tumuhairwe, J. K, Gumisiriza, C, & Tumuhairwe, F. K. Agrobiodiversity of banana (*Musa* spp.) production in Bushwere, Mbarara district, Uganda, In Agricultural biodiversity in smallholder farms of East Africa. F. Kaihura and M. Stocking (Eds.) United Nations University Press. (2003).

[49] Obiefuna, J. C. Biological weed control in plantains (*Musa* AAB) with Egusi melon (*Colocynthis citrullus* L.). Biological Agriculture and Horticulture (1989). , 6, 221-227.

[50] Oppong, F. K, Osei- Bonsu, K, Amoah, F. M, & Opoku- Ameyaw, K. Evaluation of Basta (glufosinate ammonium) for weed control in Coffee. Journal of the Ghana Science Association (1998). , 1(1), 60-68.

[51] Owen, M. D. K. The value of alternative strategies for weed management. (Abstract) III international Weed Science Congress, Foz do Iguassu, Brazil, 6- 11 June, (2000). Published by the International Weed Science Society, Oregon, U.S.A. , 50.

[52] Owen, M. D. K, & Zelaya, I. A. Herbicide- resistant crops and weed resistance to herbicides. Paper presented at the Symposium 'Herbicide- resistant crops from biotechnology: current and future status' held by the Agrochemicals Division of the American Chemical Society at the 227th National Meeting, Anaheim, CA, March, (2004). , 29-30.

[53] Pancho, J. V. Seed sizes and production capabilities of common weed species in the rice fields of the Philippines. Philippine Agriculturist (1964). , 48, 307-316.

[54] Prostko, E. P, Culpepper, A. S, Webster, T. M, & Flanders, J. T. Tropical Spiderwort identification and control in Georgia field crops. Circ. 884. University of Georgia College of Agriculture and Environmental Science / Coop. Ext. Ser. Bull., Tifton. (2005). http://pubs.caes.uga.edu/caes-pubs/pubs/PDF/c884.pdfaccessed 20 September 2006).

[55] Queneherve, P, Chabrier, C, Auwerkerken, A, Topart, P, Martiny, B, & Martie-luce, S. Status of weeds as reservoirs of plant parasitic nematodes in banana fields in Martinique. Crop Protection (2006). , 25, 860-867.

[56] Ragone, D, & Wilson, J. E. Control of weeds, nematodes and soil born pathogens by soil solarization. Alafua Agricultural Bulletin, 14. University of Hawaii, USA., (1988). , 13, 13-20.

[57] Robinson, A. F, Inserra, R. N, Caswell-chen, E. P, Vovlas, N, & Troccoli, A. *Rotylenchulus* species : identification, distribution, host ranges and crop plant resistance. Nematropica (1997). , 15, 165-170.

[58] Rodriguez, M. G, Sanchez, L, & Rodriguez, M. E. Plant parasitic nematodes associated with coffee (*Coffea arabica*) in Cajalbana, Cuba. Revista de Proteccion Vegetal (2000). , 15(1), 38-42.

[59] Shcherbakova, J. A. The effect of sowing depth of agricultural crops and fertilizers on the growth and development of *Commelina communis*. Sibirskii Vestnik Sel'skokhozyaistvennoi Nauki, (1974). , 6, 33-37.

[60] Singh, N. D. Studies on selected hosts of *Rotylenchulus reniformis* and its pathogenicity to soybean (*Glycine max*). Nematropica(1975). , 5(2), 46-51.

[61] Smith, D. Impact of natural enemies on the leaf mining fly Liriomyza commelinace, In Proceedings of the Interamerican Society for Tropical Horticulture, (1990). abstract only)., 34, 101-104.

[62] Stamps, R. H. Prodiamine suppresses spreading dayflower (*Commelina diffusa*) facilitating hand-weeding in leatherleaf fern (*Rumohra adiantiformis*) ground beds. Journal of Environmental Horticulture(1993). , 11(2), 93-95.

[63] Standish, R. J. Prospects for biological control of *Tradescania fluminensis* Vell. (Commelinaceae). Doc Science Internal Series 9, New Zealand Department of Conservation. (2001). http://www.doc.govt.nzaccessed 15 March 2006).

[64] Standish, R. J. Experimenting with methods to control Tradescantia fluminensis an invasive weed of native forest remnants in New Zealand. New Zealand Journal of Ecology (2002). , 26(2), 161-170.

[65] Sun YiHui, Zhang RongBao, Zhou JinHui, Xiao Gang, Sun YanHui and Guan HongDan.Herbicidal effects and safety of Imazethapyr AS for weeds control in soybean field. Pesticide Science and Administration (2008). in Chinese with English abstract)., 29(1), 27-29.

[66] Terry, P. J. Weed Management in Bananas and Plantains. In Weed Management for Developing Countries. Labrada, R., J.C. Caseley and C. Parker (eds.) FAO Plant Production & Protection Paper 120. (1996). , 11-315.

[67] Tian Jing and Zhao ChangShanCloransulam-methyl 84% WG controlling destructive weeds in soybean field. Agrochemicals (2009). in Chinese with English abstract)., 48(5), 376-378.

[68] Tonzani, R, Cardoso, G. V, Zonta, E, Merry, W, Parraga, M. S, & Pereira, M. G. Effects of four weed extracts on lettuce. (Abstract) III International Weed Science Congress, Foz do Iguassu, Brazil, 6-11 June, (2000). Published by the International Weed Sicence Society, Oregon, U.S.A. 2000. , 34.

[69] Tuffi Santos, L.D., Meira, R.M.S.A., and Santos, I.C.Effect of glyphosate on the morpho-anatomy of leaves and stems of *C. diffusa* and *C. benghalensis*. Planta daninha (2002)., 22(1), 101-107.

[70] Urich, R, Coronel, I, Silva, D, Cuberos, M, & Wulff, R. D. Intraspecific variability in *Commelina erecta*: response to phosphorus addition. Canadian Journal of Botany (2003)., 81, 945-955.

[71] Valverde, R. A. Brome mosaic virus isolates naturally infecting *Commelina diffusa* and *Commelina communis*. Plant Disease (1983).

[72] Van Rijin, P. J. Weed Management in the humid and sub- tropics. Royal Tropical Institute, Amsterdam, The Netherlands. (2000).

[73] Voll, E, Franchini, J. C, Cruz, R. T, Gazziero, D. L. P, Brighenti, A. M, & Adegas, F. S. Chemical interactions of *Brachiara plantaginea* with *Commelina benghalensis* and *Acanthospermum hispidium* in soybean cropping systems. Journal of Chemical Ecology (2004).

[74] Walker, S. R, & Evenson, J. P. Biology of *Commelina benghalensis* L. in south-eastern Queensland. 1. Growth, development and seed production. Weed Research UK (1985a)., 25(4), 239-244.

[75] Walker, S. R, & Evenson, J. P. Biology of *Commelina benghalensis* L. in south-eastern Queensland. 2. Seed dormancy, germination and emergence. Weed Research UK. (1985b)., 25(4), 245-250.

[76] Waterhouse, D. F. Editor). Biological control of weeds: Southeast Asian prospects. Canberra, Australia; Australian Center for International Agricultural Research (ACIAR). (1994).

[77] Webster, T. M. Weed survey- southern states: broadleaf crops subsection. Proceedings of the Southern Weed Science Society (2001)., 54, 244-259.

[78] Webster, T. M. and MacDonald, G.E. A survey of weeds in various crops in Georgia. Weed Technology (2001)., 15, 771-790.

[79] Webster, T. M, Burton, M. G, Culpepper, A. S, York, A. C, & Prostko, E. P. Tropical Spiderwort (*Commelina benghalensis*): A tropical invader threatens agroecosystems of the Southern United States. Weed Technology (2005)., 19(3), 501-508.

[80] Webster, T. M, Burton, M. G, Culpepper, A. S, Flanders, J. T, Grey, T. L, & York, A. C. Tropical Spiderwort (*Commelina benghalensis* L.) Control and Emergence Patterns in Preemergence Herbicide Systems. Journal of Cotton Science (2006a)., 10, 68-75.

[81] Webster, T, Grey, T, Burton, M, Flanders, J, & Culpepper, A. Tropical Spiderwort (*Commelina benghalensis*): the worst weed in cotton? In Proceedings of the 2006 Beltwide Cotton Conference, January 3-6, (2006). San Antonio, Texas) 2006b., 2181-2183.

[82] Webster, T, Flanders, J, & Culpepper, A. Critical period of tropical spiderwort (*Commelina benghalensis*) control in cotton. Weed Science Society of America Abstracts (2006). c , 80.

[83] WeedScience org. Group O/4 resistant spreading dayflower (*Commelina diffusa*), USA: Hawaii. (2005). http://www.weedscience.org/Case/Case.asp?ResistID=394accessed 15 March 2007)

[84] Wilson, A. K. Commelinacea- review of the distribution, biology and control of the important weeds belonging to this family. Tropical Pest Management (1981). , 27(3), 405-418.

[85] Yang YuTing, Lin ChangFu, Geng HeLi, Sun BaoXiang and William H. Ahrens.Studies on the action of bromoxynil and 2,4-D butyl ester herbicides combinations against dayflower (Commelina communis L.). Pesticides (2001). in Chinese with English abstract)., 40(7), 37-38.

[86] Zhang XiuRong, Ma Shu, Ying, Dai BingLi, X.R. Zhang, S.Y. Ma and B.L. Dai.Monophagy of *Lema scutellaris* on *Commelina communis*. Acta Entomologica Sinica (1996). , 39, 281-285.

[87] Zimmerman, A. De nematoden der koffiewortels. Deel I Mededeel's. Lands Plantentium (Buitenzorg) (1898).

Chapter 22

Integrating Herbicides in a High-Residue Cover Crop Setting

Andrew J. Price and Jessica A. Kelton

Additional information is available at the end of the chapter

http://dx.doi.org/10.5772/56142

1. Introduction

Sustainable agriculture requires the use of multiple, integrated weed management practices to ensure long-term viability. A number of cultural, mechanical, and chemical weed control options can be utilized in a production system to reduce weed interference and safeguard crop yield. The dependence on one single weed control strategy may result in short-term success; however, long-term use can lead to multiple setbacks including poor soil health, reduced crop production, and increasing herbicide resistance. In turn, employing multiple weed control tactics simultaneously may prove difficult without previous knowledge as to how best to implement an integrated weed management system. To that end, this chapter is dedicated to illustrating successful herbicide use in conjunction with cover crops and their residues, practices proven not only to suppress weed germination and growth, but also to reduce soil erosion and water runoff and build soil organic matter and thus subseqent productivity.

Use of cover crops, particularly those producing high amounts of biomass (greater than 4,500 kg ha^{-1}), can provide numerous benefits for a cropping system [1]. However, care must be taken when choosing herbicides to apply to these cover crops both prior to and after primary crop planting. This chapter provides an overview of effective herbicide choices for use prior to and within cover crop as well as efficient application methods for use after planting the primary crop(s). We also discuss herbicide interception by cover crop residue and means to control reduced efficacy due to interception. It is hoped that this summary will aid in the adoption of sustainable farming practices to ensure successful agricultural productivity for future generations.

2. Conservation agriculture

As demands are placed on agriculture to produce increasing yields for a growing global population, the need to implement systems with high productivity and sound environmental standards is key to ensuring agricultural sustainability for future generations. To this end, conservation agriculture is a systems-based approach for food, feed, and fiber production that utilizes a number of practices aimed at maintaining yields while limiting energy and chemical inputs, minimizing soil degradation and erosion, and reducing long-term, detrimental impacts to the environment [2]. Conservation agriculture is comprised of many different management practices, particularly cultural techniques such as crop rotation, planting date, and seeding rate, that can reduce dependence on chemical inputs for successful yield production. Moreover, limited tillage practices, or conservation-tillage, is essential to conservation agriculture systems to ensure soil quality, reduce runoff, and lessen energy consumption on agricultural lands.

2.1. Conservation tillage

Conservation-tillage, or reduced-tillage, has been proven to provide multiple benefits in agricultural settings. In addition to erosion and runoff control, soil health improvement, and reduced energy demands, reduced-tillage practices can produce crops yields similar to that of conventional systems [3-5]. The use of reduced-tillage, however, can alter weed communities. Seed production by annual weed species remains, in most part, on the soil surface where it is subject to increased decomposition and predation. With reduced competition and minimized soil disruption, perennial weed species can become established and dominate the weed community in conservation-tillage [6]. To aid in the control of both annual and perennial weeds, the use of cover crops for ground cover can reduce herbicide requirements in conservation-tillage settings.

2.2. Cover crops

A number of cereal and legume cover crops are utilized in various crop productions for several purposes. Currently, a large portion of cover crops are planted as a green manure which are turned under prior to sowing the primary crop [7,8]. In reduced-tillage, however, cover crops are grown as a ground cover and remain on the soil surface after cover crop termination. In addition to further reducing soil erosion, increasing soil organic matter, and improving water infiltration, cover crops can provide a level of weed suppression both prior to and during the primary growing season [9]. When compared to fallow conservation-tillage systems, cover crops offer increased weed control through direct resource competition while actively growing as well as through shading and/or allelopathy after termination. Covers grown to produce high levels of biomass, in particular, can increase shading of germinating weed species and provide greater ground cover for an extended period during the growing season. When employing cover crops, however, knowledge concerning herbicide use both during cover crop production and primary crop growth is essential.

http://dx.doi.org/10.5772/56142

2.3. Herbicide use

2.3.1. Cover crop establishment and termination

To produce substantial cover crop biomass, it is imperative to adequately manage cover crop production. Besides using correct seeding rates, early planting dates, and sufficient fertilizer applications, it is important to be aware of herbicide applications made prior to cover crop establishment. Often times, postemergent (POST) herbicides applied late season or post-harvest can have residual carryover than may be detrimental to cover crops. Rotation restrictions listed on herbicide labels should be referred to when planning POST applications and cover crop species.

To manage cover crops before cash crop planting, herbicides are typically utilized for cover crop termination. Most often, these herbicides, such as glyphosate and glufosinate, are non-selective with little to no carryover risk. However, consideration should be given to in-season chemical weed control regimes in order to limit repeated applications of a single herbicide mode of action. Moreover, care should be taken to avoid reduced herbicide rates applied for cover termination to reduce the risk of herbicide resistance [10]. Recent research has focused on mechanical termination with a roller or crimper which may reduce or eliminate the need of these herbicides for cover crop termination [11].

2.3.2. Cash crop establishment and management

Although use of in-season herbicides can be substantially reduced when using high-residue cover crops, some chemical applications are generally required to achieve the most effective weed suppression and minimize crop loss due to weed competition. While an ideal agricultural system would require no chemical inputs for sufficient weed control, practicality dictates the use of herbicides to guarantee crop yield since no system, as of yet, exists that can successfully suppress weed populations without intensive labor or mechanical requirements. To this end, cover crops are a means to minimize, rather than eliminate, herbicide inputs in crop systems. In recognizing the fact that the majority of agricultural systems will require chemical weed control measures for optimum crop production even when utilizing cover crops, it is essential to understand how cover crops affect herbicide selection and efficacy for each crop.

Primarily, the use of reduced-tillage and cover crops eliminates the ability to utilize preplant incorporated herbicides which offer residual soil activity [11]. Furthermore, cover crop residue can impede preemergent (PRE) herbicide applications from reaching the soil surface, reducing herbicide efficacy [12]. While postemergent chemical weed control can be effective alternatives in these settings, many weed species can prove to be difficult to control if not killed early in the season. Moreover, resistance concerns essentially necessitate the use of preemergent herbicides with differing mechanisms of action to avoid selection pressure for resistant weed biotypes [13].

Along with many cultural pracitces, production of crops under reduced-tillage with cover crops requires development of specific herbicide regimes to ensure minimal chemical inputs while achieving sufficient weed control to allow for successful crop production. The following sections review major crops produced globally, describe research conducted in respect to reduced-tillage production, as well as list available herbicides for use when using reduced-tillage and cover crops. These reviews are designed to provide information that can be beneficial for producers implementing conservation-tillage.

3. Wheat

Global production of wheat (*Triticum aestivum* L.) was estimated at approximately 217 million hectares in 2010 [14] representing the largest single crop, in area grown, and providing approximately 19% of the caloric intake of the world's diet [15]. In recent years, concerns have been noted over stagnant wheat yields due to drought and rising temperatures attributed to global warming [16]. Efforts to maintain current wheat production levels and identify potential measures to aid in increasing yield have led researchers to explore conservation practices in wheat systems. In addition to preserving high crop yields, long-term conservation systems are intended to protect environmental quality and reduce chemical and energy inputs necessary for crop production. Components of conservation systems such as reduced- or no-tillage can produce crop yields equal to or exceeding conventional tillage practices while reducing erosion, water runoff, and increasing water infiltration.

Much research has been conducted to evaluate wheat productivity in conservation-tillage practices. Reports reveal similar or increased grain yield for reduced-tillage compared to conventional tillage systems [17-19]. With little or no tillage operations, some chemical applications are required to achieve successful levels of weed control; however, with herbicide applications, weed species have been effectively controlled below levels that could reduce yield [20]. To offset the herbicide needs in conservation-tillage, evaluations of cover crops as ground cover have been conducted. Crops such as mustard (*Sinapis alba* L.), pea (*Pisum sativum* L.), and lentil (*Lens culinaris* Medik.) have proven to be good choices with little yield differences [21]. However, other reports show negative impacts on wheat production when implementing cover crops prior to wheat production for reasons such as increased weed competition, primarily *Bromus* spp., and reduced fertilizer uptake [22].

Like most crops produced in conservation-tillage, herbicide options may be limited to a degree whether utilizing a cover crop or not. With reduced-tillage, preplant incorporation of residual herbicides cannot be utilized. Moreover, when planting into cover crops, soil-applied pre-emergent herbicides may be less effective due to interception by crop residue. When planting wheat, preplant burndown herbicides may be necessary to control early weeds. POST herbicides are also necessary to control weeds that germinate after planting. Table 1 lists many of the herbicide options for use in conservation-tillage systems for wheat production.

4. Maize

Maize, or corn (*Zea mays* L.), is one of the most economically important grain crops worldwide with 162 million ha produced in 2010 [2]. In addition to being a staple in human and livestock

http://dx.doi.org/10.5772/56142

diets in many countries, corn is also used for bioethanol production and the manufacturing of many non-food products. Consumption of corn and products derived from corn continues to increase. Given the demand, it is imperative for sustainable production systems that produce high yields while preserving long-term productivity of the land to be implemented.

Conservation-tillage practices have been researched and utilized for several decades in some regions such as the Midwest in the United States. As with many other crops, some variability has been noted for corn yield in no-tillage systems compared to conventional tillage methods. However, many reports show at least equal corn yields can be achieved when tillage practices are reduced [3]. Adequate yield potential, coupled with the reduction of on-farm expenses, have made conservation-tillage systems a good fit for corn production.

Herbicide			
Common Name	**Trade Name**[a]	**Application Timing**	**Weed Species Controlled**
Carfentrazone	Aim® [23]	Preplant Burndown	Non-selective control of emerged broadleaves and grasses
Glufosinate	Liberty® [24]		
Glyphosate	Roundup WeatherMax®[25]		
Paraquat	Gramoxone® [26]		
Chlorsulfuron + Metsulfuron	Finesse® [27]	PRE or POST[b]	*Bromus* species, annual ryegrass (*Lolium multiflorum*), kochia (*Kochia scoparia*)
Pyrasulfotole + Bromoxynil	Huskie™ [28]	Early POST	Emerged broadleaf seedlings such as dandelion (*Taraxacum officinale*); suppression of established dandelion and henbit (*Lamium amplexicaule*)
Thifensulfuron + Tribenuron	Harmony® Extra [29]	POST	Actively growing broadleaves, wild garlic (*Allium vineale*); suppression of Canada thistle (*Cirsium arvense*)
Clearfield wheat			
Imazamox	Beyond® [30]	POST	Broadleaves henbit and chickweed (*Stellaria media*), grasses barnyardgrass (*Echinochloa crus-galli*), jointed goatgrass (*Aegilops cylindrica*), volunteer cereals (non-Clearfield types)

[a]Trade names listed are representative of available herbicides. Inclusion of particular trade names does not suggest author endorsement.

[b]PRE, preemergence; POST, postemergence.

Table 1. Herbicides for use in reduced-tillage wheat production.

A major limiting factor to adopting reduced-tillage in corn production is the concern of less effective weed control. Tillage has long been used as a means for weed seed burial which reduces the number of seeds in the upper portion of the soil, the area most favorable for germination for most species. In addition to weed seed remaining in the upper layer of soil, shifts in weed species have also been noted. With the implementation of conservation-tillage, most crop systems experience a shift in weed species from annuals to perennials dominating the weed community.

Perennial weed species, largely controlled with tillage practices, can thrive on less disturbed crop land. For effective weed control, producers implementing reduced-tillage have relied on increased herbicide applications. To curb herbicide use, cover crops have been adopted in conjunction with reduced-tillage corn systems. Research has shown that utilizing a legume or grain cover crop can reduce weed density and growth while not affecting corn yield [31,32]. For corn in particular, cover crops offer a potential benefit in addition to weed suppression. Adequate nitrogen availability is essential for corn development. The use of legume cover crops, such as hairy vetch (*Vicia villosa* Roth), red clover (*Trifolium pratense* L.), or medics (*Medicago* spp.), may provide a portion of corn nitrogen requirements and reduce fertilizer inputs into the system [33]. Some research indicates that legume covers do not reduce fertilizer requirements but improves grain production with standard fertilizer applications [34]. Other research shows that legume covers can provide some nitrogen required for successful corn production[35,36]. Selecting the right legume cover crop for maximum nitrogen contribution with timely availability for corn uptake is key for utilizing these crops as nitrogen sources.

Use of burndown herbicides prior to corn planting is critical for early season weed control when using cover crops. A residual herbicide applied in conjunction with the herbicide used for cover crop termination can broaden weed species controlled as well as extend control into the season. A number of PRE herbicides are available that can be applied without incorporation into the soil and are effective even with plant residue on the soil surface. These herbicides and POST herbicide choices that can be successfully utilized in conservation-tillage corn with cover crops are listed in Table 2.

Herbicide			
Common Name	**Trade Name**[a]	**Application Timing**	**Weed Species Controlled**
Glufosinate	Liberty® [24]	Preplant burndown	Emerged weed species
Glyphosate	Roundup WeatherMax® [25]		
Paraquat	Gramoxone® [26]		
2,4-D	Agri Star® 2,4-D [37]		
Atrazine	Aatrex® [38]	Preplant or PRE[b]	Broadleaves such as kochia (*Kochia scoparia*); suppression of foxtail (*Setaria* spp.), velvetleaf (*Abutilon theophrasti*). Can also be applied POST

http://dx.doi.org/10.5772/56142

Herbicide			
Common Name	**Trade Name[a]**	**Application Timing**	**Weed Species Controlled**
Flumioxazin	Valor® [39]		Broadleaf species such as horseweed (*Conyza canadensis*); suppression of grass species such as panicum (*Panicum* spp.) and goosegrass (*Eleusine indica*)
Pendimethalin	Prowl® [40]		Germinating, small-seeded grass and broadleaf species such as crabgrass (*Digitaria* spp.) and common lambsquarters (*Chenopodium alba*)
S-metolachlor	Dual Magnum® [41]		Grass and broadleaf species such as foxtail and *Amaranthus* spp.
Carfentrazone	Aim® [23]	POST[c]	Certain broadleaf weed control; tank mix with atrazine or dicamba
Bromoxynil	Buctril® [42]		Broadleaf weeds such as burcucumber (*Sicyos angulatus*), giant ragweed (*Ambrosia trifida*)
Dicamba	Banvel® [43]		Annual broadleaf species as well as certain perennial species such as dock (*Rumex* spp.) and wild onion (*Allium* sp.)
Mesotrione	Callisto® [44]	POST	Broadleaf species such as wild mustard (*Sinapis arvensis*), nightshade (*Solanum* spp.), and Canada thistle (*Cirsium arvense*)
Tembotrione	Laudis® [45]		Broadleaf and grass species such as common chickweed, purple deadnettle (*Lamium purpureum*), *Amaranthus* spp., and large crabgrass (*Digitaria sanguinalis*)
Ametryn	Evik® [46]	POST-directed spray	Grass species such as Texas panicum, goosegrass, and foxtail
Linuron	Lorox® [47]		Broadleaf and grass species such as dog fennel, common ragweed (*Ambrosia artemisiifolia*), velvetleaf, and annual ryegrass (*Lolium multiflorum*)
Clearfield Corn			
Imazethapyr + Imazapyr	Lightning® [48]	POST	Broadleaves, grasses, and sedges such as kochia, ragweed, quackgrass (*Elytrigia repens*), and nutsedge (*Cyperus* spp.)
LibertyLink Corn			
Glufosinate	Liberty®	POST	Broadleaf and grass species; ragweed, horseweed, johnsongrass seedlings
Roundup Ready Corn			

Herbicide			
Common Name	Trade Name[a]	Application Timing	Weed Species Controlled
Glyphosate	Roundup Weathermax®	POST	Nonselective control of some broadleaf and grass species
Glyphosate + *s*-metolachlor + atrazine	Expert® [49]	PRE or POST	Annual broadleaves and grasses; perennials such as quackgrass, dandelion (*Taraxacum officinale*), and Canada thistle

[a]Trade names listed are representative of available herbicides. Inclusion of particular trade names does not suggest author endorsement.

[b]PRE, preemergence.

[c]POST, postemergence.

Table 2. Herbicides for use in reduced-tillage maize production.

5. Rice

Production of rice (*Oryza sativa* L.) in 2010 was near 154 million ha worldwide [2]. In many regions, rice provides nearly half or more of calories consumed by humans [50] and is the most important grain crop grown. Rice yield has steadily grown in the past several decades due to breeding and fertilizer advancements; however, it is necessary for rice yield to continue to improve in order to meet increased demands by a growing world population. Given that little land exists in rice-producing countries to expand production, it is necessary for methods to be established that can continue yield improvement without depleting future soil productivity.

Wetland, transplant rice production is the dominant and highest yielding rice system in most regions [50, 51]. However, the water and energy requirements may limit rice production as competition for resources increases [52]. To reduce strain on environmental and economic resources and to ensure sustainable rice systems in the future, dry-seeded rice production has been implemented in some areas [53]. Dry-seeded rice production can be initiated in conjunction with conservation-tillage with fewer water demands, lower energy and labor requirements, and reduced soil erosion. Research has reported that dry-seeded rice in no-tillage can be a successful alternative to conventional systems [52].

A limiting factor to widespread adoption of dry-seeded, reduced-tillage rice, however, is reduced weed control. For rice, transitioning from wetland, conventional systems to a dry system with reduced-tillage can affect weed compositions in multiple ways. Standing water can reduce germinating weed seeds while the transplanted rice becomes established; removing this water barrier can increase weed numbers [54]. Additionally, reduced-tillage practices can result in an increase of weed seed germination due to less seed burial.

In dry-seeded rice, mulches have been suggested as a means to combat weed increases [51]. Little research has been conducted to fully understand the benefits of cover crops for weed control in rice; however, legume covers have been associated with increased rice yield and reduced weed biomass in upland rice [55]. Future research needs include addressing the effects of cover crops on rice production in dry-seeded rice systems.

Due to challenging weed issues in rice systems, particularly dry-seeded rice, herbicide use will continue to be necessary for effective weed suppression in both conventional and reduced-tillage systems. The implementation of cover crops into these systems may lessen the herbicide requirements but will not eliminate the use of chemicals altogether. Currently there are a number of preemergent and postemergent herbicides available for use in rice production (Table 3); however, as dry-seeded, conservation-tillage rice systems increase in popularity, more herbicide options may become available.

Herbicide			
Common Name	**Trade Name[a]**	**Application Timing**	**Weed Species Controlled**
Clomazone	Command®[56]	PRE[b]	Grass species such as barnyardgrass (*Echinochloa crus-galli*), crabgrass (*Digitaria* spp.), and panicum (*Panicum* spp.)
Halosulfuron	Permit® [57]		Broadleaf species such as dayflower (*Commelina erecta*) and kochia (*Kochia scoparia*). Broadleaf and grass species may be controlled with a POST application.
Pendimethalin	Prowl® [40]		Germinating, small-seeded grass and broadleaf species such as crabgrass (*Digitaria* spp.), foxtail, and common lambsquarters (*Chenopodium alba*)
Quinclorac	Facet® [58]		Broadleaf and grass species such as morningglory (*Ipomoea* spp.), and barnyardgrass. Can also be applied POST
Thiobencarb	Bolero® [59]		Grass and broadleaf species such as barnyardgrass, dayflower (*Commelina communis*), and eclipta (*Eclipta alba*)
Acifluorfen	Ultra Blazer® [60]	POST[c]	Grasses and broadleaves such as foxtail (*Setaria* spp.), panicum, and eclipta
Bensulfuron	Londax® [61]		Broadleaf and sedge species, particularly aquatic weeds such as ducksalad (*Heteranthera limosa*) and ricefield bulrush (*Scirpus mucronatus*)
Bentazon	Basagran® [62]	POST	Broadleaf and sedge species such as dayflower, eclipta, and yellow nutsedge (*Cyperus esculentus*)
Carfentrazone	Aim® [23]		Broadleaf species such as common cocklebur (*Xanthium strumarium*), dayflower, and *Amaranthus* spp.

Herbicide			
Common Name	**Trade Name**[a]	**Application Timing**	**Weed Species Controlled**
Propanil	Stam® [63]		Grass, rush, and broadleaf species such as barnyardgrass, spikerush (*Eleocharis* spp.), and curly dock (*Rumex crispus*)
Cyhalofop	Clincher® [64]	After Flooding	Grass species such as barnyardgrass, broadleaf signalgrass (*Brachiaria platyphylla*), and junglerice (*Echnochloa colona*)
2,4-D	Agri Star® 2,4-D [37]		Annual and perennial weed species such as cocklebur, morningglory, and dock
Clearfield Rice			
Imazamox	Beyond® [30]	POST	Grass and broadleaf species such as morningglory, barnyardgrass, and panicum
Imazethapyr	Newpath® [65]		Grass, sedge, and broadleaf species such as barnyardgrass, morningglory, and nutsedge
Imazethapyr + Quinclorac	Clearpath® [66]		Grass, sedge, and broadleaf species such as junglerice, eclipta, morningglory, and nutsedge

[a]Trade names listed are representative of available herbicides. Inclusion of particular trade names does not suggest author endorsement.

[b]PRE, preemergence.

[c]POST, postemergence.

Table 3. Herbicides for use in reduced-tillage rice production.

6. Soybean

Production of soybean [*Glycine max* (L.) Merr.], estimated at 102 million ha in 2010 [2], meets a number of livestock and human food needs as well as industrial demands for use in products such as paints, lubricants, and biofuel. Due to its diversity of uses, the soybean is an important field crop for much of the world. In light of the value of soybeans, it is essential to establish sustainable growing practices to ensure global demand continues to be met.

Implementation of conservation practices, such as reduced-tillage, can be utilized as components of alternative management systems replacing conventional systems to provide erosion and runoff control while reducing labor and cost inputs. In the United States, in fact, approximately 80% of soybeans were produced with some form of conservation-tillage by 2006 [67]. This increase in conservation-tillage can be attributed to the environmental and economic benefits achieved with reduced-tillage as well as the commercial availability of herbicide-tolerant soybeans, which have made successful chemical weed control achievable with the use of fewer herbicides.

Early work in conservation-tillage soybean have reported equal or improved yield in soybean with reduced-tillage compared to conventional systems [68, 69]. Previous research has also examined soybean systems planted behind wheat or a cover crop such as rye with improved weed control being noted when compared to a fallow system [70] and greater yield with a cover crop than with just the previous crop's stubble [71]. The inclusion of plant residue, either from a cover crop or a previous crop, provides a level of weed control by acting as a physical barrier for germinating weed seed or through allelopathic inhibition released by some cover crop species. The weed control provided by ground cover is crucial in a no-till practice due to the loss of control from tillage reduction and the shift towards more difficult to control perennial weed species.

While cover crops and plant residue have been identified as means to reduce weed emergence when implemented in reduced-tillage practices further measures are required to keep the weed population below an acceptable level [70]. Many cultural practices, such as crop rotation, row spacing, and planting date, can be manipulated in such a way as to reduce weed populations; however, herbicide use is still necessary in many systems.

As with most field crops grown in conservation-tillage systems, soybean production with reduced-tillage has heavily relied on postemergent herbicide applications. Use of cover crops in these systems may also contribute to the tendency for fewer PRE herbicides due to interception concerns. However, the increase in herbicide-resistant weed species such as Palmer amaranth (*Amaranthus palmeri* S. Wats) and horsweed [*Conyza canadensis* (L.) Cronq.] in herbicide resistant crops, like soybean, necessitates the use of multiple herbicides to slow the development of weed resitance and safeguard the effectiveness of current herbicide options for the future. Table 4 provides a partial list of herbicides that can be utilized in reduced-tillage soybean with cover crops.

Herbicide			
Common Name	**Trade Name**[a]	**Application Timing**	**Weed Species Controlled**
Glufosinate	Liberty® [24]	Preplant Burndown	Emerged weed species
Glyphosate	Roundup WeatherMax® [25]		
Paraquat	Gramoxone® [26]		
2,4-D	Agri Star® 2,4-D [37]		
Clomazone	Command® [56]	PRE[b]	Grasses and broadleaves such as crabgrass (*Digitaria* spp.), panicum (*Panicum* spp.), velvetleaf (*Abutilon theophrasti*), and Florida beggarweed (*Desmodium tortuosum*)
Dimethenamid	Outlook® [72]		Grass and broadleaf species such as foxtail (*Setaria* spp.), panicum, and *Amaranthus* spp.
Flumioxazin	Valor® [39]		Broadleaf species such as horseweed (*Conyza canadensis*); suppression of grass species such as panicum and goosegrass (*Eleusine indica*)

Herbicide			
Common Name	**Trade Name[a]**	**Application Timing**	**Weed Species Controlled**
Imazaquin	Scepter® [73]		Broadleaf and grass species such as morningglory (*Ipomoea* spp.), velvetleaf, and foxtail
Metribuzin	Metribuzin [74]		Broadleaf and grass species such as *Amaranthus* spp.and broadleaf signalgrass (*Brachiaria platyphylla*)
Pendimethalin	Prowl® [40]		Grass and broadleaf species such as panicum and *Amaranthus* spp.
S-metolachlor	Dual Magnum® [41]		Grass and broadleaves such as barnyardgrass (*Echinochloa crus-galli*), crabgrass, and Florida pusley (*Richardia scabra*)
Bentazon	Basagran® [62]	POST[c]	Broadleaf weeds such as coffee senna (*Senna occidentalis*) and velvetleaf
Chlorimuron	Classic® [75]		Broadleaf weeds such as Florida beggarweed and morningglory
Cloransulam	FirstRate® [76]		Broadleaf weeds such as common cocklebur (*Xanthium strumarium*) and velvetleaf
Fluazifop	Fusilade® [77]		Annual and perennial grass species such as crabgrass and bermudagrass (*Cynadon dactylon*)
Imazethapyr	Pursuit® [78]		Broadleaf and grass species such as morningglory and crabgrass
Lactofen	Cobra® [79]		Broadleaf species such as croton (*Croton* spp.) and Florida beggarweed
Sethoxydim	Poast® [80]		Grass species such as foxtail, crabgrass, and panicum
LibertyLink Soybean			
Glufosinate	Liberty®	POST	Broadleaf and grass species such as *Amaranthus* spp., morningglory, and goosegrass
Roundup Ready Soybean			
Fomesafen + Glyphosate	Flexstar® [81]	POST	Broadleaf and grass species such as morningglory, velvetleaf, and broadleaf signalgrass
Glyphosate	Roundup WeatherMax®	POST	Grass and broadleaf species such as Florida beggarweed, crabgrass and groundcherry

[a]Trade names listed are representative of available herbicides. Inclusion of particular trade names does not suggest author endorsement.

[b]PRE, preemergence.

[c]POST, postemergence.

Table 4. Herbicides for use in reduced-tillage soybean production.

http://dx.doi.org/10.5772/56142

7. Cotton

Cotton production around the world is estimated at approximately 23 million tonnes (lint production) [2] with China, India, and the United States being the top producers [82]. Efforts to adopt sustainable cotton practices have led producers to utilize conservation-tillage systems in cotton production. Besides environmental benefits achieved with reduced-tillage, major economic advantages can be realized due to reduced time, labor, and fuel requirements when operating with less tillage. Prior to the introduction of herbicide-resistant crops, adoption of reduced-tillage was difficult due to control of weed species required multiple and costly herbicide inputs [13]. In some instances, effective herbicides were not available to control problematic weed species such as perennials that can thrive in reduced-tillage. When glyphosate-resistant cotton was made available, reduced-tillage became practical since a broad spectrum of weed species could be controlled with a single herbicide [83].

Extensive research has been carried out in conservation-tillage cotton with positive benefits seen for cotton yield [84-86]. Moreover, with herbicide-resistant cotton varieties, weed control has been as successful as conventional tillage cotton. Because of this success, conservation-tillage practices have been widely adopted in areas such as the southeastern United States. This dependence on a single herbicide, however, has led to the appearance of herbicide-resistant weed species and now threatens the feasibility of reduced-tillage cotton production. Currently, research efforts are focused on identifying ways to ensure the long-term viability of conservation-tillage while controlling established populations of herbicide-resistant weed species and reducing the risk of future development of resistant weeds.

Multiple weed management tactics are necessary to control weed resistance development with cover crops playing an important role in resistance management. The use of cover crops, particularly high-residue crops such as rye and black oat, can reduce herbicide inputs through shading and allelopathy. The use of high-residue crops allows for maximum shading of the soil surface during the beginning of the season while also providing a ground cover for a longer period into the growing season. Cover crops, along with multiple herbicide modes of action and rotation, have been shown to effectively control weeds in reduced-tillage cotton [87, 88].

A number of herbicide choices are available for use with conservation-tillage cotton (Table 5). PRE herbicides are especially important in early-season weed control to ensure management of weed species that are difficult to control later in the season. Although concerns have been raised as to whether cover crops reduce the efficacy of PRE herbicides, it has been suggested that any loss in weed control due to herbicide interception is offset by the control provided by cover crop residue [89-91].

Herbicide			
Common Name	**Trade Name[a]**	**Application Timing**	**Weed Species Controlled**
Dicamba	Banvel® [43]	Preplant Burndown	Emerged weed species
Flumioxazin	Valor® [39]		
Glufosinate	Liberty® [24]		

Herbicide			
Common Name	**Trade Name[a]**	**Application Timing**	**Weed Species Controlled**
Glyphosate	Roundup WeatherMax®[25]		
Paraquat	Gramoxone® [26]		
Clomazone	Command® [56]	Preplant or PRE[b]	Grasses and broadleaves such as crabgrass (*Digitaria* spp.), panicum (*Panicum*spp.), velvetleaf (*Abutilon theophrasti*), and Florida beggarweed (*Desmodium tortuosum*)
Fluometuron	Cotoran® [92]		Grasses and broadleaves such as signalgrass (*Brachiaria sp.),* horseweed (*Conyza canadensis*) and sicklepod (*Senna obtusifolia*)
Pendimethalin	Prowl® [40]		Grass and broadleaf species such as foxtail (*Setaria* spp.), panicum, and *Amaranthus* spp.
Prometryn	Caparol® [93]		Annual grass and broadleaves such as groundcherry (*Physalis* sp.), Florida pusley (*Richardia scabra*), and panicum
S-metolachlor	Dual Magnum®[41]		Grass and broadleaves such as barnyardgrass (*Echinochloa crus-galli*), crabgrass, and Florida pusley
Clethodim	Select® [94]	POST[c]	Grass species such as crabgrass, panicum, and foxtail
Herbicide			
Common Name	Trade Name	Application Timing	Weed Species Controlled
Quizalofop	Assure® [95]		Annual and perennial grasses such as foxtail, goosegrass (*Eleusine indica*), and bermudagrass (*Cynodon dactylon*)
Sethoxydim	Poast® [80]	POST	Grass species such as foxtail, crabgrass, and panicum
Trifloxysulfuron	Envoke® [96]		Broadleaf and grass species such as coffee senna (*Senna occidentalis*), barnyardgrass, and Florida beggarweed
Diuron	Direx® [97]	POST-directed spray	Broadleaf and grass species such as sicklepod, velvetleaf, and crabgrass
Linuron	Linex® [98]		Broadleaves and grasses such as morningglory, Florida pusley, and panicum
MSMA	MSMA [99]		Grass and broadleaf species such as crabgrass, Florida beggarweed, and *Amaranthus* spp.

Herbicide			
Common Name	**Trade Name[a]**	**Application Timing**	**Weed Species Controlled**
LibertyLink Cotton			
Glufosinate	Liberty®	POST	Broadleaf and grass species such as *Amaranthus* spp., morningglory, and goosegrass
Roundup Ready Cotton			
Glyphosate	Roundup WeatherMax®	POST	Grass and broadleaf species such as Florida beggarweed, crabgrass, foxtail, groundcherry, and velvetleaf

[a]Trade names listed are representative of available herbicides. Inclusion of particular trade names does not suggest author endorsement.

[b]PRE, preemergence.

[c]POST, postemergence.

Table 5. Herbicides for use in reduced-tillage cotton.

8. Peanut

Groundnut, or peanut (*Arachis hypogaea* L.), was planted on approximately 21 million ha between 2011 and 2012 wordwide with top production occurring in China, India, Indonesia, the United States, and some African countries such as Nigeria, Senegal, and Sudan [100]. Besides being a nutrient rich food source, the peanut is utilized for its oil in cooking and manufacturing as well as a livestock feed. In the United States, peanuts offer an exceptional rotational crop with cotton to replenish soil nitrogen. The benefits of peanuts to a cotton system, which have been shifting toward long-term, reduced-tillage practices, have necessitated the adoption of minimum tillage practices in peanut production as well.

The increased farming costs of conventional tillage systems have spurred producers to implement conservation-tillage to reduce expenses; however, peanut growers face unique difficulties when using these systems [101,102]. Particularly, concerns over peanut response to reduced-tillage due to peanut growth habits have required research in order to identify successful means of conservation-tillage integration into peanut production [102, 103].

Peanut yield variability under reduced-tillage compared to conventional tillage has been noted as one of the greatest concerns when adopting conservation-tillage practices [101,102]. Inconsistent yield response by peanut has been noted in previous studies investigating conservation-tillage. Research has reported yields of peanut to be reduced or equal to conventionally tilled peanut [101, 104]; other studies have shown reduced-tillage peanuts to produce equally or greater than conventional tillage peanuts [103,105]. Research efforts continue to recognize the contributing factors that affect peanut response to tillage systems.

Weed management in conservation-tillage peanut is also a concern for producers. Weed control in peanut, regardless of tillage system, can be problematic due to the extended growing season and unique growth habits [106,107]. Generally, peanut production requires an incorporated residual as well as a POST herbicide to provide effective weed control under the slow-closing canopy of peanut [107]. Moreover, in-season cultivation for weed management cannot be implemented due to the potential to damage developing peanut pods [106,108].

Weed control in reduced-tillage peanuts can be even more difficult than in conventional tillage due to the loss of weed control through seed burial and the inability to utilize preplant incorporated herbicides [109]. This results in increased dependence on post emergent herbicides which may or may not control the number of perennial weed species that may predominate in a reduced-tillage setting; the loss of effective weed management can reduce peanut productivity due to weed competition [102,107].

Utilization of cover crops in peanut systems may offer beneficial weed control while reducing the need for increased postemergent herbicide applications. Research has shown effective weed control with cover crops in strip-tillage peanut systems that use a dinitroaniline pre-emergent herbicide over cover crop residue [107]. Other effective herbicides used in conservation-tillage peanut systems are listed in Table 6.

9. Herbicide interception

Preemergent, residual herbicides must reach the soil surface to be effective. When spraying over cover crop residue, herbicide applications can be intercepted and absorbed prior to reaching the soil surface. Herbicides, such as acetochlor, chlorimuron, and oryzalin have been shown to be impeded by plant stubble [113,114]. While timely rainfall can move herbicides to the soil, some portion of herbicide can be retained in the residue.

Herbicide amounts intercepted by stubble can affect weed control achieved with the herbicide; efficacy can be reduced by cover crops either through physical interception preventing soil contact or through increased microbial activity in the residue speeding herbicide degradation [115]. Increases in soil organic matter from extended conservation-tillage practices may also increase herbicide adsorption within the soil [116]. Additionally, herbicide persistence and carryover risks may be increased when applied to residue [114]. Certain crops may be susceptible to herbicides at low doses that can persist in cover crop residue that would otherwise have dissipated in bare soil. However, little research has been done to determine the extent of persistence for most herbicides.

Methods to reduce herbicide interception are limited when using cover crops. Interception could potentially be managed, particularly in strip-till operations, through banded herbicide applications over the row allowing for in-row weed control while reducing herbicide inputs. Furthermore, a water-based, microencapsulated herbicide formulation, like Prowl H_2O® (pendimethalin), may allow more herbicide to reach the soil after a rain event or irrigation.

http://dx.doi.org/10.5772/56142

Herbicide			
Common Name	**Trade Name[a]**	**Application Timing**	**Weed Species Controlled**
Glyphosate	Roundup WeatherMax® [25]	Preplant Burndown	Emerged weed species
Paraquat	Gramoxone® [26]		
2,4-D	Agri Star® 2,4-D [37]		
Diclosulam	Strongarm® [110]	PRE[b]	Broadleaf species such as eclipta (*Eclipta prostrata*) and *Amaranthus* spp.
Flumioxazin	Valor® [39]		Broadleaf species such as horseweed (*Conyza canadensis*)
Pendimethalin	Prowl® [40]		Grass and broadleaf species such as foxtail (*Setaria* spp.) and *Amaranthus* spp.
Acifluorfen	Ultra Blazer® [60]	POST[c]	Broadleaf and grass species such as coffee senna (*Senna occidentalis*) and velvetleaf (*Abutilon theophrasti*)
Bentazon	Basagran® [62]		Broadleaf species such as morningglory (*Ipomoea* spp.) and velvetleaf
Chlorimuron	Classic® [75]		Broadleaf weeds such as Florida beggarweed (*Desmodium tortuosum*) and morningglory
Clethodim	Select® [94]		Grass species such as panicum, foxtail, and crabgrass (*Digitaria* spp.)
Imazapic	Cadre® [111]		Broadleaf and grass species such as morningglory, *Amaranthus* spp. and crabgrass
Imazethapyr	Pursuit® [78]		Broadleaf, grass, and sedge species such as Florida pusley (*Richardia scabra*), crabgrass, and nutsedge (*Cyperus* spp.)
Paraquat	Gramoxone®		Grass and broadleaf species
Sethoxydim	Poast® [80]		Grass species, foxtail and panicum
2,4-DB	Butyrac® [112]		Broadleaf species such as velvetleaf and prickly sida (*Sida spinosa*)

[a]Trade names listed are representative of available herbicides. Inclusion of particular trade names does not suggest author endorsement.

[b]PRE, preemergence.

[c]POST, postemergence.

Table 6. Herbicides for use in reduced-tillage peanut.

10. Conclusion

The ever increasing demands on global agriculture dictate the use of intensive, high-yielding production practices. However, the inability to sustain these systems long-term necessitates the implementation of more energy-efficient, environmentally-sound practices that can still produce successful yields. Conservation agriculture practices seek to achieve these goals in order to ensure current and future agricultural production. While components of conservation agriculture, such as reduced-tillage and cover crops, are fundamental practices in these systems, herbicides are still valuable and necessary weed management tools within conservation systems. Integrating these management practices can be challenging and continue to warrant research to identify the most successful means of utilizes herbicides in conjunction with reduced-tillage and cover crops.

Author details

Andrew J. Price[1*] and Jessica A. Kelton[2]

*Address all correspondence to: Andrew.price@ars.usda.gov

1 United States Department of Agriculture, Agricultural Research Service, National Soil Dynamics Laboratory, Auburn, Alabama, USA

2 Auburn University, Auburn, Alabama, USA

References

[1] Reiter, M.S., D.W. Reeves, C.H. Burmester, H.A. Torbert. Cotton nitrogen management in a high-residue conservation system: Cover crop fertilization. *Soil Science Society of America Journal,* 2008; 72, 1321-1329, ISSN 1435-0661.

[2] Food and Agriculture Organization of the United Nations (FAO). FAOSTAT 2010. Available online at http://www.fao.org/ag/ca/index.html (accessed 13 August 2012).

[3] DeFelice, M.S., P.R. Carter, and S.B. Mitchell. Influence of tillage on corn and soybean yield in the US and Canada. Online. Crop Management. 2006. http://www.plantmanagementnetwork.org/pub/cm/research/2006/tillage/ (accessed 12 August 2012).

[4] Reeves, D.W. 1997. The role of soil organic matter in maintaining soil quality in continuous cropping systems. *Soil and Tillage Research,* 43, 131-167, ISSN 0167-1987.

[5] Truman, C.C., D.W. Reeves, J.N. Shaw, A.C. Motta, C.H. Burmester, R.L. Raper, and E.B. Schwab. Tillage impacts on soil property, runoff, and soil loss variations of a

Rhodic Paleudult under simulated rainfall. Journal of Soil and Water Conservation, 2003; 58,258-267, ISSN 0022-4561.

[6] Swanton, C.J., K.J. Mahoney, K. Chandler, and R.H. Gulden. Integrated weed management: Knowledge-based weed management systems. *Weed Science*, 2008; 56, 168-172, ISSN 0043-1745.

[7] Norsworthy, J.K., M.S. Malik, P. Jha and M.B. Riley. Suppression of *Digitaria sanguinalis* and *Amaranthus palmeri* using autumn-sown glucosinolate-producing cover crops in organically grown bell pepper. *Weed Research*, 2007; 47, 425-432, ISSN 0043-1737.

[8] Treadwell, D.D, N.G. Creamer, J.R. Schultheis, and G.D. Hoyt. Cover crop management affects weeds and yield in organically managed sweetpotato systems. *Weed Technology* 2007; 21, 1039-1048, ISSN 0890-037X.

[9] Brennan, E.B. and R.F. Smith. Winter cover crop growth and weed suppression on the Central Coast of California. *Weed Technology* 2005; 19, 1017-1024, ISSN 0890-037X.

[10] Clark, A., editor. Managing Cover Crops Profitably. College Park, MD, USA: Sustainable Agricultural Research and Education (SARE); 2007.

[11] Price, A.J., J.A. Kelton. Weed control in conservation agriculture. In: Soloneski S. and M. Larramendy (ed.) Herbicides: Theory and Applications. Rijeka: InTech; 2010. p. 3-16.

[12] Gaston, L.A., D.J. Boquet, and M.A. Bosch. Pendimethalin wash-off from cover crop residues and degradation in a Loessial soil. *Communications in Soil Science and Plant Analysis* 2003; 34, 2515-2527, ISSN 0010-3624.

[13] Price, A.J., K.S. Balkcom, S.A. Culpepper, J.A. Kelton, R.L. Nichols, and H. Schomberg. Glyphosate-resistant Palmer amaranth: A threat to conservation tillage. *Journal of Soil and Water Convervation* 2011; 66(4), 265-275, ISSN 0022-4561.

[14] Mitchell, D.O. and M. Mielke. Wheat: The global market, policies, and priorities. In Aksoy M.A. and J.C. Beghin (eds.) Global Agricultural Trade and Developing Countries. Washington, DC, USA: World Bank; 2005. p. 195-214.

[15] Food and Agriculture Organization of the United Nations (FAO). Conservation agriculture 2011. Available online at http://www.fao.org/docrep/013/al977e/al977e00.pdf (accesssed 13 August 2012).

[16] Zhao, H., G. Gao, X. Yan, Q. Zhang, M. Hou, Y. Zhu, Z. Tian. Risk assessment of agricultural drought usning the CERES-Wheat model: a case study of Henan Plain, China. *Climate Research* 2011; 50, 247-256, ISSN 0936-577X.

[17] Bonfil, D.J., I. Mufradi, S. Klitman, and S. Asido. Wheat grain yield and soil profile water distribution in a no-till arid environment. *Agronomy Journal* 1999; 91, 368-373, ISSN 0002-1962.

[18] De Vita, P., E. Di Paolo, G. Fecondo, N. Di Fonzo, and M. Pisante. No-tillage and conventional tillage effects on durum wheat yield, grain quality and soil moisture content in southern Italy. *Soil and Tillage Research*, 2007; 92, 69-78, ISSN 0167-1987.

[19] Gruber, S., C. Pekrun, J. Mohring, and W. Claupein. Long-term yield and weed response to conservation and stubble tillage in SW Germany. *Soil and Tillage Research*, 2012; 121, 49-56, ISSN 0167-1987.

[20] Wilson, H.P., M.P. Masgianica, T.E. Hines, and R.F. Walden. Influence of tillage and herbicides on weed control in a wheat (*Triticum aestivum*)- soybean (*Glycine max*) rotation. *Weed Science*, 1986; 34, 590-594, ISSN 0043-1745.

[21] Guy, S.O. and R.M. Gareau. Crop rotation, residue durability, and nitrogen fertilizer effects on winter wheat production. *Journal of Production Agriculture*, 1998; 11, 457-461, ISSN 0890-8524.

[22] Dao, T.H. Crop residues and management of annual grass weeds in continuous no-till wheat (*Triticum aestivum*). *Weed Science*, 1987; 35, 395-400, ISSN 0043-1745.

[23] FMC Corporation. 2012. Aim® *Herbicide Label.* Philadelphia, PA, USA: FMC Corporation Agricultural Products Group. 15 p.

[24] Bayer CropScience. 2011. Liberty® *Herbicide Label.* Research Triangle Park, NC, USA: Bayer CropScience LP. 20 p.

[25] Monsanto Company. 2009. Roundup WeatherMax® *Herbicide Label.* St. Louis, MO, USA: Monsanto Company. 54 p.

[26] Syngenta Crop Protection. 2011. Gramoxone® *Herbicide Label.* Greensboro, NC, USA: Syngenta Crop Protection, LLC. 55 p.

[27] E. I. du Pont de Nemours and Company. 2009. DuPont™ Finesse® *Herbicide Label.* Wilmington, DE, USA: E.I. du Pont de Nemours and Company, Inc. 12 p.

[28] Bayer CropScience. 2011. Huskie™ *Herbicide Label.* Research Triangle Park, NC, USA: Bayer CropScience LP. 24 p.

[29] E. I. du Pont de Nemours and Company. 2010. DuPont™ Harmony® Extra *Herbicide Label.* Wilmington, DE, USA: E.I. du Pont de Nemours and Company, Inc. 13 p.

[30] BASF Corporation. 2011. Beyond® *Herbicide Label.* Research Triangle Park, NC, USA: BASF Corporation. 23 p.

[31] Yenish, J.P., A.D. Worsham, and A.C. York. Cover crops for herbicide replacement in no-tillage corn (*Zea mays*). *Weed Technology* 1996; 10, 815-821, ISSN 0890-037X.

[32] Clark, A.J., A.M. Decker, and J.J. Meisinger. Seeding rate and kill date effects on hairy vetch-cereal rye cover crop mixtures for corn production. *Agronomy Journal*, 1994; 86, 1065-1070, ISSN 0002-1962.

[33] Fisk, J.W., O.B. Hesterman, A. Shrestha, J.J. Kells, R.R. Harwood, H.M. Squire, and C.C. Sheaffer. Weed suppression by annual legume cover crops in no-tillage corn. *Agronomy Journal*, 2001; 93, 319-325, ISSN 0002-1962.

[34] Utomo, M., W.W. Frye, and R.L. Blevins. Sustaining soil nitrogen for corn using hairy vetch cover crop. *Agronomy Journal*, 1990; 82, 979-983, ISSN 0002-1962.

[35] Wagger, M.G. Cover crop management and nitrogen rate in relation to growth and yield of no-till corn. *Agronomy Journal*, 1989; 81, 533-538, ISSN 0002-1962.

[36] Decker, A.M., A.J. Clark, J.J. Meisinger, F. Ronald Mulford, and M.S. McIntosh. Legume cover crop contributions to no-tillage corn production. *Agronomy Journal*, 1994; 86, 126-135, ISSN 0002-1962.

[37] Albaugh. 2012. Agri Star® 2,4-D Amine *Herbicide Label.* Ankeny, IA, USA: Albaugh, Inc. 36 p.

[38] Syngenta Crop Protection. 2009. Aatrex® *Herbicide Label.* Greensboro, NC, USA: Syngenta Crop Protection, LLC. 24 p.

[39] Valent U.S.A. 2010. Valor® *Herbicide Label.* Walnut Creek, CA, USA: Valent U.S.A. Corporation. 27 p.

[40] BASF Corporation. 2008. Prowl® *Herbicide Label.* Research Triangle Park, NC, USA: BASF Corporation. 24 p.

[41] Syngenta Crop Protection. 2011. Dual Magnum® *Herbicide Label.* Greensboro, NC, USA: Syngenta Crop Protection, LLC. 54 p.

[42] Bayer CropScience. 2005. Buctril® *Herbicide Label.* Research Triangle Park, NC, USA: Bayer CropScience LP. 36 p.

[43] Arysta LifeScience. 2004. Banvel® *Herbicide Label.* Cary, NC, USA: Arysta LifeScience North America, LLC. 27 p.

[44] Syngenta Crop Protection. 2012. Callisto® *Herbicide Label.* Greensboro, NC, USA: Syngenta Crop Protection, LLC. 32 p.

[45] Bayer CropScience. 2010. Laudis® *Herbicide Label.* Research Triangle Park, NC, USA: Bayer CropScience LP. 19 p.

[46] Syngenta Crop Protection. 2011. Evik® *Herbicide Label.* Greensboro, NC, USA: Syngenta Crop Protection, LLC. 8 p.

[47] Tessenderlo. 2010. Lorox® *Herbicide Label.* Phoenix, AZ, USA: Tessenderlo Kerley, Inc. 14 p.

[48] BASF Corporation. 2008. Lightning® *Herbicide Label.* Research Triangle Park, NC, USA: BASF Corporation. 10 p.

[49] Syngenta Crop Protection. 2009. Expert® *Herbicide Label.* Greensboro, NC, USA: Syngenta Crop Protection, LLC. 31 p.

[50] Fairhurst, T.H. and A. Dobermann. Rice in the global food supply. *Better Crops International,* 2002; 16, 3-6. http://www.ipni.net/ppiweb/bcropint.nsf/$webindex/0E477FFC43BD62 DA85256BDC00722F62/$file/BCI-RICEp03.pdf. (accessed 31 August 2012).

[51] Farooq, M., K.H.M. Siddique, H. Rehman, T. Aziz, D. Lee, and A. Wahid. Rice direct seeding: Experiences, challenges and opportunities. *Soil and Tillage Research,* 2011; 111, 87-98, ISSN 0167-1987.

[52] Mishra, J.S. and V.P. Singh. Tillage and weed control effects on productivity of a dry seeded rice-wheat system on a Vertisol in Central India. *Soil and Tillage Research,* 2012; 123, 11-20, ISSN 0167-1987.

[53] Chauhan, B.S. and D.E. Johnson. Influence of tillage systems on weed seedling emergence pattern in rainfed rice. *Soil and Tillage Research,* 2009; 106, 15-21, ISSN 0167-1987.

[54] Rao, A.N., D.E. Johnson, B. Sivaprasad, J.K. Ladha. and A.M. Mortimer. Weed management in direct-seeded rice. *Advances in Agronomy,* 2007; 93, 153–255, ISSN 0065-2113.

[55] Becker, M. and D.E. Johnson. Legumes as dry season fallow in upland rice-based systems of West Africa. *Biology and Fertility of Soils,* 1998; 27, 358-367, ISSN 0178-2762.

[56] FMC Corporation. 2011. Command® *Herbicide Label.* Philadelphia, PA, USA: FMC Corporation Agricultural Products Group. 19 p.

[57] Gowan Company. 2007. Permit® *Herbicide Label.* Yuma, AZ, USA: Gowan Company. 18 p.

[58] BASF Corporation. 2010. Facet® *Herbicide Label.* Research Triangle Park, NC, USA: BASF Corporation. 9 p.

[59] Valent U.S.A. 2001. Bolero® *Herbicide Label.* Walnut Creek, CA, USA: Valent U.S.A. Corporation. 4 p.

[60] United Phosphorus. 2009. Ultra Blazer® *Herbicide Label.* King of Prussia, PA, USA: United Phosphorus, Inc. 10 p.

[61] United Phosphorus. 2010. Londax® *Herbicide Label.* King of Prussia, PA, USA: United Phosphorus, Inc. 9 p.

[62] Arysta LifeScience. 2005. Basagran® *Herbicide Label.* Cary, NC, USA: Arysta LifeScience North America, LLC. 12 p.

[63] United Phosphorus. 2010. Stam® *Herbicide Label.* King of Prussia, PA, USA: United Phosphorus, Inc. 7 p.

[64] Dow AgroSciences. 2011. Clincher® *Herbicide Label.* Indianapolis, IN, USA: Dow AgroSciences LLC. 4 p.

[65] BASF Corporation. 2011. Newpath® *Herbicide Label.* Research Triangle Park, NC, USA: BASF Corporation. 12 p.

[66] BASF Corporation. 2011. Clearpath® *Herbicide Label.* Research Triangle Park, NC, USA: BASF Corporation. 10 p.

[67] Ebel, R. Soil management and conservation. In: Osteen, C., J. Gottlieb, and U. Vasavada (eds.) Agricultural Resources and Environmental Indicators, 2012. EIB-98, United States Department of Agriculture, Economic Research Service, August 2012. p 33-36. Available from http://www.ers.usda.gov/Publications/eib- economic-information-bulletin/eib98.aspx (accessed 5 September 2012).

[68] Campbell, R.B., R.E. Sojka, and D.L. Karlen. Conservation tillage for soybean in the U.S. Southeastern Coastal Plain. *Soil and Tillage Research,* 1984; 4, 531-541, ISSN 0167-1987.

[69] Robinson, E.L., G.W. Langdale, and J.A. Stuedemann. Effect of three weed control regimes on no-till and tilled soybeans (*Glycine max*). *Weed Science,* 1984; 32, 17-19, ISSN 0043-1745.

[70] Price, A.J., D.W. Reeves, and M.G. Patterson. Evaluation of weed control provided by three winter cereals in conservation-tillage soybean. *Renewable Agriculture and Food Systems,* 2005; 21, 159-164, ISSN 1742-1705.

[71] Liebl, R., F.W. Simmons, L.M. Wax, and E.W. Stoller. Effect of rye (*Secale cereale*) mulch on weed control and soil moisture in soybean (*Glycine max*). *Weed Technology,* 1992; 6, 838-846, ISSN 0890-037X.

[72] BASF Corporation. 2008. Outlook® *Herbicide Label.* Research Triangle Park, NC, USA: BASF Corporation. 17 p.

[73] BASF Corporation. 2009. Scepter® *Herbicide Label.* Research Triangle Park, NC, USA: BASF Corporation. 15 p.

[74] Loveland Products. 2008. Metribuzin *Herbicide Label.* Greeley, CO, USA: Loveland Products, Inc. 26 p.

[75] E. I. du Pont de Nemours and Company. 2010. DuPont™ Classic® *Herbicide Label.* Wilmington, DE, USA: E.I. du Pont de Nemours and Company, Inc. 15 p.

[76] Dow AgroSciences. 2011. FirstRate® *Herbicide Label.* Indianapolis, IN, USA: Dow AgroSciences LLC. 6 p.

[77] Syngenta Crop Protection. 2011. Fusilade® *Herbicide Label.* Greensboro, NC, USA: Syngenta Crop Protection, LLC. 39 p.

[78] BASF Corporation. 2011. Pursuit® *Herbicide Label.* Research Triangle Park, NC, USA: BASF Corporation. 27 p.

[79] Valent U.S.A. 2007. Cobra® *Herbicide Label.* Walnut Creek, CA, USA: Valent U.S.A. Corporation. 29 p.

[80] BASF Corporation. 2010. Poast® *Herbicide Label.* Research Triangle Park, NC, USA: BASF Corporation. 24 p.

[81] Syngenta Crop Protection. 2009. Flexstar® *Herbicide Label.* Greensboro, NC, USA: Syngenta Crop Protection, LLC. 26 p.

[82] National Cotton Council of America. Production Ranking 2012. http://www.cotton.org/econ/cropinfo/cropdata/rankings.cfm (accessed 5 September 2012).

[83] Carpenter, J. and L. Gianessi. Herbicide tolerant soybeans: Why growers are adopting Roundup Ready varieties. *AgBioForum,* 1999; 2, 65-72, ISSN 1522-936X.

[84] Schwab, E.B., D.W. Reeves, C.H. Burmester, and R.L. Raper. Conservation tillage systems for cotton in the Tennessee Valley. *Soil Science Society of America Journal,* 2002; 66, 569-577, ISSN 1435-0661.

[85] Nyakatawa, E.Z., K.C. Reddy, and D.A. Mays. Tillage, cover cropping, and poultry litter effects on cotton: II. Growth and yield parameters. *Agronomy Journal,* 2000; 92, 1000-1007, ISSN 1435-0645.

[86] Keeling, W., E. Segarra, and J.R. Abernathy. Evaluation of conservation tillage cropping systems for cotton on the Texas Southern High Plains. *Journal of Production Agriculture,* 1989; 2, 269-273, ISSN 0890-8524.

[87] Bauer, P.J. and D.W. Reeves. A comparison of winter cereal species and planting dates as residue cover for cotton growth with conservation tillage. *Crop Science,* 1999; 39, 1824-1830, ISSN 0002-1962.

[88] Reeves, D.W., A.J. Price, and M.G. Patterson. Evaluation of three winter cereals for weed control in conservation-tillage nontransgenic cotton. *Weed Technology,* 2005; 19, 731-736, ISSN 0890-037X.

[89] Johnson, M.D., D.L. Wyse, and W.E. Lueschen. The influence of herbicide formulation on weed control in four tillage systems. *Weed Science,* 1989; 37, 239-149, ISSN 0043-1745.

[90] Lindwall, C.W. Crop management in conservation tillage systems. In P. Unger (ed.) Managing Agricultural Residues. Boca Raton, FL, USA: Lewis Publishing, 1994. p. 185-210.

[91] Westerman, P.A., M. Liebman, F.D. Menalled, A.H. Heggenstaller, R.G. Hartzler, and P.M. Dixon. Are many little hammers effective? Velvetleaf (*Abutilon theophrasti*) population dynamics in two- and four-year crop rotation systems. *Weed Science,* 2005; 53, 382-392, ISSN 0043-1745.

[92] Makhteshim Agan. 2009. Cotoran® *Herbicide Label.* Raleigh, NC, USA: Makhteshim Agan of North America, Inc. 6 p.

[93] Syngenta Crop Protection. 2011. Caparol® *Herbicide Label.* Greensboro, NC, USA: Syngenta Crop Protection, LLC. 21 p.

[94] Valent U.S.A. 2007. Select® *Herbicide Label.* Walnut Creek, CA, USA: Valent U.S.A. Corporation. 30 p.

[95] E. I. du Pont de Nemours and Company. 2010. DuPont™ Assure® *Herbicide Label.* Wilmington, DE, USA: E.I. du Pont de Nemours and Company, Inc. 13 p.

[96] Syngenta Crop Protection. 2011. Envoke® *Herbicide Label.* Greensboro, NC, USA: Syngenta Crop Protection, LLC. 43 p.

[97] E. I. du Pont de Nemours and Company. 2011. DuPont™ Direx® Extra *Herbicide Label.* Wilmington, DE, USA: E.I. du Pont de Nemours and Company, Inc. 21 p.

[98] Tessenderlo. 2010. Linex® *Herbicide Label.* Phoenix, AZ, USA: Tessenderlo Kerley, Inc. 14 p.

[99] Drexel Chemical Company. 2009. MSMA *Herbicide Label.* Memphis, TN, USA: Drexel Chemical Company. 4 p.

[100] United States Department of Agriculture Foreign Agricultural Service (USDA-FAS). Peanut area, yield, and production. http://www.fas.usda.gov/psdonline/psdreport.aspx? hidReportRetrievalName=BVS&hidReportRetrievalID=918&hidReportRetrievalTemplateID=1#ancor (accessed 5 September 2012).

[101] Jordan, D.L., J.S. Barnes, C.R. Bogle, G.C. Naderman, G.T. Roberson, and P.D. Johnson. Peanut response to tillage and fertilization. *Agronomy Journal*, 2001; 95, 1125-1130, ISSN 0002-1962.

[102] Tubbs, R.S. and R.N. Gallaher. Conservation tillage and herbicide management for two peanut cultivars. *Agronomy Journal*, 2005; 97, 500-504, ISSN 0002-1962.

[103] Johnson, W.C. III, T.B. Brenneman, S.H. Baker, A.W. Johnson, D.R. Sumner, and B.G. Mullinix, Jr. Tillage and pest management considerations in a peanut-cotton rotation in the Southeastern coastal plain. *Agronomy Journal*, 2001; 93, 570-576, ISSN 0002-1962.

[104] Brandenberg, R.L., D.A. Herbert, Jr., G.A. Sullivan, G.C. Naderman, and S.F. Wright. The impact of tillage practices on thrips injury of peanut in North Carolina and Virginia. *Peanut Science*, 1998; 25, 27-31, ISSN 0095-3679.

[105] Marois, J.J. and D.L. Wright. Effect of tillage system, phorate, and cultivar on tomato spotted wilt of peanut. *Agronomy Journal*, 2003; 95, 386-389, ISSN 0002-1962.

[106] Wilcut, J.W., A.C. York, W.J. Grichar, and G.R. Wehtje. The biology and management of weeds in peanut (*Arachis hypogaea*). In H.E. Pattee and H.T. Stalker (eds.) Advan-

ces in Peanut Science. Stillwater, OK, USA: American Peanut Research Educational Society, 1995. p. 207-244.

[107] Grichar, W.J., B.A. Besler, R.G. Lemon, and K.D. Brewer. Weed management and net returns using soil-applied and postemergence herbicide programs in peanut (*Arachis hypogaea* L.). *Peanut Science*, 2005; 32, 25-31, ISSN 0095-3679.

[108] Rao, V.R. and U.R. Murty. Botany-morphology and anatomy. In J. Smartt (ed.) The Groundnut Crop: A Scientific Basis for Improvement. London: Chapman & Hall, 1994. p. 43-95.

[109] Price, A.J. and J.W. Wilcut. Weed management with diclosulam in strip-tillage peanut (*Arachis hypogaea*). *Weed Technology*, 2002; 16, 29-36, ISSN 0890-037X.

[110] Dow AgroSciences. 2010. Strongarm® *Herbicide Label.* Indianapolis, IN, USA: Dow AgroSciences LLC. 5 p.

[111] BASF Corporation. 2012. Cadre® *Herbicide Label.* Research Triangle Park, NC, USA: BASF Corporation. 9 p.

[112] Albaugh. 2010. Agri Star® Butyrac® *Herbicide Label.* Ankeny, IA, USA: Albaugh, Inc. 9 p.

[113] Banks, P.A. and E.L. Robinson. Soil reception and activity of acetochlor, alachlor, and metolachlor as affected by wheat (*Triticum aestivum*) straw and irrigation. *Weed Science*, 1986; 34, 607-611, ISSN 0043+1745.

[114] Schmitz, G.L., W.W. Witt, and T.C. Mueller. The effect of wheat (*Triticum aestivum*) straw levels on chlorimuron, imazaquin, and imazethapyr dissipation and interception. *Weed Technology,* 2001; 15, 129-136, ISSN 0890-037X.

[115] Reddy, K.N., M.A. Locke, and L.A. Gaston. Tillage and cover crop effects on cyanazine adsorption and desorption kinetics. *Soil Science,* 1997; 162, 501-509, ISSN 0038-075X.

[116] Locke, M.A., K.N. Reddy, and R.M. Zablotowicz. Weed management in conservation crop production systems. *Weed Biology and Management,* 2002; 2, 123-132, ISSN 1445-6664.

Chapter 23

Herbicide Safeners: Effective Tools to Improve Herbicide Selectivity

Istvan Jablonkai

Additional information is available at the end of the chapter

http://dx.doi.org/10.5772/55168

1. Introduction

Herbicide safeners, formerly referred to as herbicide antidotes, are chemical agents that increase the tolerance of monocotyledonous cereal plants to herbicides without affecting the weed control effectiveness. The use of safeners offer several benefits to agricultural weed control. Safeners may allow: (1) the selective chemical control of weeds in botanically related crops; (2) the use of nonselective herbicides for selective weed control; (3) the counteraction of residual activity of soil-applied persistent herbicides such as triazines in crop rotation systems; (4) an increase in the spectrum of herbicides available for weed control in "minor" crops; (5) an expansion and extension of the uses and marketability of generic herbicides; (6) the elucidation of sites and mechanism by serving as useful biochemical tools [1]. The commercial viability of safener concept is indicated by the growing number of herbicide-safener products available on the pesticide market. With the use of safeners, difficult weed control problems can be addressed and without safeners, many herbicidally active substances could have never been applied for weed control [2].

The concept to enhance crop tolerance to nonselective herbicide by using chemical agents was introduced by Otto Hoffman. In the late 1940s Hoffmann serendipitiously found that no herbicide injury symptons were developed in tomato plants previously treated with 2,4,6-T, an inactive analogue of herbicide 2,4-D when plant were exposed accidentally to vapors of 2,4-D due to the malfunction of the ventillation system of the greenhouse [3]. Following this observation Hoffmann reported later the antagonistic effects of 2,4-D against herbicidal injury by barban after foliar treatments of wheat plants [4]. Research and development in finding new safeners as well as subsequent commercialization proceeded very intensively in the 1970s. Since the patent application of the safening properties of 1,8-naphthalic anhydride (NA) intensive research on discovery of new safeners resulted in compounds with diverse chemistries (Table 1) successfully applied to alleviate injury symptoms by various classes of herbicides in cereal crops.

NA patented by Hoffmann [5] has been considered as the most versatile safener showing less botanical and chemical specificity than other safeners developed later. NA protected cereals as seed treatments against various herbicide chemistries [6]. NA was reported to be mildly phytotoxic to maize (chlorosis and growth inhibition) under some growing conditions. One problem in treating seeds with safeners prior to planting is that phytotoxicity can increase as the time the safener is exposed to the seed increases. With NA, the phytotoxicity to the crop increases with increased time the safener is in contact with the seed during storage. This problem has thus far prevented NA from being introduced to the commercial market [7].

The introduction of dichloroacetamide derivatives developed as safeners against thiocarbamates and chloroacetanilides was a breakthough in the history of the safeners since these compounds can be applied to the soil in preplant incorporated (PPI) or preemergence (PRE) technology in prepackaged tank mixture with the herbicide. Generally, prepackaged herbicide-safener mixtures offer several advantages over seed safeners. First of all, the manufacturer controls all components of the formulation secondly, the farmers buy and use a single and reliable product which allows a wider selection of crop cultivars. Dichlormid exhibited a remarkable degree of chemical and botanical specificity in protection of maize against thiocarbamates such as EPTC, butylate, vernolate but the safener was less protective to maize against chloroacetanilides. In addition to dichlormid, a number of dichloroacetylated amine derivatives were marketed. Among them AD-67, a spiro-oxazolidine compound was commercialized to protect maize plants against acetochlor while benoxaxor can be used to safen *S*-metolachlor or racemic metolachlor in maize. Furilazole, in addition to providing protection against acetochlor, has a very good safening effect on sulfonylureas particularly halosulfuron. The dichloromethyl-1,3-dioxolane MG-191 the most active member of dichloroacetal and ketal derivatives, protects maize against thiocarbamate and chloroacetanilide injuries. MG-191, similarly to dichlormid, is more effective against thiocarbamates than chloroacetanilides.

The oxime ethers such as cyometrinil, oxabetrinil, and fluxofenim were marketed as seed treatment safeners to protect sorghum plants against chloroacetanilides, in particular, metolachlor. Flurazole, a 2,4-disubstituted 5-thiazolecarboxylate is also a seed safener allowing the safe use of alachlor in sorghum. The phenylpyrimidine safener fenclorim was introduced against pretilachlor in rice and can be used in tank mixture formulated together with the chloroacetanilide herbicide.

The urea type dymron and the thiocarbamate dimepiperate are actually herbicidally active compounds that possess safening activity against pretilachlor [8] and bensulfuron [9] in rice.

Trends toward post-emergence herbicide treatments and the use of high-activity herbicide molecules have led to the development of safeners with post-emergence application in winter cereals. A new era in safener research began with the discovery of 1,2,4-triazolcarboxylates and fenchlorazole-ethyl was developed as a post-emergence safener against ACCase inhibitor fenoxaprop-ethyl in wheat in a tank mixture with the herbicide. Similarly, the dihydropyrazol dicarboxylate mefenpyr-diethyl was used against ACCase inhibitors including fenoxaprop-

http://dx.doi.org/10.5772/55168

ethyl as well as mesosulfuron and iodosulfuron in a variety of cereals. The main application of 8-quinolinoxy-acetate cloquintocet-mexyl is against clodinafop-propargyl in wheat. Dihydroisoxazole-carboxylate isoxadifen-ethyl can safen herbicides of various mode of action. First, it was applied in maize in combination with foramsulfuron but mixture with foramsulfuron and iodosulfuron-methyl is also in use. In rice, it can be used with fenoxafop-P-ethyl and ethoxysulfuron. The arylsulfonyl-benzamide, cyprosulfamide is the latest achievement in safener research. It protects maize against isoxaflutole pre-emergence and can also be used in maize with isoxaflutole plus thiencarbazone in pre-emergence and early post-emergence applications [10].

Interestingly, no successful safeners have been developed for broad-leafed crops. Recently, the non-phytotoxic microbial inhibitor dietholate (*O, O*-diethyl-*O*-phenyl phosphorothioate) [11] used to inhibit soil microbes that degrade thiocarbamate herbicides was patented as a Table 1 safener for cotton plants against injuries by clomazone [12].

Despite large amount of information published on the activity, mode of action and uses of safeners during the 50-year history of these herbicide antagonists this overview will focus on several less addressed topics such as a) relationships between the molecular structure and the safening properties; b) basis for differential chemical selectivity; and c) safener effects on detoxifying enzymes in crop plants and weeds.

Chemical class	Name	Structure[a]	logP	Herbicide	Crop	Appl. method
Anhydride	1,8–Naphthalic anhydride (NA)	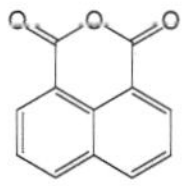	2.54	Thiocarbamates	Maize	Seed–treatment
Dichloro–acetamide	Dichlormid	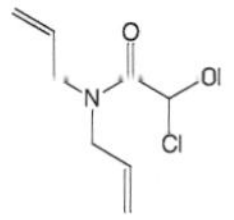	1.84	Thiocarbamates Chloroacet-anilides	Maize	PPI, PRE
	Furilazole	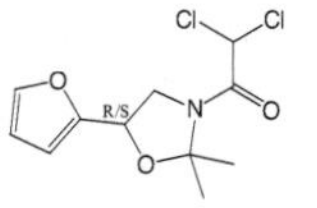	2.12	Acetochlor Halosulfuron-methyl	Maize	PRE
	AD–67	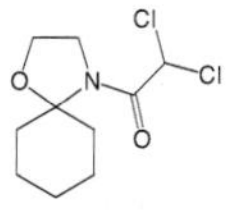	2.32[b]	Acetochlor	Maize	PRE
	Benoxacor	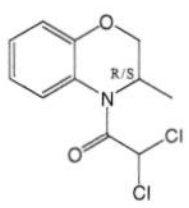	2.69	Metolachlor	Maize	PRE

Chemical class	Name	Structure[a]	logP	Herbicide	Crop	Appl. method
Oxime ether	Cyometrinil	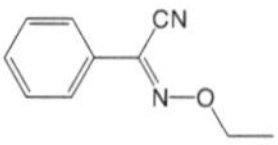	1.56	Chloroacet–anilides (metolachlor)	Sorghum	Seed–treatment
	Oxabetrinil	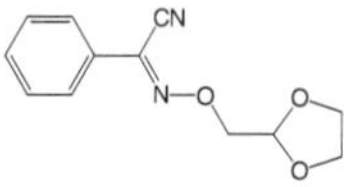	2.76	Chloroacet–anilides (metolachlor)	Sorghum	Seed-treatment
	Fluxofenim	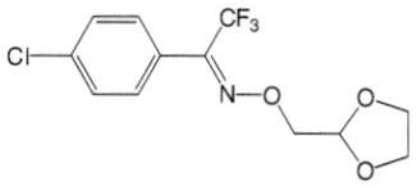	2.90	Chloroacet-anilides (metolachlor)	Sorghum	Seed–treatment
Thiazole carboxylic acid	Flurazole	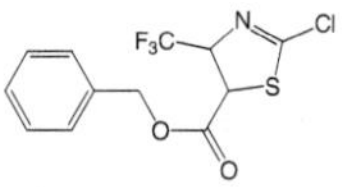	3.64[b]	Alachlor	Sorghum	Seed–treatment
Dichloromethyl-ketal	MG–191	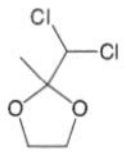	1.35[b]	Thiocarbamates Chloroacet-anilides	Maize	PRE
Phenyl–pyrimidine	Fenclorim	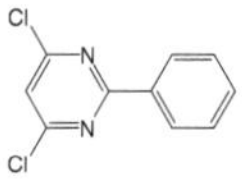	4.17	Pretilachlor	Rice	PRE
Urea	Dymron	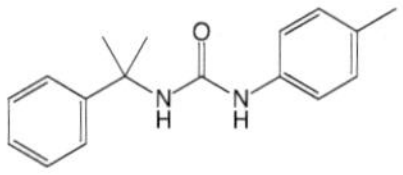	2.70	Pyributicarb Pretilachlor Pyrazosulfuron–ethyl	Rice	PRE, POST
Piperidine–1–carbothioate	Dimepiperate	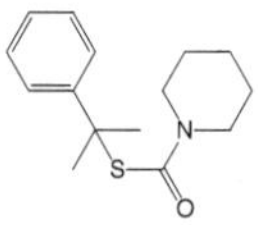	4.02	Sulfonylureas	Rice	POST
8–Quinolinoxy–carboxylic esters	Cloquintocet–mexyl	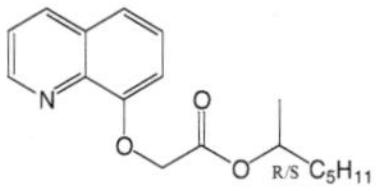	5.03	Clodinafop–propargyl	Cereals	POST
1,2,4–Triazole–carboxylate	Fenchlorazole–ethyl	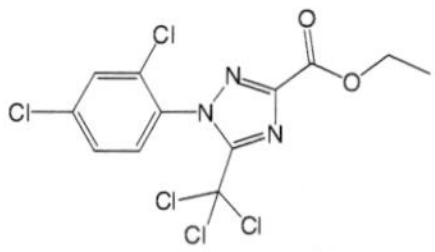	4.52	Fenoxaprop–ethyl	Cereals	POST

Chemical class	Name	Structure[a]	logP	Herbicide	Crop	Appl. method
Dihydropyrazole–dicarboxylate	Mefenpyr–diethyl	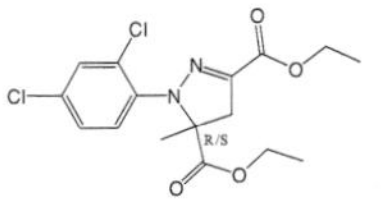	3.83	ACCase inhibitors Sulfonylureas	Wheat, Rye, Triticale, Barley	POST
Dihydroisoxazole–carboxylate	Isoxadifen–ethyl	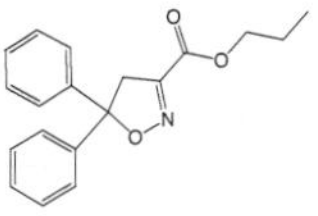	3.88[b]	ACCase inhibitors Sulfonylureas	Maize Rice	POST
Arylsulfonyl–benzamide	Cypro–sulfamide	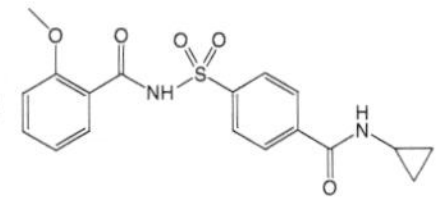	2.09[b]	Isoxaflutole	Maize	PRE, POST

[a] Safeners used as racemic mixtures are indicated by *R/S* in their structures.

[b] Log P values unavailable were calculated by ALOGPS 2.1 program available online at www.vclab.org/articles/cite.html.

Table 1. Structure, logP and application of some important safeners.

2. Structure–safening activity relationships

Structure-activity correlations are very important in the search for biological activity because they provide useful information about chemical substituents that are necessary for the required bioactivity. Published structure-activity correlation studies with safeners and analogous compounds have been limited.

Hoffmann's original patent for NA against EPTC in maize claimed only a few NA analogs such as alkyl esters, barium and tin salts as well as *N,N'*-diallyl naphthalene-1,8-dicarboxylic acid, *N,N'*-diallyloxamide, *N,N'*-dipropynyloxamide, *N,N,N',N'*-tetrapropynyloxamide and dipropynylmalonamide [5]. In addition to the original patent, the effects of other structural analogs of NA were tested against EPTC in maize as seed dressing [13]. The presence of the dicarboxylic anhydride group and at least one aromatic ring attached directly to the anhydride appeared to be essential for the protective activity of NA structural analogues. Derivatives such as acenaphthylene-1,2-dione, benzoisoquinoline-1,3-dione, 4-amino-NA, naphthalic-dianhydride, phtalic anhydride as well as diphenic anhydride showed safening effects while chlorinated NA, 2-phenylglutaric anhydride and phenalene-1-one were toxic to maize.

Detailed structure-activity correlations were conducted mainly with various amide safeners that protect maize from thiocarbamate injury. Studies with several hundred of amides revealed that the most effective safeners were *N,N*-disubstituted acetamides [14] or substituted *N*-acetyl-1,3-oxazolidines [15, 16]. Structure-activity studies with dichloroacetamides revealed that *N,N*-disubstituted derivatives were more effective than monosubstituted amides. A variety of substituents on the nitrogen atom including alkyl, haloalkyl, alkenyl and heterocyclic groups impart various degrees of protective activity. Nevertheless, mono- and trichloroacetamides exhibited less safening activity than dichloro analogues [17, 18]. Based on these SAR studies similarities between the chemical structure of the herbicide and its safener, the possible competitive antagonism between the thiocarbamate and the safener molecules for a common target site has been postulated [19]. Computer-aided molecular modeling (CAMM) studies supported this theory [20]. Superimposing of the structures of dichlormid and EPTC revealed that the two chlorine atoms of the safener do not superimpose over any functional group of the EPTC. If structure of EPTC sulfoxide, the very phytotoxic EPTC metabolite, and the dichlormid were superimposed, the two compounds were similar with functional groups in the same location on both molecules. Comparative three-dimensional quantitative structure-activity relationship studies using comparative molecular field analysis (CoMFA) also supported the competitive antagonism theory and predicted a structure of *N*-allyl-*N*-methoxyethoxymethyl dichloroacetamide as a potent highly effective safener [21].

Structure-safening activity studies with oxime ether derivatives revealed that the safening activity is affected by the number of nucleophilic sites present in the molecule. An oxime ether with two nucleophilic sites was more effective than those with only one. In addition to cyometrinil, oxabetrinil and fluxofenim pyridin-2-aldoxime *O*-ethers such as benzyl and phenylethyl ethers were protective to grain sorghum in seed treatments against metolachlor. The oxime and aldehyde derivatives tested, in terms of decreasing safening effectiveness, were dimethyglyoxime > benzophenone oxime > pyridine-2-aldoxime > benzoin--oxime > methyl thioacetohydroxamate >pyridine-2-aldoxime methiodide > 5-nitro-furancarboxyaldehyde [22]. CAMM evaluations of the oxime ether analogues cyometrinil, oxabetrinil and fluxofenim revealed that as the effectiveness of the safener increases so does its molecular similarity to metolachlor [20].

Structure-safening activity relationships for thiazol-5-carboxylic acids against acetamide herbicides were described for 60 derivatives in the original patent [23]. Thiazolecarboxylates substituted by a trifluoromethyl in the 4-position are clearly superior to those substituted in the 4-position by methyl in reducing herbicidal injury to sorghum. Another preferred group of thiazolecarboxylates contained a halogen atom at position 2 preferably chlorine.

A structure-activity relationship study to safen maize against acetochlor was carried out with the herbicide safener MG-191 and its acetal and ketal analogues at preemergence application [24-26]. Open chain acetals formed from 1,1-dichloroacetaldehyde exhibited only marginal safening efficacy. Dialkyl ketals of 1,1-dichloroacetone showed increasing effectiveness up to 3 carbon length of the alkyl group with further increases in carbon atoms resulted in loss of activity. The 5-, 6- and 7-membered 1,3-dioxacycloalkanes prepared from dichloroacetaldehyde had hardly detectable safening activity. However, introducing alkyl or aryl substitution at the 2-position of the 1,3-dioxacycloalkane ring remarkably increased the safening activity. Regarding ring size the highest activity observed was for 2-dichloromethyl-2-methyl-1,3-dioxepane. Replacing an oxygen in the 1,3-dioxolane ring for nitrogen resulted in oxazolidines with reduced safening activities but alkyl or aryl substitution on the nitrogen increased the safening activity of compounds. Replacement of oxygens by sulfur atoms leads to less active derivatives among which 1,3-dithiolane derivative showed higher activity than the oxathiolanes. Various 1,3-dioxolane-4-ones provided significant protection against the acetochlor. Benzo[1,3]-dioxoles were ineffective while benzo[1,3]dioxin-4-ones were protective in safening maize. 5-Dichloromethyl-3-substituted-isoxazoles were also active safeners.

Unfortunately, no publication has been reported for the other chemistry of safeners. However, no unifying structural motifs for compounds to be safeners can be predicted from these studies.

3. Chiral safeners

The importance of the chirality in the biological activity has long been recognized. Since biochemical processes in the cells take place in chiral environment and most enzymatic pathways are stereoselective, a high degree of enantiomeric and enantiotopic selectivity can be obtained when chiral or prochiral molecules are introduced into biological systems. About one fourth of the presently available pesticides are chiral, existing as two mirror images called enantiomers. These stereoisomers generally possess identical physico-chemical properties but widely different biological activities, such as toxicity, mutagenicity, and carciogenecity [27]. The active enantiomer of the chiral pesticide would have the desired effect on target species while the other may be inactive [28].

Among the commercially available safeners, four such as benoxacor, furilazole, cloquintocet-mexyl, and mefenpyr-dietyl are chiral compounds but used exclusively as racemic (*R*/*S*) mixtures in herbicidal compositions and no information accessible on the safening efficacy of the individual enantiomers. In one recent patent, the *R* enantiomer of furilazole is described in a herbicidal mixture as a safener [29].

Nevertheless, only a few molecules have been reported as safeners in enantiomerically pure form. The optical isomers of 4-(dichloromethylene)-2-[*N*-(α-methylbenzyl)imino]-1,3-dithiolane hydrochloride were synthesized and were tested against triallate in wheat [30]. The *R* enantiomer exhibited high safening activity and its activity exceeded that of the *S* and the racemic compound. The monoterpene *R*-carvone was found more effective than the *S* enantiomer to safen maize against acetochlor injury [31]. 2-Dichloromethyl-2-methyl-[1,3]oxathiolane 3-oxide, a structural analogue of the MG-191 safener, was prepared and the enantiomers were separated by chiral HPLC [32]. The more polar diastereomeric pair was as effective as MG-191 while the other exhibited only marginal protection against acetochlor. Inducibility of ZmGSTF1-2 from roots was more enhanced by the stereoisomers with higher safening efficacy while only one of these enantiomers was effective in shoots. The findings indicated the importance of the stereochemistry in the protective effectiveness. The safener (*S*)-3-dichloroacetyl-2,2-dimethyl-4-ethyl-1,3-oxazolidine was found to induce the GSH content and GST activity in root and shoot of maize seedlings but the effect of the *R* form was not reported in these experiments [33]. As a future prospect, the needs for broad application of the green technology in the sustainable agriculture will probably induce a shift in the use and development of enantiomerically pure safeners.

4. Prosafeners and natural compounds with safening activity

The term prosafeners refers to molecules with safening activity undergoing biotransformation to the actual safening agent prior to exhibiting their safening effect. Substituted *N*-phenylmaleamic acids and their progenitors *N*-phenylmaleimides and *N*-phenylisomaleimides exhibited safening activity against alachlor in sorghum at preemergence application [34]. Simple hydrolytic ring-opening reaction of *N*-phenylmaleimides and *N*-phenylisomaleimides results in the *N*-phenylmaleamic acid derivatives with safening activity. Two thiazolidine derivative L-2-oxathiazolidine-4-carboxylic acid (OTC) [35] and thioproline (L-thiazolidine-4-carboxylic acid) [36] have been reported to safen sorghum against tridiphane injury. OTC is converted by 5-oxoprolinase to *S*-carboxy-L-cysteine which spontaneously decarboxylates to yield L-cysteine. The conversion of thioproline to cysteine takes place in two steps, first proline oxidase yields *N*-formyl-L-cysteine from which cysteine is forming by hydrolysis. Either source of cysteine elevates the glutathione level in plants and therefore enhance herbicide detoxication.

Safening activities of natural cyclic hydroxamic acids (DIMBOA, DIBOA, and MBOA) as well as synthetic analogues such as 1,4-benzoxazin-3-ones and 1,3-benzoxazolidin-2-ones were prepared and tested to safen maize against acetochlor and EPTC injuries [37]. Cyclic hydroxamic acids were supposed to act as safeners by catalyzing hydroxylation of herbicides containing reactive chlorine in their structure and they are ineffective against herbicides not possessing leaving groups. While no safening activities of natural hydroxamic acids were detected, the synthetic analogues exhibited low to moderate activity.

Metabolism of the herbicide safener, fenclorim resulting, in a semi-natural product with safening activity has recently been described in *Arabidopsis thaliana* cell cultures [38]. The metabolism of fenclorim mediated by GSTs yielded *S*-(fenclorim)-glutathione conjugate that was sequentially catabolized to *S*-(fenclorim)-cysteine then to 4-chloro-6-(methylthio)-phenylpyrimidine (CMTP). Although the fenclorim conjugates tested showed little GST inducing activity in *Arabidopsis*, the formation of CMTP resulted in metabolic reactivation, with the product showing enhancing activity similar to that of parent safener. In addition, CMTP safened rice plants and induced rice GSTs. The formation of CMTP by metabolic bioactivation can contribute to the longevity of safener action since it was found stable 8 – 24 h after application.

Oxylipins constitute a family of oxygenated natural products which are formed from fatty acids. Safeners and reactive electrophilic oxylipins (RES oxylipins) have a common biological activity in that they both strongly induce the expression of defence genes and activate detoxification responses in plants [39, 40]. Surprisingly, the application of oxylipin A has been found to reduce the herbicidal injury [41].

5. Interaction of safeners and herbicides on the absorption and translocation

Published results on how safeners affect the herbicide absorption are rather contradictory and, therefore, no general conlusion can be drawn. In an excellent summary the effect of 15 safeners

http://dx.doi.org/10.5772/55168

toward various herbicides was reviewed [2]. Interestingly the majority of papers published report safener-enhanced herbicide uptake followed by no effect then reduced uptake results. According to a recent study mefenpyr-diethyl had no effect on the uptake of either mesosulfuron-methyl or iodosulfuron-methyl-sodium [10]. These results suggest that the influence of safeners on the herbicide uptake may not be a decisive factor in the protective action. However, the knowledge of absorbed amounts of safeners and herbicides by crops may help to determine the optimal herbicide/safener ratios applicable in the agricultural practice. In addition, determination of the site of safener and herbicide uptake can contribute to prepare the most selective herbicide-safener mixture. A suitable placement of soil-applied herbicides to roots or the emerging shoots is of great practical importance in achieving the most effective weed control and the least injury to crop plants.

Studies on how maize can differentiate in the absorbtion of herbicides and safeners were conducted with radiolabeled EPTC, acetochlor and MG-191 [42, 43]. Time-dependent uptake of root-applied [^{14}C]EPTC reached a maximum after 6h and decreased up to 3 days (Figure 1). The first measurable shoot growth inhibition appeared just after 1-day-exposition to the herbicide and 38% shoot length inhibition was observed 3 DAT. In general, the MG-191 safener had no influence on the herbicide uptake except for 1 DAT when the safener enhanced the herbicide uptake by 1.5-fold as compared to that in the unsafened plants. Nevertheless, the safener conferred a complete protection to maize throughout the study. The highest amount of herbicide uptake was 65 µg/g fresh weight.

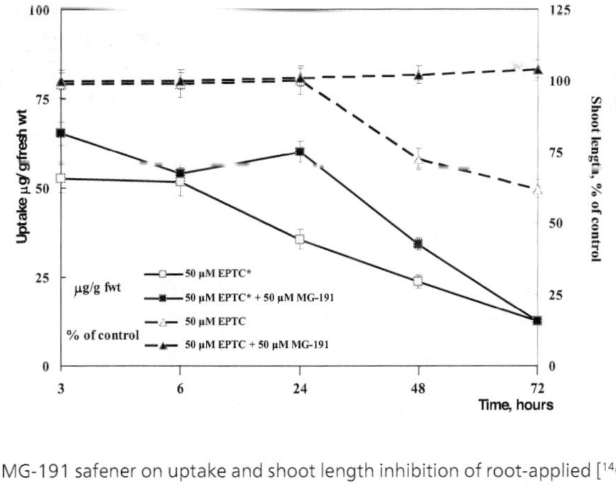

Figure 1. Influence of MG-191 safener on uptake and shoot length inhibition of root-applied [^{14}C]EPTC.

As a comparision, the amount of root-absorbed [^{14}C]acetochlor was continuously increased up to 3 days (Figure 2). As a result of increasing uptake the first detectable shoot length inhibition occurred 6h after treatment. At 3 DAT 28% shoot and 52% root (data not shown) growth inhibition by the herbicide occurred. Addition of the MG-191 safener did not affect the acetochlor absorption by maize seedlings but completely antagonized the herbicide shoot growth inhibition. The maize seedlings absorbed much higher amounts of acetochlor (377 μg/g fresh weight).

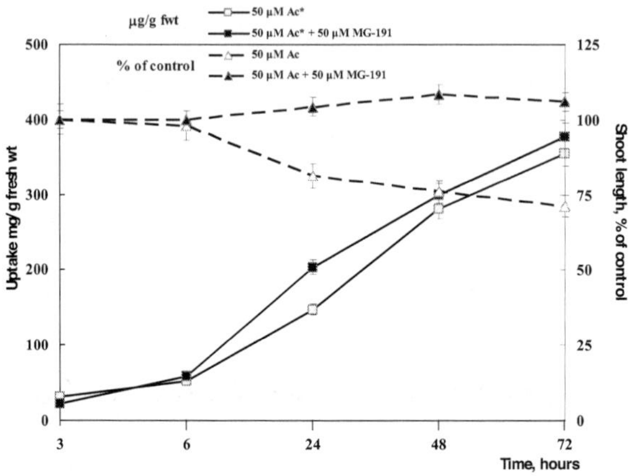

Figure 2. Influence of MG-191 safener on uptake and shoot length inhibition of root-applied [^{14}C]acetochlor.

All previous efforts to elucidate modes of action of safeners focused on the fate of various herbicides as affected by the safener treatments while no studies were conducted on how uptake, translocation and metabolism of safeners were influenced by herbicides. For a better understanding of the herbicide-safener interaction, absorption of [^{14}C]MG-191 by maize seedlings was studied as influenced by EPTC. Absorbed amount of the labeled safener following application to the roots of 5-day-old maize plants increased over the time and no influence of EPTC on uptake was observed (Figure 3). At a higher safener concentration (50 μM), plants absorbed higher amounts of radiolabel than at a lower concentration (10 μM) but plants contained low levels (3% and 1%) of the safener applied. The highest value for the safener content in the maize seedlings was less than 8 μg/g fresh weight.

These data clearly suggest that even this small amount of safener offer protection to maize. The absorbed herbicide/safener ratio (μg/μg) at 3 DAT accounted for 50 with acetochlor and 1.7 with EPTC at same concentrations of the herbicide. These results may partly explain why safening efficacy of MG-191 toward EPTC is higher than toward acetochlor under field conditions. Site of uptake can also affect the MG-191 effectiveness. In experiments using a charcoal barrier to separate shoot and root zones of maize, the influence of site of safener placement on acetochlor phytotoxicity was studied [44]. MG-191 was the most protective when both the safener and the herbicide were applied simultaneously to shoots and roots but also satisfactory protection was achieved when the safener was applied in the root zone and the herbicide to the emerging shoots. This also indicates the main site of uptake for acetochlor absorption is the coleoptile while the root-uptake is very significant in the safener performance. Under field conditions the more water-soluble MG-191 (log P, 1.35) can be more easily leached

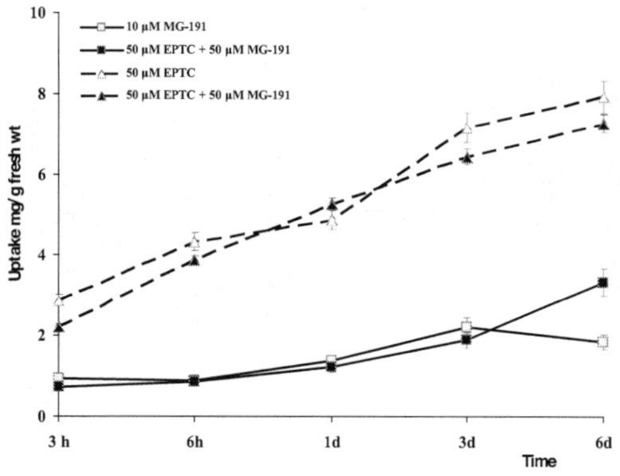

Figure 3. Time-course uptake of root-applied [^{14}C]MG-191 by 5-day-old maize seedlings and the influence of EPTC.

to the roots of maize plants than the less water-soluble acetochlor (log P 4.14). The higher logP of acetochlor also supports its higher uptake as compared to MG-191.

It is also difficult to put the results of safener affected translocation of absorbed herbicides in perspective. Reduction of translocation of herbicides such as acetochlor, methazachlor, and imidazolinones from roots of maize to the shoots following treatments with dichlormid, BAS 145,138 and NA is likely a consequence of the safener-enhanced herbicide metabolism to more polar and less mobile products [45-49]. On the other hand, no effect of MG-191 on EPTC and acetochlor translocation has been observed [42, 44]. It is interesting to note that safener MG-191 and the herbicide acetochlor exhibit different translocation patterns (Figure 4). While the majority of the absorbed radiolabel from [^{14}C] acetochlor was found in the roots and coleoptiles of maize seedlings (Figure 4a), the root-applied [^{14}C]MG-191 distributed evenly within the plants (Figure 4b) showing similar mobility and distribution as EPTC (data not shown). This may be further evidence for the higher protective efficacy of this safener against EPTC as compared to acetochlor. The similar translocation pattern of the herbicide and the safener may be a prerequisite for the high level of safening activity.

6. Action of safeners on the glutathione-mediated detoxification of herbicides

Various chemistries of safeners were found to enhance the herbicide detoxification in the safened plants by elevating the activity of the mediating enzymes such as glutathione S-transferases (GSTs), cytochrome P450 mixed function oxidases (CYPs), glycosyltransferases (UGTs) and ATP-binding casette (ABC) transporter proteins as well as a cofactor endogenous glutathione (GSH) involved in detoxification of herbicides [2, 50-52]. The best studied group

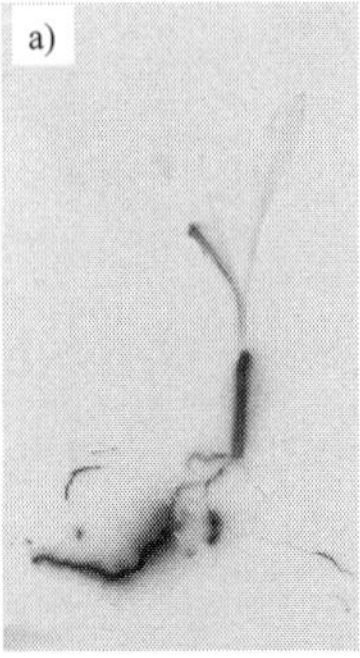

Autoradiograph of [^{14}C]acetochlor-treated (2.5 kg/ha) maize seedling 11 DAT

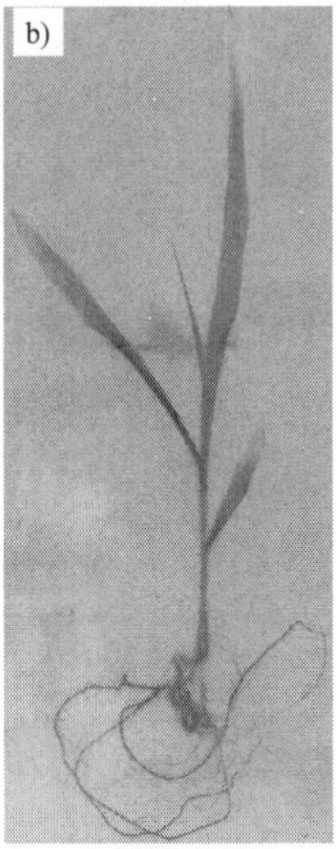

Autoradiograph of [^{14}C]MG-191-treated (50 μM) maize seedling 3 DAT

Figure 4. Distribution of root- and shoot-applied [^{14}C]acetochlor and root-applied [^{14}C]MG-191 in maize seedlings.

of plant enzymes involved in herbicide metabolism is the GSTs that mediate the conjugation of the major cellular thiol tripeptide, GSH with herbicide substrates. GSTs are multifunctional enzymes, each composed of two subunits which catalyze conjugation of a broad range of electrophilic substrates with GSH [53]. Herbicides known to conjugate with GSH include thiocarbamates, chloro-*s*-triazines, triazinone sulfoxides, chloroacetanilides, diphenylethers, some sulfonylureas, aryloxyphenoxypropionates, thiazolidines, and sulfonamides [54, 55]. Plant GSTs comprise a large and diverse group, with 54 GST genes encoded by the *Arabidopsis* genome, and have been classified on sequence similarity, genomic organization and functions into several distinct subclasses [56]. In plants, phi (F) and tau (U) classes are the most prominent GSTs involved in herbicide detoxification [57-59]. In addition to up-regulating GST expression, safeners also enhance the activity of enzymes involved in sulfate assimilation and GSH biosynthesis thereby elevating the level of GSH [50, 60].

Only two studies are available in the literature on how the safener structure affects the expression of GST isoforms. The herbicide safener MG-191 (2-dichloromethyl-2-methyl-1,3-dioxolane) and its less effective structural analogue dichloromethyl-dioxolanone (NO-17; 2-dichloromethyl-2,5-dimethyl-1,3-dioxolane-4-one) were reported to differentially enhance the expression of members of the GSTs in maize [61].

None of these safener molecules had influence on the expression of *Zm*GSTF1-2 (Figure 5a and b). However, MG-191 and, to a lesser extent NO-17 selectively enhanced the expression of tau class *Zm*GSTU1 in both root and shoot tissues after 1 day of treatment (Figure 5c and d). Addition of cycloheximide to the treatment solutions suppressed the enhancement of expres-

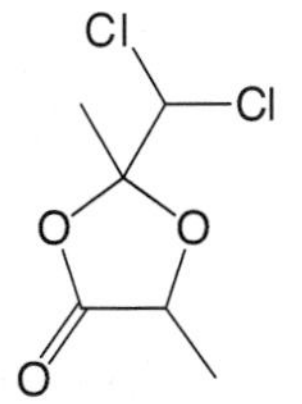

NO–17

sion of *Zm*GSTU1 only in the roots. *Zm*GSTU1 has previously been shown to play a key role in metabolism of nitrodiphenyl ether type herbicides [54].

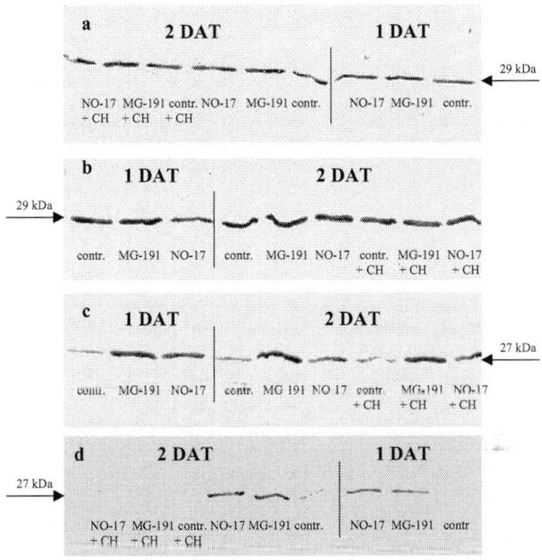

Figure 5. Western blots of crude GST extracts from maize roots and shoots (a) Analysis of GSTs from maize shoots using the anti-*Zm*GSTF1-2 serum.(b) Analysis of GSTs from maize roots using the anti-*Zm*GSTF1-2 serum.(c) Analysis of GSTs from maize shoots using the anti-*Zm*GSTU1-2 serum.(d) Analysis of GSTs from maize roots using the anti-*Zm*GSTU1-2 serum.

Analysis of isoenzyme profile of maize GSTs revealed that phi class of GSTs predominate, with ZmGSTF1 as the major subunit which is present constitutively and shows high specificity to 1-chloro-2,4-dinitrobenzene (CDNB) substrate [62]. A second phi type GST termed *Zm*GSTF2 accumulates following treatments with herbicide safeners. These subunits can dimerise together to form *Zm*GSTF1-1 and *Zm*GSTF2-2 homodimers as well as *Zm*GSTF1-2 heterodimer. In addition to these three phi GST isoenzymes a phi type GST *Zm*GSTF3 and three tau class GSTs *Zm*GSTU1, *Zm*GSTU2 and *Zm*GSTU3 are present in lower amounts [63, 64]. While the expression of *Zm*GSTF2 was enhanced by auxins, herbicides, the herbicide safener dichlormid and glutathione, the *Zm*GSTU1 subunit was induced more selectively, only accumulating significantly in response to dichlormid treatment [63]. Although *Zm*GSTF2 has been considered more active in detoxifying metolachlor and alachlor than *Zm*GSTF1 it is far less abundant [65]. The importance of *de novo* synthesis of the isoenzyme *Zm*GSTU1 in its safening action is difficult to explain. Nevertheless, these results indicate that dichloromethyl-dioxolane type MG-191 is a more specific inducer of maize GSTs than other compounds commonly used to safen thiocarbamate or chloroacetanilide herbicides in maize.

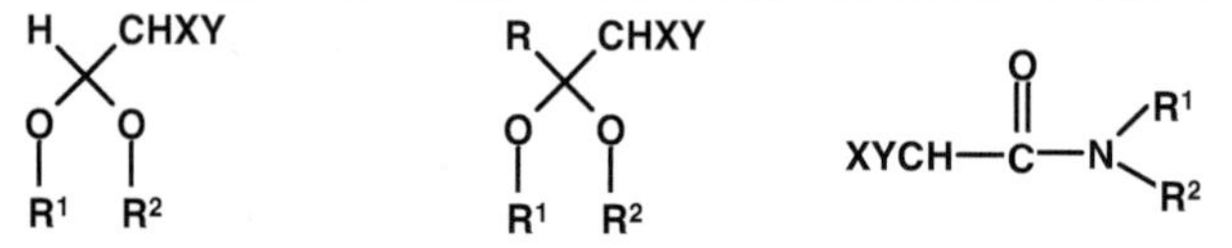

1a-l **2a-k** **3a-d**

Code	R	X	Y	R^1	R^2	Protection[a] (%)	GSH[b]	GST(CDNB)[c]	GST (Ac)[d]
							treated/control		
Ac	-	-	-	-	-	-	1.11	1.48	3.74
1a	-	H	Cl	Et	Et	24	1.49	0.69	2.03
1b	-	H	Br	Et	Et	60	1.53	0.94	3.76
1c	-	Cl	Cl	Et	Et	8	0.69	1.42	1.83
1d	-	Cl	Cl	Pr	Pr	0	0.80	0.95	1.38
1e	-	Cl	Cl	Bu	Bu	-6	1.22	0.96	0.91
1f	-	Cl	Cl	i-Bu	i-Bu	-2	1.20	0.90	1.61
1g	-	Cl	Cl	$-(CH_2)_2-$		18	0.93	0.88	1.23
1h	-	Cl	Cl	$-(CH_2)_3-$		14	0.60	0.89	0.90
1i	-	Cl	Cl	$-CH_2C(CH_3)_2CH_2-$		-3	0.91	0.88	1.33
1j	-	Cl	Cl	$-(CH_2)_4-$		11	0.95	1.03	1.33
1k	-	Cl	Cl	$-(CH_2)_5-$		0	0.98	1.24	0.58
1l	-	Cl	Cl	$-(CH_2)_6-$		3	0.82	1.32	0.83
2a	Me	Cl	Cl	Et	Et	62	1.15	1.22	0.65
2b	Me	Cl	Cl	Pr	Pr	63	0.98	1.18	0.85
2c	Me	Cl	Cl	Bu	Bu	38	0.78	0.88	3.93
2d	Me	Cl	Cl	i-Bu	i-Bu	14	0.85	1.07	4.72
2e	Ph	Cl	Cl	$-(CH_2)_2-$		41	2.00	1.94	2.23
2f	Me	Cl	Cl	$-(CH_2)_2-$		64	1.18	1.83	3.93
2g	Me	Cl	Cl	$-(CH_2)_3-$		68	1.31	1.49	1.96
2h	Me	Cl	Cl	$-CH_2C(CH_3)_2CH_2-$		66	1.62	1.77	1.44
2i	Me	Cl	Cl	$-(CH_2)_4-$		70	1.71	1.48	0.92
2j	Me	Cl	Cl	$-(CH_2)_5-$		50	1.24	1.27	1.19

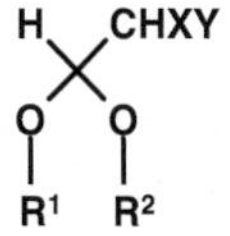

1a-l

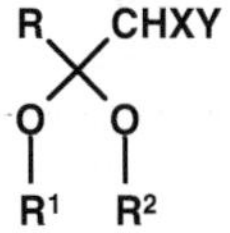

2a-k

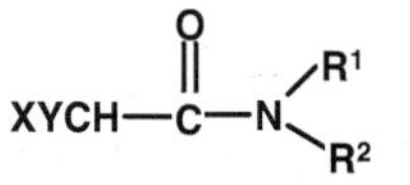

3a-d

Code	R	X	Y	R^1	R^2	Protection[a] (%)	GSH[b]	GST(CDNB)[c]	GST (Ac)[d]
							treated/control		
2k	Me	Cl	Cl	$-(CH_2)_6-$		60	1.38	1.39	1.14
3a	-	Cl	Cl	allyl	allyl	81	1.78	1.24	4.69
3b	-	H	Cl	H	allyl	48	2.25	1.25	3.60
3c	-	H	Cl	allyl	allyl	2	1.45	1.16	2.39
3d	-	H	Br	allyl	allyl	22	0.98	0.90	2.98

[a] based on shoot length; protection (%) = 100 x [(herbicide + safener)] / [control - herbicide]; shoot lengths 14 DAT: control, 27.9+5.3 cm, acetochlor, 3.1±0.3 cm;

[b] GSH content relative to that of untreated control; $GSH_{contr.}$: 0.55±0.09 μmol/g fresh weight;

[c] GST(CDNB) activity as compared to that of untreated control; $GST_{contr.}$: 3.87±0.33 nkat/mg protein;

[d] GST(Ac) activity as compared to that of untreated control; $GST_{contr.}$: 8.26±1.68 pkat/mg protein

Table 2. Safening activity and inducibility of shoot GSH content and GST activities by acetals, ketals and amides in maize

In other, structure and GST isoform expressing ability studies with acetal and ketal analogues of MG-191 as well as mono-and dichloroacetamides (Table 2) demonstrated that the safener structure affects the specific expression of GSTs mediating the detoxication of acetochlor (Matola et al., 2003). Nevertheless, no correlation was found between the degree of induction of GSH and GSTs and the safening activity as related to the structure. A higher inducibility of these GST isoforms was observed in root tissues (Figure 6a and c). In shoots, when the heterodimer *Zm*GSTF1-2 was used the expression of the constitutive *Zm*GSTF1 and inducible *Zm*GSTF2 was enhanced only by **2f** (MG-191) and its analogue **2g** having a 6-membered ring (Figure 6b). These molecules and also **2h** were the most potent inducers of the expression of tau class *Zm*GSTU1 in shoot tissues (Figure 6c). *Zm*GSTU1 has previously been shown to play a key role in metabolism of nitrodiphenyl ether herbicides [54]. These results confirm previous findings that dichloromethyl-ketal safeners are more specific inducers *Zm*GSTU1-2 than other compounds commonly used to safen thiocarbamate and chloroacetanilide herbicides in maize [61].

The exact mechanism of the safener-mediated enhancement of GST activity is not completely understood. GSTs are induced by a diverse range of chemicals and accompanied by the production of active oxygen species. Thus the connection between safener-mediated protection of crops and oxidative stress tolerance has been suggested [66]. Many GSTs are effective not only in conjugating electrophilic substrates but also function as glutathione peroxidases. Safeners may induce GST expression by mimicking oxidative insult [67]. Our results indicate that safener structure plays a decisive role in specific expression of GSTs mediating the detoxication of chloroacetamide herbicides. Since no correlation between the degree of induction of levels of GSH and GST isoforms and the safener activity was found, the mode of action of safeners is a more complex process than simply promoting the metabolism of herbicides.

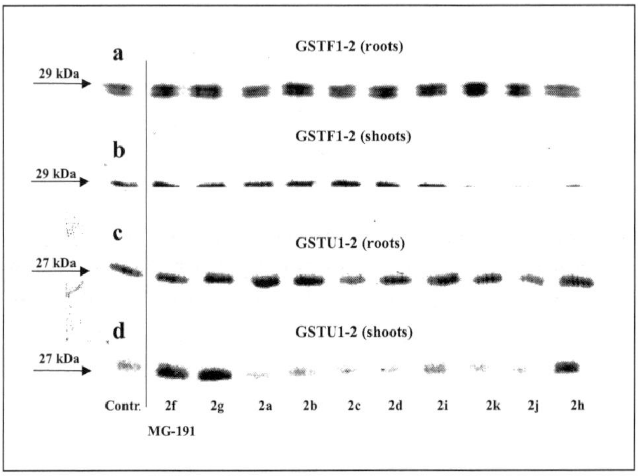

Figure 6. Western blots of crude GST extracts from maize roots and shoots; (a) and (b) analysis of GSTs using the anti-*Zm*GSTF1-2 serum from maize roots and shoots; (c) and (d) analysis of GSTs using the anti-*Zm*GSTU1-2 serum from maize roots and shoots.

7. Effect of safeners on herbicide detoxification enyzmes in weeds

Studies on the mechanism of action of safeners revealed that herbicide safeners improve crop tolerance to herbicides by regulating the expression of genes involved in herbicide metabolism [68]. It is widely accepted that safeners selectively protect crop plants against herbicide injury by stimulating the plant detoxifying mechanism at herbicide rates required for effective weed control. Nevertheless, only a few papers were published on the safener effect of GSTs and cytochrome P450 monooxygenases of various weed species. To a better understanding on why safeners do not provide protection to weeds it is essential to explore the safener action on detoxification enzymes of weeds.

7.1. Effect of safeners on weed glutathione (GSH) content and glutathione *S*-transferase enzyme (GSTs) activities

Safeners such as MG-191, dichlormid, AD-67, BAS-145138, and flurazol were reported to reduce phytotoxicity of EPTC in grassy weeds [69]. MG-191, BAS-145138 and flurazole offered

http://dx.doi.org/10.5772/55168

moderate safening to *Bromus secalinus* (bromegrass) and flurazole was also moderately protective in *Setaria glauca* (yellow foxtail) at sublethal rate of EPTC. Safener-induced elevation of GSH contents and GST activities is widely considered as key element for increased tolerance to thiocarbamates and chloroacetanilides of safened plants [50]. Tolerance of plant species such as maize, soybean and several weeds to acetochlor has been correlated with their glutathione and homoglutathione content [70]. It was also apparent that a relationship exists between the relative GST activities toward alachlor and metolachlor in maize and various weed species ([71]. GST activities toward metolachlor were found to correlate well with the selectivity of the herbicide toward the broadleaf weeds but not toward the grass weeds [72]. However, there was no correlation between total activity of cysteine biosynthesis from serine (CBS) and susceptibility to metolachlor of sorghum, maize, and various grassy weeds [73]. GST isozymes involved in herbicide metabolism is cell suspension culture of a grass weed *Setaria faberi* (giant foxtail) exhibited a similar level of complexity to those from maize cell cultures [74].

Nevertheless, much less is known about GSH or other non-protein thiol contents and GST activities of different weed species following treatments by herbicides and safeners. In order to explain differential physiological and biochemical responses of monocot and dicot weeds to these herbicides, non-protein thiol levels and GST activities were studied in selected mono- and dicot weeds species [75]. The most sensitive *Echinochloa crus-galli* (ECHCR, barnyard-grass) contained higher level of non-protein thiols than less sensitive dicot seedlings (Figure 5). Nevertheless, thiol contents in the most tolerant maize and in the least sensitive monocotyledonous *Bromus secalinus* (BROSE, cheatgrass) were comparable. In general, either herbicide or safener pretreatments did not alter thiol contents substantially. *Abuthilon theophrasti* (ABUTH, velvetleaf) was the only exception because 1 μM acetochlor and 10 μM AD-67 resulted in remarkable increases (73% and 87%, respectively) in the levels of non-protein thiols.

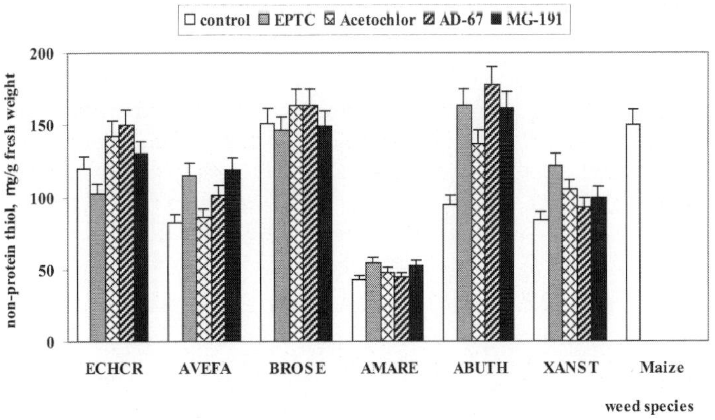

Figure 7. Effect of treatments on non-protein thiol contents of mono- and dicot weed species.

Glutathione S-transferase (GST) activities using CDNB substrate were not correlated with herbicide susceptibility of the selected weed species (Figure 6a). The GSTs extracted from monocot seedlings exhibited much higher activities than from dicot seedlings. GST_{CDNB} activity detected in *Avena fatua* (AVEFA, wild oats) exceeded that in maize. In general, elevation of GST_{CDNB} activities following pretreatments with both herbicides and safeners were more pronounced (2- to 10-fold of controls) in the highly sensitive *Echinochloa crus-galli* and *Amaranthus retroflexus* (AMARE, redroot pigweed) compared to less sensitive species.

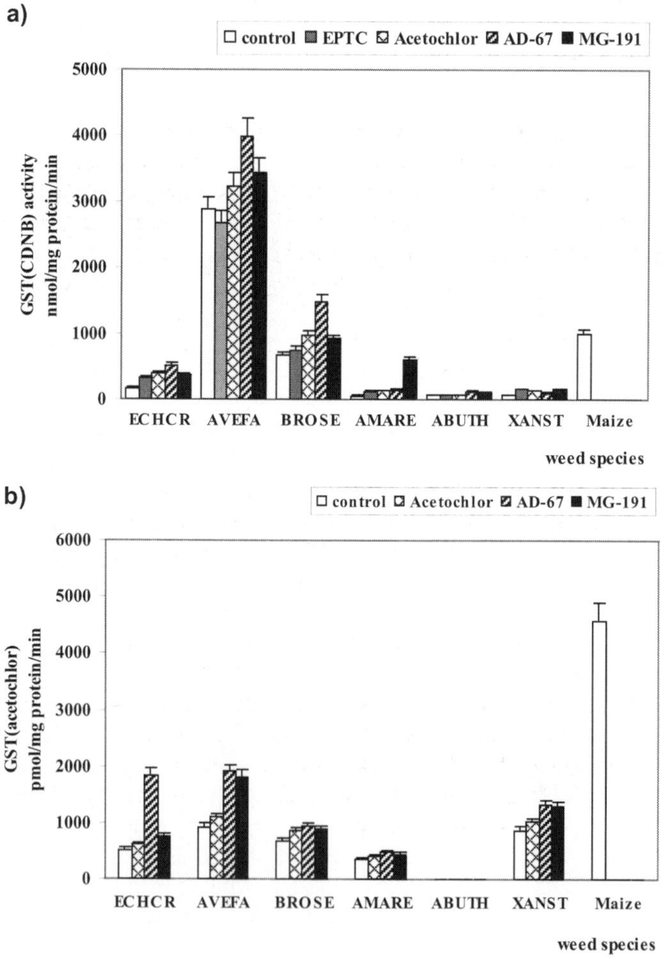

Figure 8. Effect of treatments on glutathione S-transferase activities of selected weed species. a) GST_{CDNB} activities; b) $GST_{acetochlor}$ activities of untreated and treated 6-day-old etiolated seedlings.

With [^{14}C]acetochlor substrate, $GST_{acetochlor}$ activities of both mono- and dicot seedlings were in the same range except for velvetleaf (ABUTH) (Figure 6b). Regardless of treatment, extractable GSTs from velvetleaf did not show specificity for acetochlor. Nevertheless, $GST_{acetochlor}$ activities in all weed species were less expressed than in maize. No correlation was found between enzyme activity and acetochlor susceptibilities of these weed species. In monocot seedlings higher enzyme inductions (up to 2-fold increase) were observed as compared to those in dicots following safener treatment. Nevertheless, $GST_{acetochlor}$ activity of the maize seedlings exceeded those of weed species which may indicate that the higher detoxication capability of crop plant is closely related to the herbicide tolerance. It is also noteworthy that both GSH and cysteine conjugates of chloroacetamides were found inhibitory to GSTs from maize, *Avena fatua*, and *Echinochloa crus-galli*suggesting that GSH conjugation in crops and weeds takes place in a complex manner [76].

Interestingly, *Arabidopsis* plant cultures were more responsive to induction by safeners than either maize or wheat [77]. Enhancement of GST_{CDNB} activity was greatest with fenclorim however treatment with flurazole, CMPI and benoxacor also offered significant increases. *O*-Glucosyltransferase and *N*-glucosyltransferase activities were also stimulated but to a lesser extents. Safeners mefenpyr diethyl and fenchlorazole-ethyl enhanced fenoxaprop-ethyl tolerance of weed *Alopecurus myosuroides* (black-grass) [78]. In black-grass, these detoxification pathways were only slightly enhanced by safeners, suggesting that metabolism alone was unlikely to account for increased herbicide tolerance. Instead, it was determined that safening was associated with an accumulation of glutathione and hydroxymethylglutathione and enzymes with antioxidant functions including phi and lambda glutathione transferases, active as glutathione peroxidases and thiol transferases respectively. In addition to enhanced glutathione metabolism safener treatment resulted in elevated levels of flavonoids in the foliage of black-grass plants, notably flavone-C-glycosides and anthocyanins. Safening of grass weeds was concluded as a mechanism associated with an inducible activation of antioxidant and secondary metabolism. The ability of safeners to induce GSTs of grassy weeds can be exploited in phytoremediating herbicide-contaminated soils. In recent studies safener benoxacor was used to enhance GSTs of the perennial grass *Festuca arundinancea* to establish a basis for preventing environmental herbicide pollution [79]. Further studies revealed that in addition to benoxacor cloquintocet-ethyl, fenchlorazol-ethyl, fenclorim, fluxofenim and oxabetrinil were also able to enhance GST activity in *Festuca* [80]. These results indicate that herbicide diffusion following the runoff of surface waters can be prevented or significantly reduced by vegetating buffer strips with *Festuca* and by the combination of herbicide and a suitable safener. By this way, the application of safeners can be extended by using non crop-species in phytoremediating contaminated soils.

7.2. Interaction of safeners on weed cytochrome P450 monooxygenases

The involvement of cytochrome P450 monooxygenases in herbicide detoxication and selectivity has been well demonstrated [81, 82]. The role of cytochrome P450 monooxygenases in enhanced metabolism of resistant weed species has also been documented [83, 84]. Nevertheless, only a few examples can be found in the literature as to cytochrome P450-dependent monooxygenase system in weed species [85].

Monocotyledonous (*Avena fatua, Bromus inermis, Echinochloa crus-galli*) and dicotyledonous (*Amaranthus retroflexus, Abuthilon threophrasti, Xantium strumarium*) weeds were used to study the interaction of safeners, herbicides metabolized by cytochrome P450 enzymes, and P450 inhibitors on herbicide phytotoxicity and P450 levels of weeds and maize [86]. The safener NA was slightly protective to all monocots at the reduced rate (50 g/ha) of nicosulfuron and also exhibited safening effects on dicots against all herbicides. MG-191 reduced growth inhibition of EPTC in *A. fatua* and *E. crus-galli*.

Species	Cytochrome P450, pmol/mg protein		
	Control	NA[a]	ABT[b]
A. fatua[c]	41±11	49±12	36±17
B. inermis	ND[d]	ND	ND
E. crus-galli	17±8	14±9	ND
A. retroflexus	10±4	21±8	ND
A. theophrasti	51±24	89±32	54±27
X. strumarium	ND	ND	ND
Maize[e]	67±14	73±15	96±18

[a] 0.5 %w/v; [b] 1 μM; [c] 7-day-old etiolated weed seedlings; [d]ND not detectable; [e]4-day-old etiolated maize seedlings.

Table 3. Cytochrome P450 contents of mono- and dicot weeds and influence of treatment with the safener NA and P450 inhibitor ABT.

Weed microsomal cytochrome P450 enzymes were found less stable than those from maize. Carbon-monoxide difference spectra for *B. inermis* and *X. stumarium* could not be recorded probably due to dark colors of microsomal preparations and difficulties in resuspending the microsomes. Cytochrome P450 content in the microsomal membrane fraction of *A. fatua* was 2.4-fold greater than in *E. crus-galli* (Table 3). Among dicotyledonous plants, *A. theophrasti* contained 5.1-fold higher level of the enzyme as compared to that of *A. retroflexus*. However, the P450 level was higher in maize than in weeds.

It is difficult to evaluate changes in the enzyme contents of weed species pretreated with the safener NA or the P450 inhibitor ABT due to the high values of standard deviation of the data. Following treatments with NA, a stimulating tendency could be observed for weeds except *E. crus-galli*. With maize the NA treatment had no enhancing effect on the enzyme content. However, a significant increase (43%) was found when maize seedlings treated with ABT but the P450 inhibitor was uneffective on weed P450s.

For further characterization of *in vivo* interaction of the combination of the herbicides with safeners and inhibitors microsomes isolated from etiolated maize seedlings were used (Figure 7). Treatment of maize seedlings with nicosulfuron resulted in 30% elevation in P450 level while no effect of EPTC was found. The combination of NA with either bentazon or nicosul-

http://dx.doi.org/10.5772/55168

furon decreased P450 levels by about 50% as compared to the untreated control. Interestingly, without herbicide pretreatment with NA had no influence on maize P450. The inhibitory effect of NA *in vitro* on maize P450 was reported by the formation of an enzyme-NA Type I complex [87]. Pretreatments with the combination of MG-191 and all herbicides yielded slight increases in the enzyme concentration. It is interesting to note that no binding of MG-191 to P450 was detected [88] which may indicate why MG-191 was not inhibitory to P450. The P450 inhibitor PBO simultaneously applied with bentazon and nicosulfuron substantially reduced P450 levels while the ABT was less inhibitory.

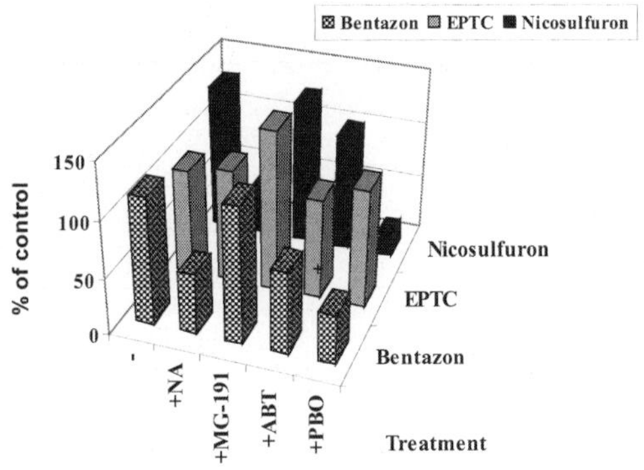

Figure 9. Interaction of herbicides with safeners and cytochrome P450 inhibitors on P450 enzymes extracted from 4-day-old etiolated maize seedlings. Treatments were as follows: bentazon, 10 μM; EPTC, 10 μM; nicosulfuron, 10 μM; NA 0.5%w/v; MG-191, 10 μM; ABT, 1 μM; PBO, 10 μM.

These results demonstrate that safeners can marginally protect weed species by stimulating the herbicide detoxifying enzymes but the lower level of these enzymes in weeds as compared to those in crops provide a basis for the botanical selectivity of safeners.

8. Mechanism of safener action

The mechanism by which safeners act is currently unknown despite the widespread agricultural use and the substantial experimental evidence accumulated on the biochemical basis of action. Safeners appear to induce a set of genes that encode enzymes and biosynthesis of cofactors involved in the herbicide detoxication [50, 52, 89, 90].

The exact mechanism of safener-mediated enhancement of GST activity is not completely understood. GSTs are induced by a diverse range of chemicals and accompanied by the production of active oxygen species. Thus the connection between safener-mediated protection of crops and oxidative stress tolerance has been suggested [66]. Many GSTs are effective not only in conjugating electrophilic substrates but also function as glutathione peroxidases. Safeners may induce GST expression by mimicking oxidative insult [67]. Herbicide safeners increase herbicide tolerance in cereals but not in dicotyledonous crops. The reason(s) for this difference in safening is unknown. Treatment of *Arabidopsis* seedlings with various safeners resulted in enhanced GST activities and expression of GSH-conjugate transporters such as *At*MRP1-4 [91]. Safeners also increased GSH content of *Arabidopsis* seedlings. However, treatment of *Arabidopsis* plants with safeners had no effect on the tolerance of seedlings to chloroacetanilide herbicides. Immunoblot analysis confirmed that *At*GSTU19 was induced in response to several safeners. These results indicate that, although *Arabidopsis* may not be protected from herbicide injury by safeners, at least one component of their detoxification systems is responsive to these compounds.

Concerning the location of safener binding site(s) of plants few studies have been conducted. A high-affinity cytosolic-binding site for the dichloroacetamide safener (*R,S*)-3-dichloroacetyl-2,2,5-trimethyl-1,3-oxazolidine was found in etiolated maize seedlings ([92]. The binding was highest in the coleoptiles and lowest in the leaves. A good correlation was shown between the safener effectiveness. Chloroacetanilide and thiocarbamate herbicides were effective inhibitors of safener binding at low concentrations. The inhibition by alachlor and EPTC was shown to be competitive. The safener binding protein (SafBP) was purified to homogeneity having a molecular mass of 39 kDa [93]. Based on the peptides obtained from proteolytic digests of SafBP a cDNA encoding SafBP was cloned and expressed in *E. coli*. The predicted primary structure of SafBP was related to a phenolic *O*-methyltransferase but SafBP did not catalyze O-methylation of catechol or caffeic acid. It was concluded that SafBP may not be the primary site of action of the dichloroacetamide safeners. Supporting the participation of *O*-methyltransferases in the safener action, treatment of wheats (*Triticum aestivum* L.) with cloquintocet-mexyl resulted in an accelerated depletion of flavone *C*-glycosides and a selective shift in the metabolism of endogenous phenolics [94]. Changes in phenolic content were associated with an increase in *O*-methyltransferase and *C*-glucosyltransferase activity toward flavonoid substrates.

Proteomic methods were used to identify herbicide safener-induced proteins in the coleoptile of *Triticum tauschii* [95]. The herbicide safener, fluxofenim, dramatically increased protein abundance in the molecular range in the molecular weight range of 24 to 30 kDa as well as a few higher molecular weight protein and overall 20 proteins were identified. Among the eighteen inducible proteins 15 were glutathione *S*-transferase subunits that fall into three subclasses: eight proteins were from the tau subclass, six proteins were from phi subclass, and one was from the lambda class. Another three safener inducible proteins showed homology to the aldo/keto reductase family with proteins that have roles in glycolysis and the Krebs cycle. One of the two constitutively expressed proteins showed the highest homology to the dehydroascorbate reductase subclass of GSTs while the other to an ascorbate peroxidase. Results indicated that the induced proteins were associated with herbicide detoxication and with general stress response. In another study with cloquintocet-mexyl safener and dimethenamid herbicide 29 safener-induced and 10 herbicide-regulated proteins were identified in *Triticum tauschii* seedlings [39]. Surprisingly, mutually exclusive sets of proteins were identified following herbicide or safener treatment suggesting a different signaling pathway for each chemical. Safener-responsive proteins were mostly involved in

http://dx.doi.org/10.5772/55168

xenobiotic detoxication whereas herbicide-regulated proteins belonged to several classes involved in general stress responses. Quantitative RT-PCR revealed that multidrug resistance-associated protein (MRP) transcripts were highly induced by safeners and two MRP genes were differently expressed.

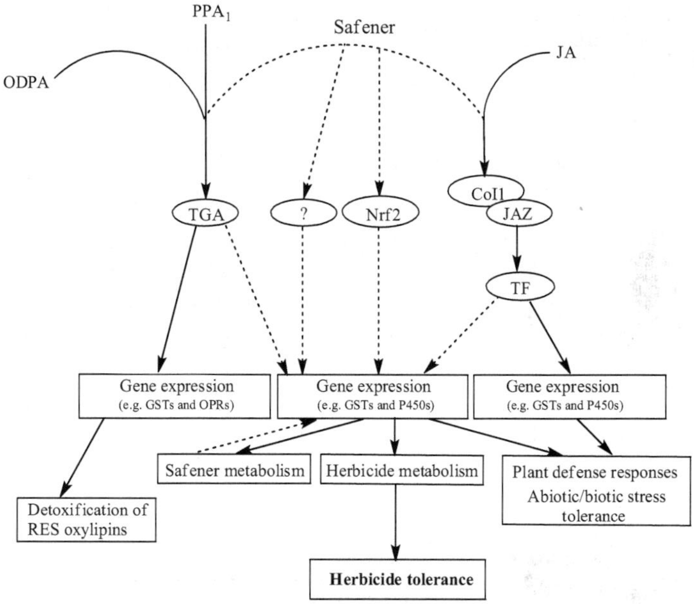

Figure 10. Suggested safener-mediated signalling pathway for regulation of defense genes and activation of detoxification pathways in plants by Riechers et al. [52]. Dashed lines indicate possible but unproven signaling pathways while solid lines indicate known signaling pathways. ODPA: 12-oxo-phytodienoic acid; OPRs: ODPA-reductases; PPA_1: A_1-type phytoprostanes; JA: jasmonic acid; TGA: TGA transcription factor; Nrf2: nuclear factor (erythroid-derived 2)-like 2; Col1: coronative insensitive protein 1; JAZ: transcriptional repressor protein; TF: transcription factor/activator.

Safeners were suggested to trigger an unidentified, preexisting signaling pathway for detoxification of endogenous toxins or xenobiotics [96]. According to a new hypothesis, safeners may be utilizing an oxidized lipid-mediated (oxylipins) or cyclopentenone-mediated signaling pathway which subsequently leads to the expression of GSTs and other proteins involved in detoxification and plant defense [52]. Some possible safener-mediated signaling pathways for the regulation of defense genes and activation of detoxification pathways have been suggested (Figure 8). Safeners may tap into a RES oxylipin-mediated signaling pathway and up-regulate TGA transcription factors, an Nrf2-Keap1-mediated as well as jasmonic acid-mediated signaling pathways. Safeners and oxylipins as reactive electrophilic species (RES oxylipins) have a common biological activity since both strongly induce the expression of defense genes and activate detoxification responses in plants [39, 40].

9. Conclusions

Fifty-year of herbicide safeners resesearch and use confirms that these molecules offered new ways to improve herbicide selectivity. Although this technology now competes with herbicide-tolerant, genetically-modified or naturally-selelected crops, safeners still comprise an important part of the herbicide market in maize, cereals and rice [10]. Many of the commercial safeners are in off-patent status offering a chance for the generic manufacturers to enter the market together with off-patent herbicides. In contrast, recent herbicide mixture patents with new herbicides still allow their exclusive usage by the patent holder [10].

Although safeners do not improve herbicide tolerance in dicot plants, but the utilization of biotechnology tools may help in extending the safener response from monocot to dicots. It was found, however that *Arabidopsis* transgenic plants did not respond to safeners at whole-plant level despite the increase of the expression of tau class protein in the roots [91]. Additionally, knowledge of critical regulatory elements in the promoters or untranslated regions of genes encoding detoxification enzymes, or a comprehensive understanding how gene expression is up-regulated by safeners might lead to the precise manipulation of transgene expression of plants [52].

The use of safeners to enhance tolerance of plants to organic pollutants such as herbicides, heavy metals or oils in the environment (soil, water) could also be a promising application of these chemicals. Phytoremediation studies with soils contaminated with oils and heavy metals and safener-treated wheat seeds have recently been reported [97]. While untreated seeds were unable to germinate on the contaminated soil, safener treatments resulted in seedlings briefly growing before succumbing to the pollutants.

Author details

Istvan Jablonkai

Institute of Organic Chemistry, Research Centre for Natural Sciences, Hungarian Academy of Sciences, Budapest, Hungary

References

[1] Hatzios K. K. (1989). Development of herbicide safeners: Industrial and university perspectives. In: *Crops safeners for herbicides*. Hatzios K. K., Hoagland R. E. (Eds). pp 3-45, Academic Press, San Diego, USA.)

[2] Davies J., Caseley J. C. (1999). Herbicide safeners: a review. *Pestic. Sci.*, 55, 1043–1058.

[3] Hoffmann O. L. (1978). Herbicide antidotes: From concept to practice. In: *Chemistry and action of herbicide antidotes*. Pallos F. M., Casida J. E. (Eds). pp 35-61, Academic Press, New York, NY, USA.

[4] Hoffmann O. L., Gull P. W., Zeising H. C., Epperly J. R. (1960). Factors influencing wild oat control with barban. *Proc. North Cent. Weed Control Conf.*, 17, 20.

[5] Hoffman O. L. (1971). Coated corn seed. *US Patent* 3,564,768.

[6] Abu-Hare A. Q., Duncan H. J. (2002). Herbicide safeners: uses, limitations, metabolism, and mechanism of action. *Chemosphere*, 48, 965-974.

[7] Monaco T. J., Weller S. C., Ashton F. M. (Eds). (2002). Herbicides and the plants, In: *Weed science: Principles and practices*. pp 98-126, Wiley, New York, NY, USA.

[8] Miyauchi N., Kobayashi K., Usui K. (2002). Differential safening activity of dymron and fenclorim on pretilachlor injury in rice seedlings in soil. *Weed Biol. Manag.* 2, 46-51.

[9] Matsunaka S., Wakabayashi K.(1989). In: *Crop Safeners for Herbicides*. Hatzios K. K., Hoagland R. E. (Eds). pp 47-61, Academic Press, San Diego, USA.

[10] Rosinger C., Bartsch K., Schulte W. (2012). Safener for Herbicides. In: *Modern Crop Protection Compounds*. Krämer W., Schirmer U., Jeschke P., Witschel M. (Eds) Vol. 1. pp 371-398. Wiley-VCH, Weinheim, Germany.

[11] Tam A. C., Behki R. M., Khan S. U. (1988). Effect of dietholate (R-33865) on the degradation of thiocarbamate herbicide by an EPTC-degrading bacterium. *J. Agric. Food. Chem.*, 36, 654-657.

[12] Keifer D. W. (2005). Method for safening crop from the phytotoxic effect of herbicide by use of phosphorated esters. US Patent 6,855,667.

[13] Hatzios K. K., Zama P. (1986). Physiological interactions between the herbicide EPTC and selected analogues of the antidote naphthalic anhydride on two hybrids of maize. *Pestic. Sci.*, 17, 25-32.

[14] Pallos F. M., Brokke M. E., Arneklev D. R. (1975). Antidotes protect corn from thiocarbamate herbicide injury. *J. Agric. Food Chem.*, 23, 821-822.

[15] Dutka F., Komives T., Marton A. F., Hulesch A., Fodor-Csorba K., Karpati M. (1979). Structure-activity relationships of herbicide antidotes. *Proc. Hung. Annu. Meet. Biochem.*, 19, 1-4.

[16] Görög K., Muschinek G., Mustardy L. A., Faludi-Daniel A. (1982). Comparative studies of safeners for prevention of EPTC injury in maize. *Weed Res.*, 22, 27-33.

[17] Pallos F. M., Reed A. G., Arneklev D. R., Brokke M. E. (1978). Antidotes protect corn from thiocarbamate herbicide injury. In: *Chemistry and action of herbicide antidotes*, Pallos F. M., Casida J. E., (Eds), pp 15-20, Academic Press, New York, NY, USA.

[18] Stephenson G. R., Chang F. Y. (1978). Comparative activity and selectivity of herbicide antidotes. In: *Chemistry and action of herbicide antidotes*, Pallos F. M., Casida J. E., (Eds), pp 35-61, Academic Press, New York, NY, USA.

[19] Stephenson G. R., Bunce J. J., Makowski R. I., Curry J. C. (1978). Structure-activity relationships for *S*-ethyl-*N,N*-dipropyl thiocarbamateantidotes in corn. *J. Agric. Food Chem.*, 26, 137-140.

[20] Yenne S. P., Hatzios K. K. (1990). Molecular comparisons of selected herbicides and their safeners by computer-aided molecular modeling. *J. Agric. Food Chem.*, 38, 1950-1956.

[21] Bordas B., Komives T., Lopata A. (2000). Comparative three-dimensional quantitative structure-activity relationship study of safeners and herbicides. *J. Agric. Food Chem.*, 48, 926-931.

[22] Chang T. S., Merkle M. G. (1982). Oximes as seed safeners for grain sorghum (*Sorghum bicolor*). *Weed Sci.* 30, 70-73.

[23] Howe R. K., Lee L. F. (1980). 2,4-Disubstituted-5-thiazolecarboxylic acids and derivatives. *US Patent* 4,199,506.

[24] Jablonkai I., Matola T. (2002). Structure–activity relationships of 2-dichloromethyl-1,3-dioxacycloalkanes and heteroanalogues in safening maize against chloroacetanilide herbicides. 10th IUPAC International Congress on Chemistry of Crop Protection, Basel, Switzerland, August 4-9, 2002, Book of Abstracts, p. 132.

[25] Matola T., Jablonkai I., Dixon D., Cummins I., Edwards R. (2003). Structure of dichloromethyl-ketal safeners affects the expression of glutathione S-transferase isoforms. *Proceedings of BCPC - Weeds*, vol 2, 527-532.

[26] Matola T., Jablonkai I. (2007) Safening efficacy of halogenated acetals, ketals and amides and relationships between the structure and effect of glutathion and glutathione S-transferases in maize. *Crop Prot.*, 26, 278-284.

[27] Lewis D. L., Garrison A. W., Wommack K. E., Whittemore A., Steudler P., Melillo J. (1999). Influence of environmental changes on degradation of chiral pollutants in soils. *Nature*, 401, 898-901.

[28] Garrison A. W. (2006). Probing the enantioselectivity of chiral pesticides. *Environ. Sci. Technol.*, 40, 16-23.

[29] Wittingham W. G. (2012). 6-Amino-4-pyridine carboxylate derivatives and their preparation, agrochemical compositions and use as herbicidal safener. *UK Patent Application*, GB 2484982.

[30] Bollinger F. G., Hemmerly D. M., Mahoney M. D., Freeman J. J. (1989). Optical isomers of the herbicidal antidote 4-(dichloromethylene)-2-[*N*-(α-methylbenzyl)imino]-1,3-dithiolane hydrochloride. *J. Agric. Food Chem.*, 37, 484-485.

[31] Jablonkai I., Matola T., Cummins I., Dixon D., Edwards R. (2010). Safening activity of carvone stereoisomers against acetochlor herbicide in maize. Royal Australian Chemical Institute's 13th National Convention in conjunction with the 12th IUPAC International Congress of Pesticide Chemistry, Melbourne, Australia, July 4-8, 2010, Abstr. No. 624.

[32] Jablonkai I., Visy J., Matola T., Cummins I., Dixon D., Edwards R. (2010). Diastereomers of a chiral safener 1,2-dichloromethyl-[1,3]oxathiolane 3-oxide exhibit differential safening activity against acetochlor in maize. Royal Australian Chemical Institute's 13th National Convention in conjunction with the 12th IUPAC International Congress of Pesticide Chemistry, Melbourne, Australia, July 4-8, 2010, Abstr. No. 623.

[33] Zhao L. X., Liu C. G., Fu Y., Xing Z. Y., Gao S. (2012). Induction of maize glutathione *S*-transferase by herbicide safeners and their effect on enzyme activity against chlorsulfuron. *Advanced Materials Research*, vol. 518-523, 5480-5483.

[34] Rubin B., Kirino O. (1989). Herbicide prosafeners: Chemistry, safening activity, and mode of action. In: *Crop Safeners for Herbicides*. Hatzios K. K., Hoagland R. E. (Eds). pp 317-351, Academic Press, San Diego, USA.

[35] Hilton J. L., Pillai P. (1986). L-2-oxathiazolidine-4-carboxylic acid protection against tridiphane toxicity. *Weed Sci.*, 34, 669-675.

[36] Hilton J. L., Pillai P. (1988). Thioproline protection against herbicide toxicity. *Weed Technol.*, 2, 72-76.

[37] Jablonkai I., Dutka F. (1996). Safening activity of natural hydroxamic acids and analogous compounds against herbicide injury to maize. *J. Environ. Sci. Health. Part B. - Pesticides, Food Contaminants, and Agricultural Wastes*, 31, 555-559.

[38] Brazier-Hicks M., Evans K. M., Cunningham O. D., Hodgson D. R. W., Steel P. G., Edwards R. (2008). Catabolism of glutathione conjugates in *Arabidopsis thaliana*: Role in metabolic reactivation of the herbicide safener fenclorim. *J. Biol. Chem.*, 283, 21102-21112.

[39] Zhang Q., Xu F. X., Lambert K. N., Riechers D. E. (2007). Safeners coordinately induce multiple proteins and MRP transcripts involved in herbicide metabolism and detoxication in *Triticum tauschii* seedling tissues. *Proteomics*, 7, 1261-1278.

[40] Mueller M. J., Berger S. (2009). Reactive electrophilic oxylipins: pattern recognition and signalling. *Phytochemistry*, 70, 1511-1521.

[41] Kreuz K., Riechers D. E., Zhang Q. (2010). The use of oxylipins as safeners and safening herbicidal compositions comprising oxylipins. *WO* 2011/134539.

[42] Jablonkai I. (1991). Basis for differential chemical selectivity of MG-191 safener against acetochlor and EPTC injury to maize *Z. Naturforsch.*, 46c, 828-835.

[43] Jablonkai I., Dutka F. (1995). Uptake, translocation and metabolism of MG-191safener in corn (Zea Mays L.), *Weed Sci.*, 43, 169-174.

[44] Jablonkai I., Repasi J., Dutka F. (1991). Effect of the site of MG-191 application on acetochlor herbicide uptake, distribution and phytotoxicity. *Pestic. Sci.*, 31, 91-93.

[45] Jablonkai I., Dutka F. (1985). Effect of R-25788 antidote on the uptake, translocation and metabolism of acetochlor herbicide by corn. *J. Radioanal. Nucl. Chem., Letters*, 96, 419-426.

[46] Barrett M. (1989) Protection of corn (*Zea mays*) and sorghum (*Sorghum bicolor)* from imazethapyr toxicity with antidotes. *Weed Sci.* 37, 296-301.

[47] Fuerst E. P., Lamoureux G. L. (1992). Mode of action of the dichloroacetamide antidote BAS 145138 in corn. II. Effects on metabolism, absorption and mobility on metazachlor. *Pestic. Biochem. Physiol.*, 42, 78-87.

[48] Little D. L., Ladner D. W., Shaner D. L. (1994). Modeling root absorption and translocation of 5-substituted analogs of the imidazolinone herbicide, imazzapyr. *Pestic Sci.*, 41, 171-185.

[49] Davies J., Caseley J. C., Jones O. T. G., Barrett M., Polge N. D. (1998). Mode of action naphthalic anhydride as a safener for herbicide AC 263,222 in maize. *Pestic. Sci.*, 52, 29-38.

[50] Hatzios K. K., Burgos N. (2004). Metabolism-based herbicide resistance: regulation by safeners. *Weed Sci.*, 52, 454-467.

[51] Coleman J. O. D., Blake-Kalff M. M. A., Emyr Davies T. G. (1997). Detoxication of xenobiotics by plants: chemical modification and vacuolar compartmentation. *Trends Plant Sci.*, 2, 144-151.

[52] Riechers D. E., Kreuz K., Zhang Q. (2010). Detoxification without intoxication: Herbicide safeners activate plant defence gene expression. *Plant Physiol.*, 153, 3-13.

[53] Marrs K. A. (1996). The function and regulation of glutathione S-transferases in plants. *Ann. Rev. Plant Physiol. Plant Mol. Biol.*, 47, 127-158.

[54] Cole D. J., Cummins I., Hatton P. J., Dixon D., Edwards R. (1997). Glutathione transferases in crops and weeds. In: *Regulation of enzymatic systems detoxifying xenobiotics.* Hatzios K. K. (Ed). pp 107-154, Wiley, Chichester, UK.

[55] Hatzios, K. K. (2001). Functions and regulation of plant glutathione *S*-transferases. In: *Pesticide biotransformation in plants and microorganisms: Similarities and divergencies.* Hall J. C, Hoagland R. E., Zablotowicz R. M. (Eds). pp 218-239, ACS Symposium Series 777, American Chemical Society, Washington, DC, USA.

[56] Dixon D. P., Hawkins T., Hussey P. J., Edwards R. (2009). Enzyme activities and subcellular localization of members of the *Arabidopsis* glutathione transferase superfamily. *J. Exp. Bot.*, 60, 1207-1218.

[57] Edwards R., Dixon D. P., Walbot V. (2000). Plant glutathione *S*-transferases: enzymes with multiple functions in sickness and in health. *Trend Plant Sci.*, 5, 193-198.

[58] Dixon D. P., Lapthorn A., Edwards R. (2002). Plant glutathione transferases. *Genome Biol.*, 3, 3004.1-3004.10.

[59] Dixon D. P., Skipsey M., Edwards R. (2010). Roles for glutathione transferases in secondary plant metabolism. *Phytochemistry*, 71, 338-350.

[60] Farago S., Brunold C., Kreuz K. (1994). Herbicide safeners and glutathione metabolism. *Physiol Plant.*, 91, 537-542.

[61] Jablonkai I., Hulesch A., Cummins I., Dixon D. P., Edwards R., (2001). The herbicide safener MG-191 enhances the expression of specific glutathione *S*-transferases in maize. *Proceedings of BCPC - Weeds*, vol 2, 527-532.

[62] Dixon D. P., Cole D. J. Edwards R (1997). Characterisation of multiple glutathione transferases containing the GST I subunit with activities toward herbicide substrates in maize (*Zea mays*). *Pestic. Sci.* 50, 72-82.

[63] Dixon D. P., Cole J., Edwards R. (1998). Purification, regulation and cloning of glutathione transferase (GST) from maize resembling the auxin-inducible type-III GSTs. *Plant Mol. Biol.* 36, 75-87.

[64] Dixon D. P., Cole D. J., Edwards R. (1999). Dimerization of maize glutathione transferases in recombinant bacteria. *Plant Mol. Biol.* 40: 997-1008.

[65] Rossini L., Jepson I., Greenland A. J., Sari Gorla M. (1996). Characterization of glutathione *S*-transferase isoforms in three inbred lines exhibiting differential sensitivity to alachlor. *Plant Physiol.* 112: 1595-1600.

[66] Theodoulou F. L., Clark I. M., He X. L., Pallett K. E., Cole D. J., Hallahan D. L. (2003). Co-induction of glutathione *S*-transferases and multidrug resistant protein by xenobiotics in wheat. *Pestic. Manag. Sci.*, 59, 202-214.

[67] Dixon D. P., Cummins I., Cole D. J., Edwards R. (1998b). Glutathione-mediated detoxication system in plants. *Curr. Opin. Plant Biol.* 1, 258-266.

[68] Davies J. (2001). Herbicide safeners – commercial products and tools for agrochemical reseseach. *Pesticide Outlook*, February 2001, 10-15.

[69] Hulesch A., Dutka F. (1993). Investigation of the safening of EPTC on several grassy crops and weeds by various safeners. *Proceedings of Brighton Crop Protection Conference – Weeds*, vol 1, 207-212.

[70] Breaux E. J., Patanella J. E., Sanders E. F. (1987). Chloroacetanilide herbicide selectivity: analysis of glutathione and homoglutathione in tolerant, susceptible, and safened seedlings. *J. Agric. Food Chem.*, 35 474–478.

[71] Hatton P. J., Dixon D., Cole D. J., Edwards R. (1996). Glutathione transferase activities and herbicide selectivity in maize and associated weed species. *Pestic. Sci.*, 46, 267-275.

[72] Andrews C. J., Skipsey M., Townson J. K., Morris C., Jepson I., Edwards R. (1997). Glutathione transferase activities toward herbicides used selectively in soybean. *Pestic. Sci.*, 51, 213-222.

[73] Hirase K., Molin W. T. (2002). Measuring cysteine biosynthesis activity from serine in extracts from sorghum, corn and grass weeds, and their metolachlor susceptibility. *Weed Biol. Manag.*, 2, 52-59.

[74] Hatton P. J., Cummins I., Price L. J., Cole D. J., Edwards R. (1998). Glutathione transferases and herbicide detoxification in suspension-cultured cells of giant foxtail (*Setaria faberi*). *Pestic Sci.* 53, 209-216.

[75] Jablonkai I., Hulesch A., Dutka F. (1995). Influence of herbicides and safeners on glutathione content and glutathione S-transferase activities of monocot and dicot weeds. Proceedings of the International Symposium on Weed and Crop Resistance to Herbicides -Cordoba (Spain), DePrado R., Jorrin J., Garcia-Torres L., Marshall G. (Eds), p. 89-91.

[76] Jablonkai I., Hulesch A, Barta I. C. (1997). Glutathione and cysteine conjugates inhibit glutathione S-transferase enzymes mediating GSH conjugation of the herbicide acetochlor. *Proceedings of the Brighton Crop Protection Conference - Weeds*, vol 2, p.801-806.

[77] Edwards R., Del Buono D., Fordham M., Skipsey M., Brazier M., Dixon D. P., Cummins I. (2005). Differential induction of glutathione transferases and glucosyltransferases in wheat, maize and *Arabidopsis thaliana* by herbicide safeners. *Z. Naturforsch.*, 60c, 307-316.

[78] Cummins I., Bryant D. N., Edwards R. (2009). Safener responsiveness and multiple herbicide resistance in the weed black-grass (Alopecurus myosuroides). *Plant Biotech. J.*, 7, 807-820.

[79] Del Buono D., Scarponi L., Espen L. (2007). Glutathione *S*-transferases in *Festuca arundinacea*: Identification, characterization and inducibility by safener benoxacor. *Phytochemistry*, 68, 2614-2624.

[80] Scarponi L., Del Buono D., Quagliarini E., D'Amato R. (2009). *Festuca arudinacea* grass and herbicide safeners to prevent herbicide pollution. *Agron. Sustain. Dev.*, 29, 313-319.

[81] Scalla R. (1991). Interaction of herbicides with safeners and synergist. In: *Pesticide chemistry: Advances in international research, development, and legislation*. Frehse H. (Ed). pp 141-150, Wiley-VCH, Weinheim, Germany.

[82] Durst F., Benveniste I., Lesot A., Salaün J.-P., Werck-Reichhart D. (1997). Induction of plant cytochrome P450. In: *Regulation of enzymatic systems detoxifying xenobiotics in*

plants. Hatzios K. K. (Ed). pp 19-34, Kluwer Academic Publishers, Dordrecht, Netherlands.

[83] Burnett M. W. M., Loveys B. R., Holtum J. A. M., Powles S. B. (1993). A mechanism of chlortoluron resistance in *Lolium rigidum*. *Planta*, 190, 182-189.

[84] Burnett M. W. M., Loveys B. R., Holtum J. A. M., Powles S. B. (1994). Identification of two mechanisms of sulfonylurea resistance within one population of rigid ryegrass (*Lolium rigidum*) using a selective germination medium, *Weed Sci.*, 42, 153-157.

[85] Burton J. D., Maness E. P. (1992) Constitutive and inducible bentazon hydroxylation in shuttercane (*Sorghum bicolor*) and Johnsongrass (*Sorghum halapense*). *Pestic. Biochem. Biochem. Physiol.*, 44, 40-49.

[86] Jablonkai I., Hulesch A. (1996). Cytochrome P450 levels of monocot and dicot weeds and influence of herbicides, safeners and P450 inhibitors on enzyme contents. Proceedings of the 2nd International Weed Control Congress - Copenhagen (Denmark), Brown H., Cussans G. W., Devine M. D., Duke S.O., Fernandez-Quintanilla C., Helweg A., Labrada R. E., Landes M., Kudsk P., Streibig J. C. (Eds), Vol 3, p. 789-794.

[87] Barta I. C., Dutka F. (1991) Interaction of maize cytochrome P450 with safeners and 1-aminobenzotriazole. *Proceedings of Brighton Crop Protection Conference – Weeds*, vol 3, 1127-1132.

[88] Jablonkai I., Hatzios K. K. (1994). Microsomal oxidation of the herbicides EPTC and acetochlor and of the safener MG-191 in maize. *Pestic. Biochem. Physiol.*, 48, 98-109.

[89] Gatz C. (1997). Chemical control of gene expression. *Ann. Rev. Plant Physiol. Plant Mol. Biol.*, 48, 89–108.

[90] Padidam M. (2003). Chemically regulated gene expression in plants. *Curr. Opin. Plant Biol.*, 6, 169-177.

[91] DeRidder B. P., Goldsbrough P. B. (2006). Organ-specific expression of glutathione *S*-transferases and the efficacy of herbicide safeners in Arabidopsis. *Plant Physiol.*, 140, 167-175.

[92] Walton J. D., Casida J. E. (1995). Specific binding of a dichloroacetamide herbicide safener in maize at a site that also binds thiocarbamate and chloroacetanilide herbicides. *Plant Physiol.*, 109, 213-219.

[93] Scott-Craig J. S., Casida J. E., Poduje L., Walton J. D. (1998). Herbicide safener-binding protein of maize. *Plant. Physiol.*, 116, 1083-1089.

[94] Cummins I., Brazier-Hicks M., Stobiecki M., Franski R., Edwards R. (2006). Selective disruption of wheat secondary metabolism by herbicide safeners. *Phytochemistry*, 67, 1722-1730.

[95] Zhang Q., Riechers D. E. (2004). Proteomic characterization of herbicide safener-induced proteins in the coleoptile of *Triticum tauschii* seedlings. *Proteomics,* 4, 2058-2071.

[96] Riechers D. E., Vaughn K. C., Molin W. T. (2005). The role of plant glutathione *S*-transferases in herbicide metabolism. In: *Environmental fate and safety management of agrochemicals.* Clark J. M., Ohkawa H. (Eds), pp 216-232, ACS Symposium Series 899, American Chemical Society, Washington, DC, USA.

[97] Taylor V. L., Cummins I., Brazier-Hicks M., Edwards R. (2012). Protective responses induced by herbicide safeners in wheat. *Environ. Exp. Botany,* doi: 10.1016/jenvexpbot.2011.12.030.

Chapter 24

Herbicides — A Double Edged Sword

Mona H. El-Hadary and Gyuhwa Chung

Additional information is available at the end of the chapter

http://dx.doi.org/10.5772/55957

1. Introduction

Weeds represent a global agronomic problem that threatens the productivity of cultivated crops. Weeds compete with cultivated crops for the available moisture, nutrients and light. Consequently, weeds significantly reduce either crop yield or quality. Control of weeds is essential to maintaining the production of economic crops. Weed control may be achieved either through manual eradication or herbicide application. Balanced usage of herbicides should be considered in controlling weeds. Low concentrations of herbicides may act as growth regulators for the main crop metabolism [1]. However, in some cases, herbicides may affect the main crop adversely by interfering with its essential biochemical processes such as respiration, photosynthesis, protein metabolism, and hydrolytic enzyme activity [1].

Herbicide interference with the morphology, physiology and biochemical pathways of treated plants varies according to the characteristic actions of the herbicide and depends upon the degree of tolerance or susceptibility of the crop plant species. Environmental factors and soil conditions affecting plant growth, as well as herbicide formulation, herbicide degradation and application method would significantly influence the effects of herbicides on treated plants. Once an herbicide reaches the site of action in the treated plants, the biochemical processes are affected. Herbicides differ in their site of action and may have more than one site of action. As the herbicide concentration increases in plant tissue, additional sites of action may become involved. The effect of herbicides on growth, productivity and different metabolic activities has been studied extensively in many investigations such as in **El-Hadary** [1].

1.1. A word from the authors

Authors intended to give some examples for commercial herbicides that were applied in agronomic systems within the past fifty years. These examples include those herbicides which may now be internationally prohibited but are still used in the developing and underdeveloping countries due to their low price and the little information available about them. References have been included that cover a long era of research concerning herbicide application in order to include those prohibited herbicides. Also, references were included that focus on research that was conducted in under- and developing countries.

1.2. Herbicides

This chapter will discuss different herbicide groups, classification, selectivity, interference with metabolic processes and hazardous action upon crop plants. Also, the relation to naturally occurring phenomena, such allelopathy and future prospects of genetic engineering in the production of plant herbicides themselves, will be mentioned.

2. Classification of herbicides (Broad lines)

There are different broad lines upon which herbicides could be classified:

2.1. Application timing

Time of application of an herbicide is so critical for getting satisfying results. Herbicides application is achieved either pre-emergence or post-emergence of the weed seedlings. Pre-emergence involves herbicide application prior to seed germination while post-emergence means application after seed germination and active growth. Moreover, post-directed application refers to targeting the treatment to a particular portion of the plant once emerged and growing.

2.2. Application method

Herbicides may be applied either as a foliar spray or a soil treatment. The application method may take either the broadcast pattern through treatment of the entire area or the spot pattern through specified area treatment.

2.3. Chemical groups

The chemical group to which an herbicide belongs indicates its mode of action. A good classification and description for herbicides is provided by "Compendium of Pesticide Common Names" at the web site of *http*://www.alanwood.net/pesticides/class_herbicides.html.

2.4. Mode of action

Herbicides poisonous action goes either by contact or systematically. Herbicides can be classified according to their mode of action into two categories; non-selective herbicides and selective herbicides. Non-selective herbicides are characterized by having a general poisonous effect to the plant cells while selective herbicides can recognize the plant which they affect and kill it by interference with its principle biochemical processes.

3. Selectivity of herbicides

Selectivity of herbicides for eradicating weeds can be achieved through employing some factors related to:

3.1. Biochemical differences

Based on the biochemical differences between weeds and crops, or even weeds between each other, selectivity can be achieved. There is a great diversity of types of weeds usually growing in one crop. When employing an herbicide based on biochemical differences, the crop plant would possess a defense mechanism that is usually absent in most of the competing weed species. Consequently, the herbicide would react with the biochemical metabolism of the weeds without any fatal interference on main crop metabolism.

3.2. Morphological differences

The selectivity which depends upon morphological differences is characteristic for post-emergence herbicides. Dicotyledonous plants have leaves spread out and exposed meristematic tissue, so that the toxin is directed to the growing point situated at the center of a rosette. While upright leaves of monocotyledonous plants enable plants to form a sheath around the meristem that protects it from receiving the herbicidal spray (Figure 1) [1]. Therefore, such morphological differences can be recruited to work with monocotyledon crops against dicotyledon weeds.

3.3. Chronological selectivity

Chronological selectivity utilizes the time period necessary for growing both weeds and crop plants. In other words, it depends upon the fact that some weeds are shallower rooted and grow more rapidly than the crop plants. In consequence, many of the potentially more competitive weeds that emerge before the crop can be sprayed by a foliage spray. The time of application of the herbicide is important for chronological selectivity to be successful. That means if the non-selective herbicides are applied too early, many of the germinating weed seedlings will escape and break through the soil surface; however, the crop may be damaged if those herbicides are applied too late (Figure 1) [1].

3.4. Positional selectivity

Positional selectivity is based upon the localization of weeds on the soil surface related to the main plant crop position. If seeds, tubers, etc., of the crops are large compared with those of the weeds, they become sown or placed quite deeply in the soil compared with the more shallow competitive weed seeds. Consequently, positional selectivity can often be achieved by spraying the soil surface with soil acting herbicides. These herbicides are able to destroy weed seeds growing in the top few millimeters of the soil, whereas the large seeds of the crop are protected by the fact that they are sown deeper in the soil. Bacteria and other microorganisms attack and inactivate most herbicides when used at economic concentrations so the potential hazard to the crop is reduced (Figure 1) [1].

3.5. Placement selectivity

Placement selectivity can be achieved for non-selective substances when it is possible to direct a foliar spray in such a way that it makes contact only with the leaves of weeds and not the crop [2].

3.6. Genetic engineering

If the mode of action of an herbicide is known and the target proves to be a protein, genetic engineering may well allow the crop gene coding for that protein to be isolated. It is then possible to alter that crop gene so that it is less affected by the herbicide [2].This will be discussed in detail at the end of the chapter.

4. Herbicide interference with physiological and biochemical processes and plant response

Mode of action of herbicides can lead to various physiological and biochemical effects on both growth and development of the emerging seedlings as well as the established plants. These physiological and biochemical effects are followed by various types of visual injury symptoms on susceptible plants. The incidental damage extent depends on the selectivity of the herbicide as well as the applied concentration. The herbicide application is always recommended at a certain dose termed as recommended dose (R), above which, a great damage to the crop plant may be obtained. Overdoses threaten not only the crop plant but also the environment and human health. Some herbicides in lower than recommended doses may act as growth regulators for crop plants [1,2].

Even recommended doses may have undesired effects upon the crop. The undesired effects might occur in the form of chlorosis, defoliation, necrosis, morphological aberrations, growth stimulation, cupping of leaves, marginal leaf burn, delayed emergence, germination failure, etc. These injury symptoms may appear on any part of the plant.

The various physiological and biochemical processes affected by herbicides are grouped under five broad categories including: respiration, mitochondrial activities, photosynthesis, protein synthesis, nucleic acid metabolism, and hydrolytic enzyme activities. Most herbicides can affect at least one or all of these processes. The following discusses their effect on various biochemical processes.

4.1. Respiration and mitochondrial activities

Cellular respiration that takes place in mitochondria involves the synthesis of ATP and the transport of electrons and protons from respiratory substances to oxygen. Herbicides affect

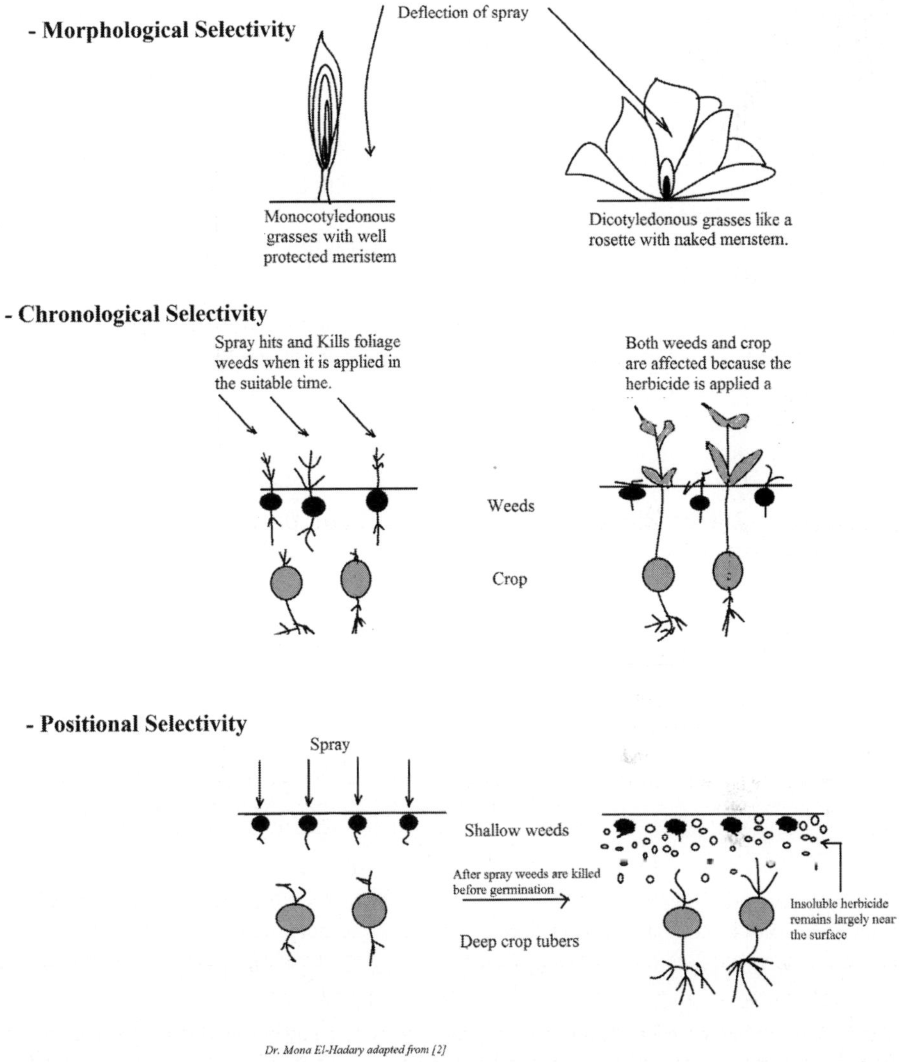

Figure 1. Factors Exploitable to Achieve Selectivity of Herbicides [1] as adapted from [2].

the mitochondrial activities by uncoupling the reaction responsible for ATP synthesis or interfering with electron transport and energy transfer. Uncouplers act on the membranes of the mitochondria in which phosphorylation takes place. Electrons leak through the membranes so that the charges that they normally separate are lost. As a result, energy is not accumulated for ATP synthesis [3].

4.2. Photosynthesis

Pigment content and photosynthetic activity are affected by herbicidal applications. The mode of action of herbicides on the photosynthesis process depends on the chemical group to which the herbicide belongs [3]. Herbicides affect chloroplast organization and pigment formation especially chlorophyll which is the principle absorbing pigment. Chlorophyll bleaching is a potent inhibitor for photosynthetic electron transport and CO_2 fixation.

Herbicides affect photosynthetic activity via different ways including photosynthetic pigments. The primary site of action is located at photosystem II (PSII) since they cause blocking of the Hill reaction. The oxygen evolution step is inhibited by interfering with the reducing side rather than the oxidizing side of PSII [4]. The inhibition of electron transfer through PSII causes a block in the whole transport chain as the inhibition of the noncyclic photophosphorylation and ATP synthesis. Consequently, the production of NADP is blocked and the function of the protective carotenoid system is prevented [5]. Urea herbicides inhibit both noncyclic and cyclic electron transport by forming a complex with oxidized form of an unknown component located in the electron transfer pathway close to PSII. This component also takes part in cyclic electron transport.

The photosystem I (PSI) also could be reduced by some herbicides but it requires much higher concentrations of the herbicide than that required for the inhibition of PSII. Since PSII precedes PSI and the former is blocked completely at concentrations which do not affect PSI.

In a study conducted by **El-Hadary** [1], it was observed that photosynthetic activity measured in wheat chloroplasts (variety Giza 163) was greatly reduced throughout the growth by using Brominal as an example for bromphenol herbicides but lower concentrations (1/4R, 1/2R and R) increased the activity. Pigment content represented as chlorophyll, a/b ratio and carotenoids showed a similar results [1]. In the same study, sulfonyl-urea herbicides such as Granstar were examined. It was observed that low Granstar concentrations stimulated the photolytic activity of chloroplasts while high concentrations reduced it. However, Granstar reduced a/b ratios throughout the growth stages, except a slight increase at the fruiting stage with 1/2R. Carotenoids were decreased only with high Granstar concentrations [1].

4.3. Protein and nucleic acid metabolism

Protein synthesis takes place mainly in three stages involving initiation, elongation and termination of the polypeptide chain. Blocking any one of these stages by the herbicide will cause inhibition of protein and nucleic acid synthesis. The herbicides that inhibit photosynthesis and ATP formation can lead to inhibition of protein synthesis as a secondary effect. The damage that is caused by an herbicide is governed by its chemical group. There are numerous studies that investigate effects of the herbicidal chemical groups upon protein and nucleic acid metabolism [2].

For instance, sulfonyl-urea herbicides block the biosynthesis of the branched chain amino acids in higher plants [6,7]. Aliphatic herbicides like Dalapon cause degradation of protein to ammonium compounds as detected in *Setaria lutescens* and sugar beets [8]. While acetamide herbicides such as propachlor, alachlor and prynaclor inhibited the protein content and RNA synthesis as reported in barley [9-12]. Also, metalachlor inhibited protein synthesis in barley [13]. RNA and protein synthesis in tomato were found to be inhibited by propanil [14].

Benzoic and phenylacetic herbicides had variable effects on protein. For example, chloramben had no effect on RNA and protein synthesis on susceptible species [15]. On the other hand, it was suggested that foliar-applications of dicamba increased RNA and protein levels in susceptible plants by removal of histone from the DNA template [16]

Carbamate herbicide groups include a large number of herbicides such as asulam, barban, chlorpropham, propham, desmedipham and phenmedipham. [17]. Barban was found to inhibit protein synthesis and the degree of inhibition was related to the susceptibility of the plant species. For example, barban increased nucleotide content of wild oat shoots associated with disruption of RNA and protein synthesis. Chlorproham and propham inhibited amino acid incorporation into protein and induced a reduction in protein synthesis [18]. DNA, RNA and protein synthesis are also inhibited at high concentrations (10-3 M) of propham [19].

Fluridone, paraquat, perfluidone and propanil treatments were found to reduce soluble protein levels in soybean [20]. Paraquat and diquat readily act on proteins, modifying their structure and function (e.g.lysozome) since they interact with dibasic and dicarboxylic amino acids like ornithine and glutamic acid [21].

Oxadiazon at high doses inhibited protein synthesis in soybean while RNA and DNA synthesis were less sensitive to oxadiazon [22]. Combination of 2,4-D and glufosinate had an additive effect on protein synthesis in both sorghum and soybean [22]. On the other hand, sethoxydim, R- 25788 [N, N dichloroacetamide] or R- 28725 at low doses did not inhibit protein or RNA synthesis in cells of both sorghum and soybean but sethoxydim significantly inhibited DNA synthesis while R-25788 stimulated it [23]. Thus, the combined effects of sethoxydim and the two Safeners (R- 25788 and R- 28725) on protein and RNA synthesis were additive while on DNA synthesis they were antagonistic.

The application of haloxyfop to *Zea mays* and soybean cell suspension, increased ^{14}C labeled free amino acids level and incorporation of ^{14}C leucine as a precursor revealed that haloxyfop did not inhibit protein synthesis [24].

Napropamide reduced DNA synthesis, RNA root cells of *Pea* and protein [25]. The inhibitory effect of napropamide on the mitotic cycle resulted from an inhibition in the synthesis of cell cycle specific protein. In contrast, 0.5 R, 1R and 1.5 R of metribuzin stimulated total and protein-N accumulation in soybean. Consequently, protein content was increased while RNA and DNA levels decreased [26]. Protein content of soybean yield was reported to be increased by application of 100 ppm GA_3 (gibbrellic acid) and 2g/L Librel separately or together [27].

Metoxuron had a remarkable inhibition on the total protein biosynthesis, while bromoxynil accelerated the biosynthesis of low molecular proteins (water-soluble proteins) and inhibited the biosynthesis of high molecular proteins (sodium hydroxid soluble proteins) in wheat (*Triticum aestivum*, var. Sakha 69) [28]. Bromoxynil at low doses (0.4 and 0.8 kg / fed) enhanced protein content and RNA synthesis in wheat plants after 30 to 60 days from foliar spraying [29].

Nitrogen in wheat grains, consequently protein, was found to be increased by treating wheat plants with Brominal at the 2-leaf stage [30]. Different bromoxynil levels increased the protein percentages in wheat grains [31]. The foliar spray with bromoxynil increased significantly the protein content in wheat grains [32]. Application of bromoxynil at the full recommended rate significantly increased grain nitrogen and proteins in both wheat and barley. The increase was evaluated by multiplying grain nitrogen by 5.7 as a factor in both wheat and barley [33]. Protein content in wheat vegetation (Giza 163) was significantly increased at the vegetative stage and flowering stage while decreased at the fruiting stage as a response to either low or high Brominal treatments [1]. In contrast, the protein content of wheat root was reduced. Also, protein profiling of grains is greatly altered with an induction for 19kDa and 25kDa but an inhibition for 66kDa, 100kDa and 110kDa was obtained [1].

The action of urea herbicides on protein and nucleic acid metabolism has been reported by many researchers. Although fluometron can cause an increase in the low molecular weight fraction of DNA, RNA and protein synthesis [34], diuron and monuron inhibited the same parameters as reported [35]. However, the monomethylated derivative of isouron [N-[5-(1,1-dimethy ethyl-3-iso) (azol]-N-methylurea] suppressed the protein synthesis in soybean[36].

Sulfonylurea herbicides were found to inhibit branched chain amino acids valine, leucine and isoleucine (e.g. Granstar; DPX- L 5300; tribenuron) [6, 7]. Aflon (urea herbicide), when sprayed at 1/2 R and R doses on *Phaseolus vulgaris,* induced a DNA increase in both shoot and root while RNA content was increased in shoot only [37]. Moreover, RNA content of roots was mostly decreased in response to R and 2R aflon treatments but increased as a result of the 1/2 R application [37]. Protein content of the wheat shoot system was increased with all Granstar concentrations at the vegetative stage and with low concentrations (1/2R and R) at both flowering and fruiting stages. In contrast, protein levels were decreased with 5/2R at the flowering stage and with 3/2R and 2R- and 5/2R at the fruiting stage [1]. Granstar treatments reduced the contents of root proteins at the vegetative stage and flowering stages but increased it at the fruiting stage. Protein profiling of grain proteins exerted an induction for 19kDa and 25kDa and complete suppression for 66kDa, 100kDa and 110KDa [1].

4.4. Hydrolytic enzyme activities

Enzymes of plants were affected greatly by herbicide treatments and their effect differs according to the chemical group to which the herbicide belongs. The following examples represent some effects of herbicides on the enzyme activities of some plant species.

One of the major metabolic processes that take place during seed germination is the production of hydrolytic enzymes such as α-, β-amylases that degrade stored carbohydrates into simple sugars. The production of hydrolytic enzymes requires the synthesis and presence of proteins, polyribosomes and nucleic acids. Thus, an effect of the herbicide on protein formation as mentioned above, would affect the synthesis of the hydrolytic enzymes [1, 3]. **El-Hadary** [1] reported that use of either Brominal or Granstar at different levels below and above the recommended rate induced stimulation for amylolytic enzyme activity (α and β-amylase); however, an incidence of a slight reduction in β-amylase activity was observed with 2R and higher doses of Granstar [1].

Dalapon, which is an aliphatic herbicide, did not affect the activity of hydrolytic enzymes like protease, α- amylase and dipeptidase in barley seeds [38]. Acetamides such as alachlor, propachlor and prynachlor which all were applied at pre-emergence caused an inhibition for seed germination in barley by reducing the synthesis of α-amylase enzyme [39].

It was reported that propaclor inhibited the gibberellic acid (GA_3) induced production of α-amylase in barley seeds [40]. Similarly, alachlor, propachlor and prynachlor were found to inhibit α-amylase as well as protease synthesis in barley seeds [41, 42]. It was suggested that these herbicides may act as repressors for gene action preventing the normal expression of the hormonal effect of GA_3 through the synthesis of DNA-dependent RNA. This was confirmed when higher levels of GA_3 overcame alachlor inhibition by removing the repressor effect [42]. In addition, the effect of these acetamide herbicides on α- amylase and protease was suggested to be secondary and these herbicides possibly act on the biosynthetic reactions (like protein synthesis) required for the synthesis of these hydrolytic enzymes.

Chloroamben and dicamba, which belong to the benzoic and phenylacetic acid herbicide groups, were found to inhibit GA_3-induced α-amylase synthesis and the development of amylase activity in barley seeds [40, 43]. This agrees with effect of trifluarlin, as an example for dinitroanilines, which was found to inhibit the *de novo* synthesis of hydrolytic enzymes such as protease [44] and dipeptidase in squash cotyledons [45], phytase in barley seedlings, squash cotyledons and maize embryos [39], and α-amylase in barley seeds [40].

Nitriles such as bromoxynil and ioxynil also inhibited proteolytic and amylolytic enzyme activities [46, 45]. Also, thiocarbamate herbicides were found to inhibit GA_3- induced α-amylase synthesis in susceptible weeds [17]. Acifluorfon was found to stimulate the activity of chalcone synthase, phenylalanine ammonia lyase and isoflavone 7-0- glucosy transferase which are responsible for the accumulation of isoflavonoids in soybean leaves [47].

The increase of galactonolactone oxidase was reported in common beans as a result of acifluorfen application; this enzyme is responsible for lipid peroxidation. Acifluorfen was found to increase the activity of galactonactone reductase, which prevented further oxidation of lipids [48]. Other herbicides, alachlor and glyphosate, were observed to inhibit 5- enolpyruvyl shikimate-3-phosphate (EPSP) synthase enzyme. This enzyme is responsible for the synthesis of all cinnamate derivatives (intermediates in flavonoids biosynthesis pathway) leading to reduced flavonoid synthesis in higher plants [49].

Sulfonylureas herbicides act by inhibiting acetolactate synthase enzymes, thereby blocking the biosynthesis of the branched chain amino acids in higher plants [7]. According to **Gronwald** [50], carbomothioate herbicides inhibited one or more acyl- CoA elongase enzymes which catalyze the condensation of malonyl CoA with fatty acid acyl-CoA substrates to form a very long chain fatty acid, used in the synthesis of surface lipids.

The effects of triazine, urea and nitroaniline herbicides on amylase and acid proteolytic activities of wheat grain cultivars, Salwa, Grana and Liwilla were studied by **Wybieralshi and Wybieralska** [51]. The studied herbicides were found to inhibit amylase activity in Salwa and Liwilla, but increased it in Grana. Acid proteolytic activity in Liwilla and Salwa was reduced especially by Igran 80 (terbutryn) and Dicuran 60 (Chlorotoluron), while the activity in Grana was not affected. In contrast, amylase, dehydrogenase, cellulase and xylanase activities were increased by application of the herbicides Pyramin (chloridazon), Ro-neet (cycloate) and Venzar (lenacil) when applied on the soil with 5% (w/w) addition of wheat straw [52]. Other studies suggested that application of SAN 9789 (norflurazon) as a metabolic inhibitor to *Sinapis alba* seedlings destroyed the chloroplasts but had no effect on α-amylase activity. This is due to the fact that α-amylase is a cytosolic enzyme [53].

The levels of leaf β-amylase and starch debranching enzyme in pea seedlings were found to slightly decrease in response to norflurazon-treatment [54]. However, inhibitors of chloroplastic functions, i.e.; diuron (DCMU), atrazine, tentoxin, paclobutrazol and San 9785 (4 -chloro-5-(dimethylamino)-2-phenyl-3 (2H)- pyridazinone) caused either no or only slight increases in α-amylase activity. In contrast were the inhibitors of plastidic protein synthesis lincomycin and chloramphenicol that cause an increase in α-amylase activity in pea seedlings. It is concluded that there was an inverse relationship between α-amylase activity and chlorophyll concentration in pea petals and stems [55]. Similarly an inhibition of α-amylase induction in barley seeds was reported [56]. Also, Li found that juglone decreased the content of total soluble protein and α-amylase activity induced by gibberellin by 74% and 78% in the aleuron cells of barley. It was concluded that juglone may be a metabolic inhibitor which prevents many (if not all) physiological and biochemical processes involving SH-groups in compounds such as amino acids, peptides and enzymes [57].

The activities of α-and β-amylases of castor bean and maize Giza 2 seedlings and adult plants supplemented with low concentration (0.5-2.5 μg/g) of metribuzin either alone or in combination with NaCl at 50 μg/g were increased significantly [58] but higher metribuzin concentration (5-10μg) had an opposite response. Application of 1.5-4.5kg/ha thiobencarb and butachlor six days after transplanting of 30-day-old rice seedlings affected the enzyme activities of the seedlings whether they were grown alone or with the competitive barnyard grass [59]. Moreover, both herbicides reduced α-amylase activity by increasing the concentration but a sharp increase in α-amylase activity was noted at 96h post-treatment in both species. In addition, protease (proteinase) activity was maximized after post-treatment at both 48h.and 24h in rice and grass, respectively.

Butachor (1000-3000 g/ha) and oxyfluorfen (100-300g/ha) effect on α-amylase activity and chlorophyll content in 46 rice cultivars was dependent on the degree of tolerance of each cultivar [60]. It was concluded that rice cultivars ADT-37, ASD-16 and ASD-18 were highly tolerant to butachor, whereas ADT-36, ADT-38 and PY-3 were highly susceptible. However, tolerance to oxyfluorfen was high in ASD-18 and AS-18696, while IR-50 was highly susceptible [60].

4.5. Lipid synthesis and oxidation

Substituted ureas, uracils, triazine, benzonitriles and bipyridyls markedly accelerated the photo-oxidations (lipids- per-oxidation) but peroxidation was completely prevented by NADH or NADPH [5]. Lipid peroxidation in higher plants (Duranta and Cassia) was induced by oxyfluorfen [61] but the peroxidative cell damage is controlled by antioxidative systems such as vitamins "C" and "E".

Lipid peroxidation and galactonlactone oxidase increased in response to the treatment of *Phaseolus vulgaris* leaves with acifluorfen [48] and the activity of glutathione reductase also increased to prevent further oxidation. Gronowald studies on herbicides concluded that the carbothioates group impaired the synthesis of surface lipids (waxes, cutin, and subrin) by inhibiting acyl- CoA elongases while chloroacetamide herbicides inhibited *de novo* fatty acid biosynthesis. Similarly, pyridazinones herbicides decreased the degree of unsaturation of plastidic galactolipids while aryloxyphenoxy pypropionic acid and cyclohexanedione herbicides inhibited *de novo* fatty acid synthesis. The target site for all these classes is the enzyme acetyl-CoA carboxylase [50].

The total lipid content as well as *gluco*-and *phospho*-lipid content of maize seedlings markedly decreased by application of perfluidone while in sunflower cotyledons total lipids were not affected but glycolipids increased at the expense of phospholipids [62]. Also, a decrease in lipid synthesis in soybean by Isouron was reported [36] but an increase in seed oil of soybean was obtained by 0.5R or 1R metribuzin application [26].

4.6. Carbohydrate content

Carbohydrate content is one of the most affected parameters in response to herbicide application. **Yakout** [28] demonstrated that treating wheat (*Triticum aestivum* var. Sakha 69) with metoxuron showed a slight reduction in the available carbohydrates with relatively no change in sucrose content while bromoxynil showed an increase in different carbohydrate levels. Also, the total reducing substances (may include sugars, phenolic substances, ascorbic acid, organic acid, etc.) were increased for both treatments [28].

Inhibition of the accumulated reducing sugars, sucrose and polysaccharides, in soybean leaves was observed in response to 1R and 1.5R metribuzin application and, consequently, seed carbohydrate content decreased with increasing metribuzin concentration [26]. Terbytryn herbicide was found to decrease starch content and increase sugar content in pre-emergence and post emergence applications [63]. On the other hand, bromoxynil was reported to significantly increase soluble and total carbohydrates at low doses while a higher dose (1.2 kg/Fed) inhibited their synthetic rate in wheat plants [29]. Similarly, the results of **El-Hadary** [1] found that *mono-, di- and poly*-saccharides and, consequently, total carbohydrates were increased with low doses but decreased with high doses of either Brominal or Granstar [1]. The incidental increase with low concentrations was attributed to that some herbicides act as growth regulators in low doses.

Urea herbicides including afalon-S at low doses of 1/2R and R increased the soluble and insoluble sugar contents of shoots at different stages of growth and development of *Phaseolus vulgaris* while a reverse situation was obtained in the case of a 2R application. The root tissue treated with various concentrations suffered from an obvious decrease in the content of the different carbohydrate fraction relative to those of the control [37].

The content of reducing sugars and sucrose of *Ricinus communis* cultivar Balada and maize cultivar Giza 2 seedlings and adult plants supplemented with low concentrations (0.5-2.5μg/g) of metribuzin either alone or in combination with NaCl at 50μg/g were increased significantly but decreased in response to higher concentrations (5-10μg). On the other hand, polysaccharide content of *R. communis* and maize seedlings as well as adult plants were significantly decreased in response to low concentrations of metribuzin and increased significantly at higher concentrations either alone or in combination with NaCl. Total carbohydrate content detected in *R. communis* treated with metribuzin were greater than those detected in presence of herbicide and NaCl combination [58].

Thiobencarb and butaclor herbicides when applied at 1.5-4.5 kg/ha after transplanting 30 days old rice seedlings and barnyard grass grown alone or with rice were found to have no effect either on total carbohydrate or starch and reducing sugars in rice and grass [59].

4.7. Plant growth response and yield

Plant growth and yield are greatly affected by herbicidal applications depending on the age, tolerance, dose and the active chemical group of the herbicide. The author in a previous work pointed that Brominal application on wheat induced an increase in the number of grains per spike with 1/4 R. 1/2R and R while higher doses caused a significant reduction [1]. Also, grain yield showed a detectable reduction in monosaccharides, disaccharides, polysaccharides and, consequently, total carbohydrate levels with all Brominal concentrations [1].

The percentage of germination and seedling growth of barley was decreased greatly by applications of bromoxynil [64]. But the same herbicide in different concentrations encouraged wheat growth [31]. Also, growth parameters such as plant height, weight and leaf area of wheat plants at 75 days after sowing were increased significantly by foliar application of bromoxynil at rate of 1.0 L/Fed [32, 65]. Moreover, a good seedling establishment of wheat was obtained by combinations of bromoxynil and fenoxaprop [66]. Low metribuzin concentrations (0.5-2.5μg/g) either alone or in combination with NaCl (50 μg/g) caused an increase in different growth parameters such as leaf area, length of shoot and root, water content and dry matter accumulation in both *Ricinus communis* cultivars, and maize cultivars Giza 2 throughout the different growth stages [58]. In contrast, the higher metribuzin concentration (5-10μg) affected the same parameters oppositely [58].

Productivity of the plant is affected in terms of 100 grains weight in response to herbicides treatment. The yield of wheat grains (var.Sakha 69) increased by bromoxynil application [28]. A dose of 1.5 kg/ha of bromoxynil brought an increase in weight of 100 grains [30,67]. The highest yield was obtained when one liter/fed bromoxynil was applied at the third-leaf stage [68]. The number of wheat grains/ear and grain yield were increased at a low dose (0.8kg/ fed.) of bromoxynil [29,69] while a higher dose of the same herbicide (1.2 kg/ fed) reduced the yield of wheat varieties; i.e. Sakha 69, Giza 157 and Giza 160 [29]. On the other hand, it was noticed that higher doses of bromoxynil resulted in a marked increase in both yield and grains/ear when crops were poorly developed at the time of spraying [70]. However, the application of 2.5, 3.0 liter bromoxynil /ha at the third-leaf and flowering stages on wheat significantly decreased the grain yield [71] as well as the number of spikes per plant, main spike length, weight of 100 grains and straw per plant [32].

Herbicidal effects may be varied when they are applied in combination. For example, a marked increase was observed in the grain yield, ears/plant and number of ears in barley by using a combination of bromoxynil, ioxnil and mercoprop [72]. An increase of about 20% was recorded in grain wheat yield when oxitril 4, which is a combination of oxitril and bromoxynil, was used at 130g/liter and applied at rates of 1.5,4 and 5 liters/ha [73]. In winter wheat a marked increase in yield was mentioned in response to half rate applications of various commercial herbicides (active ingredients bromoxynil, ioxynil, mocoprop, cyanazine, fluroxypyr, metasulfuron-methyl, and clopyralid) [74].

Urea herbicide such as Granstar (metasulfuron- methyl 75% water dispersible granules) was found to suppress the growth rate of wheat and barley by about 20% while weeds were completely destroyed[75]. Its application with a dose of 20-40 g/ha in 200-500 liter/ha prior to planting resulted in 50% suppression [76]. The author in a previous work applied Granstar at a dose of 0.5R, 1R, 1.5R and 2R on wheat at 40-days old and reported an increase in grains no./ spike [1]. However, a great decrease in monosaccharides, disaccharides, polysaccharides and, consequently, total carbohydrate levels was obtained in wheat grains with both low and high Granstar concentrations [1]. Also, chlorsulfuron was mentioned to reduce both the third leaf growth rate and shoot dry weight of wheat seedlings but not the root dry weight [76].

The urea herbicide metoxuron was reported to decrease wheat grain yield (var. Sakha 69) [28]. It was found that 100-seed weight of soybean was decreased by using metribuzin at rates of 0.5R, 1R and 1.5R [26]. Wheat yield was markedly increased by using tribenuron at a rate of 0-125g [77]. However, sulfonylurea herbicides, Chisel [Chlorosulfuron+thifensulfuron - methyl] and Granstar, significantly increased the productive tillering in some wheat varieties [78]. Application of trifluralin alone in the spring followed by some post herbicides resulted in a reduction in vegetative growth, shoot dry weight and wheat grain yield [79]. An application of 0.126 mM perfluidon herbicide was reported not only to decrease both fresh and dry weight but also shoot length of maize seedlings [62].

5. Hazardous action of herbicides in the agricultural environment and human health

Although the benefits gained from herbicides usage in weeds control, herbisides have undesired effects on man health and environment. Their residues remain in the soil for many years, affecting crops, water canals, grazing animals and human health and even the pollution of air.

Herbicides and pesticides have been suspected by the "National Cancer Research Institute" as a probable cause of certain cancers especially cancers of the brain, prostate, stomach and lip, as well as leukemia, skin melanomas and Hodgkin's lymphoma [80]. They also cause reproductive problems as well as infertility and nervous system diseases. The National Academy of Sciences reported that infants and children, because of their developing physiology, are more susceptible to the negative effects of herbicides and pesticides in comparison to adults. Herbicides may cause human poisoning since they affect humans through three mechanisms of entry: ingestion, inhalation and dermal absorption. In under-developed countries, the least expensive pesticides are utilized due the inability of farmers to purchase more expensive, safer products. As a byproduct of pesticide use, farmers and their families are affected daily with health problems directly resulting from pesticide exposure [81]. Herbicide toxicity and risks are not only limited by their direct use but can also present risks indirectly. Indirect risks are represented by herbicidal traces that remain in the edible plants themselves as well as the residues in the soil that may remain for a number of years before it can be degraded. Moreover, the leakage of these herbicides and their residues in water canals, vaporization and sublimation in air may be poisonous to the surrounding living organisms.

6. Natural herbicides

Allelopathy phenomenon serves the agricultural community so much. The following section discusses the related concepts to allelopathy and recruiting it as natural herbicides for weed management to be an alternative or to minimize conventional herbicide use.

6.1. Allelopathy term

Allelopathy is a natural biological phenomenon of interference among organisms in such a way that an organism produces one or more biochemicals that influence the growth, survival, and reproduction of other organisms. Allelopathy is the favorable or adverse effect of one plant on another due to direct or indirect release of chemicals from live or dead plants (including microorganisms).

6.2. Allelochemical term

Allelochemicals, or allelochemics, are a subset of low molecular weight secondary metabolites such as alkaloids, phenolics, flavonoids, terpenoids, and glucosinolates which are produced during growth and development but are not used by the allelopathic plant [82]. Allelochemicals may have beneficial (positive allelopathy) or detrimental (negative allelopathy) effects on the target organisms. Allelochemicals with negative allelopathic effects contribute in plant defense against herbivory. Also, allelochemicals could be recruited in weed management as alternatives to herbicides.

Allelochemicals are listed as six classes [83] that possess actual or potential phytotoxicity. The classes are namely alkaloids, benzoxazinones, cinnamic acid derivatives, cyanogenic compounds, ethylene and other seed germination stimulants, and flavonoids which have been isolated from over 30 families of terrestrial and aquatic plants. Like synthetic herbicides, there is no common mode of action or physiological target site for all allelochemicals.

6.3. Allelochemical occurrence

Allelochemics are present in different parts of the plant; leaves, flowers, fruits, stems, bark, roots, rhizomes, seeds and pollen. They may be released from plants into the environment

http://dx.doi.org/10.5772/55957

through volatilization, leaching, root exudation, and decomposition of plant residues. Rainfall causes the leaching of allelopathic substances from leaves which fall to the ground during period of stress, leading to inhibition of growth and germination of crop plants [84, 85].

6.4. Allelochemical classification and biosynthesis

According to the different structures and properties of allelochemicals, they can be classified into the following categories: water-soluble organic acids, straight-chain alcohols, aliphatic aldehydes, and ketones; simple unsaturated lactones; long-chain fatty acids and polyacetylenes; quinines (benzoquinone, anthraquinone and complex quinines); phenolics; cinnamic acid and its derivatives; coumarins; flavonoids; tannins; steroids and terpenoids (sesquiterpene lactones, diterpenes, and triterpenoids) [86]. The biosynthetic pathways of the major allelopathic substances are shown in Figure 2 [87].

6.5. Allelochemical interference and biological activity

The allelochemical interference implies their interference with each other as well the interference with other surrounding plants. Several chemicals can be released together and may exert toxicities in an additive or synergistic manner. Allelopathic interferences often result from the mixing action of several different compounds. Allelopathic plant extracts can effectively control weeds since mixtures of allelopathic water extracts are more effective than the application of single-plant extract. Combined application of allelopathic extracts and reduced herbicide dose (up to half the standard dose) give as much weed control as the standard herbicide dose in several field crops. Lower doses of herbicides may help to reduce the development of herbicide resistance in weed ecotypes [88]. Allelopathy thus offers an attractive environmentally friendly alternative to pesticides in agricultural pest management [88].

Response of the receiver plants to allelochemicals is not only concentration dependent but also controlled by the biochemical pathway in the receiver plant. Generally, low concentrations of allelochemicals are stimulatory while it is inhibitory with higher concentrations [89]. Allelochemical concentrations in the producer plant may also vary over time and in the plant tissue produced. Foliar and leaf litter leachates of Eucalyptus species, for example, are more toxic than bark leachates to some food crops. Typically, allelochemical concentration in field situations is below the required inhibitory level that can affect sensitive plants.

Receiver plant response to antagonistic allelochemicals is detected as certain signs on growth and development of the plants that are exposed to allelochemicals. The effect includes the inhibition or retardation of germination rate; seeds darkness and swelling; root or radicle reduction, curling of the root axis, lack of root hairs; increased number of seminal roots, swelling or necrosis of root tips; shoot or coleoptile extension; discolouration, reduced dry weight accumulation; and lowered reproductive capacity. These morphological effects may be secondary for primary events due to interference with different biochemical pathways of the receiver plant [90].

Biological activity of allelochemicals could be increased by some modifications so the end product could be more active, selective, or persistent. This is attributed to the potential

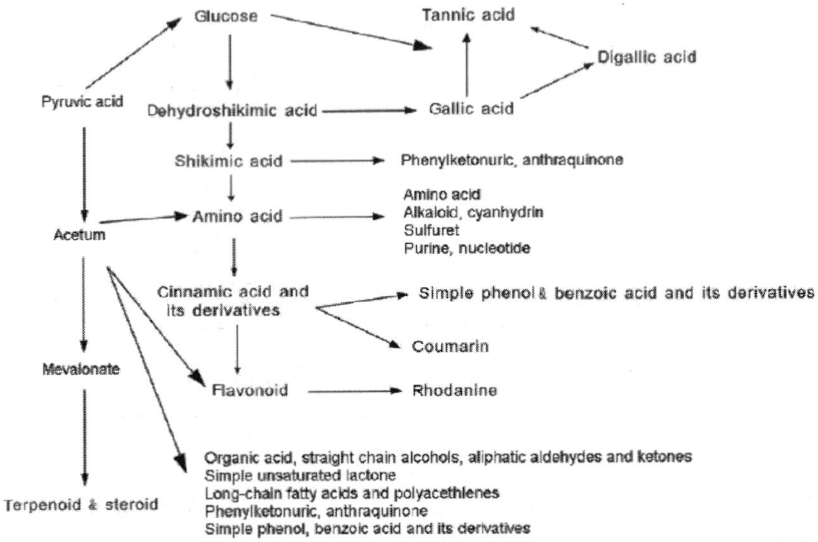

Figure 2. The Biosynthetic Pathways of the Major Allelopathic Substances [87]

phytotoxicity of alkaloids, benzoxazinones, cinnamic acid derivatives, cyanogenic compounds, ethylene and other seed germination stimulants, and flavonoids that always represent the secondary products of allelopathic plants. Biodegradable natural plant products rarely contain halogenated atoms and possess structural diversity and complexity, constituting one such class of chemicals and these can act directly as herbicides or may provide lead structures for herbicidal discovery [91]. Selection of allelopathic plants is a good and commonly used approach for identification of plants with biologically active natural products [91].

Different crops such as beet (*Beta vulgaris* L.), lupin (*Lupinus lutens* L.), maize (*Zea mays* L.), wheat (*Triticum aestivum* L.), oats (*Avena sativa* L.) and barley (*Hordeum vulgare* L.) are known to have an allelopathic effect on other crops (Rice, 1984b). For instance, some wheat cultivars were found to significantly inhibit both germination and radicle growth of annual ryegrass. The allelopathic potential of wheat cultivars was positively correlated with their allelochemical (total phenolics) content [92]. However, different allelopathic compounds of some crops important in weed management are presented in Table 1 [93].

6.6. Allelopathic plants impact

There are some examples of plants that act as natural herbicides, such as black walnut, sunflowers, sagebrush and spotted knapweed. An herbicidal chemical called catechin was extracted from the roots of spotted knapweed and can be synthesized on a larger scale and applied to a number of other invasive plants due to selectivity. Another popular species with

Crops	Scientific name	Allelochemicals
Rice	Oryza sativa L.	Phenolic acids
Wheat	Triticum aestivumL.	Hydroxamic acids
Cucumber	Cucumis sativus L.	Benzoic and Cinnamic acids
Black mustard	Brassica nigra L.	Allyl isothiocyanate
Buck wheat	Fagopyrium esculentum L.	Fatty acids
Clovers and	*Trifolium spp.*	Isoflavonoids and Phenolics
Sweet clover	*Melilotus spp.*	Phenolics
Oats	Avena sativa L	Phenolic acids and Scopoletin
Cereals	-	Hydroxamic acids
Sudangrass		Phenolic acids and Dhurrin
Sorghum	Sorghum bicolor L.	Sorgoleone

Table 1. Allelochemicals of Some Important Crops

natural herbicide abilities is the black walnut tree whose leaf extraction is often used in commercially-produced natural herbicides [94].

Other natural pre-emergent herbicides are used to control weed growth such the natural herbicide corn gluten meal. Corn gluten meal was originally developed as a medium for growing fungus, but its inhibitory effect upon the germination of weeds and grasses was detected. A cover crop of rye could work as a natural herbicide between soybean crops [94].

Herbicidal effects have been identified and quantified for more than twenty allelochemicals in *Vulpia* residues. Those present in large quantities possessed low biological activities, while those present in small quantities possessed strong inhibitory activities. Interference between different allelochemicals controls the overall phytotoxicity of *Vulpia* residues which varies according to the individual chemical structure and occurred quantity. This interference provides a pattern for suggested artificial combinations of these allelochemicals prepared in aqueous solution. Biological tests for different combinations of *Vulpia* extracts demonstrated the existence of strong synergistic effects among the identified allelochemics. Moreover, exploration of the composition of a cluster of allelochemicals, which are simple in structure, possess various biological activities and few barriers to synthesis and production; this provides an alternative option for developing new herbicides from individual plant allelochemicals [94].

Selective activity of tree allelochemicals on crops and other plants has also been reported. For example, *Leucaena leucocephala*, the miracle tree promoted for revegetation, soil and water conservation and animal improvements in India, also contains a toxic, non-protein amino acid in leaves and foliage that inhibits the growth of other trees but not its own seedlings. *Leucaena* species have also been shown to reduce the yield of wheat but increase the yield of rice. Leachates of the chaste tree or box elder can retard the growth of pangolagrass but stimulate growth of bluestem, another pasture grass. Examples that are shown in Table 2 represent some allelopathic plants and their impact as reported in published research [95].

6.7. Allelochemical modes of action

Allelochemical action goes mainly through affecting photosynthesis, respiration cell division, enzymes function and activity, endogenous hormones and protein synthesis. This suggests allelochemical action on the molecular level and gene expression [86]. Some phenolics such as ferulic acid and cinnamic acid can inhibit protein synthesis or amino acid transport and the subsequent growth of treated plants. This is attributed to the ability of all phenolics to reduce integrity of DNA and RNA [86]. A series of physiological and biochemical changes in plants induced by phenolic compounds are shown in Figure 3 [87].

Allelopathic Plant	Impact
- Rows of black walnut interplanted with corn in an alley cropping system	- Reduced corn yield attributed to production of juglone, an allelopathic compound from black walnut, found 4.25 meters from trees
- Rows of Leucaena interplanted with crops in an alley cropping system	- Reduced the yield of wheat and tumeric but increased the yield of maize and rice
- Lantana, a perennial woody weed pest in Florida citrus	- Lantana roots and shoots incorporated into soil reduced germination and growth of milkweed vine, another weed
- Sour orange, a widely used citrus rootstock in the past, now avoided because of susceptibility to citrus tristeza virus	- Leaf extracts and volatile compounds inhibited seed germination and root growth of pigweed, bermudagrass, and lambsquarters
- Red maple, swamp chestnut oak, sweet bay, and red cedar	- Preliminary reports indicate that wood extracts inhibit lettuce seed as much as or more than black walnut extracts
- Eucalyptus and neem trees	- A spatial allelopathic relationship if wheat was grown within 5 m
- Chaste tree or box elder	- Leachates retarded the growth of pangolagrass, a pasture grass but stimulated the growth of bluestem, another grass species
- Mango	- Dried mango leaf powder completely inhibited sprouting of purple nutsedge tubers.
- Tree of Heaven	- Ailanthone, isolated from the Tree of Heaven, has been reported to possess non-selecitve post-emergence herbicial activity similar to glyphosate and paraquat
- Rye and wheat	- Allelopathic suppression of weeds when used as cover crops or when crop residues are retained as mulch.
- Broccoli	- Broccoli residue interferes with growth of other cruciferous crops that follow

Table 2. Examples of Allelopathy from Published Research.

http://dx.doi.org/10.5772/55957

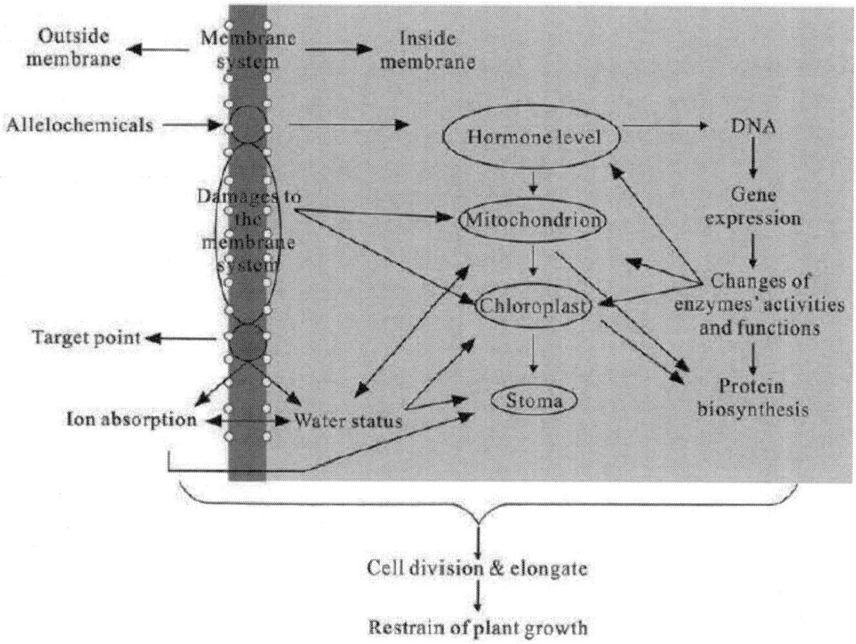

Figure 3. Mechanism of Allelochemicals [87].

6.8. Strategies of allelopathic plants application as natural herbicides

The strategy of allelochemical application is based on their antagonistic or synergistic action. Antagonistic properties of allelopathic plants are utilized in companion cropping system. Growing a companion plant which is selectively allelopathic against certain weeds and does not interfere appreciably with crop growth can greatly reduce weed establishment [96].

The interaction of weeds with crops may be positive; for instance, controlled densities of wild mustard (*Brassica campestris* L.) were interplanted with broccoli (*Brassica oleracea* var. Premium crop), crop yield increased by as much as 50% compared with broccoli planted alone [97].

Allelochemicals may be utilized as stimulators to weed seed germination before sowing the main crops, so that the germinated weeds could be eradicated easily. *Striga asiatica* is a good example for this case since it grows as a parasite to cereal grains in the southeastern United States. *Striga* normally germinates in response to compounds released from its host plants [98]. A germination stimulant, a p-benzoquinone compound from a natural host (sorghum) for *Striga* was identified. This stimulatory compound is used to induce germination of *Striga* and eradicate it before cropping its host. Ethylene was found to be a very effective germination stimulant. Also, ethylene stimulates *Striga* to germinate in the absence of a host [99] since its use as a gas at about 1.5kg/ha has been used effectively via a soil injection to trigger "suicidal" germination of *Striga* and to deplete the numbers of dormant seeds in soil [100].

6.9. Limitation of using allelopathic plants as herbicides

Recruiting allelopathy in weed management is limited by both the allelopathic plant itself and the environment. Production, release and phytotoxicity of allelochemicals are altered by biotic and abiotic soil factors [101, 102] such as plant age, temperature, light and soil conditions, microflora, nutritional status, and herbicide treatments. Toxicity of allelochemicals may be either cleared or increased after releasing into the soil by action of microbes [103] since the toxicity is influenced by soil texture. For instance, amounts of water-soluble phenolics in *P. lanceolata* leaf leachate amended soil varied depending on the soil textural classes if it is clay, sandy-loam, sand, or silty-loam [104]. Some allelopathic agents are active only under hot and dry climates as they work in the vapor phase such as monoterpenes because the high vapor density of the essential oils may penetrate into soil, affecting adversely the under growing plants [105].

High costs for synthesizing many allelochemicals stands as a limiting factor for utilizing allelochemicals. Also, the hazardous action of allelochemicals on human beings limits their use. They may be toxic [91] carcinogenic [106] or even cause thyroid, liver and kidney diseases in monogastric animals [107].

Allelopathic potentiality of some plants is influenced either by the availability or deficiency of nutrient. The deficiency of nutrients favors the production of secondary metabolites. For example in aerobic P-deficient soil, rice roots excrete organic anions, particularly citrate, to solubilize and enhance phosphorus uptake [108]. Some allelochemicals affect the growth of the plant itself, i.e., autotoxic effect as some derivatives of benzoic and cinnamic acids from the root exudates of cucumber since it inhibits root antioxidant enzymes and leaf photosynthesis, transpiration and stomatal conductance in cucumber [109].

Natural herbicides sound attractive as alternatives for herbicides but their application is still surrounded with much concern since they affect humans and environmental equilibrium. The agricultural community cannot discard the use of synthetic herbicides completely at the present time but their use can be reduced up to a certain extent by utilizing allelopathic potentiality as an alternative weed management strategy for crop production.

7. Future prospects for rationalization of herbicide usage by molecular biology

Rationalization of herbicidal use targets mainly the production of plants which are herbicidal themselves by recruiting allelopathic characters. Allelopathy is considered a genetically influenced factor [91]. Allelopathic characteristics are more likely to evolve in competitive populations such as in wild types [110]. Therefore, it is possible to enhance weed suppressive potential of crop cultivars or to transfer allelopathic characteristics from wild types or unrelated plants into commercial crop cultivars through conventional plant breeding methods or other genetic recombination strategies. There are two methods for creating herbicidal plant crops that have been suggested; regulation of gene expression related to alleochemicales biosynthesis; or insertion of genes to produce allelochemicals that are not found in the crop [88].

7.1. Gene insertion

The allelopathic phenomenon as mentioned before refers to the ability of some plant species to suppress other species by releasing allelochemicals, which are not toxic to the originating plant but toxic to surrounding vegetation. Breeding allelopathic cultivars by molecular approaches are more complicated than developing an herbicide-resistant crop. Genetic engineering of allelochemicals bases on their overexpression as valuable secondary metabolites in plants [111]. Most secondary metabolites being used as allelochemicals are products of a multi-gene system might which have to be developed and transformed into the specific crop to produce allelochemicals [112, 113].

Gene insertion targets the change of the recent biochemical pathways into another one which is able to produce new allelochemicals through the insertion of transgenes. Although there is great difficulty to satisfy this approach, it represents the promising molecular approaches available for application in the near future. Various reviews in this trend and reference book on molecular biology of weed control [112, 113] were conducted.

7.2. Regulation of gene expression related to allelochemicals

Regulation of gene expression by a biologist first requires accurate identification of the target allelochemical(s), to determine enzymes and the genes encoding them. Accordingly, a specific promoter can be inserted into crop plants to enhance allelochemical production. Allelochemicals are conditionally expressed by biotic and abiotic factors since some metabolites having allelopathic potential might be newly synthesized or highly elevated in rice plants by UV irradiation [114]. For instance, there is a differential response to UV or other environmental stresses among rice cultivars. The phenylpropanoid pathway intermediates of several allelopathic rice cultivars have the highest content of *p*-coumaric acid. The latter is a key reaction in the biosynthesis of a large number of phenolic compounds in higher plants. Phenolic compounds are derived from cinnamic acid by the catalysis of 4-hydroxylase (CA4H) enzyme. The activity of CA4H was measured to determine its response to UV irradiation in rice leaves of different varieties. *Kouketsumochi* showed induction for CA4H activity by UV after 24 h of UV irradiation for 20 min while the rice cultivar AUS 196 showed no response. The increase in CA4H enzyme activity as a required enzyme in conversion of cinammic acid into *p*-coumaric acid suggested a role for CA4H gene in the elevation of the allelopathic function in rice plants [114].

Responsiveness to environmental stresses and plant-plant interaction may be conferred by a specific promoter. A promoter which its induction is responsive to an elicitor can be used to regulate genes that are responsible for coding allelochemicals. The expression of phytoalexins and pathogenesis related genes in plants were reported in response to UV treatment and other plant defense inducers [115, 116]. UV was found to stimulate phytoalexine production in pepper. The effective motifs response to UV light was determined in tobacco by examining the expression of GUS activity of plants transformed with the constructs of various CASC (*Capsicum annuum* sesqiterpene cyclase) promoters fused into GUS gene [115]. This was followed by UV irradiation of the transgenic plants to assure the induction of the CASC promoters through examining GUS activity of the transgenic plants. The levels of GUS activity for transgenic plants with pBI121-KF1 and pBI121-KF6 were significantly elevated by UV-irradiation and had a two-to-threefold increase approximately over the untreated-transgenic plants. In contrast, GUS expression in the transgenic plants with pBI121-CaMV 35S was not changed by UV, and in the other constructs had only a very small increase [117]. The CASC promoters of both KF-1 and KF-6 were suggested to contain cis-acting elements capable of conferring quantitative expression patterns that were exclusively associated with UV irradiation. The regulation of genes associated with allelopathy could be achieved by developing a specific promoter responsive to plant-weed competition or environmental stresses. The CASC promoters of KF-1 and KF-6 obtained may be specific to UV. Thus, this promoter can be used for the overexpression of specific promoters constructed to allelochemical-producing genes [116]. To regulate the CA4H gene in the phenylpropanoid pathway, specific promoters, the CASC-KF1 and KF6, were fused to CA4H gene. The gene constructs were introduced into the binary plant expression vector pIG121-HMR with reverse primer harbouring *Bam*HI site and forward primer harbouring *Hin*dIII site as illustrated in Figure 4 [118].

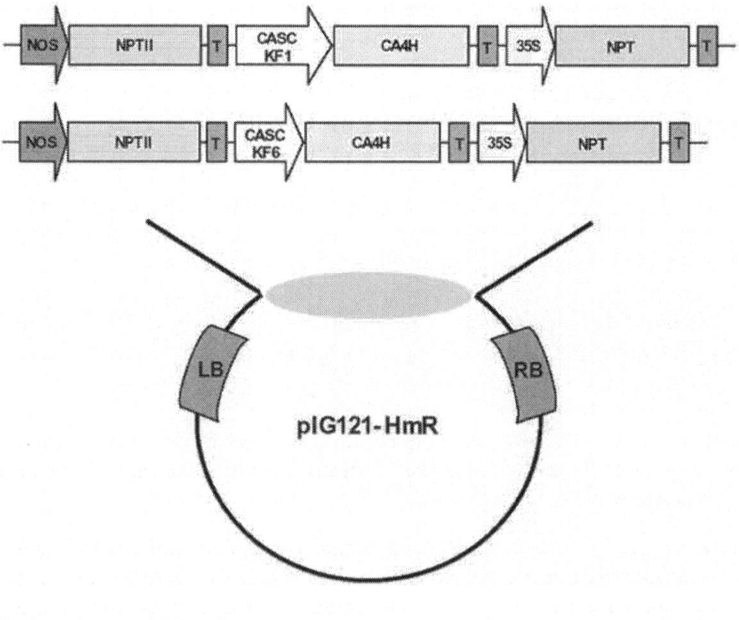

Figure 4. The Gene Cassette with Specific Promoters Responsive to UV Irradiation in pIG121-HmR [117].

8. Conclusion

Herbicides are widely used in agricultural communities on a large scale for eradicating weeds. Herbicides function by affecting different biochemical processes in weeds. Herbicides in low doses act as growth regulators for the main crop but high doses may cause crop damage. However, uncontrolled herbicide use can cause hazardous effects not only upon the main crop but also human health and the surrounding environment [80, 81]. Moreover, heavy doses of herbicides create the problem of herbicide resistance development in weeds. There is an urgent need to identify natural alternatives that can meet the demands of agrosystems without affecting the surrounding environment. Hence, the idea of recruiting the allelopathic phenomenon of some plants in inhibiting the growth of weed vegetation has been investigated. Allelopathy cannot cancel the use of herbicides completely but can minimize it. Allelopathic plant use has limitations in the application because of the potential toxicity. Thus, molecular biology can aid the agricultural community by engineering crops to be herbicides themselves through gene insertion and regulation depending on well-defined allelopathic genes or promoters, respectively. Even with well-characterized allelopathic genes, it might be very difficult to transfer genes into crops.

Author details

Mona H. El-Hadary[1,2*] and Gyuhwa Chung[3]

*Address all correspondence to: drmona3000@yahoo.com

1 Department of Molecular Biology, Genetic Engineering and Biotechnology Research Institute (GEBRI) Minufiya University, Egypt

2 Department of Botany, Faculty of Science, Damanhour University, Egypt

3 Department of Biotechnology, Chonnam National University, Korea

References

[1] El-Hadary, M. H. Effect of Brominal and Granstar Herbicides on Growth and Some Metabolic Activities in Wheat. MSc. thesis. Faculty of Science Tanta University Egypt; (1988).

[2] Hassall, K. A, Editors- Ebert, E, & Kayser, H. Staub T- Book Review: The Biochemistry and Uses of Pesticides. Structure, Metabolism, Mode of Action and Uses in Crop Protection. (2nd Edition). VCH Verlagsgesellschaft mb: Germany; (2003). DOI: 10.1002/anie.199202422.

[3] Rao, . . Principles of Weed Science (2nd ed.). India Science Publishers ISBN 1-5808-069: 1983; 125-160.

[4] Radosevich, S. R, Steinbak, K. E, & Arntsen, C. G. Effect of Photosystem II Inhibitors on Thylkaloid Membranes of Two common Groundsel (*Senecio vulgaris*) Biotypes. Journal of Weed Science (1979)., 27-216.

[5] Giannopolitis, C. N, & Ayers, G. S. Enhancement of Chloroplsat Photooxidations with Photosynthesis-Inhibiting Herbicides and Protection with NADH or NADPH. Journal of Weed Science (1978)., 26-440.

[6] Brown, H. M. Mode of Action, Crop Selectivity and Soil Relations of the Sulfonylurea Herbicides. Journal of Pesticide Science (1990)., 29(3), 263-281.

[7] Moberg, W. K, & Cross, B. Herbicides Inhibiting Branched-Chain Amino Acid Biosynthesis. Journal of Pesticide Science (1990)., 29(3), 241-246.

[8] Anderson, R. N, Linck, A. J, & Behrens, R. Absorption, Translocation, and Fate of Dalapon in Sugar Beets and Yellow Foxtail. Journal of Weeds (1962)., 10-1.

[9] Duke, W. B. An Investigation of the Mode of Action of Chloro-N-isopropylacetanilide. PhD. Thesis. Illinois Urbana University USA; (1967)., 2.

[10] Duke WB; Slife FWHanson JB. Studies on Mode of action of chloro-N-isopropylacetanilide. Abstr. Weed Science Society of America (1967)., 2.

[11] Duke, W. B, Slife, F. W, Hanson, J. B, & Butler, H. S. An Investigation on the Mechanism of Propachlor. Journal of Weed Science (1975)., 23-142.

[12] Rao, V. S, & Duke, W. B. The Effects of Acetanilide Herbicides on Polysome and Protein Formation. Abstr. Weed Science Society of America (1974).

[13] Deal LM; Reeves JT ; Larkins BA, Hess FD. Use of an *in vitro* Protein Synthesizing System to Test the Mode of Action of Chloroacetamides. Journal of Weed Science (1980)., 28-334.

[14] Hofstra, G, & Switzer, C. M. The Phytotoxicity of Propainil. Journal of Weed Science (1968)., 16-23.

[15] Gruenhagen, R. D, & Moreland, D. E. Effect of Herbicides on ATP Levels in Excised Soybean Hypocotyls. Journal of Weed Science (1971)., 19-319.

[16] Arnold, W. F, & Nalewaja, J. D. Effect of Dicamba on RNA and Protein. Journal of Weed Science (1971)., 301-305.

[17] Mann JD; Jordan LSDay BE. The Effects of Carbamate Herbicides on Polymer Synthesis. Deeds (1965a)., 13-63.

[18] Mann JD; Jordan LSDay, B.E. A survey of Herbicides for their Effect upon Protein Synthesis. Journal of Plant Physiology (1965b)., 40-840.

[19] Rost, T. L, & Bayer, D. E. Cell Cycle Population kinetics of Pea Root Tip Meristems Treated with Propham. Journal of Weed Science (1976). , 24-81.

[20] Hoagland, R. E, & Duke, S. O. Relationships Between Phenylalanine Ammonia-Lyase Activity and Physiological Response of Soybean (*Glycine max*) Seedlings to Herbicides. Journal of Weed Science (1983). , 31(6), 845-852.

[21] Szogyi, M, Cserhati, T, & Szigeti, Z. Action of Paraquat and Diquat on proteins and Phospholipids. Journal of Pesticide Biochemistry and Physiology (1989). , 34(3), 240-245.

[22] Hatzios, K. K. Effects of Glufosinate on the Metabolism of Sorghum as Influenced by 2,4-D Asparagine and Glutamine. In: Proceedings of the Southern Weed Sci. Society, 38th Annual Meeting; , 480.

[23] Hatzios, K. K, & Moon, P. A. Combined Effects of Sethoxydim and Chloroacetamide Safeners on the Metabolism of Sorghum Protoplasts and Soybean Cells. In: Proceedings of the Southern Weed Science Society, 38th Annual Meeting; , 462.

[24] Cho HY; Widholm JWSlife EW. Effects of Haloxyfop on Corn (*Zea mays*) and Soybean (*Glycine max*) Cell Suspension Cultures. Journal of Weed Science (1986). , 34(4), 496-501.

[25] Ditomaso, J. M, Rost, T. L, & Ashton, F. M. The Comparative Cell Cycle and Effects of Herbicide Napropamide on Root tip Meristems. Journal of Pesticide Biochemistry and Physiology (1988). , 31(2), 166-174.

[26] Gabr, M. A, & Shakeeb, M. A. Metabolic Changes Associated with growth of Soybean as Affected by Pre-emergence Application of Metribuzin. Canadian Journal of Botany (1988). , 66(12), 2380-2384.

[27] Salem, S. M. Effect of Some Growth Regulators and Micronutrients on Growth and Productivity of Soybean Plants. Bulletin of Faculty of Agriculture Cairo University Egypt (1989). , 40(1), 213-224.

[28] Yakout, G. A, & Soliman, E. L-S. h. a. r. a. k. y A. S. FS. The Effect of Bromoxynil and Mell oxuron Herbicides on Wheat Leaves. Alexandria Science Exchange Egypt (1987). , 8(4), 1-15.

[29] Fathi, S. F, & Shaban, A. Response of Some Wheat Cultivars to Bromoxynil. Zagazig Journal of Agriculture Research Egypt (1991). , 18(3), 729-738.

[30] Morsy, M. A, Zaitoon, M. I, Hanna, L. H, & Ibrahim, I. Z. Effect of Brominal on Yield Components and Uptake of Some Plant Nutrients in Wheat. Annals of Agricultural Science Faculty of Agricultural Science Mashtohor, Zagazig Univesty Egypt.

[31] EL-Desoky IR. The Infleunce of Some Herbicide Mixtures on Wheat and Associated Weeds. PhD thesis. Faculty of Agriculture Cairo University Egypt, 1990.

[32] Shehzad, M. A, Nadeem, M. A, & Iqba, M. Weed Control and Yield Attributes Against Post-emergence Herbicides Application in Wheat Crop. Global Advanced Research Journal of Agricultural Science Punjab, Pakistan (2012). , 1(1), 007-016.

[33] Grundy, A. C, Botman, N. D, & Williams, F. R. Effects of Herbicide and Nitrogen Fertilizer Application on Grain Yield and Quality of Wheat and Barley. Journal of Agricultural Science (1996). , 126(4), 379-385.

[34] Ali-zade, M. A, & Ismailove, A. A. The Use of Herbicides in Cotton Field and their Effect on the Nucleic Acid Content of Cotten Leaves. Weed Abst., (1979). , 28-145.

[35] Ashton, M. Y. de Villiers OT; Glenn RK, Duke W. B. Localization of Metabolic Sites of Action of Hexbicides. Pesticides. Journal of Biochemistry and Physiology (1977). , 7-122.

[36] Henry, W. T, & Hatzioz, K. K. Comparative Effects of Three Urea Herbicidal Derivatives on the Metabolism of Enzymatically Isolated Soybean Leaf Cells. Journal of Weed Researches (1987). , 27-23.

[37] El-Shafey, A. S. EL-Akkad SS. Effect of soil treatment with the herbicide Afalon-S on Certain Physiological Aspects in *Phaseolus vulgaris*. Journal of Desert Researches (1992). , 42(2), 15-18.

[38] Lotlikar, P. D, Remmert, L. F, & Freed, V. H. Effect of D and other Herbicides on Oxidetive Phosphorylation on Mitochondria from Cabbage. Journal of Weed Science (1968). , 2, 4.

[39] Penner, D. Herbicide and inorganic Phosphate on Phytase in Seedlings. Journal of Weed Science (1970). , 18-301.

[40] Moreland, D. E, Malhotra, S. S, Gruenhayen, R. D, & Shokrahii, E. H. Effects of Herbicides on RNA and Protein Synthesis. Journal of Weed Science (1969). , 17-556.

[41] Rao, . . Mechanism of action of acetanilide herbicides. PhD thesis. Cornell University Ithaca New York USA; 1974 p116.

[42] Rao, V. S, & Duke, W. B. Effect of Alachlor, Propachlor and Prynchlor on GAInduced Production of Protease and α-Amylase. Journal of Weed Science (1976). , 3.

[43] Penner, D. Herbicidal Influence on Amylase in Barley and Squash Seedlings. Journal of Weed Science (1968). , 16-519.

[44] Ashton, F. M. Relationship between Light and Toxicity Symptoms Caused by Atrazine and Monuron. In: Weeds (1965). , 13-164.

[45] Tsay, R, & Ashton, F. M. Effect of Several Herbicides on Dipeptidase Activity of Squash Cotyledons. Journal of Weed Science (1971). , 19-682.

[46] Penner, D, & Ashton, F. M. Influnce of Dichlobenil, Endothal and Bromoxynil on Kinin Control of Protolytic Activity. Journal of Weed Science (1968). , 16-323.

[47] Casio, E. G, Weissenbock, G, & Moclure, J. W. Acifluorfen-Induced Isoflavonoids and Enzymes of their Biosynthesis in Mature Soybean Leaves. Whole Leaf and Mesophyll Responses. Journal of Plant Physiology (1985). , 78(1), 14-19.

[48] Schmidt, A, & Kunert, K. J. Lipid Peroxidation in Higher Plants, the Role of Glutathione Reductase. Journal of Plant Physiology (1986). , 82(3), 700-702.

[49] Lydon, J, & Ducke, S. O. Pesticide Effects on Secondary Metabolism of Higher Plants. Journal of Pesticide Science (1989). , 25(4), 361-373.

[50] Gronwald, J. W. Lipid Biosynthesis Inhibitors. Journal of Weed Science (1991). , 39(3), 435-449.

[51] Wybieralshi, J, & Wybieralska, A. Enzyme Activity of Wheat Grain Treated with Herbicides Journal of Chemosphere (1988). , 17(1), 159-163.

[52] Pietr, S. J, & Jablonska, E. The Effect of Action of Herbicides on Some Chemical Parameters and the Enzymatic Activity of soils. Polish Journal of Soil Science (1987). , 169(2), 17-23.

[53] Manga, V. A, & Sharma, R. Lack of Functionan Interrelationship between β-amylase Photoregulation and Chloroplast Development in Mustard (*Sinapis alba* L.) Cotyledons. Journal of Plant and Cell Physiology (1990). , 31(2), 167-172.

[54] Saeed, M. Regulation of Amylolytic Enzymes in the Photosynthetic Tissues of Pea (*Pisum sativum* L.). Dissertation Abstracts International, Science and Engineer (1990).

[55] Saeed, M, & Duke, S. H. Chloroplastic Regulation of Apoplastic α-amylase Activity in Pea Seedlings. Journal of Plant Physiology (1990). , 93(1), 131-140.

[56] Yoshikawa, H, Fujimolto, E, & Doi, K. Synthesis and Biolgical Activtiy of Benzaldehyde O-alkyloximes as Abscisic Acid Mimics. Journal of Bioscience, Biotechnology and Biochemistry (1992).

[57] Li, H. H, Nishimura, H, Hasegawa, K, & Mizutani, J. Some Physiological Effects and the Possible Mechanism of Action of Juglone in plants. Journal of Weed research-Tokyo (1993). , 38(3), 214-222.

[58] Hasaneen MNAEL-Saht HM, Bassyoni FM. Growth, Carbohydrates and Associated Invertase and Amylase. Journal of Biologia Plantarum (1994). , 36(3), 451-459.

[59] Kumar, J, & Prakash, J. Effect of Thiobencarb and Butachlor on Photosynthesis, Carbohydrate Content, Amylase and Protease Activity in Rice (*Oryza sativa*) and Barnyard Grass (*Echinochloa crus galli*), Indian Journal of. Agricultural Science (1994). , 64(1), 9-14.

[60] Kathiresan, R. M, Gurusamy, A, Brown, H, Cussans, G. W, Devine, M. D, & Duke, S. O. Fernandez, Quintanilla C, Helweg A, Labrada RE, Landes M, Kudsk P, Streibig JC. Herbicide Tolerance in Rice cultivars. In: Proceedings of the Second International Weed Control Congress, Copenhagen, Denmark, 25-28 June (1996). , 1(4), 955-962.

[61] Finckh, B. F, Kunert, K, & Vitamin, C. and "E" an Antioxidative System Against Herbicide-Induced Lipid Peroxidation in Higher Plants. Journal of Agriculture and Food Chemistry (1985). , 33(4), 574-577.

[62] Valadon LRGKates M. Effect of Perfluidon on Metabolism of Lipids in Maize (*Zea mays* L.) and Sunflowe (*Helanthus annuus* L.). Journal of Plant-Growth Regulation (1984). , 1984(3), 2-111.

[63] Bansal, G. L, & Sharma, V. K. Effect of Terbutryn on Maize (*Zea mays*) and Watergrass (*Echinochloa colonum*). Changes in Chlorophyll Content and Carbohydrates. Indian Journal of Weed Science (1989).

[64] Abdou, R. F, & Ahamed, S. A. Cytological and Developmental Effects of Four Herbicides on Barely. Journal of Rachis (1989). , 8(2), 14-16.

[65] El-Bagouri, I. H, Wassif, M. M, Kadi, M. A, & Sabet, S. A. Response of Barley to Foliar Application of Some Micro Nutrients Under the Conditions of Saline Water Irrigation and Highly Calcareous Soil. Desert Intuitional Bulletin A.R.E. (1983). , 14-1.

[66] Malik, N. Meadow bromegrass and crested wheat grass forage yield response to herbicides applied during establishment. Bibliographic Citation Journal production Agric. (1991). , 4(4), 508-515.

[67] Majid, A, Hussein, M. R, & Mkhtar, M. A. Studies on Chemical Weed Control in Wheat. Pakistan Journal of Agricultural Research (1983). , 21(4), 167-171.

[68] Gonzalez, M. J, & Ferrandez, G. A. Early Weed Control in Wheat. Revista de los CREA (1987). , 124-5.

[69] Ashraf, M. Y, & Bahig, N. A. Response of Wheat (*Triticum aestivum* L.) to herbieidal Wheat Control. Biblographic citation Nucleus- Karahi. (1989).

[70] Fogelfors, H. Different Herbicide Doses in Barley Studies of the Actual Requirement. Swedish Crop Protection Conference. Weeds and Weed Control (1991). , 32-53.

[71] Montazeri, M, & Saber, H. K. Response of Golestan Wheat Cultivar to D and Bromoxynil at Different Growth Stages. Journal of Revista de los CREA Seed and Plant (1992). , 2, 4.

[72] Botman, N. D. Effects of Herbicide Use, Fungicide Use and Position in the Field on the Yield and Yield Components of Spring Barley. Journal of Agricultural Science (1992). , 118(1), 17-28.

[73] Hallgern, E. Effects of Some Herbicides or Mixtures of Herbicides on Annual Dicots as a Whole and on Grain Yield at Different Doses, Development Stages and Weed Densities. Vaxtodling, Institutionen for Vaxtodling, Sveriges Lantbruk, suniversitet, (1993).

[74] Grundy, A. C, Botman, N. D, & Williams, F. R. Effects of Herbicide and Nitrogen Fertilizer Application on Grain Yield and Quality of Wheat and Barley. Journal of Agricultural Science (1996). a(4) 379-385.

[75] SpiridonovYu. YA, Raskin MS, Samus MV, Grishakova OM, Shestakov VG, Yakovets VI, kirillova NA. Effectiveness of Preparations of Sulfonylurea Derivatives in Weed Control Communication 3. The Effectiveness of Granstar in Sowings of Cereal Crops. Journal of Agrokhimiya, (1990). , 8-116.

[76] Dong, B, Rengel, Z, & Graham, R. D. Effects of Herbicides Chlorsulfuron on Growth and Nutrient Uptake Parameters on Wheat Genotypes Differing in Zn-efficiency. Journal of Plant and Soil (1995).

[77] Stewart, V. R, & Keener, T. K. Evaluation of Four Sulfonylurea Herbicides for Broad Leaved Weed Control in Winridge Winter Wheat. Proceedings of the Western Society of Weed Science (1989).

[78] Drozd, D. Reaction of Spring Wheat Varieties to New Generation Herbicides (Chisel and Granstar). Biuletyn Instytutu Hodowli- I- Ak limatyzacji. Roslin (1995). , 194-199.

[79] Clay, S. A, Gaffney, J. F, & Wrage, L. I. Spring Wheat Cultivar Responses to Trifluralin and Post-emergence Herbicides. Journal of Weed Technology (1995). , 9(2), 352-355.

[80] eHow living healthy. The Effects of Herbicides and Pesticides on Humans., by Flint D. http://www.ehow.com/facts_5636303_effects-herbicides-pesticides-humans.html

[81] Kato, M. Elyanne Ratcliffe MPH, Rohrer WH. Agricultural Pesticide Exposure and its Negative Health Effects. Children Global Medicine. www.dghonline.org. http://www.globalmedicine.nl/index.php/global-medicine-1/agricultural-pesticide-exposure.

[82] Rice, E. L. Allelopathy." (2nd ed.) Academic Press: New York; , 421.

[83] Putnam, A. R. Weed Tech. (1988). , 2-510.

[84] Rice, E. L. Allelopathy. Academic Press: New York; (1974 3). p.

[85] Mann, J. Secondary Metabolism (2nd edi.). Clarendon Press: Oxford;(1987). p.

[86] Li, Z. H, Wang, Q, Ruan, X, Pan, C. D, & Jiang, D. A. Phenolics and Plant Allelopathy. Journal of Molecules (2010). 1420-3049Available at www.mdpi.com/journal/molecules-doi:10.3390/molecules15128933., 15-8933.

[87] Wang, Q, Ruan, X, Li, Z. H, & Pan, C. D. Autotoxicity of Plants and Research of Coniferous Forest Autotoxicity. *Sci. Sil. Sin.* (2006). , 43-134.

[88] FarooqJabran M, Cheema K, Wahid ZA, Siddique A, Kadambot HM The Role of Allelopathy in Agricultural. Journal of Pest Management Science; (2011). Available at http://onlinelibrary.wiley.com/doi/10.1002/ps.2091/abstract., 2011(67), 5-493.

[89] Lovett, J. V. Allelochemicals, Mycotoxins and Insect Pheromones and Allomones. In: Phytochemical Ecology. Chou CH and Waller GR (ed.).Taipei: ROC; (1989)., 49-67.

[90] Rice, E. L. Botany Review. (1979)., 45-15.

[91] Duke, S. O, Dayan, F. E, Romagni, J. G, & Rimando, A. M. Natural Products as Sources of Herbicides: Current Status and Future Trends. Journal of Weed Research (2000)., 40-99.

[92] Wu, H, Pratley, J, Lemerle, D, Haig, T, & Verbeek, B. Differential Allelopathic Potential among Wheat Accesions to Annual Ryegrass. In: DL Michalk Dl, Pratley JE (eds.) NSW 2650: Proceedings of the 9th Australian Agronomy Conference of the Australian Society of Agronomy: *"Agronomy, growing a greener future?"*, NSW July 1998, Charles Sturt University, Wagga Wagga; (1998). Avalible from http://www.regional.org.au., 2650, 20-23.

[93] Bhadoria PBSAllelopathy: A Natural Way towards Weed Management. American Journal of Experimental Agriculture (2011)., 1(1), 7-20.

[94] An, M, Pratley, J. E, & Haig, T. Allelopathy: From Concept to Reality. *In: DL Michalk Dl, Pratley JE (eds.)* NSW 2650: Proceedings of the 9th Australian Agronomy Conference of the Australian Society of Agronomy: *"Agronomy, growing a greener future?"*, NSW July 1998, Charles Sturt University, Wagga Wagga; (1998). Avalible from http://www.regional.org.au., 2650, 20-23.

[95] Ferguson, J. J, & Rathinasabapathi, B. Allelopathy: How Plants Suppress Other Plants. Publication series of Horticultural Sciences Department, Florida Cooperative Extension Service, Institute of Food and Agricultural Sciences, University of Florida. HS944 document; July 2003. Reviewed May (2009). EDIS Web site at http://edis.ifas.ufl.edu/hs186.

[96] Putnam, A. R, & Duke, W. B. Annual Reviewof Phytopathology. (1978). , 1978, 16-413.

[97] Jimenez-osornio, J. J, & Gliessman, S. R. In "Allelochemicals: Role in Agriculture and Forestry". Waller GR (ed.). American Chemical Society Washington DC (1987). , 262-274.

[98] Matusova, R, & Rani, K. Verstappen FWA, Franssen MCR, Beale MH,. Bouwmeester HJ. The Strigolactone Germination Stimulants of the Plant-Parasitic *Striga* and *Orobanche* spp. are Derived From the Carotenoid Pathway. Journal of Plant Physiology (2005)., 139(2), 920-934.

[99] Egley, G. H, & Dale, J. E. Ethylene, 2-Cloroethylphosphonic acid, and witched germination. Proceeding 23rd Annual Meeting Southern Weed Science Society; (1970)., 372.

[100] Eplee, R. E. Ethylene: a switched seed stimulant. Weed Science (1975)., 23-433.

[101] Huang, P. M, Wang, M. C, & Wang, M. K. Catalytic Transformation of Phenolic Compounds in the Soil. In Inderjit, et al. (ed.) Principles and Practices in Plant Ecology: Allelochemical interactions. CRC Press: Boca Raton FL;, 1999-287.

[102] InderjitCheng, H.H., Nishimura, H. Plant phenolics and terpenoids: Transformation, Degradation, and Potential for Allelopathic Interactions. In S. Inderjit, et al. (ed.): Principles and practices in Plant Ecology: Allelochemical interactions. CRC Press: Boca Raton FL;, 1999-255.

[103] InderjitAllelopathy Symposium. Soil Environmental Effects on Allelochemical Activity. Journal of Agronomy (2001)., 93-79.

[104] InderjitDakshini KMM. Allelopathic Effect of *Pluchea lanceolata* (Asteraceae) on Characteristics of Four Soils and Tomato and Mustard Growth. American Journal of Botany. (1994)., 81, 799-804.

[105] Koitabashi, R, Suzuki, T, Kawazu, T, Sakai, A, Kuroiwa, H, & Kuroiwa, T. Cineole Inhibits Roots Growth and DNA Synthesis in the Root Apical Meristem of *Brassica campestris* L. Journal of Plant. Research (1997)., 110, 1-6.

[106] InderjitBhowmik PC. The Importance of Allelochemicals in Weed Invasiveness and the Natural Suppression. In: Inderjit, Mallik, A.U. (ed.), Chemical Ecology of Plant: Allelopathy of Aquatic and Terrestrial Ecosystems. Birkhauser Verlag AG: Basal; , 187-192.

[107] Van Etten, C. H, & Tookey, H. L. In: CRC Handbook of Naturally Occurring Food Toxicant. Rechcighl M. Jr (ed.) CRC Press: Boca Raton;, 1983-15.

[108] Kirk GJDSantos EE, Santos MB. Phosphate Solubilization By Organic Anion Excretion from Rice Growing in Aerobic Soil: Rates of Excretion and Decomposition, Effects on Rhizosphere pH and Effects on Phosphate Solubility and Uptake. Journal of New Phytopathology (1999)., 142, 185-200.

[109] Yu, J. Q, & Matsui, Y. Phytotoxic Substances in the Root exudates of *Cucumis sativus* L. Journal of Chemistry and Ecololgy (1994)., 20-21.

[110] Putnam, A. R. editor.Tang CS- The Science of Allelopathy. John Wiley and Sons: New York; (1986 3). p.

[111] Canel, C. From genes to Phytochemicals: the Genomics Approach to the Characterization and Utilization of Plant Secondary Metabolism. Journal of Acta Horticulturae. (1999)., 500-51.

[112] Gressel, J. Molecular Biology of Weed Control. Taylor and Francis Publishers, London. Hahlbrock, K. and D. Scheel 1989. Physiology and Molecular Biology of Phenylpropanoid Metabolism. Annual Review of Plant Physiology Plant Molecular Biology (2002)., 40-347.

[113] Fei, F. H, Chun, L. C, Ze, Z, Zeng, H. L, Fang, Y. D, Cheng, L, & Zhong, H. X. The UDP-glucosyltransferase multigene family in *Bombyx mori*. Journal of Bio Med Central Genomics (2008). doi:10.1186/1471-2164-9-563.

[114] Kim, H. Y, Shin, H. Y, Sohn, D. S, Lee, I. J, Kim, K. U, Lee, S. C, Jeong, H. J, & Cho, M. S. Enzyme Activities and Compounds Rrelated to Self-Defense in UV-Challenged Leaves of Rice. *Korean* Journal of Crop Science (2000). , 46(1), 22-28.

[115] Back, K, He, S, Kim, K. U, & Shin, D. H. Cloning and Bacterial Expression of Sesquiterpene Cyclase, a Key Branch Point Enzyme for the Synthesis of Sesquiterpenoid Phytoalexin Capsidiol in UV-Challenged Leaves of *Capsicum annuum*. Journal of Plant Cell Physiology (1998). , 39(9), 899-904.

[116] El-Hadary, M. H. Molecular Studies on Some Pathogenesis-Related (PRs) Proteins in Tomato Plants. PhD thesis. Genetic Engineering and Biotechnology Research Institute (GEBRI) Minufiya University, Egypt; (2007).

[117] Shin, D. H, Kim, K. U, Sohn, D. S, Kang, S. U, Kim, H. Y, Lee, I. J, & Kim, M. Y. Regulation of Gene Expression Related to Allelopathy. *In:* Kim KU, Shin DH (eds.) *Proc. of the Inernational. Workshop in Rice Allelopathy.* Kyungpook National University, Taegu, Korea, August (2000). Institute of Agricultural Science and Technology, Kyungpook National University, Taegu 2000;109-124, 17-19.

[118] Kim, K. U, & Shin, D. H. The importance of allelopathy in breeding new cultivars. Agriculture and Consumer Protection. FAO Corporate Document Repository-Weed Management for Developing Countries (Addendum 1). Available at http://www.fao.org/docrep/006/Y5031E/y5031e0f.htm.